Coastal Research Library

Volume 17

Series Editor
Charles W. Finkl
Department of Geosciences
Florida Atlantic University
Boca Raton, FL 33431, USA

The aim of this book series is to disseminate information to the coastal research community. The Series covers all aspects of coastal research including but not limited to relevant aspects of geological sciences, biology (incl. ecology and coastal marine ecosystems), geomorphology (physical geography), climate, littoral oceanography, coastal hydraulics, environmental (resource) management, engineering, and remote sensing. Policy, coastal law, and relevant issues such as conflict resolution and risk management would also be covered by the Series. The scope of the Series is broad and with a unique cross-disciplinary nature. The Series would tend to focus on topics that are of current interest and which carry some import as opposed to traditional titles that are esoteric and non-controversial. Monographs as well as contributed volumes are welcomed.

More information about this series at http://www.springer.com/series/8795

Andrew D. Short • Antonio Henrique da F. Klein
Editors

Brazilian Beach Systems

Springer

Editors
Andrew D. Short
School of Geosciences
University of Sydney
Sydney, NSW, Australia

Antonio Henrique da F. Klein
Departamento de Geociências, Laboratório de Oceanografia Costeira
Universidade Federal de Santa Catarina
Florianópolis, SC, Brazil

ISSN 2211-0577 ISSN 2211-0585 (electronic)
Coastal Research Library
ISBN 978-3-319-30392-5 ISBN 978-3-319-30394-9 (eBook)
DOI 10.1007/978-3-319-30394-9

Library of Congress Control Number: 2016944928

Printed on acid-free paper

This Springer imprint is published by Springer Nature
The registered company is Springer International Publishing AG Switzerland

To ***Professors Dieter Carl Ernst Heino Muehe and João José Bigarella (in Memoriam)***
for ongoing leadership in Brazilian beach research

Foreword

It's likely that no one on Earth has visited more beaches than Andrew Short. In Australia alone, he visited 11,670, publishing seven books. Andy studied at the Coastal Studies Institute at Louisiana State University (LSU) where studies of applied coastal morphodynamics in the beach environment began. Then in the 1970s to 1980s, he was part of the University of Sydney's Coastal Studies Unit, which demonstrated the co-evolution of morphology and hydrodynamics explaining more thoroughly and completely the behavior of the beach and surf zone. In this way, a series of basic and logical parameters were defined and adapted, to provide the key elements to a global classification system of sandy beaches. These parameters that define beach systems were well-received internationally due to their simplicity and efficiency in explaining the interaction of sand and waves in beach behavior. Surprisingly, Andy and his colleagues defined beach stages and their behavior through years of morphodynamic field observation by the naked eye, a decade before the application of video monitoring of surf zones complimented their findings. Equally notable, they applied their findings to improve beach safety, focusing on the risks and dangers of the beach environment.

In reality, Andy has frequented the coast of Brazil since 1975 when he was part of one of the first morphodynamic studies conducted in Brazil on beach systems of Sergipe and Pernambuco. I met Andy in a hotel bar in Chile during the Sandy Beach'94, where we made notes and diagrams on napkins over a few cups of pisco sour. It cost us a hangover, which was only cured after a dip in the cold waters of the Valdivia beaches. Sixteen years ago, I had the privilege of showing him the beaches of my home state from Cassino beach to Chuí, RS. On this trip, we climbed up two lighthouses, Albardão and Fronteira Aberta, to observe the beach systems: fortunately the latter of the two only collapsed two weeks after our visit.

Antonio Klein began his Brazilian beach excursions in the 1990s during his undergraduate studies at the Universidade Federal do Rio Grande (FURG). At this time, Klein and I were doing beach surveys of the southern coast with a level and rod; Klein, who was new to the marine environment, would only move seaward after sounding the sea floor with the rod. Even so, he decided to study Concheiros do Albardão, a beach unique to the southern littoral of Rio Grande do Sul for its deposits of seashell fragments and quartz sand. With this research, he completed his master's at the Universidade Federal do Rio Grande do Sul (UFRGS) in Marine Geology. He continued his work as professor at the Universidade do Vale do Itajaí (UNIVALI) in the state of Santa Catarina, where he researched the application of coastal morphodynamics in beach safety, eventually developing an award-winning project that reduced the number of swimming accidents on the Santa Catarina coast. Between 1999 and 2004 in Portugal during his PhD, he pioneered some of the first notable studies on the beaches of Santa Catarina, focusing on embayed beaches limited by rocky headlands, which constitute the majority of Santa Catarina's beaches. In 2010, he moved to Federal University of Santa Catarina (UFSC), and this book is a result of his first research project at this university.

It is not surprising that the collaboration of these two beach enthusiasts, Andy and Klein, would result in a book of this scope and importance. This book is an unprecedented approach to Brazilian beach systems from Amapá to Rio Grande do Sul. The book begins by locating Brazilian beaches in a global classification model according to the relative importance of their principal variables: tide range and wave energy, as well as presenting the evolution of Brazilian beach studies, including management, erosion, and beach safety. This initial focus is followed by the classification of the Brazilian coastal provinces by geological inheritance, geomorphology, hydrodynamic regime, and climate. They assembled researchers with different areas of expertise in coastal geology and geomorphology from the seventeen Brazilian coastal states to improve our present knowledge of Brazilian beach systems. The book concludes with a summary of all that is known about Brazilian beach systems and what still needs to be investigated to improve our knowledge of the system as a whole. It recommends directions for future research and is a valuable tool for those responsible for coastal management.

This book is a unique opportunity in that it presents the physical variability of Brazilian oceanic beaches in a logical and accessible form, particularly for those passionate about the study of beach systems and their connections to other areas of knowledge. Students and professionals in areas such as oceanography, geography, geology, coastal engineering, and coastal management will find this book a valuable resource in their development and understanding of the mechanisms that govern beaches, hopefully using this knowledge in real life application to benefit their communities. This work is essential in the library of all those that are fascinated by oceanic beaches.

Institute of Oceanography
Federal University of Rio Grande (FURG)
Rio Grande, RS, Brazil
November 15, 2015

Lauro Júlio Calliari

Preface

This book is a culmination of decades of fieldwork, research, and publications on the many beaches that line the magnificent coast of Brazil. This research commenced tentatively and sporadically in the 1960s and mushroomed in the 1990s, cumulating in 2000 with the First Brazilian Sandy Beaches Symposium, which contained 67 presentations by Brazilian coastal researchers.

Today coastal and beach research is underway in every one of the 17 coastal states, as evidenced by the contents of this book. The first editor was introduced to the Brazilian coast in 1975 and has returned multiple times to visit and work on the coast from Amapá in the north to Rio Grande do Sul in the south. The second editor introduced embayed beach morphodynamics and beach hazards and risk to Brazil and has supervised 18 graduate students, most with coastal-beach topics, many of whom have gone on to form the basis of the next generation of Brazilian coastal scientists and managers.

This book is about the beaches of Brazil. These beaches are both a vital and the major component of the Brazilian coast, and a source of endless fascination and recreation for the Brazilian people. All Brazilians know about their coast and beaches and most seem to want to vacation there in the summer months. This combination of people and coast has however resulted in some problems, ranging from a personal level with beach safety, to a national level with coastal development. In order to address these problems, one must begin with a good knowledge of the beaches and how they behave. This book addresses both these problems as well as documenting our present knowledge of the Brazilian coast and its beautiful, abundant, and wide-ranging beach systems.

This book contains 20 chapters written by 58 authors, who between them know all that is presently known about the Brazilian coast and in particular its beach systems. Seventeen of those chapters provide a state-by-state assessment of the beaches in each state, together with introductory, island beaches, and final a review and overview chapter.

If you are wondering why an Australian is editing a book on Brazilian beaches, it has to do with my 40-year association with the Brazilian coast and the assistance of a wonderful group of Brazilian coastal colleagues who have taken the time to show me, talk about, and discuss their beautiful coast and its beach systems. I also have the good fortune to see and visit much of the Brazilian coast, always with my Brazilian colleagues. I would particularly like to thank the following for taking their time to show me some of the following coasts:

Amapá – Valdenira Santos; Pará – Nils Asp and Luci Pereira; Ceará – Jader Onofre Morais; Rio Grande do Notre – Helenice Vital; Natal to Recife – Rodolfo Angulo; Fernando de Noronha – Lauro Calliari; Recife to Vitoria – Pedro Pereira and Lauro Calliari; Espírito Santo – Jacqueline Albino; Rio de Janerio – Dieter Muehe; São Paulo – Michel Mahiques; Paraná – Rodolfo Angulo; Santa Catarina – Antonio Klein; Rio Grande do Sul – Lauro Calliari, Sergio Dillenburg, and Elírio Toldo.

Sydney, NSW, Australia — Andrew D. Short
Florianópolis, SC, Brazil — Antonio Henrique da F. Klein

Acknowledgments

We first and foremost thank our 58 colleagues in Brazil and Spain for their contributions to this book, as well as the 19 reviewers listed below who assisted the editors in reviewing all the chapters. Without you all, this book would not exist.

This book could also not have been achieved without the support of CNPq – *Science Without Borders Program* which provided funding for a 3 years (2013–2015) Visiting Professor Grant at UFSC – Graduate Geographer Program and Graduate Oceanographer Program, project no. 400302/2012-8 (Coastal Hazards and Headland Sediment Bypassing Assessment in Different Coastal Environments), part of which resulted in the compilation and editing of this book. This work also had the financial support of the following programs: Programa de Formação em Recursos Humanos em Geologia da Petrobras (PFRH-PB 240); Fundo Clima – Ministério do Meio Ambiente, project MMA-Riscos; and SMC Project (MMA) and Bolsa de Produtividade em Pesquisa – Nível 2 [Proc. No. 303550/2012-0]).

The following are thanked for reviewing the chapters:

Javier Alcantara-Carrió, Edward J Anthony, Jarbas Bonnetti Filho, Lauro J Calliari, J Andrew G Cooper, Paolo Ciavola, João M Alveirinho Dias, Oscar Ferreira, Duncan FitzGerald, Mauricio Gonzales, Federico Isla, Guilherme Lessa, Dieter Muehe, Pedro Pereira, Klaus Schwarzer, Tim Scott, Karl Stattegger, Allan Williams, and Colin Woodroffe.

At UFSC, we thank José Mauricio de Camargo and Charline Dalinghaus for assisting with figures and tables.

At Springer, we especially thank Petra van Steenbergen who set us off on the path to this book, and Marielle Klijn who guided us through the preparation and submission and review of the manuscript and Mrs. Rathika Ramkumar who oversaw the final editing, proofing and production of the book. We also thank Charles W. Finkl, the Coastal Research Library Series Editor, for this ongoing support and enthusiasm for the project.

Contents

Editors and List of Contributors

Editors

Andrew D. Short School of Geosciences, University of Sydney, Sydney, NSW, Australia

Antonio Henrique da F. Klein Departamento de Geociências, Laboratório de Oceanografia Costeira, Universidade Federal de Santa Catarina, Florianópolis, SC, Brazil

Contributors

Jacqueline Albino Department of Oceanography and Ecology, Universidade Federal do Espírito Santo, Vitória, ES, Brazil

Javier Alcántara-Carrió Oceanographic Institute of the University of São Paulo (IOUSP), São Paulo, SP, Brazil

Rodolfo José Angulo Departamento de Geologia, Laboratório de Estudos Costeiros – Lecost, Universidade Federal do Paraná, Setor de Ciências da Terra, Curitiba, Paraná, Brazil

Tereza Cristina Medeiros de Araújo Geological Oceanography Laboratory, Department of Oceanography, Center of Geoscience and Technology, Federal University of Pernambuco, Recife, PE, Brazil

Elisabeth Mary de Carvalho Baptista Núcleo de Estudos sobre a Zona Costeira do Estado do Piauí – NEZCPI, Universidade Estadual do Piauí – UESPI, Curso de Licenciatura Plena em Geografia, Teresina, PI, Brazil

Abílio Carlos da Silva Pinto Bittencourt Institute of Geosciences, UFBA, BA, Brazil

Jarbas Bonetti Departamento de Geociências, Laboratório de Oceanografia Costeira, Universidade Federal de Santa Catarina, Florianópolis, SC, Brazil

Carlos Alberto Borzone Laboratório de Ecologia de Praias, Centro de Estudos do Mar, Universidade Federal do Paraná, Setor de Ciências da Terra, Pontal do Paraná, Paraná, Brazil

Lauro Júlio Calliari Institute of Oceanography, Federal University of Rio Grande, Rio Grande, RS, Brazil

Marcelo dos Santos Chaves Departamento de Geografia, Centro de Ciências Sociais e Aplicadas, Universidade Federal do Rio Grande do Norte, Natal, RN, Brazil

Nery Contti Neto Department of Oceanography and Ecology, Universidade Federal do Espírito Santo, Vitória, ES, Brazil

José Maria Landim Dominguez Institute of Geosciences, UFBA, BA, Brazil

Catherine Fitzpatrick Institute of Oceanography, Federal University of Rio Grande, Rio Grande, RS, Brazil

Maria Luzineide Gomes Núcleo de Estudos de Geografia Física – NEGEO, Universidade Estadual do Piauí – UESPI, Curso de Licenciatura Plena em Geografia, Teresina, PI, Brazil

Moab Praxedes Gomes Departamento de Geologia/Programa de pós-Graduação em Geodinâmica e Geofísica, Centro de Ciências Exatas e da Terra, Universidade Federal do Rio Grande do Norte-UFRN Campus Universitário, Natal, RN, Brazil

Mauricio González Rodríguez Instituto de Hidráulica Ambiental de la Universidad de Cantabria "IHCantabria", Universidad de Cantabria, Santander, Spain

Omar Quetzalcóatl Gutiérrez Instituto de Hidráulica Ambiental de la Universidad de Cantabria "IHCantabria", Universidad de Cantabria, Santander, Spain

Antonio Espejo Hermosa Instituto de Hidráulica Ambiental de la Universidad de Cantabria "IHCantabria", Universidad de Cantabria, Santander, Spain

Zuleide Maria Carvalho Lima Departamento de Geografia/Programa de pós-Graduação em Geografia/UFRN/, Centro de Ciências Sociais e Aplicadas, Universidade Federal do Rio Grande do Norte, Natal, RN, Brazil

Flavia Moraes Lins-de-Barros Departamento de Geografia/Programa de Pós Graduação em Geografia, Universidade Federal do Rio de Janeiro, Rio de Janeiro, Brazil

Verónica Cánovas Losada Instituto de Hidráulica Ambiental de la Universidad de Cantabria "IHCantabria", Universidad de Cantabria, Santander, Spain

Arthur A. Machado Institute of Oceanography, Federal University of Rio Grande, Rio Grande, RS, Brazil

Michel Michaelovitch de Mahiques Oceanographic Institute of the University of São Paulo (IOUSP), São Paulo, SP, Brazil

Luis Parente Maia LABOMAR-Instituto de Ciências do Mar, Universidade Federal do Ceará, Fortaleza, CE, Brazil

Valdir do Amaral Vaz Manso Marine Geophysical and Geology Laboratory, Department of Geology, Center of Geoscience and Technology, Federal University of Pernambuco, Recife, PE, Brazil

Cristina Celia Martins Fundação de Estudos e Pesquisas Aquáticas, Fundação de Estudos e Pesquisas Aquáticas, Sao Paulo, SP, Brazil

Amilcar Carvalho Mendes Museu Paraense Emílio Goeldi-MPEG, Campus de Pesquisa, Coordenação de ciências da Terra e Ecologia, Belém, PA, Brazil

Jáder Onofre de Morais Laboratório de Geologia e Geomorfologia Costeira e Oceânica-LGCO, Universidade Estadual do Ceará, Fortaleza, CE, Brazil

Dieter Muehe Graduate Program in Geography, Universidade Federal do Espirito Santo, Vitória, Brazil

Lucas do Nascimento Prefeitura Municipal de Camaçari, Camaçari, Bahia, Brazil

Wellington Nascimento Trindade Instituto de Estudos Costeiros, Universidade Federal do Pará, Bragança, Pará, Brazil

Silvana Moreira Neves Department of Geographical Sciences, UFPE Campus Universitário, Recife, Pernambuco, Brazil

João Luiz Nicolodi Instituto de Oceanografia (IO), Universidade Federal de Rio Grande (FURG), Rio Grande, RS, Brazil

Mauricio Almeida Noernberg Centro de Estudos do Mar, Laboratório de Oceanografia Costeira e Geoprocessamento, Universidade Federal do Paraná, Setor de Ciências da Terra, Paraná, Brazil

Tiago Castro Alves Oliveira Environmental Engineering Department, Universidade Federal do Espírito Santo, Vitória, ES, Brazil

Jorge Eduardo de Abreu Paula Núcleo de Estudos de Geografia Física – NEGEO, Universidade Estadual do Piauí – UESPI, Curso de Licenciatura Plena em Geografia, Teresina, PI, Brazil

Luci Cajueiro Carneiro Pereira Instituto de Estudos Costeiros, Universidade Federal do Pará, Bragança, Pará, Brazil

Pedro de Souza Pereira Geological Oceanography Laboratory, Center of Geoscience and Technology, Oceanography Department, Federal University of Pernambuco, Recife, PE, Brazil

Felipe M. Pimenta Departamento de Geociências, Centro de Filosofia e Ciências Humanas, Universidade Federal de Santa Catarina, Florianópolis, SC, Brazil

Francisco Pinheiro Lima-Filho Departamento de Geologia/Programa de pós-Graduação em Geodinâmica e Geofísica, Centro de Ciências Exatas e da Terra, Universidade Federal do Rio Grande do Norte-UFRN Campus Universitário, Natal, RN, Brazil

Lidriana de Souza Pinheiro Instituto de Ciências do Mar-LABOMAR- Federal University of Ceará, Fortaleza, CE, Brazil

Yuri G. Pinheiro Institute of Oceanography, Federal University of Rio Grande, Rio Grande, RS, Brazil

Clécio José Lopes de Quadros Departamento de Geologia, Laboratório de Estudos Costeiros – Lecost, Universidade Federal do Paraná, Setor de Ciências da Terra, Curitiba, Paraná, Brazil

Leonardo Cruz da Rosa Departamento de Engenharia de Pesca e Aquicultura, Laboratório de Ecologia Bentônica, Universidade Federal de Sergipe, Centro de Ciências Agrárias Aplicadas, São Cristóvão, Sergipe, Brazil

Adeylan Nascimento Santos School of Engineering and Information Technology, UNIFACS, Salvador, Bahia, Brazil

Valdenira Ferreira dos Santos Instituto de Pesquisas Científicas e Tecnológicas do Estado do Amapá-IEPA, Núcleo de Pesquisas Aquáticas – NuPAq, Macapá, AP, Brazil

Eduardo Siegle Oceanographic Institute of the University of São Paulo (IOUSP), São Paulo, SP, Brazil

Filipe Galiforni Silva Oceanographic Institute of the University of São Paulo (IOUSP), São Paulo, SP, Brazil

Iracely Rodrigues da Silva Instituto de Estudos Costeiros, Universidade Federal do Pará, Bragança, Pará, Brazil

Iracema Miranda da Silveira Museu Câmara Cascudo Universidade Federal do Rio Grande do Norte- UFRN, Natal, RN, Brazil

Fernando C. Sobral Institute of Oceanography, Federal University of Rio Grande, Rio Grande, RS, Brazil

Paulo Henrique Gomes de Oliveira Sousa Oceanographic Institute of the University of São Paulo (IOUSP), São Paulo, SP, Brazil

Flavo Elano Soares de Souza Centro de Tecnologia, Colégio Agrícola de Jundiaí, Universidade Federal do Rio Grande do Norte, UFRN, Macaiba, RN, Brazil

Maria Cristina de Souza Departamento de Geologia, Laboratório de Estudos Costeiros – Lecost, Universidade Federal do Paraná, Setor de Ciências da Terra, Curitiba, Paraná, Brazil

Werner Farkatt Tabosa Departamento de Geologia/Programa de pós-Graduação em Geodinâmica e Geofísica, Centro de Ciências Exatas e da Terra, Universidade Federal do Rio Grande do Norte-UFRN Campus Universitário, Natal, RN, Brazil

Elírio Ernestino Toldo Jr. Centro de Estudos de Geologia Costeira e Oceânica, Universidade Federal do Rio Grande do Sul, Porto Alegre, RS, Brazil

Ana Vila-Concejo School of Geosciences, University of Sydney, Sydney, NSW, Australia

Helenice Vital Departamento de Geologia/Programa de pós-Graduação em Geodinâmica e Geofísica, Centro de Ciências Exatas e da Terra, Universidade Federal do Rio Grande do Norte-UFRN Campus Universitário, Natal, RN, Brazil

Chapter 1
Brazilian Beach Systems: Introduction

Antonio Henrique da F. Klein and Andrew D. Short

Abstract Brazil possesses one of the great national coastlines of the world, extending for approximately 9000 km between latitudes 4°N and 34°S. The Amazon, the world largest river dominates the northern 1500 km. South of the Amazon, sandy beaches increasingly dominate the shore with more than 4000 beaches occupying much of the coast. This chapter provides an overview of the range of beach systems that occupy the Brazilian coast. This is followed by a review of previous research on the Brazilian coast together with management issues facing the coast. It then provides an updated classification of the entire coast, dividing it into seven coastal regions based on coastal processes, geology and geomorphology, that include from the northern Amazon Gulf mud coast; Northern tide-dominated; Northern tide-modified; Northeast wave-dominated; Eastern wave-dominated, Southeast wave-dominated; and Southern wave-dominated.

1.1 Introduction

Brazil possesses one of the great national coastlines of the world, extending for approximately 9000 km between latitudes 4°N and 34°S. The coast is a classic trailing edge coast typified by numerous long meandering rivers, generally low gradient regressive coastal plains, an abundance of sediment and extensive beach-barrier systems. The Amazon, the world largest river in terms of discharge and sediment supply, dominates the northern 1500 km, maintaining a predominately mangrove-fringed mud-dominated shore, with scattered sandy beaches. South of the Amazon, however, sandy beaches increasingly dominate the shore with more than 4000 beaches occupying much of the coast and comprising 2 % of all coastal ecosystems

A.H. da F. Klein (✉)
Departamento de Geociências, Laboratório de Oceanografia Costeira,
Universidade Federal de Santa Catarina,
Florianópolis, SC 88040-900, Brazil
e-mail: antonio.klein@ufsc.br

A.D. Short
School of Geosciences, University of Sydney, Sydney, NSW 2006, Australia
e-mail: andrew.short@sydney.edu.au

A.D. Short, A.H. da F. Klein (eds.), *Brazilian Beach Systems*, Coastal Research Library 17, DOI 10.1007/978-3-319-30394-9_1

(82,778 ha) (Muehe 2003). The remaining coast is occupied by rocky shore, inlets and in sheltered locations mangroves, as well as salt-marsh in the south.

This book is about the beach systems that dominate most of the Brazilian coast. They include one of the world's longest beach-barrier systems, the 610 km long strip of sand and dunes that extend from Torres, south along the entire coast of Rio Grande do Sul, to the border at Chuí. This system includes the 242 km long Hermenegildo-Cassino beach and 193 km long Tavares-Tramandaí beach, the longest beaches in South America and some of the longest in the world. There are many other long beaches associated with extensive river deltas and coastal plains; together with embayed and small pocket beaches bordered by headlands and inlets; as well as numerous beaches located in lee of beachrock reefs and in some areas fringing coral reefs.

The entire coast is bordered by the Atlantic Ocean and much is exposed to easterly trade winds and to east through southerly seas and swell, which combined provide considerable energy to transport sediment and construct a wide range of beach, barrier, inlet and deltaic forms. Wave energy ranges from low to moderate along the tide-dominated Amapá and Pará coasts, where considerable wave attenuation takes place across the shallow inshore, to moderate along the Maranhão, Ceará and northern Rio Grande do Norte coast, to moderate to high energy along the long southeast-facing east coast. The waves drive predominately northerly longshore sand transport, which has been calculated to reach 1 Mm^3 $year^{-1}$ in some locations, together with some local and seasonal reversals in sediment transport. Tides also vary considerably with the north coast dominated by macro to mega-tides reaching 11 m in the mouth of the Amazon, decreasing to macro and meso north and east of the mouth. The east coast has meso-tides in the north and along parts of the central coast grading to micro-tides towards the south, with the lowest tide range (~0.5 m) along the coast of Rio Grande do Sul. Tidal currents are significant along the north coast with the flood tides trending to the west reinforcing the easterly wind and wave driven currents and the strong North Brazil current. The easterly Trade winds dominate much of the northeast and north coast, while southeast winds dominate the east coast down to Santa Catarina Island, south of which there is a shift to northerly wind dominance. From Rio de Janeiro to the south storm surges up to 1.0 m high contribute to coastal processes.

The considerable range in wave and tide energy maintains the full range of beach types and states along the open coast. In the north the beaches are generally tide-dominated to tide-modified, while along the east coast they range from tide-modified in the northeast to wave-dominated along the central and southern sectors, with tide-dominated beaches predominating in sheltered bays and estuaries. The beaches span the tropical to subtropical latitudes and in northeast are modified by coral and beachrock reefs, which induce the formation of lower energy crenulate beaches in their lee. Also in the north and east the Barreiras Formation outcrops along sections of coast forming eroding cliffs, and in the south particularly between Carbo Frio and Cabo de Sta. Marta numerous bedrock headlands produce many embayed and pocket beach systems.

Brazil's numerous sandy beaches are synonymous with Brazil. Not just because they occupy the majority of the coast, but also because of the way Brazilian life and lifestyle have evolved and adapted to this seemingly endless stretch of tropical and subtropical sand beaches. Almost 20 % of Brazil's population lives in coastal counties (Muehe 2003), and most major cities, except São Paulo and Brasília, are coastal, it total more than 40 million Brazilian's live near the coast and its beaches. While the coast has long been the location of all ports and smaller fishing communities, since the 1970s there has been a surge towards the coast. This has resulted in a rapid expansion of existing towns and cities as well as the development of extensive housing and second-house subdivisions and in favorable locations the growth of tourist centers with highrise hotels and resorts. All of this is bringing more people to the coast both permanently and on vacation. The resulting pressure on the beaches and their backing barriers and dunes is often intense as dunes are leveled for development, tall structures crowd the shore and overshadow the afternoon beach, and effluent pollutes the beaches.

At the same time these same beaches will be the most susceptible parts of the coast to the impacts of climate change. They are literally caught between rapidly increasing human pressure and more subtle but equally intense changing sea level and climate. As a consequence they are presently and will continue to experience considerable human and natural impacts.

It is the *aim* of this book for the first time to both document and assess the nature, dynamics and state of Brazil's beach systems on a state-by-state basis. While the book will primarily focus on the physical characteristics of the beaches and their morphodynamics, it will also touch on the impact these systems have on beach users, through the hazards they present, as well as briefly review the nature and level of development along the coast.

The book is arranged into 20 *chapters*. This chapter introduces the topic and reviews the history of beach studies in Brazil, the way the coast is classified, the range of beach types along the coast, the type of beach hazards and the general impact of population pressure and development along the coast. The following 19 chapters (Chaps. 2, 3, 4, 5, 6, 7, 8, 9, 10, 11, 12, 13, 14, 15, 16, 17, 18, 19, and 20) begin with an overview of Brazilian coastal processes particularly the wave, tide and wind regimes (Chap. 2), followed by a state-by-state coverage of the beach systems, starting in the north at Amapá and extending south to Rio Grande do Sul including a chapter on Brazil's ocean island beaches (Chaps. 3, 4, 5, 6, 7, 8, 9, 10, 11, 12, 13, 14, 15, 16, 17, 18, and 19). It finishes (Chap. 20) with an overview of both the nature and status of Brazil's beaches, as well as addressing areas where more research is required.

1.2 Brazilian Beaches

This section briefly reviews the range of beach types and states that occur globally and along the Brazilian coast, and that will be presented in more detail in each of the chapters. It then looks at some the major issues facing Brazilian beaches including beach management, erosion and safety; this is followed by a classification of the Brazilian coast (Sect. 1.3) into seven coastal regions, each of which is then described (Sect. 1.4).

1.2.1 Beach Types and States

Brazilian beach systems can be classified into three types based on relative tide range (RTR) (Masselink and Short 1993), where

$$RTR = TR / H_b \tag{1.1}$$

where TR = mean spring tide range (m) and H_b = breaker wave height (m). This parameter quantifies the relative contribution of waves and tides. When waves are relatively high and tides low and RTR < 3 beaches are *wave-dominated*. Between 3 and ~10 they are *tide-modified*; and when waves are very low and tide relatively high and RTR is between ~10 and ~50 they become *tide-dominated* (Fig. 1.1; Short 2006).

The Brazilian coast has tides ranging from 0.5 to 11 m and low though high waves. It therefore contains RTR's ranging from <1 to >50 and the full range of wave-dominated, tide-modified and tide-dominated beach types. The three beach types can be further classified into 13 beach states using the dimensionless fall velocity (Ω) (Gourlay 1968), where

$$\Omega = H_b / W_s T \tag{1.2}$$

where W_s = sediment fall velocity (m s^{-1}) and T = wave period (s).

Ω quantifies the relative contribution of wave height and period and sediment grain size (expressed as sediment fall velocity) to beach morphodynamics. Using Ω *wave-dominated beaches* can be classified into six beach states (Fig. 1.1). When waves are relatively low, periods long and sand coarse Ω < 1, the beaches are narrow and barless and called *reflective*. When waves are moderate to high (Ω = 2–5) the beaches become rip-dominated *intermediate* with usually one or two bars cut by rip channels and currents. When wave are high and sand is fine Ω > 6 the beaches become wide and *dissipative* with often multiple (2–4) shore parallel sand bars. Figure 1.1 illustrates the six wave-dominated beach states.

Tide-modified beaches go though a similar transition with the addition of a usually wide low tide bar and consist of three beach states (Fig. 1.1). The lower energy reflective state (Ω < 1) consists of a reflective high tide beach plus a wide (~100 m+) low tide terrace. The intermediate state contains a reflective high tide beach and low tide bar cut by rip channels (Ω = 2–5) on its outer low tide sector; while the higher energy dissipative state features a very wide (>200 m) low gradient featureless concave ultradissipative beach when Ω > 6 (Fig. 1.1).

The *tide-dominated beaches* consist of four states each fronted by wide intertidal sand and/or mud flats (100's–1000's m wide). They range from a low energy high tide beach fronted by ridged sand flats under higher waves, through to very low energy sand flats, tidal sand flats and finally tidal mud flats (Fig. 1.1). For a full description of the beach types and states see Short (1999, 2006).

In addition two other beach states can occur along the coast, these are high tide reflective sandy beaches fronted by intertidal rocks flats or beachrock reefs, and high tide beaches fringed by intertidal coral reef flats (Fig. 1.1). Both are common along parts of the Brazilian coast, particularly where beachrock and coral reefs fringe the shore.

In total the three beach types and rock/reef flat beaches account for 15 different beach states ranging from the high energy wave-dominated multi-bar dissipative with surf zones 300–500 m wide, to barless reflective beaches; to with increasing tide range the tide-modified beaches with surf; to the very low energy tide-dominated beaches fronted by tidal flats. Table 1.1 list the beach types and states, their abbreviations and general relationship to RTR, Ω and H_b. Note that the actual relationship

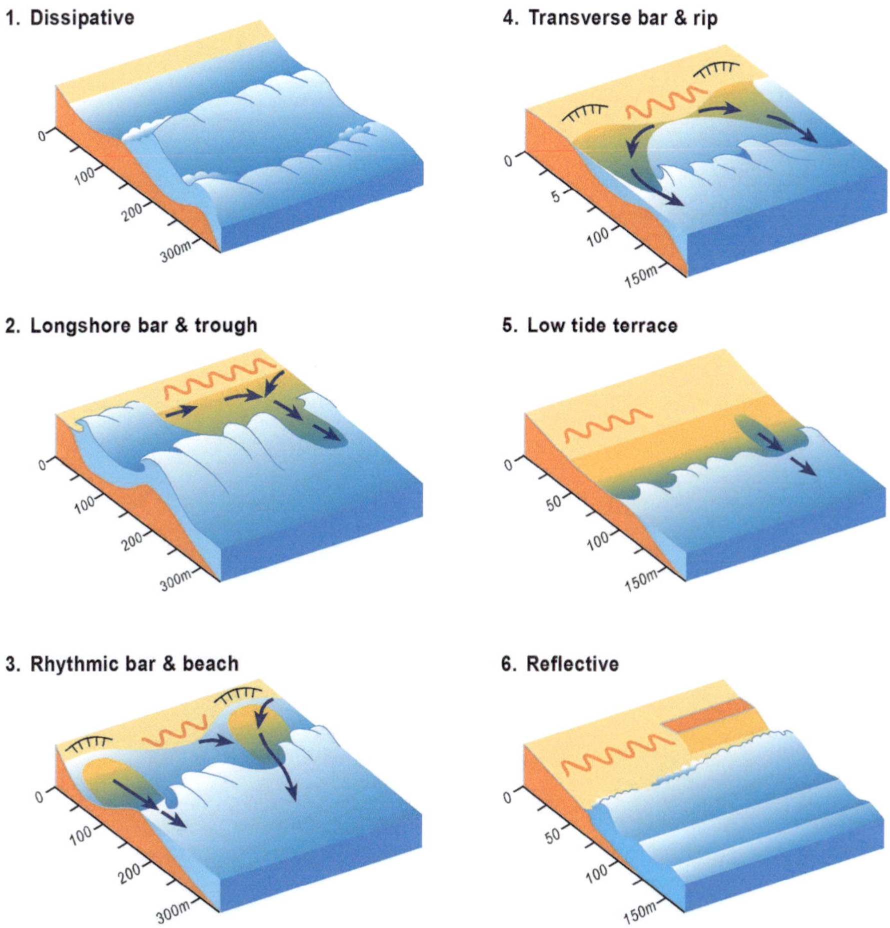

Fig. 1.1 Schematic sketch of wave-dominated (1–6), tide-modified (7–9) and tide-dominated beaches states (10–13); and beaches fronted by rock or reefs flats (14 and 15) (Source: Short and Woodroffe 2009)

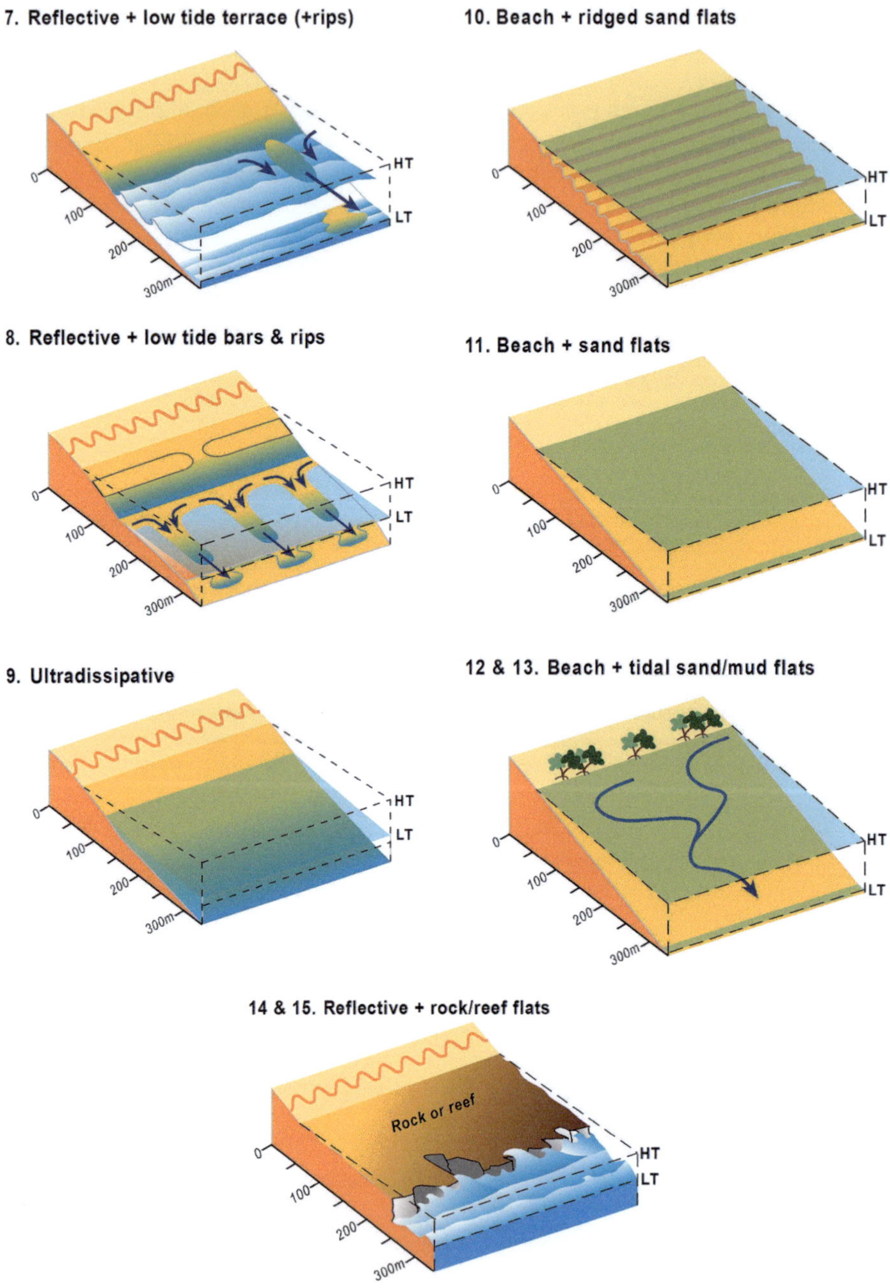

Fig. 1.1 (continued)

Table 1.1 List of the three beach types and 15 beach states and some of their environmental characteristics

No.	Abbreviation	Beach state	RTR	Ω	~H_b (m)
Wave dominated			<3	1–6	
1	D	Dissipative	<1	>6	>2
		Intermediate	<3		
2	LBT	Longshore bar & trough	<3	~5	<2
3	RBB	Rhythmic bar & beach	<3	~4	>1.5
4	TBR	Transverse bar & rip	<3	~3	~1.5
5	LTT	Low tide terrace	<3	~2	~1
6	R	Reflective	<3	~1	<1
Tide-modified			3 ~ 10	1–6	
7	R + LTT	Reflective + low tide terrace		~1	<1
8	R + LTR	Reflective + low tide bar & rips		~3	~1
9	UD	Ultradissipative		>5	~1
Tide-dominated			~10 ~ 50	<1	<<1
10	B[#] + RSR <0.5	Beach + ridged sand flats		<1	<0.5
11	B + SF	Beach + sand flats		<1	<0.3
12	B + TSF	Beach + tidal sand flats		<1	<0.2
13	B + TMF	Beach + tidal mud flats		<1	<0.2
*Rocks/reef**					
14	R + RF	Reflective + rock flats	–	–	–
15	R + CF	Reflective + coral reef flats	–	–	–

The RTR, Ω and H_b are all approximate and will vary between wave environments, while 14 and 15 are independent of waves and tides (Short 1999). Also see Fig. 1.1.
[#]Beach indicates a very low energy strip of high tide sand
[*]Rock and reef fronted beaches form independently of RTR, Ω and H_b

will vary with wave environments and need to be determined locally. The beaches fronted by rocks flats or coral reefs are independent of waves and tides.

The relationship between the wave-dominated, tide-modified and tide-dominated beach states and H_b, W_s and RTR is also presented in Fig. 1.2. Figure 1.2a plots the impact of increasing Ω and RTR on beach type and state while Fig. 1.2b plots the relation between beach state and H_b, sand size and RTR. It shows how wave-dominated beaches have the highest waves and lowest RTR with fine to medium sand, with a coarsening towards the reflective end. Tide-modified beaches have moderate waves, increasing RTR and medium more poorly sorted sand; while tide-dominated beaches have low waves, high RTR and the coarsest material, which is very poorly sorted. The figures are based on Australian data and made need modification in other coastal environments.

The forgoing applies to beaches in general and is largely based on Australian studies. However Brazil, like Australia, has tides ranging from micro to mega, and waves from low to high, together with beach sand ranging from fine to coarse. One would therefore expect all the above beach types to be found along the Brazilian coast, which is in fact the case, as will be presented in the following chapters.

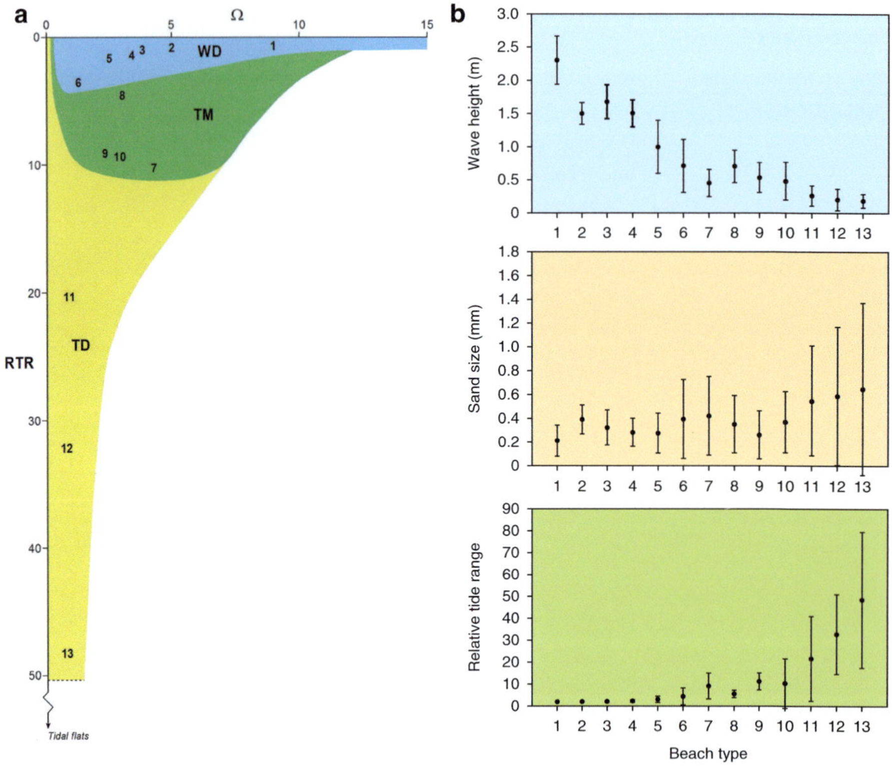

Fig. 1.2 Relationship between Ω and RTR in controlling beach type and state (*WD* wave-dominated, *TM* tide-modified, *TD* tide-dominated) (Source: Short and Jackson 2013); and (**b**) the relationship (mean and standard deviation) between the wave-dominated (1–6), tide-modified (7–9) and tide-dominated (10–13) beach states and H_b, sand size and RTR (Source: Short and Woodroffe 2009). All data are based on Australian beaches and may need modification for other coastal environments. See Table 1.1 for number legend

1.2.2 Brazilian Beach Studies

The scientific study of Brazilian beaches commenced indirectly with the publication by Darwin (1881) on the '*bar of sandstone*' (beachrock) at Recife. A century later these reefs were further studied and dated by Mabesoone (1964). During the 1960s studies of beach sediment were made in southern Brazil by Bigarella et al. (1966) and Martins and Eichler (1969). While the extensive reefs and beach sediments continue to warrant considerable study, the first publication examining the actual beaches were beach profile studies undertaken by Ottman et al. (1959) in Recife; and Kowsmann (1970) at Copacabana Beach; while Muehe (1979) presented more regional beach studies of the coast of Rio de Janeiro. The first beach morphodynamic investigations were undertaken by Suhayda et al. (1977). This team from

Louisiana State University's Coastal Studies Institute, in collaboration with PETROBAS, conduced beach experiments on a dissipative beach at Aracaju (SE), and measured the impact of beachrock reefs on wave refraction, attenuation and currents at Suape (PB). The second author (ADS) was a member of this team.

It was not until the 1990s however that beach studies became more widespread with Calliari and Klein (1993, 1995) and Toldo et al. (1993) examining beach morphodynamic states in Rio Grande do Sul state, followed by Hoefel and Klein (1998) conducting the first investigations into beach hazards and beach safety in Santa Catarina state. Since then there has been a considerable increase in beach research and publications throughout Brazil. An important reference is the Hofel's book – "*Morfodinâmica de praias arenosas oceânicas – uma revisão bibliográfica*" (Hoefel 1999). The author presented a literature review of beach morphodynamics based on the Australian coastal geomorphology school. At the end of book she presented a list of important papers published until 1995, and comments about the main results of these papers. She cited almost 50 references, with the number of publication decreasing from RJ, RS, PR, BA, PE, CE, RN, SC, PA, PI. She did not find citations for AM, AL, SE or ES.

The *First Brazilian Symposium on Sandy Beaches* (BSSB) in 2000 (Klein et al. 2003) contains 67 papers of which 21 dealt with beaches and beach change; 20 with beach ecology; 14 with coastal development and human impact on beaches; and five with dunes and barriers. A paper presented in the meeting by Muehe (2003) referenced 72 Brazilian beach papers published between 1993 and 2000. In relation to Brazilian marine geology research the proportion of publication related to beach profile and morphodynamic represents only ~2 % of a total of registered paper from 1941 to 1999 (Tessler and Mahiques 1996; Muehe 2003). Another important result of the meeting, was a document making suggestions for future studies on Brazilian Beach Systems (Finkl et al. 2003), including improving the knowledge about the coastal processes that shape the Brazilian beaches (waves, tides, bathymetry, wind, etc.), to permit a better characterization of process and/or erosive trends along the coastline. The same issues and challengers highlighted by BSSB were proposed earlier by Neves and Muehe (1995) and Hoefel (1999), and during the "*I Simpósio Nacional sobre Erosão Costeira*" (Recife-2008), and "*I and II Simpósio sobre Mudanças Climáticas da Zona Costeira*" (INCTCilma) held in Rio Grande (2010) and Salvador (2011).

In 2004, the *International Coastal Symposium* (Klein et al. 2006) was held at Itapema (SC). At the symposium 402 papers were presented, with half (207) devoted to the Brazilian coastal environments, with 38 papers on beach morphodynamic, beach changes, shoreline changes and beach management.

In the same year (2004) the Ministry of Environment, established the ORLA Project, aimed to optimize planning of coastal areas by integrating environmental, urban, and national heritage policies, and providing special attention to beaches under national jurisdiction (Oliveira and Nicolodi 2012). The coastal and marine management in Brazil was defined in specific legislation, such as 'Territorial Waters', 'Coastal Zone', among others. From 2004, Decree 5300, which regulated the Coastal Management Act in Brazil, established a new geographical area of land management: the seashore.

Table 1.2 Brazilian coastal states and their research groups that contributed to this book

Amapá	Instituto de Pesquisas Científicas e Tecnológicas do Estado do Amapá-IEPA
Pará	Grupo de Estudos Marinhos e Costeiros, Centro de Geociências, UFPa
Maranhão	Curso de Geografia, Departamento de Geociências, UFMa
Piauí	Núcleo de Estudos de Geografia Física, UESPI
Ceará	Instituto de Ciências do Mar, UFC; Laboratório de Geologia e Geomorfologia Costeira e Oceânica-LGCO, UEC
Rio Grande do Norte	Departamento de Geologia, UFRGN
Paraíba	Instituto de Geociências, UFBA??
Pernambuco	Laboratório de Geologia e Geofisica Martinha and Departamento de Oceanografia, UFPe
Alagoas	Departamento de Geografia e Meio Ambiente, UFAl
Sergipe	Laboratório de Ecologia Bentônica and Núcleo de Geologia, UFSE
Bahia	Instituto de Geociências, UFBA
Espírito Santo	Programa de Pós-Graduação em Geografia and Departamento de Oceanografia e Ecologia, UFES
Rio de Janeiro	Departamento de Geografia, Laboratório de Geografia Marinha, UFRJ
	Laboratório de Geologia Marinha (LAGEMAR – UFF)
	Instituto de Oceanografia – UERJ
São Paulo	Instituto de Oceanografia – USP
Paraná	Setor de Ciências da Terra, UFPr
Santa Catarina	Laboratório de Oceanografia Costeira, UFSC http://loc.ufsc.br
Rio Grande do Sul	CECO – Centro de Estudos Costeiros e Oceânicos UFRGS; Laboratório de Oceanografia Geológica, UFPG http://www.ufrgs.br/igeo/ceco/
	LOG – Laboratório de Oceanografia Geológica, FURG

In 2010 a Special Issue of Coastal Engineering addressing Headland Bay Beaches (Hsu and Klein 2010) presented 15 papers with five from Brazil. In this same year, the SMC Brazil project utilized 60 years of wave, tide and storm surge to hindcast Brazilian wave climates, as well as compiling a shoreline and bathymetry database (Oliveira 2012). In the 15 years since BSSB there has been ongoing and considerable growth in beach studies, particularly at universities and supported by government funding (e.g.: CNPq – National Science and Technology Institutes – INCT Clima), with coastal and beach research now underway in every coastal state (Table 1.2), and as indicated in the following chapters.

In 2006, Muehe (2006a) under the auspice of PGGM (Brazilian Geophysics and Geology Program) published "*Erosão e Progradação do Litoral Brasileiro*", followed by Dillenburg and Hesp (2009) book on the Quaternary coastal barrier systems along the Brazilian coast. These two book were the first to provide broad unified coverage of the coast. In the 2000 year the Brazilian Enviroment Ministerium edited a series titled the Brazilian Coastal Atlas ("*Atlas na escala da União, Vulnerabilidade costeira, Senssibilidade ao derramamento de oleo,…*"). This book is designed to build upon these solid foundations to provide the first coverage of the beach systems of every coastal state.

1.2.3 Issues Facing Brazilian Beaches

1.2.3.1 Coastal Management

There is considerable variation in the length, type, location, access and level of development of Brazil's many beaches. All are however public places where everyone has the right of access and use (MMA 2006). This right should always be guaranteed, making the beaches available for everyone on equal terms, as in the "Constitution". However, in Brazil there is in places an overlap of the administration of beach management, leading to possible conflicts and inefficient management, as well as private use of the public space (Scherer 2013). As a consequence there are many challenges in coastal and beach management in Brazil including: land ownership; regulating tourism and urban projects; regulating beach bars and restaurants; ensuring accessibility, biodiversity conservation and cultural maintenance; and erosion control (Santana 2003; Scherer 2013). Another major problem is that Brazilian legislation does not consider sand dunes as part of the beaches, leading to a fragmented management of beach ecosystems (Scherer 2013). Even tools designed for Brazilian beach and shoreline management, such as Projeto Orla – a governmental project (Oliveira and Nicolodi 2012), and Blue Flag Program – an instrument of non-governmental initiative (Scherer 2013), have been difficult to implement due to the lack of a solid institutional and administrative basis and lack of financial support (Scherer 2013). However there have been some good examples such as the development of ORLA projects during the last decade (Oliveira and Nicolodi 2012).

There is an urgent need to improve legislation in order to achieve efficient and democratic beach management. In general terms Brazil should guarantee free public access to all beaches (MMA 2006) and introduce more efficient standardized beach rules. It would also be beneficial to initiate a beach classification and certification scheme that includes water and sand quality (Scherer 2013). Regulations to control uses and activities on the Brazilian coastline is urgent, together with an acknowledgment that there are overlapping areas of responsibilities, which can lead to conflicts (Fig. 1.3, Table 1.3). The management of the beaches should be shared among multiple structures and users, applying a governance process (Table 1.4) (Scherer 2013).

1.2.3.2 Coastal Processes

Recent trends in **sea level** along the Brazilian coast were reviewed by Neves and Muehe (1995), Mesquita (2003) and Muehe (2006b). More recently Losada et al. (2013) examined sea level changes together with tides, storm surges and extreme events to derive a 'total sea level' trends and forecasting. In Brazil they found sea level was rising, storm surges were increasing towards the south, and El-Niño events were positively effecting sea level (Table 1.5).

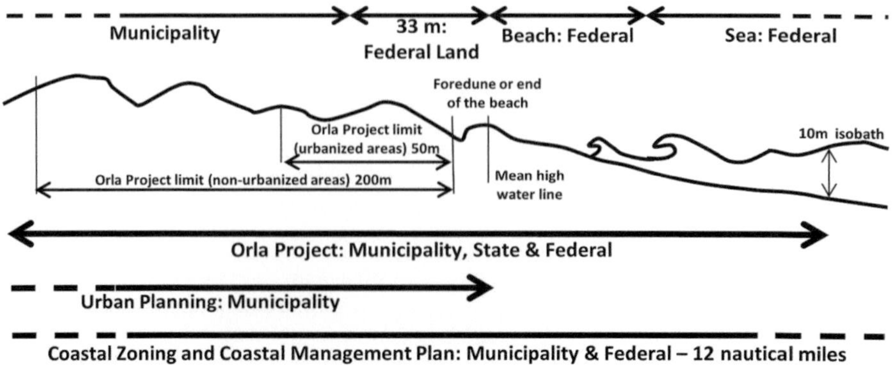

Fig. 1.3 Coastal profile, tools and public administration (Source: Scherer 2013). Based in Orla project (MMA 2006 p. 28)

Table 1.3 Beach regions and public management in Brazil (Scherer 2013)

Beach sector (from offshore to inland)	Beach use predominant	Management skill
Sea	Public (with permits required for private use)	Union (Federal government)
Beach	Public (with the occasional private use permits)	Union (central government), city hall (local government)
33 m/"*Terrenos de Marinha*"	Public (with allocation of public and private use under different regimes: use permission and concession of the right of resolvable use, lease, rental, sale, occupation, assignment and tenure)	Union (central government), city hall (local government)
Beyond 33 m Inland	Private (with public areas such as parks and streets)	City hall (local government)

Table 1.4 Administration levels and planning and management tools

	Administration level		
Planning instruments management	Federal government	State government	Local government
Master plan			X
Ecological Economic Zoning Coastal (ZEEC), Coastal Zone Management Plan (PGZC)		X	X
Orla project	X	X	X

Source: Scherer 2013

X = when the public administration focuses directly on the instrument

Table 1.5 Mean sea-level change along the Brazilian coast

Author	Place	Rate	Period
Pirazolli (1986)	Fortaleza (CE) and Belém (PA)	a sea-level fall	Tidal Gage Period 20 year (1950–1970)
	Recife (PE)	3.7 mm $year^{-1}$	
	Salvador (BA)	1.6 mm $year^{-1}$	
	Canavieiras (BA)	3.1 mm $year^{-1}$	
	Imbituba (SC)	0.55 mm $year^{-1}$	
Aubrey et al.(1988)	Fortaleza (CE)	0.3 mm $year^{-1}$	Tidal Gage Period 20 year (1950–1970)
	Belém (PA)	3.4 mm $year^{-1}$	
	Recife (PE) -	0.2 mm $year^{-1}$	
	Salvador (BA)	2.7 mm $year^{-1}$	
	Canavieiras (BA)	4.1 mm $year^{-1}$	
	Rio de Janeiro (RJ)	3.6 mm $year^{-1}$	
	Imbituba (SC)	0.7 mm $year^{-1}$	
Silva (1992)	Rio de Janeiro (RJ)	12.6 mm $year^{-1}$	Tidal Gage 1965–1986
Harari and Camargo (1994)	Recife (PE)	5.6 mm $year^{-1}$	Tidal Gage 1946–1988
França (2000), Mesquita (2003)	Equatorial Atlantic	4.0 mm $year^{-1}$	Altimetry
IHCantabria	Salvador (BA)	~2 mm $year^{-1}$	Hindcast/Forcasting
Losada et al. (2013)			Tide gauges (1950–2009)

Continuous records of **wave height** along the Brazilian coast are limited and discontinuous, with a network of wave gauges only initiated in the last few years. To overcome the lack of field data Pianca et al. (2010) used WWWIII hindcasting to present an analysis of the Brazil wave climate along the shelf-break. They suggested six wave regions (north, northeast, east, central, southeast and south). Melo Filho et al. (1995) reported that the northern region is also exposed to northern hemisphere swell. Neves Filho (1992) examined the increase in deviations between predicted and measured tides in Rio de Janeiro (Ilha Fiscal) and São Paulo (Cananéia) during the period from 1965 to 1986 and inferred an increase in storminess; while Reguero et al. (2013) found that since 1950 there has been a gradual increase in H_s and storm wave height and a positive shift in wave direction on the eastern Brazilian coast. However without coastal monitoring the impact of these changes is unknown.

Beach Erosion can be a result of sea level rise, changing wave climate and a deficient sediment budget as well being aggravated by uncontrolled urbanization. The book "*Erosão e Progradação do Litoral Brasileiro*" (Muehe 2006a) provides broad review of beach erosion in Brazil. Although widespread and in some segment severe it is not yet a serious threat considering the coast as a whole (Muehe 2006b). Major problems are most frequently associated with human intervention with the sediment flux or associated with the morphodynamic of river mouth (Muehe 2006b). But, regional vulnerability also occurs in areas with a permanent loss of sediments and where exposed to tectonic subsidence (Neves and Muehe 1995).

Critical Beach Erosion is occurring in Fortaleza and Recife-Olinda as well as several other places along the coast and in the vicinity of river mouths. Problems arise as houses are constructed too close to the beach or inlet. This problem has already been recognized and specific guidelines to leave a buffer zone between the beach and the front of the urbanized area has been recommended by governmental agencies. Nevertheless these guidelines have largely not been adopted. Also the already urbanized areas cannot be removed. While the coastline as a whole is not seriously endangered there are many areas of concern. The North and Northeast region which have very low gradient shoreface-inner shelf, such as the Amazon region, will experience substantial shore recession in response to sea level rise (Muehe 2001). Some Northeast areas are also losing sediments to the dune field resulting in a negative sediment budget. Along the Southeast and South coast beach barriers that prograded seaward during the mid Holocene during a positive sediment supply are now transgressing landward with wave overwash and localized erosion, owing to a negative sediment supply.

Under a rising sea level scenario beachrock outcrops which front long stretches of the Northeast coast will have their protection reduced as sea levels rises and wave height increases due to the increase in water depth. This will result in variable shoreline response as the shoreline is realigned to the changing waves. The sedimentary cliffs of the North, Northeast and part of the Southeast regions provide good protection as their recession, although increasing, will still be slow. However, Muehe (2006b) reported that the lack of long-term monitoring of shoreline mobility, wave climate and sea level makes difficult to distinguish between short-term erosion events and long-term recession trends.

1.2.3.3 Beach Safety

All beaches are inherently hazardous as they contain a range of permanent and variable features and processes that pose a risk to the beach-going public. Variable hazards include variable water depth, breaking waves and surf zone currents, the most hazardous of which are rip currents (Short and Hogan 1994; Short 1999). Permanent hazards include deepwater seaward of the beach, beachrock reefs, groynes, jetties and inlets. All these hazards are prominent along the Brazilian coast, with *rip currents* dominating most of the wave-dominated and some of the tide-modified beaches, together with numerous topographic rips associated with the many beachrock reefs and headlands. Along the lower wave energy, but higher tide range northern coast the variable water depth and numerous tidal inlets and their strong tidal currents pose considerable hazards.

The first Brazilian study to address the issue of beach hazards and public safety was Hoefel and Klein (1998) on Santa Catarina (SC) beaches. Klein et al. (2003) presented an analysis of 15,000 rescues and applied the model presented by Short and Hogan (1994) to the SC beaches. They founded that social factors such as the perception of hazard signs by beach users is the main issue. Calliari et al. (2010)

reviewed beach safety management studies in the south (RS, SC, PR) and northeast (CE). They found ~83–86 % of rescues occurred in rip currents associated with intermediate beach states.

Beach safety is a major public issue along the Brazilian coast and while lifeguards are provided at many beaches, there is much that can be done to improve public safety, as will be discussed in Chap. 20.

1.3 Brazilian Coastal Classification

The long length of the Brazilian coast together its wide range of climate, coastal processes and sediment availability requires the coast be divided into a series of regions and subsequently in provinces and/or sectors in order to understand the impact these factors have on the coast and its beach, dune and barriers systems.

There have been several descriptive classification of the coast (Table 1.6) with the first by Silveira (1964) followed by Zembruscki et al. (1972) and Schaffer-Novelli et al. (1990). The more recent Dominguez (2006, 2009) and Muehe (2010) classifications are based on the geological structure, as well as Quaternary sedimentation and contemporary coastal processes. While they are in general agreement there is some variation in the location of boundaries between Sergipe and São Paulo, and while Dominguez focuses more on geological influence, Muehe focuses more on barrier types. Both classifications are however very useful in understanding the regional variations along the coast.

This book is concerned with a finer scale of morphological features, that of beaches and their morphodynamics, which are in turn related to the larger scale geological control and Quaternary sedimentation. However at the finer scale wave climate, tide regime and sediment size are also all very influential, together with coastal geology (headlands, beachrock, reefs, islands and shelf gradient), accommodation space, orientation and more recently human intervention and impact.

Table 1.6 presents a summary all the above classifications, together with the one developed for this book, focusing on the variation in beach systems. This Table is designed to provide a national overview, with considerably more detail provided in each of the following chapters. The beach classification is based on RTR (Eq 1.1), which defines wave-dominated, tide-modified and tide-dominated beaches; together with the dominant barrier form/length. Based on these parameters the coast is divided into seven regions, which are discussed in the following section.

1.4 Brazilian Coastal Regions

As indicated in Table 1.6 there is a range of ways of classifying the Brazilian coast. At a national and geological scale the large northern Amazon and Parnaiba basins, and southern Parana basin, together with the smaller Potiguar and Reconcavo-Camamu basins are all dominated by near continuous sedimentary shores (Fig. 1.4a).

Table 1.6 Classifications of the Brazilian coast

The Brazilian coast division

State	**This chapter:** Klein & Short (2016)	Silveira (1964)	Schaeffer-Novelli et al. (1990)	Dominguez (2009)	Muehe (2010)
AM	**1. Amazon Delta mud coast** – Tide-dominated beach + sand/mud flats 1900 km	Amazon or equatorial region	Coast of Amapá	The tidal embayment of the Amazon	North coast dominated by tides and mangroves
PA	**2. Northern tide–dominated** barriers & estuaries – Tide-modified-tide-dominated (tide-dominated in estuaries) 1400 km		Gulf of Amazonas		
MA	**3. Northern tide–modified**, transgressive barriers – Tide-modified to wave-dominated (RN Tide-modified in lee of reefs & in bays) 1300 km		Reentrâncias Maranhenses		
PI		Northeast region or coastal barriers	East cost of Maranhão to the Calcanhar Cape (RN)	The sediment starved coast of Northeastern Brazil	Northeast coast with deficit sedimentary
CE	**4. Northeast wave–dominated** beaches & transgressive barriers – wave-dominated (tide-modified in lee of reefs & in bays) 800 km				
RN			Northeast coast from the Calcanhar Cape to the Recôncavo Baiano		
PB	**5. Eastern wave–dominated** regressive deltas & barriers – wave-dominated (RJ – tidal-modified in bays) 2000 km				
PE					
AL					Mixed coast with sedimentary cliffs and wave dominated deltas
SE				The dip-fed wave-dominat coast of Eastern Brazil	
BA		East coast	Recôncavo Baiano to the Frio Cape		
ES					
RJ	**6. Southeast wave–dominated** embayed barriers – Wave dominated (tide-modified in bays) 1700 km			The high-relief coast of Southeastern Brazil	Coast with lagoons associated to the double sandy barriers
SP		Southeast region or escarpment crystalline coast	Frio Cape to the torres		
PR				The strike-fed sandy coast of Rio Grande do Sul	Southeast coast dominated by rocky cliffs
SC	**7. Southern wave–dominated** transgressive barriers & beaches 750 km				
RS		Southern region, southern or subtropical coastline	Rio Grande do Sul Coast		Sandy coast of Rio Grande do Sul with multiple sandy barriers

Modified from PBMC 2014, pg 101-2

Name of States: *AM* Amapá, *PA* Pará, *MA* Maranhão, *PI* Piaú, *CE* Ceará, *RN* Rio Grande do Norte, *PB* Paraíba, *PE* Pernanmbuco, *AL* Alagoas, *SE* Sergípe, *BA* Bahia, *ES* Espírito Santo, *RJ* Rio de Janeiro, *SP* São Paulo, *PR* Paraná, *SC* Santa Catarina, *RS* Rio Grande do Sul

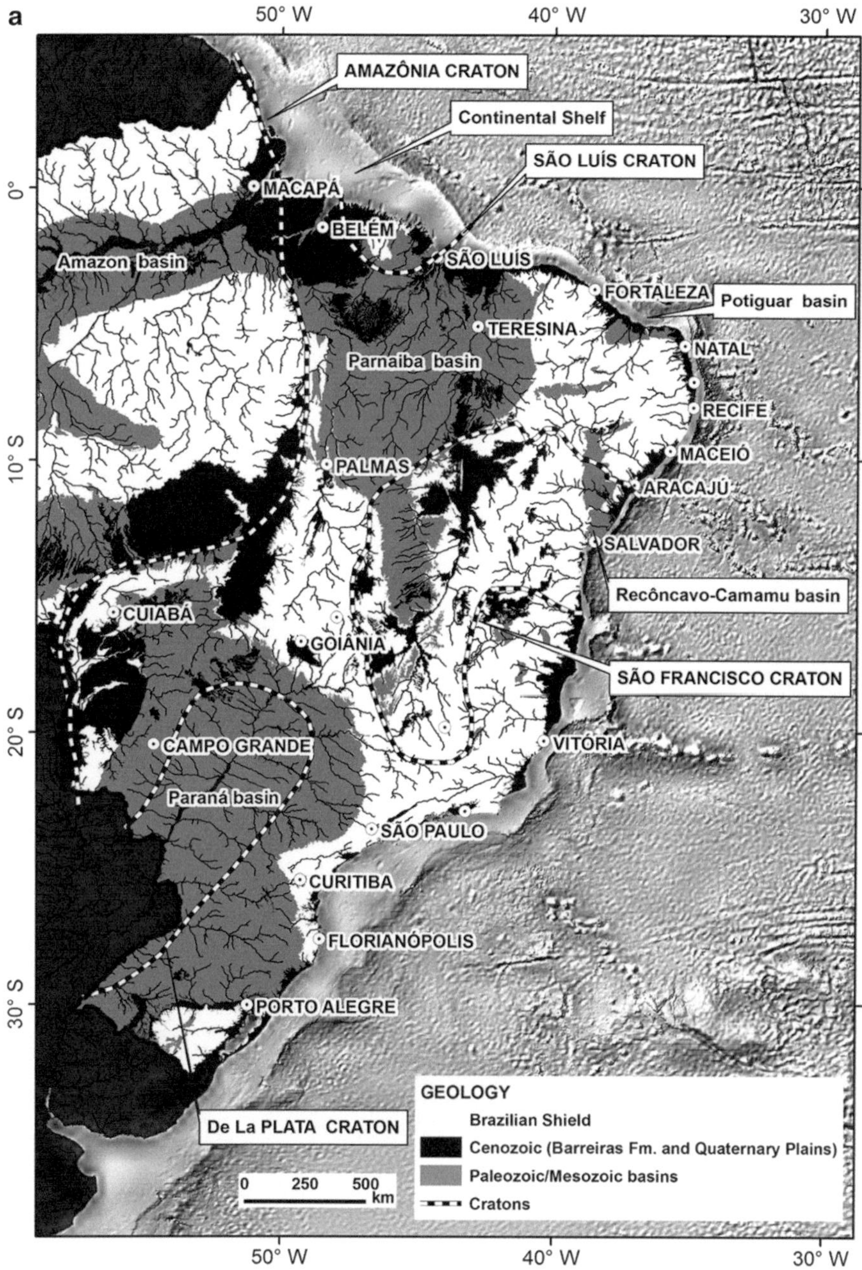

Fig. 1.4 (**a**) Simplified geology of Brazil with location of the state capitals (Source: Dominguez 2009); (**b**) Major processes acting on the Brazilian coastal zone (Source: Dominguez 2009)

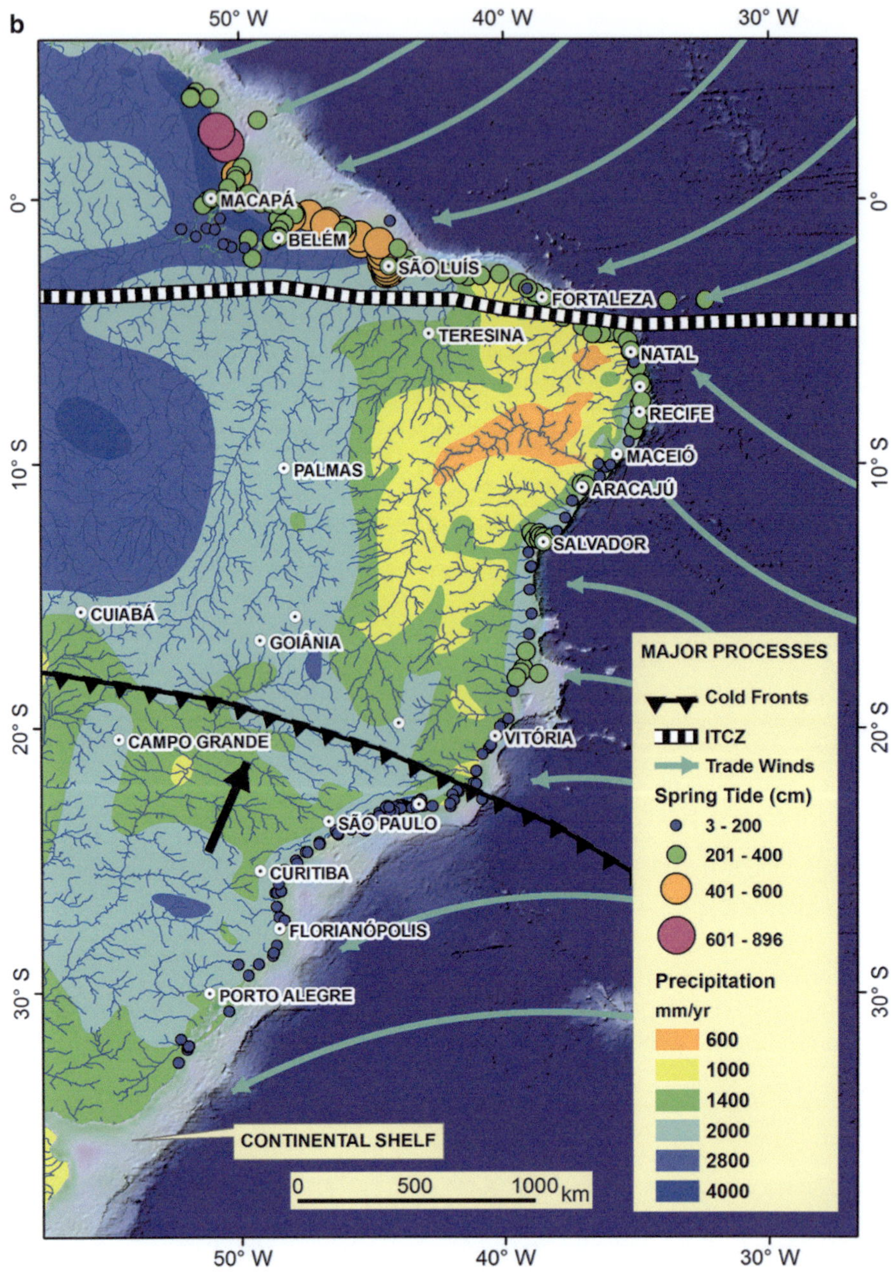

Fig. 1.4 (continued)

Elsewhere the coast is dominated by high-grade metamorphic rocks of the Brazilian Shield associated with the Sao Francisco, São Luis, Rio de la Plata and Amazonia cratons, which are most prominent along the southeast coast between Cabo Frio and Santa Marta. In places the basement rocks are covered by Tertiary sediments known as the Barreiras Formation, which forms the coastal tablelands of northeastern and eastern Brazil (Maranhão, Ceará, Rio Grande do Norte, Paraiba, Pernambuco, Alagoas, Sergipe, Bahia and Espírito Santo). In addition neo-tectonic and faulting along parts of the northeast coast has produced numerous coastal bays and offsets. The coastal geology therefore provides a first order division of the coast into sedimentary drainage basins and bedrock-influenced, the latter including metamorphic and the Tertiary sediments. The geological influence also extends to the nature and quality of sediment delivered to the coast.

Brazilian coastal sediments range from mud to sand and from abundant to deficient. While the numerous rivers between the Amazon and La Plata are the ultimate source of much of the sediment, most of this sediment was deposited on the shelf by these rivers at low sea level and reworked onshore during Quaternary sea level transgressions and highstands. Today the coastal sediments are derived from these shelf sources and from those rivers that deliver sediment directly to the coast, such as the Amazon (mud) and São Francisco (sand). In general the Amazon mud dominates the northern most coast, the Paraná-La Plata and Itajaí rivers the southern coast, and a number of rivers, including the large São Francisco, the central east coast. As a consequence much of the coast has an abundance of sediment apart from the more arid northeast coast, which with only a few small rivers is sediment deficient.

There is considerable variation in coastal processes between the tide-dominated north coast to a wave-dominated east and south coast (Fig. 1.4b). Tides range from meso-mega in the Amazon and along the north coast decreasing to meso along the northeast coast and micro along the southeast coast. In contrast waves are low to moderate in the north increasing to moderate in the northeast and moderate to high along the east coast. Winds are predominately trade wind easterlies, apart from the south where northerly winds dominate. Waves and wind combine to transport considerable sediment, with wave processes generating overall northerly sand transport, while the trade wind maintain generally north-trending transgressive dunes (Fig. 1.4b).

Based on the geology, sediments, coastal processes and beach types the coast has been divided in this chapter into seven coastal regions (Table 1.6, Fig. 1.5), regions, which largely follow those of Dominguez (2006, 2009). The coastal length refers to the open coast, apart from the Amazon where it includes parts of the mouth and its distributaries. The following sections briefly discuss the characteristics of each of the seven coastal regions. More detailed descriptions of the regions and their beaches are provided in the following chapters.

Fig. 1.5 Brazil's seven coastal regions and their boundaries

1.4.1 The Amazon Delta Mud Coast

The Amazon mud coast (Region 1) extends from the border at Cape Orange for 1900 km through the intricate Amazon mouth to Cape Maguari at the northeastern tip of Marajó Island (Fig. 1.6). This is a wide, low gradient sedimentary coast dominated by the massive delivery of mud to the coast and its predominately northerly transport as far as the Orinoco River. The tide reaches a maximum of 11 m in the Amazon mouth and along the northern shores of Marajó Island decreasing northward to 4 m at Cape Orange. Deepwater waves are generated by the easterly trade winds and are low to moderate, decreasing to low to zero along the very low gradient shore. The mud is transported to the northwest by the strong North Brazilian Current, which flows between 60 and 100 cm s^{-1}, assisted by the easterly trade wind, the easterly waves and the flooding tides of the Amazon (Anthony et al. 2010).

The coast and shoreline is dominated by the mud, which blankets the coast forming wide inter-sub-tidal mud flats, in places several kilometer wide, which grade into a very low gradient nearshore and 100–300 km wide continental shelf, all of which attenuates wave energy, which at the shore is low to zero. Landward the flats are vegetated by wide mangrove and várzea forests. Because of the absence of upland drainage systems the coast is reasonably continuous along the Amapá coast, interrupted only by small rivers and minor tidal creeks. The mud is transported westerly in the form of mud banks and migratory mud waves (Anthony et al. 2010), which have developed large mud capes such as capes Orange, Caciporé and do Norte and Maracá Island.

North of the Amazon mouth the shoreline has several long sandy beaches fronting by wide sand to mud tidal flats. The large distributary mud islands of the Amazon River dominate the river mouth and its distributary channels. Several tide-dominated sand beaches are located along the eastern shores of the islands, while small pocket beach are scattered along some of the river and distributary channels, particularly the North and South channels. The large Ilha de Marajó is part of the Pós Barreiras Formation, and occupies the southwestern area of the region. The island's northern shore faces into the Amazon's South channel and is exposed to the trade winds and generally low easterly waves which have developed a series of low beach ridges and westerly migrating recurved spits along the 200 km long northern coast all fronted by wide intertidal sand flats. This region terminates at Cape Maguari, the northeastern tip of the island, and part of a 10 km wide regressive beach ridge barrier system.

1.4.2 Tide-Dominated Barriers & Estuaries of Pará-Maranhão

The eastern updrift sector of the Amazon embayment (Region 2) extends for 1400 km from Cape Maguari down into and including both side of the 200 km long Marajó Bay across eastern Pará and western Maranhão as far as Baleia Island, on

Fig. 1.6 Coastal region 1 – the tide-dominated Amazon delta and mud coast

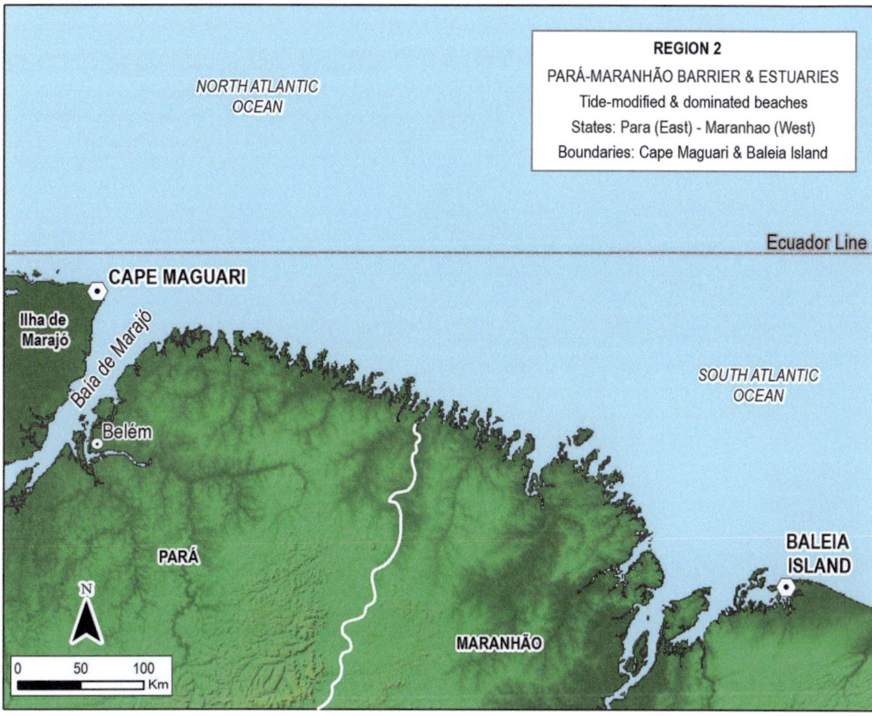

Fig. 1.7 Coastal region 2 – the highly indented tide-dominated Pará-Maranhão barriers and estuaries

the eastern side of São Marcos Bay (Fig. 1.7). Tide range peaks at 10 m at the mouth of Marajó Bay decreasing to 3 m at Belém and to 6 m in the east. Souza Filho (1995) has demonstrated that this coastal zone has experienced active progradation during the last 5000 years as a result of a combination of excess sediment supply, and a drop in sea level, in a tropical tide-dominated environment. The trade winds blow from the northeast in summer and southeast in winter generating low to moderate waves which are attenuated across wide intertidal sand and mud flats, with zero wave energy at low tide and only low waves reaching the beaches at high tide.

The east coast of Marajó Iisland extends for more than 200 km south of Cape Maguari into the Marajó Bay (Fig. 1.7), decreasing in wave energy as the bay narrows to the south. Numerous generally small low sandy beach ridges and tidal creeks shoals and spits fronted by sandy tidal flats occupy parts of both sides of the bay as far as Ponta de Pedras, the remainder being dominated by mangroves. The beaches are predominately tide-dominated, with low narrow sandy high tide beaches fronted by wide sand to mud tidal flats, with some bordered by bluffs of the Barreiras Formation. Extending east of Marajó Bay the coast is characterized by more than 30 funnel-shaped, tide-dominated muddy estuaries, that are several kilometer wide at their mouth and extends on average about 20 km inland. The drowned estuarine

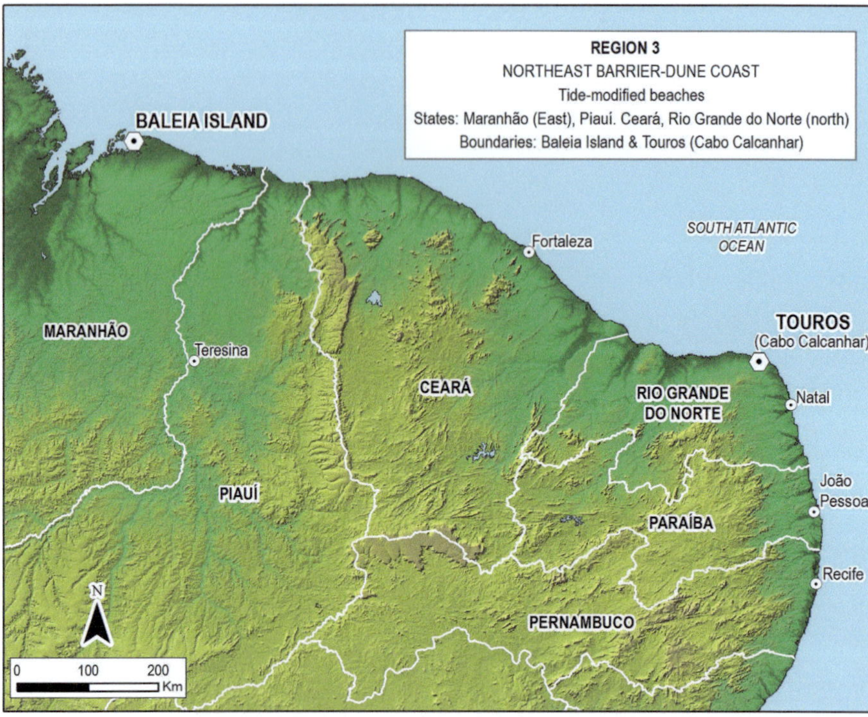

Fig. 1.8 Coastal region 3 – the northeast tide-modified barrier-dune coast

valleys were initially carved into the Barreiras Formation and possibly amplified and perpetuated by the action of the tides (Dominguez 2009), with the macro to mega-tides resulting in strong flows within the estuaries. Muddy shores and mangroves dominate the estuaries, with mangroves occupying 7500 km^2 of the coast (Souza Filho et al. 2009). The estuaries are separated by low peninsulas of the Barreiras Formation, which at their seaward end are capped by prograding generally short (<10 km) convex tide-dominated sandy '*drumstick*' beach ridge-barriers (Hayes 1979). In places the shoreline has prograded 30 km. The prograding barriers are separate by wide mangrove-filled swales. The waves, wind and tide generate a net westerly transport of sediment, which recurves into the wide inlets, which act as sediment sinks. Beaches are tide-dominated consisting of a low high tide beach fronted by intertidal sand flats up to a few hundred meter wide and more extensive tidal sand shoals near the inlets (Fig. 1.5c).

1.4.3 Northern Tide-Modified Barrier Coast

The east coast of Maranhão (east of São Luís), Piauí, Ceará and the northern coast of Rio Grande do Norte extends for 1300 km between Baleia Island and Touros (Fig. 1.8). In sharp contrast to the western Maranhão coast, it has a relatively smooth continuous coastline, interrupted by occasional rivers and headlands. This is a north to northeast facing shore exposed to the persistent easterly trades, which arrived from the northeast in summer with strong southeasterly winds in winter. The trades generate waves to 2 m and drive substantial westerly longshore sediment transport and numerous extensive west-trending transgressive dune fields, with both headland bypassing and overpassing prevalent. Tides range from 6 m in the west to 2.5 m in the east maintaining tide-modified beach systems which range from R+LTT to R+LT rips to UD (see Table 1.1), some with multi-bars, widening to tide-dominated tidal sand flats around creek and inlet mouths. The Barreiras Formation dominates the hinterland and outcrops along the coast as eroding sea cliffs, which together with beachrock and coral-algal reefs contribute hard substrates on this otherwise sandy coast. This region includes the Potiguar basin, where faulting has lead to the formation of two of the largest estuaries/bays of northeastern Brazil: The Açu and the Apodi.

1.4.4 North eastern Wave-Dominated Beachrock Coast

At Touros the northern Brazilian coast terminates and turns to face east for the remainder of the coast. The easterly orientation exposes it directly to the southeast trades, as well as higher southerly waves and meso-tides (2.5 m). The coast however has an arid interior (see Fig. 1.4b) and this aridity and small drainage basins result in a sediment deficient region that extends for 800 km from Touros to Coruripe (including Rio Grande do Norte (east), Paraiba, Pernambuco and Alagoas) (Fig. 1.9).

Fig. 1.9 Coastal region 4 – the northeast wave-dominated beachrock coast

The coast is dominated by northerly sediment transport including headland bypassing and overpassing, tide-modified beaches, northerly dune transgression usually set in embayed beaches, with headlands consisting of either bedrock and/or beachrock reefs, with the reefs becoming more prevalent to the south. The Barreiras Formation dominates the hinterland and parts of the coast

While the coast is relatively straight and uniform in its easterly orientation, the beach systems vary considerable along the coast depending on wave exposure and sand size. This variation is a product of the numerous headlands of the Barreiras Formation; the occurrence of shore parallel beachrock and in places coral reefs, both exposed and submerged; with the beachrock becoming near continuous for the southern 400 km between Baía da Traição and Coruripe; numerous creeks and inlets; and variation in sediment size. As a consequence the beaches range from long straight fine-grained wave-dominated fully dissipative multi-bar systems, though rip-dominated intermediate with both wave-dominated and tide-modified rip systems, and in lee of the reefs short to crenulate reef-sheltered tide-modified beaches.

1.4.5 Eastern Wave-Dominated Deltaic Coast

The São Francisco delta marks the northern boundary of this 2000 km long region, which extends south to Cape Frio and includes Sergipe, Bahia, Espírito Santo and northern Rio de Janeiro states (Fig. 1.10). This region is dominated by substantial fluvial input and beachrock, all exposed to moderate to high waves and meso-tides. The Barreiras Formation dominates the hinterland but only occasionally outcrops at the coast forming both active and inactive cliffs. The humid climate however maintains several large rivers with the São Francisco, Jequitinhonha, Doce and Paraíba do Sul delivering a sediment load of 52×10^6 t year^{-1} (Syvitski et al. 2005). At the coast these form protruding wave-dominated regressive barrier-deltas, which supply sand for generally northerly sand transport, with some southerly transport on the southern side of the deltas. Dune transgression is limited to the São Francisco delta. Between the rivers the Barreiras Formation and some Cretaceous and Precambrian rocks reach the coast as sea cliffs.

Tides are meso (2.5 m) along most of the coast and the moderate to high deepwater waves are moderated by the wide shallow shelf and locally by beachrock reefs resulting in beach systems which range from longer wave-dominated dissipative beaches on exposed sections with fine sand, to rip-dominated intermediate (both wave-dominated and tide-modified), to reflective in sheltered areas, with smaller embayed beaches in the sections dominated by bedrock.

Within this region the Camamu-Recôncavo rifted basins had resulted in faulted blocks and basins, which have formed the large Todos os Santos and Camamu bays. Lesser faulting also influences the coast to the north forming offsets and headlands.

Fig. 1.10 Coastal region 5 – the eastern wave-dominated deltaic coast

1.4.6 Southeast Wave-Dominated Rocky-Embayed Coast

This southeast region extends for 1700 km from Cape Frio to Cape Santa Marta and includes southern Rio Janeiro, São Paulo, Paraná and much of Santa Catarina state (Fig. 1.11). The horst and garbens of the Serra da Mantiqueira and Serra do Mar mountain ranges dominate much of the coast of Rio de Janeiro and São Paulo resulting high relief and numerous bays, embayments and islands. The larger Guanabara and Paranaguá bays are flooded rifts. In Santa Catarina pre-Cenozoic basement granites outcrop along the center of the state including Santa Catarina Island. The entire region is typified by rocky sections separating embayed beaches, inlets and estuaries. Barriers range from regressive in protected embayments to transgressive with extensive dunes in exposed embayments, with northerly sand transport including headland bypassing and overpassing, particularly in Santa Catarina.

At Cape Frio the coast turns and faces south to southeast exposing it to the dominant southerly waves, while the tide decreases down the coast from 2.5 m to 0.5 m, resulting in both tide-modified and increasingly wave-dominated beaches. These processes combined with numerous rocky-embayed sections, occasional rock reefs

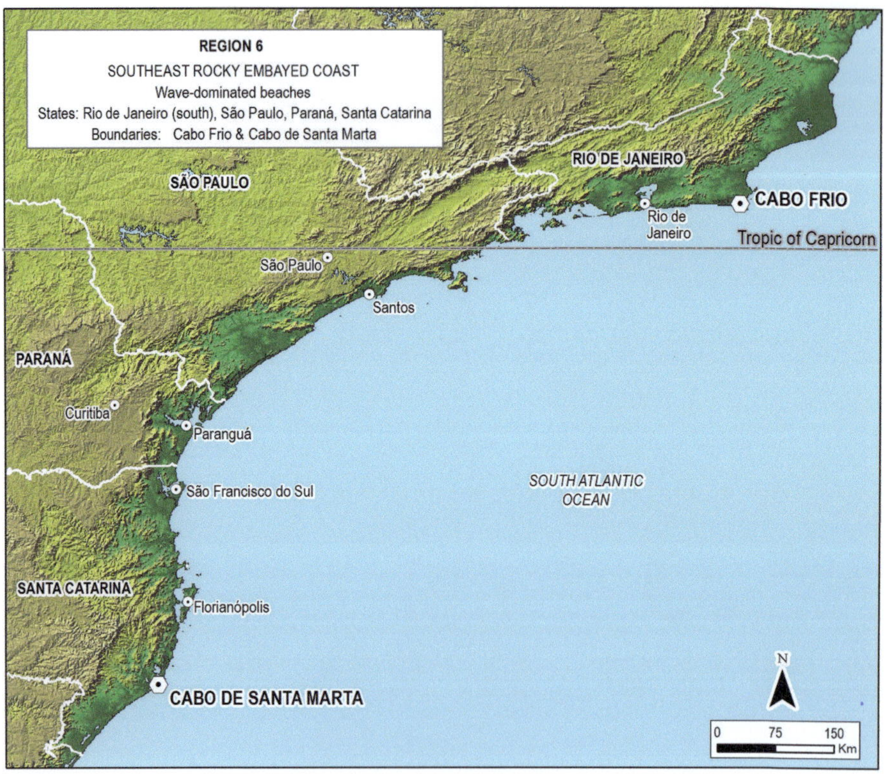

Fig. 1.11 Coastal region 6 – the southeast wave-dominated rocky embayed coast

and beachrock have produced a full range of wave-dominated and tide-modified beach systems, ranging from fully dissipative to reflective, including lower energy tide-modified, and even tide-dominated cheniers in the very sheltered Tijucas embayment.

1.4.7 Southern Wave-Dominated Barrier Coast

The 750 km long coast from Cape Santa Marta to the Uruguayan border at Chuí consists of a near continuous high energy dissipative beach and backing barrier dominated by south-trending transgressive dunes, which in turn are backed by large freshwater lakes and lagoons and three early Quaternary barrier systems, the Quaternary coastal plain having an area of 33,000 km^2 (Fig. 1.12). This is a high-energy wave-dominated coast, which includes the southern part of Santa Catarina and all of Rio Grande do Sul. Tides are micro (<1 m) and only two major inlets the

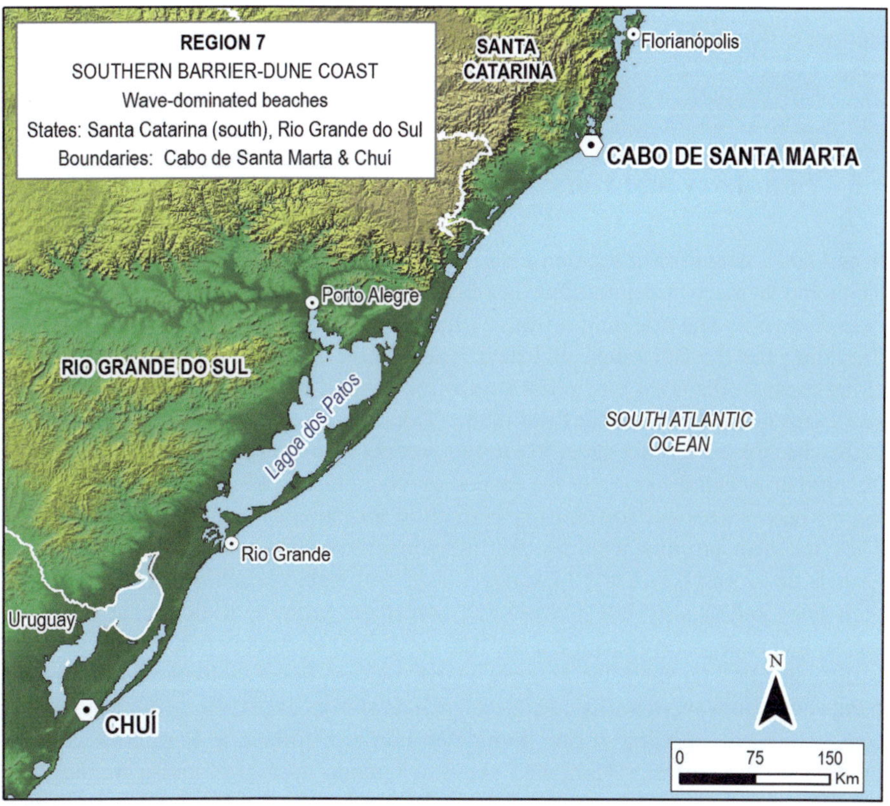

Fig. 1.12 Coastal region 7 – the southern wave-dominated barrier coast

Rio Grande and Tramandaí reach the sea. The long barrier system forms two broad arches separated by the Rio Grande Inlet and tends to diverge and increase in width northwards, following the dominant sediment transport.

Sediment supply came probably from de La Plata River located to the south with additional contribution from local rivers (Jaguarão, Camaquã, Jacuí, and Tubarão) at low sea level. The La Plata is the second largest river system in South America (sediment load: 91×10^6t year^{-1}) (Syvitski et al. 2005), and has delivered large volumes of sediments to the coastal zone during the Quaternary. These sediments were predominantly dispersed northwards during lower sea levels (Milliman 1975; Rocha et al. 1975; Urien et al. 1978) and later moved onshore by rising sea level and wave action. Dillenburg and Hesp (2009) have shown that the Holocene barrier has migrated landward from an offshore position during Early to Middle Holocene (10–6 ka).

The wave-dominated beach system is essentially fully dissipative along its entire length, the only variation being the sand size, which ranged from 0.125 mm to 1.4 mm and consequent number of bars. The coarser sand (0.35–1.4 mm) maintains two shore parallel bars, while areas of finer sand (0.125–0.2 mm) result in three to occasionally four bars. The surf zone average 300–400 m in width increasing to up to a kilometer during high waves. The 242 km long stretch of beach between Hermenegildo and Cassino is the longest beach in South America and one of the longest beaches in the world (https:geoprospectiva.worldpress.com/).

1.5 Summary and Conclusions

Brazil has a magnificent 9000 m long coast that extends from the tropical Amazon River and its macro-tidal muddy coast to the sub-tropical long sandy beaches of Rio Grande do Sul. The tide ranges from 11 m to <1 m and waves from <<1 m to 1.5 m providing for the full range of beach types. Wave-dominated beaches are found along most of the open east coast south from Natal, where exposure to southerly swell and micro to meso-tides dominate. Tide-modified beaches occur in areas of increasing tide range and generally lower waves (<1 m). They are also found along the northern-central section of the east coast where meso-tides occur and in particularly where waves are lowered by refraction and attenuation over inshore reefs. They are also prominent along the meso to macro-tidal coasts of northern Rio Grande do Norte, Ceará and Maranhão. Tide-dominated beaches occur in areas of high tides and/or very low waves. They are most prominent along the macro to mega-tidal coasts of Amapá, Pará and Maranhão, but also occur wherever waves are sufficiently low for the RTR to exceed ~10, and can occur as in sheltered embayments even on wave-dominated coasts, such as at Tijucas, SC. In general terms the coast can be divided into a tide-dominated northern Amapá to Pará coast, a tide-modified coast between Maranhão and Rio Grande do Norte, and a mixed tide-modified and wave-dominated east coast, becoming more wave-dominated as tide range decreases to micro south from Espirito Santo.

More than 40 million Brazilian live in the coastal zone, a number that is increasing, as more Brazilians wish to live, work and vacation at the coast. The response to this increase over the past 40 years has been a substantial increase in coastal population and development along beachfronts and in coastal dune fields. Coastal development is continuing seemingly unabated and in places uncontrolled or planned.

All this is taking place on or in lee of the full range of dynamic beach types and states, on a coast in places experiencing massive longshore sand transport, massive dune transgression, headland sand bypassing and overpassing, together with the natural oscillation and rotation in beach behavior. In other words a very dynamic coast and shoreline. Unfortunately most development takes no notice of the type of beach system, its sediment budget, its natural dynamics and hazard zone, its ecosystems nor the impact of the development on the beach system. As a consequence there are hundreds of sites where poorly sites structures have been undermined and eroded by the sea, in place leading to poorly designed beach armoring, thereby replacing the sandy beach with large rocks and further degrading the natural system.

All of this is also occurring at a time of changing climate, which is already leading to sea level rise, and is also expected to lead to changes in wave climate, tidal range and river discharge. These natural changes will place more pressure on the beaches leading to accelerated shoreline erosion, and where the beaches are developed to either destruction of beachfront property and/or massive armoring of the shoreline, all at great cost to the community and the State.

This book sets out to describe the beach systems of Brazil from Amapá to Rio Grande do Sul, their nature, behavior, stability and issues. It is doing this in the hope of raising understanding and awareness of both the nature and plight of Brazil's 4000 beach systems. They are a major national natural resource, which can be exploited to considerable economic and social benefit to the community and the nation. However in order to optimize this benefit the beaches themselves need to be taken into account, so that in the process of development they are not smothered and degraded. Beaches need room to be beaches to respond to storms and calms, to move sand alongshore, to allow sand to blow into the dunes, to maintain their rich ecosystems. To date most development on beaches is ignorant of or has ignored this need, with the consequent degradation of the beach and often associated coastal dune systems. As developed beaches are degraded so too will the potential benefit from these beaches be degraded, as money is spent and wasted fixing problems (pollution, seawalls, congestion), and ultimately holidaymakers seek more suitable locations for vacation leaving abandoned or degraded beach resorts. The increasing beach population, particularly during vacations, combined with the inherently hazardous beach systems (rip currents, waves, reef, etc.), is also leading to an increase in the number of beach incidents (rescues, first aid, etc.) and fatalities by drowning. This also must be addressed in order to maintain public beach safety.

Attractive 'natural' beaches and development can go hand in hand, but only if the development is planned, controlled and sympathetic with the needs of the entire beach system. The following chapters present the beaches of each state, their nature, dynamics, stability and level of existing development, together with problems facing the beaches. It is hoped that this book will highlight the nature and needs of Brazil's beach systems, and that these needs be taken into account as the coast continues to develop.

References

Anthony JE, Gardel A, Gratiot N, Proisy C, Allison MA (2010) The Amazon-influenced muddy coast of South America: a review of mud-bank–shoreline interactions. Earth Sci Rev 103:99–121. doi:10.1016/j.earscirev.2010.09.008

Aubrey DG, Emery KO, Uchupi E (1988) Changing coastal levels of South America and the Caribbean region from tide-gauge records. Tecconophys 154:269–284

Bigarella JJ, Freire SS, Salamuni R, Viana R (1966) Contribuição ao estudo dos sedimentos praias recentes, II Praias de Matinhos e Caiobá, Bol Universidade Federal Paraná Geog Física 6., pp 1–109

Calliari LJ, Klein AHF (1993) Características morfodinâmicas e sedimentológicas das praias oceânicas entre Rio Grande e Chuí, RS. Pesquisas 20(1):48–56

Calliari LJ, Klein AHF (1995) Variabilidade das praias oceanicas entre Cassino e Chui, RS. Pesquisas 11:55–63

Calliari LJ, Klein AHF, Albuquerque MG, Morcellin O (2010) Beach safety management in Brazil. In: Leatherman S, Fleremeyer J (eds) Rip currents. CRC Press, Boca Raton, pp 215–223

Darwin C (1881) On the remarkable bar of sandstone of Pernambuco on the coast of Brazil. Phil Mag J Sci, London 19:257–261

Dillenburg SR, Hesp PA (2009) Geology and geomorphology of Holocene coastal barriers of Brazil. Springer, Berlin, p 380

Dominguez JML (2006) The coastal zone of Brazil: an overview. J Coast Res SI39:16–20

Dominguez JML (2009) The coastal zone of Brazil. In: Dillenburg SR, Hesp PA (eds) Geology and geomorphology of holocene coastal barriers of Brazil. Springer, Berlin, pp 17–51

Finkl CW, Klein AHF, Calliari LJ (2003) Reports of meetings: Brazilian sandy beaches symposium: morphodynamics, ecology, uses, hazards and management. J Coast Res 16(4):1188–1191

França CAS (2000) Contribuição ao Estudo da Variabilidade do Nível do Mar na Região Tropical Atlântica por Altimetria por Satélite TOPEX/POSEIDON e Modelagem Numérica. PhD thesis. São Paulo, p 274

Gourlay MR (1968) Beach and dune erosion tests. Delft Hydraulics Laboratory, Report No.M935/M936

Harari J, Camargo R (1994) Tides and mean sea level in Recife (PE): 8o 3.3'S 34o 51.9'W: 1946 to 1988. Boletim do Instituto Oceanográfico

Hayes MO (1979) Barrier Island. Morphology as a functionof tide and wave regime. In: Leatherman SP (ed) Barrier islands from the Gulf of St Lawrence to the Gulf of Mexico. Academic, New York, pp 223–304

Hoefel FG (1999) Morfodinamica de Praias Arenosas Oceanicas: Uma Revisao Bibliografica, vol 1, 1st edn. Editora da Univali, Itajai, p 92

Hoefel FG, Klein AHF (1998) Environmental and social decision factors of beach safety in the central northern coast of Santa catarina, Brazil, vol 2. Notas Técnicas da FACIMAR, Itajaí, pp 155–166

Hsu JC, Klein AHF (eds) (2010) Hydrodynamics and applications of headland-bay beaches. Coast Eng, 57 (2): 75–240

Klein AHF, Short AD (2016) Brazilian beach systems. In Short AD, Klein AHF (eds), Brazilian beach systems, Coastal research library. Springer, Dordrecht

Klein AHF, Santana GG, Diehl FL, Menezes JT (2003) Analysis of hazards associated with sea bathing: results of five years work in oceanic beaches of Satna Catarina State, Southern Brazil. J Coast Res SI 35:107–116

Klein AHF, Menezes JT, Diehl FL, Abreu JGN, Polette M, Sperb RM, Sperb RC (2006) Santa Catarina. In: MMA (ed) Erosão e progradação do litoral brasileiro. Ministério do Meio Ambiente, Brasília, pp 401–436

Kowsmann RO (1970) Variações de curto e longo prazo de um perfil de praia em Copacabana, Rio de Janeiro. Publicação do Instituto de Pesquisas da Marinha 39:1–5

Losada IJ, Reguero BG, Méndez FJ, Castanedo S, Abascal AJ, Mínguez R (2013) Long-term changes in sea-level components in Latin America and the Caribbean. Glob Planet Chang 104:34–50

Mabesoone JM (1964) Origin and age of sandstone reefs of Pernambuco (northeastern Brazil). J Sediment Pet 34:715–726

Martins LR, Eichler BB (1969) Propriedades texturais dos sedimentos litorâneos de Santa Catarina. 11: areias de praia e dunas, trecho Laguna-Imbituba. Notas e Estudos. Escola de Geologia, Universidade Federal do Rio Grande do Sul 2(1):27–44

Masselink G, Short AD (1993) The effect of tide range on beach morphodynamics and morphology: a conceptual model. J Coast Res 9:785–800

Melo Filho E, Alves JHGM, Jorden V, Zago F, Dias WSS (1995) Instrumental confirmation of the arrival of North Atlantic swell to the Ceará coast. In: International Conference on Port Coastal Engineering in Developing Countries (COPEDEC IV), 4, 1995, Rio de Janeiro. Proc, Rio de Janeiro, RJ 3:1984–1996

Mesquita AR (2003) Sea level variations along the Brazilian coast: a short review. J Coast Res SI35:21–31

Milliman JD (1975) A synthesis. Contributions to sedimentology. Upper continental margin sedimentation of Brazil. Stuttgart 4:151–175

MMA (2006) Projeto Orla: fundamentos para a gestão integrada. 74p., Ministério do Meio Ambiente (MMA)/Ministério do Planejamento, Orçamento e Gestão, Braslília, DF, Brasil. ISBN: 8577380297. http://www.mma.gov.br/estruturas/orla/_arquivos/11_04122008111238.pdf

Muehe D (1979) Sedimentology and topography of a high energy coastal environment between Rio and Cabo Frio. Anais Da Academia Brasileira De Ciencias 51(3):473–481

Muehe D (2001) Critérios morfodinâmicos para o estabelecimento de limites da orla costeira para fins de gerenciamento. Revista Brasileira de Geomorfologia, Uberlandia – MG 2(1):35–44

Muehe D (2003) Beach morphodynamics research in Brazil: evolution and applicability. J Coast Res SI35:32–42

Muehe D (2006a) Erosão e Progradação do Litoral Brasileiro, vol 1. Ministério do Meio Ambiente, Brasília, p 475

Muehe D (2006b) Erosion in the Brazilian coastal zone: an overview. J Coast Res SI39:43–48

Muehe D (2010) Brazilian coastal vulnerability to climate change. Pan-Amer J Aq Sci 5(2):173–183

Neves Filho SC (1992) Variações da maré meteorológica no litoral sudeste do Brasil. Dissertação de Mestrado. Universidade Federal dório Rio de Janeiro, COPPE, Programa de Engenharia Oceânica, Rio de Janeiro, RJ, Brasil

Neves CF, Muehe D (1995) Potential impact of sea-level rise on the metropolitan region of Recife, Brazil. J Coast Res SI 14:116–131

Oliveira MR (2012) Diálogos Brasil-Espanha: Sistema de Modelagem Costeira. Instituto Ambiental Brasil Sustentável – IABS/Agência Espanhola de Coopera- ção Internacional para o Desenvolvimento – AECID/Ministério do Planejamento – MP/Ministério do Meio Ambiente – MMA/Editora IABS, Brasília, p 72. ISBN ISBN 978-85-64478-15-2

Oliveira MR, Nicolodi JL (2012) A Gestão Costeira no Brasil e os dez anos do Projeto Orla. Uma análise sob a ótica do poder público. Revista da Gestão Costeira Integrada 12(1):91–100

Ottmann F, Nóbrega R, Coutinho PN, Oliveira SPB (1959) Estudo topográfico e sedimentológico de um perfil da praia de Piedade – Recife – Pernambuco. Trabalhos do Instituto de Biologia Marinha, Recife 1:19–37

PBMC (2014) Impactos, vulnerabilidades e adaptação às mudanças climáticas. In: Assad ED, Magalhães AR (eds) Contribuição do Grupo de Trabalho 2 do Painel Brasileiro de Mudanças Climáticas ao Primeiro Relatório da Avaliação Nacional sobre Mudanças Climáticas. COPPE/ Universidade Federal do Rio de Janeiro, Rio de Janeiro, p 414. ISBN ISBN: 978-85-285-0207-7

Pianca C, Mazzini PLF, Siegle E (2010) Brazilian offshore wave climate based on NWW3 reanalysis. Braz J Oceanogr 58:53–70. doi:10.1590/S1679-87592010000100006

Pirazzoli PA (1986) Secular trends of relative sea level (RSL) changes indicated by tide-gauge records. J Coast Res SI 1:1–26

Reguero BG, Méndez FJ, Losada IJ (2013) Variability of multivariate wave climate in Latin America and the Caribbean. Glob Planet Chang 100:70–84

Rocha JM, Milliman JD, Santana CI, Vicalvi MA (1975) Southern Brazil. Contributions to sedimentology. Upper continental margin sedimentation of Brazil. Stuttgart 4:117–150

Santana GG (2003) Tourism development in coastal areas – Brazil: economic, demand and environmental issues. J Coast Res SI 35:85–93

Schaeffer-Novelli Y, Cintrón-Molero G, Adaime RR, Camargo TM (1990) Variability of mangrove ecossystems along the Brazilian coast. Estuaries 13(2):204–218

Scherer MEG (2013) Gestão de Praias no Brasil: Subsídios para uma Reflexão. Revista da Gestão Costeira Integrada 13:3–13, 10.5894/rgci358

Short AD (1999) Beach hazards and safety. In: Short AD (ed) Beach and shoreface morphodynamics. Wiley, Chichester, pp 293–304

Short AD (2006) Australian beach systems – nature and distribution. J Coast Res 22:11–27

Short AD, Hogan CL (1994) Rips and beach hazards, their impact on public safety and implications for coastal management. J Coast Res SI 12:197–209

Short AD, Jackson DJW (2013) Beach morphodynamics. In: Shroder JF (ed) Treatise in geomorphology, vol 10. Academic, San Diego, pp 107–129

Short AD, Woodroffe CD (2009) The coast of Australia. Cambridge University Press, Melbourne, p 288

Silva GN (1992) Variação de Longo Período do Nível Médio do Mar: Causas, Consequências e Metodologia de Análise – Tese de Mestrado, Programa de Engenharia Oceânica. COPPE/ UFRJ, Rio de Janeiro

Silveira JD (1964) Morfologia do litoral. In: Azevedo A (ed) Brasil: a terra e o homem. Companhia Editora Nacional, São Paulo, pp 253–305

Souza Filho PWM (1995) A planície costeira bragantina (NE do Pará): influência das variações do nível do mar na morfoestratigrafia costeira durante o Holoceno. Dissertação de Mestrado, Belém, Pará, p 123

Souza Filho PWM, Gonçalves FD, Rodrigues SVP, Costa FR, Miranda FP (2009) Multi-sensor data fusion for geomorphological and environmental sensitivity index mapping in the Amazonian mangrove coast, Brazil. J Coast Re SI 56:1592–1596

Suhayda JN, Hsu SA, Roberts HH, Short AD (1977) Documentation and analysis of coastal processes, northeast coast of Brazil, Technical report no. 238. Coastal Studies Institute, Louisiana State University, Baton Rouge, p 98

Syvitski JPM, Vörösmarty CJ, Kettner AJ, Green P (2005) Impact of humans on the flux of terrestrial sediment to the global coastal ocean. Science 308(5720):376–380. doi:10.1126/ science.1109454

Tessler MG, Mahiques MM (1996) Levantamento Bibliográfico Sobre A Geologia Marinha do Brasil, 1841–1992. PGGM/MMA/CIRM/FEMAR/CNPq/v. único, São Paulo, p 320

Toldo EE Jr, Dillenburg SR, Cunha LOBP, Almeida LESB, Tabajara LL, Martins RR (1993) Parametros morfodinâmicos da praia de Imbé. RSPesquisas 19:27–32

Urien CM, Martins LR, Martins I (1978) Modelos deposiónales em la Platafroma Continental de Rio Grande do sul, Uruguay y Buenos Aires. Congresso Geológico Argentino 8. Actas 2:639–658

Zembruscki SJ, Barreto HT, Palma JJC, Milliman JD (1972) Estudo preliminar das províncias geomorfológicas da margem continental brasileira. In: Congresso Brasileiro de Geologia, 26, 1971, Belém Anais Belém: v. 2, p. 187–209

Chapter 2
Brazilian Coastal Processes: Wind, Wave Climate and Sea Level

Mauricio González Rodríguez, João Luiz Nicolodi, Omar Quetzalcóatl Gutiérrez, Verónica Cánovas Losada, and Antonio Espejo Hermosa

Abstract The coast of Brazil has substantial physical and environmental diversity, constituting a constant challenge for coastal management. This diversity is characterized by heterogeneity regarding the morphology of the coast and its hydrodynamic components, such as waves, tides, sea level changes and atmospheric pressure gradients. In this chapter an historical review regarding the existing observed data is presented.

Also a detailed description of the wave climate, astronomical tide and storm surge along the Brazilian coast is provided, based on the SMC-Brasil databases: Downscaled Ocean Waves (DOW), Global Ocean Tides (GOT) and Global Ocean Surges (GOS). Finally a briefly description of the SMC-Brasil is provided focusing on it is principal characteristics and an example of it is application to thc Massaguaçu beach is shown on Appendix.

2.1 Introduction

The coast of Brazil has substantial physical and environmental diversity, constituting a constant challenge for coastal management. This diversity is characterized by heterogeneity regarding the morphology of the coast and its hydrodynamic components, such as waves, tides, sea level changes and atmospheric pressure gradients. In

M. González Rodríguez (✉) • O.Q. Gutiérrez • V.C. Losada • A.E. Hermosa
Instituto de Hidráulica Ambiental de la Universidad de Cantabria "IHCantabria", Universidad de Cantabria, Parque Científico y Tecnológico de Cantabria. C/ Isabel Torres, N° 15, 39011 Santander, Spain
e-mail: mauricio.gonzalez@unican.es; omar.gutierrez@unican.es; veronica.canovas@unican.es; antonio.espejo@unican.es

J.L. Nicolodi
Instituto de Oceanografia (IO), Universidade Federal de Rio Grande (FURG), Av. Italia, km 8 S/N, Rio Grande, RS, Brazil
e-mail: joaoluiznicolodi@gmail.com

A.D. Short, A.H. da F. Klein (eds.), *Brazilian Beach Systems*, Coastal Research Library 17, DOI 10.1007/978-3-319-30394-9_2

general, there is increasing energy of incident waves from north to south, a reverse trend occurring in tides, with tidal range decreasing from north to south.

Another relevant factor is the storm surge, which can be defined as the difference between the observed tide and astronomical tide. The storm surge is responsible for the increase or decrease of sea level in relation to predicted astronomical tides. This phenomenon usually involves intrusion of seawater into low areas, causing flooding. However, when negative, the meteorological tide can adversely affect coastal activities, such as the dynamic of ports.

The combination of positive storm surge with the wave setup may result in extreme values of run-up (maximum vertical excursion of swash on the shoreline), which can result in inundation, destruction of dunes and coastal defences (i.e. seawalls), especially when these storms coincide with astronomical spring tide.

In Brazil, the impact of these processes varies along the coast: The northern Amazon gulf region is highly dynamic, with interaction between the massive water and sediment discharge, the high annual precipitation, wave action, and especially the meso to mega-tidal regime (Pereira et al. 2010). The trade wind generated southeast and northeast waves are generally less than 1.5 m (Innocentini et al. 2000), reaching 3 m in summer off the Amapá coast (Cachione et al. 1995), however they remain low at the shore after crossing the low gradient shelf-nearshore. The January-May wet season raises water level along the coast, while the high tides dominate most of the year.

Along the Northeast coast, the wave climate is maintained by the trade winds generated by Tropical South Atlantic Anticyclone, with waves arriving from east, northeast and southeast. The highest incidences of waves occurs between June and September, arriving from the south and southeast with heights 1–2 m and average period between 7 and 12 s (Tessler and Goya 2005). The tidal range is about 4 m (meso-tidal), decreasing to 3 m in estuarine areas.

The subtropical Southeast coast between Cabo Frio (RJ) and Chui (RS) receives southerly waves generated by high latitudes storms in the South Atlantic (Pianca et al. 2010). The northeast trade winds are also influential during much of the year. The predominance of southerly waves generates northerly longshore sediment transport, which can be observed through the orientation of river mouths and morphology of ebb deltas. The south to southeast waves have an average period of 10–16 s and average height 1–4 m (Tessler and Goya 2005). Tides are 2 m in the north decreasing to 0.5 m in the south.

The action of these hydrodynamic variables, especially when combined with the storm surge, can result in damage to coastal communities, through the deposition of fluid mud on the beach (Calliari and Faria 2003), loss of land, destruction of properties and natural habitats, devaluation of property and tourism, lowering of tax collection and the loss of lives (Teixeira 2007).

In Brazil, coastal erosion exceeds shoreline progradation, with higher rates of erosion on beaches, cliffs and estuaries, in that order, while in the estuaries, the reports of erosion and progradation are equivalent, though in some states the erosion is mainly concentrated in the vicinity of river mouths, such as the coast of Santa Catarina and Paraná (Muehe 2006). Severe erosion is occurring at the mouth of the São Francisco and Paraíba do Sul rivers, mainly due to damming of the rivers, which traps sediment upstream and changes the sedimentary balance of coastal zones

(Muehe 2006). Along the Rio Grande do Sul coast severe beach erosion is occurring due to the seabed topography refracting and concentrating wave energy on sections of beach (Calliari and Speranski 2006). In Rio Grande do Norte between Guamaré and Macau severe erosion is affecting long stretches of coast, which threatens the oil pumping installations (Costa Neto 2001). The erosion is a product of the large tidal range and the consequent high velocity tidal currents as well as the constant wave-driven westerly sediment transport (Muehe 2006).

Within this context, it is important to understand the interactions between oceans and coastal zones and the climate change-related variables. Moreover, it is essential to build a strategic vision for the coastal zone, so that steps may be taken in response to existing hazards as well as the new scenarios of global warming, rising sea levels and coastal erosion (Nicolodi and Pettermann 2011). It is worth emphasizing that while there is in Brazil a reasonable degree of knowledge on the subject, the systematic monitoring of waves and tides is still incipient.

The main initiative for the wave and tide data measurements has been the *Global Ocean Observing System* (GOOS). The aim of GOOS/Brazil is the implementation, expansion and consolidation of a system of oceanographic, meteorological and climatological information in the South Atlantic for the purpose of producing knowledge, to enable oceanographic and meteorological forecasting in the maritime area of national responsibility, and thereby reduce vulnerabilities and risks from extreme events and climate change.

It is in this context that the major networks for monitoring waves and tides of Brazil are:

- *Buoys National Program* (PNBOIA): drifting and anchored buoys in the coastal region to provide real time meteorological and oceanographic data to the scientific community. In 2014 five buoys anchored near the shelf break were operational: Recife (PE), Porto Seguro (BA), Cabo Frio (RJ), Santos (SP) and Rio Grande (RS).
- *Monitoring Network Waves in Shallow Water*: a network of buoys anchored in shallow waters along the Brazilian coast, to monitor the real-time wave climate. This aims to provide data for understanding the interactions between the continent and the ocean, coastal engineering projects, port and ocean, marine mining, navigation, studies of variations of the shoreline and coastal processes. In 2014 there was one operational buoy (Recife) and two in maintenance (Rio Grande and Santos)
- *Global Sea Level Observing System* (GLOSS-BRAZIL): monitoring sea level to support environmental research that will improve the social and economic planning of the country. In 2014 eleven tide gauges were operational along the Brazilian coast.

Even with the development of these networks, there is a serious shortage of long and reliable coastal data series in Brazil. The reason is the small number of operational systems, the short operational time of these networks and logistical difficulties of maintenance and support. This situation has resulted in only short series and data gaps owing to long periods of inactivity. More detailed information about these networks can be obtained in http://www.goosbrasil.org/.

A considerable part of the data on Brazilian waves and tides was obtained through research projects, with inherently limited periods of monitoring and private data. A study of existing publications (which used wave data) in major databases available on the Internet indicates that most of these publications (60 %) obtained information about wave parameters from visual estimation. Data from buoys occur in only 21 % of papers, and in most cases the buoys were anchored only during the period of the scientific project.

In this context, this chapter is intended to present and discuss data from waves and tides existing in Brazil. Based on these data it is possible to analyze the current coastal dynamics, especially with the use of tools that combine numerical models and measured data. An example is the *Coastal Modeling System* (SMC-Brasil, http://smcbrasil.ihcantabria.com/), which is an initiative of the Ministry of Environment in partnership with the Spanish government to transfer methodologies and tools to support the Brazilian coast management.

This chapter is structured as follows: a data review section, followed by a section on wind hindcast, in which results of atmospheric reanalysis are presented, as well as wind characterization sources and temporal variability of the wind. This is followed with the results of wave climate hindcasting, with emphasis on Global Ocean Waves (GOW) reanalysis and the wave height calibration. The next topic at issue is a Downscaled Ocean Waves (DOW), where a hybrid downscaling methodology to transfer the wave climate to coastal areas has been used, especially with the SWAN model (Booij et al. 1999), version 40.85 together with an overview on the spatial and temporal variability of the wave climate. Sea level hindcast data is analyzed, focusing on Global Ocean Tides (GOT) and Global Ocean Surges (GOS), with specific emphasis on the characterization of these variables along the Brazilian coast. In the Appendix a description of SMC-Brazil, the coastal modeling and management system developed for Brazilian coast, together with a study case to illustrate the application of the system is presented.

2.2 Data Review

In Brazil, there are just a few records of wave and tide measurements, especially with a long and reliable data series. This chapter summarize the key initiatives and briefly describe the available data.

2.2.1 Waves

Waves can be measured by three main methods: visual estimations, in situ techniques and using satellite sensors. Visual estimates are still used today, especially in smaller projects, and have been validated by several researchers (Jardine 1979; Bryant 1979; Guedes Soares 1986; Plant and Griggs 1992).

Visual observations, especially wave heights, are fairly reliable if carried out by experienced observers who follow specific instructions. Observations should be recorded as they are often the only source of information (Holthuijsen 2007).

In situ instruments may be located at the sea surface (e.g. floating surface buoy) or below the sea surface from a fixed location (e.g. pressure transducer, ADCP). All are used to acquire time records of the up-and-down motion of the surface (Holthuijsen 2007).

Measurements from satellite sensors are a more recent alternative, particularly for global scale, and they have been used to calibrate global models of wave propagation (Aage et al. 1998).

The most common way of measuring waves is to follow the three-dimensional motion of the water particles at the sea surface. This can be done with buoys, which measure the vertical acceleration with an onboard accelerometer, while modern buoys can also acquire directional information. The data is transmitted by radio communication to a land based receiving station and more recently with satellite communication and position detection by the Global Positioning System (GPS).

As mentioned in the previous section, waves monitoring programs in Brazil are currently under the coordination of GOOS-Brazil. The history of Brazilian wave monitoring dates back to the 1960s, with the pioneering buoy measurements of Wainer (1963), on the coast of Rio Grande do Sul (Tramandaí Beach) for 9 months. This was one of the first series of wave data using this type of equipment in Brazil and the significant wave height (Hs) and period (Ts) can be observed in Fig. 2.1. From these data, it was possible to predict significant and maximum waves for return periods of 30 and 100 years (Strauch et al. 2009).

While more wave buoys have been used in recent years, they are not popular because of the high cost of equipment and maintenance and the high risk of accidents.

However, the urgent need for wave data in Brazil, motivated researchers from the Federal University of Rio de Janeiro (UFRJ), with the aid of the Brazilian Navy, to create an alternative project 'the Sea Sentinels'. The aim of this project is to monitor nearshore wave conditions in Brazil by means of visual observations. In order to

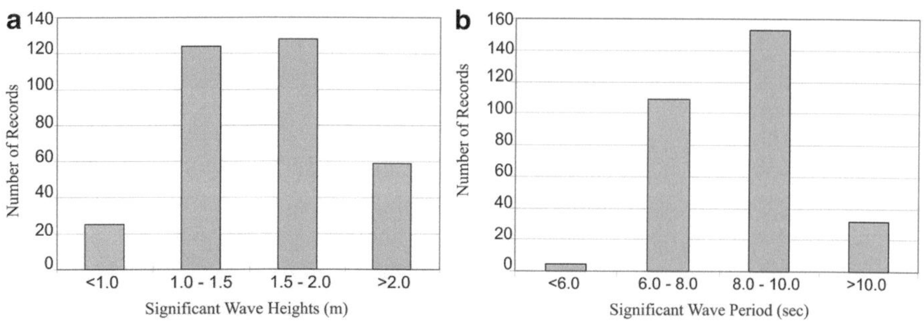

Fig. 2.1 Distribution of significant heights (**a**) and periods (**b**) in Tramandaí between 1962 and 1963 (Wainer 1963) (Adapted from Strauch et al. 2009)

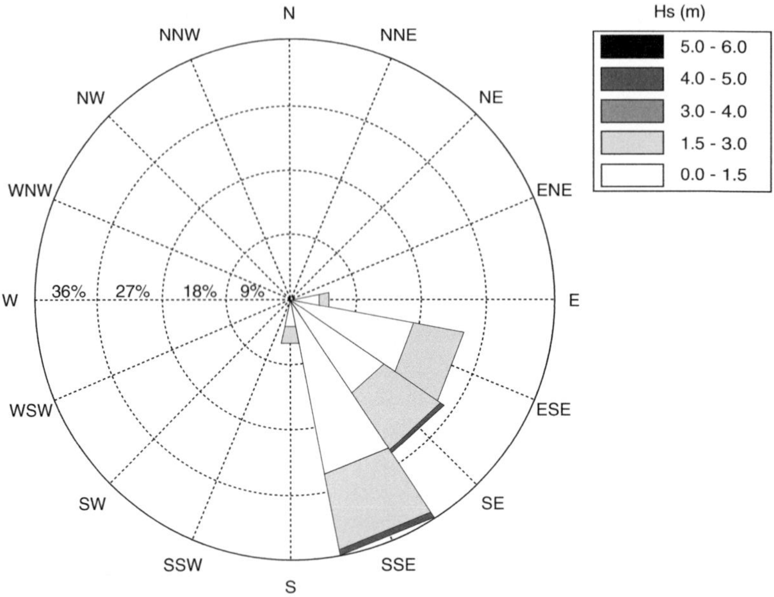

Fig. 2.2 Directional distribution of the significant wave height at the mouth of Lagoa dos Patos (RS) (Strauch 2001)

obtain the observations, a group of 25 surfer volunteers from the eight coastal states were trained by the Brazilian Navy and UFRJ to act as the sentinels. This project monitored wind direction and velocity, wave height, wave period, wave direction and nature and direction of the alongshore current (Melo 1993).

The qualitative results of this project contributed to the definition of three coastal regions and their wave climates: A southern region exposed to South Atlantic southern swell; an eastern region dominated by locally generated sea waves; and a northern region, under the constant action of southeast trade wind waves and also exposed to northern hemisphere swell (Melo 1993).

In southern Brazil, the Federal University of Rio Grande (FURG), in partnership with the Center for Development of Nuclear Technology (CDTN), anchored a non-directional waverider at the mouth of the Lagoa dos Patos, in Rio Grande. This equipment worked for 28 months, contributing to the understanding of wave climate of the region, as shown in Fig. 2.2

In Santa Catarina state, researchers from the Federal University of Santa Catarina (USFC) implemented the Coastal Information Program (PIC online), which maintained a buoy anchored off Santa Caterina island of between 2001 and 2003. The program monitored waves, sea surface temperature and indirect information about currents on the Santa Catarina platform. The data was freely available in real time via the internet. A summary of the program and the data analysis can be found in Melo (2004) and Pimenta et al. (2004). Data from this program are available on UCSD website (http://cdip.ucsd.edu/?nav=historic&sub=map&xmap_id=24). The significant height, peak period, mean direction and temperature for 2003 are shown in Fig. 2.3.

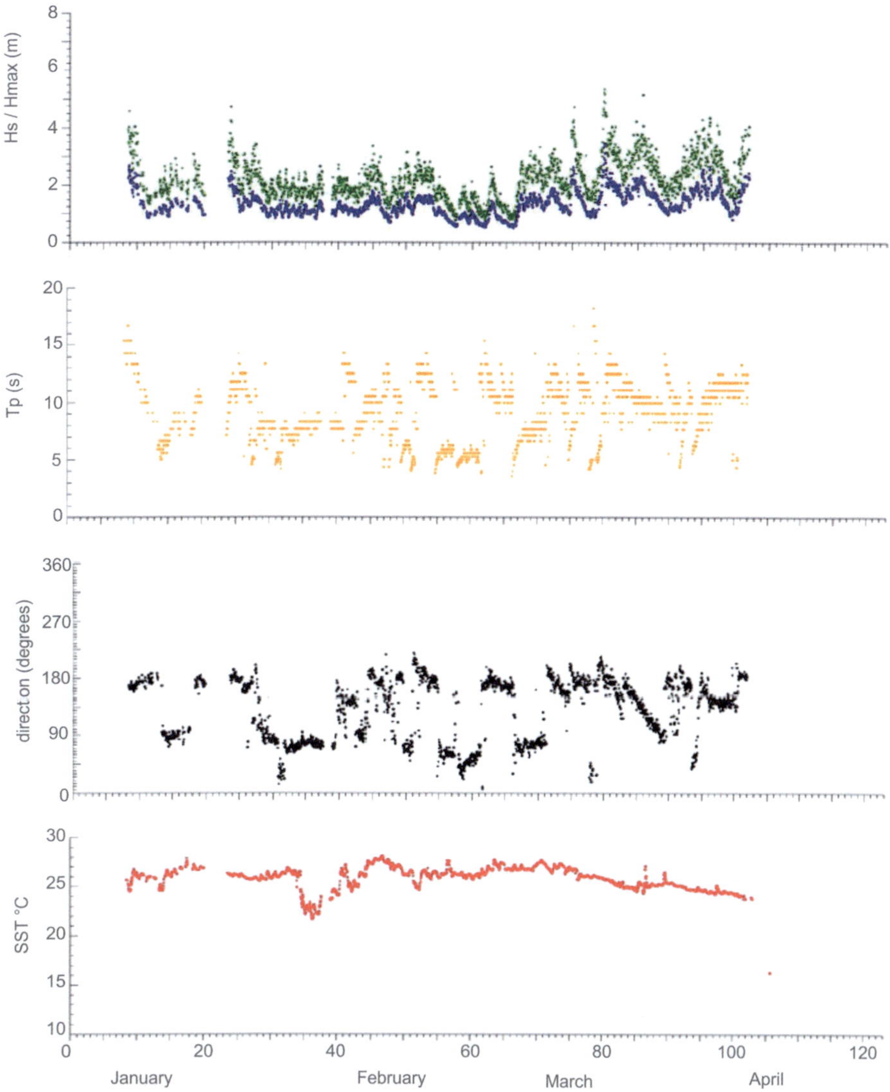

Fig. 2.3 Time series of parameters measured in the first quarter of 2003 off Santa Catarina island. Significant heights ($H_{1/3}$ – *blue*) and maximum (H_{max} – *green*); peak period (T_p); Direction corresponding to T_p; and sea surface temperature (*SST*)

At governmental level, the GOOS-Brazil Program maintains two networks that measure wave parameters: the National Program Buoys (PNBOIA) and the Monitoring Network Waves in Shallow Water (REDE ONDAS).

The PNBOIA have a network of drifting and anchored coastal buoys to provide meteorological and oceanographic data in real time to the scientific community. In 2014 five buoys were operational: Recife (PE), Porto Seguro (BA), Cabo Frio (RJ), Santos (SP) and Rio Grande (RS). These buoys are anchored near the shelf break.

Wave data is also received via satellite through the Argos system. The Argos program is jointly administered by the American agency, NOAA, and the French CNES. The data from this program can be obtained from: http://www.goosbrasil.org/tiki-index.php?page=PNBOIA%20Data#.

The WAVES NETWORK is a network of buoys anchored in shallow waters along the Brazilian coast, to monitor the wave climate through real-time knowledge of sea conditions. In 2015 there were five operational buoys (Recife, Praia do Forte, Paranaguá, Tramandaí and Rio Grande) and one in maintenance (Santos). The data from this government-university joint operation can be obtained at: http://redeondas.herokuapp.com/.

2.2.2 *Tides*

The Brazilian initiative of measuring the sea level can be grouped into three distinct periods. The first basically involved the setup and maintenance of tide gauges, focusing on obtaining information for navigation and harbor applications, elaboration of nautical charts and altimetric surveys, which did not require accurate estimates. The second phase, from the 1990s to date, is marked by an improvement in the establishment of reference levels (either local or the Vertical Datum) and the creation of PTNG (Permanent Tide Network for Geodesy) along with more precise and accurate estimates using continuous GPS (CGPS), gravimeters and altimetry (Lemos and Ghisolfi 2011).

As a very detailed description of first and second periods can be found in Lemos and Ghisolfi (2011) the following briefly reviews the three periods.

2.2.2.1 1st Period (1910–1980)

Between 1910 and 1920 the DHN and INPH (National Institute of Hydrologic Research) made the first sea level measurement, initially focusing on navigation, harbour applications, elaboration of nautical charts and altimetry surveys (Neves 2005).

Between 1919 and 1920, the Brazilian General Chart Commission (now extinct) operated a tide gauge in the city of Torres in the Rio Grande do Sul state.

During the 1970s, 281 sites throughout the Brazilian coast were sampled, with most sites lasting less than a month.

The first geographic readjustment of the Torres datum was established in 1952. By that time, more than 5000 reference levels had been set near the Brazilian tide gauges.

In 1959, after 9 years of observations (1949–1957), by the Inter-American Geodetic Survey the Datum was readjusted and moved to the city of Imbituba (SC). However during the 1960s other important tide-gauge stations were deactivated

2.2.2.2 2nd Period (1990s)

In 1994, the IBGE took over the tide gauge located at Porto de Imbetiba in the city of Macaé (Rio de Janeiro) and upgraded the station to become a pilot station for the future PTNG. The PTNG was established in 1997 and stations were located at Imbituba (SC), Macaé (RJ), Salvador (BA), Fortaleza (CE), and Santana (AM). The network became operational effectively from 2001, after the installation of digital equipment in Macaé and Imbituba.

Unfortunately, the majority of the tide gauges, once active, were either not maintained or destroyed. An exception is the Cananéia station, where the time-series is more than 50 years old. According to Pirazzoli (1986), who analyzed the data of long-term variations of MSL measurement from a data set available in the Permanent Service for Mean Sea Level (PSMSL), the rate of variation of the MSL had a period of 20 years. Hence, the studies on long-term tendency should have at least 50 years of data.

2.2.2.3 3rd Period: The Present

Currently, the most important initiative for monitoring tides is the GLOSS-BRAZIL. It is a network to monitor sea level, managed by the Brazilian Navy. The GLOSS-Brazil Implementation Plan was finalized in October 2004, and in 2014 eleven tide gauges were operational along the Brazilian coast.

As an example, we highlight the oldest tide gauge station in Brazil, which is located on the south coast of São Paulo, in Cananeia. This station has the longest time series, with measurements carried out since 1954 by the same AOTT tide gauge. Costa (2007) analyzed the data between 1954 and 2004 (Fig. 2.4). In this study, the authors estimate a sea level rise on the order of 4.2 mm $year^{-1}$.

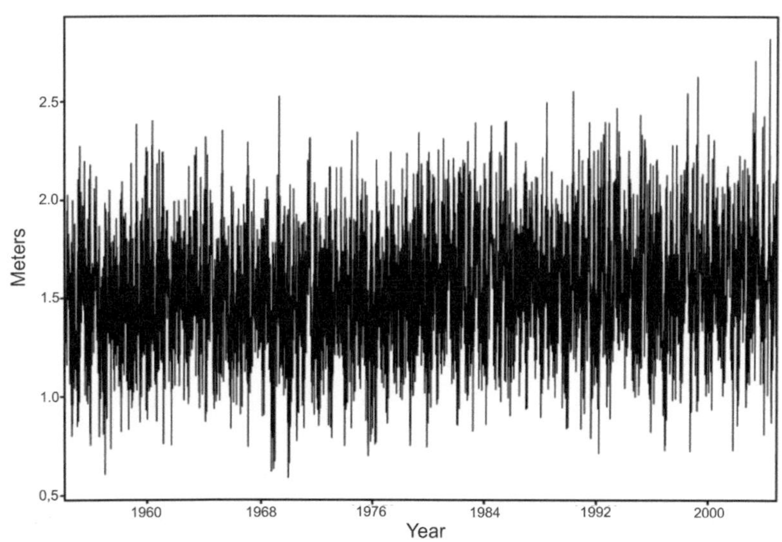

Fig. 2.4 Relative sea level in the region of Cananeia, São Paulo

2.3 South Atlantic Wind Systems

2.3.1 Spatial and Temporal Variability

South Atlantic atmospheric circulation is dominated by anticyclonic flow around the subtropical gyre, which is strongly influenced by interoceanic connections. At the Brazil-Malvinas Confluence (between approximately 30–50°S), there are sharp thermal contrasts between the cold circumpolar and South Atlantic warm air masses that form strong east-tracking lows east of the Drake Passage (Garcia et al. 2004). These lows are the leading source of the southerly swells that dominates Brazil's southeast facing coasts.

Further to the north, in the ITCZ region (between approximately 30°N–30°S) the easterly trade winds flow onto the north and east facing coast of Brazil. These extensive and persistent winds, while not as powerful as the mid-latitude westerly lows, are able to generate short period seas of moderate height. Seas are also be generated locally by thermal sea-breezes at the coast. Both the westerlies and trade winds, have seasonal fluctuations as indicated in Fig. 2.5.

Regarding trade winds, which are the leading source of most of the northern Brazilian wave energy, it is worthwhile to differentiate between North and South Hemispheres. Whereas the Northern Hemisphere northeast trades, are stronger during winter and spring, the Southern Hemisphere, southeast trades are almost constant, with minor seasonal fluctuations. The most striking seasonal feature (austral spring-summer) is the directional shift occurring between 10°S and 20°S, in which the winds turn their direction from the southeast to the northeast readjusting to the land contours.

2.4 Brazilian Wave Climate: Spatial and Temporal Variability

Description of wave climate along the Brazilian coast was based on the SMC-Brasil waves database. This database was obtained by improving the spatial resolution by means of a downscaling (DOW) of a global ocean waves hindcast (GOW). The database consists of 60 year temporal series along the Brazilian coast.

GOW, have been simulated using the Wave Watch III model (WWIII, Tolman 2009). Simulations are computed on a global grid with a spatial resolution of 1.5° in longitude and 1° in latitude. More details about GOW can be found on Reguero et al. (2012). GOW have been calibrated by means of a directional methodology proposed by Mínguez et al. (2011). The procedure is based on altimetry measurements. Although the comparison is only made when there is coexistence of both data sources, GOW and satellite, the correction is applied for the full period of wave hindcast. The preliminary validation using both buoy and satellite altimetry data showed a good agreement between the different satellites and, as a consequence,

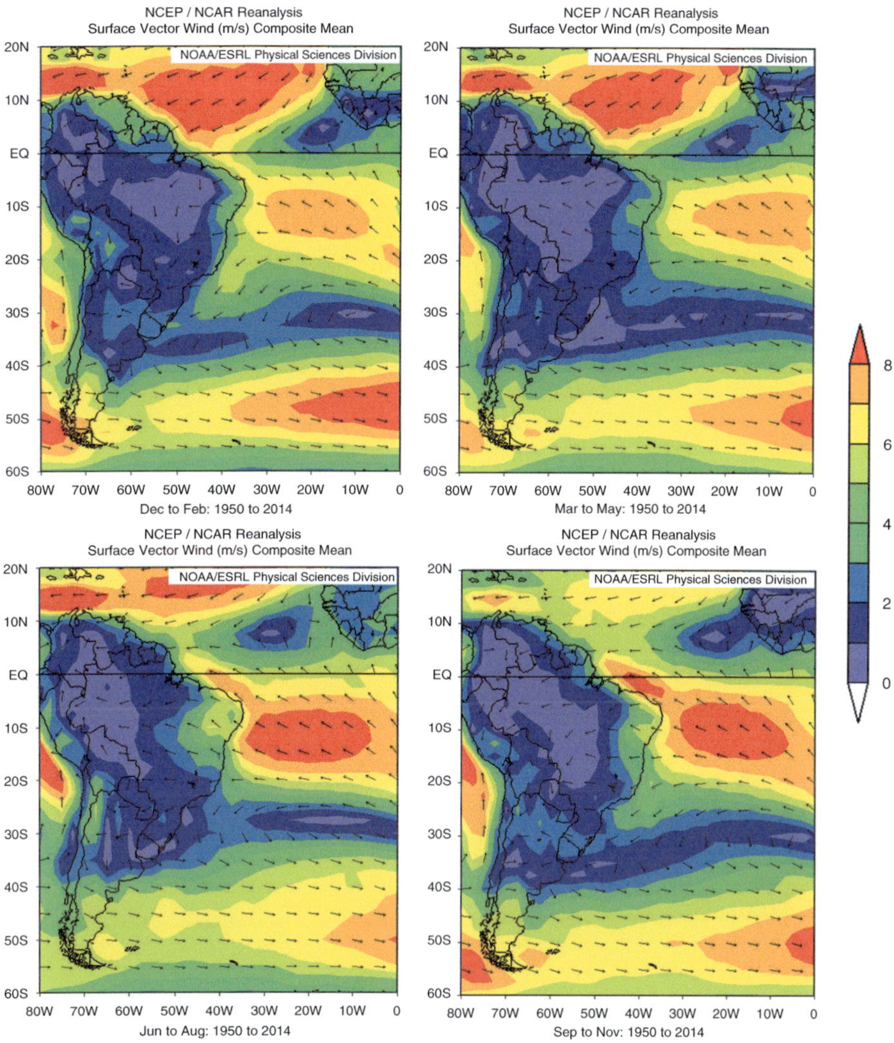

Fig. 2.5 Seasonality of the South Atlantic winds (Data from NCEP/NCAR reanalysis)

they are combined for comparison with reanalysis results. Temporal coverage of the satellite data extends from 1992 to 2008. The satellite wave height calibration procedures summarized in Cotton (1998) and Woolf et al. (2002), and later updated by Hemer et al. (2010). More information about this topic can be found in IHCantabria (2013a).

GOW database does not offer appropriate description of waves in coastal areas, therefore a hybrid downscaling methodology, described in Camus et al. (2011), and SWAN (Booij et al. 1999) simulations were applied to increase the resolution of wave climate on shallow areas. This improved database was named DOW.

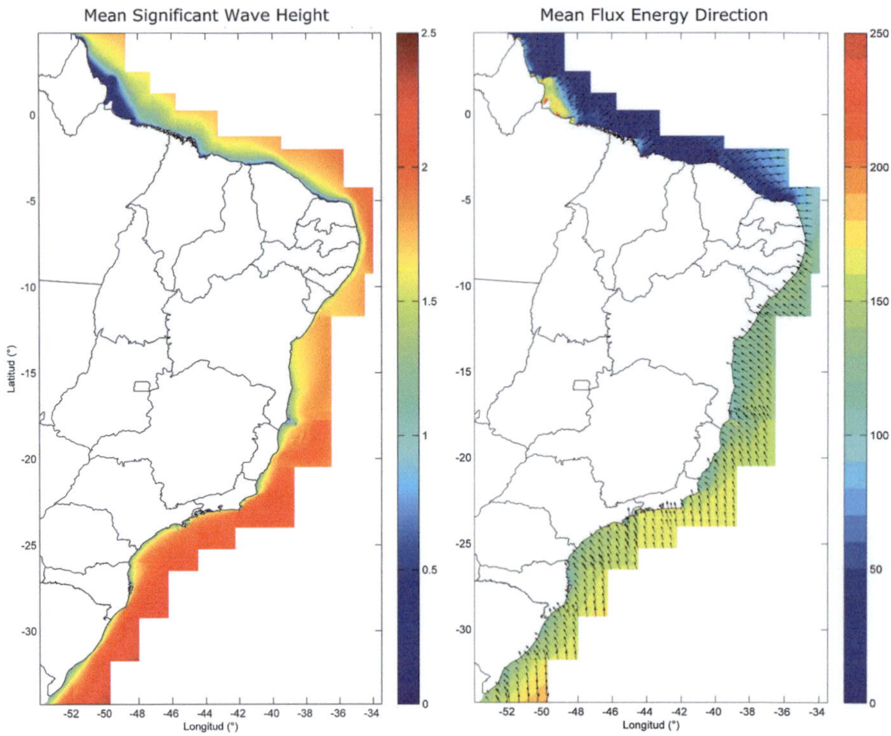

Fig. 2.6 Annual mean H_s (*left*) and FE (*right*) along the coast of Brazil from DOW database for the 1948–2008 period

2.4.1 Mean Waves

In order to characterize mean coastal wave conditions along the Brazilian coast, the annual mean significant wave height H_s and the mean wave energy flux (FE) have been determined making use of the DOW from 1948 to 2008.

Figure 2.6 illustrates the H_s and FE along the Brazilian coast based on GOW. Surprisingly to some extent when compared with the wind maps shown in Fig. 2.5, larger waves are found along the southern coast (down to 20°S) due to the action of the extratropical lows that are able to produce longer period waves than the trade winds, which dominate the northern coast. Longer periods means more propagation capacities of swells when compared with seas. This fact can be corroborated when observing the FE mean direction, presenting a dramatic shift around 5°S.

Seasonality of the coastal wave climate can be observed in Fig. 2.7. As expected, wave climate fluctuations are the response to the regional winds seasonality. The north coast receives more energetic waves during summer due to the intensification

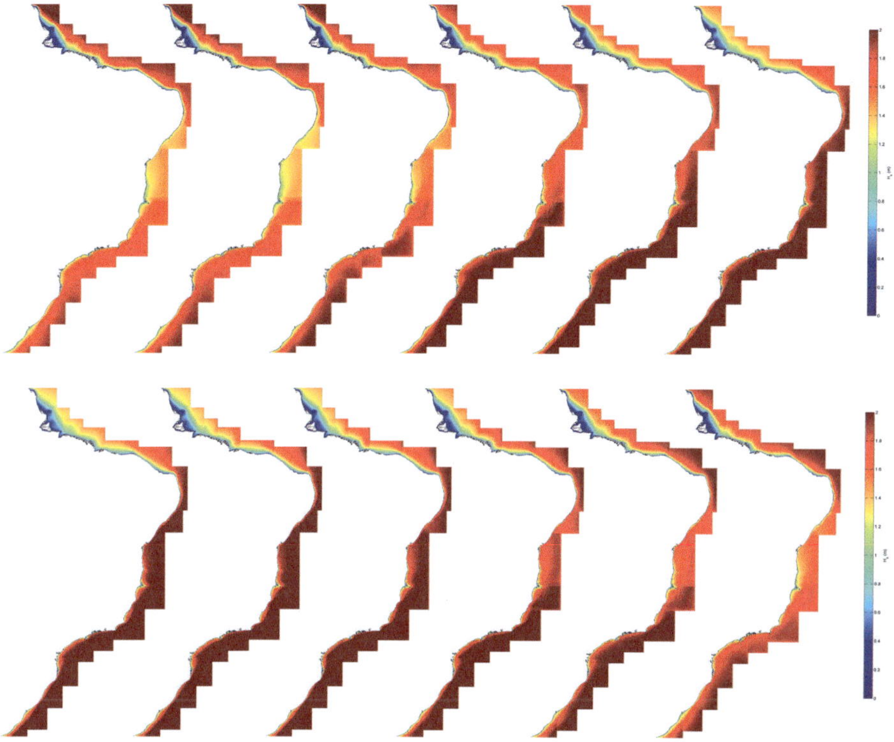

Fig. 2.7 H_s seasonality maps (from January to December) (in meters) on Brazilian coast from DOW database for the 1948–2008 period

of the Northern Hemisphere trades. Between 10 and 15°S H_s peaks during winter owing to the strength of Southern Hemisphere trades. Further to the south larger H_s during winter months are due to the more frequent and deeper lows forming over the Drake Passage.

2.4.2 *Extreme Waves*

In order to characterize extreme significant wave height values at the coast, the H_{s12} map is shown in Fig. 2.8. H_{s12} is the significant wave height exceeded for about 12 h a year; this value is used characterize the stormiest event throughout an averaged year. As expected, only the most southern latitudes receive H_{s12} waves up to 5 m. Nevertheless, these waves tend to decay rapidly due to attenuation across the continental shelf. As can be observed, sea states up to $H_{s12}=3$ m are extremely rare on the north facing coast of Brazil.

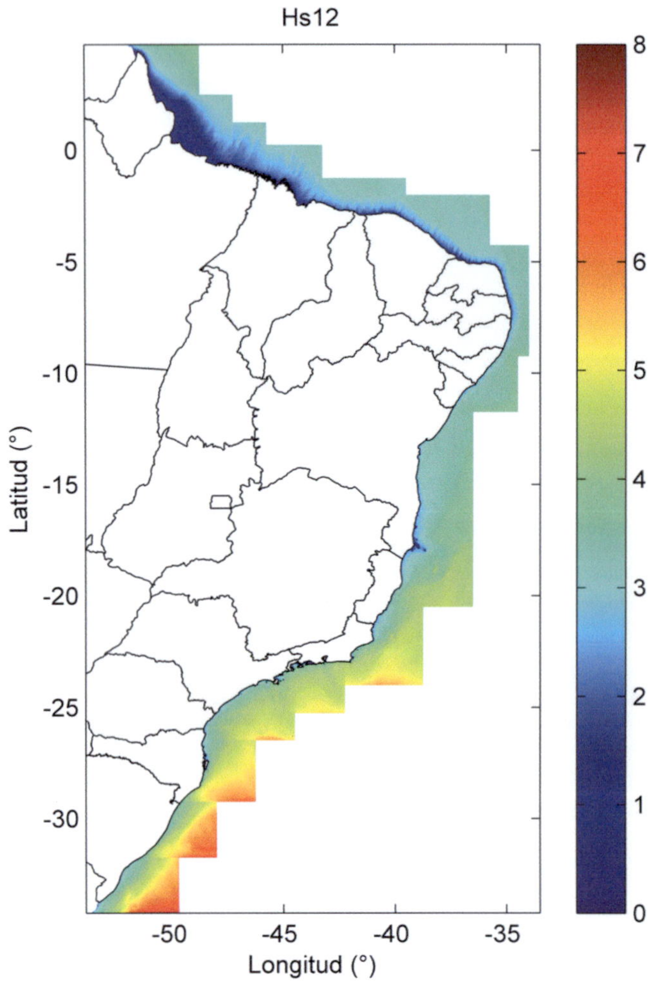

Fig. 2.8 Hs_{12} values (in meters) along the Brazilian coast from DOW database for the 1948–2008 period

2.5 Astronomical Tide Range Along the Brazilian Coast

The characterization of the astronomical tide along the Brazilian coast is based on the Global Ocean Database (GOT) of SMC-Brasil. GOT database is composed by a set of 60 year hourly temporal series along the Brazilian coast. This database was created based on the harmonic constants of TPXO Global Tidal Solution developed by Oregon University (Egbert et al. 1994; Egbert and Erofeeva 2002). The database consists of a 60-year time series of astronomical tide elevation from 1949 to 2009.

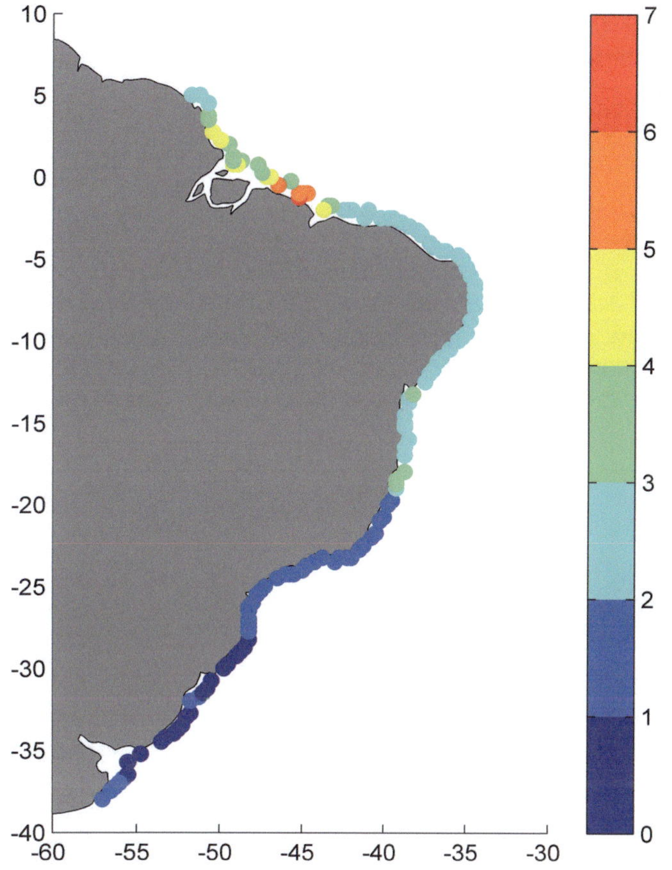

Fig. 2.9 Maximum tidal range along the Brazilian coast from the GOT database along the Brazilian coast for the 1948–2008 period

Database was validated with tide gauges located along the Brazilian coast from University of Hawaii Sea Level Center (UHSLC), the Instituto Nacional de Pesquisas Hidroviarias (INPH) and the Marina do Brasil (MB). Details about the database and validation can be found on IH-Cantabria (2013b).

Figure 2.9 shows the maximum tidal range values, with highest (meso to mega) tides around the Amazon Gulf (Amapá, Pará and Maranháo states), meso-tides along the northeast coast, decreasing to micro-tides down the southeast coast.

2.6 Storm Surge Along the Brazilian Coast

Storm surges are primarily caused by low pressure and strong onshore winds that produce an increase in the sea level on the coast. It is therefore essential to consider storm surge in the calculation of flooding levels on the coast.

The storm surge level variation can be obtained by subtracting from tide gauges the predicted astronomical tide. Unfortunately, real data provided by measurement networks are scarce and present several limitations, both in terms of spatial and temporal coverage (Cid et al. 2014). In order to overcome these limitations numerical models have become useful tools for the generation of long term and high resolution databases. The Storm Surge database (GOS) was obtained by means of numerical simulation and was validated using tidal gauge series. It was simulated using the Regional Ocean Model System (ROMS) developed by Rutgers University (Haidvogel et al. 2000; 2008; Shchepetkin and McWilliams 2005).

The model is set-up for Southern Atlantic covering the area 65°S to 35°N and 20°W to 125°W with a horizontal resolution of 0.25°. The bathymetry is extracted from the ETOPO 2 (NOAA 2006) database, a global topography of 2 min resolution derived from depth soundings and satellite gravity observations (Smith and Sandwell 1997).

GOS database was validated using tide gauge records from UHSLC, INPH and Marina do Brasil. Details about the database and the validation can be found on IH-Cantabria (2013b).

2.6.1 Spatial Variability

In order to characterize the GOS database along the Brazilian coast the following variables were calculated: historical maximum elevation, mean and standard deviation of the elevation, and elevation exceeded the 50 %, 10 % and 1 % of the time.

Figure 2.10 shows the historical maximum elevation along the Brazilian coast. The minimal elevations are found on the northern coast (<0.5 m), while in the south they reach the maximum values (>2 m), where the coast is exposed to extra tropical storms. Parise et al. (2009) found that highest storm surges in the south are related to south-westely winds. Machado et al. (2010) found four patterns of these synoptic situations, the first three confirmed the Parise et al. (2009) observations and the fourth is related to a high pressure system.

The standard deviation (Fig. 2.11) gives a spatial variability measurement of the storm surge. In the northern part it is less than 15 cm meanwhile in the south it could reach the 40 cm.

2.7 Coastal Flooding

The determination of the flooding level on a beach requires taking into account several meteorological and oceanographic variables and the interaction between them. These variables vary spatial and temporally and also interact with the beach morphology. The most important variables are astronomical tide, storm surge and highest level reached by the wind waves over the beach slope.

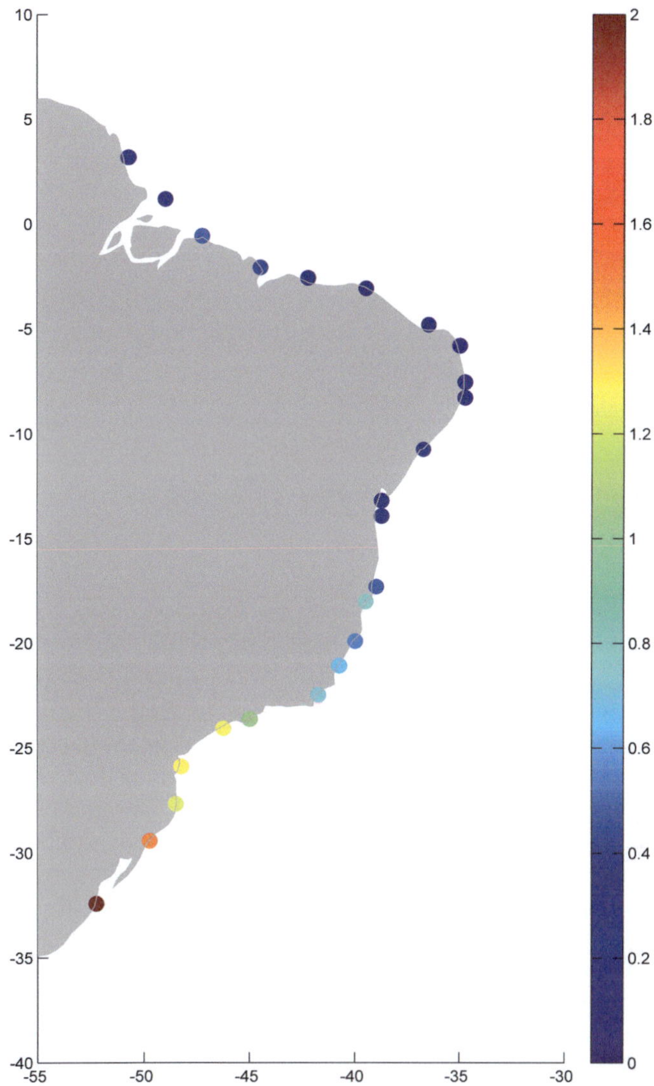

Fig. 2.10 Historical maximum storm surge (meters), obtained from GOS database for the 1948–2008 period

The flooding stage at one instant could be defined as the elevation over a reference level (RL) produced by the combination of the instant tidal level and the run up (Ru). The tidal level is also the combination of the astronomical tide elevation (AT) and storm surge elevation (SS), as is shown in Fig. 2.11. Some of the factors affecting the storm surge and run up are random and have a probability of occurrence.

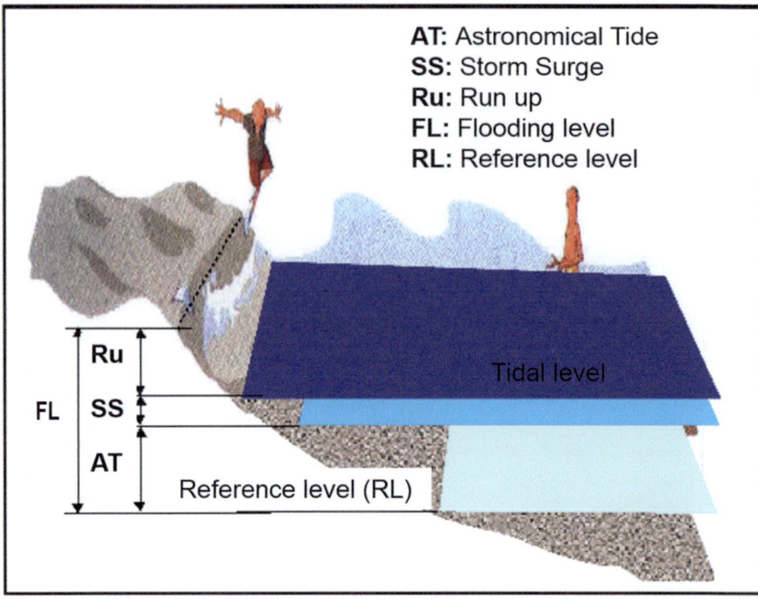

Fig. 2.11 Outline of the factors that define the flooding stage (GIOC 2003)

Therefore, the flooding stage determination is a stochastic problem. The consequences are that it is impossible to calculate a deterministic flooding stage and therefore a probability must be associated to each specific storm. The association of the flooding stage with a probability is fundamental for structures designs, delimitation of public domains, protection of coastal ecosystems, etc.

A simplified approach to this problem, that provides general information about the flooding on coastal areas and can be used to obtain pre-design or rough values of the inundation in a fast and simply way is presented on a simplified flooding Atlas for the Brazilian coast, summarizes these results (‘*Uma Proposta de Abordagem para o Estabelecimento de Regimes Probabilísticos de Inundação Costeira do Brasil*’, Ministério do Meio Ambiente 2014). A second and complex method that uses the astronomical tide, storm surge and wind waves databases described previously for the calculations, as source of forcing is implemented on SMC-Brasil (see Apendix 1). This methodology requires the use of detailed bathymetries, and numerical models to propagate wind waves from deep to shallow water, and find the breaking point. This method must be applied to a specific beach for local flooding assessment. In both cases the flooded area can be determined using the flooding stages as inputs for model simulations.

2.7.1 Methodology Differences

Although both methodologies can be used to produce flooding charts, and to define flooding contours associated to return periods, there are differences in the character of results, which lead to different approximations and application of the results. The simplified methodology gives a rough idea of the flooding in a coastal area, and was used to generate a flooding assessment document for immediate use; while the detailed methodology provides a fine approach to the coastal flooding taking into all the morphological and dynamical characteristics of a specific beach. In this case, the flooding charts must be obtained through numerical tools implemented on *SMC-Brasil*.

The main differences between both methodologies are on the characteristics of the beach and the wave propagation. Table 2.1 lists the differences between the methodologies.

Table 2.1 Summary of the differences between the simplified methodology, used on 'Uma Proposta de Abordagem para o Estabelecimento de Regimes Probabilísticos de Inundação Costeira do Brasil' (Ministério do Meio Ambiente 2014), and detailed methodology, used on SMC-Brasil, to calculate the flooding level

Elements		Simplified methodology	Detailed methodology (SMC-Brasil)
General	Approximation	Global Brazilian Coast	Local applications
	Spatial approach	For coastal regions (~100 km)	For a specific beach
Levels	Astronomical tide	60 year GOT series	60 year GOT series
	Storm surge	60 year GOS series	60 year GOS series
Wind Waves	Database	60 year DOW series	60 year GOW series
	Wave propagation	Snell's approximation. Using straight and parallel bathymetry	OLUCA-Sp numerical model. Using high resolution bathymetry
	Propagated cases	All of them one by one (hourly)	Propagation of a small number of representative cases
	Breaking wind waves criteria	Defined as $H_b/h_b=0.8$	OLUCA-Sp (Spectral wave breaking):
			Battjes and Janseen 1978
			Thornton and Guza 1983
			Winyu and Tomoya 1998
	Run-Up	Nielsen and Hanslow (1991)	Nielsen and Hanslow (1991) or flooding numerical model
Beach	Bathymetry	Nautical charts + topography	Nautical charts + local surveys + topography
	Beach slope	Just for reflective and dissipative beaches	Real profile
	Beach orientation	For different Beach orientations in the region	Not relevant

2.8 SMC-Brasil: Hindcast Dynamic Databases and a Coastal Process Numerical Tools

2.8.1 Introduction

Shoreline plays an important role in human life from a socio-economic, commercial, recreational, residential and touristic point of view, as well as being important for its biodiversity and environmental wealth. During recent decades, man has realized the importance of preserving this biodiversity and the need of a good knowledge of dynamics and effects on the shoreline to ensure the coast stability and its conservation through sustainable actions and a proper environmental management.

In order to properly respond to these needs, from 1995 to 2002, the Environmental Hydraulics Institute "IHCantabria" from University of Cantabria developed a Beach Nourishment and Protection Manual, which included a design and evaluation methodology collected in some technical documents and a user-friendly system called Coastal Modeling System (SMC), which took into account all those methodologies.

Since then IHCantabria has been working on the development of new methodologies, databases and coastal models to improve the knowledge of dynamics and effects on the coast and, since 2009 up to now, some of those scientific advances have been integrated in a new advanced version called SMC-Brasil (http://smcbrasil.ihcantabria.com/), which has been developed by The Environmental Hydraulics Institute "IHCantabria", the Coastal Oceanography Laboratory of the Federal University of Santa Catarina (UFSC) and the Oceanographic Institute of the Sao Paulo University (USP), with the support of the International Spanish Cooperation Agency (AECID), the Brazilian Ministério do Meio Ambiente (MMA) and Ministério do Planejamiento, Orçametno e Gestao/ Secretaria de Patrimonio da União (MP-SPU).

The main objective of the SMC-Brasil is to provide a coastal numerical tool and a series of reference documents, that help technicians in the design, execution and monitoring of coastal projects; to establish a strategy in order to prevent coastal erosion and estimate flooding risks of Brazilian littoral zones.

2.8.2 SMC-Brasil

SMC-Brasil is a user-friendly system specifically designed to assist coastal designers and managers in the analysis of marine and littoral dynamics to understand the changes in coastal caused by those dynamics.

The SMC-Brasil is composed of methodological documents for specific coastal topics and two user-friendly systems (Fig. 2.12).

- Thematic documents: these technical documents collects all the database descriptions and the methodologies used to estimate and analyze waves, sediment

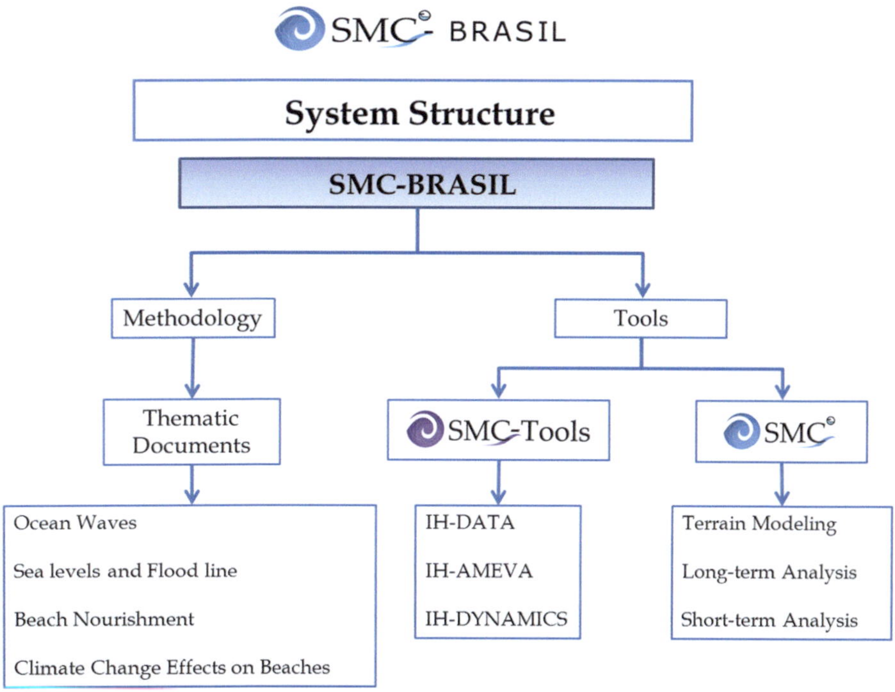

Fig. 2.12 The SMC-Brasil is composed of four methodological thematic documents and two numerical tools

transport, beach stabilization, climate change effects on beaches, etc. All this information has been summarize in four technical documents: Ocean Waves (IHCantabria 2013a); Sea levels and flood line (IHCantabria 2013b); Beach nourishment (IHCantabria 2013c); and Climate Change Effects on Beaches (IHCantabria 2013d).

- Numerical tools: a set of numerical models and statistical tools are used develop coastal projects and manage the coast using the methodology and databases, described in the thematic documents. There are two principal modules: SMC-Tools and SMC. The first one covers bathymetry and marine dynamics databases, obtained by reanalysis. The second one includes numerical model and statistical tools and permits the application of those methodologies and formulations proposed in the cited above documents in Brazilian coastal projects.

SMC-TOOLS was specially developed for SMC-Brasil version in order to help coastal designers and managers to obtain input data (wave and sea level series, bathymetries and images), marine and littoral dynamics on the beach and to readily process all this information to carry out coastal projects and managements.

It is important to highlight that the SMC-Brasil is not a static system, but allows for the incorporation of new databases, methodologies and morphodynamic models, at different time and spatial process scales.

2.8.3 System Skills

SMC-Brasil permits the study of marine and littoral dynamics required to understand the coastal response to those dynamics at different spatial and time scales (short, medium and long-term, or even changes in the near future). This makes it suitable for many coastal engineering applications: sediment transport studies, coastal evolution and stabilization studies, coastal flooding assessment, beach nourishment projects, etc.

In order to show some skills of SMC-Brasil, some results obtained in a study of Massaguaçu Beach (north coast of Sao Paulo) are shown in the Appendix 1.

2.9 Summary and Conclusions

Brazil has a large shoreline (approximately 9000 km), which spans from 5°N to 35°S parallel, resulting in considerable physical and environmental diversity.

This diversity is characterized by heterogeneity regarding the coastal morphology and the marine climate (waves, sea level, etc) and meteorology (atmospheric pressure gradients, winds, etc).

Previous studies reveal wave climate and sea level vary along the coast and there are some specific zones with a large seasonal variability.

In general, wave energy increases from north to south, while tidal range decreases from north to south. However, some zones have a regional behavior and morphology due to the complexity of the dynamics affecting the coast.

For example, Amazon gulf region is conditioned by the interaction of different meteorological and oceanographic agents, owing to the intense water and sediment discharge, the high annual precipitation, wave action, and especially the meso-mega tidal regime (Pereira et al. 2010). These agents also have significant seasonal variability and generate a challenging morphology to study.

In this chapter the diverse behavior has been confirmed by analyzing the reanalysis wave, tide, wind, coastal flooding and storm surge databases generated for Brazilian coast, following the methodologies described in IHCantabria (2013a, b). These databases have been calibrated and validated and reflect a good fit to previous studies, except in Amazon region, due to its complexity.

During the last few decades, the interest in Brazilian coast has increased owing to major environmental and coastal erosion problems. For example, Massaguaçu beach (SP) has an erosion problem in its central part, leading to shoreline retreat and the exposure of littoral infrastructures (see Appendix 1). In order to help in the study of coastal dynamics and their effects on the coast, a new version of Coastal Modeling System (SMC-Brasil) has been developed. It is a user-friendly system specifically designed to assist coastal designers and managers in the analysis of marine and littoral dynamics to understand the changes in coastal caused by those dynamics.

SMC- Brasil includes new reanalysis databases (more than 60 years, every hour), methodologies and tools to analyze local dynamics and its response on the coast in the short, medium and long-term, or even estimate their changes in the future to analyze the potential effects and impacts of those changes on the coast. These characteristics make this system suitable for many coastal engineering applications. For example, this system can carry out studies of sediment transport along the coast in order estimate the potential sedimentation or erosion areas to estimate the gravity of the problem and propose coastal works to reduce the problem.

Coastal flooding is another important topic worldwide, because sea level rise and extremely water levels are flooding some areas that were not previously flooded or are inundated more frequently. In order to help in the estimation of potential inundation zones, a flooding assessment document (Ministério do Meio Ambiente, 2014) was elaborated. This document provides general information about the flooding in coastal areas and can be used to obtain pre-design or rough values of the inundation. However, a more reliable estimation can be obtained by using SMC-Brasil, because it has implemented the required databases and methodologies to consider all the local dynamical processes and morphological variables in the study. This system takes into account the potential climate change effects on the flood line, and consequently, it helps in the estimation of potential flooding risk along Brazilian coast.

Appendix 1: Massaguaçu Beach SMC-Brasil Case Study

Massaguaçu beach is an embayed beach located on the north coast of Sao Paulo in lee of several islands that affect the wave propagation toward the coast and, consequently, the beach morphology. It has an erosion problem along its central part (Fig. 2.13).

Fig. 2.13 Erosion problems in central part of Massaguaçu Beach

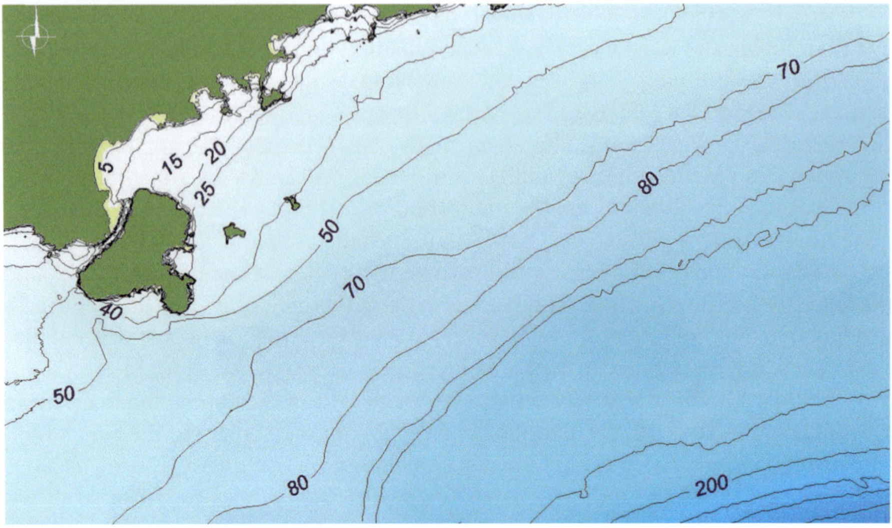

Fig. 2.14 Study area bathymetry obtained from SMC-Brasil (Database generated with General Bathymetric Chart of the Oceans, Brazilian nautical charts and some local bathymetries)

Bathymetry and wave climate data needed to carry out this study were obtained from SMC-Brasil database, which contains offshore and local bathymetry (Fig. 2.14) and wave climate information (Fig. 2.15).

The SMC-Brasil wave climate database, predicts 85 % of the waves arriving from the east-south, with the most energetic waves coming from the south to south-southeast.

Once the offshore wave climate is characterized, it is possible to propagate the wave climate toward the coast using SMC-Brasil. Figure 2.16 shows a southerly storm propagation. As can be seen in this figure, although southern waves are very energetic offshore, the islands provide considerable shelter to Massaguaçu Beach and significantly reduce the wave energy that reaches the beach (approximately 70 % in this case).

The analysis of current patterns for all the wave directions in the study area revealed that there are three main zones along the beach (Fig. 2.17):

- In the south, currents are irregular, with direction depending on the wave direction as well as transverse currents.
- In the center, there is a reduction in current magnitude and a change in direction.
- In the north, currents generally increase toward the northeast, except at the end of the beach, where an offshore rip current is generated for some wave directions.

These wave dynamics and currents generate a net sediment transport from the central part of the beach toward the extremes, with a net longshore sand transport toward the north, resulting in an erosional "hot spot" in the center of the beach,

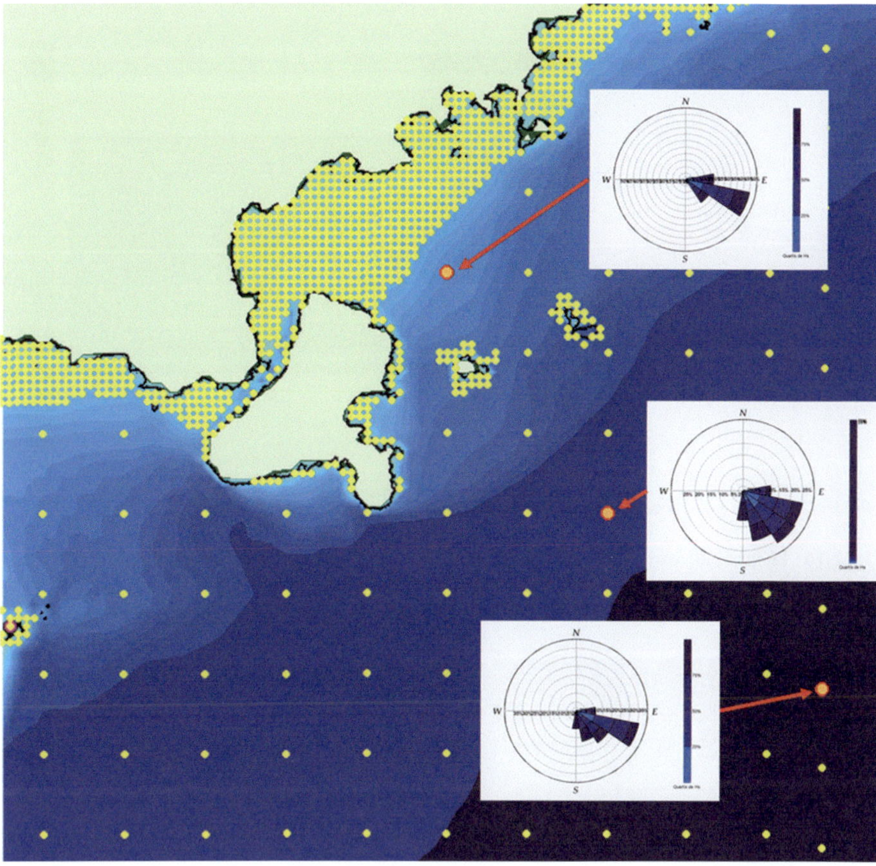

Fig. 2.15 Downscaled Ocean Waves (DOW) points near the study site are shown. Wave climate direction of three points are shown to indicate the variations of wave direction in the area of study

where historically the beach has had erosion problems. In order to check the beach stability, the equilibrium planform of Massaguaçu was obtained by using SMC-Brasil. This system fits different equilibrium planform models based on the wave climate at the control point and the energy flux direction. Figure 2.18 shows the long-term equilibrium planform in blue and the shoreline in 2006 in black, and confirms sediment transport toward the north is responsible for shoreline retreat in central part of the beach.

Once the wave-beach morphodynamic are analyzed, coastal works can be proposed to reduce the erosion problem in the study area. For example, one of the proposed solutions in this study was the construction of a detached breakwater in the north (Fig. 2.19). This solution generates a static equilibrium planform in the central-north zone that could reduce the present littoral drift toward the north. The proposed detached breakwater could generate a 60 m width dry beach in the north and predicts a 40 m shoreline advance at central part. However, it requires a large amount of sand nourishment (approximately 1,400,000 m^3) and there is a lack of

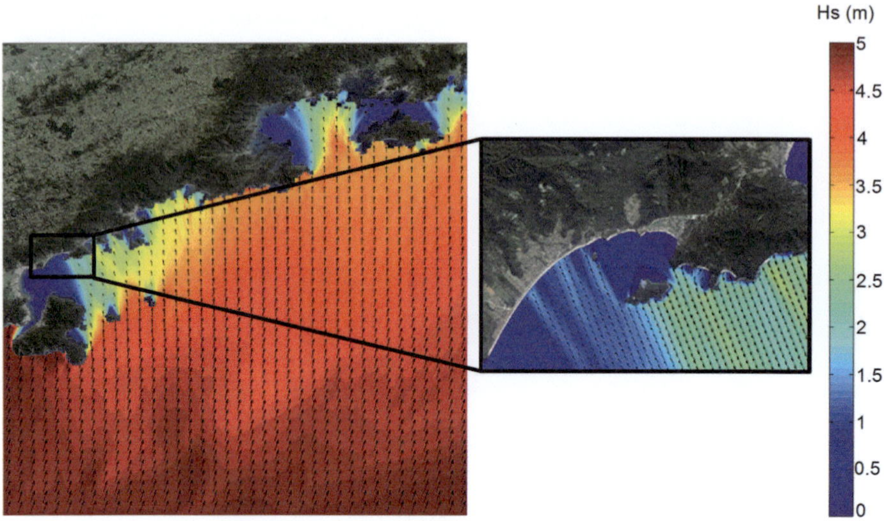

Fig. 2.16 Height and wave direction maps obtained for a southern storm ($H_s = 5$ m and $T_p = 15$ s, approximately). Massaguaçu Beach insert to right

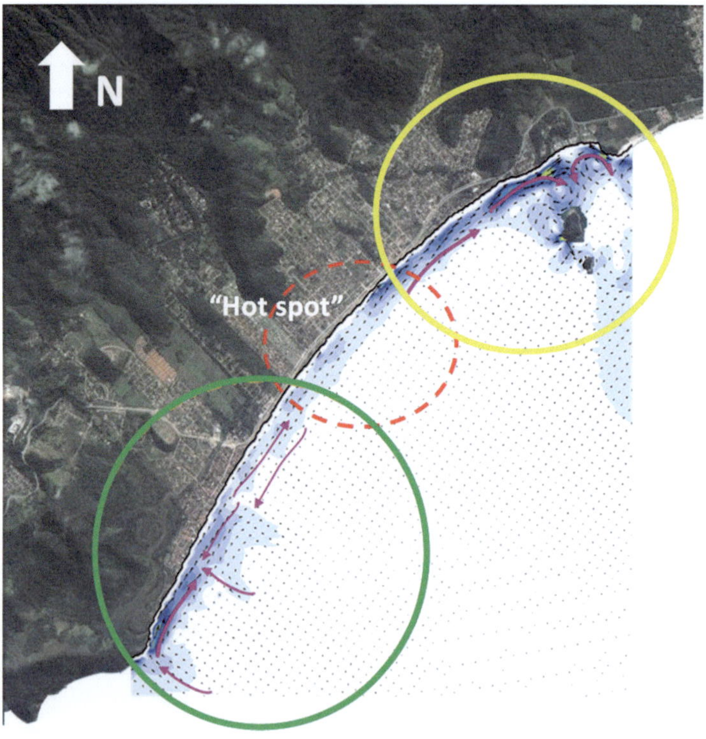

Fig. 2.17 Analysis of current patterns in the three zones of the study area using SMC-Brasil

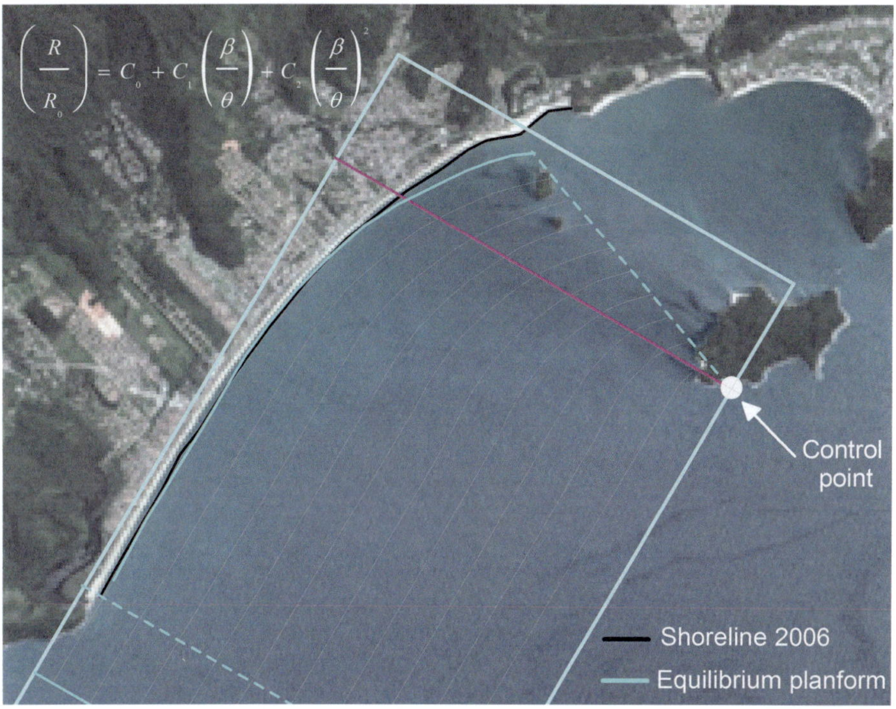

Fig. 2.18 Massaguaçu equilibrium planform

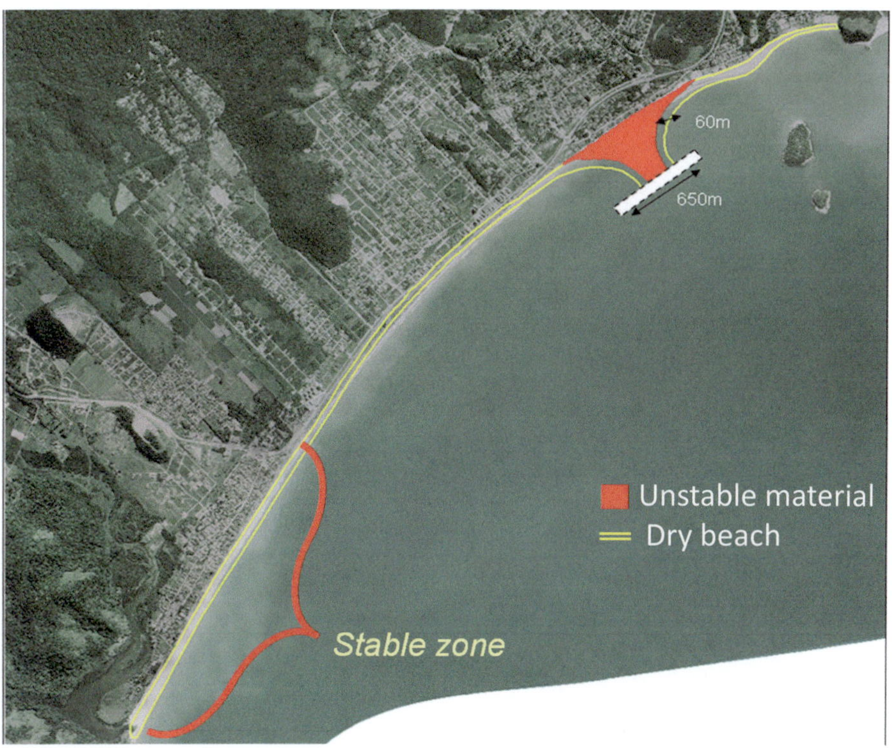

Fig. 2.19 Location of the dettached breakwater and predicted equilibrium planform proposed to solve the central beach erosion

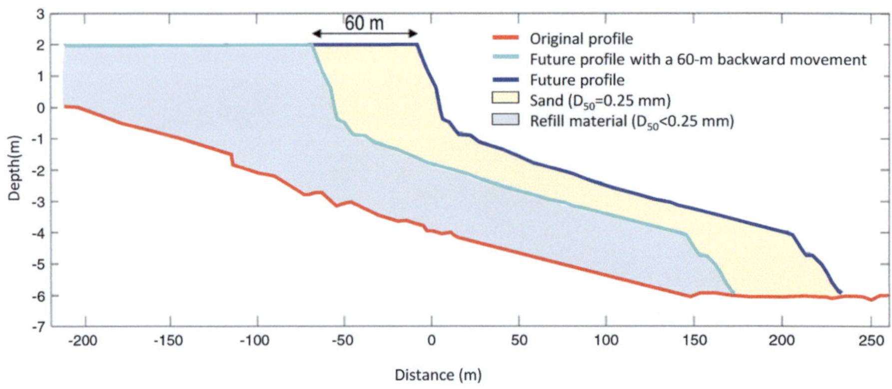

Fig. 2.20 Future equilibrium profile proposed to solve the erosion problem at central part

natural sand sources near the study site. In fact, there is a minimum sand volume required to stabilize the beach, which depends on the dynamics, their variability and the nourishment sand size; but the rest of the refill material can come from other sources, even artificial. In order to reduce the sand volume, it was proposed to nourish the active profile (beach profile affected by dynamics and their variability) with a sediment that permit stabilizing the beach in the middle and long-term ($D_{50}=0.25$ mm), with the non-active profile filled with another material ($D_{50}<0.25$ mm) because it is not affected by the dynamics and their variability. Figure 2.20 shows the associated equilibrium profile.

Finally, the SMC-Brasil can assess the impacts of global climate change impacts in future solutions and can take into account measures to mitigate negative impacts in the present design.

References

Aage CTD, Allen DJT, Carter G (1998) Oceans from space. Édition Ifremer, Plouzané, 162 p

Battjes JA, Janseen JPFM (1978) Energy loss and set-up due to breaking of random waves. In: Proceedings of the 16th international coastal engineering conference, ASCE, pp 569–587

Booij N, Ris RC, Holthuijsen LH (1999) A third-generation wave model for coastal regions. Part I: model description and validation. J Geophys Res 104(C4):7649–7666

Bryant E (1979) Comparison of computed and observed breaker wave heights. Coast Eng 9:39–50

Cachione DA, Drake DE, Kayen RW, Sternberg RW, Kineke GC, Tayle GB (1995) Measurements in the bottom boundary layer on the Amazon subaqueous delta. Mar Geol 125:235–239

Calliari LJ, Faria AFG (2003) Bancos de lama na praia do Cassino: formação, implicações geomorfológicas, ambientais e riscos costeiros, estudo de caso: maio de 2003. In: Abstracts of the IX Congresso da Associação Brasileira de Estudos do Quaternário. Universidade Federal de Pernambuco, Recife

Calliari LJ, Speranski N (2006) Padrões de refração de ondas para a do Rio Grande do Sul e sua relação com a erosão costeira. In: Muehe D (ed) Erosão e progradação do litoral brasileiro, 1st edn. Ministério do Meio Ambiente, Brasilia, pp 446–454

Camus P, Méndez F, Medina R (2011) A hybrid efficient method to downscale wave climate to coastal areas. Coast Eng 58(9):851–862. doi:10.1016/j.coastaleng.2011.05.007

Cid A, Castanedo S, Abascal AJ, Menéndez M, Medina R (2014) A high resolution hindcast of the meteorological sea level component for Southern Europe: the GOS dataset. Clim Dyn 43:2167–2184. doi:10.1007/s00382-013-2041-0

Costa DS (2007) Variação do nível médio do mar – técnicas para a avaliação. Dissertation, University of São Paulo, USP

Costa Neto LX (2001) Transporte longitudinal de sedimentos ao largo do município de Macau, litoral Norte do RN. XIX Simpósio de Geologia do Nordeste. Natal, SBG, p 65–66

Cotton P (1998) A feasibility study for a global satellite buoy intercalibration experiment, Southampton oceanography centre resolution consultation report 26. Southampton Oceanography Centre, Southampton, p 73

Egbert GD, Erofeeva SY (2002) Efficient inverse modelling of barotropic ocean tides. J Atmos Ocean Technol 19(2):183–204. doi:10.1175/1520-0426(2002)019<0183:EIMOBO>2.0.CO;2

Egbert GD, Bennett AF, Foreman MGG (1994) TOPEX/POSEIDON tides estimated using a global inverse model. J Geophys Res 99(C12):24821–24852. doi:10.1029/94JC01894

Garcia CAE, Sarma YVB, Mata MM, García VMT (2004) Chlorophyll variability and eddies in the Brazil–Malvinas Confluence region. Deep-Sea Res 51:159–172. doi:10.1016/j.dsr2.2003.07.016

GIOC (2003) Documento Temático, Atlas de Inundación del Litoral Peninsular Español. Ministerio de Medio Ambiente – Dirección General de Costas and Universidad de Cantabria, Spain

Guedes Soares C (1986) Calibration of visual observations of wave period. Ocean Eng 13(6):539–547

Haidvogel DB, Arango HG, Hedstrom K, Beckmann A, Malanotte-Rizzoli P, Shchepetkin AF (2000) Model evaluation experiments in the North Atlantic Basin: simulations in nonlinear terrain-following coordinates. Dyn Atmos Oceans 32(3–4):239–281. doi:10.1016/S0377-0265(00)00049-X

Haidvogel DB, Arango H, Budgell WP, Cornuelle BD, Curchitser E, Di Lorenzo E, Fennel K, Geyer WR, Hermann AJ, Lanerolle L, Levin J, McWilliams JC, Miller AJ, Moore AM, Powell TM, Shchepetkin AF, Sherwood CR, Signell RP, Warner JC, Wilkin J (2008) Ocean forecasting in terrain-following coordinates: formulation and skill assessment of the regional ocean modeling system. J Comput Phys 227(7):3595–3624. doi:10.1016/j.jcp.2007.06.016

Hemer M, Church J, Hunter J (2010) Variability and trends in the directional wave climate of the southern hemisphere wave climate of the southern hemisphere. Int J Climatol 30:475–491. doi:10.1002/joc.1900

Holthuijsen LH (2007) Waves in oceanic and coastal waters. Cambridge University Press, New York, 388 p

IHCantabria (2013a) Documento Temático de Onda

IHCantabria (2013b) Documento temático do Nível do Mar e Cota de Inundação

IHCantabria (2013c) Documento Temático de Recomposicao das praias

IHCantabria (2013d) Documento Temático dos Efeitos de Mudança Climática em Praias

Innocentini V, Prado SCC, Pereira CS, Arantes FO, Brandão IN (2000) Marulhos no litoral norte do Brasil geradas por furacões: Caso 24 de outubro de 1999. In: Abstracts of the XI Congresso Brasileiro de Meteorologia, Rio de Janeiro, October 2000

Jardine TP (1979) The reliability of visually observed wave heights. Coast Eng 9:33–38

Lemos AT, Ghisolfi RD (2011) Long-term mean sea level measurements along the Brazilian coast: a preliminary assessment. Pan-Am J Aquat Sci 5(2):331–340

Machado AA, Calliari LJ, Melo E, Klein AHF (2010) Historical assessment of extreme coastal sea state conditions in southern Brazil and their relation to erosion episodes. Panam J Aquat Sci 5(2):277–286

Melo E (1993) The sea sentinels project: watching waves in Brazil. In: Proceedings of the 8th symposium on coastal ocean manage, ASCE, pp 505–517

Melo E (2004) Programa de Informação Costeira on-line: um balanço dos primeiros dois anos de funcionamento. In: Proceedings of the 1 seminario e workshop em engenharia oceânica, FURG, pp 1–23

Mínguez R, Espejo A, Tomás A, Méndez FJ, Losada IJ (2011) Directional calibration of wave reanalysis databases using instrumental data. J Atmos Ocean Technol 28:1466–1485. doi:10.1175/JTECH-D-11-00008.1
Ministério do Meio Ambiente (MMA) (2014) Uma Proposta de Abordagem para o Estabelecimento de Regimes Probabilísticos de Inundação Costeira do Brasil. Ministério do Meio Ambiente, Brasília, No prelo
Muehe D (2006) Erosão e progradação do litoral brasileiro. Ministério do Meio Ambiente, Brasilia, 476 p
Neves CF (2005) O nível médio do mar: uma realidade física ou um critério de engenharia? Vetor 15(2):19–33
Nicolodi JL, Pettermann RM (2011) Vulnerability of the Brazilian coastal zone in its environmental, social, and technological aspects. J Coast Res SI 64:1372–1379
Nielsen P, Hanslow DJ (1991) Wave run-up distributions on natural beaches. J Coast Res 7(4):1139–1152
NOAA, National Geophysical Data Center (2006) Etopo2v2 global gridded 2-minute database
Parise CK, Calliari LJ, Krusche N (2009) Extreme storm surges in the south of Brazil: atmospheric conditions and shore erosion. Braz J Oceanogr 57(3):175–188
Pereira PS, Calliari LJ, Barletta RC (2010) Heterogeneity and homogeneity of Southern Brazilian beaches: a morphodynamic and statistical approach. Cont Shelf Res 30:270–280
Pianca C, Mazzini PL, Siegle E (2010) Brazilian offshore wave climate based on NWW3 reanalysis. Braz J Oceanogr 58:53–70
Pimenta FM, Melo E, Franco D, Zavialov PO (2004) Assessment of santa catarina shelf currents through the analysis of indirect measurements. J Coast Res SI 39:651–655
Pirazzoli PA (1986) Secular trends of relative sealevel (RSL) changes indicated by tidegauges records. J Coast Res SI 1:1–26
Plant NG, Griggs GB (1992) Comparison of visual observations of wave height and period to measurements made by an offshore slope array. J Coast Res 8(4):957–965
Reguero BG, Menéndez M, Méndez FJ, Mínguez R, Losada IJ (2012) A Global Ocean Wave (GOW) calibrated reanalysis from 1948 onwards. Coast Eng 65:38–55. doi:10.1016/j.coastaleng.2012.03.003
Shchepetkin AF, McWilliams JC (2005) The regional ocean modeling system: a split-explicit, free-surface, topography following coordinates ocean model. Ocean Model 9:347–404
Smith WHF, Sandwell DT (1997) Global sea floor topography from satellite altimetry and ship depth soundings. Science 277:1956–1962
Strauch JC (2001) Caracterização do Estado do Mar em Rio Grande – XIV Simpósio Brasileiro de Recursos Hídricos. Aracajú, SE
Strauch JC, Cuchiara DC, Toldo EE Jr, Almeida LESB (2009) O Padrão das Ondas de Verão e Outono no Litoral Sul e Norte do Rio Grande do Sul. Rev Bras Recur Hidr 14(4):29–37
Teixeira PS (2007) Subsídios para o gerenciamento costeiro por erosão da falésia no balneário Hermenegildo. Dissertation, Federal University of Rio Grande, FURG
Tessler MG, Goya SC (2005) Processos costeiros condicionantes do litoral brasileiro. Revista do Departamento de Geografia, USP 17:11–23
Thornton EB, Guza RT (1983) Transformation of wave height distribution. J Geophys Res 99(c10):5925–5938
Tolman H (2009) User manual and system documentation of Wavewatch-III version 3.14. NOAA/NWS/NCEP Technical Note (276) p 194
Wainer IJ (1963) Technical report. Análise e Previsão das Alturas de Onda em Tramandaí. Federal University of Rio Grande do Sul, Porto Alegre. 30 p
Winyu R, Tomoya S (1998) Energy dissipation model for regular and irregular breaking waves. Coast Eng 40(4):327–346
Woolf D, Challenor PG, Cotton PD (2002) Variability and predictability of the North Atlantic wave climate. J Geophys Res 107(C10):9-1–9-14. doi:10.1029/2001JC001124

Chapter 3
Beaches of the Amazon Coast: Amapá and West Pará

Valdenira Ferreira dos Santos, Andrew D. Short, and Amilcar Carvalho Mendes

Abstract The Amazon river flows into the funnel-shaped Amazon Gulf with its shoreline stretching for hundreds of kilometers in either direction. The river delivers massive amounts of mud and some sand to the coast, which is dominated by the river flow, macro to mega-tides, general low easterly waves and the strong Brazilian Current. While most of the coast consists of wide intertidal mud flats and muddy riverbanks, there are approximately 500 sandy beaches located in five Amazon sub-provinces. These are the northern Amapá coast with some longer beaches; the Amazon river Amapá and Pará shores with beaches located on the outer interdistributary islands and some small beaches along river banks; and the northern coast of Pará's Marajó island of which more than half is tide-dominated sandy beaches.

3.1 Introduction

Brazil's northernmost coastal province is dominated by the Amazon River and its equatorial location between 4°N to 1.5°S. The Amazon shelf and river mouth were flooded during the postglacial marine transgression forming the 200 km wide Amazon funnel-shaped estuarine-river mouth occupied by the two major channels together with numerous inter-distributary channels and islands. The high discharge and sediment load, combined with macro to mega-tidal ranges, in a hot wet tropical

V.F.d. Santos (✉)
Instituto de Pesquisas Científicas e Tecnológicas do Estado do Amapá-IEPA, Núcleo de Pesquisas Aquáticas – NuPAq, Rodovia JK, km 10, Fazendinha Campus, 68903-280 Macapá, AP, Brazil
e-mail: valdeniraferreira@gmail.com

A.D. Short
School of Geosciences, University of Sydney, Sydney, NSW 2006, Australia
e-mail: andrew.short@sydney.edu.au

A.C. Mendes
Museu Paraense Emílio Goeldi-MPEG, Campus de Pesquisa, Coordenação de ciências da Terra e Ecologia, 1901, Perimetral Avenue – Terra Firme, 66077-830 Belém, PA, Brazil
e-mail: amendes@museu-goeldi.br

A.D. Short, A.H. da F. Klein (eds.), *Brazilian Beach Systems*, Coastal Research Library 17, DOI 10.1007/978-3-319-30394-9_3

Table 3.1 Shoreline characteristics of the Amazon coast

	Amazon coastal sub-provinces	Coast length (km)[a]	Beach number	Beach length (km)
1[b]	northern Amapá coast (inc. Ilha de Maracá 125 km)	675	16	142
2[b]	southern Amapá Amazon (ocean beaches)	495	3	44
	Canal do Norte do Rio Amazonas (pocket)		14	4
3[b]	Pará Amazon Coast (inc. Canal do Sul do Rio Amazonas)	1500	~50	~110
4[b]	Pará: Marajó Island (north & east coast)	390	168	121
	Sub-total (this chapter):	3060	~260	~420
5[c]	Pará: east coast	790	247	n/a
	Total:	3850	~510	n/a

[a]Note distances in the river mouth are approximate and minimal
[b]This chapter
[c]Chapter 4

environment have all contributed to the present coastal environment, which is dominated by river channels with muddy shorelines backed by mangrove forests and inundated varzea lowlands. The entire coast has undergone substantial Holocene progradation on the order of kilometres, together with considerable northerly transport of the Amazon mud as far as Venezuela. Sandy beaches with wide sand flats and mud flats, only occur in more exposed locations, the latter mainly along the northern Amapá coast and on Pará river mouth islands.

The Amazon Coastal Zone (ACZ) is a physiographic feature extending for 2700 km from the Orinoco Delta in the Venezuela to Baia de São Marcos in Maranhão State, and includes the coasts of the Republics of Guyana and Suriname, French Guyana and northern Brazil. This chapter will focus on the Brazilian section of the ACZ, specifically the coasts of Amapá and western Pará.

The Amapá-Pará section of the ACZ coast extends for approximately 3850 km from the border with French Guyana at Cabo Orange to Pará's eastern border near Cabo Gurupi. This section can be divided into five sub-provinces (Table 3.1 and Fig. 3.1): (1) the northern open Amapá coast including Maracá island; (2) the southern Amapá Amazon coast extending along the Amazon's North Channel and including Bailique archipelago and other islands; (3) the Pará Amazon mouth which includes Canal do Sul do Rio Amazonas, together with of several inter-distributary channels and river mouth islands; (4) Pará's Marajó island north and east coasts; and (5) the eastern Pará coast. This chapter is concerned with the Amapá and western Pará coast, including the Amazon river mouth and the northern and eastern shores of Marajó Island, a coastline of approximately 3060 km. The eastern Pará coast, commencing at Belém, is covered in Chap. 4.

Although some studies were undertaken in the late twentieth century, it is only since 2002 that scientists from Instituto de Pesquisas Científicas e Tecnológicas do Estado do Amapá-IEPA, Museu Paraense Emílio Goeldi-MPEG and Universidade Federal do Pará-UFPA, commenced detailed studies of the Brazilian Amazon coast and shoreline as part of a research network concerned with environmental monitor-

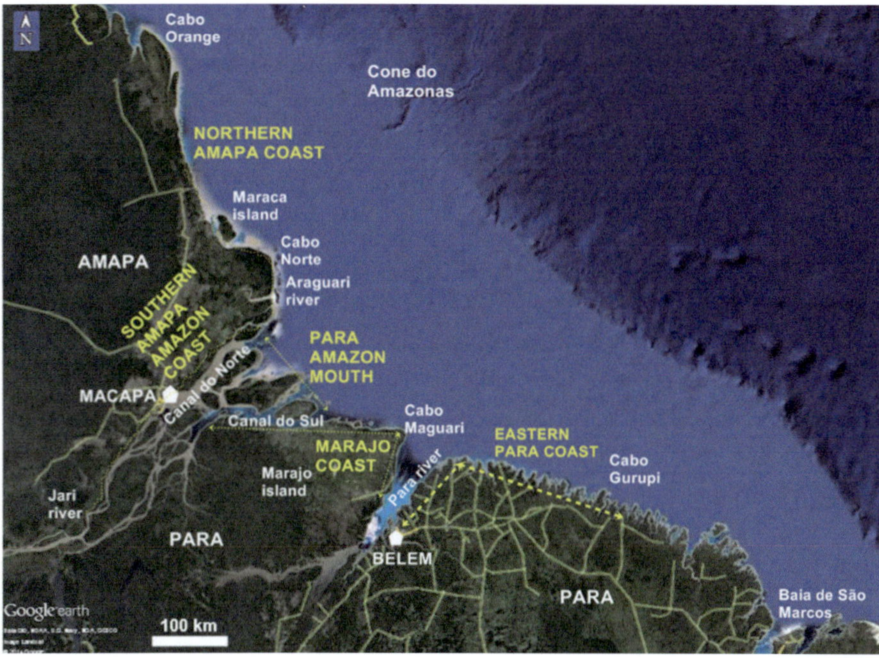

Fig. 3.1 The Amazon coast of Amapá and Pará states (Source: Google earth)

ing of areas affected by the oil industry (Rede 05-PETROMAR, PIATAM Mar, PIATAM Oceano and Cartas SAO FZA projects) together with monitoring and modelling of erosion and coastal occupation (Millennium Institute Project, INCT-Ciências do Mar) and other projects to support the management plans for coastal conservation units. This chapter will review our growing knowledge of this coast and in particular the sandy beaches that occur along this predominately muddy mangrove-lined, tide-dominated coast at the mouth of the world's largest river. Our present knowledge however remains preliminary, as most beach investigations have been superficial with detailed studies yet to be undertaken.

3.2 The Amazon Coastal Environment

3.2.1 Geology and Physiography

This coast is located in the outer part of the Amazon basin where the sediments of the ACZ have developed a wide Quaternary plain in Amapá and Pará states (Fig. 3.2). The plain is backed in the north by pre-Cambrian rocks of the Amazonian craton (crystalline basement) described by Tassinari and Macambira (1999), while in the south of Amapá state and southeastern of Pará state the sequences from the Amazonas and Parnaiba sedimentary basins outcrop with sediments from the

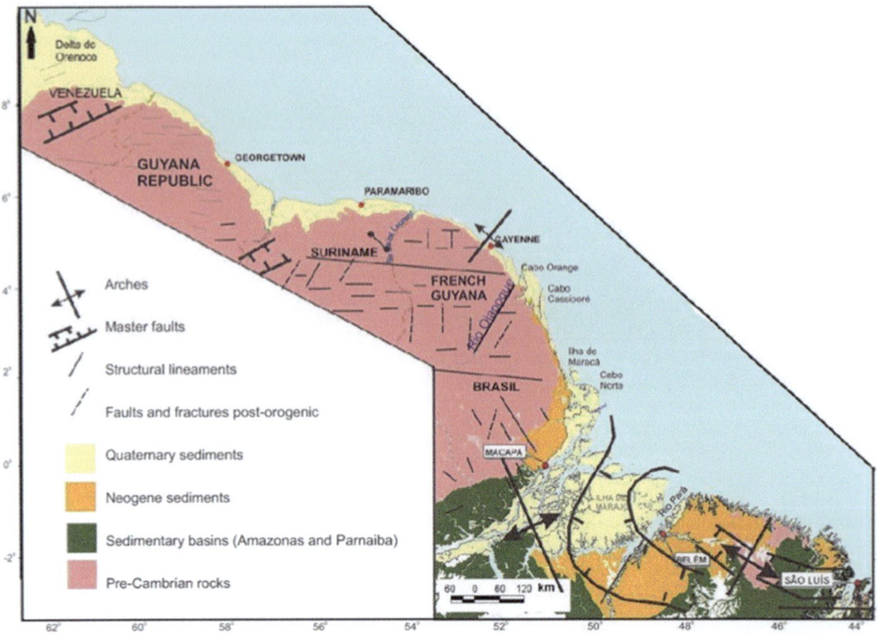

Fig. 3.2 Geological structure of the Amazon coast (Source: Modified from Santos 2006)

Barreiras Formation and Pós-Barreiras, which reaches the coast in Pará state (Fig. 3.2). Several estuaries cross the coastal plain, aligned by the regional structure. The most representative estuaries are, from the north to south: Oyapock, Cassiporé, Cuñani, Calçoene, Araguari, and the Amazon itself, and the estuary of the Pará River. Most of the Amapá and western Pará shoreline consists of intertidal mud or sand flats up to a few kilometer wide on the open coast, but narrowing along the river banks, all backed by wide dense varzea forests with transition to mangrove towards the Amazon mouth. Two nearly continuous zones of sandy sediments occur along the northern coast of Amapá south of Cabo Cassiporé (see 3.3.1), and along the northern and eastern coast of the island of Marajó (see 3.3.4 and 3.3.5), as well as on some of the more exposed sections of the river mouth island, particularly Mexiana Island (see 3.3.3). Elsewhere, small pocket beaches are located along some of the river channels (see 3.3.2). Sandy sediments have also been found elsewhere along the coast in the mouths of estuaries and deposited in cheniers, sand bars, sand waves, sand plains and sand ridges (Augustinus 1989; Prost 1989; Mendes 1994; Silveira 1998; Allison et al. 1995, Souza-Filho and Paradella 2003, Anthony et al. 2014).

The present coast is a product of the Holocene sea level rise, which drowned the Amazon mouth, together with Quaternary sea-level changes, the large fluvial sediment supply and the reworking of relict sediments on the continental shelf. The Amazon shelf has experienced massive sedimentation and is presently more than 300 km wide, gently sloping, with an overall gradient of 1:2240 until the shelf break at 100 m depth (Milliman 1979). The most prominent feature observed in the adjacent Amapá shelf is the Amazon Cone (Fig. 3.1), situated north of the Amazon river mouth. It is located on the outer shelf and is about 700 km long and 250 km wide, reaching 650 km in the northernmost sector with an area of approximately 162,000 km^2 (Damuth and Kumar 1975).

The four coastal regions covered in this chapter commence with the northern Amapá coast. North of the Capo Cassiporé is an extensive zone of pelitic deposition, the largest on the planet, forming mud banks that extend along of 1600 km of coast including French Guyana-Suriname-Guyana coasts and reach the Orinoco Delta in Venezuela. These mud banks originate from the Amazon sediments and migrate shoreward and then alongshore by a combination of wave forcing, tidal currents, easterly wind-induced coastal currents and proximity to the coast of the North Brazil Current (NBC) (Augustinus 1978; Wells and Coleman 1978; Allison and Lee 2004; Anthony et al. 2013, 2014; Nittrouer and DeMaster 1996). This coast is predominately muddy north of Cape Cassiporé, while sandy sediments with intercalation of mud plains dominate the shoreline to the south until the mouth of the Amazon river.

The eastern Amapá coast has experienced substantial sedimentation with the Cabo Norte area, up to 80 km wide, dominated by three lacustrine belts. According to Silveira (1998) these lakes have largely infilled over the past few hundred years resulting in a palaeo-drainage network with ox-bow lakes such as Lago Duas Bocas, Lago Novo, Lago Comprido de Dentro, Lago dos Botos, and Lago Mutuco forming the Meridional Lacustrine belt. The lakes of the oriental belt are now occupied by extensive mangrove ecosystems over consolidated muddy deposits, which are more than 20 km in width with a predominance of *Rizhophora* species. The tallest mangrove trees and widest mangrove extension in Brazil occur on the wide coastal plain at Cabo Norte with trees up to 40 m high and along the Amapá coast. This area has ideal conditions to maintain mangroves, with a large amount of suspended sediment transported by the Amazon River, high temperatures, large tidal ranges and saline to brackish water (Costa Neto et al. 2006, Rabelo et al., 1994).

The Pará Amazon coast includes the Amazon mouth, which is influenced by the massive Amazon River and dominated by a 300 km wide funnel-shaped estuarine mouth. The river divides into two major channels, the Canal do Norte (North Channel) and Sul (South Channel), together with numerous smaller interconnected channels. The North Channel flows northeast along the Amapá coast and is 14.5 km wide at its mouth adjacent to Curua Island. The South Channel flows east-southeast between Mexiana and Marajó islands and is 17 km wide at its mouth. The broad coastal plain at the mouth of the Amazon and along northern Amapá contains several paleochannels and lines of paleocoast that record the Holocene evolution of the region (Boaventura and Narita 1974; Silveira 1998; Rossetti and Goes 2008; Santos et al. 2009; Jardim et al. 2013) (Fig. 3.3). The cities of Macapá and Santana, located right on the equator, are the principal settlements on this section of coast. These two adjoining cities and their adjacent communities are spread along about 30 km of the Canal do Norte of the Amazon coast. They are located where the first exposures of the higher elevation of Pleistocene deposits occur on the riverbanks enabling part of the cities to be located above the annual flood levels. The Portuguese built a substantial fort on the Pleistocene bluffs at Macapá in 1764.

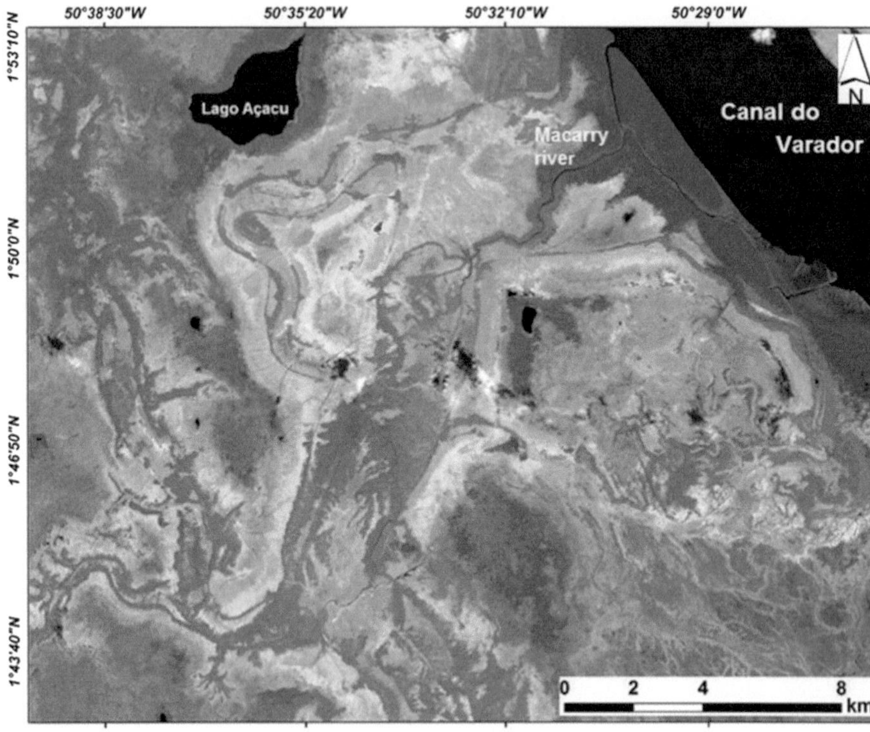

Fig. 3.3 Paleochannel along the Amazon coast. Paleo-Macarry in front of Ilha de Maracá (Source: Landsat 7 ETM+, band 4, 18 November 2000)

The Marajó Island coast is dominated in part by the elevated Pleistocene deposits which form most of the island and bluffs and headlands where exposed at the coast, particularly along the eastern shore of the island. Holocene deposits form a narrow coastal plain along the northern coast, which widens to about 10 km at Cabo Maguari on the northeastern tip of the island. The coastal deposits then narrow south into Marajó Bay until the Paracauari river, with the older deposits being exposed south of the river forming the coastal plateau (França and Souza-Filho 2006). Sandy beaches occupy about one third of the coast the remainder dominated by mangroves.

3.2.2 Climate

The Amazon coast straddles the equator and is exposed to a hot, wet tropical climate (Koppen Af), with temperatures averaging between 25 and 27 °C. The weather is modulated by the position of the Intertropical Convergence Zone (ITCZ), which shifts from around 14°N in August to 2°S in March-April. Trade winds blowing towards the ITCZ dominate the wind regime and arrive predominantly from the north through east, with wind velocities averaging 2–4 m s^{-1}, increasing up to 10 m s^{-1} during September. Rainfall averages 2600 mm at Macapá increasing both northward to reach 3750 mm at Cayenne, and eastward reaching 2900 mm at Belem. Most of the rain arrives during the January to May wet season, while humidity is high year round and usually above 80 %.

Climatic variations are produced by La Nina and El Nino events, causing precipitation to be higher during La Nina and lower during periods of El Nino. These conditions are also reflected in river discharge (Richey et al. 1989; Santos et al. 2010).

3.2.3 *Fluvial Processes*

The Amazon River mouth is part of a vast and complex fluvial-estuarine system that includes the Amazon River and dozens of estuaries that contribute to form the largest run-off of water, sediments, dissolved nutrients and organic material found anywhere in the world (Meade et al. 1979, 1985; DeMaster and Pope 1996; Geyer et al. 1996). The enormous fluvial discharge of the Amazon River into the Atlantic Ocean is approximately 20 % of the total discharged on the whole planet. The mean water discharge is around 180,000 m^3 s^{-1} with peaks of up to 220,000 m^3 s^{-1}, occurring in May–June and a minimum of 100,000 m^3 s^{-1} in November–December (Richey et al. 1986). It is assumed the water discharge of the smaller rivers that border the Amazon will have seasonal patterns that are more or less synchronous with that of the Amazon river. Accord Martinez et al. 2009, the water discharge has a weak correlation with the suspended sediment concentration, which presents a significant increase in the budget of the Amazon with a suspended sediment discharge 20 % higher between 1996 and 2007.

Towards the coast the discharge is affected by the macro to mega-tides that are amplified in crossing the broad shelf and within the funnel-shaped mouth and distributaries, with the tide extending up to 800 km upstream to the town of Óbidos (Beardsley et al. 1995; Nittrouer et al. 1995).

The river level rises and falls with the tides in the lower reaches, with asymmetrical flow reversals – a longer ebb and shorter flood, the latter producing tidal bores in some channels, called *pororoca* (Santos 2006). The river level also rises and falls seasonally peaking during the wet season and reaching up to 5 m higher (at Óbidos) during the wet season, before falling during the dry season. El Nino and La Nina events also influence the height of the rises and falls of the river level.

3.2.4 *Sediment Discharge*

The Amazon River delivers a massive sediment load estimated at 754 Mt $year^{-1}$ (+/− 9 %), which is deposited on the inner shelf and then 30 % is transported to the northwest as large migratory mud banks for 1600 km northward towards French

Guyana, Suriname and Venezuela (Anthony et al. 2010, 2014). According to Muller-Karger et al. (1988) the net effect of the Amazon river discharge, the North Brazil Current (NBC) and the trade winds and waves is to mix the river water with the saline water of Atlantic Ocean, forming an immense turbid aqueous plume, with reduced salinity, which extends at least 300 km northwest.

While the suspended load is well documented (Gibbs 1967; Meade et al. 1979, 1985; Nittrouer et al. 1986), there is also a small (~1 % of load) but substantial bedload transported by the Amazon river, estimated at 0.01–0.05 Mt day^{-1} or 3.65–18.25 Mt year^{-1} (Dunne et al. 1998). The bedload consists of medium to fine sand and is deposited as small sandy beaches and sand flats along some riverbanks, and at the mouth as more extensive tide-dominated beaches and extensive sand flats.

Along the northern coast of Amapá tide-dominated sandy beaches are backed in places by beach ridges and/or cheniers. Silveira 1998 pointed out three clusters of chenier plains on the north coast of Amapá with the source of sediments being intertidal sandflats, in addition to material from the crystalline basement which also contributes to the sandy beach sediments (Mendes 1994),

3.2.5 *Coastal Processes*

Amazon coastal processes are related to the persistent easterly trade winds and associated easterly waves, the macro through mega-tides and their associated currents, coupled with the massive water and sediment discharge from the Amazon, and the northwest flowing NBC.

3.2.5.1 Tides

The 300 km wide shallow Amazon shelf amplifies the tidal range, with the entire coast exposed to semi-diurnal macro-megatides. The presence of the extensive shoal seaward from Cabo Norte influences the tidal behavior (Beardsley et al. 1995; Nittrouer et al. 1995). North of the shoal, the open Amapá coast, including Ilha do Maracá, receives tides as a standing wave in near-resonance with the embayment formed by the shoal and shoreline. In this area the tides range from 4 m in the Ponta dos Índios (Oiapoque river) increasing southwards to 10 m near Ilha do Maracá, the highest in Brazil, then decreasing to 5 m at Ponta Guará in Araguari river mouth. South of the shoal, the tides behave as a progressive wave decreasing from 5 m at the Amazon mouth to 4 m at the Jari River mouth, approximately 300 km upstream. The irregular Pará-Marajó coast is bordered by a series of distributaries of the Amazon, Pará and Tocantins rivers, including Marajó Bay. Tides increase from 6 m to 10 m along the northern shore of Marajó Island, then decrease into Marajó Bay falling to 3 m at Belem. Because of the high tides, strong tidal currents flow along all shores, increasing in strength in the confined river mouths and distributaries. According to Nittrouer et al. (1991a) tidal effects on the Amazon river propagate 800 km upstream to the town of Obidos.

Fig. 3.4 Tidal bore (pororoca) at Araguari river-mouth. Date: 30 October 2008 (Photo: Valdenira Santos, AMASIS Project)

Strong tidal currents (~2 m s^{-1}) are manifest as semi-diurnal tidal currents oriented across the shelf and into the river and its distributaries. The wide shelf also amplifies the tidal wave resulting in tidal resonance in the river mouth (Beardsley et al. 1995; Geyer et al. 1996).

The ideal physiographic conditions associated with the macrotidal regime including a shorter flooding tide, result in the presence of tidal bores called "*pororoca*" (Fig. 3.4) along much of the northern coast of Amapá coast and the Amazon mouth. The phenomenon is stronger at equinox spring tides, when the heights of the waves reach up to 2.5 m (Santos et al. 2005). The velocity of the wave can exceed 23 km h^{-1} in the Araguari river mouth.

3.2.5.2 Waves

The northern Amapá and open Pará coast are exposed to waves generated by the easterly trade winds. They produce easterly ocean waves 1–2 m high, with periods <10 s, 95 % of the year, with the largest waves arriving from January to March when the winds are stronger (Schaeffer-Novelli and Cintron-Moler 1988, Kineke 1993). Very short seas with $H_b \sim 0.1$–0.5 m and $T \sim 2$–5 s accompany the persistent trades and occur both along the coast and within broad sections of the rivers (Fig. 3.5a). While the waves are higher during the January to March wet season they are more effective in periods of low river discharge from September to November.

At the shore however wave energy is very low to zero and extremely tide dependent. Ocean waves are first attenuated by the wide low gradient inner shelf and then by the often kilometres wide intertidal mud and sand flats, as well as tidal mud and sand river mouth shoals, with lowered waves only reaching the shore at

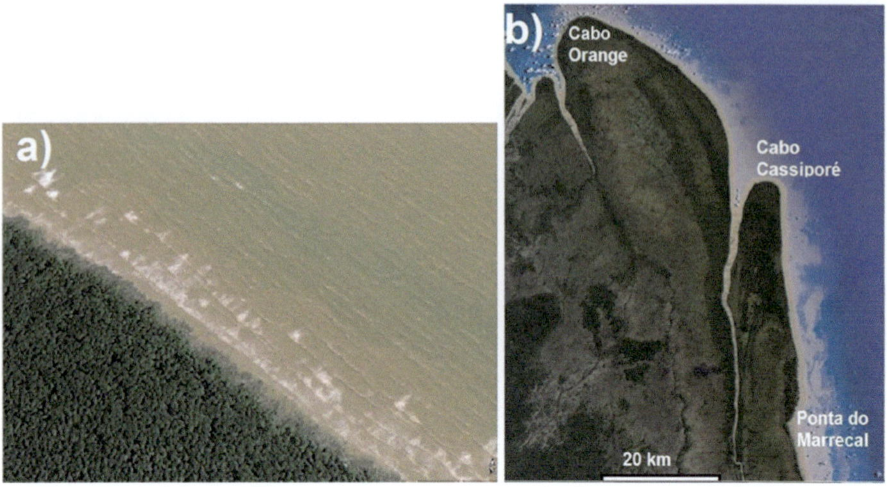

Fig. 3.5 (**a**) Waves breaking across tidal flats near the mouth of the Amapá Grande river (see Fig. 3.6). (**b**) The 85 km long Cabo Cassiporé. (*right*) and Cabo Orange (*left*) coasts have been formed from prograding mud deposits.(Source: Google earth)

high tide, while at low tide they are dissipated across the wide flats often breaking kilometers seaward of the high tide shore. Short seas accompany the swell and are the only waves inside the river and its distributaries. As a consequence wave energy is low to very low for the entire coast, which combined with the high tide range maintains a tide-dominated coast and beaches.

The highest energy section of coast is located between the south of Cabo Cassiporé and Ilha de Maracá where the waves maintain a longshore current that transports sand northward until south of Cabo Cassiporé (Fig. 3.5b). From Cabo Cassiporé northward to Cabo Orange, the shoreline is accreting with prograding mud deposits dominating the shoreline from 85 km south of Cabo Cassiporé to the border with French Guyana. East of Cabo Orange the inter-mudflats are 2–5 km wide (Nittrouer et al. 1991b; Silveira 1998.)

3.2.5.3 Coastal Currents

A strong northwest current prevails along the coast, particularly along the northern Amapá coast. The current is reinforced by the easterly trade winds, wave driven currents and the flooding tidal currents, which move perpendicular to the isobaths of the continental shelf resulting in strong currents across the shelf with speeds exceeding 2 ms^{-1} during spring tides (Beardsley et al. 1995; Geyer and Kineke 1995; Nittrouer and DeMaster 1996). On the muddy plains of the northern Amapá coast, the tidal current velocity can exceed 0.35 ms^{-1} (Allison et al. 1994). Similar velocities have been recorded along the Guyana coasts (Anthony et al. 2010).

The dominance of the easterly winds, waves, tidal and ocean currents all result in east to northeasterly sediment transport along the Amapá coasts. North of the

Amazon mouth, between the Araguari river and Cabo Cassiporé, considerable sand has been deposited as beaches and sand flats. North of the Amapá coast mud dominates the coasts of French Guiana, Suriname, Guyana and part of eastern Venezuela coast, extending in total for 1600 km and forming the longest muddy coast in the world (Allison et al. 1995; Anthony, et al. 2010, 2014).

3.3 Amazon Beach Systems

The Amapá-west Pará-Amazon coast has an open and riverine coastline of approximately 3150 km (Table 3.1), the latter including numerous interdistributary river mouth islands. This section will discuss the sandy beaches located in each of the four sub-provinces listed in Table 3.1.

3.3.1 Northern Amapá Coast (Including Ilha de Maracá)

The northern Amapá coast extends from the Oiapoque river, which forms the border with French Guiana to the southern side of the Araguari river mouth, a distance of 550 km (675 km including Ilha de Maracá) (Fig. 3.6). Sandy beaches occupy 142 km (21 %) of this coastline. Two types of beaches occur on this coast: beach ridge systems and high tide sandy beaches with tidal flats.

Beach ridge systems occupy 109.2 km alongshore of south of Cabo Cassiporé and Amapá Grande river mouth and are arranged in 13 segments along the coast, which is experiencing erosion at rates up to 78.5 my^{-1} (Fig. 3.7) (Silva 2010). Strong tidal currents perpendicular to the coast (with tidal range up 9 m) and the easterly waves maintain the beach ridges (Fig. 3.7a, b), which are up to 50 m wide and have recurved spits at the end of each segment. Beach ridges are migrating to landward with the continuous retreat of the coast, marked by exhumed and overturned Avicennia mangrove. According Mendes (1994) sandy sediments are derived from proximal sources of pre-Cambrian basement and Tertiary sediments carried from the estuaries that cut this coast. Allison et al. (1995) and Nittrouer et al. (1991b) suggest that these areas were formed by ephemeral sandy deposits overlying over-consolidated ("relict") mud on the shoreface deposited during a progradational phase of Amazon mudflats between 500 and 1300 years BP. Chenier clusters deposited during progradational events are also preserved on the coastal plain (Silveira 1998). Anthony, et al. (2014) reported similar eroding chenier systems along the Surinam and Guyana coasts.

Tide-dominated high tide sand beaches fronted by wide tidal sand flats occur in discontinuous accretionary deposits for 32.8 km north of Sucuriju (Fig. 3.7c) and north and south of Araguari river (Araqueçaua and Paratur channels and Vitória island) (Fig. 3.7d). The beaches are composed of fine to very fine 3–5 m thick sand deposits, covering discordantly fine sediments, and are intersected by meandering channels (tidal channels and tidal creeks) that extend from the mangrove vegetation

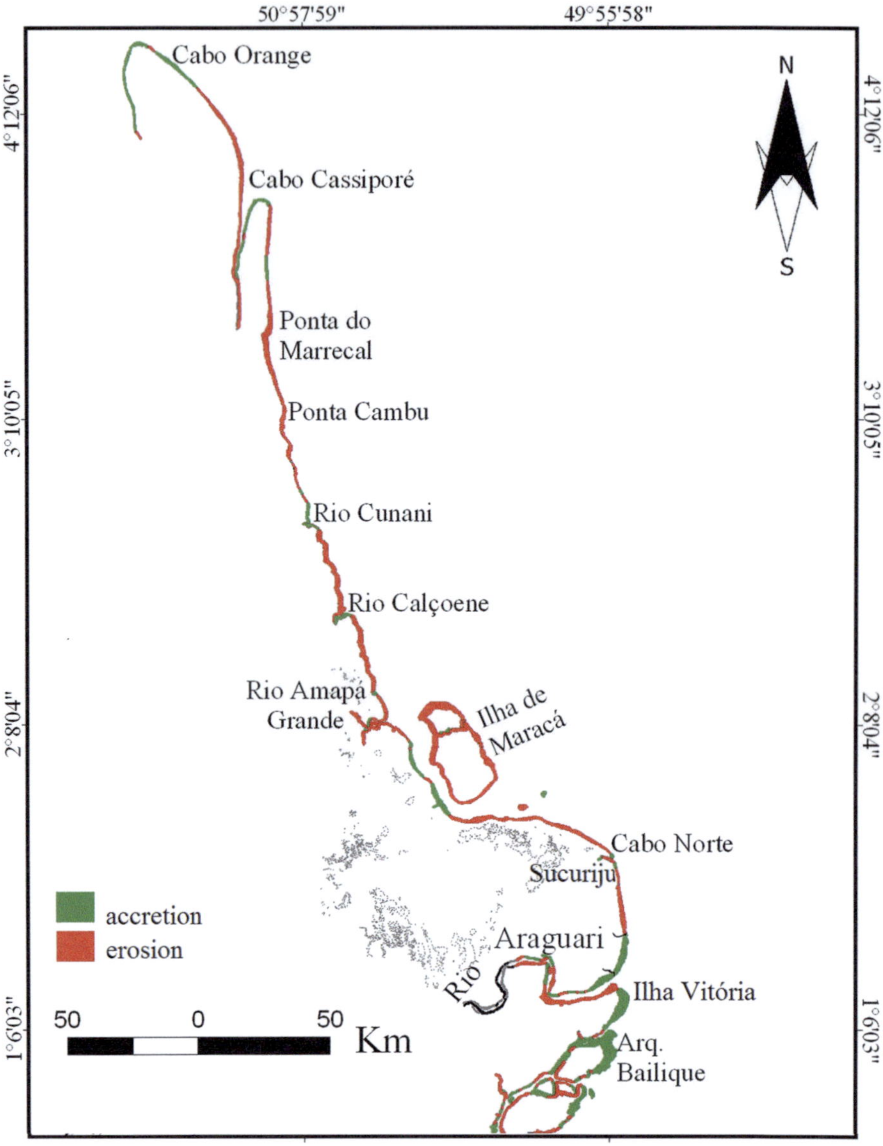

Fig. 3.6 Northern coast of Amapá (Modified from Silva 2010)

to the intertidal zone. The high tide beaches are up to 60 m wide and of low to moderate slope (1–3°).

The upper intertidal zone usually has surface wave ripples but with few physical or biogenic structures. In the mid-tide zone wavy flaser bedding is associated with tidal currents (Mendes 1994). The very low gradient tidal flats up to 3 km wide occupy the mid-lower intertidal. The flats contain wave ripples and patches of muddy sediments.

Fig. 3.7 Beaches on the northern coast of Amapá between Cabo Cassiporé and Araguari River. (**a**) Sand ridges north of Goiabal Beach; (**b**) Goiabal Beach, located near Rio Calçoene; (**c**) Sucuriju beach; (**d**) Ilha Vitoria Beach (Photos: Valdenira Santos, September 2013, Cartas SAO FZA Project)

Sucuriju beach, at Cabo Norte, is a small beach 2.6 km long and 200 m wide, located 1.2 km from Sucuriju village. To the south Paratur and Araqueçaua beaches extend for 22 km but have a narrower high tide beach. Mangroves dominate the landscape behind these beaches and young mangroves grow on the tidal flats. The beach on Vitória island is 650 m long with a tidal channel aligned north-south separating it from the muddy intertidal zone. Behind the beach, a "*varzea*" forest ecosystem dominates with a few mangroves. All of these beaches are exposed directly to tidal bores, which fluidize structures deposited at the limit of the bore.

Goiabal beach, near Calçoene, is one of the few beaches accessible by road and is popular during the summer vacation. It is backed by low vegetated beach ridges and comprises a 70 m wide high tide beach composed of fine sand and sloping at 2°, and then low-gradient intertidal sand flats that extend seaward for at least 2 km (Fig. 3.8).

Mendes (1994) examined the heavy mineral assemblages in the sandy deposits and found they are derived from proximal sources, notably the pre-Cambrian basement which, according to Allison et al. (1995) supplies ephemeral deposits that accumulate at estuarine mouths. Santos (1994) also found bryozoan shells and siliceous spicules in sandy sediments at the mouth of the Araguari river indicating that some sediment is derived from the inner continental shelf. A few pelecypod shells are also found in the beach deposits. During the wet season the ripple troughs are covered by a thin layer of mud and/or organic material, forming flaser deposits.

Due to the difficulty of access, all 16 ocean beaches are in natural state of preservation. Goiabal beach (Figs. 3.7b and 3.8) is the most famous and most popular

Fig. 3.8 View north along Goaibal beach showing the narrow high-tide beach and start of the wide intertidal sand flats (Photo: A D Short)

beach of Amapá State while many of the beaches are used for cattle grazing, with cattle ranches occupying the backing coastal plain.

3.3.2 Southern Amapá Amazon River Coast

The southern Amapá Amazon coast is strongly influenced by the Amazon river with its high discharge and variable water levels and the macro-megatides. This sector has 495 km of mainland shore, including Bailique archipelago, Pedreiras, Cajari and Santana islands, and extends from the Amazon mouth to Jari river mouth, located 350 km upstream, which is also the border with Pará state (Fig. 3.1). The main islands of Bailique archipelago are Franco (109 km^2), Bailique (159 km^2), Faustino (34.5 km^2), Brigue (22.5 km^2), Curuá (345 km^2) and Parazinho (2.64 km^2), which are all part of the Parazinho Biological Reserve.

Two types of beaches occur on this coast: high tide sandy beaches fronted by wide tidal sand flats, and smaller pocket beaches usually fronted by river-truncated mud flats. High tide beaches and wide sand flats (Fig. 3.9b) occupy 43.3 km (8.8 %) of this coastline and are most frequent on the more exposed eastern and southeastern shores of Bailique Archipelago. The best-known high-tide sand beach is located in Parazinho island (Fig. 3.9a). It is 4 km-long and 50 m-wide grading into wide tidal flats (Fig. 3.9a). The pocket beaches occupy 4 km (0.8 %) of coast and occurs along the banks of Canal do Norte of Amazon river located between 120 and 150 km inside the Amazon river mouth, including on either side of Macapá city (Fig. 3.10c).

Fig. 3.9 High-tide beach on Amapa's Parazinho island at the Amazon river mouth (**a**); and (**b**) wide exposed sand flat with mud hollows on Amapá's Farol Praia on Ilha Vitória (Photos: A D Short)

They are all located on the northern shores and represent small narrow strips of high tide sand fronted by intertidal mud flats (Fig. 3.10a) of variable width, and then the deeper river channel, which ranges in width from 16 km in the north decreasing to 13 km in the south. The beaches are all short, the longest being 0.38 km in length decreasing to just 0.1 km long. Most are fringed by varzea forest together with scattered mangroves along the shore (Fig. 3.10b).

From the mouth of the Amazon river to inside Canal do Norte, extensive sand waves occur in the intertidal and subtidal zone, as well as in the tidal bars (Fig. 3.11). The crests are orthogonal or almost orthogonal to the direction of the tidal current, which is predominantly northeasterly, the same direction as that of the ebb tide current. Sand flats in the Bailque archipelago are up to 4 km wide and more than 20 km long and some are accreting at the mouth of the Amazon a rate of 272 myr^{-1} (Silva et al. 2011). The beaches, tidal bars and sand flats are composed of silt and fine to very fine sand derived from the Amazon river bedload discharge (Martinez et al. 2009).

3.3.3 Pará-Amazon Riverine-Island Shores

The western Pará coast commences at the border of Jari river and extends through the Amazon mouth islands towards the Marajó archipelago and the southern shores of Marajó Bay. Its length is dependent on how many of the Amazon interdistributary mouth islands are included and how they are measured. For this study the coast was measured essentially northeast from the mouth of Jari river along the southern shores of Canal do Norte and including the islands of Grande of Gurupá, do Pará, de Serraria, Jurupari, Caviana de Dentro, Caviana de Fora, Canivete, Janaucu and Mexiana and then east following the southern shores of Canal do Sul along the north (~1500 km) and then the east coast of Marajó Island (390 km), providing a total length of 1890 km (Table 3.1, Fig. 3.12).

The riverine shorelines of the islands are dominated by fluvial processes, as well as tidal fluctuations and low fetch-limited wind waves. The shores are dominated by

Fig. 3.10 Beaches of the Amazon river coast. (**a**) Pocket beaches fronted by intertidal mud flat; (**b**) pocket beach bordered by varzea forest; and (**c**) river pocket beach near Fazendinha (Macapá) (Photos: Valdenira Santos, July 2014, Cartas SAO FZA Project)

Fig. 3.11 Sand waves in the intertidal zone at Canal do Norte of the Amazon river (Photo: Valdenira Santos, October 2013, Cartas SAO FZA Project)

Fig. 3.12 (**a**) 4 km wide sand flats on Janauçu island; note the backing series of vegetated beach ridges; (**b**) 130 m long pocket beach on river bank on Mexiana Island; and (**c**) 1.3 km long high-tide beach and flats on the exposed eastern tip of Mexiana island (Source: Google earth)

mud banks and mud flats backed by inundated lowlands, with mangroves increasing towards the Amazon river mouth.

Based on a view of the coast using Google Earth, about 50 beaches were detected along the Pará-Amazon coast. They had a total length of 109 km, occupying 7.2 % of the shore, with an average length of 2.1 km. Many are however poorly expressed and highly irregular; some appeared as eroding and others were littered with vegetation debris, usually large logs. They are all located on five river mouth islands including Caviana de Dentro (6 beaches), Janauçu (3), Caviana de Fora (7), and in particular the east-northeast coast of Mexiana (30). The beaches are of three types: wide sand flats possibly backed by a high-tide beach or beach ridges and grading into varzea forest, with the flats extending up to 20 km longshore (Fig. 3.12a); small pockets of sand on the banks of the river channel (Fig. 3.12b); and well developed high-tide beaches fronted by sand flats (Fig. 3.12c).

3.3.4 *Marajó Island – North Coast*

The northern coast of Marajó Island extends east-west for 200 km between Chaves and Cabo Maguari. The first 80 km faces north across the 7–16 km wide Canal do Sul, the southern channel of the Amazon River, while the remainder are exposed to the Atlantic. The shoreline is predominately sandy with at least 39 low-energy tide-dominated beach systems (Figs. 3.13 and 3.14) that average about 1 km in length and occupy at least 40 km of the coast. In profile the beaches consist of a steep high-tide sandy beach, some backed by eroding bluffs of the Pos Barreiras Formation, others by low beach ridges or dunes, while they are fronted by intertidal sand flats up to 1 km wide.

The remainder of the coast consists of forest-fringed sandy tidal flats, 20 large tidal creeks and numerous smaller creeks, together with several low islands between 100 km to 150 km (east of Chaves). The islands are mangrove-fringed with some containing sandy beaches on their more exposed north to eastern shores. The easterly wind waves and swell produce westerly longshore sediment transport manifested by several west-trending recurved spits (Fig. 3.15).

The eastern 100 km of coast has advanced with former beaches, recurved spits and creek channels extending up to 10 km inland, particularly in the lee of the northeastern tip of the island at Maguari Cape. This low-lying section of coast is also rich in mangrove forests grading into varzea forests.

Fig. 3.13 The beach at Chaves is backed by steep eroding bluffs, partly protected by a seawall. The jetty extends 200 m out across the sand flats (Photo: Valdenira Santos, Cartas SAO FZA Project)

Fig. 3.14 Narrow high-tide beach on the north coast of Marajó island (Photo: José Roberto Pantoja, January 2014, Cartas SAO FZA Project)

Fig. 3.15 A 2 km long beach-spit on the northeastern coast of Marajó island. The beach is fronted by 400 m wide ridged sand flats and backed by older beach ridges and spits (Source: Google earth)

Fig. 3.16 Tide-dominated beaches consisting of a high-tide beach fronted by 500 m-wide ridged sand flats adjacent to creek mouths at Cajuuna (Source: Google earth; photos: A G Miranda, Cartas SAO FZA Project)

3.3.5 Marajó Island – East Coast (Marajó Bay)

The eastern coast of Marajó island extends for 190 km between Cabo Maguari and Ponta do Malato. While it faces into the easterly trade winds, it also faces across Marajó Bay, which narrows from 60 km at its mouth to 15 km in the south. This is still sufficient for Atlantic swell to impact the first 50 km of coast and for the trade winds to generate short seas along the entire shore. The large tide range (10 m decreasing south to 3 m) generates strong shore-parallel tidal currents, as well as cross-shore currents associated with the numerous small inlets and tidal creeks along this section of coast. The coast consists of mangroves, tidal creeks and generally short tide-dominated beaches fronted by intertidal sand ridges, sand flats and tidal sand flats, some of the latter containing transverse to sinuous sand shoals. The sand flats average 200–300 m in width widening to several hundred meters in front of creek mouths. The ridged sand flats contain low (<10 cm amplitude) shore-parallel evenly spaced sand ridges with spacing averaging about 80 m (Fig. 3.16) This is similar to sand ridge spacing on tide-dominated Australian beaches (Short 2006).

The east coast can be divided into two sectors either side of the Paracauari river. To the north is the 60 km long 'low' coast, which is exposed to higher waves (Atlantic swell and seas) and has longer, more continuous, beaches separated by tidal inlets and composed of fine sand resulting in low gradient high-tide beaches and sand flats. The beaches average 2.5 km in length and occupy 72 % of this sector. This section also has numerous tidal creeks and inlets and associated ebb tide shoals, with mangroves extending 2–3 km inland. It has experienced shoreline regression of between 1 and 10 km, with the former shorelines marked by stranded beaches and recurved spits (Fig. 3.17).

As a consequence there is little development along this section of coast. South of the river is the 130 km long lower-energy ‘high’ coast backed by steep bluffs and slopes, with fewer tidal channels and mangroves. The beaches are composed of medium to coarse sand (França and Souza-Filho 2003) and tend to be embayed between protrusions in the bluffs, forming small pocket and embayed beaches. They average only 0.33 km in length and occupy only 30 % of the shore. The remainder of the shore is composed of the bluffs, and in the south, mangroves. Because of the coarser sand and decreasing wave height the high-tide beaches are steeper but still fronted by both ridged sand flats and tidal flats.

In total, beaches occupy 82.5 km (43 %) of the shore, and are generally short with a mean length of 0.64 km ($\sigma = 0.94$ km), and the longest only 4.9 km in length. Overall, there tends to be a net sand transport to the south on at least 20 of the beaches, as manifested by southerly spits, while at some creek mouths the beaches are convex with spits curving to the north and south, and on seven beaches just to the north. The spits are maintained by both the easterly waves and strong tidal flows. Thirteen of the beaches between Jubim, Joanes and Monsarás and around Santana and Santa Maria, are backed by steep bluffs and slopes and are ‘embayed’. Mangroves increase in occurrence to the south and some grow on some of the tidal flats along the beaches, with the southernmost 15 beaches all fronted by some mangroves. Overall this is a very low wave-energy shoreline, with generally low narrow high-tide sandy tide-dominated beaches fronted by wide sand flats. Most beaches consist of a single ridge, some overwashed and/or backed by low dunes, with only a few beaches indicating shoreline regression leaving a series of beach ridges.

3.4 Discussion and Conclusion

The Amazon coast is a large complex coastal system involving the interaction of the world’s largest river in terms of discharge and sediment in a broad funnel-shaped gulf dominated by macro to megatides, generally low easterly waves, strong coastal currents, and dispersion of sediments along the coast, deflected in the ocean by the NBC. While 99 % of the sediment load is suspended, the 1 % bedload is still estimated at a massive 3.65–18.25 Mt year^{-1} (Dunne et al. 1998). The transport of much of the mud northward along the Amapá coast and beyond as far as Venezuela is well-documented (Anthony et al. 2010, 2014); however there has been little investigation of the bedload and its contribution to channel, river mouth and coastal sedimentation, particularly in the form of tide-dominated beaches and sand flats.

Recent investigations undertaken around the Amazon mouth as part of the Research Network 05-N/NE (AMASIS and AMASTRAT subprojects) and Cartas SAO FZA project, have visited a number of the approximately 260 beaches that have been identified along this predominately muddy coast. These beaches occupy approximately 12 % of the 3150 km long Amapá and western Pará Amazon coast.

Fig. 3.17 Regressive shoreline at Soure showing the inner beach ridges, mangrove-filled interbarrier depression and modern coast with beaches fronted by 200–300 m wide tidal sand flats (Source: Google earth; photo: A G Miranda, Cartas SAO FZA Project)

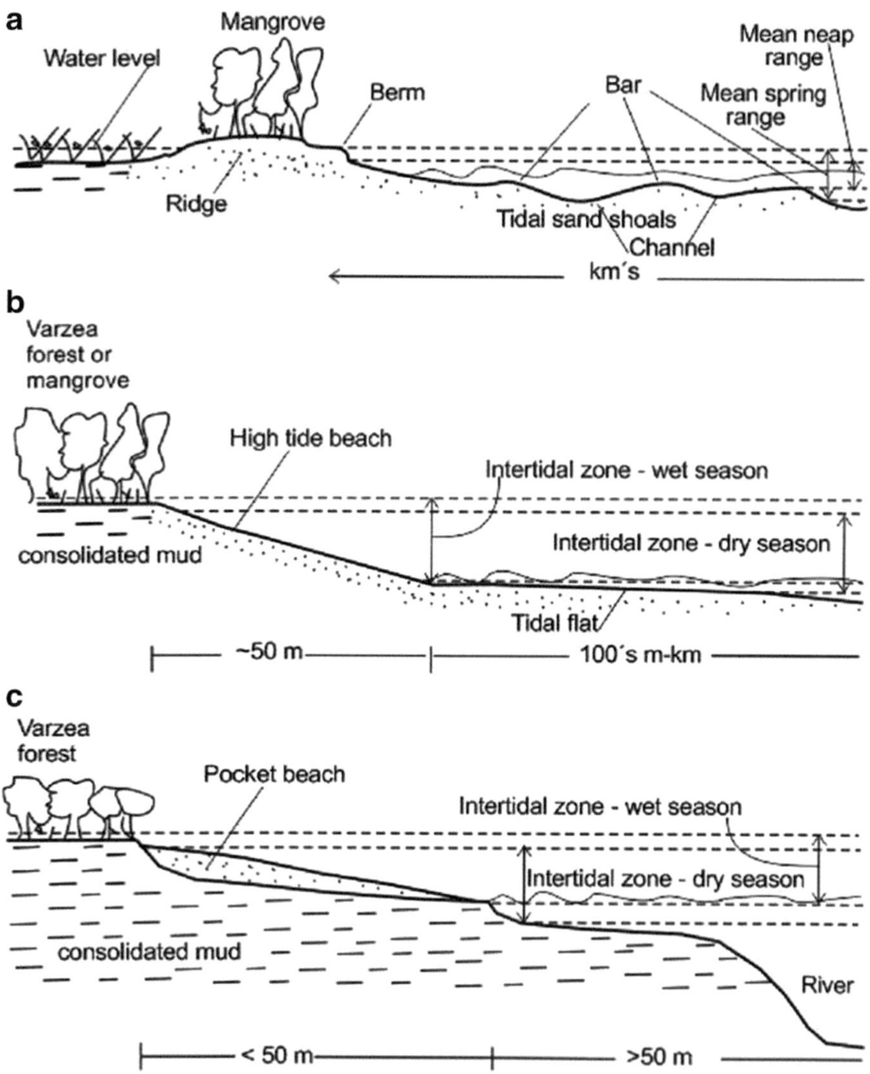

Fig. 3.18 The three typical beach types found along the Amazon coast. (**a**) low beach ridge grading into intertidal tidal sand shoals, which may be kilometers wide (See Figs. 3.7, 3.9 and 3.12); (**b**) steeper high-tide beach fronted by intertidal sand flats (See Figs 3.15 and 3.16); and (**c**) small (and short) pocket beach found along some river banks (See Figs. 3.10)

The beaches are of three types (Fig. 3.18). At the river mouth are extensive areas of wide intertidal sand flats extending several kilometers along the coast and up to a few kilometers in width, which may or may not have a small high-tide beach and/or beach ridge (Fig. 3.18a), similar to those reported by Anthony, et al. 2014. These flats are often host to mangroves and grade landward into older beach ridges and varzea forest or swamp. Second are the best developed beaches which occur on the most exposed shores, including the northern Amapá coast; the exposed eastern side of some of the river mouth islands; and in particular on the north and east coasts of Marajó island, where they occupy 30 % of the shore. They usually consist of a well-developed steeper high-tide beach usually at least 50 m wide, then intertidal sand flats from a few meters (on river channels) to kilometers wide. The flats may be ridged, featureless or contain tidal drainage features, and may grade seaward into mud flats. Mud pockets can also be found on some of the inner flats (Fig. 3.18b). These beaches contain the four tide-dominated beach states (Short 2006). The third type is found along the river banks where there are occasionally small pockets of sand usually less than 200 m in length deposited at high tide on the riverbanks and often composed of medium to coarse fluvial sand, usually overlying mud deposits (Fig. 3.18c). The most accessible of these are at Macapá city where they are used for recreation.

This chapter is the first to report on preliminary inspections of some of these beaches. There is still much to be done in order to fully understand the transport pathways of the Amazon bedload, and the role the beaches and sand flats play in the overall sedimentation process. Many of the beaches are also unstable and experiencing erosion and/or associated longshore sand transport, while others are bordered by dynamic tidal channels and associated tidal shoals. Vegetation debris including large logs litters many of the upper beaches. The beaches are just a small part of a highly dynamic Amazon coast, the whole of which requires more thorough investigation.

Acknowledgements To FINEP (grants 01.07.0143.00, AMASTRAT Subproject) and CNPq (552967/2011-4, Cartas SAO FZA Project) for the financial support for the fieldwork and for flights over the study area. To Kerly Jardim for the help with the pictures.

References

Allison MA, Lee MT (2004) Sediment exchange between Amazon Mudbanks and shore-fringing mangroves in French Guiana. Mar Geol 208:169–190. doi:10.1016/j.margeo.2004.04.026

Allison MA, Nittrouer CA, Kineke GC, Sternberg RW (1994) Use of an instrumented tripod system to examine sediment dynamics and fine-scale strata formation in muddy surfzone and nearshore environments. J Coast Res 10:488–496

Allison MA, Nittrouer CA, Faria LEC Jr (1995) Rates and mechanisms of shore face progradation and retreat downdrift of the Amazon river mouth. Mar Geol 125:373–392. doi:10.1016/0025-3227(95)00020-Y

Anthony EJ, Gardel A, Gratiot N, Proisy C, Allison MA (2010) The Amazon-influenced muddy coast of South America: a review of mud-bank–shoreline interactions. Earth Sci Rev 103:99–121. doi:10.1016/j.earscirev.2010.09.008

Anthony EJ, Gardel A, Proisy C, Fromard F, Gensac E, Peron C, Walcker R, Lesourd L (2013) The role of fluvial sediment supply and river-mouth hydrology in the dynamics of the muddy, Amazon-dominated Amapá–Guianas coast, South America: a three-point research agenda. J S Am Earth Sci 44:18–24. doi:10.1016/j.jsames.2012.06.005

Anthony EJ, Gardel A, Gratiot N (2014) Fluvial sediment supply, mud banks, cheniers and the morphodynamics of the coast of South America between the Amazon and Orinoco river mouths. In: Martini, IP, Wanless HR (eds) Sedimentary coastal zones from high to low latuitudes: similarities and differences. Geololical Society, London, Special Publication, 388:533–569. DOI.org/19.1144/SP388.8

Augustinus PGEF (1978) The changing shoreline of Surinam (South America). Found Sci Res Surinam and the Neth Antilles, Utrecht, 95, pp 232

Augustinus PGEF (1989) Cheniers and cheniers plains: a general introduction. Mar Geol 90(4):219–229. doi:10.1016/0025-3227(89)90126-6

Beardsley RC, Candela J, Limeburner R, Geyer WR, Lentz SJ, Castro BM, Cacchione D, Carneiro N (1995) The M2 tide on the Amazon shelf. J Geophys Res 100(C2):2283–2319

Boaventura FMC, Narita C (1974) Geomorfologia. Folha Na/NB.22-Macapá. In: DNPM-Projeto RADAM, 6 Levantamento de Recursos Naturais, MME/DNPM, Rio de Janeiro, pp II/1-II/27

Costa Neto SV, Senna CSF, Coutinho RS (2006) Vegetação das Áreas Sucuriju e Região dos Lagos no Amapá. In: Neto C (ed) Inventário Biológico das Áreas do Sucuriju e Região dos Lagos, no Amapá. PROBIO/IEPA, Macapá, pp 41–79

Damuth JE, Kumar N (1975) Amazon cone: morphology, sediments, age and growth pattern. Geol Soc Am Bull 86:863–878

DeMaster DJ, Pope RH (1996) Nutrient dynamics in Amazon shelf waters: results from AMASSEDS. Cont Shelf Res 16(3):263–289. doi:10.1016/0278-4343(95)00008-O

Dunne T, Mertes LAK, Meade R, Richey JE, Forsberg BR (1998) Exchanges of sediment between the floodplain and channel of the Amazon in Brazil. Geol Soc Am Bull 110(4):450–467. doi:10.1130/0016-7606(1998)110<0450:EOSBTF>2.3.CO;2

França CF, Souza-Filho PWM (2003) Análise das Mudanças Morfológicas Costeiras de Médio Período na Margem Leste da Ilha de Marajó (PA) em Imagem Landsat. Rev Bras Geosci 33(2-suplemento):127–136

França CF, Souza-Filho PWM (2006) Compartimentação Morfológica da Margem Leste da Ilha de Marajó: Zona Costeira dos Municípios de Soure e Salvaterra – Estado do Pará. Rev Bra Geomorfologia 7(1):33–42

Geyer WR, Kineke GC (1995) Observations of currents and water properties in the Amazon frontal zone. J Geophys Res 100:2321–2339. doi:10.1029/94JC02657

Geyer WR, Beardsley RC, Lentz SJ, Candela J, Limeburner R, Johns WE, Castro BM, Soares ID (1996) Physical oceanography of the Amazon shelf. Cont Shelf Res 16:575–616. doi:10.1016/0278-4343(95)00051-8

Gibbs RJ (1967) The geochemistry of the Amazon River system: part I. The factors that control the salinity and the composition and concentration of the suspended solids. Geol Soc Am Bull 78:1203–1232. doi:10.1130/0016-7606(1967)78[1203:TGOTAR]2.0.CO;2

Google Earth pro V 7.1.2.2041. (2013). http://www.earth.google.com

Google Earth V 7.1.2.2041 (2012) High resolution images from Digital Globe 2014 http://www.earth.google.com. (December 22, 2012)

Jardim KA, Santos VF, Silveira OFM (2013) Mapeamento morfológico da região do Cabo Norte utilizando imagens ALOS PALSAR, Amapá, costa amazônica. In: (eds) XIII Simpósio de Geologia da Amazônia (SGA), Belém. DATA. Anais do X Congresso ABEQUA, ABEQUA, Guarapari, 6 p (CD-ROM)

Kineke GC (1993) Fluid muds on the Amazon continental shelf. PhD dissertation, University of Washington, Seattle WA, 249 pp

Martinez J-M, Guyot JL, Filizola N, Sondas F (2009) Increase in suspended sediment discharge of the Amazon River assessed by monitoring network and satellite data. Catena 79(3):257–264. doi:10.1016/j.catena.2009.05.011

Meade RH, Nordin CF Jr, Curtis WF, Rodrigues FMC, do Vale CM, Edmond JM (1979) Sediment loads in the Amazon River. Nature 278:161–163. doi:10.1038/278161a0

Meade RH, Dunne T, Richey JE, Santos UM, Salati E (1985) Storage and remobilization of suspended sediment in the lower Amazon River of Brazil. Science 228:488–490. doi:10.1126/science.228.4698.488

Mendes AC (1994) Estudo sedimentológico e estratigráfico de sedimentos holocênicos da costa do Amapá, setor entre a Ilha de Maracá e o Cabo Orange. Dissertação Mestrado em Geologia, Universidade Federal do Pará, Centro de Geociências, Curso de Pós-graduação em Geociências, Belém, 247 p
Milliman JD (1979) Morphology and structure of Amazon upper continental margin. Am Assoc Pet Geol Bull 63(6):934–950
Muller-Karger FE, McClain CR, Richardson PL (1988) The dispersal of the Amazon's water. Nature 333:56–59. doi:10.1038/333056a0
Nittrouer CA, DeMaster DJ (1996) The Amazon shelf setting tropical, energetic, and influenced by a large river. Cont Shelf Res 16(5–6):553–574. doi:10.1016/0278-4343(95)00069-0
Nittrouer CA, Curtin TB, DeMaster DJ (1986) Concentration and flux of suspended sediment on the Amazon continental shelf. Cont Shelf Res 6:151–174
Nittrouer CA, DeMaster DJ, Figueiredo AG, Rine JM (1991a) AMASSEDS: an interdisciplinary investigation of a complex coastal environment. Oceanography 4:3–7
Nittrouer CA, Kuehl SA, Rine JM, Figueiredo AG, Faria-Junior LEC, Dias GTM, Silva MAM, Allison MA, Pacioni TD, Segall MP, Underkoffer EC, Borges HV, Silveira OF (1991b) Sedimentology and stratigraphy of the Amazon continental shelf. Oceanography 4:33–38
Nittrouer CA, Kuehl SA, Sternberg RW, Figueiredo AG Jr, Faria LEC Jr (1995) An introduction to the geological significance of sediment transport and accumulation on the Amazon continental shelf. Mar Geol 125:177–192. doi:10.1016/0025-3227(95)00075-A
Prost MT (1989) Coastal dynamics and chenier sands in French Guiana. Mar Geol 90:259–267. doi:10.1016/0025-3227(89)90128-X
Rabelo BV, Souza CB, Quintas DFP, Ávila JES, Chagas MAA, Gibson VM (1994) Abordagens sobre os Manguezais do Amapá: contribuições para debate. Anais do II Encontro do SILVOLAB, Macapá. March 1994, 20 p
Richey JE, Meade RH, Salati E, Devol AH, Nordin CF Jr, Santos UDM (1986) Water discharges and suspended sediment concentrations in the Amazon River 1982–1984. Water Resour Res 22:756–764. doi:10.1029/WR022i005p00756
Richey JE, Nobre C, Deser C (1989) Amazon river discharge and climate variability: 1903 to 1985. Science 246:101–103. doi:10.1126/science.246.4926.101
Rossetti DF, Goes AM (2008) Late Quaternary drainage dynamics in northern Brazil based on the study of a large paleochannel from southwestern Marajó Island. An Acad Bras Cienc 80(3):579–593. doi:10.1590/S0001-37652008000300017
Santos VF (2006) Ambientes Costeiros Amazônicos: Avaliação de Modificações por Sensoriamento Remoto. Tese de Doutorado, Universidade Federal Fluminense/Instituto de Geociências, Niterói, p 306
Santos VF. dos (1994) Análise Textural e Mineralógica dos Sedimentos Arenosos do Baixo Curso do Rio Araguari, Cabo Norte, Amapá. Trabalho de Conclusão de Curso. Universidade Federal do Pará, Belém p 71
Santos VF, Figueiredo AG, Silveira OFM, Polidori L, Oliveira DM, Dias MB, Santana LO (2005) Processos Sedimentares em áreas de macro-marés influenciados pela pororoca-estuário do rio Araguari-Amapá-Brasil. Resumos do X Congressso da Associação Brasileira de Estudos do Quaternário (ABEQUA), Guarapari. DATA. Anais do X Congresso ABEQUA, ABEQUA, Guarapari, 6 p (CD-ROM)
Santos VF, Polidori L, Silveira OFM, Figueiredo AG (2009) Aplicação de dados multisensor (SAR e ETM+) no reconhecimento de padrões de uso e ocupação do solo em costas tropicais: costa Amazônica, Amapá, Brasil. Rev Bras Geof 27(suppl1):39–55. doi:10.1590/S0102-261X2009000500004
Santos VF, Figueiredo AG, Silveira OFM, Polidori L (2010) The ENSO Influence on the Amazonian Coastal Plain, Brazil. EOS Trans. AGU, 91(26) Meet Am Suppl, Abstract OS13B-07
Schaeffer-Novelli Y, Cintron-Moler G (1988) Expedição aos manguezais do Amapá – Ilha de Maracá. IO/USP, São Paulo, p 99 (Rel. Tec)

Short AD (2006) Australian beach systems – nature and distribution. J Coast Res 22:11–27. doi:10.2112/05A-0002.1

Silva MV (2010) Análise Multitemporal Quantitativa da Linha de Costa Amapaense aplicando Dados de Sensores Remotos Óticos e Radar (1992–2000). Trabalho de Conclusão de Curso, Universidade Federal do Amapá, Macapá, 122 p

Silva MV, Santos VF, Silveira OFM (2011) Análise Multitemporal de Mudanças nas Ilhas Costeiras do Estado do Amapá aplicando Dados de Sensores Remotos. In: Epiphanio JCN, Galvão LS (eds) XV Simpósio Brasileiro de Sensoriamento Remoto, Curitiba. 30 April – 05 May 2011. Anais do XV SBSR, INPE, São José dos Campos, pp 8614–8622. DVD. ISBN: 978-85-17-00057-7

Silveira OFM (1998) A Planície Costeira do Amapá. Dinâmica de Ambiente Influenciado por Grandes Fontes Fluvais Quaternárias. Tese Doutorado, Universidade Federal do Pará/Centro de Geociências, Belém p 215

Souza-Filho PWM, Paradella WR (2003) Use of synthetic aperture radar for recognition of Coastal Geomorphological Features, land-use assessment and shoreline changes in Bragança coast, Pará, Northern Brazil. An Acad Bras Cienc 75(3):341–356. doi:10.1590/S0001-37652003000300007

Tassinari CCG, Macambira MJB (1999) Geochronological provinces of the Amazon Craton. Episodes 22(3):174–182

Wells JT, Coleman JM (1978) Nearshore suspended sediment variations, central Surinam coast. Mar Geol 24:47–54. doi:10.1016/0025-3227(77)90069-X

Chapter 4
Coastal Morphodynamic Processes on the Macro-Tidal Beaches of Pará State Under Tidally-Modulated Wave Conditions

Luci Cajueiro Carneiro Pereira, Ana Vila-Concejo, and Andrew D. Short

Abstract Amazon coast is a vast and complex estuarine–marine system with highly dynamic coastal environments driven by an abundance of sediment reworked by fluvial, tidal and wave processes. The coast of Pará occupies the eastern Amazon Gulf and is more than 2000 km long, with more than 400 sandy beaches. It can be divided into five coastal sectors: the Amazon river-distributary coast; the north coast of Marajó Island; the west and east coasts of Marajó Bay; and the highly indented eastern barrier island coast between Marajó and Gurupi bays. This chapter will examine the eastern Pará coast, between Belém in Marajó Bay and Gurupi Bay, a macro to mega-tidal coast dominated by tide-modified and tide-dominated sand beaches. There is however considerable spatial and temporal variability in beach state, dependent on the amount of nearshore sheltering by inter to sub-tidal sandbanks, with *tide-modified* beaches ($3 < \text{RTR} < 15$) on the more exposed sections and *tide-dominated* ($\text{RTR} > 15$) on the more sheltered sections, together with tide-modified beaches at high tide and tide-dominated at low tide. The impact of this tidal modulation on wave energy and beach morphodynamics is examined on three beaches, together with their associated beach hazards. These are then combined with seasonal tourism population to assess the level of risk. The large seasonal beach population coupled with unplanned beach development generates addition hazards including eroding buildings, traffic congestion and accidents on some beaches, and sewerage and litter pollution. Coastal planning and regulation is urgently required for this coast to enable it to develop a safe and sustainable tourist industry and to reduce the beach hazards and risk.

L.C.C. Pereira (✉)
Instituto de Estudos Costeiros, Universidade Federal do Pará,
Alameda Leandro Ribeiro, sn, Aldeia, Bragança, Pará 68600-000, Brazil
e-mail: cajueiro@ufpa.br; luci.pereira@pq.cnpq.br

A. Vila-Concejo • A.D. Short
School of Geosciences, University of Sydney, Sydney, NSW 2006, Australia
e-mail: ana.vilaconcejo@sydney.edu.au; andrew.short@sydney.edu.au

A.D. Short, A.H. da F. Klein (eds.), *Brazilian Beach Systems*, Coastal Research Library 17, DOI 10.1007/978-3-319-30394-9_4

4.1 Introduction

Coastal processes are primarily controlled by wave and tidal forcing, mainly as a consequence of local and regional-scale processes. Most beach morphodynamic studies have been undertaken in sub-tropical and temperate coasts with limited studies of macro-tidal beaches and particularly equatorial beaches. This chapter reports on the eastern Amazon Pará coast beaches, which are located near the massive Amazon river/estuarine systems and exposed to macro-tides which modulate the low to moderate wave energy as it crosses the inter to sub-tidal sandbanks (Anthony et al. 2013; Pereira et al. 2014).

Brazilian Amazon beaches straddle the equator between 4°N and 4°S centered on the Amazon mouth at ~0° latitude and represents 35 % of Brazil's 8500 km long coastline (Isaac and Barthem 1995). The Amazon mouth and associated distributaries, estuaries and tidal channels, contain one of the largest mangrove systems in the world (Kjerfve and Lacerda 1993). This mangrove-dominated shoreline also contains extensive tidal flats, salt marshes, cheniers, beach ridges, deltas and coastal dunes (Anthony et al. 2013; Souza-Filho et al. 2003). Due to its low latitude, Coriolis acceleration is weak and coastal circulation is more affected by factors such as riverine discharge, wind and wave forces, and the macro to mega-tides that generate strong shore parallel and shore normal tidal currents (Lentz 1995; Nittrouer and DeMaster 1986, 1996).

This vast and complex estuarine–marine system is a highly dynamic environment that includes the Amazon River and dozens of estuaries that contribute to form the world's largest run-off of sediments, dissolved nutrients and organic material (DeMaster and Pope 1996; Geyer et al. 1996). The enormous fluvial discharge of the Amazon River into the Atlantic Ocean is approximately 20 % of the total discharged on the whole planet. The mean fluvial discharge is around 180,000 m^3s^{-1} with peaks of up to 220,000 m^3s^{-1}, occurring in May–June and a minimum of 100,000 m^3s^{-1} in November–December (Richey et al. 1986). The massive suspended sediment discharge estimated at 754×10^6 M tons year^{-1} (+/– 9 %) is primarily transported northwest as large migratory mud banks towards Guinea (Anthony et al. 2010). While only 1 % of the sediment discharge is bedload, it still represents a substantial 3.65–18.25 M tons year^{-1} (Dunne et al. 1998), sufficient to supply numerous sandy beaches and extensive sand flats.

This chapter reviews the coastal processes and beach systems of eastern Pará state, which extends east of the Amazon mouth (Fig. 4.1). This section of coast includes the 265 km long eastern shores of Marajó bay and the 500 km long Pará open coast which consists of about 15 regressive tide-dominated barriers and barrier islands clusters, each separated by elongate funnel-shaped estuaries extending more than 20 km inland. The barriers recurve into the estuarine mouths with extensive inter to sub-tidal sand shoals dominating the mouths. This chapter will also discuss the impact of macro-tides and moderate waves on three beaches typical of the eastern Pará coast. The aim is to provide both an overview of the coast as well as insight into beaches behavior and beach usage in this equatorial, river and tide-dominated coast.

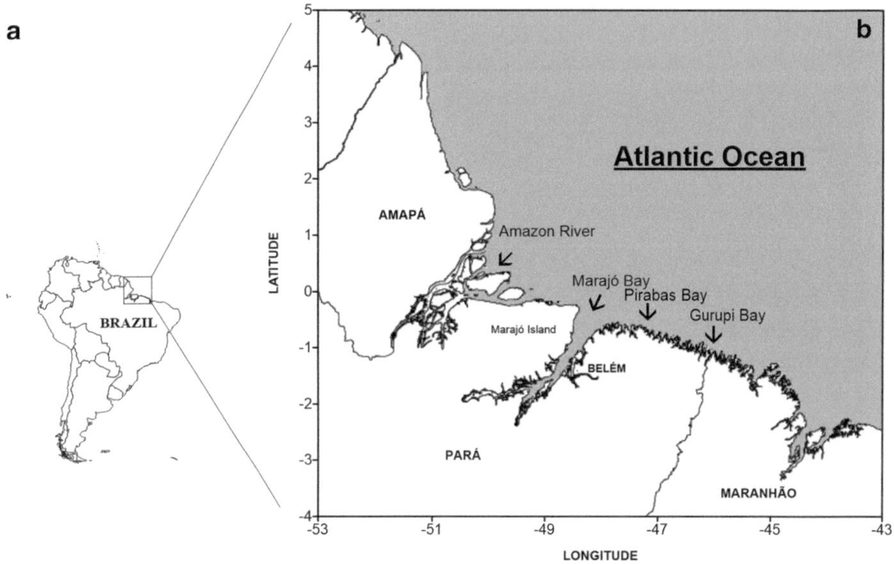

Fig. 4.1 (**a**) South America; (**b**) Brazilian Amazon Coast, indicating the location of the Amazon River mouth, Marajó Island and the eastern Pará coast between Belém and Gurupi Bay

4.2 Coastal Geomorphology

The coast of Pará commences 350 km inside the mouth of the Amazon River and extends east to Gurupi Bay, a linear distance of 650 km (Fig. 4.1). The length of the coast however is far more difficult to define owing in the west to its long boundary with the Amazon River and the numerous river mouth islands and distributaries including the large Marajó Island, the elongate Marajó Bay and in the east the highly irregular and indented coast between Marajó and Gurupi bays.

Franzinelli (1992) divides the Pará coast into two primary geomorphological zones: (i) the "emergence coast" represented by Marajó Island, with a relatively straight coastline; and, (ii) the "submergence coast" between Marajó and Gurupi bays. This second sector is then subdivided in two subsections: (i) from Marajó Bay to Pirabas Bay; and, (ii) between Pirabas Bay and Gurupi Bay. Between Marajó and Pirabas bays, the coastal plateau reaches the coastline forming terraces and active cliffs composed of Tertiary sediments from Barreiras and Pirabas formations. The coastal plain is narrow with a width of less than 2 km while the estuaries extend up to 60 km inland. The second subsection (Pirabas Bay and Gurupi Bay) comprises the Bragança-Viseu basin with well-developed mangrove areas up to 30 km wide and estuaries and tidal creeks extending up to 80 km inland. Coastal sedimentary environments dominate east from Pirabas Bay while the coastal plateau extends southward as low inactive cliffs.

Table 4.1 provides a very approximate length for each coastal section following the more open coast and avoiding many creeks and inlets. The actual length of the shoreline would be far greater than that provided in the table particularly for the Amazon River mouth, Marajó Bay and the indented eastern coast. A summary of these coastal sections is provided in Sect. 4.4.1.

Table 4.1 Pará coastal sections: length and number of beaches

Sections	Coast length (approx.) (km)	Number beaches
1. Amazon River mouth	~1500	~50
2. Marajó Island (N)	245	40
3. Marajó Bay (W)	185	128
4. Marajó Bay (E)	265	157
5. East coast (Marajó-Gurupi bays)	525	90
Total	~2720	~465

4.3 Coastal Processes

4.3.1 Climate

The Amazon coastal climate is equatorial humid with well-defined dry and wet seasons (Köppen classification=Am). The main meteorological system responsible for local rainfall patterns is the Intertropical Convergence Zone- ITCZ (Marengo 1995). The rainy season occurs when the ITCZ shifts southward (between January and June) to the coast of Maranhão and Pará (~4°S latitude) when total rainfall often exceeds 2000 mm. During the dry season (July–December), the ITCZ shifts to the Northern Hemisphere (>2°N latitude), resulting in higher air temperatures and rainfall normally less than 100 mm (Figueroa and Nobre 1990; Marengo 1995).

Figure 4.2 shows the longer-term monthly rainfall as well as rainfall level, air temperature and wind speeds, between November 2008 and October 2009 from the INMET, when total rainfall exceeded 3000 mm (Fig. 4.2a). Air temperature are normally high and stable with mean values around 26 °C during the rainy season, rising to around 28 °C during the second half of the year (Fig. 4.2b). During the dry season, fair weather conditions prevail with moderate wind speeds (Fig. 4.2b) reaching a maximum monthly average of 6.0 $m.s^{-1}$.

4.3.2 Waves

The waves on the Amazon equatorial coast are generated by the prevailing trade winds and arrive from northeast and east. Data from CPTEC-INPE (Weather Forecasting and Climatic Studies Center of the Brazilian Space Agency) show that winds blowing perpendicular to coastline ($40° < \theta < 60°$N) during the wettest months (February, March and April) generate the highest offshore significant wave heights (1.2–1.6 m) with wave periods between 5 and 9 s, arriving from $35° < \theta < 50°$N (Fig. 4.3). During the driest months (September, October and November), intermittent high wind velocities (8–9 ms^{-1}) blow from northeast and east ($80° < \theta < 90°$N), generating local waves from $65° < \theta < 85°$N with heights between 1.0 and 1.3 m with periods between 4 and 5 s (Fig. 4.3). During transitional period (June/July, rainy-dry season), the weaker southeast winds generate the lowest offshore wave energy (normally, > 1500 Nm^{-2}).

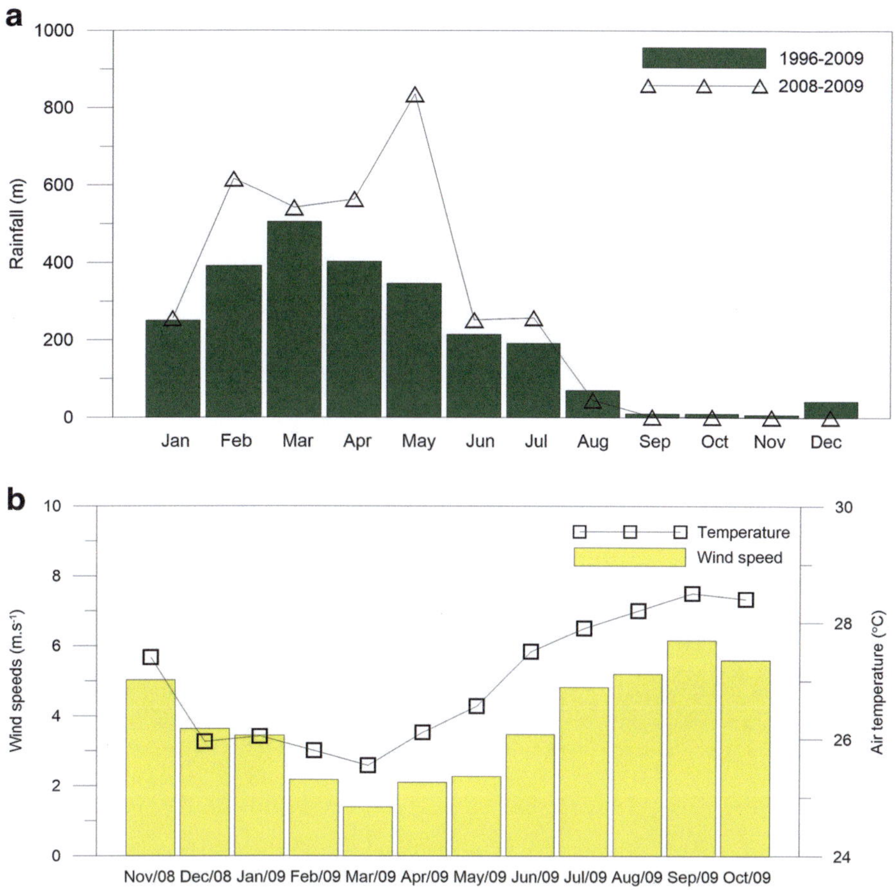

Fig. 4.2 (**a**) Total monthly rainfall (mm) observed during long term and between 1996 and 2009 from INMET; and (**b**) air temperature and wind speed values recorded between November 2008 and October 2009 from INMET (Source: www.inmet.gov.br)

The generally moderate energy deepwater waves are further modulated by inter and sub-tidal sandbanks and bars maintained by the wave, tide and fluvial processes (Anthony et al. 2013; Pereira et al. 2013b; Pereira et al. 2014).

4.3.3 Tides

The 300 km-wide shallow Amazon shelf amplifies the tidal range along the coastline reaching a maximum of 11 m in the Amazon mouth, decreasing to 7 m then 4 m northward along the Amapá coast and to 5 m along the Pará coast, as well as decreasing to 3 m at Belém (www.dhn.mar.mil.br). Strong tidal currents affect the entire coast. Their velocity is controlled by two components: (i) strong semi-diurnal tidal

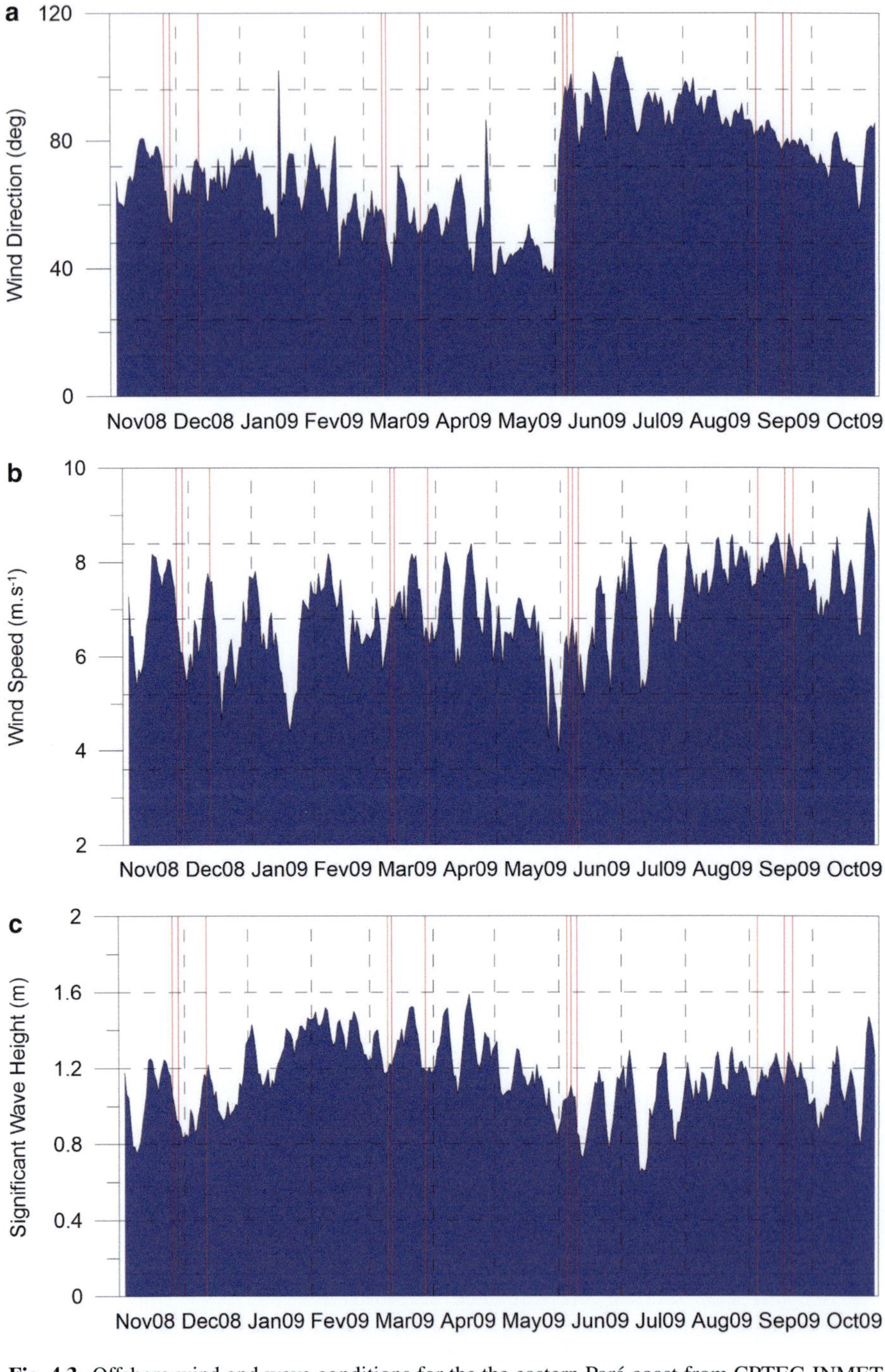

Fig. 4.3 Offshore wind and wave conditions for the the eastern Pará coast from CPTEC-INMET data, between November 2008 and October 2009. The *red lines* represent the dates when the nearshore surveys were undertaken

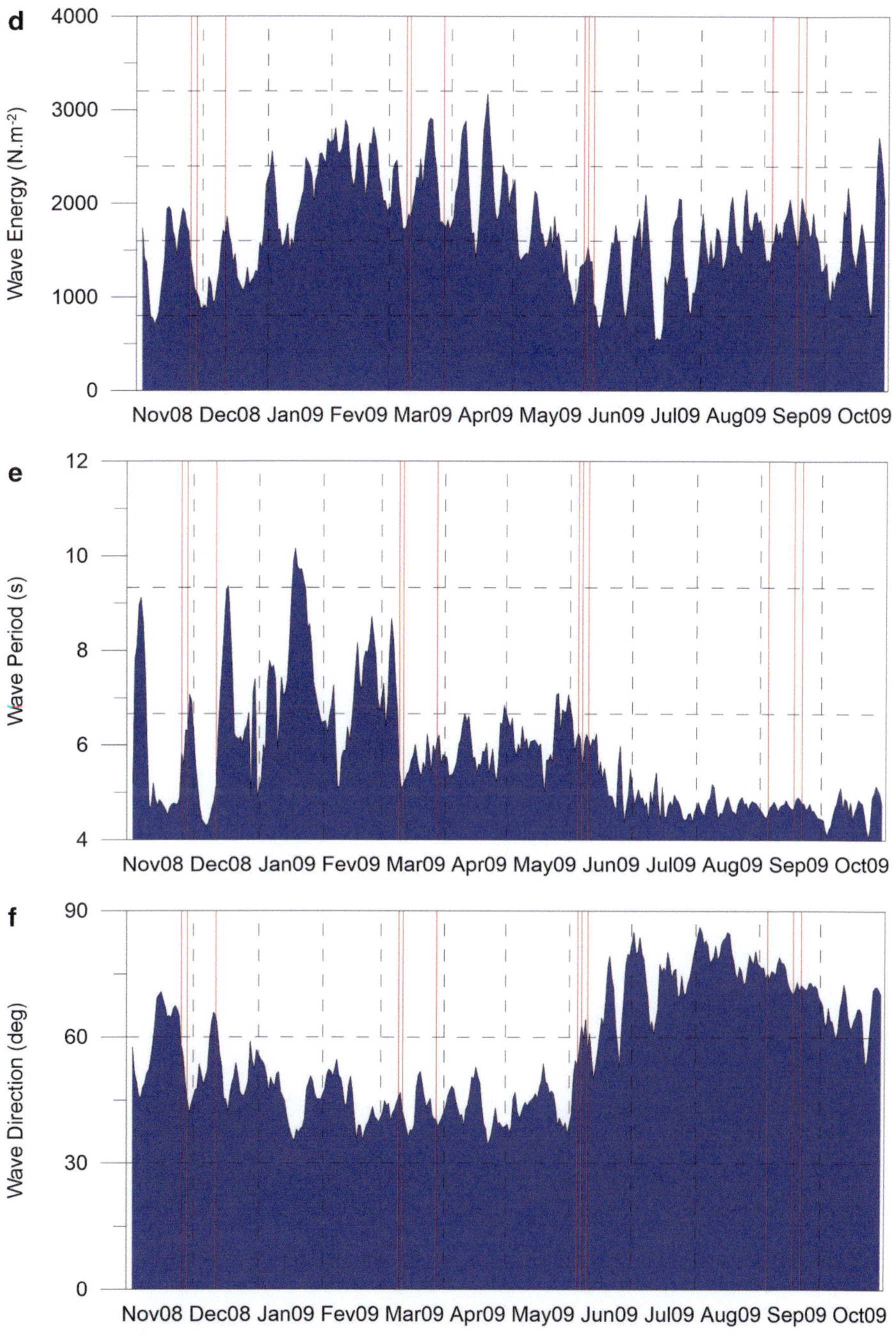

Fig. 4.3 (continued)

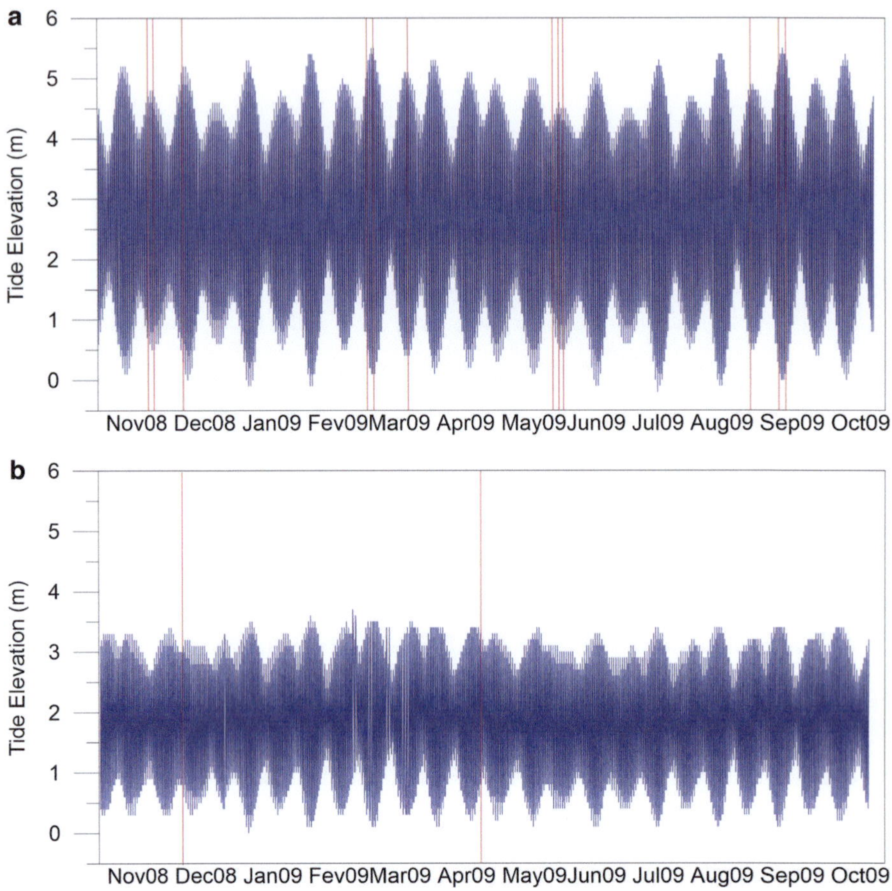

Fig. 4.4 Tide elevation from DHN (Directorate of Hydrography and Navigational of the Brazilian Navy). Fundeadouro de Salinópolis station between November 2008 and October 2009 (**a**) and Belém Harbour station from November 2008 and October 2009 (**b**). The red lines represent the dates when the nearshore surveys were undertaken in Ajuruteua, Atalaia and Princesa beaches (**a**), and Paracauari (**b**)

currents oriented across the shelf, and (ii) a shelf current, which maintains a strong northwesterly surface currents (Beardsley et al. 1995; Geyer et al. 1996).

Macro-tides dominate the eastern Pará coast. Figure 4.4a shows DHN data with spring tidal ranges (sTR), typically ranging from 4.0 to 5.5 m in Fundeadouro de Salinópolis and from 3.0 to 3.5 m in Belém Harbour (Fig. 4.4b). Data obtained *in situ* along the northeastern coast of the Pará shows that the largest sTR occur during equinoctial spring tides (>5.0 m), but the effect of strong winds can generate large meteorological tides during non-equinoctial conditions. On Amazon coast, the effect of the equinoctial can normally be observed at the end of February, March and/or in the beginning of April (first semester), as well as in the end of August, September and/or at the beginning of October (second semester). Tidal attenuation occurs when tidal wave propagates into the Marajó estuary, with the reduction in

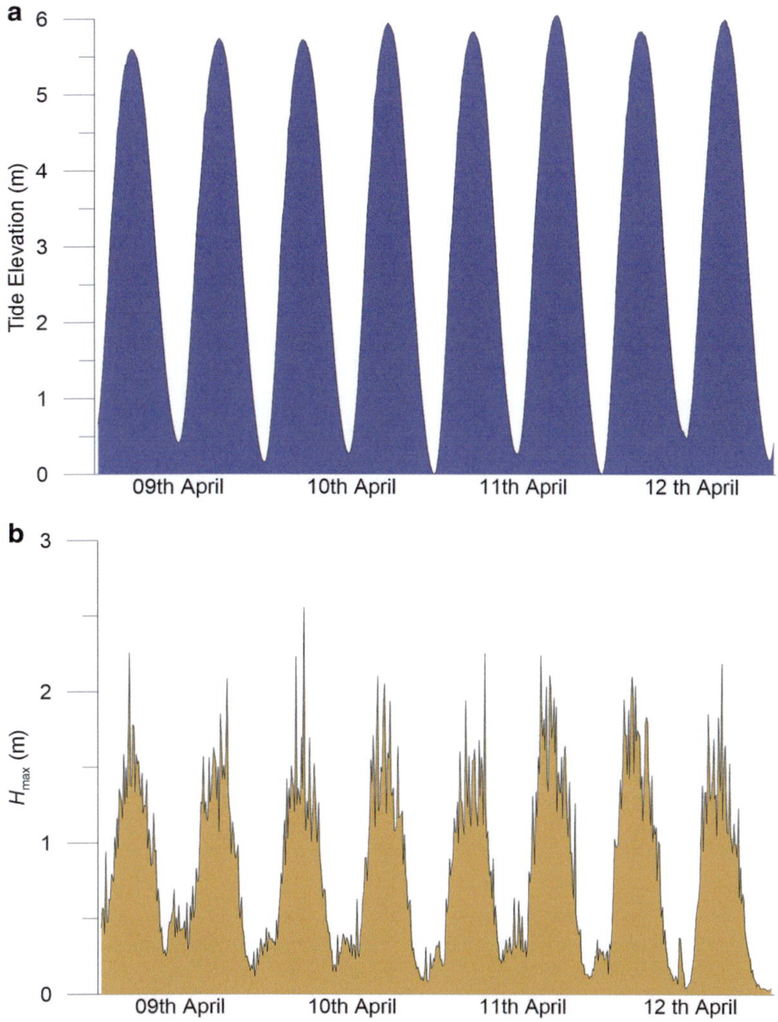

Fig. 4.5 Hydrodynamic condition: (**a**) tide elevation and (**b**) maximum wave height at Ajuruteua beach, during an equinoctial period

Belém (Belém Harbour station) is on the order of 35 % when compared to northeastern coast (Fundeadouro de Salinópolis station).

4.3.4 *Wave-Tide Interactions*

As the tide falls along the eastern Pará coast the shoreline becomes increasing sheltered by extensive inter and sub-tidal sandbanks or bars resulting in a tide-modulation of the breaker wave climate. Figure 4.5 illustrate the impact of tide elevation on wave height at the sheltered Ajuruteua beach field site during the equinoctial period, with tides to 6 m and waves reaching 2.5 m at high tide, then dropping to <0.5 m at low tide. Significant wave height in the sub-tidal zone under spring tide conditions vary from 0.1 m (ebb/low tide) to 1.5 m (flood/high tide) and the wave energy values range from 200 $N.m^{-2}$ (ebb/low tide) to 2200 $N.m^{-2}$ (flood/high tide) in the most sheltered beach (Ajuruteua) and between 600 $N.m^{-2}$ (ebb/low tide) and 2800 $N.m^{-2}$ (flood/high tide) on the more exposed beach (Atalaia). These data show the influence of the tidal elevation and the presence of sandbanks in attenuating most wave energy during low tide, when waves break on the offshore sandbanks. However, when the tide rises, waves start to propagate over the banks, reaching the maximum wave energy at the shoreline during high tide. The wider role of tide in modulating wave height, period and tidal currents is discussed in Sect. 4.5.

The sandbanks also delay of the flood tide inside the banks generally leading to longer ebb tide periods (~7 h) and shorter flood tide (~5 h) periods particularly during equinoctial period at the most sheltered beach (Ajuruteua). On the more exposed Atalaia beach the lag is less pronounced because it has fewer sandbanks and consequently less influence on the ebbing tide.

Similar to wave height, wave period is also modulated by tide elevation and range. During lower tides, the waves break on the offshore sandbanks and bars and only locally-generated short period wind waves reach the shore; as the tide rises, the longer period ocean waves start to propagate over the sandbanks reaching the peak period values during higher tides.

Within the inshore areas, the highest current velocities correspond to tidal currents and occur along the beaches located near the mouth of the larger estuaries and bays. As a consequence of the local tide asymmetry, the highest velocities are observed during the flood tides, with higher velocities also coinciding with the highest sTR, while the lowest velocities occur in June. The directions of the tidal currents are strongly influenced by the estuarine morphology with ebb currents flowing seaward and flood currents flowing into the rivers. The bi-directional circulation pattern is more evident on the more sheltered beaches. On the more exposed beaches the northeast winds and waves drive a northwestward current that can dominate over the tidal flows.

4.4 Coastal Sections

The entire Pará coast extends from more than 2700 km from 300 m upstream in the Amazon River, through the Amazon mouth and its distributaries, to Marajó island and bay and finally along the indented eastern coast. This section briefly describes the three western sections, which are covered in detail in Chap. 3, then provides an overview of the two eastern sections that are the subject of this chapter, followed by a more detailed overview of the eastern barrier island coast.

4.4.1 The Pará Coastal Sections

The Pará coast can be divided into five morphological sections based on shoreline morphology and relative contribution of fluvial, tide and wave processes.

Section 1 The *Amazon River mouth* coast consists of the Amazon River banks and its numerous distributaries and islands. It is a river and tide-dominated coast with a shoreline dominated by *varzea* forest and mangroves fronted by tidal mud flats and containing about 50 sandy beaches which occupy about 100 km of the 1500 km of mud and mangrove-dominated shoreline (see Sect. 3.3.3).

Section 2 *Marajó Island* is a low emergent coast and has a relative straight north coast containing a mix of narrow low energy tide-dominated beaches and mangroves that for much of its length faces into the 10–15 km wide South Canal (Canal do Sul), a major Amazon distributary. The eastern 120 km is more open to the Atlantic though the coast remains mangrove-dominated with small scattered tide-dominated beaches and tidal sand and mud flats (see Sect. 3.3.4).

Section 3 *Marajó Bay* divides the coast in two. The 150 km long bay is 60 m wide at its mouth narrowing to 15 km wide at its base where it joins the junctions of the Pará and Tocantins rivers. Tide-dominated beaches line both sides on the bay. On the *western Marajó Bay* coast there are 125 beaches spread along the 180 km of shoreline. The shoreline is relatively straight and continuous except where cut by several small rivers and many smaller creeks. This coast has a 60 km long 'low' northern sector and a 120 km long 'high' southern sector where the Barreiras Formation has been eroded to form coastal bluffs in places. Most of the beaches are low, short (<5 km) and many lie adjacent to creeks and rivers with recurved spits common as well as evidence of both north and south longshore transport. Most of the beaches face east into the prevailing trade winds receiving fetch-limited waves generated within the bay, as well as being affected by the strong shore-normal and shore-parallel tide and river flows. Many of the beaches are fronted by intertidal sand flats and then wider mud flats (see Sect. 3.3.5).

Section 4 The *eastern side of Marajó Bay* (Fig. 4.6) has approximately 265 km of shoreline spread along a more discontinuous shoreline dominated by mangroves with some larger rivers and creeks cutting the coast into large bays, distributary islands and extensive tidal shoals. Waves are restricted to low short wind waves generated within the bay, while tides reach up to 10 m at the bay entrance decreasing to 3 m at Belém (Fig. 4.4). The combination of low short waves (<0.5 m) and meso to mega-tides produces predominately tide-dominated beaches. While mangroves and mud flats dominate most of the coast, particularly in the north, there are over 150 generally short, low energy tide-dominated beaches along the coast. Most face west or northwest into the bay. They commence just north of Abaetetuba, 45 km southwest of Belém, and extend along more than 240 km of shore to the bay mouth, near Vigia. The beaches from at creek mouths, at the base of and between

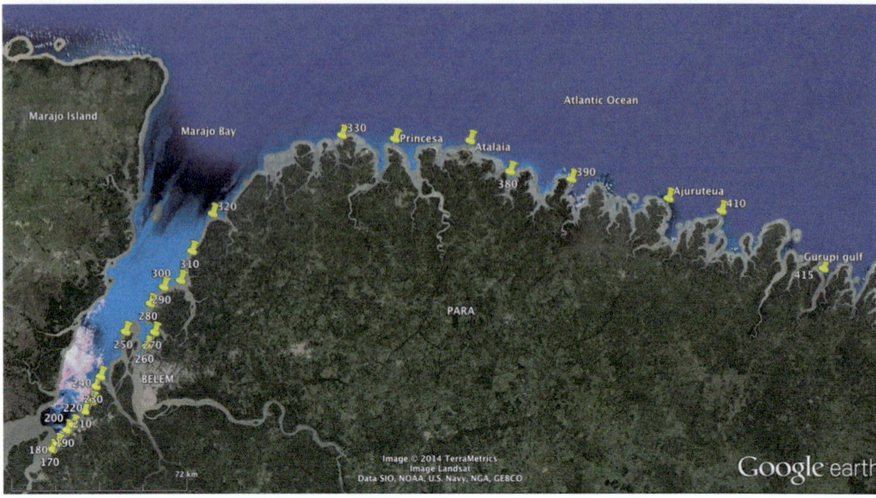

Fig. 4.6 The eastern Pará coast extends from near Belém in Marajó Bay to the border at Gurupi gulf. *Place marks* indicated roughly every tenth 10 beach as well as beaches discussed in text (Source: Google earth)

low headlands where the low coastal plateau (Barreiras Formation) reaches the shore, and where mangroves are dominant as beaches ridges or cheniers along the mangrove shore. No work has been published on these beaches and only a few beaches near Belém have been inspected by the authors. The Belém beaches each consisted of a medium quartz sand high tide beach that grade abruptly into tidal sand or mud flats around mid-tide (Fig. 4.7). They were all backed and or bordered by the Barreiras Formation, which at one beach forms 10 m high bluffs. Some of the Belém beaches are popular recreation sites and are backed by seawalls and restaurants, all of which impinge on the backbeach.

Section 5 *Marajó-Gurupi bays*. The eastern coast commences at Romana Beach on the eastern entrance to Marajó bay and extends east for approximately 525 km to Gurupi Bay (Fig. 4.6). This is a highly indented shoreline consisting of approximately 15 major bays and associated tidal creeks and their distributaries, which from funnel-shaped estuaries, averaging 5–10 km in width at their mouths and extending up to 20 km inland, with tidal creeks extending further inland. The bays are separated by the drowned plateau which grades seaward into low regressive barriers and mangroves terminating at highly dynamic recurved tide-modified barrier islands (Hayes 1979) all less than 10 km in length. The bays are filled with extensive mangroves, sandy tidal shoals and channels. The beaches face into the Atlantic waves, however, with the 4–6 m tidal range and strong tidal currents most are tide-dominated and fronted by extensive intertidal sand flats and in places tidal sand shoals that extend for up to 2 km seaward. Only the most exposed may be tide-modified with rip currents forming at low tide. Section 4.5 will examine three beaches located along the eastern Pará coast in greater detail.

Fig. 4.7 Do Amor beach at Belém (Located near 270 in Fig. 4.6) consist of a narrow steep high tide beach grading abruptly into intertidal sand flats (Seawalls are common on the popular Belém beaches)

4.4.2 The Barrier Island Beaches

The eastern coast between Pirabas and Gurupi bays is dominated by barrier islands and estuaries. There are approximately 15 clusters of barrier islands separated by the funnel-shaped estuaries, with the clusters containing a total of 90 beaches. All the beaches are tide-modified to tide-dominated and tend to be relatively short (1–10 km) and convex in plan from often recurving into the major or secondary bays and inlets, many terminating at migrating recurved spits. Beach orientation therefore ranges from east to north to west.

Beach processes are dominated by macro-tides (4–6 m), high velocity flood dominated tidal currents (>1.0 m.s^{-1}) and moderate wave energy (H_s up to 2.0 m) particularly during the rising and high tides. During falling to low tide the shore is partly sheltered by the sandbanks permitting only low short period seas to be generated under fetch-limited conditions. In contrast, during the rising and high tide, higher and longer offshore waves move across the flooded sandbanks and reach the shoreline (Fig. 4.5) (Pereira et al. 2012a, b, 2013a, b). Incident wave energy on these beaches is therefore dependent on both deepwater wave conditions and on degree of shelter provided by the presence and distribution of sandbanks.

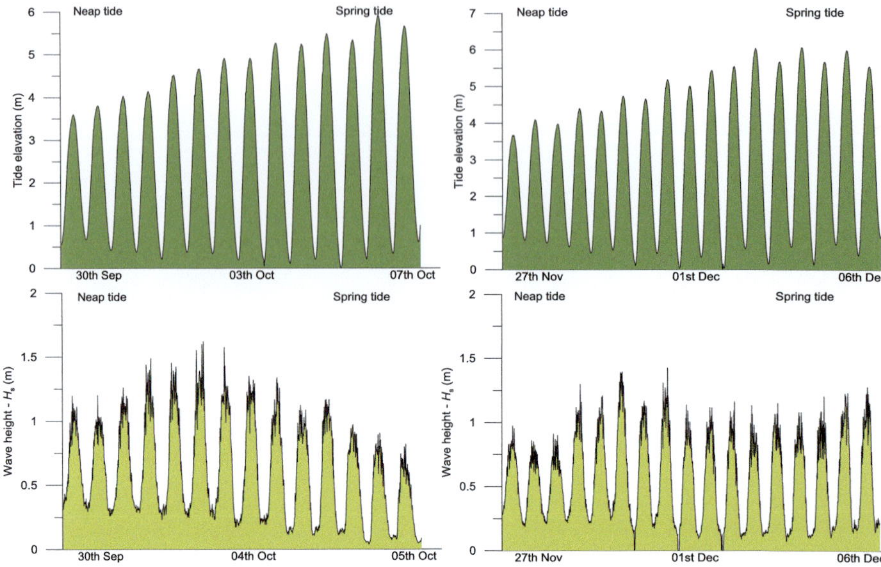

Fig. 4.8 Tidal modulation of wave height (H_s) during neap and spring tide under varying offshore wave conditions at Ajuruteua beach

The morphodynamic processes on three sandy beaches along the eastern Pará coast under influence of different tidally modulated wave energy are presented in Sect. 4.5. Physical hazards are also considered in Sect. 4.6 as this coast contains the most popular tourist beaches in the State, in particular Ajuruteua, Atalaia and Princesa beaches.

4.5 Beach Morphodynamics

4.5.1 Tidal Modulation of Processes

Beach morphodynamics on all Pará beaches is modulated by the seasonal variation in wave height together with semi-diurnal tidal fluctuation and lunar though seasonal tidal variation in tide range. Figure 4.8 shows the impact of semi-diurnal and spring to neap tides over a 7-day period at Ajuruteua beach. Water depth across the beach increases by up to 6 m at spring high tide permitting the longer period ocean waves to cross the sandbanks and surf zone and reach the beach, while at low tide exposure of the sandbanks eliminates ocean waves and only permits short fetch limited waves to reach the base of the low tide beach. As a consequence breaker wave height varied from a 1–1.5 m at high tide to a of 0.1–0.3 m at low tide, with the lowest height during spring low tide when the sandbanks are most exposed, resulting in considerable variable in wave height and beach-surf zone morphodynamics with each tidal cycle.

Fig. 4.9 (**a**) Princesa, (**b**) Atalaia and (**c**) Ajuruteua beaches (See Fig. 4.6 for location) (Source: Google earth)

Beach morphodynamics in Pará have been investigated on three of the eastern barrier island beaches: Princesa, Atalaia and Ajuruteua beaches (Fig. 4.9). These beaches illustrate the roles of waves and tides in controlling both spatial and temporal variability in beach morphodynamics along the eastern Pará coast. Ajuruteua is bordered by Caeté river and is the most sheltered by sandbanks. Princesa is moderately sheltered and is bordered by Maracanã and Marapanim rivers, while Atalaia is the most exposed and is bordered by Sampaio river. All three beaches are composed of uniform fine sand (mean grain size 0.129–0.160 mm). The beaches have a moderately sloping (2–5°), narrow high tide beach (<20–30 m) and a low gradient intertidal zones (1–2°) that ranges in width from 150 to 300 m. Figure 4.10 shows the three beaches during an ebbing tide.

Offshore wave data for the beaches were obtained from CPTEC-INPE based on WaveWatch III wave modeling. Nearshore-surf zone wave data were collected at the surf zone of the beaches during four field campaigns of 25 h duration using a bottom-mounted mooring with wave and tide data-logger (TWR 2050). Wave statistics were based on 512 samples at a burst rate of 4 hz of 10 min duration. Tidal water level were acquired every 2 s and mean values obtained every 10 min.

Figure 4.11 illustrate the role of tide in modulating beach processes and morphodynamics at two of the beaches. Breaker wave height, period and energy all experience their greatest modulation on the most sheltered Ajuruteua beach, where the extensive sandbanks filter out all ocean waves at low tide, while least modulation occurs on the more exposed Atalaia (see Fig. 4.11b, c). During lower tides, the waves break on the offshore sandbanks and bars and only locally-generated short

Fig. 4.10 The three studied beaches: (**a**) Princesa, (**b**) Atalaia and (**c**) Ajuruteua, showing the low intertidal gradient during an ebbing tide

period wind waves reach the shore; as the tide rises, the longer period ocean waves start to propagate over the sandbanks reaching the peak period values during higher tides.

Tidal currents, as expected reverse with the tide (Fig. 4.11e). The sandbanks also delay of the flood tide inside the banks generally leading to longer ebb tide periods (~7 h) and shorter flood tide (~5 h) periods particularly during equinoctial period at the most sheltered beach (Ajuruteua). On the more exposed Atalaia beach, the lag is less pronounced because it has fewer sandbanks and consequently less influence on the ebbing tide.

Within the inshore areas, the highest current velocities correspond to tidal currents and occur along the beaches located near the mouth of the larger estuaries and bays. As a consequence of the local tide asymmetry, the highest velocities are observed during the flood tides (Fig. 4.11d), with higher velocities also coinciding with the highest sTR, while the lowest velocities occur in June. The directions of the tidal currents are strongly influenced by the estuarine morphology with ebb currents flowing seaward and flood currents flowing into the rivers. The bi-directional circulation pattern is more evident on the more sheltered Ajuruteua and Princesa beaches. On the more exposed Atalaia beach the northeast winds and waves drive a northwestward current that can dominate over the tidal flows.

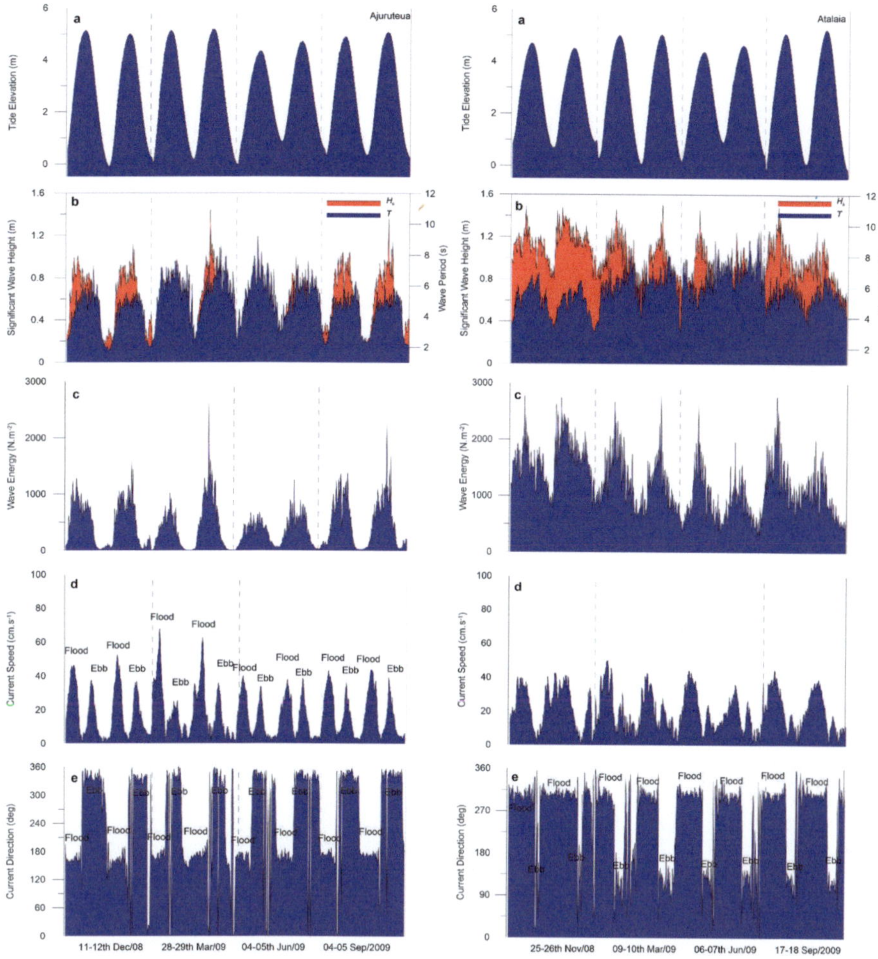

Fig. 4.11 Nearshore condition at the most sheltered (Ajuruteua, *left*) and the most exposed (Atalaia, *right*) beaches (*a* tidal elevation, *b* wave height and period, *c* wave energy, *d* current speed and *e* current direction)

4.5.2 *Tidal Modulation of Beach Morphodynamics*

The role of tides is further illustrated in Fig. 4.12 which plots the impact of the both wave attenuation and tidal oscillation on the beach morphodynamics. In general, when wave energy is highest ultradissipative beach conditions prevail while when wave height decreases (June, transitional period) low-tide bar characteristics prevail. The central column shows how the higher offshore waves have the potential to maintain tide-modified conditions with a relatively low RTR (3–6.3) and high Ω

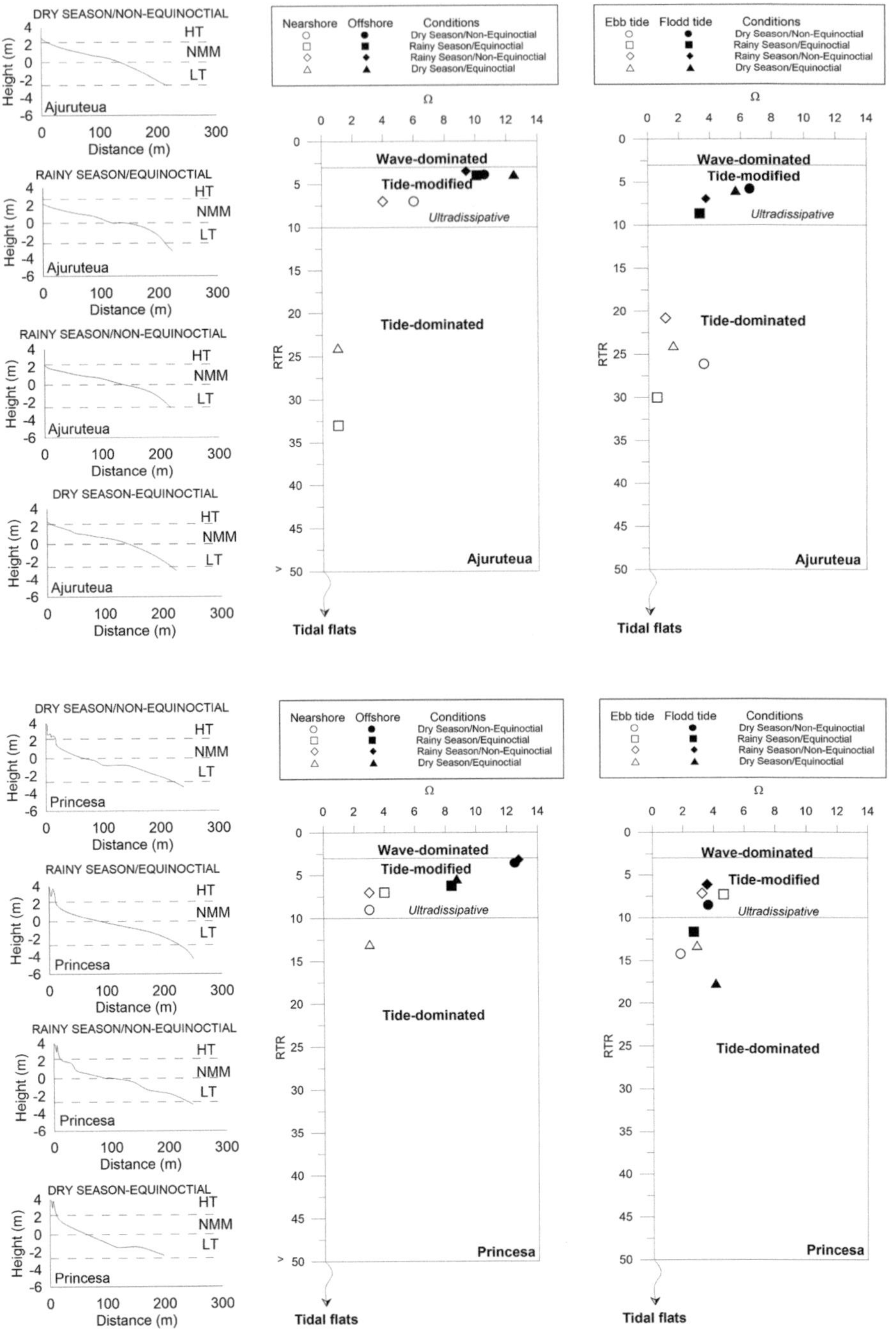

Fig. 4.12 Crosshore and seasonal variation in beach morphodynamics expressed by variation in the relative tidal range (*RTR*) and dimensionless fall velocity (Ω) at Ajuruteua, Princesa and Atalaia beaches. Columns show seasonal beach profiles (*left*); predicted beach state based on offshore and nearshore-surfzone conditions (*center*); and predicted beach state during flood and ebb tides (*right*)

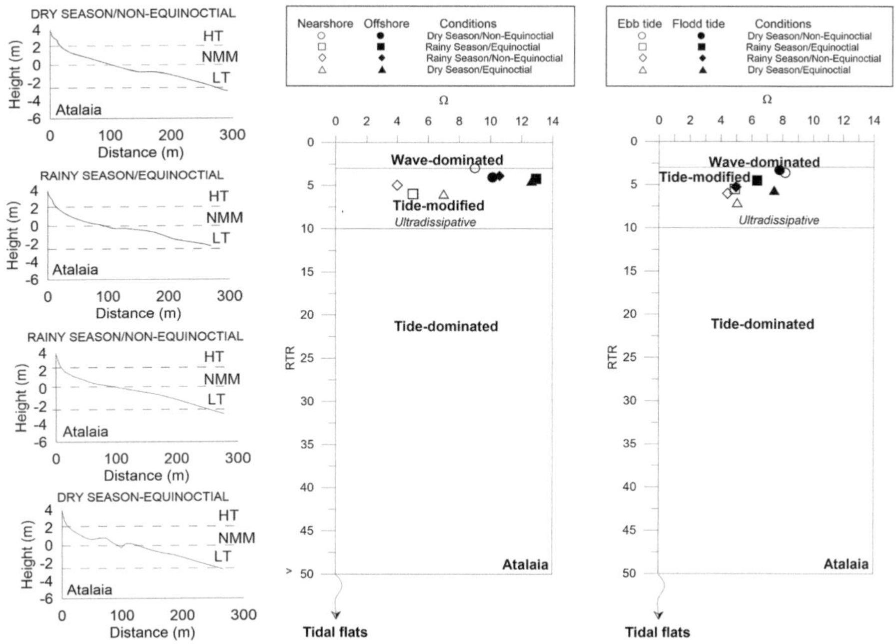

Fig. 4.12 (continued)

(8–13). In the nearshore-surf zone however on Ajuruteua and Princesa beaches wave attenuation across the banks and bars lower breaker wave height resulting in a higher nearshore RTR (7–33) and lower Ω (1–6), producing in a shift from tide-modified to tide-dominated conditions (Short 2006), the greatest shift on the most sheltered Ajuruteua. The right-hand column plots the impact of rising flood tide and falling ebb tides on these parameters. During flooding tides, higher waves maintain both lower RTR and higher Ω, particularly on the more sheltered beach resulting in a tide-modified condition, with falling ebb tides resulting in higher RTR and lower Ω and tide-dominated conditions. However as the beaches are essentially inactive at low tide, it is the high tide condition that control and determine the beach state. In addition, upper foreshore erosion occurs during the equinoctial tides, while upper foreshore sedimentation during non-equinoctial conditions (Pereira et al. 2013a).

The impact on each of the three beaches can be summarized as follows: The most sheltered Ajuruteua beach receives the lowest breaker height ($H_b < 0.2$) during equinoctial periods (March and September), due to longer ebb tide period, which occurs during the highest tidal ranges which coincide with the lowest modal H_b values, resulting in characteristics of tide-dominated tidal sand flats. During the non-equinoctial period (June and December) when modal H_b reaches higher values the beach shifts to an ultradissipative state. The ultradissipative beach consists of a narrow sloping high tide beach and 200 m wide low gradient intertidal zone.

The second most sheltered beach (Princesa) also maintains an ultradissipative profile with a narrow high tide beach and a 200 m wide low gradient intertidal.

This agrees with the morphodynamic conditions, which are predominately tide-modified with RTR ranging from 6 to 18 and Ω values varying between 2 and 5. The highest RTR values are recorded normally during dry season (September) when H_b reaches minimal values. Like Ajuruteua beach, conditions during the flooding tide conditions maintain an ultradissipative state during non-equinoctial period.

On the more exposed beach (Atalaia), the higher waves maintain a lower RTR between 3 and 6 at high tide only rising slightly to 4–7 at low tide. During the higher energy dry season when RTR is 3 the beach tends to be wave-dominated, while tide-modified condition occur when wave energy is lower with RTR ranging from 4 to 7 and Ω values varying between 4 and 5 during ebbing tide and from 4 and 8 during flooding tide. When $\Omega > 5$ the beach tends to be ultradissipative, while during the lower energy when $\Omega = 2–5$ intermediate states with low tide bars prevail with rip channels and currents forming in the shallow subtidal at low tide.

The results from these three beaches indicate that along the eastern Pará coast there is considerable spatial and temporal variability in beach state dependent on the amount of nearshore sheltering by intertidal sandbanks, with tide-modified beaches on the more exposed sections and tide-dominated on the more sheltered section.

Temporal variability in beach morphodynamics is therefore controlled by the combination of offshore wave conditions, the neap to spring tidal cycle and the semi-diurnal tidal cycle. The impact of lunar and semi-diurnal tidal cycles on wave height is illustrated in Fig. 4.8 and the impact of the semi-diurnal tides on wave height and beach state in Figs. 4.11 and 4.12.

4.6 Physical Hazards

Physical hazards associated with Pará beaches include the presence of deep tidal channels with strong tidal currents (>1.0 $m.s^{-1}$), rip currents, intertidal rocks, moderate to high hydrodynamic energy including wind-waves with wave > 1 m during high tide and macro-tides (4–6 m) which cause considerable variation in beach width and water depth. The presence of beachfront development (Fig. 4.13) and the large numbers of vehicles that drive onto the most popular beach (Atalaia) are other elements of risk discussed in this section. The lack of local coastal management plans together with the large number of beachgoers, many with low swimming skills and poor perception of physical hazards, results in a high level of beach risk and incidents.

4.6.1 Beach Population

Beach risk level increases during the busy vacation periods when thousands of people occupy the beaches. During these periods lifesavers record an increase in the number of drownings and other injuries (personal communication by local lifesavers). The risk level increases most during July (school vacation) when beach-goer numbers dramatically increase even though hydrodynamic energy is lower, as well as during the busy March-April (Easter) and September (Brazilian Independence Day) when equinoctial tides occur and hydrodynamic energy is higher.

4.6.2 High Waves

Hazardous high-energy events typically occur: (i) at the end of the dry season (November/December) when to moderate wind-wave energy is generated by the strongest trade winds of the year; (ii) during the mid-rainy season (March) coinciding with the higher equinoctial tides wind-wave energy; and, (iii) during the mid-dry season (September) marked by other equinoctial period with strong winds (Fig. 4.2). During these periods and particularly during high tide the offshore waves can propagate to the shoreline with negligible dissipation of energy producing the most hazards bathing conditions including rip currents on the high energy Atalaia beach.

4.6.3 Macro-Tides

The meso to macro-tides are another source of hazards. The tide rises (5–6 m during spring tides) in less than 6 h (due to tide asymmetry) rapidly covering the wide intertidal zone. Because of the tidal tide asymmetry flooding tides are shorter and stronger. The maximum hazard to beachgoers occurs 2–3 h after low tide, when tidal current velocities are at a maximum. High velocity tidal currents adjacent to the shore are especially dangerous to beachgoers on the most sheltered beaches (Ajuruteua and Princesa) and, also in the tide channels located adjacent to the beaches.

4.6.4 Intertidal Rocks

The presence of intertidal rocks (Fig. 4.13d) and intertidal channels and troughs in the surf zone are particularly hazardous when the tide is flooding and are most dangerous for children, who often play in the intertidal sectors of the beach. Waves breaking on intertidal rocks can cause serious injuries when beachgoers may be knocked down and then washed into the sea, while strong currents in the channels can result in drowning with many incidents being reported by local residents and lifesavers (personal communication).

On all beaches, wave energy and current speeds are greatly reduced during the low tide, resulting in a less hazardous bathing environment.

Fig. 4.13 Hazards caused by high wave energy during the equinoctial period (*a–c*) and presence of intertidal rock at Princesa beach (*d*)

4.6.5 Hazards to Property

The beachfront properties consist mostly of rudimentary wooden structure (Figs. 4.13 and 4.14a), except at Atalaia where there are also luxurious beach houses (Fig. 4.14b). Beachfront properties are often affected by beach erosion occurring during equinoctial tides when waves can propagate to the properties causing extensive damage (Fig. 4.13). Injuries to beachgoers can occur because the local civil defense authorities seldom ban the access to areas where there is a risk of building collapse. Beachfront buildings are typically built on stilts to avoid the effects of the macro-tides and flooding. However, when they are located in the intertidal zone they can represent a great risk to beachgoers, particularly as the tide quickly rises during spring tides, and can trap beachgoers between the water and the stilts. This occurs mainly during the equinoctial period when the tides are the highest.

4.6.6 Beach Pollution

Beach and water pollution on the popular beaches is a major issue because each of the beachfront buildings has a rudimentary cesspit (Fig. 4.15) located within the intertidal zone, backshore or dunes. This situation has negative repercussions on

Fig. 4.14 Erosion problems exacerabated by structures built in the active zone at the most popular beaches (**a**) Ajuruteua and (**b**) Atalaia

Fig. 4.15 Rudimentary cesspit located within the intertidal zone at (**a**) Ajuruteua and (**b**) Atalaia

beach water quality particularly during the busy annual vacation period. Contamination occurs primarily during high tide and the beginning of the ebb tide, when the water level reaches the area in which the cesspits are located, in particular on days when the numbers of beachgoers are at their highest levels (e.g. July vacation period). The presence of cesspits from houses and commercial establishments (bars, restaurants, hotels, and guesthouses) result in unpleasant sights and odors, as well as contamination by bacteria, such as fecal coliforms.

4.6.7 Beach Vehicles

Another anthropogenic hazard is the presence of vehicles on Atalaia beach (Fig. 4.16). Cars, buses and trucks are driven onto the beach and serious problems have been recorded including car accidents, submerged cars and traffic jams. During vacation period, mainly in July, some vehicles have been trapped by the quickly rising tide. Cars become stuck in the intertidal zone due to the congestion caused as

Fig. 4.16 (**a**) Traffic conjection on Atalaia beach during the vacation period; (**b**) bogged truck caught by the flooding tide on Atalaia; and (**c**) unique wheeled transport (donkey cart) on Princesa beach

many of vehicles trying to get off the beach to higher ground (personal communication). Another anthropogenic annoyance is sound pollution, which is a consequence of the high volume and large number of car stereos.

4.7 Recreational Use

Beach tourism in Pará is heavily biased towards the July school vacation period. This is different from other parts of Brazil as the usually popular January school vacation period coincides with the region's long and intense rainy season. This unfavorable aspect of the climate, together with the relative isolation of the beaches, and the low quality of their services and infrastructure, combine to limit the development of the local tourism industry.

Atalaia is the most popular oceanic beach. This beach has also received the most investment from local authorities, and offers facilities, which attract more affluent visitors. This is reflected in the large number of luxurious beach houses in the northwest sector, and the much larger number of vehicles on the beach during peak visitation periods.

By contrast, island-bound Princesa beach, like the rest of Maiandeua Island, provides only rudimentary services and infrastructure. In fact, local residents and public authorities have worked together to guarantee the preservation of the island's natural resources and landscapes, which has included the establishment of Environmental Protection Area of Algodoal-Maiandeua, decreed by state law 5621 of November 27th, 1990. These initiatives have limited the island's infrastructure, including a prohibition on motor vehicles (the only wheeled transport is provided by donkey carts, see Fig. 4.16c), and only 22 bars along the 14 km of the beach. In comparison with Atalaia, Princesa attracts visitors with a very different profile, in particular, younger people interested in enjoying the natural beauty of the area.

At Ajuruteua beach, local residents rely on traditional or commercial fishing and/or tourism for their livelihood. Despite the rudimentary facilities available at this beach, it receives some 90,000 visitors during the vacation (Pereira et al. 2007; Oliveira et al. 2011). The cultural and historical importance of the nearby city of Bragança and the beauty of the local coastline tend to attract visitors with a range of ages and economic profiles.

In terms of recreational use and local topographic characteristics, these beaches were divided by Sousa et al. (2011) into three zones:

(i) *Zone 1* encompasses the most landward part of the beach, where the infrastructure that provides services for beachgoers – bars, restaurants and hostelries – is installed. This zone is normally located on the dune fields or intertidal zone. During the high tide, this area accommodates practically all the beachgoers, as the rest of the beach is covered by seawater;
(ii) *Zone 2* is subdivided into upper, intermediate and lower subzones, the area of which at any given time depends on the tidal conditions (spring/neap, high/low). This zone is used primarily for leisure activities, such as sports (football, volleyball), sunbathing, and the consumption of food and drink at bar tables set up during ebb and low tides. At Atalaia, cars, buses and trucks are driven onto the beach and parked in this zone. Cars are prohibited on Maiandeua Island and Ajuruteua;
(iii) *Zone 3* is on the waterline. At Ajurutea and Atalaia, this zone is used for bathing and nautical sports (kite surf, jet ski, surf). Nautical sports are prohibited on Princesa beach.

Studies carried out by Sousa et al. (2011) show that during a vacation period (July), the average number of visitors per time is similar at all three beaches. Peak visitation times coincide with the period of most intense insolation, i.e., normally between 10 am and 2 pm. For all beaches, the peak days are Saturday and Sunday and zone 1 is the most visited, while the occupation of zone 2 is determined by the tide cycle (Fig. 4.17).

4.8 Beach Development and Management

Pará has good potential for beach tourism, however the development of this industry is limited particularly by the lack of adequate access and infrastructure. Initiatives have been implemented at municipal, state, and federal government levels,

Fig. 4.17 Recreational dynamics is influenced by the tidal cycle during the varation periods at Ajuruteua beach, shown here at (**a**) low, (**b**) mid and (**c**) high tide

providing economic incentives aimed at improving services and infrastructure while maintaining the natural beauty of the region's beaches, but the results have been slow in coming. The unregulated occupation of land, and the lack of services and infrastructure are the principal factors underpinning the social and environmental problems that characterize this sector of the Amazon region.

There are a number of issues that need to be addressed to improve beach management and public safety on these beaches.

Inappropriate Beachfront Development and Pollution In general, the waterfront of these beaches is dominated by a number of precarious wooden structures built on stilts on mangroves, dunes and the intertidal zone. This is compounded by the lack of street lights, public telephones and garbage collection, together with no public water supply or sanitation system, as toilets are only available in the bars, and are restricted to customers, and there are no public toilets. In addition, street cleaning is intermittent. As a consequence of the lack of a public sanitation system or refuse collection, sewage and solid waste are deposited directly onto the beach. During the vacation period (July and a few other holiday periods) there is an increase in the discharge of sewage and the disposal of solid waste directly onto the beach as well as considerable increase in the presence of plastics, paper and metal, food leftovers, and human and animal excrement.

Beach Safety Resources The presence of lifeguards occurs mainly during the July vacation period, however the low number of lifeguards is often inadequate for the large number of beach users (see Fig. 4.17) particularly if these holidays coincide with equinoctial tides (e.g. Easter-March/April and Independence-September days), when the hydrodynamic energy is the highest. Also surfing, kite surfing and jet skis are aquatic activities practiced on the popular beaches of Atalaia and Ajuruteua and can be dangerous for other beach users. While public rescue equipment (life buoys, lifejackets and life rafts) can be found on these beaches, first aid equipment is not available. As first-aid stations or hospitals are normally located dozens of kilometers inland from these beaches local first aid equipment is essential.

Beach Traffic At the popular Atalaia beach vehicles are driven onto the beach and parked in the intertidal zone, mainly during the peak season (Fig. 4.16). This has caused traffic jams and accidents, as well as the degradation of dunes.

Beach Erosion Along the most of the tourist beaches, some of the buildings and infrastructure have been partially or completely destroyed by erosive processes (Figs. 4.13, 4.14, and 4.15). This process occurs during the equinoctial spring tides in March and September, which result in the loss of street lights, bars, restaurants and hostels. When the erosion become too great the wooden structures are simply moved to new locations on the dunes. While all these beaches receive a large number of tourists during holiday periods, there has been no investment to protect the coast and/or replace the sand lost from the coastline.

Coastal Management Princesa beach is the least developed with sparse infrastructure, and only few bars built on the dunes and intertidal zone. Under the Environmental Protection Area (Algodoal/Maiandeua) management, the construction of hotels or other buildings, except bars and fishermen's huts, is strictly prohibited. The island can only be reached by boat and motorized vehicles are prohibited on the island, with the only wheeled transport provided by donkey carts. There are no campgrounds on the island, although Algodoal village offers a selection of lodgings. Sewage disposal is based on the few precarious cesspits built in the dunes or intertidal zone, while refuse is dumped in the dunes.

The implementation of a Coastal Management Plan and/or programs considering conflicts/losses and benefit/profit are necessary for the beaches of the Amazon coast of Brazil. We suggest several courses of actions that could be implemented to improve the current status of the studied beaches, including:

(i) Appropriate land use planning to improve the local landscape and to accommodate morphodynamic changes, avoid destruction of natural resources (dunes, mangrove forest, lagoons, etc.), avoid risks and conflicts between the economic sectors and the key areas of environmental conservation. This will require:

(a) planning development and restricting it to behind the foredune;
(b) prohibition of cars accessing the intertidal zone;
(c) zonation of water activities in swimming areas delimited by flags, buoys and other signs, together with more lifeguards;
(d) beach safety education program for beachgoers with regards to the hazards, including maps and information board (permanent and/or provisory) along the beaches showing the physical hazards.

(ii) Implementation of programs that promote the sustainable exploitation of natural resources (fish, crabs, mangrove forest, sediment beach, *etc.*) to allow the stability of the local population who live on these resources. This step will require:

(a) development of extension works with the local inhabitants through the implementation of environmental education programs;
(b) incentives for a sustainable tourism development to increase job opportunities and improve the local economy without affecting the natural environmental;
(c) incentives to form of associations to consolidate and to fortify the local economic sectors, for example, the fishermen class;

(iii) Investment in local infrastructure and services (roads, school, medical assistance, basic water distribution, public illumination, *etc.*) to improve life quality of local population and the safety of seasonal beachgoers;
(iv) Coastal zone monitoring (physical, ecological and social) to evaluate the current status and avoid future social and environmental problems.

In the context of Integrated Coastal Zone Management these actions should be undertaken with the participation of all stakeholders. It is important to include the participation of the local people so their needs are fulfilled and there is a sustainable plan for their traditional way of live.

4.9 Conclusions

The coast of Pará is approximately 2700 km long and has more than 450 sandy beaches. It is located just south of the equator adjacent to the Amazon River mouth, an area of numerous rivers/distributaries and islands and tidal creeks exposed to macro-mega tides and moderate energy trade wind waves. These factors combine to produce five coastal sectors: the Amazon river coast; the north coast of Marajó Island; the west and east coasts of Marajó Bay; and the highly indented eastern barrier island eastern coast between Marajó and Gurupi bays.

The 800 km long eastern Pará coast discussed in this chapter includes the eastern shore of Marajó bay where low energy tide-dominate beaches abut the low scarps of the Barrieras Formation, and the barrier island-estuary dominated east coast, where both tide-modified and tide-dominated beaches prevail.

A study of three beaches of varying wave exposure along the eastern barrier island coast is used to highlight the role of the large tides in modulating wave energy at daily though neap-spring and seasonal time scales. The high tides and low to moderate energy waves combine with the usually fine sand to maintain tide-modified to tide-dominated beaches, characterized by wide intertidal zones, often accompanied by inter to sub-tidal sandbanks, shoals and deep channels. During the tidal cycle there are substantial changes in tidal elevation, current direction and velocity, breaker wave height and beach morphodynamics.

The impact of these beach states and processes on the beach-going population has been assessed, with identification of many physical beach hazards: waves, tides and tide state, tidal currents and intertidal rocks. When these are combined with the seasonal tourism population they result in a dramatic increase in risk leading to incidents and drowning. In addition the unplanned beach development and large seasonal beach population generate addition hazards including eroding buildings, traffic congestion and accidents on some beaches, and sewerage and litter pollution.

Coastal planning and regulation is urgently required for the coast to enable it to develop a safe and sustainable tourist and fishing industry and to reduce the hazards discussed above and risk to the beach users.

References

Anthony JE, Gardel A, Gratiot N, Proisy C, Allison MA (2010) The Amazon-influenced muddy coast of South America: a review of mud-bank–shoreline interactions. Earth Sci Rev 103:99–121

Anthony JE, Gardel A, Proisy C, Fromard F, Gensac E, Peron C, Walcker R, Lesourd S (2013) The role of fluvial sediment supply and river-mouth hydrology in the dynamics of the muddy Amazon-dominated Amapá-Guianas coast South America: a three-point research agenda. J S Am Earth Sci 44(No Spécial):18–24

Beardsley RC, Candela J, Limeburner R, Geyer WR, Lentz SJ, Castro BM, Cacchione D, Carneiro N (1995) The M2 tide on the Amazon shelf. J Geophys Res 100(C2):2283–2319

DeMaster DJ, Pope RH (1996) Nutrient dynamics in Amazon shelf waters: results from amassed. Cont Shelf Res 16(3):263–289

Dunne T, Mertes LAK, Meade R, Richey JE, Forsberg BR (1998) Exchanges of sediment between the floodplain and channel of the Amazon in Brazil. Geol Soc Am Bull 110(4):450–467. doi:10.1130/0016-7606(1998)110<0450:EOSBTF>2.3.CO;2

Figueroa SN, Nobre CA (1990) Precipitations distribution over central and western tropical South America. Climanálise 5:36–45

Franzinelli E (1992) Evolution of the geomorphology of the coast of the State of Pará Brazil. In: Prost MT (ed) Évolution des littoraux de Guyane et de la Zone Caraïbe Méridionale pendant le Quaternaire. ORSTOM, Paris, pp 203–230

Geyer WR, Beardsley RC, Lentz SJ, Candela J, Limeburner R, Johns WE, Castro BM, Soares ID (1996) Physical oceanography of the Amazon shelf. Cont Shelf Res 16:575–616

Hayes MO (1979) Barrier island. Morphology as a functionof tide and wave regime. In: Leatherman SP (ed) Barrier islands from the Gulf of St Lawrence to the Gulf of Mexico. Academic Press, New York, pp 223–304

Isaac VJ, Barthem RB (1995) Os Recursos pesqueiros da Amazônia brasileira. PR-MCT/CNPq Museu Paraense Emílio Goeldi

Kjerfve B, Lacerda LD (1993) Mangroves of Brazil. In: Lacerda LD (ed) Conservation and sustainable utilization of mangrove forests in Latin America and Africa regions. International Society for Mangrove Ecosystems, Okinawa, pp 245–272

Lentz SJ (1995) The Amazon River plume during AMASSEDS: subtidal current variability and the importance of wind forcing. J Geophys Res 100:2377–2390

Marengo J (1995) Interannual variability of deep convection in the tropical South American sector as deduced from ISCCP C2 data. Int J Climatol 15:995–1010

Nittrouer CA, DeMaster DJ (1986) Sedimentary processes on the Amazon continental shelf: past present and future research. Cont Shelf Res 6:5–30

Nittrouer CA, DeMaster DJ (1996) The Amazon shelf setting tropical, energetic and influenced by a large river. Cont Shelf Res 16:553–573

Oliveira SMO, Pereira LCC, Vila-Concejo A, Gorayeb A, de Sousa RC, Souza-Filho PWM, da Costa RM (2011) Natural and anthropogenic impacts on a macrotidal sandy beach of the Brazilian Amazon (Ajuruteua): guidelines for coastal management. J Coast Res 64:1385–1389

Pereira LCC, Guimarães DO, da Costa RM, Souza-Filho PWM (2007) Use and occupation in Bragança Littoral Brazilian Amazon. J Coast Res 50:1116–1120

Pereira LCC, da Silva NIS, Costa RM, Asp NE, da Costa KG, Vila-Concejo A (2012a) Seasonal changes in oceanographic processes at an equatorial macrotidal beach in northern Brazil. Cont Shelf Res 43:95–106

Pereira LCC, Pinto KST, Costa KG, Vila-Concejo A, da Costa RM (2012b) Oceanographic conditions and human factors on the water quality at an Amazon macrotidal beach. J Coast Res 285:1627–1637

Pereira LCC, de Oliveira SMO, Costa RM, da Costa KG, Vila-Concejo A (2013a) What happens on an equatorial beach on the Amazon coast when La Niña occurs during the rainy season? Estuar Coast Shelf Sci 135:116–127

Pereira LCC, Vila-Concejo A, Short AD (2013b) Influence of subtidal sand banks on tidal modulation of waves and beach morphology in Amazon macrotidal beaches. J Coast Res 65:1821–1826

Pereira LCC, Pinto KST, Vila-Concejo A (2014) Morphodynamic variations of a macrotidal beach (Atalaia) on the Brazilian Amazon Coast. J Coast Res 70:681–686

Richey JE, Meade RH, Salati E, Devol AH, Nordin CF Jr, Santos U (1986) Water discharges and suspended sediment concentrations in the Amazon River 1982–1984. Water Resour Res 22:756–764

Short AD (2006) Australian beach systems – nature and distribution. J Coast Res 22(1):11–27

Sousa RC, Pereira LCC, Silva NIS, de Oliveira SMO, Pinto KST, da Costa RM (2011) Recreational carrying capacity of three Amazon macrotidal beaches during the peak vacation season. J Coastal Res 64:1292–1296

Souza-Filho PWM, Tozzi HAM, El-Robrini M (2003) Geomorphology landuse and environmental hazards in Ajuruteua macrodital sandy beach northern Brazil. J Coast Res 35:580–589

Chapter 5
Maranhão Beach Systems, Including the Human Impact on São Luís Beaches

Luci Cajueiro Carneiro Pereira, Wellington Nascimento Trindade, Iracely Rodrigues da Silva, Ana Vila-Concejo, and Andrew D. Short

Abstract The 1200 km of coastline of Maranhão is the second longest amongst Brazilian states, and is located just south of the equator between the Gurupi River and the Parnaíba Delta. Maranhão represents a transition zone from the tide-dominated Amazon-Pará coast to the more wave-dominated Ceará coast. Morphologically, this coast can be divided in three provinces: (i) Western coast – heavily indented macro-tidal coast composed of funnel-shaped, tide-dominated estuaries, separated by low mangrove-dominated peninsulas capped by dynamic barrier islands and tide-modified to tide-dominated beaches; (ii) Central coast – occupied by the four arms of the funnel-shaped, tide-dominated Maranhão Gulf, with tide-modified and tide-dominated beaches, occupying most of the more exposed outer estuarine shoreline; and (iii) Eastern coast – a relatively straight coast with tides decreasing towards the east from 7 to 3 m, resulting in wave-dominated beaches backed by extensive Holocene and Pleistocene trasgressive dune fields. In the Central Maranhão Gulf is São Luís, the most urbanized zone on the Maranhão coast. Its tide-modified beaches present many beach hazards to the beachgoers. In addition, the beach quality, considering natural (climatic, morphodynamic, and hydrological processes) and social (occupation of the land and types of services and infrastructure) conditions has been adversely affected in different ways. Local beaches are being increasingly occupied by urban development, which is affecting the equilibrium of depositional systems, primarily through the occupation of dunes and cliffs. In addition, the local water quality has been affected by the lack of an adequate public sanitation system and the presence of numerous illegal sewers, which contribute to the elevated bacteriological contamination. The lack of local urban planning has caused serious social and environmental problems and effective management plans for this sector of the Amazon coast are urgently needed.

L.C.C. Pereira (✉) • W. Nascimento Trindade • I.R. da Silva
Instituto de Estudos Costeiros, Universidade Federal do Pará,
Alameda Leandro Ribeiro, sn, Aldeia, Bragança, Pará 68600-000, Brazil
e-mail: cajueiro@ufpa.br; luci.pereira@pq.cnpq.br; etobio17@yahoo.com.br; iracely@ufpa.br

A. Vila-Concejo • A.D. Short
School of Geosciences, University of Sydney, Sydney, NSW 2006, Australia
e-mail: ana.vilaconcejo@sydney.edu.au; andrew.short@sydney.edu.au

A.D. Short, A.H. da F. Klein (eds.), *Brazilian Beach Systems*, Coastal Research Library 17, DOI 10.1007/978-3-319-30394-9_5

5.1 Introduction

The state of Maranhão has a 1230 km long coast that faces generally north into the Atlantic. It is located just south of the equator between the Gurupi River (1°S) and the Parnaíba Delta (2.5°S) and represents a transition zone from the tide-dominated Amazon-Pará coast to the more wave-dominated Ceará coast. It is part of one of the most extensive and well-preserved areas of tropical coastline anywhere in the world. The coast ranges from the eastern heavily indented macro-tidal western coast; to central funnel-shaped mangrove-lined Maranhão Gulf; to the relatively straight meso-tidal Eastern coast.

According to Souza-Filho (2005), the mangroves along the macrotidal northeastern Pará and northwestern Maranhão, named by authors as Amazon Macrotidal Mangrove Coast (AMMC), encompasses a total area of 7591 km^2, representing the largest continuous mangrove belt in the world and contains 56.6 % of Brazil's mangroves. In Maranhão state there are 5414 km^2 of mangroves between Gurupi and São José bays (Fig. 5.1), where they extend up to 40 km inland especially in the irregular macrotidal coast between Gurupi and Turiaçu. The mangroves are less extensive between Turiaçu and Cumã, particularly around Alcântaras where the coast has exposed cliffs of the Itapecuru Formation (Souza-Filho 2005). While mangroves dominate the many estuaries tide-modified and tide-dominated beaches dominate the open coast and outer more exposed estuarine shores.

The anthropogenic landscape varies from the metropolitan regions of São Luís with a total population of one million people (www.ibge.gov.br) to isolated areas that are either uninhabited or sparsely occupied by traditional populations. The urban centers encompass a wide economic base, which includes, shipping, fisheries, tourism, commercial trading, etc (Andrade et al. 2010; Silva et al. 2013). By contrast, in small coastal communities the local economy is based primarily on fishing (Dias et al. 2009). Tourism is supported by the presence of historic buildings, natural resources like beaches and manifestations of local culture, such as local music and dance festivals, handicrafts and cuisine (Silva et al. 2011). In the principal ports, the risks of the accidental spillage of petroleum or its derivatives, or other chemical products, represent a constant threat to the local environment including the beaches (Andrade et al. 2010).

In São Luís and many coastal towns, unregulated occupation of the land, in particular of mangroves, dunes and cliffs, and the lack of basic public services, such as sanitation, have an adverse impact on both the quality of life and the beaches on São Luís Island (Silva et al. 2013). In addition public organisations responsible for the management of the natural resources of Maranhão coasts are far from satisfactory patricularly close to urbanized areas. The lack of local urban planning has caused serious social and environmental problems and effective management plans for this sector of the Amazon coast are urgently needed.

This chapter will commence with a review of the geology, climate and coastal processes along the Maranhão coast, followed by a brief overview of the entire Maranhão coast, its coastal provinces and their beach systems. It will then present

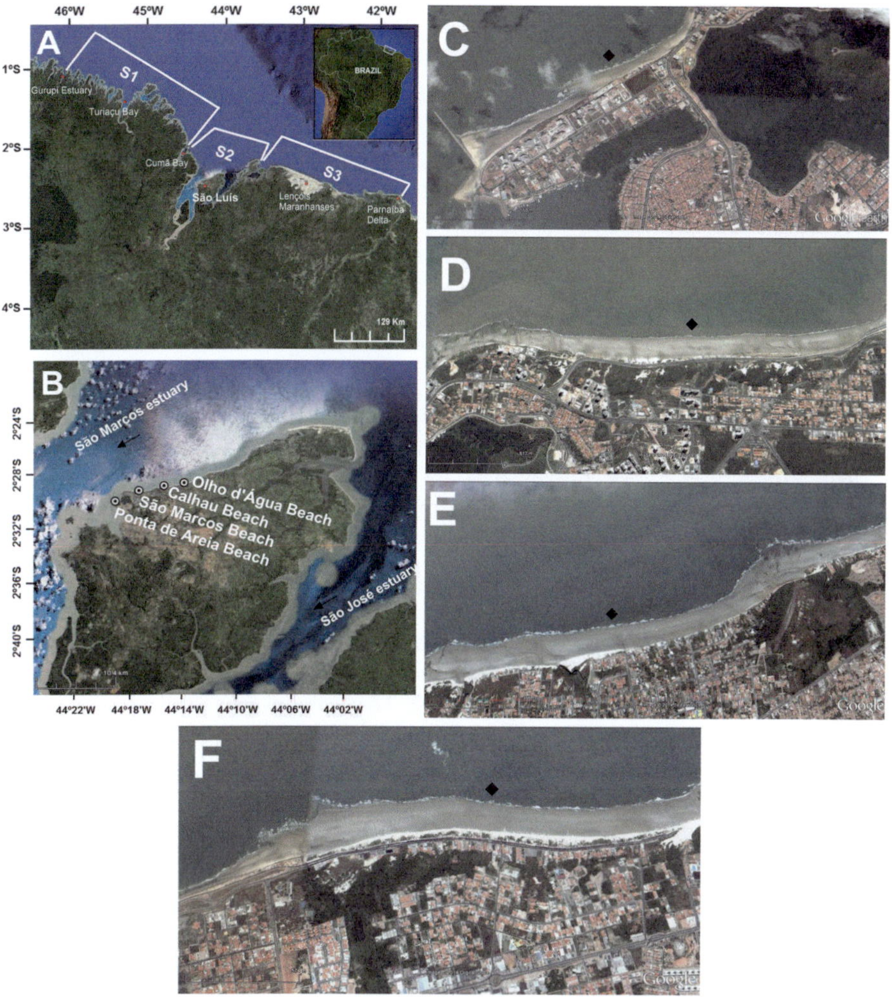

Fig. 5.1 The coast of Maranhão (**a**) showing the three sub-provinces (S1, S2, S3), a detail of São Luís Island with the location of the urban beaches (**b**) and a detail (**c–f**) of these heavily developed beaches (Source: Google earth)

a more detailed examination of the beaches of São Luís and their use and management.

5.2 Coastal Environment

The Maranhão has a long and highly variable coastline, which can be divided into three distinctive provinces (Fig. 5.1).

5.2.1 Coastal Provinces

(i) The *western province* (S1) (known as Reentrâncias Maranhenses) extends for 520 km from the eastern border at the Gurupi River mouth to the Maranhão Gulf. This heavily indented macro-tidal coast is composed of about 20 tide-dominated, funnel-shaped mangrove-lined estuaries, separated by low mangrove-dominated peninsula capped by dynamic barrier islands, and tide-modified to tide-dominated beaches. The barrier shores are highly dynamic being affected by macrotides (4–7 m), with strong tidal currents and substantial tidal modulation of the waves (Pereira et al. 2011).

(ii) The *central province* (S2) consists of the four arms of the tide-dominated funnel-shaped Maranhão Gulf with 490 km of outer shoreline that encompass Cumã, São Marcos, São José and Arraial bays. These bays are feed by the Mearim, Pindaré, Munim, and Itapecuru rivers. Tide-modified and tide-dominated beaches occupy most of the more exposed outer estuarine shoreline inlcuding São Luís Island, together with a section of transgressive dunes along the most exposed northeastern corner of the island.

(iii) The *eastern province* (S3) extends for 240 km, as a relatively straight coast, to the border at the Parnaíba Delta. Tides decrease towards the east from 7 m to 3 m where the coast becomes occupied by longer primarily wave-dominated beaches, backed by extensive Holocene and Pleistocene dune transgression, which extend tens of kilometers inland. These dunes are located in the large Lençois Maranhenses National Park with an area of 155,000 ha (Abakerli 2001).

5.2.2 Geology

Maranhão's coast is dominated by low relief plateaus composed of Miocene to Pliocene Barreiras Formation together with extensive Holocene and Pleistocene marine and aeolian deposits the latter both onlaping and overlaping the Barreiras Formation. In the east the formation has been heavily dissected by fluvial erosion that extends out onto the continental shelf, with the valleys flooded during the Holocene sea level rise to form the present highly irregular shoreline (Dominguez 2009).

The central Maranhão Gulf, is approximately 90 km in width, and is the largest re-entrant on the northern Brazilian coast, second only to the Amazon River mouth. The gulf extends 150 km inland, as well as seaward across the continental shelf. The gulf has apparently been eroded into the sedimentary rocks of the Parnaíba basin during sea-level low stands and flooded by the sea level rise (Dominguez 2009).

The Barreiras Formation extends to the east where it has been blanketed for tens of kilometers inland by both Pleistocene and Holocene dune transgression. The continental shelf narrows to the east from 200 km off the western province, to 150 km off the gulf and 60 km off the eastern province (Dominguez 2009).

5.2.3 Climate

The Maranhão's coast is located between two climate zones. In the western sector the climate is hot and wet tropical (As, Köppen classification) with a longer and more intensive rainy season marked by high rainfall levels (normally above 2000 mm $year^{-1}$), typical of the Amazon coast. The eastern sector has a hot and semi-wet tropical climate of the type (Aw), indicating a hot wet summer and a drier winter (www.ambientebrasil.com.br). The main meteorological system regulating the rainfall patterns is the Intertropical Convergence Zone (ITCZ) (Marengo 1995). The rainy season along the Brazilian Amazon coast occurs when the ITCZ shifts southward (between January and June) to the coast of Maranhão and Pará, resulting in lower air temperatures and heavy rain. During the dry season (July–December) the ITCZ shifts north of the equator, resulting in higher temperatures and lower rainfall rates, mainly due to the influence of the instability lines associated with the sea breeze (Figueroa and Nobre 1990; Marengo 1995).

According to historical data (1970–2010) from INMET (www.inmet.gov.br), the rainy season on São Luís Island typically occurs between January and June, with February to May representing 70 % of the total annual precipitation (2200 mm). The easterly trade winds are typically less than 5 $m.s^{-1}$ and the average temperature is 26.8 °C.

The dry season normally extends from July to December, with the lowest rainfall rates (20 mm per month) occuring between September and November. Northeast trade winds with speeds higher than 6 $m.s^{-1}$ and air temperature above 27 °C are characteristic of the dry season. Figure 5.2 shows the mean air temperature, the total monthly rainfall and maximal wind speed and direction values observed between 1970 and 2010 in São Luís.

5.2.4 Coastal Processes

Semidiurnal macrotides (> 6.0 m) are characteristic of the western and gulf provinces, whereas semidiurnal mesotides (> 3 m) occur along the eastern province (FEMAR 1997). Highest and lowest tide ranges on the open coast are 7 m in Maranhão Gulf and 3 m at the Parnaíba Delta, respectively. Figure 5.3 shows 1 year of tidal records obtained at São Luís harbour (central province) and at Tutóia harbour (eastern province) from the Brazilian Hydrographic Agency. In São Luís the spring tides are just over 6 m, and the neaps less than 2.5 m, while at Tutóia the springs are 3 m and neaps 2 m.

On the open coast the relative tide range (RTR) varies from < 3 on the most exposed sections of the eastern coast ($TR = 3$ m, $H_b = 1.4$ m, $RTR = \sim 2$) with wave-dominated beaches, to > 3 on exposed sections of the gulf and western coast ($TR = 6$ m, $H_b < 1.4$ m, $RTR > 4$) with tide-modified beaches, and finally to > 10 on the sheltered section of the Gulf and western coast ($TR = 6$ m, $H_b < 0.5$ m, $RTR > 10$) with tide-dominated beaches.

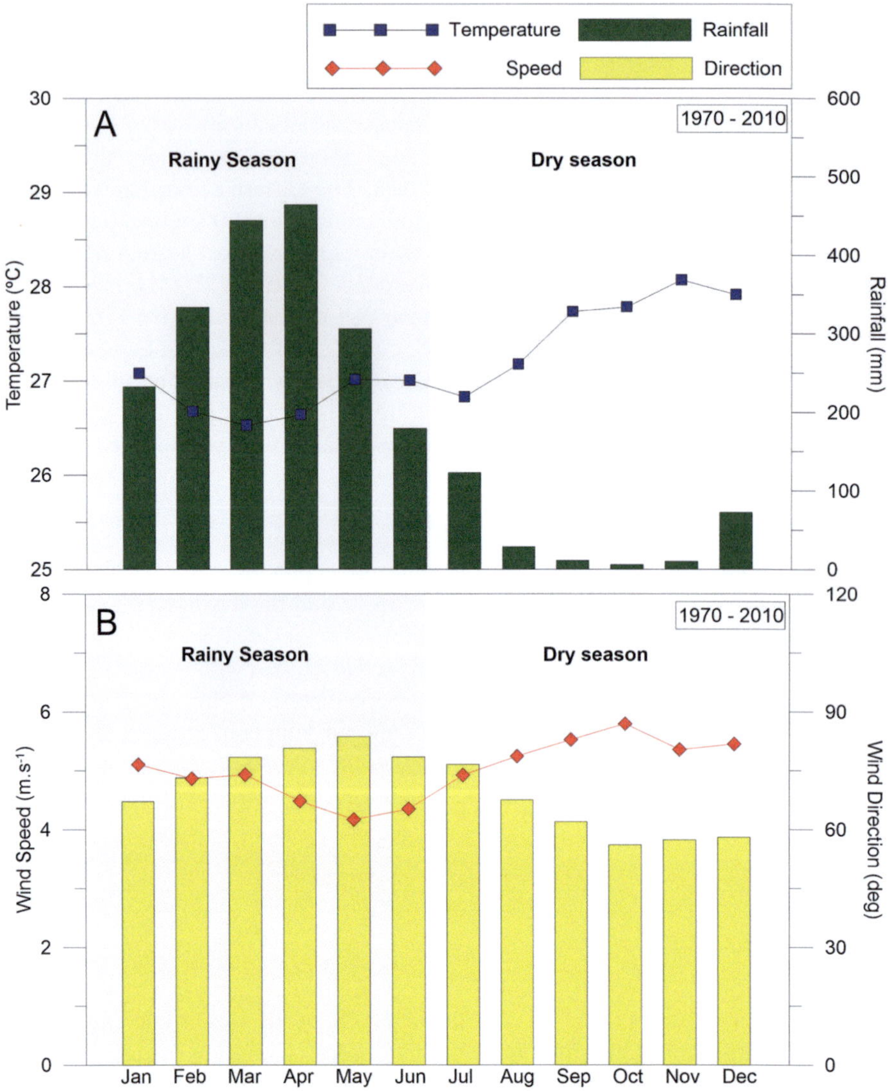

Fig. 5.2 Average of total monthly (**a**) air temperature and rainfall, and (**b**) wind direction and maximal wind speed values recorded between 1970 and 2010 from INMET (São Luís Station)

Surf Zone Processes Figures 5.4, 5.5, and 5.6 show water level, current and wave data during a tidal cycle on the tide-modified beaches in São Marcos Bay (Fig. 5.1). Nearshore-surf zone data was undertaken using a bottom-mounted mooring (1.7 m depth at low tide) housing a current meter, wave and tide data-logger (TWR 2050). Tidal water oscillation were acquired every 2 s and mean values obtained every 10 min. Wave statistics were based on 512 samples at a burst rate of 4 hz of 10 min duration. The red lines in Fig. 5.3a represent the dates when the nearshore surveys were undertaken in São Luís beaches, with each field campaign 13 h in duration.

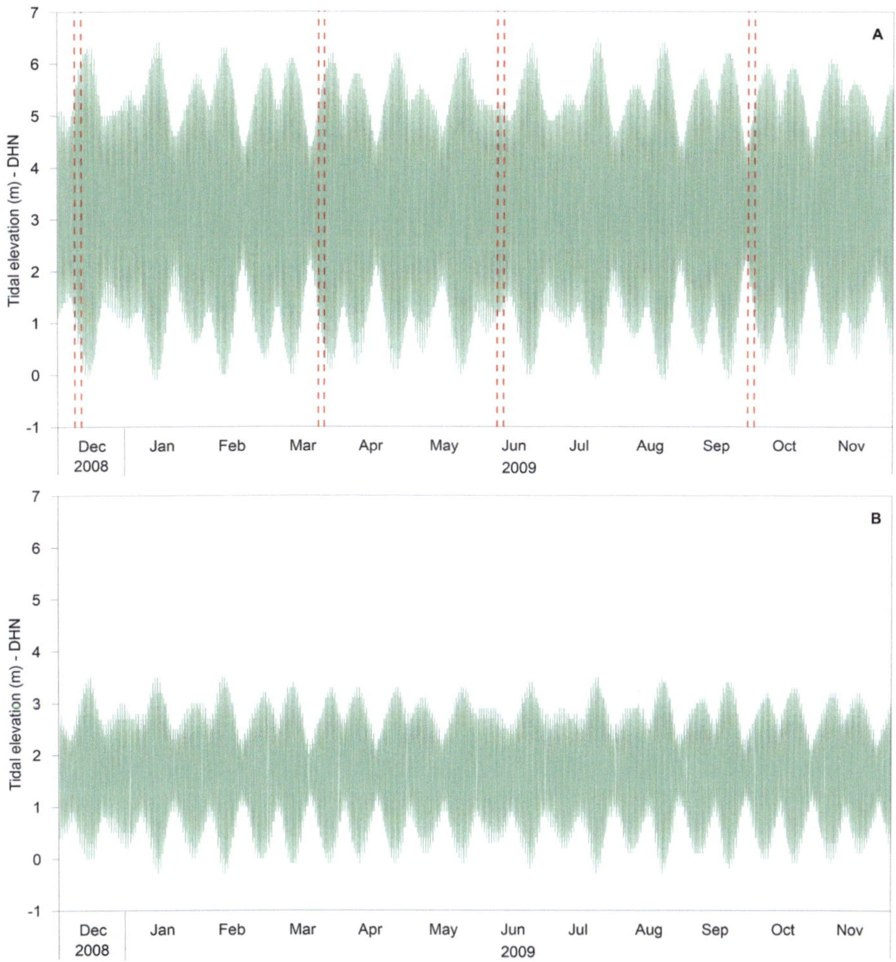

Fig. 5.3 Tide elevation from DHN (Hydrographic and Navigational Department of the Brazilian Navy). São Luís (**a**) and Tutoia (**b**) stations between December 2008 and November 2009. The *red lines* in **a** represent the dates when the nearshore surveys (Figs. 5.4–5.6) were undertaken in São Marcos, Calhau and Olho d'Água

The figures show that tidal currents are asymmetric, with flood tidal currents stronger than the ebb tidal currents. Maximum current speeds were 0.45 $m.s^{-1}$ at São Marcos beach, 0.5 $m.s^{-1}$ at Calhau beach and 0.4 $m.s^{-1}$ at Olho d'Água beach. In São Marcos and Calhau, the highest velocities were recorded in March coinciding with the equinoctial period and the highest tide range, whereas the lowest values were found in June (transitional season, rainy to dry season). Morais (1977) and Feitosa (1989) report tidal currents reaching 2.5 $m.s^{-1}$ in tide channels at São Marcos Bay, however in the outer part of this litoral, northeast winds and waves drive a northwestward current that can dominate over the tidal flows reducing the southeastward current during ebbing tides. The current speeds in the outer part of São Marcos Bay oscillate from 0.4 to 1.5 m s^{-1} (Feitosa 1989).

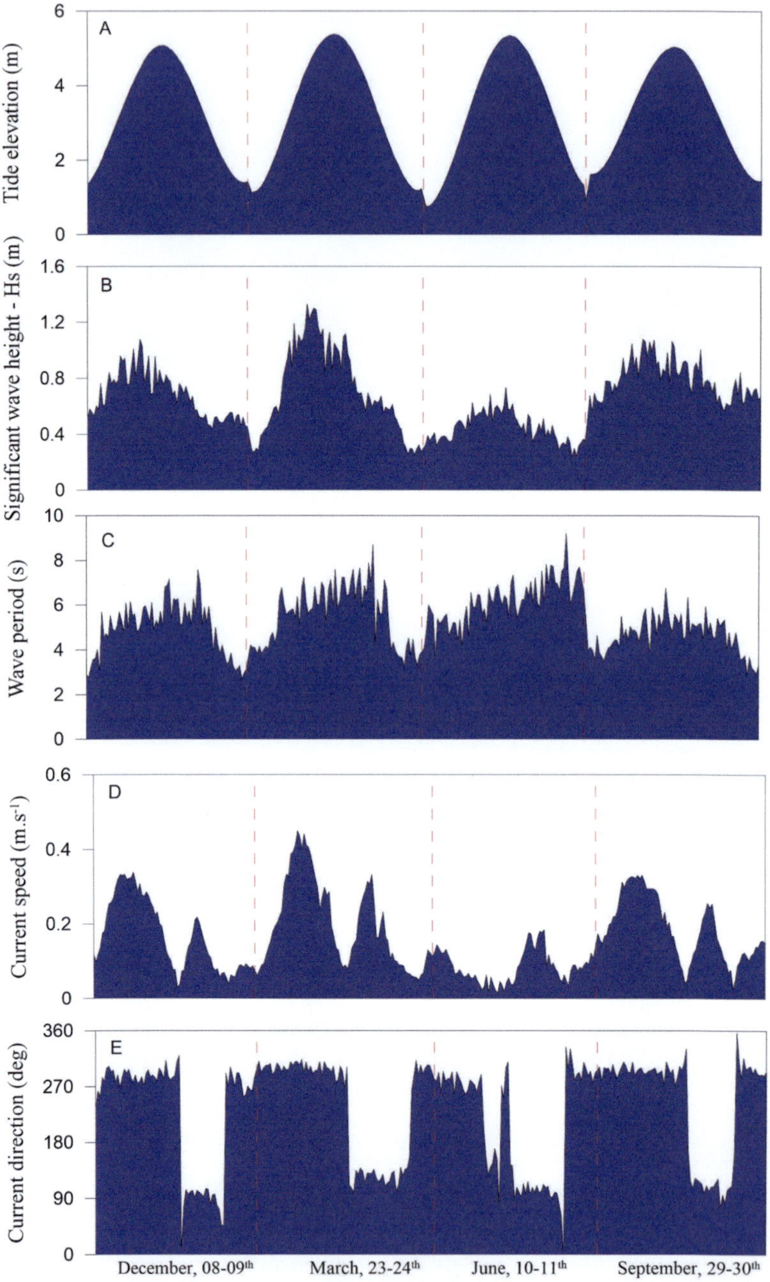

Fig. 5.4 Nearshore condition (tidal elevation-(**a**), wave height-(**b**), wave period-(**c**), current speed-(**d**) and current direction-(**e**) on the sheltered São Marcos beach

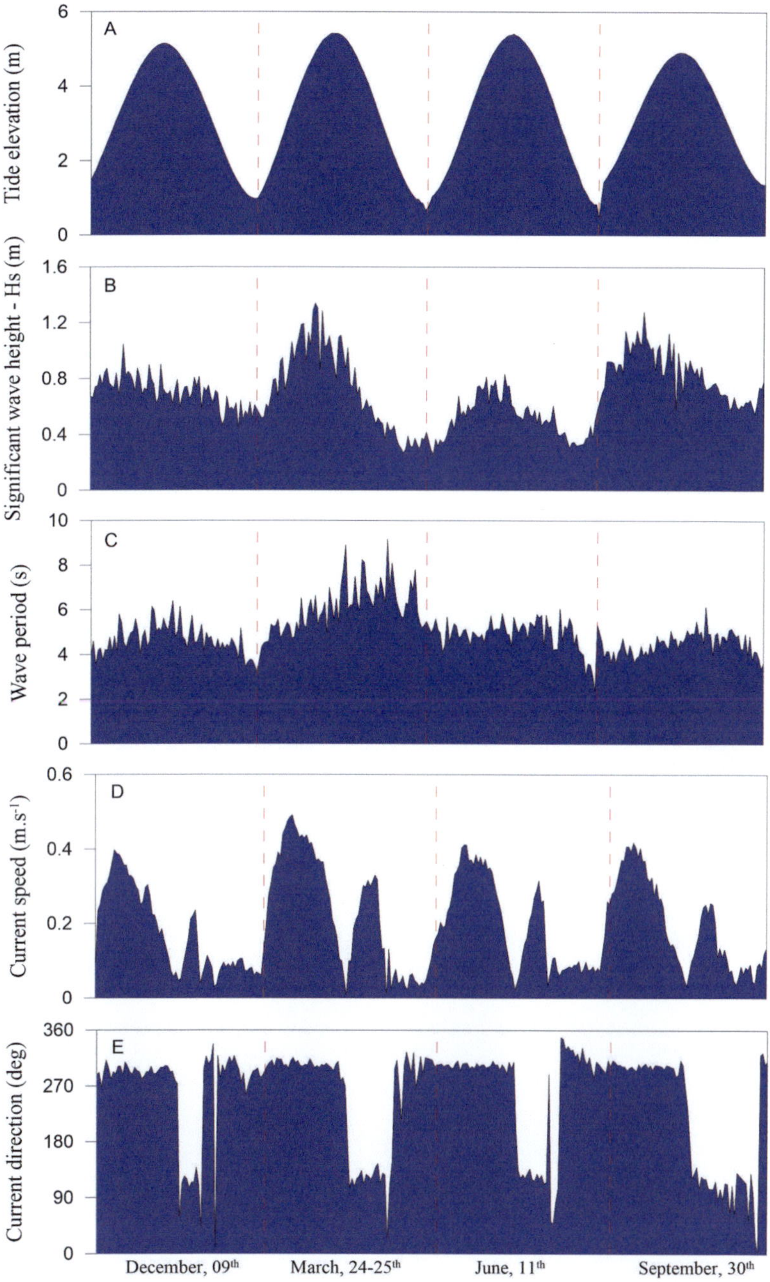

Fig. 5.5 Nearshore condition (tidal elevation-(**a**), wave height-(**b**), wave period-(**c**), current speed-(**d**) and current direction-(**e**) on Calhau beach

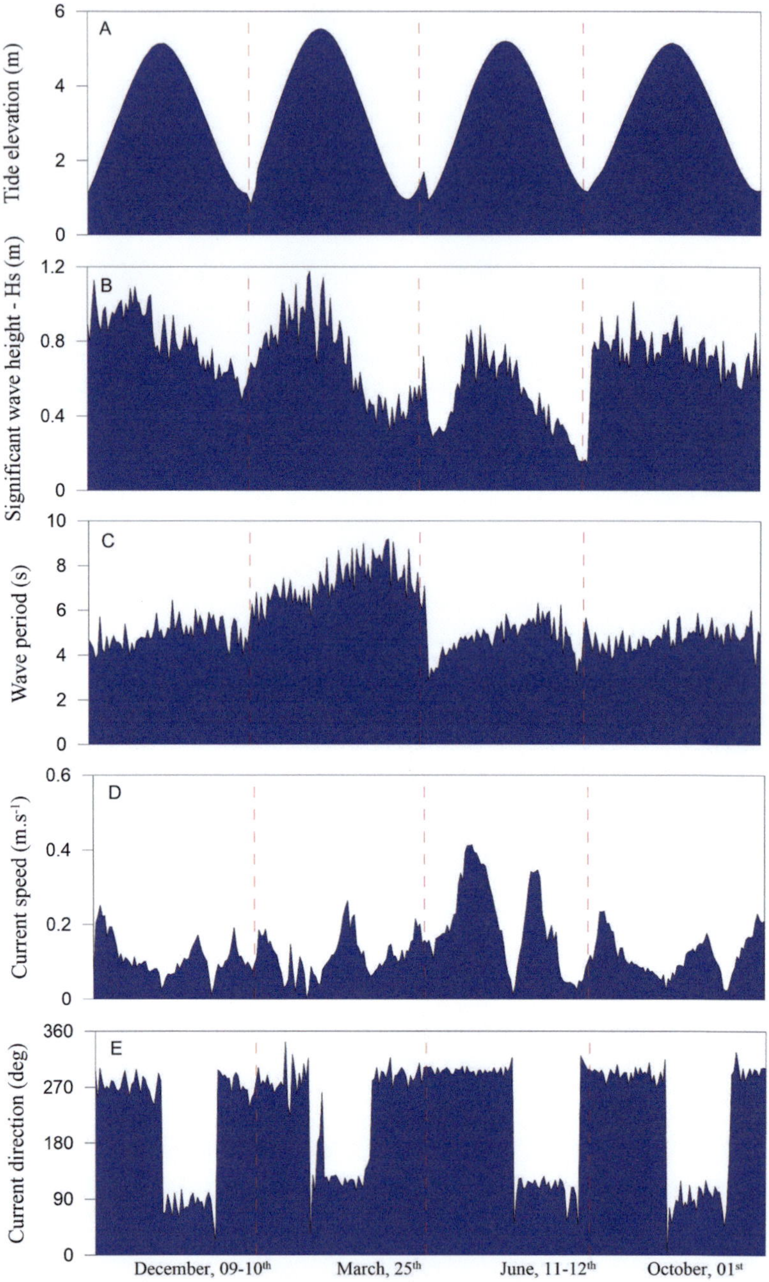

Fig. 5.6 Nearshore condition (tidal elevation-(**a**), wave height-(**b**), wave period-(**c**), current speed-(**d**) and current direction-(**e**) on Olhod'Água beach

Nearshore significant wave heights in São Luís ranged from 0.6 to 1.4 m. They are, however, tide-modulated particularly in locations where sand shoals prevent wave propagation during low tide. As a consequence H_s values (Figs. 5.4, 5.5, and 5.6) varied from a maximum at high tide (> 1.0 m) to minimun (< 0.5 m) at low tide. Again, the highest H_s were registered in March, while the lowest were recorded in June.

5.3 Maranhão Beach Systems

The Maranhão coast is dominated by fine sand exposed to meso- through macro-tides and moderate easterly trade wind waves. The three provinces presented in Sect. 5.2.1 are used to provide an overview of the beach systems. As there have been no previous studies of the Maranhão beach systems, this overview is based largely on an examination of each of the beaches using Google Earth, together with field-work on some of the São Luís beaches.

5.3.1 West Coast – Reentrâncias Maranhenses (Province S1)

The western coast of Maranhão is essentially a continuation of the eastern coast of Pará with heavily dissected Barreiras Formation dominating the coast forming deeply incised valley and ridges, now occupied by tide-dominated funnel-shaped, estuaries in the flooded valleys and regressive barrier islands and mangroves on the exposed ridges. This indented shore is known as Reentrâncias Maranhenses with approximately 20 estuaries between the Pará border and the Maranhão Gulf, ranging from 5 km to 15 km in width at their mouths and extending 15 km to 25 km inland. The outer shoreline of the estuaries and peninsulas totals 520 km and contains 42 sandy beaches. The entire western coast has been regressing during the Holocene, with abundant evidence of barrier regression in the form of inner beach and foredune ridges many terminating as spits. Barriers average 5 km wide, but reach up to 15 km on Mutuoca Island. The convex, recurved barriers and beaches at their tips range from a few kilometer to a maximum of 10 km in length. Easterly waves and strong tidal currents build west-trending spits that recurve into the estuaries.

Souza-Filho (2005) divides this coastal province into two sub-provinces. The first extends from Gurupi Bay to Turiaçu Bay, where mangroves attain their maximum extension, extending approximately 40 km inland and covering an area of 1846 km^2. The second sub-province extends between Turiaçu and Cumã bays, where the mangroves narrow to around 26 km and are continuous apart from an 8 km section of shore. This sub-sector has 1945 km^2 of mangroves.

Like the east Pará beaches described in the previous Chap. 4, the west-coast beaches are highly dynamic systems affected by the substantial westerly longshore transport, strong boundary tidal currents and associated tidal sand shoals that extend several kilometers off- and alongshore, and substantial tidal modulation of the waves. The latter results in wind-wave/swell only reaching the beach at high tide, while the tidal shoals effectively filter out most or all of the waves at low tide.

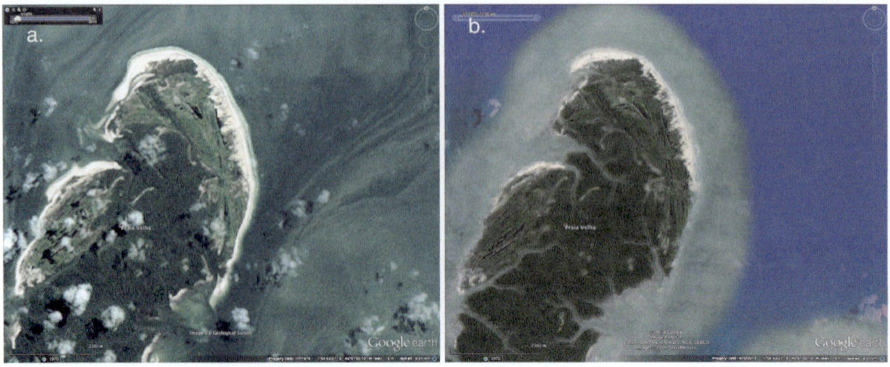

Fig. 5.7 Google images of Irmãos Island taken in 1970 (*left*) and 2013 (*right*) indicating substantial easterly erosion and westerly accretion of the barriers and beaches (Source: Google Earth)

Figure 5.7 shows a typical regressive barrier and beaches at the head of Irmãos Island. Between 1970 and 2013 the eastern barrier's distal spit was eroded approximately 2 km, while the northeastern spit extended by more than 1 km and the 5 km long western barrier/beach grew substantially both westward and into the estuary. All these changes indicate a substantial sediment transport resulting in major changes in beach and barrier form over periods of years to decades.

The west-coast beaches range from tide-modified, in more exposed locations, to tide-dominated, in more sheltered locations and into the estuaries. The tide-modified beaches consist of a high tide beach fronted by a low gradient intertidal zone up to a few hundred meters wide. These often grade laterally and seaward (in some places) into tidal-sand shoals and channels. The tide-modified beaches tend to be crenulate in plan form owing to pulsative sand transport into the estuaries. Low waves only reach the beaches at high tide, while strong tidal flows contribute to longshore sand transport.

Of the 40 beaches along the west coast 15 are located on the tip of the peninsulas and tend to face north to northeast and are more exposed and tide-modified, while the 27 beaches are sheltered and/or located inside the wide estuarine mouths are tide-dominated and may face east or west. These recurved estuarine beaches attest to the dominance the tidal currents have on sediment transport and the beach/barrier systems.

All but one (Pericaua, in the east) of the west-coast beaches is difficult to access due to tidal creeks and mangroves dominating the peninsulas and backing all the beaches. As a consequence the beaches remains undeveloped in a natural state.

5.3.2 *Maranhão Gulf (Province S2)*

Maranhão Gulf is a broad (90 km wide), coastal reentrance that extends over 160 km inland. It includes Cumã Bay in the west, the central São Marcos and São José bays and the eastern Tubarão Bay. This coastal segment has 490 km of open coast and contains 144 generally low-energy, tide-modified and tide-dominated beaches. Most of these beaches are located along the outer 50 km of the estuarine shores.

Sixty of these beaches are located on São Luís Island, which will be discussed in more detail in section 5.4. Mangroves cover an area of 1623 km^2 (Souza-Filho 2005) and dominate much of the bay's western and eastern shore, lying in lee of the outer beach and barrier islands.

The beaches are affected by ocean waves close to the gulf's mouth and by fetch-limited low seas inside the gulf where macro-tides and strong tidal currents prevail. The beaches can be classified into two systems: barrier beach systems backed by mangrove, and mainland beaches systems backed by the bluffs sculpted on the Barreiras Formation.

5.3.2.1 Barrier Island Beaches

The mangrove-backed barrier islands and beaches tend to occur along the shores of the gulf, particularly in the western Cumã Bay and on the cluster of mangrove-dominated islands located at the eastern entrance to the gulf, including Cotindiha, Carrapatal and Santana islands. Most of these sectors have experienced shoreline regression and the beaches are backed by a mix of mangroves, inner beach ridges and sand spits. Most of these beaches are undeveloped and only accessible by boat. Figure 5.8 illustrates a typical east facing estuarine beach on Cajual Island. Note the

Fig. 5.8 The eastern tip of Cajual Island, showing from top the mangrove shoreline, the dynamic tidal sand shoals and long south migrating spit, with a cluster of regressive recurved spits and mangroves backing the beach (Source: Google Earth)

Fig. 5.9 View of barrier islands in the northeast of the Gulf, showing the highly dynamic nature of their evolution and their beaches, which are exposed to southern longshore sand transport (Source: Google Earth)

complex tidal sand shoals in the north, the 3 km long spit in the center and smaller embayed beaches in the south.

The barrier islands at the northeast entrance are even more dynamic. Figure 5.9 shows two small barrier island that have undergone up to 2 km of regression as well as southerly migration on the order of 2 km. Their beaches are tide-modified and consist of a high tide beach and 200 m wide low tide ridge and runnels with the runnels skewed to the south.

5.3.2.2 Mainland Beaches

The mainland beaches are backed by the Barreiras Formation (described by Rossetti 2000), that dominates the protruding coast between Cumã and São Marcos bays and the center of the gulf particularly the exposed outer shores of São João and São Luís islands. The beaches commence in Cumã Bay where there is the first occurrence of the Itapecuru Formation at the coast along a 10 km long section of its western shore, which also forms the break in the near continuous mangroves (Fig. 5.10). Souza-Filho (2005) measured the mangrove area between Turiaçu and Cumã bays at 1945 km^2. The Barreiras-Itapecuru Formation dominates the center

Fig. 5.10 Alcântara beach in São João Island is attached to the Barreiras Formation in the north, then extends as a south migrating spit together with south skewed bars and channels (Source: Google Earth)

of the gulf being exposed around the outer shores of São João and São Luís islands. It forms 10 m high bluffs fronted by tide-modified and tide-dominated beaches and tide-dominated beaches and sand flats. The more exposed beaches are tide-modified with a high tide beach, in places backed by bluffs, and fronted by a wide low tide terrace. The beaches are generally accessible by vehicle but only those in São Luís Island (discussed in more detail in Sect. 5.4) have been impacted by intense development (Fig. 5.11).

5.3.3 East Coast (Province S3)

At Baleia Island there is a marked change in the orientation and nature of the coast. The 12 km long island consist of a 4 km wide series of regressive spit-like strandlines and mangroves that connect to the Pleistocene transgressive dunes that dominate the remainder of the coast (Fig. 5.12).

Fig. 5.11 The popular and heavily developed Olhod'Água beach, in São Luís Island, onlaps the Barreiras Formation and is fronted by a 200 m wide low tide terrace (Source: Google Earth)

Fig. 5.12 Balaia Island and Ponta dos Mangues form the boundary between the tide-dominated mangrove-lined shore to the west and sandy wave-dominated shore to the east (Source: Google Earth)

Fig. 5.13 The wave-dominated dissipative double bar Lençóis Maranhenses beach is 65 km long and backed by Brazil's largest active transgressive dune system (Source: Google Earth)

One kilometer east of the island at Ponta dos Mangues the near continuous sandy beaches and dunes commence and extends east for 230 km to the border at the Parnaíba Delta. The extensive dune area is known as "Lençóis Maranhenses" (Machado et al. 2000; Palma 1979), the largest active dune field in Brazil (Hesp et al. 2009) that extends up to 30 km inland. The dunes are comnposed of fine and very fine sand (Santos 1996) are largely protected as part of the Lençóis Maranhenses National Park, the largest coastal national park in Brazil. There are 26 fine-sand (Santos 1996), wave-dominated dissipative beaches containing two (Fig. 5.13) and three (Fig. 5.14) longshore bars along this relatively straight coast, the longest section running continuouly for 65 km. Only along the eastern 60 km of coast, where a series of barrier islands are formed close to the Parnaíba River mouth, are some sheltered lower energy tide-dominated and tide-modified beaches present. Tutóia beach (Fig. 5.15), grades with increasing wave energy, from a tide-dominated, high-tide beach to a 100 m wide tide-modified low tide terrace. The tide-dominated sector has 400 m wide sand flats containing up to 10 sand ridges.

Fig. 5.14 Preguiça beach is 25 km long, composed of fine sand and consists of a continuous wave-dominated triple bar system, with backing transgressive dunes (Source: Google Earth)

5.4 São Luís Beach Systems

This section briefly reviews the coast and coastal processes on the northern portion of São Luís Island, in the São Marcos Bay area (Fig. 5.1), which forms a 12 km-long waterfront between the inner Ponto d'Areia and São Marcos, central Calhau and outer Olho d'Água. This coastal area is strongly affected by a number of different anthropogenic activities and coastal management initiatives are virtually non-existent.

The city of São Luís is located on São Luís Island, 20 km inland from the mouth of the gulf (Fig. 5.1). The shoreline is 35 km long and located within a mangrove-dominated fluvial marine ecosystem exposed to a high-energy hydrodynamic environment (Pereira et al. 2011; Trindade et al. 2011).

São Luís was declared a World Cultural Heritage site by UNESCO in 1977 in recognition of its rich historical, architectural and cultural heritage. The city is one of the most important artistic centers in northern Brazil, although its main economic activities are industry, commerce, and services. Its coast has been traditionally used for fishing and leisure activities, however during the past few decades, considerable

Fig. 5.15 Tutóia beach is sheltered in lee of barrier islands and spits of the Parnaíba Delta. It grades from a tide-dominated beach with ridged sand flats either side of the creek mouth, to a tide-modified beach plus low tide terrace in the east (Source: Google Earth)

social and environmental problems have developed as a result of the intense exploitation of its natural resources and rapid urban growth, together with a complete lack of town planning (Silva et al. 2013). This development has altered the natural environment to fit different social and economic interests.

The urban development of the coastal zone of northern São Luís is characterized by the construction of buildings on the dune systems that run parallel to the shoreline and on the backing bluffs and cliffs. These buildings were erected without council permits or controls. Up until the 1960s, the coastal strip between Ponta de Areia and Araçagi beaches had very few inhabitants, but beginning in the early 1970s, the city underwent rapid and unregulated expansion (Espírito Santo 2006). While the city's economy is based on industry, commerce, and services, tourism has become increasingly important in recent years, focusing primarily on the city's cultural heritage and surrounding natural environments. Nowadays, few tourists and locals visit the city beaches which are overdeveloped and where bathing is prohibited owing to pollution.

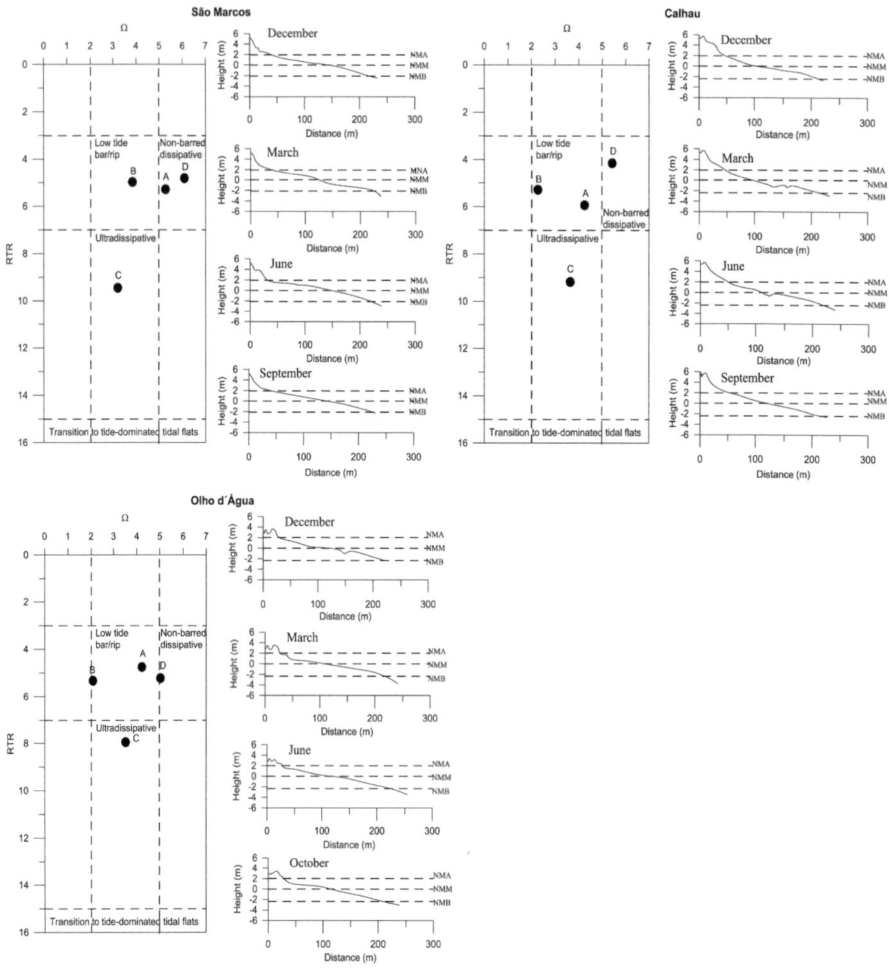

Fig. 5.16 Morphodynamic classification and seasonal beach profiles for São Marcos, Calhau and Olhod'Água beaches. *A* December, *B* March, *C* June, *D* September/October

5.4.1 Morphological Aspects

São Luís area is most urbanized zone on the Maranhão coast, and unregulated occupation on the dunes and intertidal zones has contributed to erosion problems (Silva et al. 2009). The beaches are composed of uniform fine sand (D_{50}=0.17–0.16 mm). They are tide-modified and consist of a moderately sloping (2–5°) high tide beach face, with a 250–300 m, wide low gradient-intertidal beach and surf zone (~2°). Upper foreshore erosion occurs during the high-energy events with bar-trough morphology forming during higher wave energy conditions (Fig. 5.16).

Based on Short (2006), these beaches are tide-modified with a 5–7 m tide range and exposed to low to moderate wave heights (H_b=0.4–0.9 m) with short wave period (T=4–8 s.) The RTR ranges from 3 to 15. Figure 5.16 shows ultradissipative characteristic occuring during a lower energy period ('C' June) with RTR varing between 8 and 10 and the dimensionless fall velocity (Ω) from 3 to 4. During equinoctial period ('B' marked by a higher wave period and weaker wind speed), intermediate low-tide bar/rip beach state is observed (RTR=4–6; Ω=5–6). However, during equinoctial tides with shorter wave periods (stronger wind speed), non-barred dissipative characteristic is recorded (Fig. 5.16). The highest modal H_b values are also recorded along the more exposed outer sector.

São Luís's urban beach has experienced 32,000 m^3 of erosion in 1 year (Fig. 5.17a, b), however there is considerable spatial and temporal variability in beach behavior (Pereira et al. 2011). Severe erosion usually occurs during the rainy season, with the exception of the inner section of the São Marcos Bay, which experienced accumulation. During the dry season, with the strong northeast winds, aeolian sediment transport causes accumulation in the middle section while erosion occurs in the inner (Ponta de Areia) and outer (Olho d'Água) sections (Fig. 5.1). As a consequence, three seasonal patterns are evident: (i) March (the most energetic period) results in erosion events along the middle (Calhau) and outer (Olho d'Água) sections, with sediment accumulation along the inner (São Marcos) section; (ii) June has accumulation in the inner and middle sections, with erosion in the outer section; and, (iii) September experiences accumulation in the middle section, but erosion in the inner and outer sections.

5.4.2 *Territorial Occupation and Facilities*

São Luís beaches are readily accessible with some located near downtown São Luís, one of the most urbanized coastal zones in northern Brazil. While the occupation and construction on dunes, cliff or intertidal zones is not permitted by the State and Federal laws (number 4669 Oct 2006, www.saoluis.ma.gov.br and number 7.661, 16 May 1988), houses, hotels, restaurants, bars, residential and commercial buildings, roads and walkways have been built on the dunes, intertidal and/or cliff zones (Fig. 5.17). Silva et al. (2013) estimated that the unoccupied area on Calhau and Olho d'Água beaches decreased 11 % between 2008 and 2010 (Table 5.1), while that in São Marcos and Ponta de Areia reduced by 9 %, and 6 %, respectively. In addition to the illegal occupation, the high buildings hinder the local air circulation, and shade the beaches during the afternoon (Fig. 5.17).

The beachfront development offers a variety of privately-owned facilities, which provide beachgoers with amenities and leisure infrastructure (Fig. 5.17). The number of private establishments – bars and restaurants, hotels and guesthouses, and shower stalls –has increased rapidly in recent years. On the other hand, public services available on the São Luís beaches are insufficient to satisfy the demand, resulting in a lack of an adequate public sanitation system (Fig. 5.17).

Fig. 5.17 Images of some São Luís beaches showing erosion problems (**a** and **b**); occupation on dunes (**c** and **d**); presence of building at Ponta de Areia (**e**); cars on Olhod'Água (**f**); and lack of basic sanitation (**g** and **h**)

Table 5.1 Percentage (%) of the territory occupied by urban development on the beaches of São Luís in 2008 and 2010

Land occupation	Ponta de Areia	São Marcos	Calhau	Olho d'Água
		2008		
Dunes	69	64	18	59
Dune and intertidal zone	0	0	41	10
Bluffs and-or Cliffs	0	15	0	0
Unoccupied	31	21	41	31
		2010		
Dunes	78	70	25	70
Dune and intertidal zone	0	0	45	10
Bluffs and-or Cliffs	0	15	0	0
Unoccupied	22	15	30	20

Silva et al. (2013)

The better and more abundant facilities are available at the more affluent São Marcos and Calhau beaches, and this unbalanced distribution reflects the economic differences in the area and in the profile of the beach users. Using lamp posts as an index of security, São Marcos and Calhau beaches can be considered a safer zone. Free public services such as phones and parks are also concentrated in this area, reflecting the "degree of interest" of the administration in one specific part of the São Luís littoral and, on the other hand, the high affluence of people with better economic profile. The best private services such as showers, bars/restaurants, water sport renting and ground sports are also located in São Marcos and Calhau, clearly reflecting the spatial distribution of beach users number and their economic profile, because the main objective of these services is to obtain an economic profit.

5.4.3 Human Influence

The population of São Luís city has grown from 696,371 inhabitants in 1991 to 1,014,837 in 2010, resulting in increasing environmental problems as a consequence of the lack of adequate infrastructure and services as well as unregulated and often illegal developments. This scenario persists, despite the city was being a World Cultural Heritage site.

The city's beaches are used for fishing, leisure, and tourism, but the presence of sewage from local bars and other buildings flowing directly onto the beaches have affected the water quality closest to the city center, impeding their use (Silva et al. 2011). The excessive disposal of sewage onto the beaches of São Luís (Fig. 5.17) has resulted in a complete ban on bathing since 2009. This prohibition has had a considerable impact on recreational activities in the area, and consequently, on its economy. The effects of this discharge on water quality was evaluated by Silva et al. (2009, 2011, 2013) at four of the city's beaches – Ponta de Areia and São Marcos, located in the inner bay, and Calhau (middle) and Olho d'Água, in the outer bay. More than 100 sewage outlets were detected, and as a consequence the thermotolerant coliform concentrations have exceeded the limit recommended by CONAMA (> 1100 MPN.100 ml^{-1}).

5.4.4 Beach Use

On all four beaches, zone 1 (backshore zone) offers food and drink services; while zone 2 (intertidal zone) is used primarily for leisure activities, such as football, volleyball, and sunbathing, as well as for the consumption of food and drinks at bar tables, except during the high tide. Zone 3 (surf zone) is rarely used for bathing and aquatic sports (surf, raid-surf, kit surf, windsurf, and others), because of the official prohibition of bathing all along the coast.

A study undertaken during the school holidays, at the beginning of the summer season, showed that the recorded density of beachgoers was much lower than expected (Silva et al. 2011). Poor water quality has adversely affected the recreational carrying capacity in São Luís beaches.

5.4.5 Hazard and Risks

5.4.5.1 Beach Hazards

Along the São Luís beaches, the highest energy conditions occur during spring tides (>6.0 m), which have strong tidal currents (>1.0 $m.s^{-1}$), and periods of moderate waves (H_b>1.0 m, T: 3–8 s). Ponta de Areia beach, which is located in the innermost sector of the bay, is the most sheltered, while Olho d'Água is the most exposed. The strongest tidal currents are associated with the equinoctial spring tides, which occur in the mid rainy season (March/April) and mid dry season (September/October), and are considered dangerous for bathers. On the other hand waves are of moderate energy during the second semester (September and December), when winds are stronger, and are suitable for surfing on the most exposed beaches (Calhau and Olho d'Água). The lowest energy conditions occur in June (transitional period), when the lowest velocity winds are recorded and tidal currents and wave energy are also at their lowest. This is the period of lowest hazard for the bathers.

Calhau and Olho d'Água receive the highest number of visitors, followed by São Marcos and Ponta de Areia, with, peak visitation times are between 10:00 h and 14:00 h, and on weekends (Silva et al. 2011). These represent the periods of highest risk. The beaches are however patrolled by the lifeguards.

5.4.5.2 Erosion Hazards

The occupation and development on the sand dunes, bluffs and intertidal areas contributes to the erosion issues on the São Luís coast. Between September 2007 and April, 2008, erosion rates were 0.1 $m^3 m^{-1}$ at Ponta de Areia, 2.4 $m^3 m^{-1}$ at São Marcos, the most occupied beach, 1.1 $m^3 m^{-1}$ at Calhau, and 1.5 $m^3 m^{-1}$ at Olho d'Água. This contributed to erosion and destruction of the promanade pavement along Ponta de Areia, São Marcos, and Calhau beaches (Silva et al. 2009).

5.4.5.3 Pollution Hazards

The inadequacies of the city's public sanitation system and its rapid urban growth has contributed to an ever-increasing number of unauthorized sewage outlets that discharge effluents directly onto the local beaches. This situation affects not only the carrying capacity, but also the spatial distribution of beachgoers. Beach use of the intertidal zone is also modulated by the rise and fall of the tides.

Other risks are related the bacteriological contamination. One third of the samples analyzed by Silva et al. (2011) had thermotolerant coliform concentrations of above 1100 MNP.100 ml^{-1} during vacation period. Similar results were obtained on the same beaches during the low visitation season in 2007–2008 (Silva et al. 2009). These results are presumably the reason for the prohibition of bathing by the local authorities. Signs have been erected, prohibiting bathing all along the coast.

Strong tidal currents and moderate wave energy contribute to the rapid circulation of the water, but even though the authorities have shut some sewage outlets down, the number of outlets is still intolerably high and the beaches remain polluted and closed for bathing.

5.4.6 Beachgoer Perceptions and Classification of the Beaches

The poor water quality also influenced the type of beachgoers that visit the beaches. Affluent local residents or tourists staying in waterfront hotels do not appear to use these beaches to bathe. A study made by Silva et al. (2011) showed that the majority of beachgoers in Ponta de Areia (worse infrastructure and services) are young and with low income. An opposite pattern can be found in Calhau (best infrastructure and services) that is mainly visited by mature beachgoers (31–69 years) with higher income.

Silva et al. (2013) applied a Urban Quality Index, using the indices of beach security (wind exposure, beach slope, tide current speeds, tidal elevation and wave height), comfort (air temperature, beach sediment water clarity, precipitation level, beach width at low tide, beach vegetation cover) and service and infrastructure (availability of potable water, public sanitation system, street lighting, public transport distance, availability of waste bins, safety, availability of accommodation, recreational facilities, and urban development of the seafront). The index ranges from excellent (0.903–1) to good (0.805–0.902), to regular (0.804 and 0.707).

At Ponta de Areia the lowest score was recorded during the dry season (0.773), when tidal currents were weaker, the vegetation was reduced, public sanitation was poor, bacteriological contamination of the water was rampant and a vast number of unregulated occupation was observed. The score tended to increase during the transition period, when supervision improved. At São Marcos, the lowest scores (0.744) were recorded during the rainy season due to the increased wave height, high rainfall rates, lack of public sanitation, and bacteriological contamination. At Calhau, the lowest score (0.763) was recorded during the rainy season, and was related to the strong currents and tides, high precipitation levels, the lack of public sanitation, and bacteriological contamination. While this beach has relatively good services and infrastructure, low scores were also recorded during the dry season, due primarily to the currents and tides. The score improved (0.781) during the transition period, influenced by the latter parameters and also comfort factors.

The lowest scores recorded during this study were from Olho d'Água (0.702), which is the most exposed to wind, has the worst-quality services and infrastructure, and very little vegetation. The score improved during the transition period (0.718), due to the more favorable tides and currents (lower hydrodynamic period).

In general, increasing rainfall in the first half of the year has a negative effect on the quality of the beaches. Moderate wave energy and relatively strong tidal currents also present hazards to bathers. However, the lack of adequate services and infrastructure, in particular the lack of public sanitation and the resulting contamination of the water by fecal coliforms, represent the principal factors determining beach quality and usage.

5.5 Conclusion

The coast of Maranhão is about 1200 km long and has more than 200 sandy beaches, and represents a transition zone from the tide-dominated Amazon-Pará coast to wave-dominated Ceará coast. It is located near to equator (1–2.5° S) and is morphologically divided in three distinct provinces: (i) an heavily indented macro-tidal western coast with 42 beaches and 520 km of coast, similar to the eastern Pará coast and composed of tide-dominated estuaries, separated by low mangrove-dominated peninsulas capped by dynamic barrier islands and tide-modified to tide-dominated beaches; (ii) central province with 161 beaches along 495 km of coast and dominated by the tide-dominated Maranhão Gulf with tide modified and lower energy tide-dominated beaches; and (iii) eastern province with 18 longer beaches along 240 km of coast: a straight coast that extends from east of the gulf to the border at the Parnaíba delta. In this sector, tides decrease from 7 m to 3 m, and the coast is occupied by wave-dominated beaches backed by extensive Holocene and Pleistocene dune transgression, with a transition to tide-modified toward the Parnaíba Delta.

A study in the most urbanized zones on the Maranhão coast, São Luís in Central segment, investigated the nearshore wave, tide and current conditions, as well as natural and anthropogenic hazards. The tide-modified beaches are composed of fine sand and have a moderate slope. The strongest tidal currents are associated with the equinoctial spring tides, which occur in the mid rainy season (March/April) and mid dry season (September/October), and are considered dangerous for bathers. Waves are of moderate energy during the second semester, mainly in September and December when winds are stronger, and are appropriate for surfing on the most exposed beaches (Calhau and Olho d'Água). The lowest energy conditions occur in June (transitional period), when the lowest velocity winds were recorded and tidal currents and wave energy were also at their lowest. This is the period of lower hazard for the bathers. Erosion occurs mainly during high hydrodynamic event.

Unregulated occupation on the dunes and intertidal zones has contributed to the erosion processes. The beachfront development offers a variety of privately-owned facilities, which provide beachgoers with amenities and leisure infrastructure. On the other hand, public services are insufficient to satisfy the demand from beachgoers, and the principal problem is the lack of an adequate public sanitation system. The excessive disposal of sewage onto the beaches of São Luís has resulted in a universal ban on bathing since 2009. This prohibition has had a considerable impact on recreational activities in the area, and consequently, on its economy.

A number of measures should be taken by local authorities to improve the current status of the city's shoreline, including: (i) immediate prohibition of further urban development on unoccupied dunes, bluffs and sea cliffs; (ii) construction of a city-wide drainage and sewage treatment system; (iii) immediate shutting down of all the sewage outlets that discharge effluents directly onto the beaches or into the rivers that drain into São Marcos Bay; (iv) establishment a permanent monitoring system for the water quality of the beaches used for recreation by the local population and tourists; and (v) installation of a system of daily waste collection, which should include an increase in the number of waste bins and disposal points on all the beaches.

References

Abakerli S (2001) A critique of development and conservation policies in environmentally sensitive regions in Brazil. Geoforum 32(4):551–565

Andrade MMN, Szlafstein CF, Souza-Filho PWM, Araújo AR, Gomes MKT (2010) A socioeconomic and natural vulnerability index for oil spills in an Amazonian harbor: a case study using GIS and remote sensing. J Environ Manage 91:1972–1980

Silva IR, Pereira LCC, Guimarães DO, Trindade WN, Asp N, Costa RM (2009) Environmental status of urban beaches in São Luís (Amazon coast). J Coast Res 56:1301–1305

Silva IR, Pereira LCC, Sousa RC, Oliveira SMO, Guimarães D, Costa RM (2011) Amazon beaches (São Luís, Brazil): recreational use, environmental indicators, and the perception of beachgoers. J Coast Res 64:1287–1291

Silva IR, Pereira LCC, Trindade WN, Magalhães A, Costa RM (2013) Natural and anthropogenic processes on the recreational activities in urban Amazon beaches. Ocean Coast Manage 76:75–84

Dias JA, Polette M, Carmo JA, Pereira LCC (2009) The Brazilian Amazon coast zone. Preface. J Int Coast Zone Manage 9(2):3–7

Dominguez JML (2009) The coastal zone of Brazil. In: Dillenburgh S, Hesp P (eds) Geology and geomorphology of Holocene Coastal Barriers of Brazil. Springer, Amsterdam, pp 17–51

Espírito Santo JM (2006) São Luís: Uma Leitura da Cidade. Instituto de Pesquisa e Planificação da Cidade (IPLAM), São Luís

Feitosa AC (1989) Evolução Morfogenética do Litoral Norte da Ilha do Maranhão. Rio Claro, UNESP: 210 p. Dissertação de Mestrado

FEMAR (1997) Catálogo das Estações Maregráficas do Brasil. FEMAR, Rio de Janeiro

Figueroa SN, Nobre CA (1990) Precipitations distribution over Central and Western Tropical South America. Climanálise – Boletim de Monitoramento e Análise Climática 5(6):36–45

Hesp P, Maia LP, Sales VC (2009) The Holocene Barriers of Ceará, Piauí and Maranhão States, Northeast of Brazil. In: Dillenburgh S, Hesp P (eds) Geology of holocene Brazilian barriers. Springer, Amsterdan, pp 326–347

Machado MRB, Silva ZA, de Castro ACL (2000) Estudo da biologia reprodutiva de *Rhizoprionodon porosus* POEY, 1861 (Condrichthyes: Carcharhinidae) na plataforma continental do estado do Maranhão, Brasil. Boletim Lab Hidrobiologia 13:51–65

Marengo JA (1995) Interannual variability of deep convection over the tropical South American sector as deduced from ISCCP C2 data. Int J Climatol 15(9):995–1010

Morais JO (1977) Processos de sedimentação na baia de São Marcos, estado do Maranhão (Brasil). Arq Cienc Mar 17(2):153–164

Palma JJC (1979) Geomorfologia da Plataforma Continental norte brasileira. In: Projeto REMAC, Geomorfologia da margem continental brasileira e das áreas oceânicas adjacentes. Rio de Janeiro, PETROBRAS/CEPENE/DINTEP. (Série Projeto REMAC. 7), 177 p

Pereira LCC, Vila-Concejo A, Trindade WN, Short AD (2011) Influence of high- energy conditions on beach changes in tide-dominated (Amazon, Brazil) and wave-dominated (NSW, Australia) coastal environments. J Coast Res 64:115–119

Rossetti DF (2000) Influence of low amplitude/high frequency relative sea level changes in a wave-dominated estuary (Miocene), São Luís Basin, northern Brazil. Sediment Geol 133:295–324

Santos JHS (1996) Análise por Geoprocessamento da Ocupação na Franja Costeira ao Norte da Cidade de São Luís – MA. Disertação de Mestrado, UFRJ, p 149

Short AD (2006) Australian beach systems – nature and distribution. J Coast Res 22(1):11–27

Souza-Filho PWM (2005) Costa de manguezais de macromaré da Amazônia: cenários morfológicos, mapeamento e quantificação de áreas usando dados de sensores remotos. Rev Bras Geofísica 23(4):427–435

Trindade WN, Pereira LCC, Guimarães D de O, Silva IR, Costa RM da (2011) The effects of sewage discharge on the water quality of the beaches of São Luís (Maranhão, Brazil). J Coast Res 64:1425–1429

Chapter 6
Piauí Beach Systems

Jorge Eduardo de Abreu Paula, Jáder Onofre de Morais, Elisabeth Mary de Carvalho Baptista, Maria Luzineide Gomes, and Lidriana de Souza Pinheiro

Abstract Coastal plain of Piauí State is composed by variety of geomorphological distinct features including low sandy beaches, sandstone reefs, estuarine and marine complexes systems, coastal lagoons, tidal plains, sandy spits, extensive sand dunes, deltaic plain and others features. These features result from the combined action of waves, coastal currents, fluvial action, sea level variation and climatic and meteorological factors. Piauí State is located in the northeast Brazil covering an area of 251,578 km^2 with a population of 3,118,360 (IBGE 2010). The territory lies between 02°44′ and 10°55′S and extends for 900 km in a north-south direction, but has a coastline of only 66 km in the north aligned in an east-west direction located between Ceará and Maranhão States. The range of geo-environmental conditions represents most of the Brazilian biomes. This chapter presents the coastal and morphological dynamics and the features of Piauí Beach Systems. It discusses the variables that control the active coastal processes, with the primary objective to understand the relationship between these variables and human interventions.

J.E.d.A. Paula (✉) • M.L. Gomes
Núcleo de Estudos de Geografia Física – NEGEO, Universidade Estadual do Piauí – UESPI, Curso de Licenciatura Plena em Geografia,
Rua João Cabral, 2231 – Pirajá, 64.002-095 Teresina, PI, Brazil
e-mail: jorgeabreupaula@yahoo.com.br; luzineidegomes@bol.com.br

J.O. de Morais
Laboratório de Geologia e Geomorfologia Costeira e Oceânica-LGCO, Universidade Estadual do Ceará, Av. Dr. Silas Munguba, 1700, Campus do Itaperi, 60.714.903 Fortaleza, CE, Brazil
e-mail: jaderonofre@gmail.com

E.M.d.C. Baptista
Núcleo de Estudos sobre a Zona Costeira do Estado do Piauí – NEZCPI, Universidade Estadual do Piauí – UESPI, Curso de Licenciatura Plena em Geografia, Rua João Cabral, 2231 – Pirajá, 64.002-095 Teresina, PI, Brazil
e-mail: elisabethbaptista@bol.com.br

L.d.S. Pinheiro
Instituto de Ciências do Mar-LABOMAR- Federal University of Ceará, Av.Abolição No. 3207, Meireles, 60165–081 Fortaleza, CE, Brazil
e-mail: lidriana.lgco@gmail.com

A.D. Short, A.H. da F. Klein (eds.), *Brazilian Beach Systems*, Coastal Research Library 17, DOI 10.1007/978-3-319-30394-9_6

6.1 Introduction

Piauí State is located in the northeast Brazil covering an area of 251,578 km^2 with a population of 3,118,360 (IBGE 2010). The territory lies between 02°44′ and 10°55′S and extends for 900 km in a north-south direction, but has a coastline of only 66 km in the north aligned in an east-west direction located between Ceará and Maranhão States. The range of geo-environmental conditions represents most of the Brazilian biomes.

The settlement of this territory is considered to have been taken place from the interior toward the coast, which distinguishes Piauí from other northeastern Brazilian states and explains the more recent settlement of the coast. The coast has an area of approximately 1200 km^2, which represents 0.85 % of the Brazilian coastline and 0.29 % of the Brazilian border (Fig. 6.1). The coast is dominated by the Parnaíba river delta in the west together with rivers and streams of the Portinho and Camurupim basin. The beaches are predominantly sandy, framed by typical vegetation and drainage network. In this context, the physical-geographical and environmental characteristics that encompasses Piauí coast and its beaches are presented in this chapter.

The coast displays unique features throughout its area (Baptista 1981, 2010). In 1996, it became part of the Environmental Protection Area (APA) of the Parnaíba River Delta, established by federal decrees/n° of 28/08/1996. It covers the entire coast of Piauí as well as the coastal parts of neighboring states of Maranhão and Ceará, totaling 313,809 ha (Reis et al. 2012).

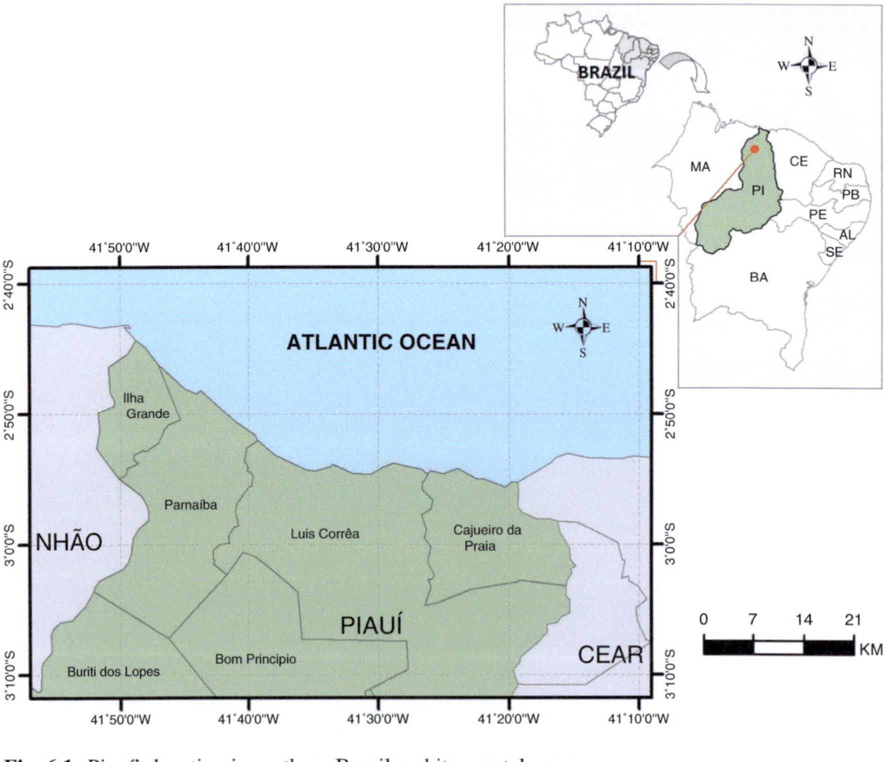

Fig. 6.1 Piauí's location in northern Brazil and its coastal zone

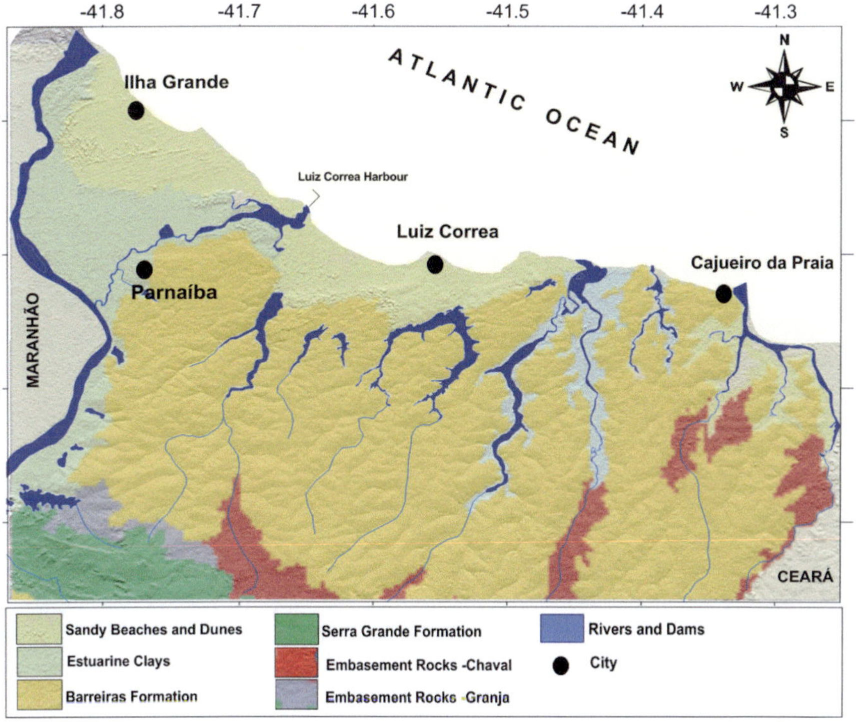

Fig. 6.2 Geological Map of Piauí coastal zone (Source: Geobank-CPRM, 2009)

6.1.1 *Geology*

The geology of the Piaui's coastal zone is related to changes in sedimentation processes of the Parnaíba river basin, also named either of Maranhão – Piauí basin or Meio Norte sedimentary basin (Baptista 1981). The sedimentation occurred in four cycles in which the sediment deposition varied from marine transgressions and regressions associated with coastal subsidence and uplift, which occurred from Paleozoic (Lower Silurian) to Tertiary/Pleistocene (Fig. 6.2).

Barreiras Formation is the dominant geological feature along the coast. It averages 23 m in thickness and is characterized by Tertiary fluvial sediments, consisting predominantly of sand and sandy-clayey sediments with large variation in grain size. The sediments are usually poorly sorted and of variegated colors indicating close relationship with the coastal plain as well as river systems. Barreiras sediments form gently sloping ramps in a tabular interfluvial drainage system.

The Quaternary deposits are represented by Holocene beach systems, aeolian dune fields, fluvial-lacustrine and estuarine deposits and are still are being deposited along the coast.

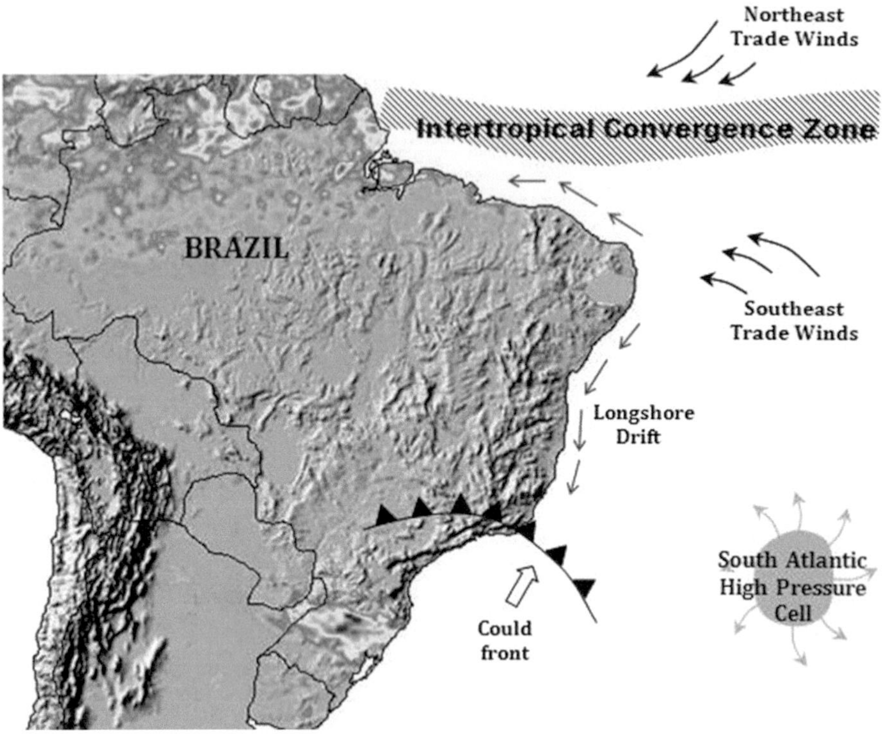

Fig. 6.3 Atmospheric circulation in the eastern – northeast coastal region of Brazil (Source: Dominguez et al. 1992)

Ferruginous sandstone of the Tertiary Barreiras Formation and Quaternary beachrock crop out on beaches and coastal plain (Baptista 2010). In some areas are exhumed erosion surfaces represented by crystalline basement rocks (Chaval and Granja basement rocks formations).

6.1.2 Climate

This coast is subject to Equatorial Atlantic Air Mass with strong northeast and southeast trade winds blowing most of the year and which average 4.4 m s^{-1}. The tropical climate is moderated by proximity to the ocean, with a hot and humid summer rainy season and drier winter. The average temperature is 25 °C and ranges between 30 and 20 °C.

The Intertropical Convergence Zone (ITCZ) is the main seasonal rainfall system in the Brazilian Northeast, mainly occurring from March to April, and sometimes expanding from February to May (Fig. 6.3). The annual average rainfall exceeds 1200 mm, with 80 % between January and May, with low rainfall from August to October (Brazil 2002).

Fig. 6.4 Delta of the Parnaíba river located between Piaui and Maranhão states (Source: Google earth)

6.1.3 Drainage

The most important drainage network of Piauí State is the Parnaíba River basin, which drains almost the entire state. The Parnaiba River Delta is the dominating feature of the coast (Fig. 6.4) and forms the Piau/Maranhão state boundary on west side together with the coastal basin of Portinho – Camurupim rivers (Baptista 1981). The Timonha river that drains the western part of Ceará, flows into the Ubatuba river with a joint estuary shared between Ceará and Piauí's Cajueiro da Praia district (Morais et al. 2004, 2005). Other streams and small tributaries occur along the littoral.

The coastal lagoons of Piauí are either drowned valleys or in lee of barriers and exposed to marine influence. Only Sobradinho and Santana lakes are registered as coastal lakes. Other important ones are Portinho, Jaboti, Mutucas and Camelo lakes, which have formed either from rivers, streams, rainwater supply or derived from the surface water flow caused by dune field. There are also lakes of phreatic origin (Baptista 2010).

The coastal sediment are derived from the Parnaíba River deltaic complex and the other of rivers and streams that flow to the coast, together with marine sediments transported by waves, tides and currents. Aeolian dune transport is also significant. All sediments are transported to the west by both waves and wind.

The beach sediments are composed of fine to medium quartz sand together with heavy minerals, shell fragments and micaceous minerals. They are moderately sorted and predominantly whitish in color. In the eastern Cajueiro da Praia and Barrinha beaches (Cajueiro da Praia) poorly sorted medium sand predominates. Further westwards Carnaubinha beach (Luis Correia) consists predominantly of moderately sorted medium sand. Morro Branco and Barra Grande (Cajueiro da Praia) have poorly sorted fine and coarse sands, respectively. Itaqui and Coqueiro

(Luis Correia) beaches have moderately sorted coarse and very fine sand, respectively (Baptista 2010). Atalaia beach (Luis Correia) has fine to very fine sand (Bittencourt et al. 1992) and Macapá beach (Luis Correia) follows the general characterization of Piaui beaches with moderately well sorted fine and medium sands.

6.1.4 Coastal Provinces and Geomorphology

The coast of Piauí starts in the west at the main Parnaíba river mouth (Fig. 6.5) followed by the Ilha Grande, with a series of beach embayments extending in the southeast-northwest direction between main Parnaíba river mouth and Igaraçu distributary. A crystalline rock outcrop occurs at Pedra do Sal's beach. Thereafter the coast has a slight bend to the east coming to the mouth of Igaraçu River. Then, it follows the beaches of Atalaia, Coqueiro and Itaqui to the mouths of San Miguel and Camurupim rivers, continuing to the mouth of the Ubatuba river and then to the tidal delta of Timonha river, where the Ceará's coast initiates (Baptista 1981).

The Piauí coast is dominated by sandy beaches, stable and transgressive sand dunes and fluvial-marine plains, all of which compose the coastal plain. In addition, there are also lacustrine and fluvial-lacustrine plains, and fluvial plains in the lower courses of the rivers that flow to the coast. The Barreirias Formation is usually located 30–50 km inland but extends to the coast in the east, where it is known as the coastal tableland. It merges further inland with crystalline basement rocks or with sandstone formations of the Parnaíba River (Maranhão – Piauí) sedimentary basin (Baptista and Horn Filho 2012). The typical littoral vegetation is dominated by mangroves that grow in the mouths of rivers and streams, as well as the

Fig. 6.5 Piauí's coastal geomorphological features, which include sandy dunes migrating to the Parnaiba river; basement rocks of the Chaval Formation; and the Barreiras Formation (Coastal Tableland)

formation of "tidal wetlands" or "salt marshes" in floodplain terraces, consisting of palustrine sediments of fluvial-marine plains which are a major feature of this coast (Baptista 2010).

Piauí's main beaches have a similar morphological configuration and include the nearshore, foreshore and backshore zones which are differentiated by the local characteristics of sediments accumulation, water drainage and the presence of rocks. Outstanding features include the micro-cliff aggregate of ferruginous sandstone reefs on Cajueiro da Praia beach, sandy spit at Itaqui beach and rocky promontory on Pedra do Sal beach.

6.2 Piauí Beach Systems

6.2.1 Coastal Processes and Parameters

Coastal processes determine the nature and evolution of the coastal area and consist of a range of parameters including wave climate, tides, currents, morphodynamic behavior, spatial variation of the volume of transported sediments, and climatic behavior. These processes and their impact on the Piauí coast and beaches are described in the following sections.

6.2.2 Waves, Tides and Currents

The wave climate is based on Paula (2013) along with analyses of the data obtained from Luis Correia (Piauí) harbor and Pecém (Ceará) harbor, and analysis of data available from the Weather Forecasting and Climate Studies Center (CPETC) of the National Institute of Space Research (INPE). During the monitoring period (2010–2012) waves averaged (H_b) 0.83 m during the dry season and 0.64 m in the rainy season. The highest values in the dry season are related to the higher winds velocities during this period.

The deepwater waves arrive predominantly from the north-northeast (340–10°). Near the coast they are affected by submerged rocks, such as the Pedra do Sal promontory (composed by crystalline basement rocks) and sandstone reefs, which cause wave attenuation and refraction-diffraction. At high tide higher waves reach the beach face, causing the medium and long term retreat of the shoreline, which forms small embayments. These areas of shoreline retreat induced by diffraction waves are characteristic in all beaches with reefs or rocky headland presence such as Cajueiro, Sardim, Itaqui, Coqueiro and Cotia beaches (east-west direction).

The peak wave period (T_p) varies between 5.8 and 12.4 s. Sea waves with T_p between 6 and 9 s prevailed by more than 70 % of the time during the rainy season and 90 % of the time during the dry season. Swell waves (>9 s period) were more significant during the rainy season, arriving approximately 25.7 % of the time (Fig. 6.6).

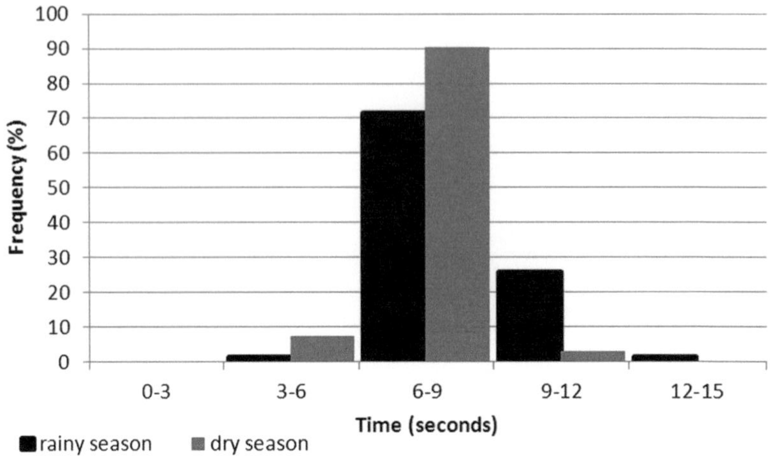

Fig. 6.6 Histogram of frequency of peak wave period (T_p) for the rainy and dry seasons (2010–2012) in the Piauí coast (Source: Paula 2013)

6.2.3 *Tides*

The Piauí coastline has semi-diurnal, meso-tides with maximum range of 3.7 m. The meso tide range is responsible of significant changes in coastal sedimentation process (Tessler 2008). At high tide, the coastal currents are stronger and variation in beach profiles and volume is greater.

6.2.4 *Surfzone Currents*

Wave-driven longshore current are the main component of sediment transport in the coastal zone. Data on incidence and waves heights in the surf zone were used to estimate longshore current velocities between 2010 and 2011 (Paula 2013). The average longshore current velocity during the period was 1.1 m s^{-1}, occasionally reaching 1.6 m s^{-1} in association with rip currents. These currents were observed along most of the coastal beaches of Piauí, including Atalaia beach (Luis Correia District) and in other areas where wave refraction around promontories and submerged beachrock.

6.2.5 *Beach Types and States and Spatial Variation in Beach State*

Piauí beaches are dominated by tide-modified reflective and low-tide terrace beaches. The morphodynamic classification of Piauí beaches is based on field work using: transverse beach profiles; variation of the beach profile inclination; analysis

Table 6.1 Beach characteristics along the Piauí coast (see Fig. 6.5 for cell location)

Cells	Beaches	Description	Beach type/state[1]
01	Profile 1 – Mangue Profile 2-Cajueiro Morro Branco and Barrinha beaches	Marine and fluvial interaction area (Ubatuba river mouth) (Fig. 6.7a) strongly influenced by river discharge and the contact with marine waters generating pelitic sediments deposit forming a tidal flat system colonized by various algae. The substrate is usually composed of very fine sand with high percentages of silt and clay, there is still sparse gravel with biodetritic sediments (of marine and vegetal origin). The beach is reflective with a 550 m wide low tide terrace (Fig. 6.7b)	Reflective + low tide terrace
02	Profile 3 – Barra Grande Profile 4 – Ponta do Anel (to the Cardoso and Camurupim mouth rivers)	Beaches with diffraction waves process generating changes in coastal current and formation of shore parallel bars, which are submerged at high tide. The beach width is 500 m at low tide (Fig. 6.7c)	Reflective + low tide terrace
03	Macapá, Carnaubinha, Arrombado Profile 5 – Itaqui Profiles 6, 7 – Coqueiro	Urbanized beaches frequented by tourists, visitors and vacationers. Three factors are responsible for beach erosion: (a) occupation and seawalls trapping backshore sediments; (b) diffraction wave process; and (c) the steep slope of the beach (Fig. 6.7d). It has a narrow beach face, which reaches 100 m at low tide	Reflective + low tide terrace
04	Profiles 8, 9, 10 – Atalaia	This is the most popular and urbanized beach of entire Piauí coast (Fig. 6.7e). It is an exposed beach with low slope and up to 450 m wide at low tide. There is no indication of erosion. There is considerable aeolian sand transport into the dune fields	Reflective + low tide terrace
05	Ostras (after the Igaraçu river mouth) Profile 11 – Eólica Profile 12 – Pedra do Sal Profile 13 – Cotia Profile 14 – Maceió Barra das Canárias	It is an exposed beach with a steep gradient resulting in a narrow beach face, up to100 m wide at low tide. This area is influenced by Pedra do Sal headland that causes diffraction downstream and retreat of the shoreline (Fig. 6.7f, g)	Reflective + low tide terrace

[1]Based on RTR parameter (Masselink and Short 1993)

of wave climate; and beach sediment characteristics. The coastline was divided into five monitoring cells and the beaches classified using the RTR model (Masselink and Short 1993) (Table 6.1).

Fig. 6.7 (**a**) Mangue beach bordering Ubatuba River estuary; (**b**) Cajueiro beach; (**c**) Barra Grande beach with algae on the shoreface; (**d**) eroding foredune on Macapá beach; (**e**) Atalaia beach; (**f**) Ostras beach; and (**g**) Eólicas beach near Parnaiba river

6.2.6 Beach-Dune Interactions

Conventional meteorological data and experimental field surveys were used to investigate the beach-dune interaction in the region. The dominant winds are the trade winds (resulting from the macroscale atmospheric dynamics) and the sea breezes (resulting from the meso-microscale atmospheric dynamics).

The trade winds are directly controlled by the mobility of the ITCZ. The south-easterly trade winds are more intense when the ITCZ moves farther north during the dry season (between August and October). When the ITCZ is south of the equator (between April and May) the trade winds are weaker. The strongest winds are also related to local/regional circulation, as when the sea breeze and trade winds coincide they generated the highest velocity winds from the east-northeast which increase both wave height and aeolian sand transport.

The direction and intensity of the winds in contact with the beach face determines the degree of sediment mobilization from the intertidal zone to the sand dunes fields. According to the Agro-meteorological Bulletin of the EMBRAPA/INMET (2000–2010) the prevailing winds are from north-northeast and east-southeast. This is reflected in the records of the wind acting on the coast, and the direction of dune migration. (Fig. 6.8).

The maximum velocity of winds on the coast of Piauí reaches 10.7 m s^{-1}, which generally exceeds the minimum value necessary to transport sediment (3.05 m s^{-1}) less than 0.25 mm diameter (corresponding to medium and fine sand). Bittencourt et al. (1990) characterized the textural changes in Atalaia beach sediments under the action of the winds confirming the competence in the sediment transport.

Fig. 6.8 The Piauí coastal zone showing the E-NE Trade winds (*red arrows*) and direction of dune movement (Source: Paula 2013)

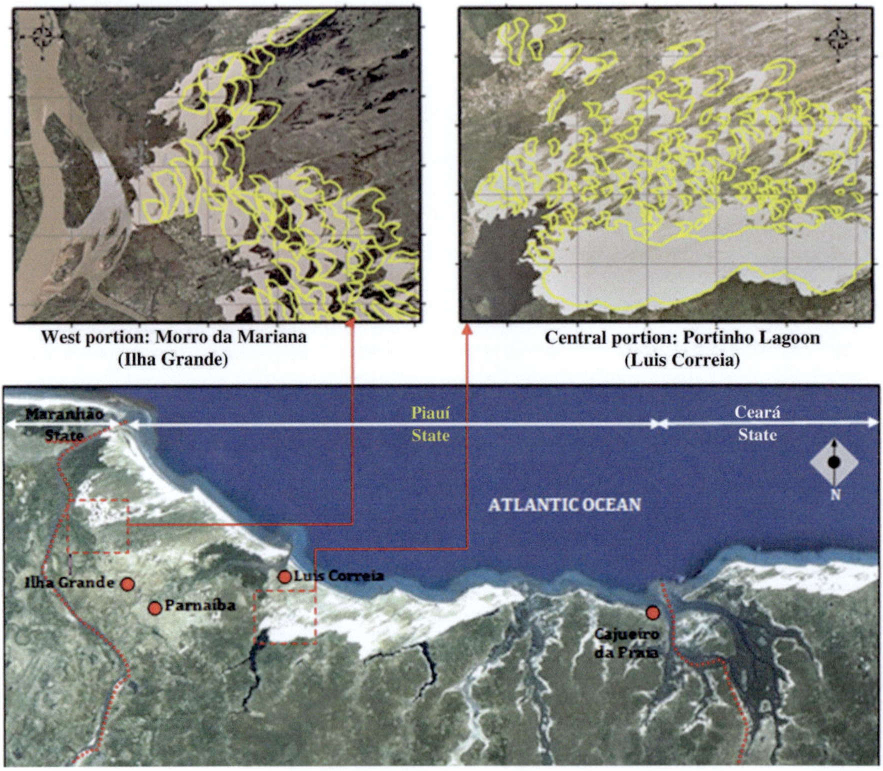

Fig. 6.9 Migration of sand dunes between 1987 and 2006. The *yellow* polygons represent the vectorization of the dune front in the Landsat image of 1987 superimposed in the orthophotos from 2006 (Source: Paula 2013)

The dominant northeast trade winds accumulate sand in foredunes, which protect the backshore from storm wave action as the dune-stored sediments are returned to the beach during periods of erosion. This sand can also be released to migrate inland as transgressive dunes (Paula et al. 2005), during the Pleistocene and at present. However recent occupation and degradation of the Pleistocene dunes and their vegetation cover has generated instability of superficial layer of the soil causing them to migrate. In addition contemporary aeolian deposits are also migrating covering houses, silting up fluvial courses and urban areas.

Paula (2013) analyzed the dunes morphology and migration and estimated migration in two sectors of the coastline using multi-temporal images (orthorectified aerial image from 2006 and Landsat satellite image from 1987). The average dune migration in the central and west portion of Piauí coast was 21.1 and 21.6 m yr^{-1} respectively (Fig. 6.9).

The orientation of much of the coastline is east–west to southeast-northwest exposing it to the east-northeast winds permitting maximum landward aeolian sand transport. As a consequence this stretch of coast acts as a sink for littoral sediments, with no effective renewal of the beach system by dune sediments.

6.2.7 Beach-Hardrock Interactions

Rocks exposed on Piauí beaches are composed of three distinct types: sandstone reef of Barreiras Formation, beachrock and granitic rock outcrops (Fig. 6.10).

Barreiras Formation rocks dominate the rock outcrops occuring at six locations, with beachrock and granite on one beach each. All these rocks have significant interactions with the beaches as described below:

Barreiras Formation Sandstone Reefs (Cajueiro, Morro Branco, Barra Grande, Carnaubinha, Itaqui and Coqueiro beaches) contribute to the beach morphology as sand accumulates in lee of the reefs. They also attenuate the incident waves and tides, lowering wave energy and reducing coastal erosion. They constitute evidence of ancient consolidated beaches, indicating possible changes in the coastline. They are also important for marine and coastal biodiversity, because at their lee side, in

Fig. 6.10 Rock outcrops on Piauí's coastline. (**a**) sandstone reef, Barreiras Formation; (**b**) beachrock; (**c**) granite outcrop (Source: Baptista 2004)

so-called "strips", numerous species of animals and plants make use of them as a support substrate and food source (Baptista 2010).

Beachrock Reefs (Barrinha Beach) provide a sea-level indicator. They primarily act to protect the coast by minimizing ocean and wave energy and trapping wind-blown sand. They also provide an habitat for marine organisms (Baptista 2010).

Granitic Outcrop (Pedra do Sal beach) Granitic rocks form a promontory with beaches to either side. The eastern beach is exposed to waves and not suitable for bathing, while the western beach is calm with low waves, and used for tourism, leisure and mooring of small fishing vessels, a traditional local activity. Granitic coasts also constitute significant marine ecosystem (Baptista 2004).

6.2.8 Longshore Sand Transport, Sediment Cells and Beach Stability/Erosion

The variation in beach profile volume was calculated by profiling in each of the coastal monitoring cells. The beach profiles were then used to calculate both spatial and temporal variation in beach subaerial volume. The beach volume (V) is defined using three variables (X, Y and Z). The sectional area within the range of X is the length of the beach between the marine (X1) and landward (X2) boundaries. The Y axis extends vertically and the Z axis extends horizontally representing width of the section. The origin of coordinates is located at the lowest sea level to a fixed reference point. The volumes were calculated with SURFER 10 software that integrates data X, Y, and Z. The sediment volumes (between 2010 and 2011) behaved differently but maintained relative morphological balance with usually positive volume. Cell 2 (Barra Grande beach), Cell 4 (Atalaia beach) and Cell 5 (Pedra do Sal beach) all showed an increase in the sediment volume (Table 6.2). From the repeated morphological measurements of beach profiles the temporal variation in beach subaerial volume was obtained.

The average volume increase for the period was about 4900 m^3. The average interannual variation in beach volume is much lower suggesting beach stability. Profile 4 had the largest increase contributing 40 % of the increase in volume, which can be explained in the west by the contribution from the Parnaiba Delta which has accreted the beach profile by up to 700 m in the last decade. In contrast profiles 11 and 12 had a negative balance, which can be attributed to the influence of Pedra do Sal promontory and the occupation of intertidal zone by the beach huts.

Between 07/10 and 09/10 there was a clear negative balance, which can be explained by the high tides (3.2–3.6 m) during this the period. The coastline near the river mouths is another good example of areas of extreme short term shoreline dynamics, independent of, but potentially accelerated by, human intervention. Macapá beach (Luis Correia-PI) exemplifies this process. The Cardoso/Camurupim river mouth has changed substantially over the past three decades, with erosion of

Table 6.2 Volumetric sediment variation of beach profiles in the Piauí Coast (see Fig. 6.5). Volumes are given in m³

Cell	Beach profiles	Beach profile volumes over the months/years monitored (m³)								Interanual balance (m³)	Sediment volume situation
		02/10	04/10	07/10	09/10	01/11	04/11	09/11	11/11		
1	Profile 1	2044.8	4831.8	5652.8	3524.6	1407.3	2438.1	6559.4	6316.3	4271.4	Accretion
	Variation*	–	(+)2787.0	(+)820.9	(–)2128.1	(–)2117.3	(+)1030.7	(+)4121.3	(–)243.1		
	Profile 2	1541.5	5446.4	6281.8	5323.2	–	7275.5	5860.6	7901.4	6359.9	Accretion
	Variation*	–	(+)3904.9	(+)835.3	(–)958.5	–	(+)1952.3	(–)1414.9	(+)2040.8		
2	Profile 3	2112.7	7037.3	6968.6	7241.8	2579.5	4206.6	6991.6	6173.9	4061.1	Accretion
	Variation*	–	(+)4924.6	(–)68,7	(+)273.1	(–)4662.2	(+)1627.0	(+)2784.9	(–)817.6		
	Profile 4	4588.5	28935.1	4265.6	963.8	16812.6	5892.5	7519.9	21875.4	17286.8	Accretion
	Variation*	–	(+)24346.6	(–)24669.5	(–)3301.8	(+)15848.8	(–)10920.2	(+)1627.4	(+)14355.5		
3	Profile 5	6277.3	4874.7	5619.1	5154.6	5599.7	6182.8	5871.4	8528.7	2251.4	Accretion
	Variation*	–	(–)1402.5	(+)744.3	(–)464.5	(+)445.1	(+)583.0	(–)311.3	(+)2657.3		
	Profile 6	952.3	1260.5	2026.9	1370.5	1392.0	1903.1	2107.8	6316.3	5363.7	Accretion
	Variation*	–	(+)308.1	(+)766.3	(–)656.3	(+)21.5	(+)511.0	(+)204.6	(+)4208.5		
	Profile 7	289.7	471.5	527.9	715.8	919.2	514.7	610.2	825.3	535.5	Accretion
	Variation*	–	(+)181.8	(+)56.3	(+)187.9	(+)203.4	(–)404.5	(+)95.5	(+)215.1		
4	Profile 8	5742.8	5384.5	4378.9	4340.8	4392.5	4468.0	3996.5	3350.8	–2391.9	Erosion
	Variation*	–	(–)358.3	(–)1005.5	(–)38.1	(+)51.6	(+)75.4	(–)471.4	(–)645.6		
	Profile 9	3040.9	4563.3	4361.1	2935.9	2819.9	3224.2	5801.3	5311.1	2270.3	Accretion
	Variation*	–	(+)1522.4	(–)202.2	(–)1425.1	(–)115.9	(+)404.2	(+)2577.1	(–)490.2		
	Profile 10	10836.6	6290.3	11996.2	7014.7	5977.7	6366.8	9762.9	10013.5	–823.1	Erosion
	Variation*	–	(–)4546.3	(+)5705.9	(–)4981.5	(–)1036.9	(+)389.0	(+)3396.1	(+)250.6		

(continued)

Table 6.2 (continued)

Cell	Beach profiles	Beach profile volumes over the months/years monitored (m³)								Interanual balance (m³)	Sediment volume situation
		02/10	04/10	07/10	09/10	01/11	04/11	09/11	11/11		
5	Profile 11	6089.5	4724.2	8715.2	6432.7	6583.4	12190.5	9018.5	4654.9	−1434.7	Erosion
	Variation*	–	(−)1365.3	(+)3990.9	(−)2282.5	(+)150.7	(+)5607.0	(−)3171.9	(−)4363.6		
	Profile 12	370.1	269.7	260.4	438.1	154.7	270.4	228.5	230.4	−159.8	Erosion
	Variation*	–	(−)100.4	(−)29.3	(+)177.6	(−)283.4	(+)115.7	(−)41.9	(+)1.9		
	Profile 13	2539.6	2818.0	2872.0	3154.8	2622.4	4001.2	5132.2	3277.4	737.7	Accretion
	Variation*	–	(+)278.3	(+)53.9	(+)282.8	(−)532.3	(+)1378.7	(+)1131.0	(−)1854.7		
	Profile 14	5399.5	5164.3	7650.7	5721.6	4556.9	3744.4	8322.2	5769.0	369.8	Accretion
	Variation*	–	(−)235.2	(+)2486.4	(−)1929.0	(−)1164.6	(−)812.5	(+)4577.8	(−)2553.1		
Over all balance		–	30245.7	**−10515**	**−17244**	6808.5	1536.8	15104.3	12761.8	–	–

*– Variation from previous month (Source: Paula 2013)

up to 550 m between 1987 and 2010. Other areas that are very dynamic are located in lee of headlands and reefs where they are affected by wave diffraction. While the headlands and reefs protect the beaches from direct wave attack, they also generate the diffraction process that can produce erosion within the embayments, as it occurs at Pedra do Sal beach (Ilha Grande) where the shoreline on the downdrift sectors eroded by 180 m. In addition headland bypassing at Itaqui and Cotia beaches and overpassing at Carnaubinha generates considerable variation in the width of the downdrift beach located in lee of the headland.

Coastal Vulnerability The vulnerability of the shoreline to coastal erosion can be assessed by examining the relationship of the natural and social variables. Sea-level rise has been estimated at 4 mm yr^{-1} (Mesquita et al. 2011). If this is compared to the maximum tidal range of the region (which reaches 3.7 m at spring tide), it can be stated that this value does not presently represent great risk to the Piauí's coast zone (even in the short or medium term). This is because most of the coast has a low occupancy rate, characterized by predominately unoccupied beaches.

However in Piauí coastal urban centers where the intertidal areas are already occupied, like Coqueiro beach and Pedra do Sal beach, the level of vulnerability will be high. Atalaia beach, which has relatively consolidated urbanization, and a wide flat beach presently shows no indicators of erosion.

Thus, the present sites of coastal occupation appear to be safe if the current use and occupation are maintained, and if there are no changes in the supply of sediment from the shoreline. However, this is generally not the case as the shoreline is becoming increasingly attractive for human activities.

So, we tried to integrate data based on the indicators of coastal erosion and its causes, as proposed by Souza et al. (2005), using a presence/absence matrix indicative of the intensity of processes along the Piauí coastline, as well as of indicators of progradation or retrogradation of the shoreline based on its position in 1987 and 2010 respectively.

In an overall assessment of Piauí beaches and considering the risk of erosion, it can be stated that 50 % of the coastal beaches that can be classified as high risk to erosion associated primarily to natural causes. 40.9 % are classified as medium erosion risk, mainly the beaches of low occupyancy and seasonal use. About 4.5 % of Piauí's beaches presents very low risk of erosion, which are the most secluded beaches and almost never visited, such as the Ostras beach near the Igaraçu river mouth, and the beaches of Maceió and Canárias beaches, situated near Parnaíba river mouth. However, 4.5 % of the beaches already have a very high erosion risk, represented mainly by the more urbanized beaches such as Pedra do Sal (Fig. 6.11a) and Coqueiro (Fig. 6.11b). These are mainly associated with anthropogenic factors together with a contribution from natural factors. The stability and maintenance of these beaches is dependent on the input sediment from the rivers in the eastern sectors, from the Coreaú, Timonha, Ubatuba, São Lourenço and Camurupim rivers. Decisions will be required soon on how to mitigate the erosive processes.

Fig. 6.11 (**a**) Pedra do Sal beach, east of the promontory (Cell 5). Steep beach profile and wave attack on this high energy beach eroding the berm and destroying improvements; and (**b**) Coqueiro beach (Cell 3). Steep beach profile and wave attack eroding the beach face and damaging human structures (Source: Paula 2013)

6.3 Beach Use and Abuse

6.3.1 Beach Development and Management

The use and occupancy of beach areas (urban or rural) need to be planned so as to take account of the nature and susceptibilitiy of the surrounding environment. The same applies to the coastal zone, where the environment is extremely dynamic and unstable, requiring specific actions for its use and occupation.

Question of occupancy limits of the coast has been discussed by Muehe (2006). In areas of higher population density, a minimum set-back of 50 m from the most landward limit of the active beach should be required.

The zoning of areas to be protected must be in harmony with the management and planning of the coastal area. They should have “minimum necessary boundaries” so that the natural environmental systems remain minimally altered by human action and can operate naturally and without human impediment.

Coastal developers (seen here as all forms of intervention that cause urbanization: infrastructure installation, creation of allotments, tourist facilities installation) do not always consider the “limits” of environmental systems. Human intervention therefore must be planned considering its dynamics and risk of vulnerability.

However, despite the existence of a proposal for integrated coastal management (through the Coastal Management-GERCO) established in 1987, not all coastal states have advanced with their State Plans for Coastal Management and the development of the Orla Project 2004 (Brasil 2004).

Even though Piauí State (including the Environmental Protection Area of Parnaíba Delta River and the municipalities of Luis Correia, Parnaiba, Cajueiro da Praia and Ilha Grande de Santa Isabel) was strategically chosen for implementation and validation of methodology that provided the basis for the development of Orla Project, little has been done to implement the strategy in recent years.

Nicolodi and Oliveira (2012) also emphasize that the Piauí coast is one of the most paradoxical regions of Brazil, as while it has great potential for tourism and conservation of nature, it has very high levels of social risk. It is a coast with extremely contrasting situations of social and economic conditions. Thus, management actions on the coast of Piauí aimed at an appropriate coastal management must also aim for the reduction of social conflicts.

6.3.2 Beach Hazards and Safety

Piauí's beaches have are range of hazards and usage that need to be considered in the assessment of beach risk The hazards are related to a range of physical factors including the high tide range; bars, troughs and rip currents; river dischange; steep beach faces and muddy tidal flats; intertidal rocks and structures; and deep water at high tide (Table 6.3). The number of beach users and hence the level of risk varies both spatially and seasonally. Location and ease of access to the Coqueiro, Atalaia and Pedra do Sal beaches result in large numbers of beach users. The Atalaia beach receives the greatest number of visitors especially on weekends, due to the close proximity to Parnaíba city (urban regional center) and also because it has better infrastructure in relation to others beaches, with beach huts, hotels and inns that service all users and visitors. It must also be noted that the number of visitors increases during the high season and long holidays, often doubling the population of the Luís Correia's city.

There is a high level of risk on Piaui beaches owing to a lack of knowledge by visitors and users in relation to beach morphodynamics and potentially hazardous (rocks, headlands, river mouths and the presence of rip currents), together with the lack of effective signaling, the need for more lifeguards (concentrated mainly in Atalaia's beach). This risk has been verified by the high number of accidents involving swimming. In addition, these beaches, with the exception Atalaia beach, consistently receive waste water from sewers, beach tents and toilets without proper treatment which compromise the environmental quality and increase the risk of bacterial contamination.

6.4 Summary and Conclusions

Coastlines are dynamic regions, with their natural processes influenced by ocean-land-atmosphere interactions, which determine their main characteristics and features. Although small in extent (66 km), Piauí's coastline has unique aspects in the northeastern coastal context, with significant attributes in regard to its physiographic elements.

In this area we have beaches with distinct morphological characteristics: Extensive and wide beaches at Atalaia, Maceió and Barra das Canárias have a low-gradient slope and wide intertidal zones. But there are also beaches that are subject

Table 6.3 Hazards on Piauí's coastal beaches

Cells	Localization	Beaches	Hazards for users
01	Easternmost coast	Profile 1 – Mangue beach	Strong tidal-river discharge during low tide. Presence of oysters in mud
		Profile 2 – Cajueiro beach	Bars caused by river discharge are submerged during high tide. The rising tide quickly covers the bars, isolating them from the beach and posing risk for elderly and young children
		Morro Branco and Barrinha beaches	Intertidal beachrock and bars are submerged during high tide (Fig. 6.12)
02	Eastern coast	Profiles 3, 4 – Barra Grande	Intertidal beachrock, steep beach face, bars, high variation of low tide terraces and deep water close inshore
		Ponta do Anel (to the Cardoso and Camurupim mouth rivers)	Steep beach face, bars, high variation of low tide terraces and deep water close inshore
03	Central coast	Macapá (after the Cardoso and Camurupim mouth rivers) and Maramar	Strong river discharge during low tide. Bars caused by rivers discharge are submerged during high tide. Steep beach face, bars, deep water close inshore and rip currents
		Profiles 5, 6 and 7 – Itaqui and Coqueiro, Carnaubinha, Arrombado	Intertidal beachrock, bars are submerged during high tide and deep water close inshore
04	Central coast area	Profiles 8, 9 and 10 – Atalaia	Variation of low tide terraces and transverse bars that are submerged during high tide and rip currents
05	Western coast	Ostras (after the Igaraçu river mouth)	Strong river discharge at low tide. Bars caused by river discharge are submerged high tide and rip currents
		Profile 11 – Eólica	Steep beach face, bars and deep water close inshore
		Profile 12 – Pedra do Sal	Steep beach face, deep water close inshore, rip currents and rocky outcrop
		Profile 13 – Cotia	Steep beach face, deep water close inshore and rock structures in the intertidal zone
		Profile 14 – Maceió	Steep beach face, bars and deep water close inshore
		Barra das Canárias	Bars caused by river discharge are submerged during high tide. The rising tide quickly covers the bars, isolating them from the beach and posing risk for elderly and young children. Very unstable tidal flats due the presence of mud

Source: Paula 2013

Fig. 6.12 Beachrocks located in the intertidal zone and bars in the high tide at Barra Grande beach (Source: Paula 2013)

to interference of diffraction waves resulting beach erosion. There are also dynamic beaches such as, Mangue, Cajueiro, Ponta do Anel and Macapa, which are influenced of mobile river mouths.

The integration of natural elements and processes reflect the morphodynamic behavior of Piauí coastal beaches and the risks associated with their use. Additionally it also reflected in human activities in the form of coast occupation and socioeconomic activities which increase vulnerability. Considering morphodynamics and applying the RTR Parameter (Masselink and Short 1993), the beaches are classified as tide-modified reflective with low tide terrace, highlighting a predominant intermediate behavior.

Beach monitoring has shown that this region has high rates of longshore sediment transport and shoreline variability, and that sites prone to erosion with unsystematic urbanization like Coqueiro and Pedra do Sal beaches need to be monitored. The practice of spontaneously development without following proper plans or policy is a major problem. Monitoring of both natural and developed beach systems is therefore required to assess their impacts. More studies are therefore necessary to establish forms of consistent integrated use and responsible coastal management.

References

Baptista JG (ed) (1981) Geografia Física do Piauí 2. COMEPI, Teresina

Baptista EMC (2004) Caracterização e importância ecológica e econômica dos recifes da zona costeira do Estado do Piauí. MSc. Dissertation, Univ Fed do Piauí

Baptista EMC (2010) Estudo morfossedimentar dos recifes de arenito da zona litorânea do estado do Piauí, Brasil. Florianópolis. Ph.D. thesis, Univ Fed de Santa Catarina

Baptista EMC, Horn Filho NO (2012) Geologia e geomorfologia da planície costeira adjacente aos recifes de arenito do litoral do estado do Piauí. In: Abstract of the IX Simpósio Nacional de Geomorfologia-SINAGEO, Univ Fed do Rio de Janeiro, Rio de Janeiro

Bittencourt ACS, Dominguez JML, Moita Filho O (1990) Variações texturais induzidas pelo vento nos sedimentos da face da Praia de Atalaia, Piauí. Rev Bras Geociênc 20:201–207

Bittencourt ACSP, Boas GSV, Farias FF (1992) Variações direcionais nos parâmetros granulométricos: um indicador apropriado para o sentido da deriva litorânea. Rev Bras Geociênc 22:100–106

Brasil (2002) Ministério do Meio Ambiente. Programa Zoneamento Ecológico-Econômico do Baixo Rio Parnaíba: Subsídios Técnicos – Relatório Final. Brasília, 92p, ilust

Brasil (2004) Ministério do Meio Ambiente e Ministério do Planejamento, Orçamento e Gestão. Projeto Orla: Subsídios para um projeto de gestão. http://www.mma.gov.br/estruturas/orla/_arquivos/11_04122008110506.pdf

Dominguez JML, Bittencourt ACSP, Martin L (1992) Controls on quaternary coastal evolution of the east-northeastern coast of Brazil: roles of sea-level history, trade winds and climate. Sediment Geol 80:213–232

IBGE (2010) Instituto Brasileiro de Geografia e Estatística-IBGE. Censo Demográfico de 2010: Retratos do Brasil e do Piauí. http://www.ibge.gov.br/home/presidencia/noticias/pdf/censo_2010_piaui.pdf

Masselink G, Short AD (1993) The effect of tide range on beach morphodynamics and morphology: a conceptual model. J Coast Res 9:785–800

Mesquita R, Harari J, França CAS (2011) Nível do Mar no litoral do Brasil. III Conferência Regional sobre Mudanças Climáticas: América do Sul. Conferência Instituto Oceanográfico da Universidade de São Paulo, São Paulo – Brasil

Morais JO, Pinheiro LS, Dias CB (2004) Aspectos Hidrológicos e Hidrodinâmicos do Sistema Estuarino Timonha / Ubatuba- Ceará-Piauí, Brasil. In: Abstracts do Congresso Brasileiro de Oceanografia Itajaí, UNIVALI, Itajaí

Morais JO, Pinheiro LS, Dias CB (2005) Dinâmica do sistema estuarino Timonha/Ubatuba (Ceará-Brazil): Considerações ambientais. Dissertação de Mestrado, Universidade Federal do Ceará - UFC

Muehe D (2006) Definição de limites e tipologias da orla sob os aspectos morfodinâmico e evolutivo. In: Projeto orla: fundamentos para gestão integrada/Ministério do Meio Ambiente, Ministério do Planejamento, Orçamento e Gestão. Brasília: MMA. http://www.mma.gov.br/estruturas/orla/_arquivos/11_04122008110506.pdf

Nicolodi JL, Oliveira MRL (2012) A Gestão Costeira no Brasil e os dez anos do Projeto Orla. Uma análise sob a ótica do poder público. Rev Gestão Costeira Integrada 12:89–98

Paula JEA (2013) Dinâmica morfológica da planície costeira do estado do Piauí: Evolução, comportamento dos processos costeiros e a variação da linha de costa. PhD Thesis, Universidade Federal do Ceará – UFC

Paula JEA, Gomes ML, Morais JO (2005) Aeolian dynamic actuation on river barrage processes and the creation of coastal lagoon environments, Catú Lagoon (Aquiraz) and Cauípe Lagoon (Caucaia) – Ceará. In: Proceedings of the of the international symposium on coastal dunes. First Brazil symposium on dunes systems: "Management and Sustainable Development", Fortaleza

Reis EB, Sousa MGS, Baptista EMC (2012) Caracterização Sócio-Ambiental do Litoral Piauiense: uma Síntese Bibliográfica e Documental. UESPI (Relatório de Pesquisa), Teresina

Souza CRG, Souza Filho PWM, Esteves LS, Vital H, Dillenburg SR, Patchineelam SM, Addad JE (2005) Praias arenosas e erosão costeira. In: Souza CRG et al (eds) Quaternário do Brasil. Associação Brasileira de Estudos do Quaternário. Holos Editora, Ribeirão Preto, 382p

Tessler M (2008) Potencial de Risco Natural. In: MMA (ed) Macro Diagnóstico da Zona Costeira e Marinha do Brasil. (Ed.) Ministério do Meio Ambiente, Brasília

Chapter 7
The Beaches of Ceará

Lidriana de Souza Pinheiro, Jáder Onofre de Morais, and Luis Parente Maia

Abstract The coastal zone of Ceara State, in Northeastern Brazil, extends for 573 km in a SE- NW direction. It is predominantly composed of sandy sediments of Upper Tertiary-Quaternary age with several generations of Pleistocene transgressive dunes, together with beaches, estuarine plains and localized occurrences of cliffs. The Precambrian rocks also occur on some beaches. The climate is semi-arid tropical and the rivers only flow into the sea during the rainy season. The dunes, beaches and estuary margins have experienced serious problems derived from the loss of sand material that have been taken to be used in coastal engineering, edifications and natural environmental degradation processes concerned to the sediment budget. In this context, this chapter focuses the description of the morphodynamics characteristics of the beaches of Ceara State, highlighting the factors controlling spatial and temporal processes, as well as discussing the impacts, potential uses and limitations of these areas. The results indicate that today this coastal erosion, either natural or induced by man, is perceived as the most significant threat to maintaining services in areas that depend upon tourism, traditional economic activities, housing and other pertinent uses of the beaches. It has been pointed out that the natural and human impacts represent the main and outstanding challenges for dwellers and the coastal managers who need to find out the new ways of living and occupation, includes redesigning the processes of house constructions and keeping up with minimizing the impacts. Then, the main actions that deal with the use and abuse of this littoral are discussed in this chapter. The environmental problems have to be strictly controlled, particularly stabilization of the mobile dune and preventing development of cliffs and headlands.

L.d.S. Pinheiro (✉)
Instituto de Ciências do Mar-LABOMAR- Federal University of Ceará,
Av.Abolição No. 3207, Meireles, 60165-081, Fortaleza, CE, Brazil
e-mail: lidriana.lgco@gmail.com

J.O. de Morais
Laboratório de Geologia e Geomorfologia Costeira e Oceânica-LGCO,
Universidade Estadual do Ceará, Av. Dr. Silas Munguba, 1700,
Campus do Itaperi, 60.714.903, Fortaleza, CE, Brazil
e-mail: jaderonofre@gmail.com

L.P. Maia
LABOMAR-Instituto de Ciências do Mar, Universidade Federal do Ceará,
Av.Abolição No. 3207, Meireles, 60165-081, Fortaleza, CE, Brazil
e-mail: parente@ufc.br

A.D. Short, A.H. da F. Klein (eds.), *Brazilian Beach Systems*, Coastal Research Library 17, DOI 10.1007/978-3-319-30394-9_7

7.1 Introduction

The State of Ceará lies between 2°S–7°S and 37°W–41°W and has an area of 146,000 km^2. It extends from the border of Piauí in the west to Rio Grande do Norte and Paraiba in the east; with Pernambuco forming the southern border (Fig. 7.1). The northern coastal zone is aligned in a general west-east direction from the estuary of the Timonha river to Itarema beach, it then trends northwest to southeast to Icapui municipality, at the Rio Grande do Norte border.

The coastline extends for 573 km, with the coastal zone having an area of 146,000 km^2 which represents 14.38 % of the state (IPECE 2013). Municipalities bordering the sea have the highest population densities in the Ceará state, ranging between 200 and 2000 inhabitants per km^2, peaking in the city of Fortaleza, the State Capital (IPECE 2013).

The coastal zone extends inland from 2 to 5 km to the landward limit of the dune fields. Wide sandy beaches, paleodunes, transgressive dunes, sand spits, estuaries with mangroves, canals, lagoons, and bedrock headlands make up the framework of the coast.

The fishermen, native people and maroon represent the traditional coastal communities. In the nineteenth century the production of sea salt was one of the main activities. Today the main economic activities are coastal tourism, aquaculture in estuaries, deep-sea fishing and industrial activities associated with the Pecem and Mucuripe harbours, and industrial complexes. Urban and coastal sprawl commenced in the 1980s, to support leisure and tourism activities with visitors from Ceará State, as well as from all over Brazil and abroad.

Property speculation and the land subdivision in the vicinity of towns and small fishing have been responsible for village stress and the coastal urbanization. This

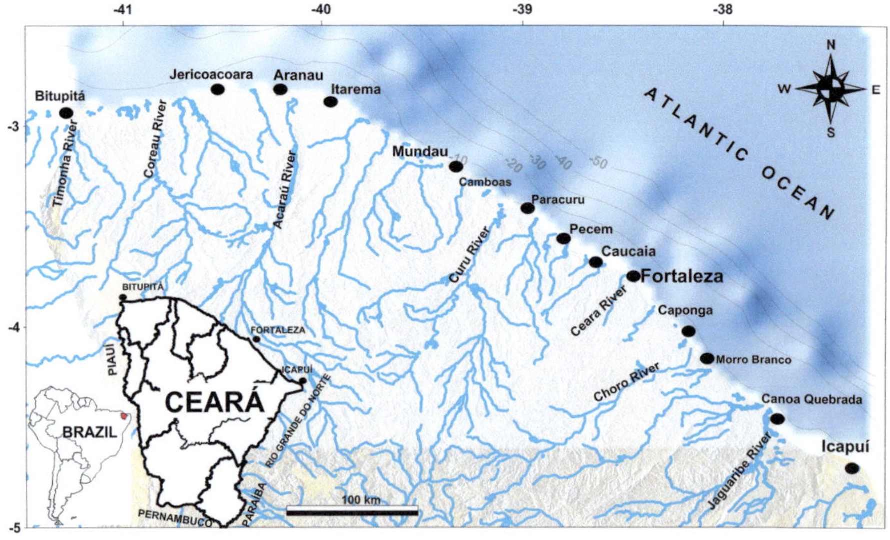

Fig. 7.1 The Ceará coastal zone

has resulted in a series of impacts which led to serious beach erosion, loss of urban infrastructure, cultural heritage, uncertainties in investment, as well as to economic depreciation of property and seaside services (Morais et al. 2006a, 2008).

The purpose of this chapter is to describe the main characteristics of the beaches of Ceará State, highlighting the factors controlling spatial and temporal processes and systems, as well as discussing the potential use and limitations of these areas.

7.1.1 Geology

The Ceará coastal zone is predominantly composed of sandy sediments of Upper Tertiary-Quaternary and contains several episodes of dunes, beaches, estuarine plains and active and paleo-sea cliffs (Fig. 7.2), all activated during period of high sea level. The Precambrian only occurs as headlands on the beaches of Pecem, Meireles, Iguape and Jericoacoara (Morais 2000).

The basal layer of the Barreiras Formation has been eroded to form rock platforms, cliffs and headlands at Morro Branco, Caponga, Bitupita, Taiba and Camocim, as well as elsewhere in northeast Brazil (Bittencourt et al. 2005). The sediments are supposed to be continental colluvial-alluvial origin and occur throughout the Northeast coast (Bigarella 1975). Rossetti et al. (2013) also discuss the marine origin of these deposits in some parts of the Northeastern Brazilian coast. The Formation also occurs along the coast in parallel sequences, as well as a sequence of transgressive episodes in Jericoacoara, Barra Nova and Canoa Quebrada (Irion et al. 2012; Morais et al. 2009). Where submerged, these cause wave refraction, diffraction and reflection of waves in the shadow zone of headlands and thereby play an important role in wave energy distribution, sediment transport and shoreline

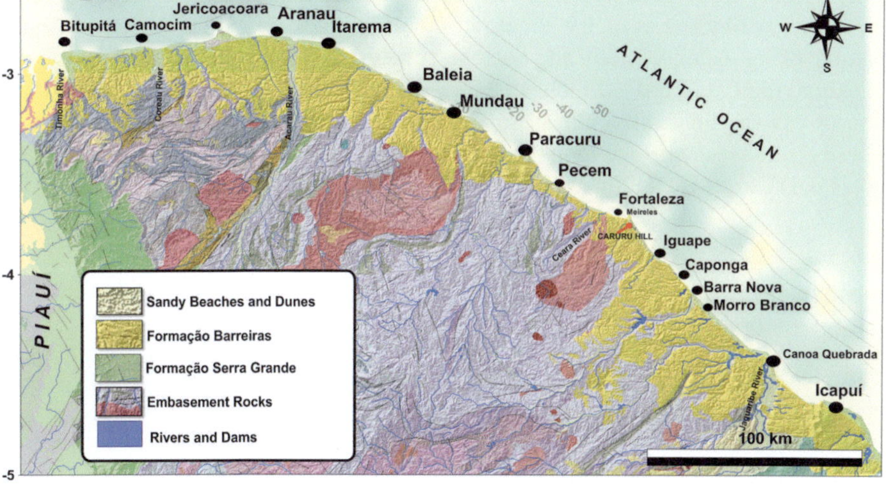

Fig. 7.2 Geological Map of Ceará coastal zone (Source: CPRM 2009)

alignment and erosion (Carvalho et al. 2007). Beachrock, is aligned parallel to the shoreline at the depth of 23 m, 30 km from the coast of Icapui on the eastern coast of the state, and appears to represent old shorelines (Monteiro and Maia 2010).

Consolidated cliffs are carved in limestones of the Jandaira Formation and sandstone of the Açu Formation of Cretaceous age at Icapuí and extend as far as Rio Grande do Norte. At Caruru hill near the mouth of the Pacoti estuary, volcanic rocks have filled areas of tension in the old Oligocene host rocks (28.6–9 Ma), with chemical and petrographic similarity to the phonolite of Fernando de Noronha (Vandoros and Oliveira 1968). This has been confirmed by Morais (1969) using a seismic profile from the archipelago of Fernando de Noronha to Fortaleza coastline.

7.1.2 Climate

The climate is warm semi-arid tropical (IPECE 2013). However, the Baturite, Maranguape and Pacatuba mountains (altitude > 700 m), influence the climate between the cities of Caucaia and Pindoretama, resulting in a tropical warm sub-humid climate. The intensity and frequency of the rainy season is dependent on the position of the Intertropical Convergence Zone (ITCZ). The ITCZ migrates in annual cycles bringing the rainy season to Ceará between March and May.

From July to November the ITCZ moves gradually to the north of the equator, bringing dry conditions, with approximately 91 % of the rainfall concentrated in the first half of the year. Campos and Studart (2003) observed that the annual rainfall rate decreases from Fortaleza (1338 mm) toward Icapui (949 mm). Toward the west coast there is also a decrease of the total rainfall, with annual averages of 650 mm.

The temperature ranges from 22 °C to 33 °C, with an average of 27 °C with the lowest temperatures during the rainy season. The wind regime is also strongly seasonal with lower wind velocities prevailing during the rainy season (average velocity 5.47 $m.s^{-1}$) and higher velocities during the dry season (average velocity 7.75 $m.s^{-1}$) (Jimenez et al. 1999). Wind direction does not show a clear seasonal pattern, being mainly easterly all year round due to the dominance of the trade winds.

Along the Ceará coast wind speed increases towards the northwest, with direction slowly varying from southeast to northeast towards the north (Maia et al. 2005). These changes may be due to the latitudinal position of each site with respect to the average ITCZ position (Jimenez et al. 1999).

7.1.3 Drainage

The rivers are intermittent and only reach the coast during the rainy season (Cavalcante and Cunha 2012). Six watersheds are located in the coastal zone: Coreaú Basin (annual run off: 1625 $hm^3\ yr^{-1}$), Acaraú Basin (annual run off: 17.4 $hm^3\ yr^{-1}$), Curú Basin (annual run off: 1.1 $hm^3\ yr^{-1}$), Litoral Basin (annual run off: 1.2 $hm^3\ yr^{-1}$), Metropolitan Basin (1.5 $hm^3\ yr^{-1}$) and Lower Jaguaribe Basin (705.6 $hm^3\ yr^{-1}$)

(established by COGERH, Hydric Resources Management company). Although the river headwaters are located in areas of higher rainfall than the coast, they flow through large areas of semi-arid climate, which causes a reduced freshwater to the coast due to evaporation and blockage by dams. There is a flow regularization during dry season to supply cities and agriculture needs. There are no perennial rivers.

Ceará State has more than 3500 dams larger than 5 ha (COGERH 2008), which means that across the state, the river flows are virtually blocked by reservoirs. Along the 299 km river course between the Castanhão Dam and Jaguaribe River mouth, in Aracati city, there are 15 dams to supply the population (Cavalcante and Cunha 2012). The dams have reduced the flow of fresh water during the rainy season by 80 % for the Timonha, Malcozinhado and Acaraú estuaries (Morais and Pinheiro 2011). The decrease in flow has resulted in silting of valleys, and prevents the transport of sediments to the coast during the rainy season thereby increasing coastal erosion processes. Freshwater from coastal sand dunes forms coastal lagoons and flow directly onto the beaches and form sandy cliffs with ephemeral channels.

7.1.4 *Waves and Tides*

Tides in Ceará are semidiurnal mesotidal with a maximum amplitude of 3.2 m (Morais 1981; Maia 1998). Waves are 80 % sea waves ($1 \leq T \leq 9$ s) and 20 % swell, with periods > 10 s (Carvalho et al. 2007). The swell waves reaching the Ceará coast, with peak period greater than 10 s are frequent between December and April (Silva et al. 2011).

In Pecem Port, on the west coast of Ceará, the significant wave height is (H_s) is predominately (75 %) between 1.1 and 1.6 m, with higher values from August to December (Silva et al. 2011). The highest recorded H_s was 2.3 m in October (Silva et al. 2011). In the Port of Fortaleza, waves with $0.8 \leq H_s \leq 1.7$ m account for 85 % of the occurrences (Maia 1998). At Aracati on the east coast of Ceará State, the most frequent H_s is between 1.3 and 1.73 m (Maia et al. 2005).

At Caponga, Futuro, and Paracuru beaches, the most frequent (75 %) breaking wave height (H_b) ranged 0.90–1.05 m, 0.50–1.1 m and 0.4–0.9 m, respectively (Pinheiro et al. 2003; Albuquerque et al. 2010; Souza 2005). The wave height is attenuated in embayed beaches and crossing longitudinal bars and rock outcrops. The predominant wave direction is east-southeast, but the most common swell direction of the first half of the year is north-northeast (Maia 1998; Carvalho et al. 2007).

7.2 Coastal Sediments

Beach sediments are predominantly medium bimodal quartz sand (Morais et al. 2006a). Figure 7.3 shows the distribution of D_{50} (mm) for 216 beaches. Fine sand dominates 32 % of the beaches, very fine sand and gravels 15 % and 8 %, respectively, while medium size sand is the most dominant on 45 % of Ceará beaches (Fig. 7.3).

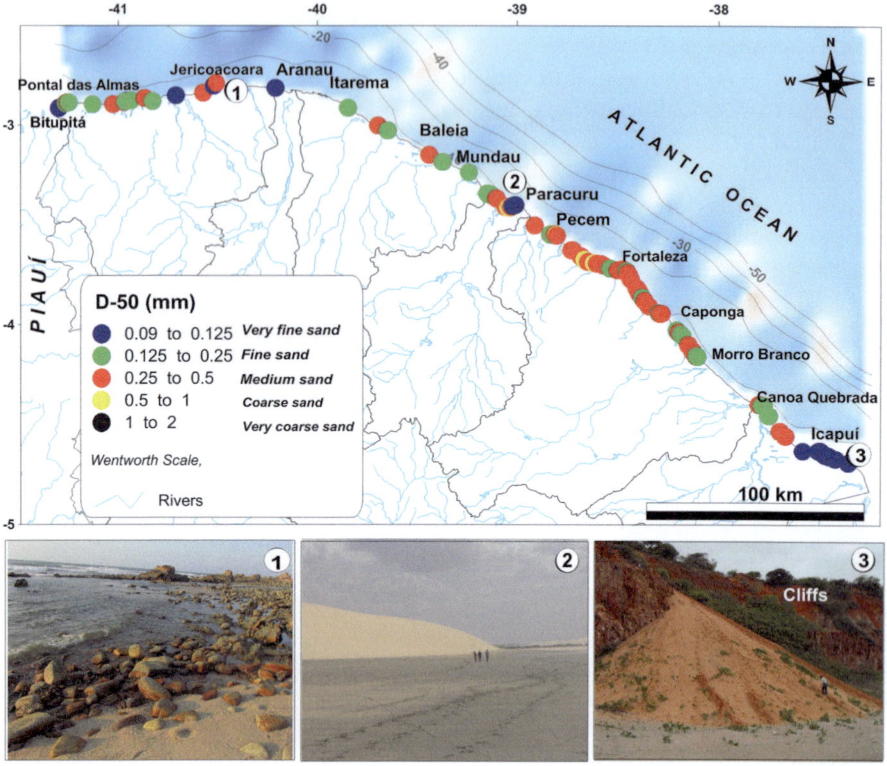

Fig. 7.3 Sediment size distribution (modal value of D_{50} (mm)) along the coast of Ceará. (*1*) Pebbles on the Jericoacoara beach; (*2*) Dunes migration towards the Paracuru beach; and (*3*) sand running down from the dunes on the cliffs top (Source: Data from the Coatal and Oceanic Systems Group coordinated by Morais, J.O- CNPq)

Longshore sand transport is to the west driven by east-southeast waves that occur 80 % of the year. Dunes also play an important role in sediment supply in some sectors of the coast, as observed in the headland overpassing at Jericoacoara, Paracurú, Icapui. Overwashing occurring during periods of high swell and contributes to a reduction in D_{50} at the Cascavel and Aquiraz beaches (Lima 2012; Maia 2014).

In areas with sandy-clay cliffs, that predominate along the east coast of the state, the rainfall in the first half of the year and groundwater discharge contributes to the input of fine sediments (silt and clay) to the beaches of Redonda and Peroba (Barros 2013; Oliveira 2012). On the west coast the presence of fine sediments is associated with proximity to the mouths of river estuaries and lagoons. Lithoclastic gravels are observed near the outcrop of conglomeratic rocks on the beach of Taiba, while the beachrock at Boi Choco, Iparana and Pacheco has resulted from marine abrasion.

Bioclastic gravels are found on the far west coast of Ceará and are cemented by calcite (Morais and Pinheiro 2005). The Fortaleza city coastline has predominantly medium siliciclastic sand out to the 15 m isobath. The inner shelf environments are characterized by the occurrence of terrigenous sediment that extends from the low-water mark to a depth of 10–20 m (Coutinho and Morais 1970; Kempf et al. 1967; Morais 2000; Freire and Cavalcante 1997). At the contiguous shelf break are thin bioclastic sands derived from the fragmentation of algae *Halimeda incrassata* and *Lithothamium* (Kempf et al. 1967; Freire and Cavalcante 1997).

7.2.1 Coastal Provinces and Geomorphology

Morais (2000) and Bensi (2006) divide the Ceará coast into two macro-compartments located east and west of Fortaleza. The eastern sector is characterized by a rocky coast with abrasion platforms and tabular sandy-clay deposits derived from cliffs and paleocliffs. Marine and rain erosion of the cliffs delivers fine-grained sediments (silt and clay) to the beach as can be seen at Morro Branco, Praia das Fontes and Redonda beaches (Morais et al. 2006c; Oliveira 2012). The west coast is dominated by mobile dunes, barrier islands backed by lagoons and mangrove areas (Bensi 2006; Morais et al. 2006a). The predominance of aeolian deposits is favored by the topography and angle of arrival of easterly waves that supply of sediments to the beaches and dunes.

The headlands are composed of pre-Cambrian rocks, beachrock and cliffs of consolidated Barreiras Formation, and Tibau, Camocim and Cretaceous rocks that comprise the sedimentary Potiguar basin (Morais 2000). These form a sequence of embayments with a flat concave shape facing the sea, resulting wave alignment following wave refraction, diffraction and reflection around and in lee of the headland area.

7.3 Ceará Beach Systems

Morais et al. (2006a) divided the 573 km long coast into five sectors, based on watershed boundaries, sediment delivery and the morphological characteristics of the beaches. Within each sector beach morphology, type of occupation of the beach and homogeneity in the modal values of H_b, T, D_{50} (mm) and tidal range, based on 216 data points, were used to characterize the sector. The modal morphodynamic feature of beaches was assigned based on Ω and RTR parameters described by Short (1999). The sectors were divided as follows: (a) Sector I (SI): Pontal das Almas – Aranaú (≈98 km); (b) Sector II (SII): Aranaú – Paraipaba (≈155 km); (c) Sector III (SIII): Curú river mouth – Pecém Port (43 km); (d) Sector IV (SIV): Cumbuco – Choró river mouth (≈162 km); and (e) Sector V (SV): Rio Choró – Requenguela (≈115 km) (Fig. 7.4). These sectors are used in the following sections to describe the beach systems and their behavior.

7.3.1 Beach Types and Morphodynamic States

Ceará beaches are predominantly tide-modified (RTR > 3). This is to be expected in a moderate wave energy mesotidal environment with a spring tide range of 3 m (Masselink and Short 1993). Considering the modal values of the sediments,

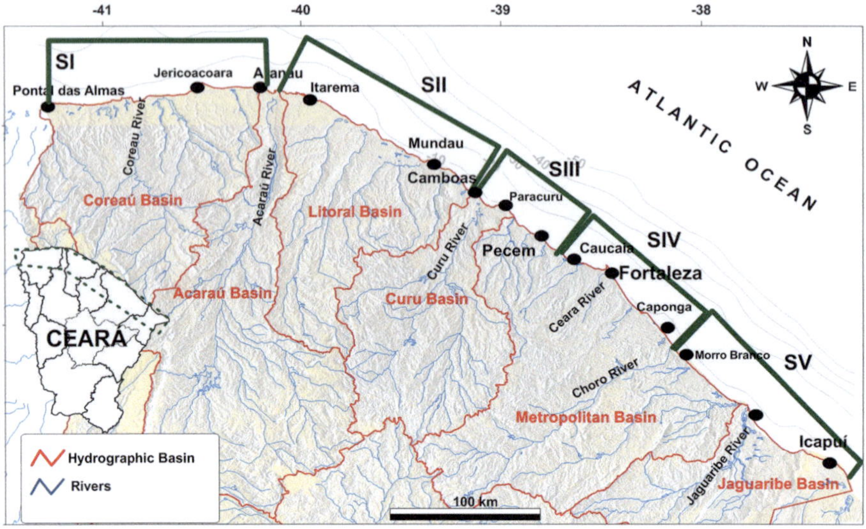

Fig. 7.4 The five coastal sectors, major drainage basins and the name of beaches mentioned in the text

Table 7.1 Ceará beach morphodynamics state

	Beach state	Number beaches	%	Beach (width) (average – m)	SD (m)
	Wave dominated ($\Omega = Hb/WsT$)[a]				
1	D (Dissipative)	3	1.3	183.3	23.6
2	LBT (Longshore bar & trough)	22	10.1	207.3	5.4
3	RBB (Rhythmic bar & beach)	1	0.4	150.0	–
4	TBR (Transverse bar & rip)	2	0.9	200.0	–
5	LTT (Low Tide Terrace)	2	0.9	200.0	–
	Sub-Total	30	13.9	–	–
	Tide-modified (RTR)[b]				
6	R+LTT (Reflective plus Low Tide Terrace)	47	21.8	352.4	197.7
7	R+LTR (Reflective plus Low Tidal bars and rips)	56	26.0	158.5	106.8
8	UD (Ultradissipative)	1	0.46	>600	–
	Sub-Total	104	48	–	–
9	Beach plus rock/reef flats	82	38	150	32.7
	Total	216	100		

[a, b]Modal Values of the Hb (m), T (s), Ws and tide range, SD = standard deviation

together with the waves and tidal characteristics at the 216 monitored points, 48.15 % of the beaches were classified as tide-modified beaches, 13.85 % as wave-dominated beaches and 38 % beaches with rock-flats (Table 7.1). The tide-modified beaches are reflective with low tidal bars and rips (R+LTR) (26 %), reflective and low tide terrace (R+LTT) (21 %) with only 0.5 % ultradissipative (UD). The wave-dominated beaches are primarily longshore bar and trough state (LBT) (10 %). Beach width varies along the coast, with the widest observed on UD (mean>600 m), followed by R+LTT (mean=352.4 m, sd ±197.65 m) and R+LTR beaches (mean=158.5 m, sd±106.84). The narrowest beaches occur at Caucaia and Fortaleza. Praia do Futuro in Fortaleza was rated as transverse bar and rip (TBR), and is characterized by the sequences of bars and rip currents that can lead to accidents and drowning.

A number of these wide flat beaches including Redonda, Bitupita and Maceió beaches are eroded during spring tides, while low frequency swell waves from north-northeast with periods greater than 10 s are responsible for flooding along several beaches.

7.3.2 Coastal Processes

The morphological characteristics of the shoreline depend on coastal hydrodynamic processes across the inner shelf and shoreface, such as wave attenuation, refraction and diffraction across outcrops, spits and sand banks and within embayments. The **sector I** between the Pontal das Almas beach and Aranaú (SI) is aligned east-west and is approximately 98 km long. The average width of the beach profile is 530 m (±119 m). Sediment size ranges from very fine to medium sand and the predominant state of the 17 beaches in this sector was R+LTT (Figs. 7.5 and 7.6).

At high tide the waves pass over the terrace and only break on reaching the base of high tide beach, similar to the reflective wave-dominated beach. As the tide falls,

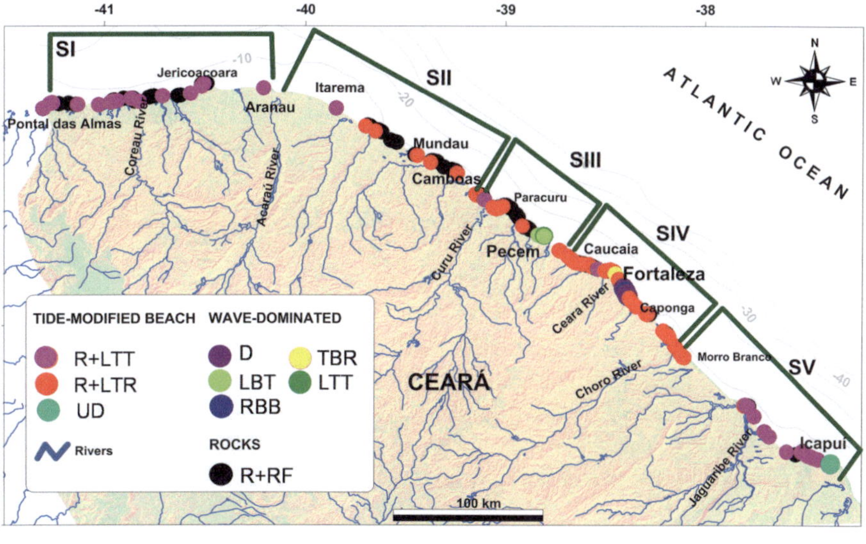

Fig. 7.5 Map showing the predominant morphodynamic states along the coast

Fig. 7.6 Sector I beaches: Reflective plus Low tide terrace beach at Camocim (**a**) and Bitupita (**b**)

waves begin to break across the terrace and at low tide they break on the outer edge producing a wide, shallow dissipative surf zone across the terrace. If rips are present, they will cut a channel across the terrace and are only active at low tide (Short 1999). Low tide also exposes the steep coarser beach face that abruptly connects to a flat low tide terrace with fine sediments, that extends for tens of meters seaward (Masselink and Short 1993; Short 1999). This region receives the discharge of the Coreaú and Acarau rivers during the rainy period, and other small drainages. Coastal lagoons formed by the closure of river mouths occur along with the dunes and sand spits.

Fig. 7.7 Sector II beaches: (**a**) Low tide terrace beach R+LTT at Almofala Beach; and (**b**) Low tide with rips with sand bars at Mundau Beach

Sector II extends from Aranaú to Camboas with the R+LTR beach state the prevalent. In the northwest, between Almofala and Aranaú, the beaches are characterized by west-trending barrier spits dominated by the R+LTT state (Fig. 7.7a), which at Itarema beach is 900 m in width. From Almofala to the mouth of the river Acaraú, the coast is characterized by sequences of coves with rocky spits dominated by the R+LTR state. High tide cusps are common and the low tide terrace may be dissected by small rip channels (Masselink and Short 1993). At high tide, surf zone processes are similar to reflective beaches with waves breaking or generally surging up the steeper beachface whereas during low tide the surf zone will be dissipative across the terrace with several lines of spilling breakers (Short 1999). The beach sediments range from fine sand to medium sand, with a predominance of fine sand suitable for aeolain transport and the average width of the beach profile is 396 m (sd±216.6 m). Aeolian transport is driven by east-southeast winds, which remobilization and transport of beach sediment into the dunes, with transgressive dune fields extending up to 4 km inland.

Also during the low tide, groundwater outcrops on the beach at the saturation point on the low tide terrace. The dissipative beaches are fed by fine sediments from sand dunes and the wave height is reduced by attenuation over the beachrock reefs which parallel the shoreline at the Mundaú and Fleicheiras beaches (Fig. 7.7b).

Sector III between the mouth of the Curú River and Pecem Harbor is 43 km long and aligned northwest-southeast. The morphology is similar to the previous Sector with sequences of spits, inlets and transverse dunes. About 73 % of the beaches are tide-modified with RTR > 3, with an average beach width of 216 m (sd ± 29 m). Pecém harbor is located offshore of Pecém beach through which are exported the raw material and products generated in the industrial complex located in the municipality of São Gonçalo do Amarante. This Port occupies a strategic position because it is equidistant from Argentina, Africa, Europe and North America for commerce and transport. The port also exports fish, leather and fruit produced in the Brazilian Northeast (IPECE 2013).

From the Curú river mouth to the Taiba beach R + LTT predominate (21 %). The sediments in this sector range from very fine sand to coarse sand with modal H_b of 0.90 m. There are also 15 km of beachrock coast and beaches (Fig. 7.8). Taiba's beach is predominantly R + LTR (Dias and Rocha-Barreira 2011). From the Taiba beach to Cauípe river's mouth, the wave-dominated LBT beach state (RTR < 3 and $\Omega \approx 5$) occurs on 66 % of the beaches, followed by LTT (9 %).

Sector IV is located between Cumbuco (Barraca das Velas) and the mouth of Choró river and has the highest population density on the Ceará coast. About 80 % of the beaches have RTR > 3, with a predominance of tide-modified beaches, the remainder are wave-dominated. The coast between Cumbuco and Iparana beaches is aligned northwest-southeast with beach sediments ranging from medium to coarse sand and a predominance of coarse sand (D_{50} > 0.65). R + LTR beach state was predominant on 12 beaches on the basis of data collected between the years 2002 and 2012.

Fig. 7.8 Sector III: Beachrock at Taiba beach

The presence of a reflective high tide beach and a more dissipative beach at low and intermediate tides is very common on all the R+LTR beaches. Beach face scarping commonly occurs at high tide but with minimal changes in the beach face. Pacheco and Iparana beaches have suffered environmental degradation owing to their proximity to the Fortaleza coastal zone and its adjoining coastal areas, which has resulted in a narrower beach backed by seawalls and buildings (Fig. 7.9). The beaches range from 60 to 120 m wide at low tide (Paula et al. 2013). Between the Iparana beach and the mouth of the Ceará River rock seawalls are prevalent with the sector between the Ceará River mouth and Mucuripe Port dominated by a sequence of 22 breakwaters, including the Titan at Mucuripe Port.

The presence of a reflective high tide beach and a more dissipative beach at low and intermediate tides is very common on all the R+LTR beaches. Beach face scarping commonly occurs at high tide but with minimal changes in the beach face. Pacheco and Iparana beaches have suffered environmental degradation owing to their proximity to the Fortaleza coastal zone and its adjoining coastal areas, which has resulted in a narrower beach backed by seawalls and buildings (Fig. 7.9a and b). The beaches range from 60 to 120 m wide at low tide (Paula et al. 2013). Between the Iparana beach and the mouth of the Ceará River rock seawalls are prevalent and the sector between the mouth of the Ceará River and the Mucuripe Port is dominated by a sequence of 22 rock groins (Fig. 7.9c).

The coast is aligned east-west with a predominance of medium sand and R+LTR beaches. Overwash episodes have been observed at Diarios beach when there is an increase of the occurrence of longer period (>10 s) north-northeast swell waves (Paula 2012; Guerra 2014,). Between the Titan jetty beach and Barra Nova beach in Cascavel, 60 % of the beaches have an RTR>3 and are R+LTR. R+LTR and R+LTT types are prevalent on Caponga and Aguas Belas beaches, respectively (Pinheiro et al. 2003; Morais et al. 2006b).

Ten percent of the beaches have an R<3, particularly Praia do Futuro, Cofeco, Porto das Dunas and Prainha (Albuquerque et al. 2009). Over a period of 24 months the 8 km long Futuro beach was wave-dominated with TBR conditions in the first half of the year and LBT in the second half.

The **sector V** is characterized by the presence of sandy and rocky cliffs bordering and backing the beaches, that extend to the Rio Grande do Norte border (SV) (Fig. 7.10). R+LTT prevails on 96 % of the beaches with an average beach width of 208 m (s.d. 94 m). The coastline between the mouth of the Choró River and Requenguela is 115 km in length and predominantly composed of very fine to medium sandy sediments. Very fine sand is prevalent all year between Retirinho and Requenguela beaches (Oliveira 2012; Barros 2013). Between the Choró and Pirangi river mouths there is a straight cliff coast.

West of Requenguela is an UD beach (Fig. 7.10a). The beach is 800 m wide at low-water spring tide, and is composed of very fine grain size sand and coarse silt, with some mud banks colonized by mangroves in the intertidal zone. Canoa Quebrada is a very well known and popular resort, receiving visitors from many regions of Brazil and abroad. It has experienced active erosion in the Barreiras Formation either from waves or from surface and groundwater action (Fig. 7.10b). It is now being protected by rockwalls.

Fig. 7.9 Sector IV beaches: (**a**) Erosion at Icaraí's beach; (**b**) Low tide terrace at Nautico beach – Fortaleza-City; and (**c**) Disispative beach at Serviluz beach

Fig. 7.10 Sector V beaches: (**a**) Cliffs and ultradissipative beach at Ponta Grossa beach; and (**b**) Low tide bars and rips at Canoa Quebrada beach

7.3.3 Beach-Barrier

The first studies on barriers islands in Ceará were made by Morais and Fonteles (2000), together with the Quaternary evolution of the coastal plain and barrier islands of Itarema. Hesp et al. (2009) provide an excellent description of the barriers in the state of Ceará. They divided them into three forms: attached barriers, barrier spits and prograded foredune plains barriers. Attached barriers occur across the mouths of rivers and estuaries and when the direction of the shoreline changes to

Fig. 7.11 Examples of sandy spit and wind farm on the the coast of Ceará

east-west. The presence of promontories traps sediment transported by longshore drift, which accumulates and is subjected to wind blown transport, forming large transgressive dunes, such as barchans observed in Jericoacoara and other parts of the coast.

Barriers spits are well developed from the mouth of Acaraú river, at Itarema and along the western coast. They occur on the east coast in lee of Ponta Grossa Iguape beach. Between 2001 and 2004 Morgado and Volta do Rio (Acaraú) spits migrated 18 m and 124 m, respectively, eroding the shoreline (Hesp et al. 2009). Wind farms have been built on sandy spits at Itarema beach. Groins were also built to contain waters stock for the wind turbines (Fig. 7.11).

Prograded foredunes plain barriers are characteristic of the coast of Icapui village where the waves arrive at an angle of 45° to the shoreline which is aligned northwest-southest. This results in a reduction in sediment transport and favors the formation of regressive foredunes ridges. (Hesp et al. 2009).

7.3.4 Beach-Dune Interactions

The dunes in Ceará extend inland from beaches composed of medium and fine quartz sands. Storm waves deliver the sediment, which is blown inland from the beach face as occurs at Paracurú, Icapui, Jericoacoara, Caponga, Iguape, and others beaches. In some cases the aeolian transport acts as an erosive agent, moving sediment inland form the beach to the dune fields (Fig. 7.12a). Geological records show that the formation and migration of dunes on the North-Northeast coast has occurred since the Pleistocene (Claudino-Sales 2002). During this period, the wind has acted

Fig. 7.12 (**a**) Barchans at Jericoacoara beach; (**b**) Aeolianite exposed at Baleia beach; (**c**) Paleodunes near the Choró River; (**d**) small barchan dune migrating across a beach at Itarema; and (**e**) dune slipface moving due to gravity at Icapuí

either as an erosive agent, transporting sediment from the beaches via headland overpassing, or indirectly by the input of sediments in waterways, and more recently advancing as dunes into urban areas.

The Ceará coast has paleodunes, and both fixed and mobile dunes extending inland of the deflation zone (Maia 1998; Claudino-Sales 2002). The subaerial exposure of the wide tide-modified beach face is a characteristic of Ceará beaches that favours aeolian sand transport, with sediment transported inland from the beach face (Fig. 7.12d). Dune forms include barchans, isolated barchans, longitudinal, parabolic, parabolic semi-fixed, hairpin, aeolianite, sand sheet and nebkas (Maia 1998; Maia et al. 2012) (Fig. 7.12b, c). The barchans are most common in the north-eastern coastal dunes, from Maranhão to the border with Parnaíba. They are composed of fine sand dunes and form where there are changes in roughness of the terrain or aerodynamic fluctuations (Pye and Tsoar 1990). They are up to several hundred meters width and length, with an average height of 20 m, while on Iguape beach they reach 30 m. The dunes dominate the seaward edge of the coastal plains and are migrating at approximately 6 m year^{-1} in the case of sand sheets without defined shapes and from 9 to 20 m year^{-1} in the case of fields of barchan dunes (Maia 1998; Jimenez et al. 1999; Levin et al. 2008).

The dunes are responsible for obstructing drainage of medium to small channels resulting in the transformation of estuarine systems into estuarine-lagoonal, lakes and ponds, such as can be seen at Jijoca, Catu, Mundaú and Malcozinhado (Pinheiro et al. 2006; Morais and Pinheiro 2011). Tsoar et al. (2009) identified fixed dunes on the Ceará coast with ages of about 132 ka, while mobile transverse dunes were dated 1320 ± 50 BP by Castro and Ramos (2006).

On beaches with dissipative characteristics and extensive terraces ebb barchans, longitudinal dunes and sand sheets predominate (Fig. 7.12d). On beaches dominated by waves, with intermediate characteristics, the frontal dunes, with a height of approximately 1.5 m, predominate. In sectors with cliffs, dunes are located on the clifftops, forming slipfaces that feed the downwind beaches. These dunes also move by both gravity effects and aeolian transport (Fig. 7.12e).

7.3.5 Longshore Sand Transport and Sediment Cells

The orientation of the coast in a northwest-southeast alignment, together with persistent easterly waves and trade winds blowing between 5 and 8 m s^{-1}, combine with sediment availability and a wide surf zone to generate substantial westerly longshore transport along the coast. In Sector I between Pontal das Almas and Acaraú where the coastline has an east-west direction, the wind takes an offshore component, and the flow of water also has a tendency to move away from the coast creating optimal conditions for sedimentation.

Modelling of wind driven surface flow performed by Bensi (2006) found that the longshore currents parallel isobaths with an average velocity of 0.13 m s^{-1} and ranging from 0.16 to 0.24 m s^{-1} between the Baleia and Fortaleza beaches (Sectors II and III) and 0.15–0.20 m s^{-1} between Morro Branco and Icapui (Sectors IV and V) (Fig. 7.13).

On the coast between the beaches of Cumbuco and Future (Sector II) surface velocity of coastal currents varies from 0.18 to 0.68 m $year^{-1}$ (Maia 1998; Silva et al 2009). Jimenez et al. (1999) estimated sediment transport on the order of 860,000 m^3year^{-1} on the coast of Fortaleza. Using the empirical model of the CERC Department of the Army (Shire Protection Manual 1984), with breaker wave height of 0.94 m, period of 6 s and arriving at an angle of 70 ° Pitombeira and Aquino (1998) obtained a rate of transport in Pecem of 854,307 m^3year^{-1}.

In the experiments on the Barra do Ceará, Meireles and Futuro beaches using ADCP over the spring tide cycles average rates of transport of 1.3 million m^3year^{-1} have been calculated for the period 2011 and 2012 (Alves 2012). In Caponga (Section IV), the velocities of currents varied from 0.3 to 0.6 ms^{-1}, with transport rates ranging from 280,000 m^3year^{-1} per year to 1.4 million m^3year^{-1} (Pinheiro et al. 2001; Morais et al. 2006b).

In the regional context of current velocities and rates of sediment transport on the coast, macro-compartments are homogeneous and reflect the behavior of local winds and waves. The velocity of the currents are modulated by changes in coastal orientation and morphology of the continental shelf and slope similar to what observed in the Sector I. On a smaller scale headland bypassing and trapping occurs with accumulation of sediments to east and erosion to the west. The sediment

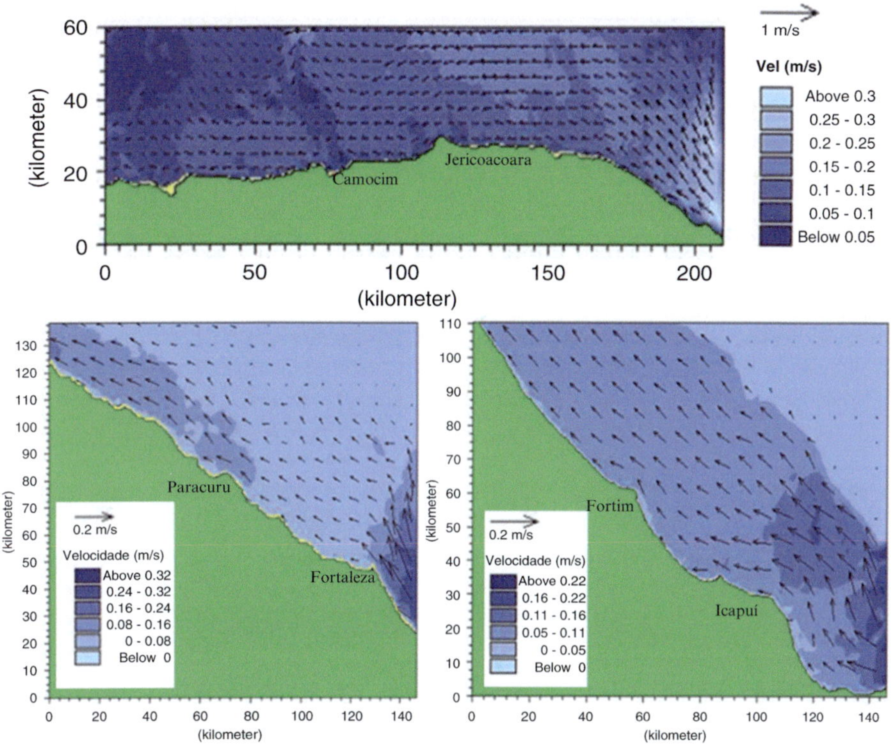

Fig. 7.13 Simulation of the surface current velocities along the coast of Ceará State (Source: Bensi 2006)

trapped by the headlands is also a source of sand for dune development and headland overpassing.

7.4 Beach Use and Abuse

7.4.1 Beach Development and Management

In recent years the Brazilian coast zone has suffered environmental degradation, generated by increasing anthropic pressures and a low capacity to absorb these impacts. The occupation of the Ceará coastal area is primarily for the industry and tourism. On the Ceará beaches, second holiday residences were the main driver of coastal development. During the seventies and eighties the coast was sought after by tourism investors for the construction of lodgings and hotels to accommodate the tourism season (Paula et al. 2012).

The dunes, beaches and estuary margins were transformed and affected by the urbanization. Today, coastal erosion is locally perceived as the most significant threat to tourism income as well as traditional economic activities. There are 20 locations experiencing coastal erosion Ceará (Morais et al. 2006a), with some locations experiencing very high rates of erosion which is resulting in serious damage to urban infrastructure as can be seen at Bitupita, Maceio, Taíba, Icaraí, Caponga, Iguape and Redonda beaches.

The Fortaleza Mucuripe Harbour impacts are a classical example of the lack of knowledge of coastal processes on Ceará's coast. When the harbour was built in 1939, the jetty changed the wave refraction, stopped the longshore transport trapping sediment at Serviluz beach. At the same time there was a lack on sediment transport downdrift to the Fortaleza coast. As a consequence the Fortaleza's beaches eroded about 200 m between Mucuripe Harbour and Ceará estuarine-river mouth (Morais 1981). In response groins were built which effected the Caucaia's coast. A seawall was been built at Iparana and Pacheco beaches to combat to the erosion. At Icaraí, Pacheco, Caponga, Mundaú, Canto Verde and Requenguela there has been damage to roads, sidewalks, kiosks, sheds, and entertainment areas (Fig. 7.14). There are also problems like rocks, beach erosion, spoils, groins, garbage and difficult beach access (Pinheiro and Rocha 2007; Medeiros 2012; Medeiros et al. 2014).

The recent "Vila do Mar" project aims to rehabilitate the 18 km of coast between the Rio Ceará mouth and Nossa Senhora das Graças neighbourhood (Fortaleza City). It includes new groins, fishing ports, sidewalks, parking lots and places for sports and leisure for beach users. A similar project occurred along the central coast of Fortaleza, with rehabilitation of the Iracema and Diarios beaches.

Among the major subjects discussed at the Orla project (Brasil 2006) were coastal erosion and the health of the beaches. It turns out, however, that many municipalities do not have the Orla project or other legal obligations to protect the coast. Constant pressure is required to prevent and mitigate the negative impacts of coastal erosion and the adoption of measures to maintain or restore the beaches as a place of recreation and economic opportunities. A more strategic and proactive approach to coastal erosion is needed for the sustainable development of vulnerable coastal zones.

7.5 Beach Hazards and Safety

Ceará has a major tourism industry, attracting many tourists in the high season periods of July, December and January, with the coastal zone accounting for about 80 % of the tourists. These tourists face a number of hazards along the coast including beach rip currents, topographic rips, groins and rock walls. Praia do Futuro, the most popular touristic beach in Fortaleza is a R + LTR beach and consequently hazardous for swimming (Albuquerque et al. 2010). The high level of risk is reflected in an average of 314 people rescued annually by lifeguards without fatalities

Fig. 7.14 Severe erosion at Icaraí beach hitting urbanization

between 2002 and 2010, while during 2013 there were 456 rescues and seven fatalities at the beach. Albuquerque et al. (2010) confirmed that the main type of seasonal hazard associated with Futuro is the rip currents, which are responsible for 86 % of rescues in sectors with TBR and RBB morphology. The main victims of drowning were men (66 %) aged between 21 and 28 years, about 50 % of those not living in Fortaleza. The high number of beach users is related to the good bathing waters compared to the other beaches in the city, and the provision of services and facilities at Futuro beach.

Among the main factors that increase the risk of accidents and drowning are: the reduced number and poor siting of lifeguard towers and number of lifeguards in relation to the number of users on the beach; there are also no signals using flags or other information in public places and in the media to warn the public about the beach hazards and risk.

Caucaia town and its municipality were identified as having the highest level of beach risk. The shore platforms, coastal protection works, the beachrock at Parcheco and Iparana and house debris that are submerged in high tide, all present substantial permanent hazards to beach users according to the methodology proposed by Short and Hogan (1994). On tide-modified beaches the main risks are associated with breaking waves at high tide and the formation of rip currents and currents in the surf zone, particularly around low tide.

7.6 Summary and Conclusions

The Ceará coastal zone is predominantly composed of sandy sediments of Upper Tertiary-Quaternary age with several generations of Pleistocene transgressive dunes, together with beaches, estuarine plains and localized occurrences of cliffs. The Precambrian rocks also occur as headlands on some beaches. The climate is semi-arid tropical, while the rivers only flow into the sea during the rainy season. The temperature ranges from 22 °C to 33 °C, with an average of 27 °C. The easterly trade winds prevail year round.

Tide is semi-diurnal with a maximum range of 3.2 m. Sea waves arrive about 80 % of the time with a modal wave period of 6.5 s, while swell occurs for 20 % of the year, with periods between10 to 20s. The dominant waves (75 %) are between 1.1 and 1.6 m, with seas arriving from the east-southeast, and swell from the north-northeast. The sediments are predominantly medium sized bimodal quartz sand.

Ceará beaches are predominantly tide-modified (RTR > 3), with approximately 48.15 % of the beaches classified as tide-modified, 13.85 % as wave-dominated and 38 % beach plus rock/reef flat. Wave-dominated are primarily longshore bar and trough (LBT) occurring on 10 % of the beaches. Beach width varies along the coast with the widest beaches associated with R + LTT and UD beaches.

Coastal dunes include paleodunes, fixed and mobile dunes, which occur beyond the deflation basins. The rates of migration of sand dunes is approximately 6 m $year^{-1}$ in the case of sand sheets without defined shapes and from 9 to 20 m $year^{-1}$ in fields of barchans dunes. Sediment transport is on the order of 860,000 m^3 $year^{-1}$ at Fortaleza. Inbalance in the sediment budget occur where urban construction obstructs the headland overpassing of sediment and the longshore transport on exposed beaches.

Today, coastal erosion is perceived as the most significant threat to maintaining income in many areas that depend on tourism and traditional economic activities. Dunes, beaches, dams and estuary margins have been transformed and affected by construction without concern for the sediment budget. This has led to an increase in beach erosion over the years. This is a challenge for both coastal dwellers and the coastal managers who need to find new ways of living with the coast, including redesigning coastal occupation as well as minimizing the impacts of the sea.

Living with coastal erosion should be on the agenda of public policies in the state, with guaranteed resources to implement the most appropriate works, maintenance and long-term monitoring. The engineering works need to look beyond just blocking the advance of the sea and need to be redesigned to accommodate different types of beach uses. While the demand for industrial enterprises, growth of cities and the development of tourism near the sea continues, there is an information gap that requires Ceará to invest in long-term studies of coastal dynamics and management.

References

Albuquerque MG, Calliari LJ, Correa ICS, Pinheiro LS (2009) Morfodinamica da Praia do Futuro, Fortaleza-CE: uma síntese de dois anos de estudos. Quat Environ Geosci 1:49–57

Albuquerque MG, Calliari LJ, Pinheiro LS (2010) Analise dos principais riscos associados ao banho de mar na praia do Futuro, Fortaleza-Ceará. Braz J Aquat Sci Technol 14:1–8

Alves RT (2012) Estudo da deriva litorânea em Fortaleza/Ce: análise quantitativa do transporte longitudinal para gestão costeira e portuária. Dissertation, Federal University of Ceará

Barros EL (2013) Caracterização Faciológica da Plataforma Continental Interna de Icapuí, CE. MSc. Dissertation, Federal University of Ceará

Bensi M (2006) Aplicação do correntômetro acústico ADCP em ambientes marinhos e estuarinos do Ceará e Paraíba – nordeste do Brasil. Dissertação de Mestrado, 134p

Bigarella JJ (1975) The barreiras group in northeastern Brazil. Anais Acad Bras Cienc 47:365–393

Bittencourt ACSP, Dominguez JML, Martin L, Silva IR (2005) Longshore transport on the northeastern Brazilian coast and implications to the location of large scale accumulative and erosive zones: an overview. Mar Geol 219:219–234. doi:10.1016/j.margeo.2005.06.006

Brasil – Ministério do Meio Ambiente (2006) Projeto Orla-Fundamentos para a gestão integrada. http://www.mma.gov.br/estruturas/orla/_arquivos/11_04122008111238.pdf. Accessed 05 Nov 2013

Campos JNB, Studart T (2003) Climatologia. In: Campos AA (ed) A zona costeira do Estado do Ceará: diagnóstico para a Gestão Integrada. Aquasis, Fortaleza, pp 51–53

Carvalho AM, Maia LP, Dominguez JML (2007) A deriva e o transporte litorâneo de sedimentos no trecho entre Cumbuco e Matões-Costa noroeste do estado do Ceará. Arq Cienc Mar 40:43–51

Castro JWA, Ramos RRC (2006) Idade das dunas móveis transversais do litoral setentrional do Nordeste brasileiro. Arq Mus Nacion 64:361–368

Cavalcante AA, Cunha SB (2012) Morfodinâmica Fluvial em Areas Semiaridas: discutindo o Vale do rio Jaguaribe-CE-Brasil. Rev Bras Geomor 13:39–49

Claudino-Sales V (2002) Les littoraux du Ceará – evolution morphologique de la zone cotiere de l'Etat du Ceará, Nord-est du Bresi, du long terme au court terme. Ph.D. thesis, Universite Paris-Sorbonne

Cogerh-Hydric Resources Management company (2008). http://portal.cogerh.com.br/downloads/category/21-seguranca-de-obras. Accessed 12 May 2013

Coutinho PN, Morais JO (1970) Distribucion de los sedimentos en la plataforma continental Norte y Nordeste del Brasil. Arq Cienc Mar 10:79–90

CPRM-Geological Brazilian Survey (2009) Geobank. http://geobank.sa.cprm.gov.br/. Accessed 02 Dec 2013

Dias ICC, Rocha-Barreira CA (2011) Comportamento morfodinâmico da praia da Taíba, Nordeste do Brasil. Rev Gest Cost Integ 11:421–431. doi:10.5894/rgci256

Freire GSS, Cavalcante VMM (1997) Cobertura Sedimentar da Plataforma Continental do Estado Ceará. DNPM, Fortaleza, pp 1–45

Guerra RGP (2014) Vulnerabilidade costeira a eventos de alta energia no Litoral de Fortaleza, Ceará. 2014. MSc, Dissertation, Federal University of Ceará

Hesp P, Maia LP, Sales VC (2009) The holocene barriers of Ceará, Piauí and Maranhão States, Northeast of Brazil. In: Dillenburgh S, Hesp P (eds) Geology of holocene Brazilian barriers. Springer, Amsterdan, pp 326–347

IPECE-Instituto de pesquisa e estrategia econômica do Ceará (2013) Ceará em números. http://www.ipece.ce.gov.br/categoria5/ceará-em-numeros. Accessed 10 Nov 2013

Irion G, Morais JO, Bungenstock F (2012) Holocene and Pleistocene sea-level indicators at the coast of Jericoacoara, Ceará, NE Brazil. Quat Res 40:1–15. doi:10.1016/j.yqres.2011.10.007

Jimenez J, Maia LP, Raventos JS, Morais JO (1999) Aeolian dune migration along the Ceará coast, north-eastern Brazil. Sedimentology 46:689–701. doi:10.1046/j.1365-3091.1999.00240.x

Kempf M, Coutinho PN, Morais JO (1967) Plataforma Continental do Norte e do Nordeste do Brasil: nota preliminar sobre a natureza do fundo. Trab Oceanogr 11:9–26

Levin DN, Tsoar H, Maia LP, Claudini-Sales V, Herrmann H (2008) Lagoas costeiras interdunares e branqueamento de dunas do Nordeste Brasileiro. Mercator – Rev Geografia 14:43–166

Lima RS (2012) Vulnerabilidade da linha de costa a eventos de alta energia na praia da caponga, Cascavel-Ceará. Dissertation, Federal University of Ceará

Maia LP (1998) Proceso costeros y balance sedimentario a lo largo de Fortaleza (NE Brasil): implicaciones para una gestion adecuada de la zona litoral. PhD thesis, University of Barcelona

Maia GG (2014) Vulnerabilidade e riscos naturais a eventos de alta energia nas praias smi-urbanas e naturais do litoral de Aquiraz, Ceará. PhD thesis, Federal University of Ceará

Maia LP, Freire GSS, Lacerda LD (2005) Accelerated dune migration and eolian transport during El Niño events along the NE Brazilian Coast. J Coast Res 6:1121–1126. doi:10.2112/03-702A.1

Maia LP, Verde RL, Pinheiro LS, Cunha SEM, Serra J (2012) Gestão de dunas costeiras e conflitos de usos nos estados do Ceará e Rio Grande do Norte- Nordeste do Brasil. In: Rodríguez-Perea A et al (ed) La gestión integrada de playas y dunas: experiencias en Latinoamerica y Europa. Sociedad Historia Natural Balears 19:149–158

Masselink G, Short AD (1993) The effect of tide range on beach morphodynamics and morphology: a conceptual beach model. J Coast Res 9:785–800

Medeiros ECS (2012) Capacidade de carga e percepção ambiental da Praia do Icaraí – Caucaia – CE. MSc. dissertation, Federal University of Ceará

Medeiros ECS, Pantalena AF, Miola B, Lima RS, Soares MO (2014) Percepção ambiental da erosão costeira em uma praia no litoral do Nordeste do Brasil (Praia da Taíba, CE). Rev Gest Cost Integ 14:471–482. doi:10.5894/rgci488

Monteiro LHU, Maia LP (2010) Uso de veículos remotos no estudo do fundo do mar. Ciencia Hoje 46:38–43

Morais JO (1969) Processo de Assoreamento do Porto do Mucuripe. Arq Cienc Mar 12:139–149

Morais JO (1981) Evolução Sedimentar da Enseada do Mucuripe. Arq Cienc Mar 21:20–32

Morais JO (2000) Compartimentação territorial e evolutiva da zona costeira do Estado do Ceará. In: Lima LC et al (eds) Compartimentação Territorial e Gestão Regional do Ceará. FUNECE, Fortaleza, pp 106–180

Morais JO, Fonteles ML (2000) Fluxos Interativos na elaboração da planície costeira de Itarema – Ceará. Rev Cienc Technol 1:63–68

Morais JO, Pinheiro LS (2005) Rochas e Minerais Industriais do Mar e em Zonas Costeiras. In: Vidal FWH et al (eds) Rochas e Minerais Industriais do Estado do Ceará. Realce, Fortaleza, pp 148–161

Morais JO, Pinheiro LS (2011) The effect of semi-aridity and damming on sedimentary dynamics in estuaries- Northeastern region of Brazil. J Coastal Res SI 64:1540–1544

Morais JO, Freire GSS, Pinheiro LS, Souza MJN, Carvalho AM, Pessoa PRS (2006a) Ceará. In: Muehe D (ed) Erosão e Progradação do Litoral Brasileiro. MMA, Brasília, 1:132–134

Morais JO, Pinheiro LS, Medeiros C, Pitombeira ES (2006b) The gabions for the protection of Caponga Beach, Ceará/Brazil: hazards and management. J Coastal Res SI 39:848–851

Morais JO, Irion G, Pinheiro LS, Kasbohm J (2009) Preliminary results on Holocene sea-level changes on Ceará coast-Brazil. J Coastal Res (SI 56):646–649

Morais JO, Tintelnot M, Irion G, Pinheiro LS (2006c) Pathways of clay mineral trasport in the coastal zone of the Brazilian continental shelf from Ceará to the mouth of the Amazon River. Geo-Mar Lett 26:16–22. doi:10.1007/s00367-005-0011-1

Morais JO, Pinheiro LS, Cavalcante AA, Paula DP, Silva RL (2008) Erosão Costeira em Praias Adjacentes as Desembocaduras Fluviais: O Caso de Pontal de Maceió, Ceará, Brasil. Rev Gest Cost Integ 8:61–76. doi:10.5894/rgci132

Oliveira MMN (2012) Aspectos morfológicos e sedimentares associados a dinâmica do litoral oeste de Icapuí, Ceará. Dissertation, Federal University of Ceará

Paula DP (2012) Análise dos riscos de erosão costeira do litoral de Fortaleza em função da vulnerabilidade aos processos geogênicos e antropogênicos. PhD thesis, University of Algarve

Paula DP, Dias JMA, Ferreira O, Morais JO (2012) High-rise development of the sea-front at Fortaleza (Brazil): perspectives on its valuation and consequences. Ocean Coast Manage 77:14–23. doi:10.1016/j.ocecoaman.2012.03.004

Paula DP, Dias JMA, Souza MAL et al (2013) Monitorização de curto prazo da praia do Icaraí (Caucaia, Ceará, Brasil) após construção de estrutura rígida do tipo bagwall para controle da erosão. In: Abstracts of the 11 Simpósio de Hidráulica e Recursos Hídricos para Países de Língua Portuguesa/ VII Congresso sobre Planejamento e Gestão das Zonas Costeiras dos Países de Expressão Portuguesa, Moçambique

Pinheiro LS, Rocha GC (2007) Percepção Dos Usuários Quanto Aos Problemas Relacionados À Erosão Na Praia Da Caponga-Cascavel-Ce. In: VII Encontro Nacional da Pós-Graduação em Geografia, Universidade Federal Fluminense, Niteroi

Pinheiro LS, Morais JO, Medeiros C (2001) Mudanças da linha de praia e feições morfológicas em Cascavel- Ceará. Arq Cienc Mar 34:117–130

Pinheiro LS, Morais JO, Pitombeiras ES (2003) Caponga shoreline rehabilitation assessment. In: Proceedings of the Brazilian symposium on sandy beaches: morphodynamics, ecology, uses, hazards and management, SI 35:564–572

Pinheiro LS, Medeiros C, Morais JO (2006) Erosive processes monitoring linked to the estuarine evolution systems nearby Aguas Belas, Cascavel, Ceará, Brazil. J Coastal Res SI 39:1403–1406

Pitombeira ES, Aquino MD (1998) Erosão Costeira em Áreas Urbanizadas. In Abstract of the VI Congresso Nacional de Controle da Erosão, UNESP, Presidente Prudente, 5–8 June 1998

Pye K, Tsoar H (1990) Aeolian sand and sand dunes. Unwin Hyman, London, p 400

Rossetti DF, Bezerra FHR, Dominguez JML (2013) Late Oligocene-Miocene transgressions along the equatorial and eastern margins of Brazil. Earth Sci Rev 123:87–112. doi:10.1016/j.earscirev.2013.04.005

Shore Protection Manual (1984) 4th edn, v 2. U.S. Army Engineer Waterways Experiment Station, U.S. Government Printing Office, Washington, DC

Short AD (ed) (1999) Beach and shoreface morphodynamics. Wiley, Chichester, 379 pp

Short AD, Hogan CL (1994) Rip Currents and beach hazards: their impact on public safety and implications for coastal management. J Coastal Res SI 12:97–209

Silva AC, Pinheiro LS, Pinheiro L et al (2009) Estudo Hidrodinâmico, Climático e Bacteriológicos Associados às Fontes Pontuais de Poluição ao Longo do Litoral de Fortaleza. Rev Bras Recur Hidro 14:1–10

Silva AC, Façanha P, Bezerra C, Araujo AS, Pitombeira E (2011) Características das ondas "Sea" e "Swell" observadas no litoral do Ceará-Brasil: variabilidade anual e inter-anual. Trop Ocean 39:17–26

Souza PHGO (2005) Dinâmica Ambiental e Transformações Socioespaciais da Orla Marítima de Paracuru- CE. Dissertation, Universidade Estadual do Ceará

Tsoar H, Levin N, Maia LP, Herrmann HJ, Sales VC (2009) The effect of climate change on the mobility and stability of coastal sand dunes in Ceará State (NE Brazil). Quat Res 71:217–226. doi:10.1016/j.yqres.2008.12.001

Vandoros P, Oliveira MAF (1968) O fonólito Mecejana-CE. Acad Bras Cienc 1:203–206

Chapter 8
Beaches of Rio Grande do Norte

Helenice Vital, Iracema Miranda da Silveira, Werner Farkatt Tabosa, Zuleide Maria Carvalho Lima, Francisco Pinheiro Lima-Filho, Flavo Elano Soares de Souza, Marcelo dos Santos Chaves, Felipe M. Pimenta, and Moab Praxedes Gomes

Abstract The Rio Grande do Norte coast extends 410 km, and consists of sandy beaches (72 %), active sea cliffs carved into Cenozoic sediments of the Barreiras and Tibau formations (26 %), and transgressive dune fields and beachrock. It comprises two different sectors: the northern (equatorial) coast trends east for 244 km, while the eastern (oriental) coast trends south for 166 km. The eastern sector is characterized by wave-dominated and some tide-modified beaches that are mainly reflective to intermediate states. In contrast, the northern sector has resulted in tide-modified and tide-dominated beaches that range from reflective (the dominant state) to intermediate. In general the Refletive + Low tide terrace (R + LTT) is present along the entire coast for most of the year, while wave-dominated Longshore bar and trough (LBT), Rhythmic bar and beach (RBB), Transverse bar and rip (TBR), Low tide terrace (LTT) and reflective (R) occur along the eastern sector, and tide-dominated Beach + tidal sand flats (B + TSF) occurs along parts of the northern sector. R + rock flats and coral reef flats are present in both sectors, where bedrock and beachrock reefs are prevalent. Beachrock reefs are very common along the Rio Grande do Norte shore, occurring in both the offshore and onshore zones. Beach morphodynamics is modified due the presence of the beachrocks. Sea level variability is dominated by tides (up 98 % of the energy spectra). The sea level subtidal

H. Vital (✉) • W.F. Tabosa • F. Pinheiro Lima-Filho • M.P. Gomes
Departamento de Geologia/Programa de Pós-Graduação em Geodinâmica e Geofísica, Centro de Ciências Exatas e da Terra Universidade Federal do Rio Grande do Norte-UFRN Campus Universitário, 1639 Lagoa Nova, Natal, RN 59072-970, Brazil
e-mail: helenicevital2004@yahoo.com.br; werner@geologia.ufrn.br; pinheiro@geologia.ufrn.br; gomesmp@geologia.ufrn.br

I.M. da Silveira
Museu Câmara Cascudo Universidade Federal do Rio Grande do Norte- UFRN, Av Hermes da Fonseca, 1398 Tirol, Natal, RN 59020-650, Brazil
e-mail: iracema_miranda@hotmail.com

Z.M.C. Lima • M.d.S. Chaves
Departamento de Geografia/Programa de Pós-Graduação em Geografia, Centro de Ciências Humanas Letras e Artes Universidade Federal do Rio Grande do Norte-UFRN/Campus Universitário, 1596 Lagoa Nova, Natal, RN 59072-970, Brazil
e-mail: zuleide@ufrnet.br; mschaves2007@gmail.com

A.D. Short, A.H. da F. Klein (eds.), *Brazilian Beach Systems*, Coastal Research Library 17, DOI 10.1007/978-3-319-30394-9_8

component is well correlated with the winds, but demonstrate very low amplitudes. Longshore currents are wind modulated, while cross-shore currents are primarily modulated by tides and secondarily by winds. Erosional hotspots and, both natural and anthropogenic hazards, are present along the Rio Grande do Norte coast.

8.1 Introduction

The Rio Grande do Norte coast lies between 04°49′53″ and 06°29′18″ S. It extends for 410 km, and consists of sandy beaches (72 %), active sea cliffs carved into Cenozoic sediments of the Barreiras and Tibau formations (26 %), and transgressive dune fields and beachrock. Owing to its location on the northeast "corner" of Brazil, Rio Grande do Norte comprises two different sectors: the northern (equatorial) coast trends east for 244 km, while the eastern (oriental) coast trends south for 166 km (Fig. 8.1). Along the semi-arid northern coast, tide-modified to tide-dominated beaches dominate, together with extensive ebb tidal deltas, active dune fields, barrier islands, and spits. Along the humid tropical eastern coast, wave-dominated to tide-modified beaches dominate, with active sea cliffs carved into tablelands alternating with vegetated dune-barrier sections. Beaches with rock flats and fringing reefs occur in both sectors. According to the Instituto Brasileiro de Geografia e Estatística (IBGE 2013), there were an estimated 3,408,510 million people residing in Rio Grande do Norte in 2014, with 1,456,065 (43 %) residing near the coast, which represents a coastal population density of ~220 inhabitants km^2.

8.1.1 Geology

Rio Grande do Norte is located in the eastern part of the northeastern South American Platform (i.e. Borborema Province). This province was defined by Almeida et al. (1977) as a complex mosaic-like folded region where major Neoproterozoic tectonic, thermal, and magmatic events were associated with the Brasiliano Cycle. The area contains three main groups of rocks (Fig. 8.2): (1) Precambrian units (3.45 Ba to 542 Ma); (2) Cretaceous units of the

F.E.S. de Souza
Centro de Tecnologia, Colégio Agrícola de Jundiaí, Universidade Federal do Rio Grande do Norte, UFRN RN 160 Km 03 Distrito de Jundiaí, Escola Agrícola de Jundiaí, Macaiba, RN 59280-000, Brazil
e-mail: flavogeo@ig.com.br

F.M. Pimenta
Departamento de Geociências, Centro de Filosofia e Ciências Humanas, Universidade Federal de Santa Catarina, UFSC/Campus Universitário, Trindade, Florianópolis, SC 88040-900, Brazil
e-mail: felipe.pimenta@ufsc.br

Fig. 8.1 Location of Rio Grande do Norte, Brazil, including the northern and eastern coastal sectors

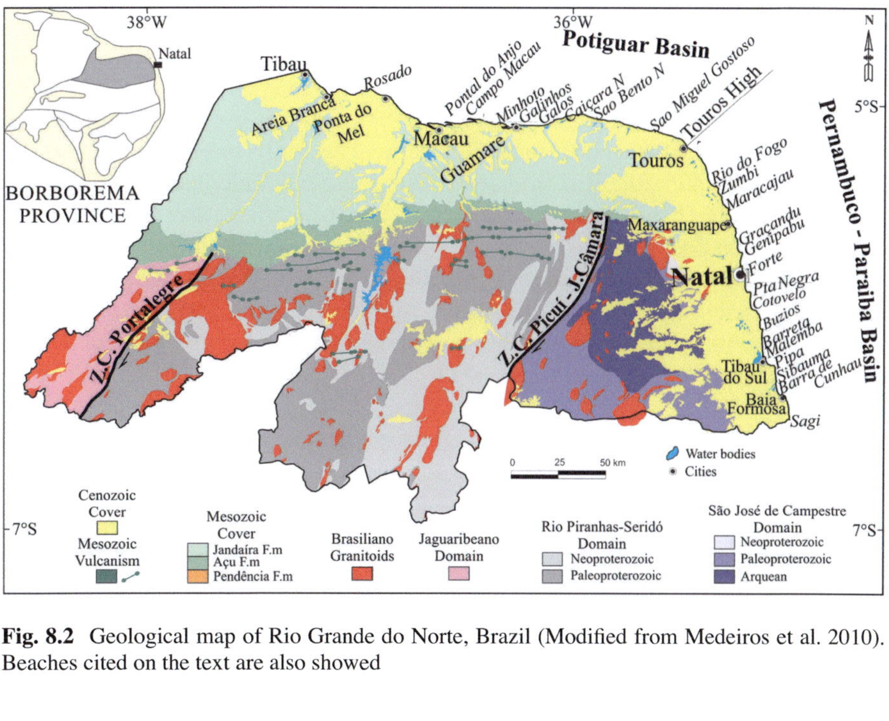

Fig. 8.2 Geological map of Rio Grande do Norte, Brazil (Modified from Medeiros et al. 2010). Beaches cited on the text are also showed

Pernambuco-Paraiba and Potiguar basins (145 to 65 Ma), and associated volcanic deposits; (3) Cenozoic sedimentary cover (65 Ma to present).

The northern and eastern Rio Grande do Norte coastal sectors are located in the Potiguar and Pernambuco-Paraíba basins, respectively. These basins, which are separated by the Touros high (Fig. 8.2), developed during the Upper Cretaceous post-rift phase of the Atlantic Ocean formation. The Potiguar Basin includes an offshore segment with an area of ~27,000 km^2, and an onshore segment covering 22,000 km^2; the latter being related to failed rift basins (Milani and Thomaz Filho 2000). The basin underwent a complex evolution, merging elements from both the Equatorial and the Southern Atlantic tectonic zones. The Pernambuco-Paraíba Basin represents the northernmost segment of the Occidental, the extensional margin of the South American continent. It lies mainly offshore and has a total area of ~35,000 km^2, of which just ~9000 km^2 are located on land. This region was the last to experience rifting owing to the nature and high rigidity of the Precambrian basement cratonic rocks (Matos 1998).

Milani and Thomaz Filho (2000) reported that tectonic events have occurred in the Potiguar Basin since the Oligocene. East-west compression along pre-existing northeast–southwest trending faults makes this one the most seismically actives regions of Brazil. Instrument and historical data also support the theory that both the Pernambuco-Paraíba and Potiguar basins are located in one of the most seismically active intraplate areas of South America (e.g. Ferreira et al. 1998, 2008).

Both basins contain well-exposed Cenozoic sandy-clayey sediments of the Barreiras Formation, which were deposited by fluvial and marine systems. The age of the Barreiras Formation has long been a source of debate, with proposed dates extending from the Miocene to the Pliocene (Salim et al. 1975; Suguio et al. 1986). Lima (2008) dated weathering profiles and found ages of 12–7 Ma. Rossetti et al. (2013) related the deposition of the Barreiras Formation to a transgressive episode during the early/middle Miocene. The Barreiras Formation is overlain by Quaternary deposits related to Pleistocene and Holocene. The most important Quaternary coastal deposits along the Rio Grande do Norte coast include dune fields, barrier island-spits, tidal channels with small tidal deltas, beachrock, and lagoonal /tidal sediments (Vital 2009).

The adjacent continental shelf represents a modern, highly dynamic mixed carbonate-siliciclastic system characterized by reduced width and shallow depths, as compared with other parts of the Brazilian shelf. It has an average width of 40 km, and the shelf-break lying at a depth of 60–70 m (Vital et al. 2010; Gomes et al. 2014; Vital 2014). The shelf is subject to the full strength of the westerly South Equatorial current, along with high winds and moderate–high tides and waves.

8.1.2 Climate

The Rio Grande do Norte climate varies from tropical dry–semi-arid (Köppen type Bs) on the northern coast, to tropical humid (Köppen type Af) on the eastern coast (Nimmer 1989), and is subjected to the movement and location of the Inter Tropical

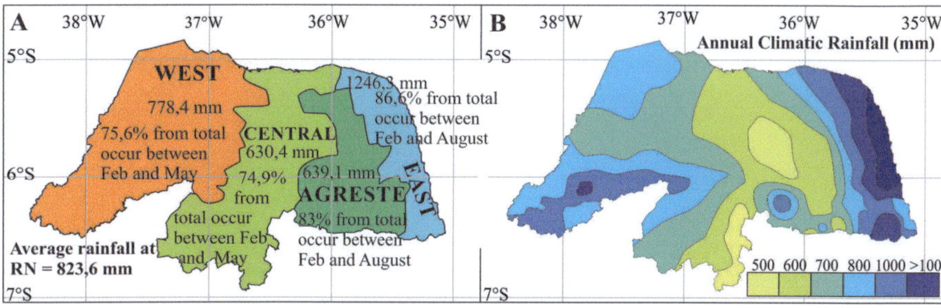

Fig. 8.3 Pluviometry in the sub-regions of Rio Grande do Norte, Brazil: (**a**) mean rainfall (mm) and the main rainfall seasons in the West, Central, Agreste, and East regions; (**b**) annual precipitation (mm) across Rio Grande do Norte (Figure after Pinheiro et al. 2010). Feb = February

Convergence Zone (ITCZ). Isohyets are generally parallel to the coast, with annual precipitation decreasing rapidly toward the interior and to the west (Fig. 8.3).

The Rio Grande do Norte coast has three precipitation zones: the West and Central regions of the northern coast, and the East region of the eastern coast (Fig. 8.3). On the northern coast, precipitation is between 600 and 800 mm year^{-1} (or lower), while the eastern coast experiences up to 1600 mm year^{-1} (Pinheiro et al. 2010). Maximum precipitation occurs during the austral spring and is strongly linked to the maximum zonal intensity of the trade winds. Higher precipitations and reduced wind speeds are associated with the ITCZ.

The dry period extends from June to January, while the rainy period extends from February to May. The mean annual air temperature is ~27 °C, with minimum temperatures (~25 °C) occurring at the end of winter (July) and maximum temperatures (~29 °C) experienced in February.

Though El Nino events are popularly believed to be associated with droughts in northeast Brazil (Kane 2001), approximately 40 % are likely to be ineffective. This is because conditions in the Atlantic may be favorable for droughts in northeast Brazil in some years, while excess rain occurs in other years. In the latter case, the effects of excess rain are due to Atlantic conditions that may reduce or even obliterate the drought effects of El Nino.

8.1.3 Drainage

The two most important hydrographic basins in the Rio Grande do Norte, the Piranhas-Açu and Apodi-Mossoró, have their mouths located on the northern coast, where they are subject to anthropogenic impacts from the oil and salt industries, shrimp farms, and tourism (Fig. 8.4). Similar to other rivers in semi-arid northeastern Brazil, they have an intermittent flow. The Piranhas-Açu River is the largest source of freshwater in this region, and has the largest hydrographic basin. It is dammed by the Armando Ribeiro Gonçalves Dam, and has a total maximum discharge of 1750 m^3 s^{-1} during ebb tides and 324 m^3 s^{-1} during flood tides (Rocha and Vital 2009). The east coast has a greater number of basins, although they are smaller

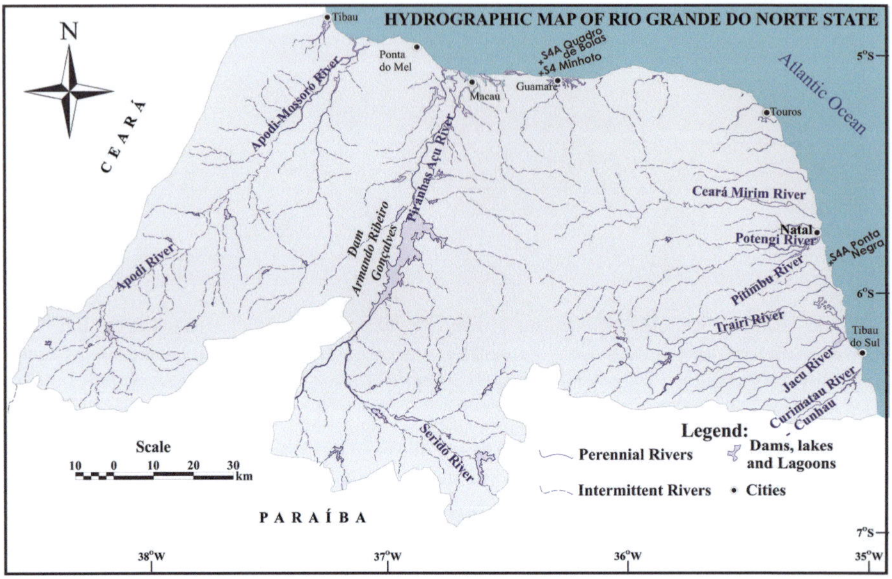

Fig. 8.4 Hydrographic network of Rio Grande do Norte, Brazil. S4 and S4A current meter are also located adjacent to Ponta Negra and Minhoto beaches

in size (e.g. the basins of the Ceará-Mirim, Potengi, Trairi, Jacu, and Curimataú rivers), which contribute to the reduced fluvial discharge and sediment deposition in the region (Fig. 8.4).

8.1.4 Tides, Currents, and Waves

The Rio Grande do Norte coast has a semi-diurnal meso-tidal regime with low amplitude subtidal variability. Sea level records for the northern shelf (Minhoto Beach, Guamaré) in June 2003, and for the eastern shelf (Ponta Negra Beach, Natal) in March of 2014, illustrate the dominant semidiurnal character of sea level variability (Fig. 8.5), with spring tide range varying from 2.5 to 3.0 m and neap-tides from 1.5 to 2.0 m. Wind forcing is relatively strong, but the associated sea level variability is very low amplitude. Tides constitute ~98 % of the sea level energy spectra (Ribeiro 2014). Currents along Minhoto Beach have a dominant semidiurnal signal with low velocities (~10 cm s^{-1}) and a residual component towards the northwest (i.e., Vnorth>0, Veast<0). For Ponta Negra Beach, tidal current amplitudes are larger (~20 cm s^{-1}), but the influence of wind in the along-shelf circulation dominates and residual currents flow north.

Relatively little is known about the Rio Grande do Norte wave climate, as compared with southern Brazil, and the knowledge we do have is based on modeling and occasional short-term observations. Pianca et al. (2010) provided the first assessment

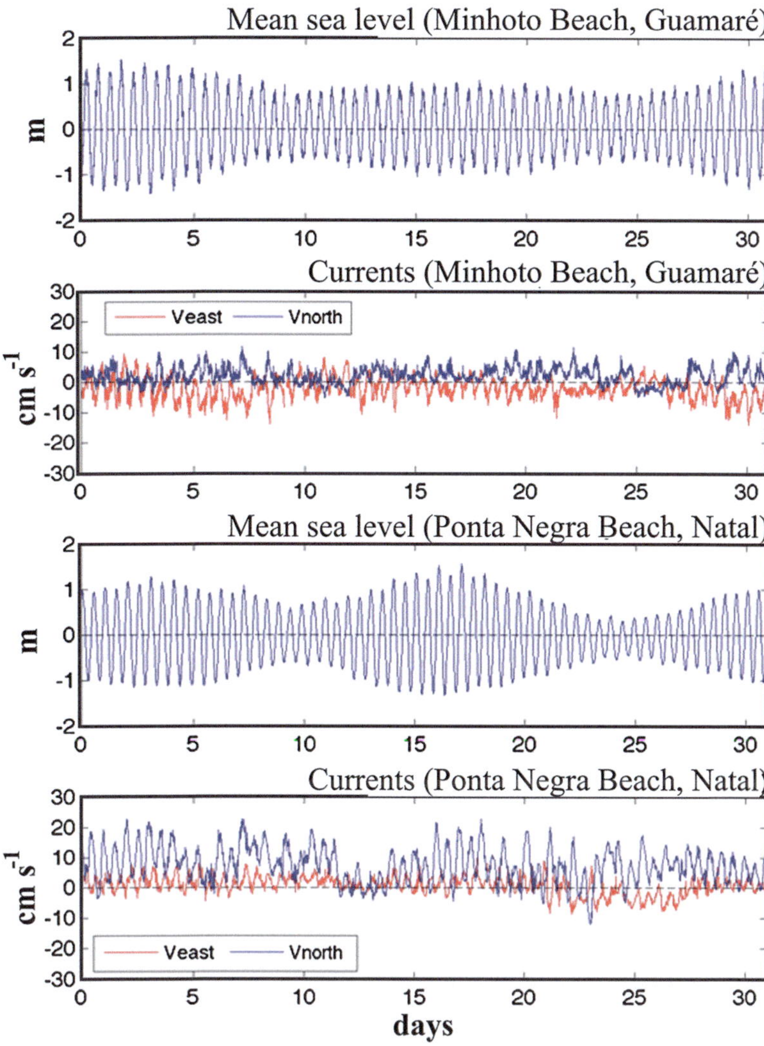

Fig. 8.5 Time series (days) of sea level (m) and coastal currents (cm s^{-1}) along Minhoto Beach, Guamaré, and along Ponta Negra Beach, Natal. Minhoto Beach records represent the northern shelf and cover the period 1 June to 17 July 2004. Ponta Negra Beach measurements represent the eastern shelf and cover the period 14 March to 15 April 2014. Mean sea level was computed from pressure gauge measurements. The eastern and northern current components are denoted as Veast (*red line*) and Vnorth (*blue line*), respectively (Modified from Araujo et al. 2004 and Ribeiro 2014)

of the wave climate for Brazilian waters using an 11-year time series (Jan 1997–Dec 2007) of the National Oceanic and Atmospheric Administration (NOAA) Wave Watch 3 (NWW3) model reanalysis. We used their results for two grid locations on the eastern (9 °S 33.75 °W) and northeastern (3 °S 37.5 °W) regions to represent the Rio Grande do Norte wave climate and to allow correlation with in situ measurements of the Rio Grande do Norte inner shelf.

Table 8.1 Rio Grande do Norte northern sector wave statistics

Variable*	Spring (SON)	Summer (DJF)	Fall (MAM)	Winter (JJA)
$\mathbf{D_p}$	E (57.5 %)	N (36.5 %)	NE (52.7 %)	SE (72.6 %)
$\mathbf{H_s}$	1–2 m (37.9 %)	1–2 m (21.3 %)	1–2 m (43.4 %)	2–3 m (51.6 %)
$\mathbf{T_p}$	6–8 s (46.1 %)	8–10 s (11.7 %)	6–8 s (36.9 %)	6–8 s (59.3 %)
$\mathbf{H_s}^{max}$	3.2 m from SE	3.1 m from N	2.8 m NE	3.4 m from SE
$\mathbf{T_p}^{max}$	21 s from N	18 s from N	16 s from N	16 s from N

Modified from Pianca et al. (2010)

*D_p spectral peak direction of the most energetic components of the wave spectra, H_s mean significant wave height (trough to crest), T_p peak period of the most energetic components of the wave spectra; max superscript = maximum observed values in 11-year time series, *N, S, E, W* north, south, east, and west, respectively. Percentages indicate the level of occurrence

Northern Sector Pianca et al. (2010) reported that the dominant wave directions (D_p) in the northern Rio Grande do Norte are from the east and north during the spring and summer, and from northeast and southeast during the fall and winter, with heights (H_s) that vary from 1 to 2 m between spring and fall, and 2–3 m during the winter (Table 8.1). The peak periods for the most energetic components of the wave spectra (T_p) are 6–8 s for all seasons, except during the summer when values are between 8 and 10 s. The highest waves (H_s^{max}) are typically between 3.1 and 3.4 m, and the longest peak periods vary between 16 and 21 s (T_p^{max}).

In situ measurements along the northern coast, close to Guamaré (Araujo et al. 2004), indicate mean H_s of 2.0 m and periods of 7.0 s for November 2003 (representing the summer dry period), and a less intense wave-field (mean height of ~1.8 m and a period of 8.3 s) from May to June 2004 (winter or rainy period). These seasonal differences in currents and waves were attributed to remote forcing and the position of the ITCZ. Stronger southeast trade winds induce more intense coastal currents and waves fields, while weaker currents and smaller and less frequent waves are observed during the winter (rainy period), when the southeast trade winds are weaker.

Eastern Sector Pianca et al. (2010) reported that for the eastern Rio Grande do Norte coast, dominant wave directions are from the east during the spring, summer, and fall, with height ranging between 1 and 2 m with periods of 6–8 s (Table 8.2). During the winter, the dominant direction is southeast. The highest waves are observed from the southeast and generally have heights of 2–3 m, with a maximum wave height of 4.3 m. The longest periods vary from 16 s in the fall to 21 s in spring.

More recently, Almeida et al. (2015) evaluated the Ponta Negra Beach wave climate by investigating propagation of deep-water spectral wave scenarios under low and high tides. The eastern sector exhibited small variability in wave direction, with the predominant direction east-southeast (75 %) in all seasons, followed by waves

Table 8.2 Rio Grande do Norte eastern sector wave statistics

Variable*	Spring (SON)	Summer (DJF)	Fall (MAM)	Winter (JJA)
$\mathbf{D_p}$	E (60.1 %)	E (50.2 %)	E (42.1 %)	SE (52.2 %)
$\mathbf{H_s}$	1–2 m (47.6 %)	1–2 m (44.6 %)	1–2 m (25.4 %)	2–3 m (35.5 %)
$\mathbf{T_p}$	6–8 s (53.8 %)	6–8 s (43 %)	6–8 s (33.2 %)	8–10 s (25 %)
$\mathbf{H_s^{max}}$	3.8 m from SE	2.6 m from N	4 m from SE	4.3 m from SE
$\mathbf{T_p^{max}}$	21 s from N	19 s from N	16 s from S	17 s from S

Modified from Pianca et al. (2010)

*D_p spectral peak direction of the most energetic components of the wave spectra, H_s mean significant wave height (trough to crest), T_p peak period of the most energetic components of the wave spectra; max superscript = maximum observed values in 11-year time series, *N, S, E, W* north, south, east, and west, respectively. Percentages indicate the level of occurrence

from the east (20 %), southeast (3 %), and east-northeast (2 %). Significant mean wave heights varied between 0.5 and 2.8 m, and waves were below 1.6 m in 75 % of sea states. Waves higher than 2.6 m had a return period of approximately 10 years. Peak period values ranged from 4 to 20 s, and were below 8 s in 75 % of sea states. Waves with periods of greater than 18 s showed a return period of more than 10 years. The analysis of the joint distribution of mean wave height to peak wave period, and mean wave height to direction, showed that the most frequent waves are those between 1.3 and 1.7 m in height, have a period of ~8 s, and are from a 110° (east-southeast) direction.

8.1.5 Coastal Sediments

The Rio Grande do Norte coast is located along the sediment-starved coast of northeastern Brazil (Dominguez 2009). The rivers of the study area are small and do not contribute a significant amount of bedload sediment to the coast. Moreover, rivers with the highest discharge (e.g. the Piranhas-Açu and Apodi-Mossoró rivers) are dammed, and reservoirs prevent sediments from reaching the ocean. As a result, the river waters that discharge into the ocean do not form large sediment plumes. Loss of sediments towards the land by dune field and spit-barrier island formation, tectonic setting, and longshore sediment removal and transport also contribute toward this negative sediment budget (Vital 2006; Vital et al. 2006).

The beaches are dominated by siliciclastic sands, with muddy sediments (mainly coarse silt) restricted to river mouths. Beach sediments are mainly moderately sorted fine–medium grained quartz. Carbonate sediments are observed only when biodetrical gravel is present (e.g. Lima et al. 2006; Chaves et al. 2006; Vital 2009). Heavy minerals, which are normally associated with high-energy periods, are found in the form of ilmenite, zircon and rutile, and are derived from rivers mouths and Barreiras Formation outcrops (Vital and Guedes 2000).

8.1.6 Coastal Provinces and Geomorphology

Most of the present-day relief along the Rio Grande do Norte coast resulted from the deformation and erosion of preexisting topography and it is characterized by coastal plains, coastal tableland, fault-controlled valleys, beachrock, and coastal dune fields (mainly barchans and barchanoids). The Touros platform, representing a structural high from the Potiguar Basin, separates the eastern and northern sectors.

Along the northern coast, dunes are mostly barchans and barchanoids. Barrier islands and sandy spit systems are restricted to the northern sector, between Ponta do Mel and Ponta dos Tres Irmãos, where cliffs from the Barreiras Formation are not found. Of the 244 km of the northern coastline, which represents 59 % of the Rio Grande do Norte coast, 194 km (80 %) is sandy beaches, 10 km (4 %) is muddy beaches, and 40 km (16 %) is active cliffs.

The eastern coast has a length of 166 km, representing 41 % of the Rio Grande do Norte coast, of which 101 km (61 %) is composed of sandy beaches, and 65 km (39 %) is composed of active cliffs of the Barreiras Formation, together with extensive parabolic or blowout dune fields controlled by vegetation. The dominant morphological signature in the eastern sector is a horst-graben structure along the margin, which has driven the erosion associated with waves refraction patterns, and explains the differential erosion on Barreiras Formation rocks. The horst-graben structure controls the morphology of the coastal tablelands (e.g. Bezerra et al. 2001), with these beaches variably referred to as zeta curved bays (e.g. Carter 1988), headland embayed beaches, structurally controlled beaches (Short and Masselink 1999), or headland bay beaches (Klein 2004).

8.2 Rio Grande do Norte Beach Systems

The Rio Grande do Norte coast is exposed to easterly and southeasterly waves and meso-tides, which has resulted in a predominantly tide-modified coast and beaches on the open shore (Fig. 8.6). Rio Grande do Norte hosts 100 beach systems along its open coast, which are tide-modified to tide-dominated along the northern sector, and tide-modified to wave-dominated along the eastern sector. Reflective and rock or coral reef flats are present in both sectors.

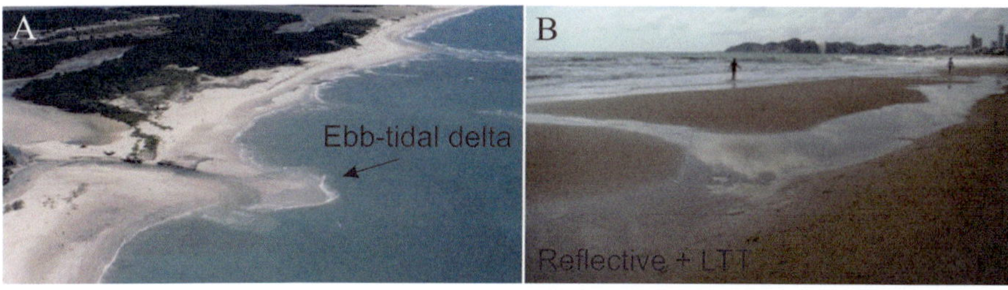

Fig. 8.6 Tide-modified beaches along the coast of Rio Grande do Norte, Brazil: (**a**) Diogo Lopes Beach, northern Sector; (**b**) Ponta Negra Beach, eastern Sector (Photos courtesy of H.Vital)

8.2.1 Coastal Processes and Parameters

The Rio Grande do Norte region experiences high-energy, coastal and shelf parallel currents driven by combined flows related to oceanic, tidal, and wave processes. Since strong winds are almost constant, water masses are well mixed with no stratification.

The breaker wave height during spring high tides, acquired monthly from different beaches along the Rio Grande do Norte coast, indicate that wave heights are greater along the eastern sector. Along the northern sector, waves have a maximum height of 0.8 m during the summer, and 0.7 m during the rainy (winter) season (Tabosa et al. 2001; Silveira et al. 2006; Lima et al. 2006; Chaves et al. 2006), with a medium wave period of 7.5 s. In the eastern sector, waves have a maximum height of 1.85 m during the summer, and a maximum height of 0.85 m during the rainy (winter) season (Chaves 2000; Souza 2004; Frazão 2003).

8.2.2 Beach Types and States

Rio Grande do Norte beaches range from wave-dominated to tide-modified and tide-dominated, with beach type controlled by the relative tidal range (RTR):

$$RTR = TR / H_b$$

where TR is the spring tide range (m) and H_b is the breaker wave height (m). Rio Grande do Norte beaches have RTR values ranging from 1 to 35 (Fig. 8.7). Those with RTR < 3 are wave-dominated, those with RTR of 3–10 are tide-modified, and those with RTR > 10 are tide-dominated. The dimensionless fall velocity Ω, can be expressed as:

$$\Omega = H_b / W_s T$$

where W_S is the sediment fall velocity (cm s^{-1}) and T is the wave period (s). Reflective beaches tend to occur when $\Omega < 1$, intermediate rip-dominated beaches

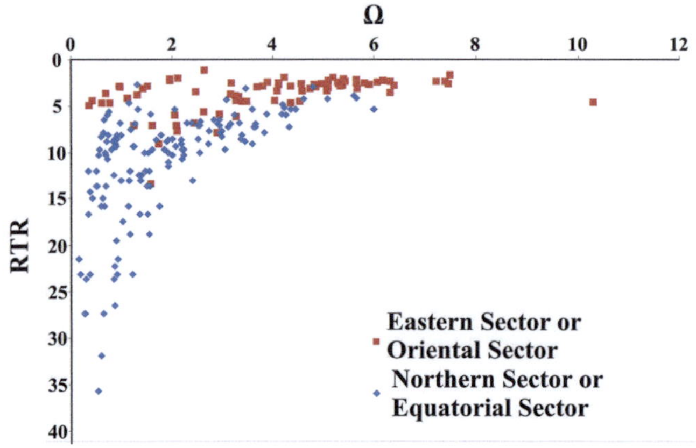

Fig. 8.7 Relationship between Ω (dimensionless fall velocity) and relative tidal range (*RTR*) of beaches along the eastern (or oriental) sector (*red squares*) and northern (or equatorial) sector (*blue diamonds*) of the Rio Grande do Norte coastline, Brazil

occur when Ω ~2–5, and dissipative beaches occur when Ω>6. The Rio Grande do Norte coast contains the full range of beach states (Fig. 8.7), with the higher energy eastern sector characterized by wave-dominated and some tide-modified beaches (predominately in reflective and intermediate states), while the lower energy and higher tidal range northern sector is characterized by tide-modified and tide-dominated beach states, with beaches ranging from reflective to intermediate (with reflective state beaches dominant).

Tide-modified beaches are by far the dominant type along the Rio Grande do Norte coast. The Reflective + low tide terrace (+ rips) (R + LTT) and Reflective + low tide bars & rips (R + LTR) rips are the most frequent along the northern and eastern coastlines, respectively. Wave-dominated beaches (RTR < 3) occur only along the eastern coast, where each of the six wave-dominated state types have been observed, although they are very few in number. Despite Ω > 6, most waves are lower than 1 m and the dissipative beaches state is the most scarce, while the Longshore bar & trough (LBT) state is the most frequent among the wave-dominated beaches. The beaches on the eastern sector have fine–medium sand, which coarsens where cliffs are present.

The northern sector contains both tide-modified and the only tide-dominated beaches, which occur where waves are very low and the tide is higher than 3 m (e.g. Galinhos, Pontal do Anjo). Sediments range from fine to coarse sand, and are moderately to poorly sorted.

8.2.3 Spatial Variations in Beach State

Spatial variations in beach state (Fig. 8.7) are primarily driven by changes in the breaker wave height, which is controlled by regional orientation (e.g. northern and eastern sectors) and bedrock, which influences wave attenuation and breaker wave height. Although the mean wave height on the inner shelf is 2.0 m, breaker waves are considerably lower owing to attenuation across the shelf and around headlands. Furthermore, TR is between 2.3 and 2.7 m on the eastern sector, but increases to 3 m on the northern sector. Along both coasts, sediments are predominately fine–medium sand, coarsening due to carbonate shells on the northern sector, and near cliffs of the Barreiras Formation along the eastern sector.

The higher waves and lower tides of the eastern sector combine with sediments to maintain both wave-dominated (in most exposed locations) and tide-modified beaches. Dissipative beaches are favored in exposed areas of finer sand, while reflective beaches are favored in areas of coarser sand. Along the northern sector, the lower waves and higher tides have favored the formation of tide-modified beaches and, in the areas with the lowest waves, tide-dominated beaches.

In general the R+LTT is present along the whole coast for most of the year, while wave-dominated LBT, Rhythmic bar & beach (RBB), Transverse bar & rip (TBR), LTT, and R occur along the eastern sector, and tide-dominated Beach+tidal sand flats (B+TSF) occurs along parts of the northern sector. Reflective+rock flats (R+RF) or Reflective+coral reef flats (R+CF) are also present in both sectors, where bedrock and beachrock reefs are prevalent.

8.2.4 Temporal Variations in Beach State

Temporal variations in the states of the Rio Grande do Norte beaches occur as a result of the lower summer and higher winter waves (Fig. 8.8a). On the eastern sector the Ω is highest during winter when waves are higher (Table 8.2), with lowest values accompanying the lower summer waves. Likewise along the northern sector the lower summer waves lead to higher RTR (more tide-dominated) compared to winter (Table 8.1). This results in some beaches shifting to higher energy states, and possibly from tide-modified to wave-dominated states between summer and winter.

Figure 8.8b plots the response of three beaches during the lunar tidal cycle. The beaches range from wave-dominated (Campo Macau in winter) to tide-modified (Pontal do Anjo-Macau in summer; Fig. 8.9), to tide-dominated (Galinhos in summer). Beach profiles were measured during the four phases of the moon. Moreover, the Galinhos Beach topographical profile was measured five times during the full moon phase, at intervals of 1 h between each profile, beginning 2 h before the low tide (for details see Lima et al. 2006 and Chaves et al. 2006). The data set

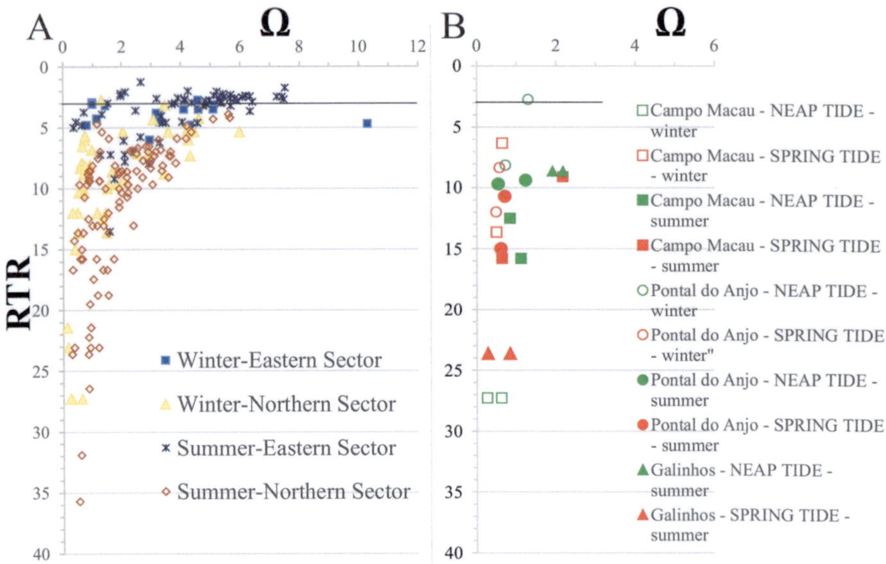

Fig. 8.8 Temporal relationship between Ω (dimensionless fall velocity) and relative tidal range (*RTR*) of beaches along the Rio Grande do Norte coastline, Brazil: (**a**) seasonal variations, with data sorted by winter and summer for the eastern (*blue squares* and *blue* x, respectively) and northern (*yellow triangles* and *brown diamonds*, respectively) sectors; (**b**) variations with the lunar cycle, with data sorted by neap tides (*green colors*) and spring tides (*red colors*), winter and summer seasons at Campo Macau Beach (*squares*), Pontal do Anjo Beach (*circles*), and Galinhos Beach (summer only; *triangles*)

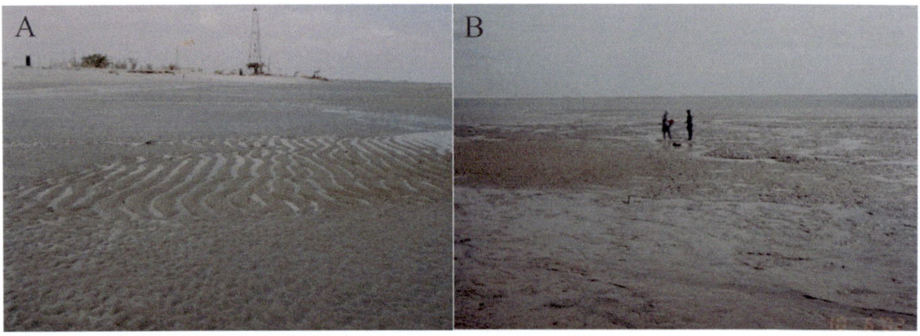

Fig. 8.9 Pontal do Anjo Beach (Northern sector, Rio Grande do Norte coastline, Brazil) during a spring low-tide: (**a**) summer of 2001; (**b**) winter (rainy season) of 2002

collected across a lunar cycle showed that deposition is highest during spring tides accompanying the full moon. The changes observed during the tidal cycle also showed that modification of the beach profile begins soon after the slack tide, when the tide begins to flood.

8.2.5 Beach-Dune Interactions

Along almost the entire Rio Grande do Norte coast, regardless of the presence of cliffs, significant aeolian sedimentation records show that past climatic and geological conditions were more favorable than at present to the accumulation of aeolian deposits.

Extensive coastal dune deposits are present along the Natal coast (eastern sector), just landward of the beach, and are termed the Barrier Dune System (Melo 1995). This barrier dune system is very important to Natal City because it hosts part of the aquifer supplying water to the city, as well as regulating groundwater distribution and the supply of water to coastal lakes (Medeiros et al. 2001). The predominant aeolian transport direction is from southeast to northwest. Where active dunes occur they are usually the reworking older inactive dunes, generally in the form of blowout. Most inactive dunes are vegetated parabolics that no longer have an active sediment source.

The northern sector has few cliffs, some with elevations of less than 5 m, which were generated by the erosion of aeolianite rocks, beachrock, and the Tibau and Barreiras formations. There are occasional dune blowouts, and more frequent active dune fields (e.g. the Zumbi and Rosado dune fields). Where there are active cliffs along the eastern sector, and in the region of Pipa, the cliff top dunes which cover coastal tablelands probably predate the erosion of the cliffs.

Coastal sediment supply and aeolian transport from beaches to the mainland are not yet fully understood along the Rio Grande do Norte coast. Both could be associated with lower relative sea level that may have exposed the shelf to wind action, although this is unlikely. It is more likely that they are related to transgressive and higher relative sea level, which allowed for the erosion of the cliffs and deposition of more sediments on the beaches; thus, generating a surplus sediment budget. Conclusively identifying the correct process requires systematic dating of the various dune fields, using a well-designed chronostratigraphic framework for correlation with the changing sea level curve.

The dunes of the Rio Grande do Norte coast are classified as recent (Holocene) dunes and paleodunes (Pleistocene), according their stability, the age of stabilization, and to morphological, sedimentological and biological criteria (Barreto et al. 2004; Angelim et al. 2006). Regardless of classification or age, the dune fields have a maximum thickness of 50 m, with the sand color white or yellow/red. In some dunes the yellowish/reddish color be attributed to pedogenetic processes, and these are considered "paleodunes"; however, these colors are also found in active dunes, resulting from the erosion of red rocks of Barreiras Formation that are rich in iron oxide cement (e.g. the Rosado dune field). Therefore, color cannot be used as a reliable criterion for characterizing "paleodunes".

One of the most important differences between the eastern and northern sectors relates to the wind transport pattern. Along the eastern sector, the direction of migration is consistently southeast-northwest, ranging from N31 °W near Paraiba Border to N64 °W at Touros. In the northern sector, at least two dominant wind transport directions are evident: southeast-northwest and northeast-southwest. In both sectors, wind transport from southeast to northwest is associated with the southeasterly trade winds, which are most active during the winter (rainy) season. Along the northeastern coast during the summer, the Atlantic Equatorial Mass provides the dominant northeast winds.

Costa Neto (2009) monitored the direction and speed of winds along the northern sector and found that between October and April, the velocities of winds from the northeast and east-northeast range between 20 and 30 km h^{-1}, but exceed 30 km h^{-1} in 5–15 % of the observations. February has the highest percentage (~15 %) of high wind speeds, while in December only 10 % of the observed wine speeds are greater than 30 km h^{-1}. The differences in the dune migration direction between the eastern and northern sectors can be attributed to the orientation of the shoreline, and to the action of the Atlantic Equatorial Mass, which has a dominant wind direction of northeast and east-northeast in the eastern sector; although, southeast winds also present significant speed values for at least 2 months a year (during the drought period). Therefore, the eastern sector has a more effective capacity for wind transport. The migration of dunes (active or inactive) is limited to ~20 km inland. However, some dunes are contained close to the shoreline owing to interception by water bodies, in particular perennial rivers of different sizes.

An integrated view of the geological framework of the Rio Grande do Norte coastal region can be seen in Ground Penetrating Radar (GPR) records (Fig. 8.10). In GPR radargrams we can interpret saline wedges and stratigraphic stacking of sedimentary packages, which emphasizes recent beach deposits (indistinct foreshore and backshore) of ~2.5 m in maximum thickness. The sandy rocks of Tertiary Barreiras Formation are interpreted as fluvial (Araujo et al. 2006) and serve as a substrate for younger sedimentary deposits, including the Potengi Formation of unknown age, which is up to 10 m thick, and Quaternary sediments of active dunes, where two generations of dunes can be identified (Fig. 8.10a). A radargram showing a northwest–southeast profile ~60 m in length and 6 m depth (Fig. 8.10b), shows foredune and sand sheets deposits. In the frontal dunes, two 2nd order surfaces are observed, recording three main stages of dune migration at Malemba Beach. Figure 8.10c shows wind-driven sedimentary packages along an 820 m section of Buzios Beach (collected with a GPR antenna of 50 MHz) that are limited by 2nd and 3rd order surfaces and by a complex arrangement of layers. To ensure that ages have geological significance, dating of these deposits will rely on the development of a chronostratigraphic framework to accurately define sampling sites.

8.2.6 Beach-Beachrock Interactions

Beachrock reefs are common along the Rio Grande do Norte shore, occurring in both the offshore and onshore zones (Fig. 8.11). Owing to high ocean temperatures, beachrock can form in a few decades, cementing the intertidal beach sands. This cementation may lead to substantial modification of Holocene coastal processes and

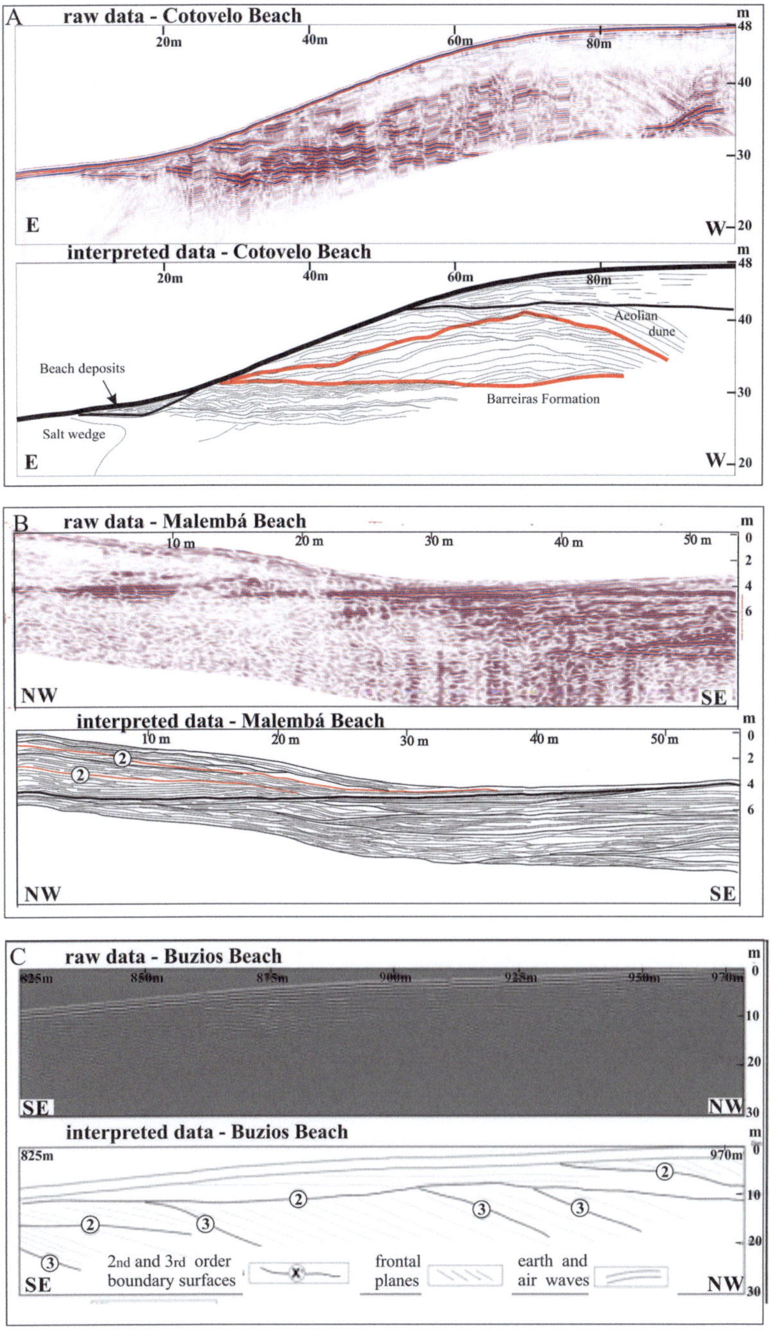

Fig. 8.10 Ground Penetrating Radar (*GPR*) radargrams of dunes along the Rio Grande do Norte coastline, Brazil. In each sub-plot, the top plot shows raw data and the bottom plot shows the interpreted cross-section: (**a**) Cotovelo Beach, including a salt wedge, beach deposits and two generations of dunes; (**b**) Malemba Beach, including a foredune with two 2nd order surfaces, and sand sheets deposits; (**c**) Buzios Beach, including 2nd and 3rd order surfaces, and a complex arrangement of layers observed in dunes

Fig. 8.11 Beachrock along the Rio Grande do Norte coastline, Brazil: (**a**) Graçandu Beach at low-tide; (**b**) Barreta Beach at high-tide; (**c**) Galos Beach; (**d**) submerged beachrock at 25 m depth, Urca do Minhoto; (**e**) sedimentary structures observed in the beachrock of Praia do Forte Beach; (Photos courtesy of H. Vital (**a**, **c**, **e**), L.H.O. Caldas (**b**), and A. Schimanski (**d**)

subsequent Holocene beach and barrier formation (Vousdoukas et al. 2007; Cabral Neto et al. 2014). Reefs are better developed on the eastern coast because of their continuity (e.g. the Barreta, Forte, and Graçandu beaches), but are also present on the northern coast (e.g. the Galinhos and Ponta do Mel beaches). Submerged beachrock reefs are also reported along the littoral region at different depths (Vianna et al. 1991; Testa and Bosence 1998, 1999; Vital 2006, 2009; Vital et al. 2008a; Santos et al. 2007; Cabral Neto et al. 2014). The most continuous structure is situated along the 20–25 m depth isobaths, but small structures are also found along the 10 and 40 m isobaths. Elevations reach 2.5–5 m above the sea floor, and the widths vary between 500 and 1000 m.

Beachrock laminations are oriented sub-horizontally and dip gently seaward (<10°). They are composed of siliciclastic (70–75 %) and bioclastic components (30–25 %), with quartz the main mineral (up to 68 %), followed by feldspar and heavy minerals. Bioclastic components can reach up to 30 % and are mainly red algae and bivalves. Beachrock composition is usually similar to modern sands of the adjacent beaches (Branner 1904; Oliveira et al. 1990; Vieira et al. 2007; Cabral Neto et al. 2014), and are characterized by swash-cross stratification in the foreshore zone (top) and trough-cross stratification in the shoreface zone (base).

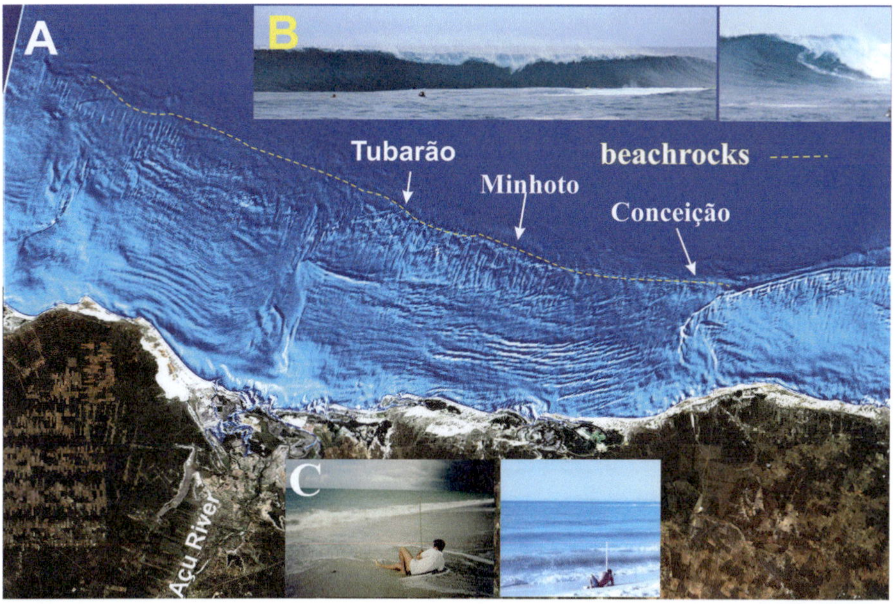

Fig. 8.12 Wave height vs. submerged beachrock along the northern sector of the Rio Grande do Norte coastline, Brazil: (**a**) Landsat 7 Enhanced Thematic Mapper plus (ETM+) image of the Brazilian tropical northeast shelf (Northern sector), running adjacent to Rio Grande do Norte and showing the almost continuous submerged beachrock (*yellow dashed line*) and different seabed features. Land is shown in *yellow–white colors*, and the shallow subaqueous shelf is shown in *blue colors* (Image modified from Gomes and Vital 2010); (**b**) offshore waves (Image courtesy of C. Bandeira); (**c**) break zone waves

Relict submerged beachrock reefs on the Rio Grande do Norte shelf (Figs. 8.11d and 8.12) are aligned parallel to the present-day coast at 20–25 m depth, and can be tracked from at least Natal to Areia Branca. The submerged beachrock lines modify beach morphodynamics by reducing and redistributing the wave energy impacting on the coastline. As a result, offshore waves on the border of the medium and outer shelf (25 m depth) are higher (2–5 m) than on the inner shelf (2 m) and breaker zone (Fig. 8.12 b, c).

8.2.7 Beach-Barrier Islands

Spit-barrier island systems on the Rio Grande do Norte northern coast range from barrier spits (e.g. Galinhos, Diogo Lopes) to barrier islands (e.g. Ponta do Tubarão, Amaro). They are composed of sandy sediments and often capped by dunes. The evolution of these barrier systems has been cyclic (Xavier Neto et al. 2001; Lima et al. 2001, 2002; Silveira et al. 2006; Silva et al. 2011) indicating an ancient system of barrier islands that has developed into the current spits, and spits that have recently detached to form barrier islands (Vital et al. 2008b, 2011; Rocha et al. 2009). Studies of modern coastal environments and sediments in this area (e.g. Vital et al. 2003a) show that barrier spits and barrier islands occur only along the east-west northern coast, confined between two important fault systems: the Carnaubais and Afonso Bezerra systems (Fig. 8.13). However, in the past, barrier islands were abundant along the entire northern coast (e.g. Caldas et al. 2006; Vital 2009; Vital et al. 2013).

The prograding nature of this coastal plain was first presented in the Silva (1991) model, in which lagoons, tidal flats, and fluvial sediments were deposited behind a barrier spit, while tidal inlet, tidal flat, and small secondary spit sediments were deposited seaward of the barrier spit. A restricted microfaunal assemblage occurred behind the barrier spit, while an open marine fauna occurred in front of the spit. The environments seaward of the barrier spit represented a tidal inlet sub-facies of the ebb-tidal delta facies preserved in the subsurface, with the ebb-tidal delta complexes attached to the mainland and promoting shoreline progradation.

8.2.8 Longshore Transport and Beach Stability/Erosion

Coastal current observations are scarce for the Rio Grande do Norte coast. The longest measurements were obtained for the eastern sector (Ponta Negra Beach, Natal), between October 2013 and April 2014. A S4A current meter was moored at 10 m depth and positioned 1.5 m from the bottom, returning more than 212 days of sea level and current observations (Ribeiro 2014). For the northeastern sector, measurements have been collected offshore of Guamaré City by the Quadro de Boias station, an S4A current meter moored at 20 m depth and positioned 10 m from the bottom, during November 2003; and nearshore, off Minhoto Beach, using an S4 current meter moored at 4 m and positioned 2 m from the bottom, during November 2003 and May–June 2004 (Vital et al. 2008a).

The Ponta Negra observations demonstrate that inner shelf currents mainly flow north, with low directional variability during the observed seasons (Fig. 8.14a). Stronger currents (>0.20 m s^{-1}) occur during summer and fall. Longshore currents mainly flow north, but ocassionally reverse to the south. Trajectories computed for virtual drifters illustrate northward residual currents, parallel to the coastline and with trajectories that range from 150 to 320 km in different seasons[1] (Fig. 8.14b). Coastal currents change direction owing to tides in the cross-shelf direction, which steer the flow towards and against the coast. The sea level subtidal component is well correlated to winds, but demonstrate very low amplitudes. Spectral analyses indicate that alongshore currents have more energy in the meteorological (~15 %) and low frequency (>50 %) bands, while cross-shore currents are dominated by semi-diurnal (>28 %) tides (Fig. 8.14c).

[1] Virtual trajectories are found by integrating the velocity field in time and further assuming a horizontally uniform flow.

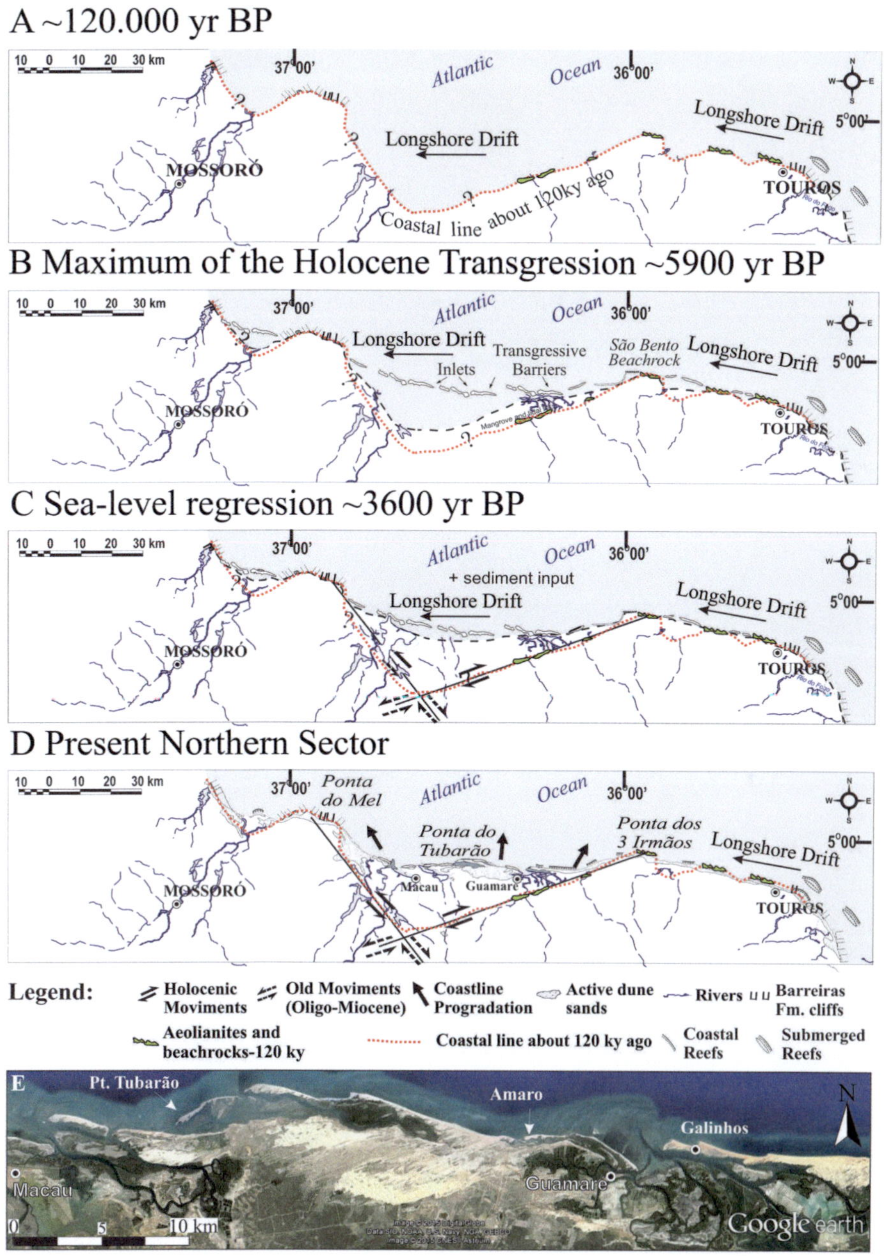

Fig. 8.13 Evolutionary model for the coastal barriers of the northern Rio Grande do Norte State coastline, Brazil (based on Silva 1991, Fonseca 1996 and Caldas 2002): (**a**) Shoreline at ~120 ka before present (BP); (**b**) shoreline at ~5.9 ka BP; (**c**) shoreline at ~3.6 ka BP; (**d**) present-day shoreline (Modified from Vital 2009 and Vital et al. 2013); (**e**) Google Earth view of the barrier-spit system (image©2015 DigitalGlobe, image date 03/24/2014)

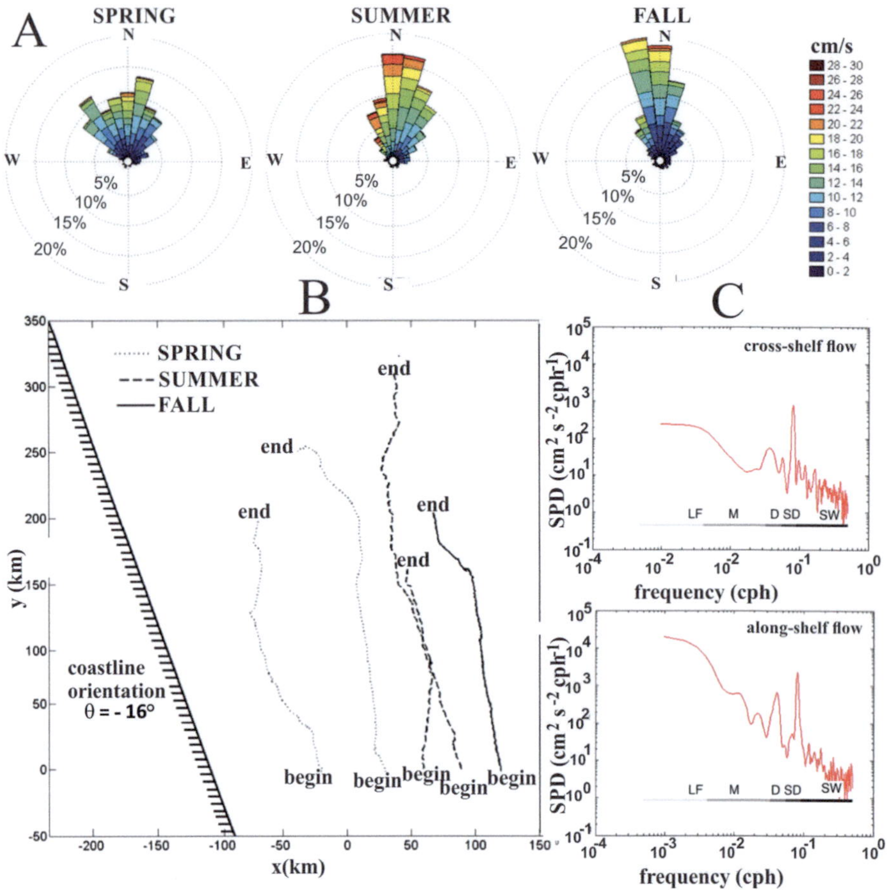

Fig. 8.14 Coastal currents for the eastern sector (Ponta Negra Beach, Natal), modified from Ribeiro (2014). For location see Fig. 8.4. (**a**) Direction and speed distributions for inner shelf currents for the summer, spring and fall. The bars follow the oceanographic convention and point towards the direction that the current flows. The bar color indicates the current speed, while the radius indicate the percentage of occurrence. (**b**) Trajectories for virtual drifters computed from inner-shelf currents records. A *straight line* illustrates the average orientation of the coastline, while trajectories are shown in kilometers. (**c**) Spectral density estimates (*SPD*) for Ponta Negra cross- and alongshelf currents during the fall. Note that the semi-diurnal band (*SD*) is dominant for the across-shelf flow, while meteorological (*M*) and lower frequency (*LF*) bands dominate the alongshelf flow. Other symbols represent the diurnal (*D*) and shallow water (*SW*) components

Along the northern sector, longshore currents flow towards the west-northwest (oblique to the coast) with a maximum of 97 cm s^{-1} during rising tides, and towards the north (perpendicular to oblique to the coast) with a maximum of 50 cm s^{-1} during falling tides. These are by far the dominant contributors to net sediment transport along the coast of Rio Grande do Norte. Owing to the obliquity of the strongest winds, alongshore wind-driven currents increase sediment transport rates, while the relatively small tidal currents (~5–60 cm s^{-1}) have only a small transport capacity.

On the northern coast, the impact of the currents on sedimentation processes is clear. Extensive west-trending spits generated by nearshore currents occur parallel to the coast (e.g. Silveira et al. 2006; Lima et al. 2006), while smaller perpendicular spits (Silva et al. 2011; Vital 2009) are generated by tidal currents. Nearshore current measurements show tidal currents up to 130 cm s^{-1}, flowing southwest, west, and northwest during flood tides, and north-northeast during ebb tides (Vital et al. 2011). The predominance of the ebb-tide is indicated by the higher values of the average currents (30–70 cm s^{-1}) when compared with the flood-tide (20–60 cm s^{-1}). The calculated longshore sediment transport is between 100 and 110 m^3 d^{-1} (Chaves et al. 2006; Vital et al. 2006).

The seven most common indicators of coastal erosional along the Rio Grande do Norte coast are (e.g. Vital et al. 2003b, 2006; Vital 2006): (1) general and progressive landward shoreline displacement (retrogradation) during the last six decades; (2) severe erosion of the Tertiary Barreiras Formation, as well as erosion of Quaternary aeolian and/or marine deposits along the coastline; (3) destruction and burial of mangroves adjacent to the beach; (4) subaerial exposure of peat bogs from ancient lagoonal or mangrove deposits on foreshore and upper shoreface surfaces; (5) persistent destruction of engineering works; (6) concentrated heavy minerals in foreshore zones; and (7) the development of beach embayments.

Furthermore, the most important factors and causes of coastal erosion along the Rio Grande do Norte coast are related to (e.g. Vital et al. 2003b, 2006; Vital 2006): (1) coastal circulation dynamics: beachrock along both the northern and eastern sectors, which are aligned parallel and intermittent to the beach, change wave energy and cause accentuated erosion and beach embayments. Where beachrock reefs are continuous, they protect extensive stretches of coast; in contrast, accentuated erosion take place where in lee of gaps in the beachrock (Fig. 8.11b). When beaches are eroded, the beachrock remain as natural breakwaters to modify wave energy, and thereby control the shape of the retreating shoreline; (2) Holocene evolution of the coastal plain: sedimentation during the Holocene has mainly been controlled by variations in sea level, longshore currents, and the advance of active dunes along the coast (Caldas et al. 2006). Intensive erosion along some stretches of the coast can be related to intense longshore drift (northerly along the eastern sector, and westerly along the northern sector) associated with a negative sediment budget and sediment loss towards the land during dune field and spit-barrier island formation; (3) naturally insufficient sediment supply: long-term trends of coastal erosion in northeastern Brazil relate to an insufficient sediment supply (Dominguez and Bittencourt 1996). The rivers in the study area are small and do not contribute significant sediments. Moreover, the largest rivers (e.g. the Açu and Mossoro-Apodi rivers) are dammed, which prevents sediments from reaching the ocean. The relationships between the small size of drainage basins, low intra-basinal relief, and low precipitation values in Rio Grande do Norte have resulted in the small sediment volume from the hinterland to the shelf (the so-called sediment-starved coast: Dominguez 2009); (4) construction of hard interface structures: hard structures prevent the further erosion of beaches or impede the motion of sand along beaches. Inappropriate structures can exacerbate the situation and harm adjacent beaches. Along the Rio Grande do Norte coast, groin fields were constructed on different beaches (e.g. Caiçara do Norte, Macau, Touros), likely because they are traditionally used to prevent erosion on shorelines with significant alongshore transport. However, these structures were built without sufficient background-knowledge of the most important aspects and mechanism impacting on coastal erosion; and (5) tectonic factors: along the northern sector, erosional areas are likely linked to large-scale bottom morphology, while along on the eastern sector, large-scale coastal morphology is the driving force. These differences are mainly due to divergent longshore drift and the negative sediment budget.

8.3 Beach Hazards and Safety

The Rio Grande do Norte beaches contain both natural and anthropogenic hazards. Natural hazards include strong rip and longshore currents, breaking waves, and variable topography associated with beachrock reefs and headlands. Anthropogenic hazards are mainly related to intense exploitation of beaches by tourism, shrimps farms, and the energy industry (wind, oil, and gas exploration). We currently possess a relatively poor understanding of how geological hazards (e.g., storms, active tectonics) impact on the coast and coastal shelf.

Most hazards reported from the eastern coast relate to rip currents (e.g. Buzios Beach), with the majority of accidents occurring when the number of people swimming increases (particularly summer tourists) and when places are not well signed. This is highlighted by preventive work guidance and education, including the placement of road signs at urban beaches with a higher incidence of drowning, which was intensified following the 2014 Fédération Internationale de Football Association (FIFA) World Cup. Because of this, since the beginning of 2015, beaches where flags or plates were placed, or where lifeguards are present, have not registered any fatal drownings.

Along the northern coast, hazards are mainly related to the higher tidal range and to the presence of tidal inlets (especially between the Galinhos and Porto do Mangue beaches) with their strong tidal, nearshore, and oceanic currents. Anthropogenic hazards relate to shrimps farms, the oil industry, and wind-energy projects in extremely fragile and dynamic environments where seabed erosion, strong currents, and sediment transport dominate.

8.4 Summary and Conclusions

The Rio Grande do Norte coast is dominated by sandy beaches backed by cliffs of the Tertiary Barreiras Formation, dune fields, barrier island-spits, and tidal channels. It is exposed to easterly and southeasterly waves and meso-tides that have resulted in a predominantly tide-modified coast and beaches on the open shore. The coast comprises two sectors: the northern, (equatorial) coast trends eastward for 244 km, and the eastern (oriental) coast trends southward for 166 km. The higher energy eastern sector is characterized by wave-dominated and some tide-modified beaches that are mainly reflective to intermediate states. In contrast, the lower energy and higher tidal range of the northern sector has resulted in tide-modified and tide-dominated beaches that range from reflective (the dominant state) to intermediate. In general the R+LTT is present along the entire coast for most of the year, while wave-dominated LBT, RBB, TBR, LTT and R occur along the eastern sector, and tide-dominated B+TSF occurs along parts of the northern sector. R+rock flats and coral reef flats are present in both sectors, where bedrock and beachrock reefs are prevalent.

Beachrock reefs are very common along the Rio Grande do Norte shore, occurring in both the offshore and onshore zones. The reefs modify beach morphodynamics by reducing and redistributing the wave energy impacting the coastline. Longshore currents are mainly towards the north along the eastern sector, and to the west along the northern sector. Sea level variability is dominated by tides (up 98 % of the energy spectra). The sea level subtidal component is well correlated with the winds, but demonstrate very low amplitudes. Spectral analyses indicate that longshore currents have more energy in the meteorological and low frequency bands, while cross-shore currents are dominated by semi-diurnal tides. Longshore currents are wind modulated, while cross-shore currents are primarily modulated by tides and secondarily by winds.

Different indicators of coastal erosion are observed along the Rio Grande do Norte coast. Areas of erosion along the northern sector are likely linked to large-scale bottom morphology, but along the eastern sector erosion is related to large-scale coastal morphology. Natural hazards along the coast are associated with strong rip and longshore currents, breaking waves, and variable topography associated with beachrock reefs and headlands. Anthropogenic hazards mainly relate to intense exploitation of Rio Grande do Norte beaches by tourism, shrimps farms, and the energy industry (wind, oil, and gas exploration).

This chapter has presented a brief overview of our existing knowledge regarding the beaches of Rio Grande do Norte. Systematic morphodynamic studies along all beaches in the region are necessary for a more accurate and complete picture of this remote corner of Brazil.

Acknowledgements The authors express their thanks to the editors (AD Short and AHF Klein) for their invitation to write this chapter and for their helpful comments on the initial version. We also thank the Navy of Brazil (SECIRM / SSN-3) with the support of boats and ships.This work was funded by CNPq (PQ 303481/2009-9, Universal 486451/2012-7), CAPES (Ciências do Mar 207/2010), FINEP/PETROBRAS (PRH-ANP22, PLAT N-NE). This is a contribution from GT 1.1 & GT 2.1 to INCT AmbTropic (CNPq/FAPESB/CAPES).

References

Almeida FFM, Hasuy Y, Brito Neves BB, Fuck RA (1977) Provincias Estruturais Brasileiras. Annals of the 7 Simpósio de Geologia do Nordeste, Campina Grande, October 1977, p 363–391

Almeida LR, Amaro VE, Marcelino AMT et al (2015) Avaliação do Clima de Ondas da Praia de Ponta Negra (Rio Grande do Norte-Brasil) através do Uso do SMC-Brasil e sua contribuição à gestão costeira. Geostão Costeira Integrada http://www.aprh.pt/rgci/pdf/rgci-532_Almeida.pdf,| DOI:10.5894/rgci532

Angelim LAA, Medeiros VC, Nesi JR (2006) Geologia e recursos minerais do estado do Rio Grande do Norte. Escala 1:500.000. Texto explicativo dos mapas geológico e de recursos minerais do estado do Rio Grande do Norte. CPRM, Recife, p 119

Araujo M, Vital H, Amaro VE, Silva MA (2004) Caracterização das Forçantes Hidrodinâmicas – Correntes, Ondas e Marés – na Região Costeira de Guamaré-Rio Grande do Norte. In: Proceedings of the III workshop PETRORISCO, Belém, September 2004 (CD-ROOM)

Araujo VD, Perez YAR, Lima RO et al (2006) Fácies e Sistemas Deposicionais da Formação Barreiras na Região da Barreira do Inferno, Litoral Oriental do Rio Grande do Norte. Geologia USP. Série Científica 6:43–49

Barreto AMF, Bezerra FHR, Tatumi SH et al (2004) Geologia e Geomorfologia do Quaternário costeiro do Estado do Rio Grande do Norte. Rev Instit Geoc – USP. Série Científica 4:1–12

Bezerra FHR, Amaro VE, Vita-Vinzi C et al (2001) Pliocene-quaternary fault control of sedimentation and coastal plain morphology in NE Brazil. J S Am Earth Sci 14:61–75

Branner JC (1904) The stone reefs of Brazil, their geological and geographical relations with a chapter of the coral reefs. Mus Comp Zool Bull, Harvard Coll, Cambridge 44:275

Cabral Neto I, Cordoba VC, Vital H (2014) Beachrocks do Rio Grande do Norte, Brasil, vol 1, 1st edn. EDUFRN, Natal, p 156

Caldas LHO (2002) Late quaternary coastal evolution of the Northern Rio Grande do Norte coast, NE Brazil. PhD thesis, Christian Albrechts University, Kiel

Caldas LHO, Stattegger K, Vital H (2006) Holocene sea-level history: evidence from coa stal sediments of the northern Rio Grande do Norte coast, NE Brazil. Mar Geol 228:39–53

Carter RW (1988) Coastal environments: an introduction to the physical, ecological and cultural systems of coastlines. Academic, London, p 617

Chaves MS (2000) Sedimentologia, Morfologia Praial eVulnerabilidade Costeira entre as Praias da Redinha e Genipabú Natal/Rio Grande do Norte. MSc dissertation, Universidade Federal de Pernambuco

Chaves MS, Vital H, Silveira IM (2006) Beach morphodynamics of the serra oil field (NE, Brazil). J Coast Res 39:594–597

Costa Neto LX (2009) Caracterização Geológica, Geomorfológica e Oceanográfica do Sistema Pisa Sal, Galinhos-Rio Grande do Norte. PhD thesis, Universidade Federal do Rio Grande do Norte

Dominguez JML (2009) The coastal zone of Brazil. In: Dillenburg SR, Hesp PA (eds) Geology and geomorphology of holocene coastal barriers of Brazil. Springer, Berlin, pp 17–51

Dominguez JML, Bittencourt ACSP (1996) Regional assessment of long-term trends of coastal erosion in Northeastern Brazil. An Acad Bras Cienc 68:355–372
Ferreira JM, Oliveira RT, Takeya MK et al (1998) Superposition of local and regional stresses in northeast Brazil: evidence from focal mechanisms around the Potiguar marginal basin. Geophys J Int 134:341–355
Ferreira JM, Bezerra FHR, Sousa MOL et al (2008) The role of Precambrian mylonitic belts and present-day stress field in the coseismic reactivation of the Pernambuco lineament, Brazil. Tectonophysics 456:111–126
Frazão EP (2003) Caracterização hidrodinâmica e morfo-sedimentar do estuário Potengi e áreas adjacentes: subsídios Pará controles e recuperação ambiental no caso de derrames de hidrocarboneto. MSc dissertation, Universidade Federal do Rio Grande do Norte
Gomes MP, Vital H (2010) Revisão da compartimentação geomorfológica da Plataforma Continental Norte do Rio Grande do Norte, Brasil. Braz J Geol 40(3):321–329
Gomes MP, Vital H, Bezerra FHR et al (2014) The interplay between structural inheritance and morphology in the equatorial continental shelf of Brazil. Mar Geol 355:150–161
IBGE (2013) Atlas do Censo Demográfico 2010. http://www.cidades.ibge.gov.br/
Kane RP (2001) Limited effectiveness of El Niños in causing droughts in NE Brazil and the prominent role of Atlantic parameters. Braz J Geophys 19:231–236
Klein AHF (2004) Morphodynamics of headland-bay beaches: examples from the coast of Santa Catarina State, Brazil. Universidade do Algarve, Faro, p 218
Lima MG (2008) A História do Intemperismo na Província Borborema Oriental, Nordeste do Brasil: implicações Paleoclimáticas e Tectônicas. PhD thesis, Universidade Federal do Rio Grande do Norte
Lima ZMC, Alves AL, Amaro VE (2001) Coastline evolution of the Galinhos Spit (NE Brazil) using air photography and landsat TM images. Pesq em Geoc 28:497–507
Lima ZMC, Vital H, Xavier P et al (2002) Sand spits from NE Brazil: high resolution Quaternary analogous for reservoir model. AAPG ACE 2002, Extended Abstract
Lima ZMC, Vital H, Tabosa WF (2006) Morphodynamic variability of the Galinhos Spit, northeastern Brazil. J Coast Res 39:598–601
Matos RMD (1998) The transversal zone: a key feature between NE Brazil and W Africa. In: Proceedings of the Rio'98 AAPG international conference and exibition. Rio de Janeiro, Brazil, p 426–427
Medeiros WE, Nascimento da Silva CC, Jardim SAF et al (2001) Integração de dados estruturais, de sensores remotos e de geofísica na prospecção hidrogeológica em terrenos cristalinos: estudo de casos no Estado do Rio Grande do Norte., Nordeste do Brasil. Simp Hidrogeologia do NE 1:445–454
Medeiros VC, Nascimento MAL, Sousa DC (2010) Geodiversidade do Estado do Rio Grande do Norte: 2.Geologia. In: Pfaltzgraff PAS, Torres FAM (eds) Geodiversidade do Estado do Rio Grande do Norte, 1st edn. CPRM, Recife, pp 15–38
Melo JG (1995) Impacto do desenvolvimento urbano nas águas subterrâneas de Natal/Rio Grande do Norte. PhD thesis, Universidade de São Paulo
Milani EJ, Thomaz Filho A (2000) Sedimentary basins of South America. In Cordani UG, Milani EJ, Thomaz Filho A, Campos DA (eds) Tectonic evolution of South America. 31 international geological congress, p 389–449
Nimmer E (1989) Climatologia do Brasil. IBGE, Rio de Janeiro
Oliveira MIM, Bagnoli E, Farias CC, et al. (1990) Considerações sobre a geometria, petrografia, sedimentologia, diagênese e idade dos "beachrocks" do Rio Grande do Norte. 36 Congresso Brasileiro de Geologia, Natal, v 2, p 621–634
Pianca C, Mazzini PLF, Siegle E (2010) Brazilian offshore wave climate based on NWW3 reanalyis. Braz J Oceanogr 58(1):53–70

Pinheiro JU, Bristot G, Lucena LRF (2010) Clima do Estado do Rio Grande do Norte. In: Pfaltzgraff PAS, Torres FAM (eds) Geodiversidade do Estado do Rio Grande do Norte, 1st edn. CPRM, Recife, pp 93–98
Ribeiro F (2014) Caracterização das Correntes da Plataforma Continental Interna da Praia de Ponta Negra. Trabalho de Cocnlcusão do Curso de Oceanografia. Universidade Federal de Santa Catarina
Rocha AKR, Vital H (2009). High resolution geophysical investigations of the Açu estuary, Northeastern Brazilian Cost. In: Amorosi A (ed) 27th IAS meeting of sedimentologists, medimond international proceedings, Monduzzi Editore, p 91–97
Rocha GR, Vital H, Rocha AKR (2009) Coastal erosion monitoring of the Arrombado tidal inlet, Macau-Rio Grande do Norte (NE Brazil). In: Amorosi A (ed) 27th IAS meeting of sedimentologists. Amorosi A (ed) Medimond international proceedings, Monduzzi Editore, p 65–71
Rossetti DF, Bezerra FHR, Dominguez JM (2013) Late oligocene-miocene transgressions along the equatorial and eastern margins of Brazil. Earth-Sci Rev 123:87–112
Salim J, Souza CS, Muniz GCB et al (1975) Novos subsídios Pará a elucidação do episódio Barreiras no Rio Grande do Norte. 7 Simpósio de Geologia do Nordeste, Fortaleza, p 149–158
Santos CLA, Vital H, Amaro VE et al (2007) Mapeamento de Recifes Submersos na Costa do Rio Grande do Norte. Rev Bras Geof 25(1):27–36
Short AD, Masselink G (1999) Embayed and structurally controlled beaches. In: Short AD (ed) Handbook of beach and shoreface morphodynamics. Wiley, New York, pp 230–249
Silva CG (1991) Holocene stratigraphy and evolution of the Açu River Delta, Rio Grande do Norte State, Northeastern Brazil. PhD thesis, Duke University
Silva AGA, Vital H, Silveira IM (2011) Presidio's barrier island migration and adjacent coastal erosion, through satellite images analysis. In: Wang P, Rosati JD, Roberts TM (eds) Coastal sediments 2011, vol 2, 1st edn. World Scientific Publishing Co. Pte. Ltd, Danvers, pp 1190–1202
Silveira IM, Vital H, Amaro VE, Chaves MS (2006) The evolutionary study of environmetal conditions of the coastal area of the municipal district of Guamaré – RN. J Coast Res 39:237–241
Souza FES (2004) Evoluçao morodinamica da região de influencia estuarina do rio Curimatau/RN. PhD thesis, Universidade Federal do Rio Grande do Norte
Suguio K, Bidegain JC, Mörner N-A (1986) Dados preliminaries sobre as idades paleomagnéticas do Grupo Barreiras e da Formação São Paulo. Rev Bras Geoc 6:171–175
Tabosa WF, Lima ZMC, Vital H et al (2001) Monitoramento Costeiro das Praias de São Bento do Norte e Caiçara do Norte – NE Brasil. Pesq em Geoc 28:383–392
Testa V, Bosence DWJ (1998) Carbonate-siliciclastic sedimentation on high-energy, ocean-facing, tropical ramp, NE Brazil. In: Wright VP, Burchette TP (eds) Carbonate ramps, Geol Soc London SP 149:55–71
Testa V, Bosence DWJ (1999) Physical and biological controls on the formation of carbonate and siliciclastic bedforms on the north-east Brazilian Shelf. Sedimentology 46:279–301
Vianna ML, Solewicz R, Cabral A, Testa V (1991) Sandstream on the northeast Brazilian Shelf. Cont Shelf Res 2:509–524
Vieira MM, Ros LF, Bezerra FHR (2007) Lithofaciology and palaenvironmental analysis of holocene beachrocks in northeastern brazil. J Coast Res 23:1535–1548
Vital H (2006) Erosão e Progradação no Litoral do Rio Grande do Norte. In: Muehe D (ed) Erosão e progradação do litoral brasileiro. Ministério do Meio Ambiente, Brasilia, pp 159–176
Vital H (2009) The mesotidal barrier of Rio Grande do Norte. In: Dillenburg SR, Hesp PA (eds) Geology and geomorphology of holocene coastal barriers of Brazil. Springer-Verlag Berlin Heildelberg. Lecture notes in Earth Sciences 107:289–324
Vital H (2014) The north and northeast Brazilian tropical shelves. In: Chiocci FL, Chivas AR (eds) Continental shelves of the world: their evolution during the last glacio-eustatic cycle, vol 41. Geol Soc Mem, London, pp 35–46

Vital H, Guedes IMG (2000) Heavy minerals of the inner continental shelf offshore the Açu River's delta. NE Brazil. Rev Bras Geosc, Rio de Janeiro 30(3):453–455

Vital H, Lima ZMC, Silveira IM et al (2003a) Barrier-spit system from the Northern Rio Grande do Norte State Coast, NE Brazil: a tectonic control? In: Proceedings of the 3th Latin-American congress on sedimentology, p 113–115

Vital H, Stattegger K, Tabosa WF et al (2003b) Why does erosion occur on the Northeast of Brazil. J Coast Res SI 35:525–529

Vital H, Amaro VE, Silveira IM (2006) Coastal erosion on the Rio Grande do Norte State (Northeastern Brazil): causes and factors versus effects and associated processes. J Coast Res SI 39:1307–1310

Vital H, Stattegger K, Amaro VE et al (2008a) Inner continental shelf off Northern Rio Grande do Norte, NE Brazil: a modern high-energy siliciclastic-carbonate platform. In: Hampson G, Dalrymple R (eds) Recent advances in shoreline–shelf Stratigraphy. SEPM SP 90:175–188

Vital H, Santos Neto F, Plácido Jr JS (2008b) Morphodynamic of a tropical tidal inlet: case study on the Rio Grande do Norte Coast, NE Brazil. Revista de Gestão Costeira Integrada 8:113–123

Vital H, Gomes MP, Tabosa WF et al (2010) Characterization of the Brazilian continental shelf adjacent to Rio Grande do Norte State, NE Brazil. Braz J Oceanogr 58(SI 1):43–54

Vital H, Rocha GR, Placido JS Jr (2011) Morphodynamics of Arrombado tidal inlet, Macau-Rio Grande do Norte (NE Brazil). In: Wang P, Rosati JT, Roberts TF (eds) Coastal sediments '11, vol 1. World Scientific, New Jersey, pp 327–338

Vital H, Tabosa WF, Souza ZS et al (2013) Carta Geológica Folha Jandaíra SB.24-X-D-III. Relatório Tecnico + 1 mapa, color. Escala 1:100.000. CPRM-Serviço Geológico do Brasil. Programa de Geologia do Brasil

Vousdoukas MI, Velegrakis AF, Plomaritis TA (2007) Beachrock occurrence, characteristics, formation mechanisms and impacts. Earth-Sci Rev 85:23–46

Xavier Neto P, Lima ZMC, Andrade PRO et al (2001) GPR images of the Galinhos Península, NE Brazil: the register of a geologic evolution from paleochannel to a sand spit? In: Proceedings of the Ist symposium GPR in sediments. Geol Soc and UCL, London 324

Chapter 9
Sandy Beaches of the State of Paraíba: The Importance of Geological Heritage

José Maria Landim Dominguez, Silvana Moreira Neves, and Abílio Carlos da Silva Pinto Bittencourt

Abstract The coastline of the State of Paraíba is approximately 145 km long. The sediment deficient character of this coast, with reduced sediment input, has resulted in long-term shoreline retreat. As a consequence, erosion of local sediment sources, such as the coarse grained Barreiras Formation, provided most of the sediment to the shoreline. These aspects, associated with a tropical climate, favoured the development of coral-algal reef structures and beachrocks which sheltered the shoreline from wave energy. The combination of medium to coarse sands and wave sheltering, has resulted in a dominance of reflective and low-energy intermediate beaches (i.e. ridge-runnel or low-tide beach terrace).

9.1 Introduction

The coastline of the State of Paraíba is approximately 145 km long, and has similar characteristics to the Alagoas coast (Chap. 11). The sediment deficient character of this coast, with reduced sediment input, has resulted in long-term shoreline retreat. As a consequence local sediment sources (e.g. erosion of cliffs) are dominant in the area and result in the exposure of older geological units along the coast. These aspects, associated with a tropical climate, favor the development of coral-algal reef structures and cementation of the beach sands. These cemented beach deposits are subsequently exhumed by the erosive retreat of the coastline, exposing beachrock reefs. The shoreline is thus sheltered from wave energy by the reef structures associated with the eroded remains of Cretaceous sedimentary rocks and resistant lateritic beds of the Barreiras Formation and also by the beachrocks. The combination of

J.M.L. Dominguez (✉) • A.C.S.P. Bittencourt
Institute of Geosciences, UFBA, Campus Ondina, Salvador, BA, Brazil
e-mail: landim@ufba.br; abilio@pq.cnpq.br

S.M. Neves
Department of Geographical Sciences, UFPE Campus Universitário, 50740-550 Recife, Pernambuco, Brazil
e-mail: silvanamneves@gmail.com

A.D. Short, A.H. da F. Klein (eds.), *Brazilian Beach Systems*, Coastal Research Library 17, DOI 10.1007/978-3-319-30394-9_9

these factors has resulted in dominance of reflective and low-energy intermediate beaches (i.e. ridge-runnel or low-tide beach terrace).

The characterization of the beaches of Paraíba was performed using a similar approach to that used for the states of Bahia (Chap. 12), Sergipe and Alagoas (Chap. 11). The data on sediment texture of the beach-face deposits were compiled from Neves (2003). These samples were collected at regular intervals (2–3 km), along with measurements of beach-face slope and notes regarding the type and number of waves breaking in the surf zone during the sampling day. Samples were dry sieved and did not undergo any preliminary treatment for removing bioclasts and organic matter. This information was complemented by observations made during overflights.

9.2 Geology and Geomorphology of the Coastal Zone

The geological framework of the Paraíba coastal zone is controlled by the following pre-Quaternary units (Fig. 9.1):

(i) Crystalline basement – composed of deformed Proterozoic rocks that integrate the Borborema structural province (Almeida et al. 1981; Brito Neves et al. 2000). They are limited to the north and south by two great crustal lineaments: Pernambuco and Patos- Paraíba, resulting in a mosaic of shear zones oriented east–west and north-northeast.

(ii) Rocks from the Paraíba sedimentary basin – limited southwards by the Pernambuco Lineament and northwards by a ramification of the Patos-Paraíba Lineament (Mamanguape Fault). The Paraíba Basin experienced late development during the separation of South America and Africa, not having developed a typical rift phase. Its infilling mainly resulted from the accumulation of sediments in marine and transitional environments, forming a homocline that plunges seaward (Mabesoone and Alheiros 1991). The development of the basin began during the late Jurassic and early Cretaceous. Of the sedimentary rocks that fill the sedimentary basin, only the uppermost units outcrop in the coastal zone: Maria Farinha (Paleocene-Eocene) and Gramame (late Campanian-Maastrichtian) formations (Beurlen 1967a, b; Barbosa et al. 2007). Both consist of detrital limestones, calciferous sandstones and marls.

    These calcareous rocks outcrop along the coastline at the foot of the Barreiras Formation, in the southern portion of the coastal zone (Compartment I) (Fig. 9.2a). These outcrops are responsible for the sinuous aspect of the coastline, which associated with the unidirectional littoral drift and the reduced input of sediments, results in its indented characteristic.

(iii) Barreiras Formation – comprises marine-transitional siliciclastic deposits formed during the Middle-Early Miocene transgression (Rossetti et al. 2013) (Fig. 9.2b). This formation forms coastal tablelands dissected by flat-floored valleys, and forms sections of active cliffs along the coast.

The Quaternary deposits include the Post-Barreiras Sediments (Rossetti et al. 2011), which where deposited over the Barreiras Formation and which thicken towards the coast. These deposits are sandy and conglomeratic, and generally

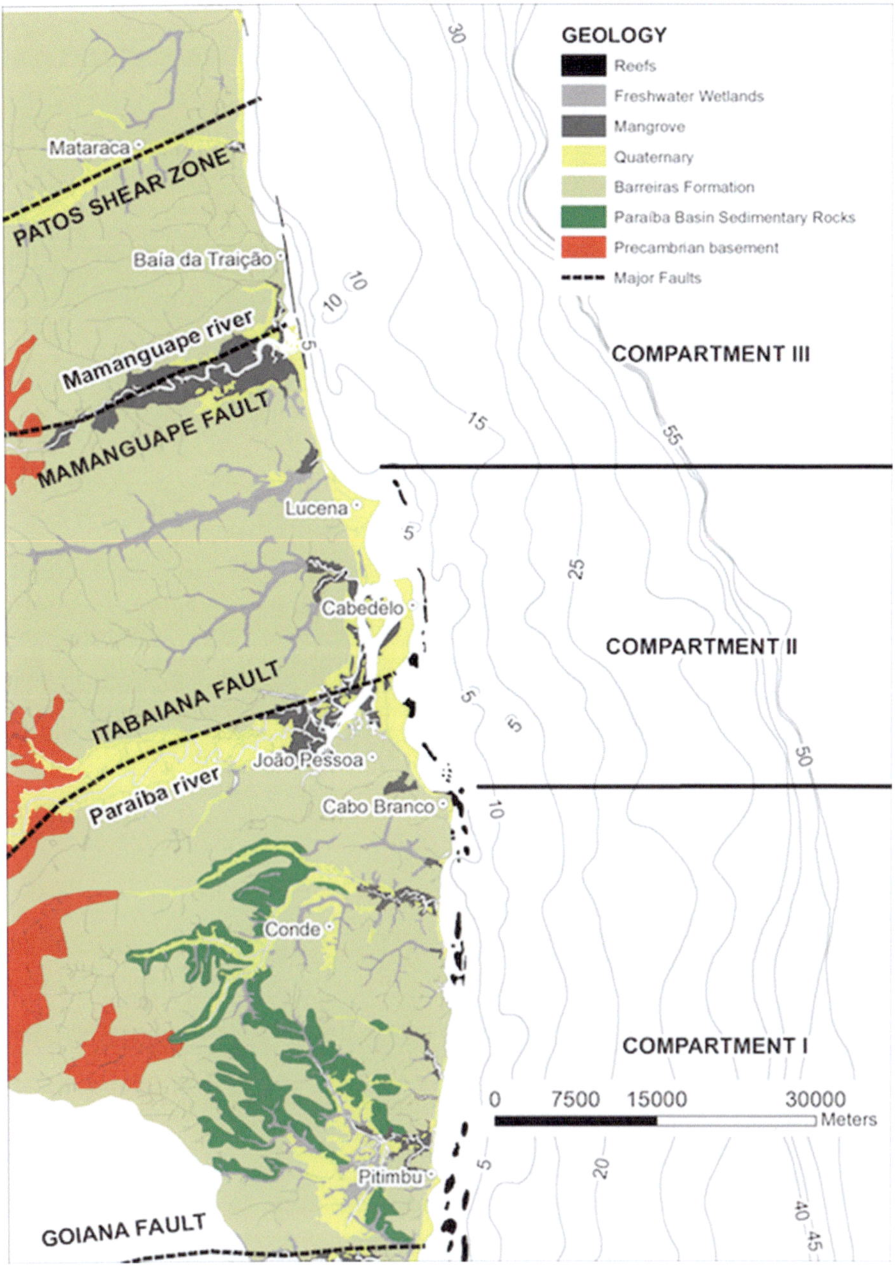

Fig. 9.1 Simplified geological map of the Paraíba State coastal zone and extent of the three coastal compartments discussed in text

Fig. 9.2 (**a**): Sedimentary rocks of the Paraíba sedimentary basin exposed at the base of sea cliffs. (**b**): Active cliffs carved into the Barreiras Formation; (**c**): Large strandplain associated with the Paraíba river mouth; (**d**): Talus deposits at the foot of active sea cliffs; (**e**): Beachrock reef at Baía da Traíção; (**f**): Reef buildup at Cabo Branco locality (*dark patches* in the offshore area are reefs at depths 10 and 15 m) (See Fig. 9.5 for location of photographs)

massive. In the northern portion of the coastal zone, there are aeolian sediments, which cover the Barreiras Formation.

The Quaternary coastal plain in Paraiba is practically absent, with the most extensive development in the central area (Compartment II) (Fig. 9.2c). Both southwards and northwards from this central compartment, the shoreline is bordered by active cliffs, at the foot of which are narrow sandy terraces, the beach prism or talus deposits (Fig. 9.2d).

Coral-algal reef formations and beachrock reefs border the coastline. The reef formations probably developed on top of limestones reefs of the Paraíba Basin

(Marinha Farinha and Gramame formations) and are concentrated in the southern portion of the coastal zone (Compartment I). In the northern portion there are no reef formations, except for a 37 km long beachrock reef that protects the entrance of the Traição Bay (Fig. 9.2e).

The width of the Paraíba continental shelf ranges from 15 to 35 km, with the shelf break located between 50 and 60 m depth. The seafloor has a very irregular topography due to the presence of the reefs (Fig. 9.2f) and incised-valleys associated with the Paraíba and Mamanguape rivers (Neves 2003). The shelf sediments consists essentially of bioclastic material, with predominance of coralline algae fragments (Barbosa 1983). Siliciclastic sediments occupy a narrow strip that borders the shoreline out to a depth of 10 m. In more sheltered areas, fragments of Halimeda are more predominant.

9.3 Sediment Supply

There are three sources of coastal sediment (Fig. 9.3):

(i) river-borne siliciclastic sediments delivered by the Paraíba River, the largest river flowing into the Paraíba coastal zone, with a watershed of 20,071.83 km^2. The historical mean net flow is approximately 14.45 m^3s^{-1}, which is representative of approximately 85 % of the area (Xavier et al. 2012). The highest daily flow rate was observed in 1985, with approximately 1600 m^3 s^{-1}, which is considered an extraordinary flood (Xavier et al. 2012). The other rivers that flow to this coastal zone region have minor discharges.

(ii) siliciclastic sediments derived from erosion of the Barreiras Formations, which outcrops north and south of the Paraíba River mouth.

(iii) bioclastic sediments eroded from the coral-algal reefs and incorporated in the nearshore-beach sediments, mainly in the southern portion of the state (Compartment I).

A 3–4 m fall in sea level during the past 5700 years, probably did not contribute to coastal progradation, as there were no important sources of siliciclastic sediments on the continental shelf.

9.4 Coastal Processes (Waves, Tides, Winds and Precipitation)

The Paraíba coast is exposed to waves arriving from the east and northeast, with heights between 1 and 2 m and periods between 6 and 8 s, and from the southeast, with heights between 2 and 3 m and periods between 8 and 10 s (Pianca et al. 2010). The effective longshore drift is predominately from south to north (Bittencourt et al. 2002, 2005), but with local reversals (Neves 2003) (Fig. 9.3). The coast experiences a semidiurnal meso-tide regime, with a tidal range on the order of 2.2 m (Salles et al. 2000) (Fig. 9.3).

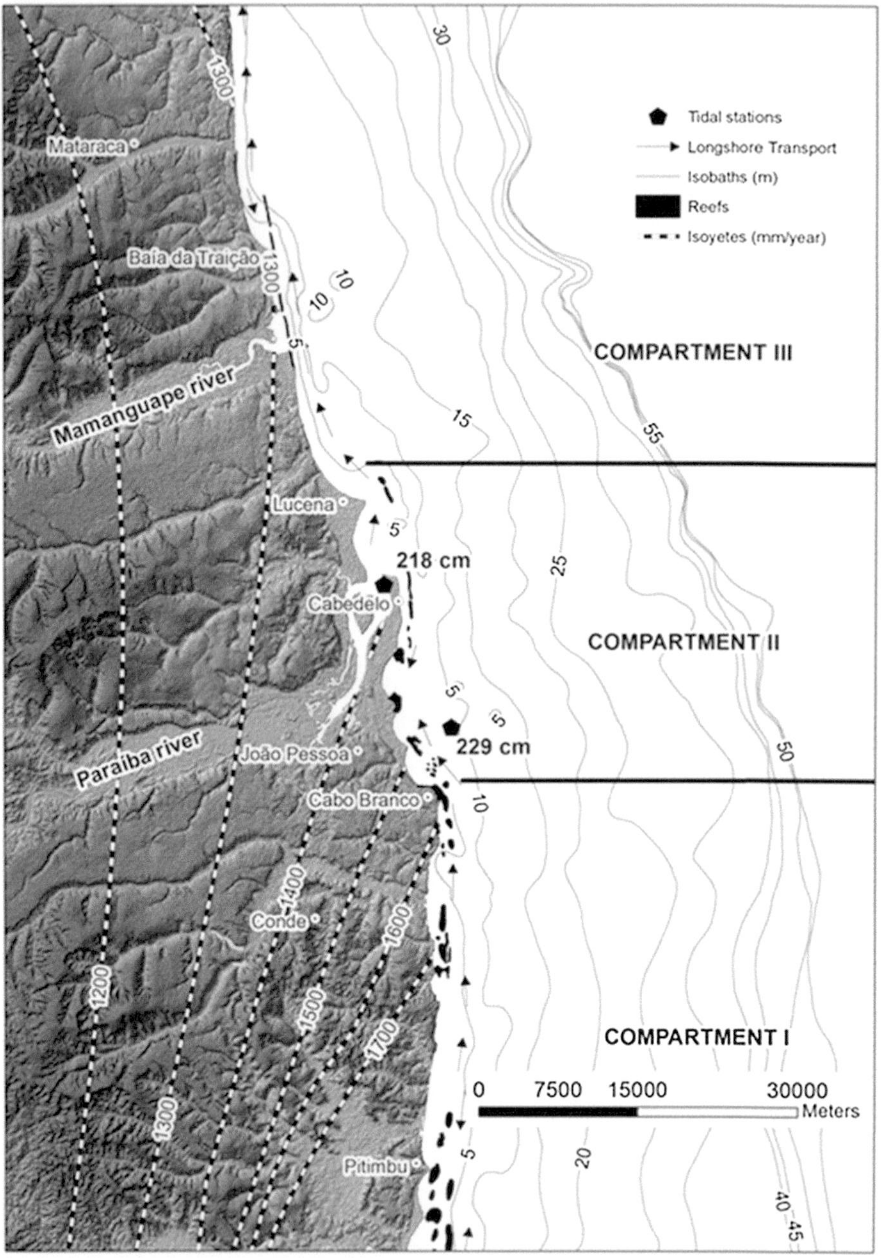

Fig. 9.3 Shaded relief of the coastal zone also showing isoyhets (mm/year), dominant longshore transport (*small arrrows*), reefs, tidal stations and respective tidal ranges, and limits of coastal compartments

Rainfall in the coastal zone ranges between 1300 and 1700 mm year^{-1} (Fig. 9.3), with the rainy season extending from March to August and the dry season, between September and February. The maximum precipitation between April and July is related to sea breeze activity, which advects medium cloud bands towards the continent, as well as the action of cold-fronts which propagate along the coast (Kousky 1979).

The area is under the influence of the South Atlantic anticyclone, with the southeast trade winds dominating throughout the year. They have moderate velocity (2 and 7 m s^{-1}) with the strongest winds between August and October. Wind intensity increases from the south to the north, associated with a decrease in rainfall, which favors the development of dunes in the northern area of the state (Compartment III) (Fig. 9.4).

9.5 Coastal Compartments and Beach Types

The Paraíba coast can be subdivided into three compartments based on coastal geomorphology (Figs. 9.1 and 9.5). Despite the fact that the beaches are predominantly reflective, there are small variations in their features due to the proximity to the cliffs of the Barreiras Formation, which supplies coarser sand, and the degree of protection provided by the inshore reefs, which affects breaker wave height and direction. These modulations in wave height, direction and sand size impact shoreline morphology and beach state.

Compartment I extends for 50 km from the Pernambuco border to Cabo Branco and has a range of features illustrated in Fig. 9.6. These include: fringing reefs bordering the shoreline (Figs. 9.1 and 9.6a); active cliffs of the Barreiras Formation and outcrops of the Cretaceous units of the Paraíba Sedimentary Basin (Fig. 9.6b) which are often protected from waves by a narrow sand terrace, often vegetated, or even the beach prism itself (Fig. 9.6b). However, this does not prevent pluvial erosion from producing large gullies on these units (Fig. 9.6c), which supply the beaches with sediment. In parts of this compartment, especially where the cliffs are active or close to the shoreline the beach sediments coarsen (Fig. 9.7). The beaches are typically reflective, with occasional low tide terraces (low-energy intermediate beaches, i.e., ridge-runnel or low-tide beach terrace), which might also be a result of higher RTR values (i.e. R+LTT beaches) (Masselink and Short 1993).

Although there are several coral-algal reefs that are easily seen in satellite images and aerial photos, they rarely emerge even during low tide. In the southern extremity of this compartment, the greater availability of finer sediment (Fig. 9.7), has resulted in the formation of salients in the back reef area, such as in the settlement of Pitimbu (Fig. 9.6d). Cabo Branco, the northern boundary of this compartment is, an iconic point of the state of Paraíba (Fig. 9.6e).

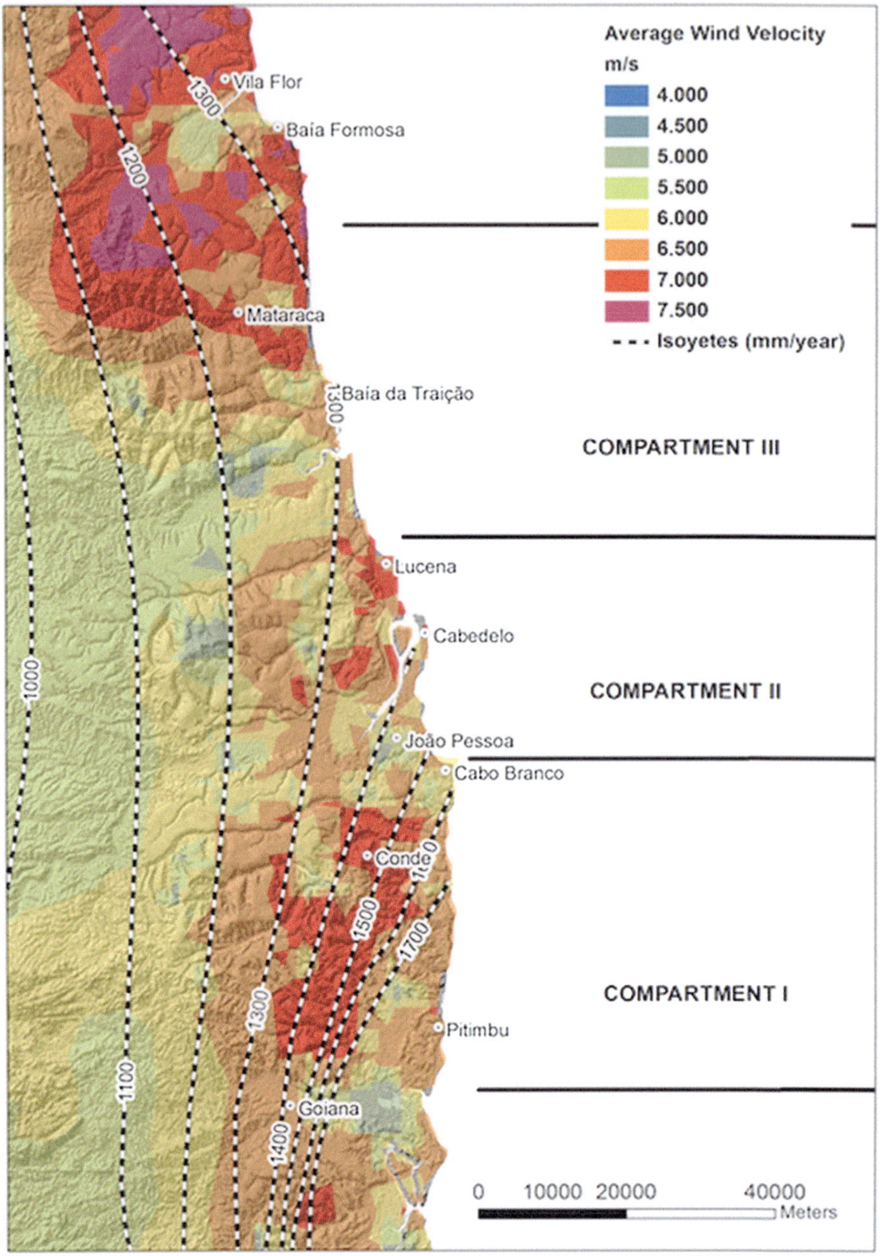

Fig. 9.4 Winds velocities at 10 m altitude, isoyhets and limits of coastal compartments. Northward increase in wind velocities is associated with a decrease in precipitation, and favors development of coastal dunes in the northern portion of the state (Wind speed data from Amarante et al. (2001))

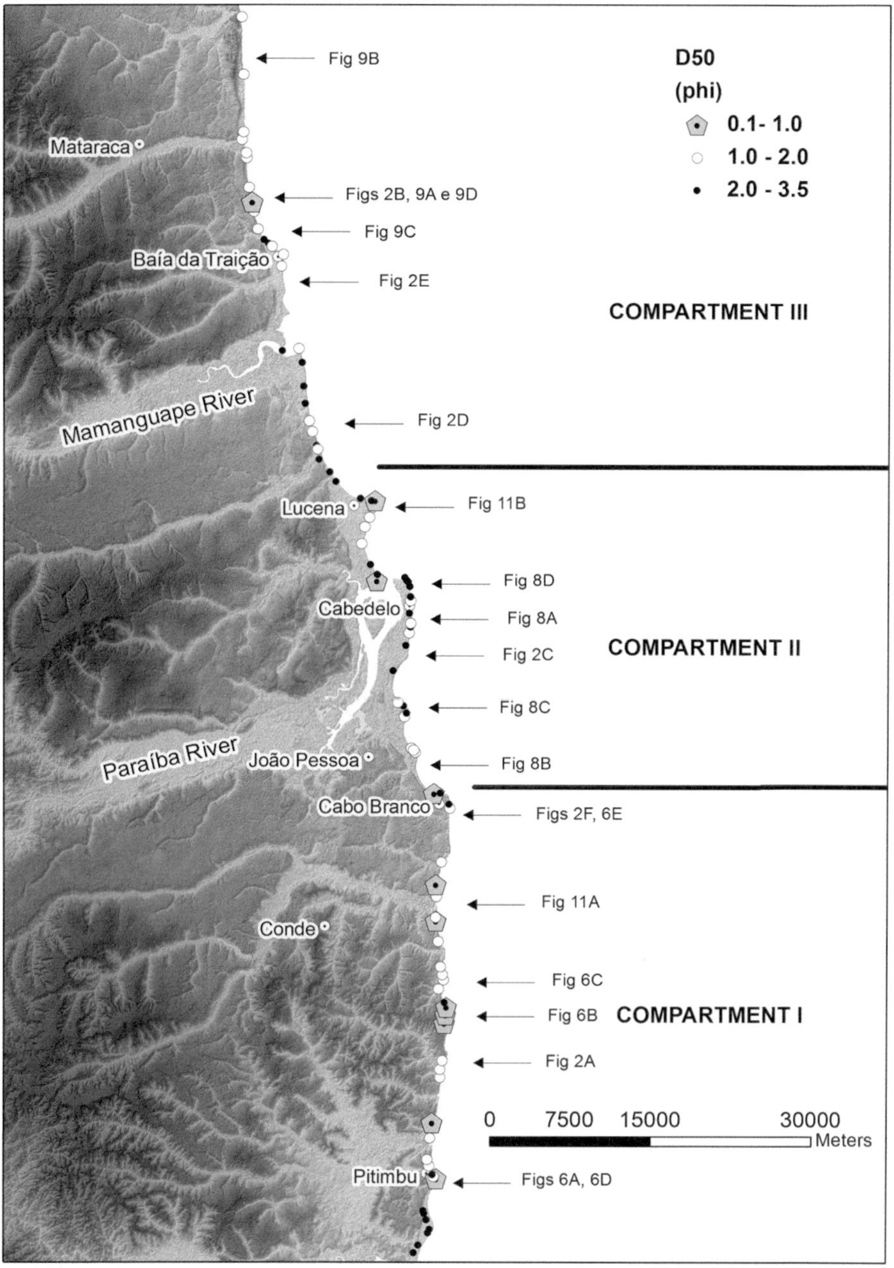

Fig. 9.5 Median (D_{50}) grain size of beach sands along the Paraíba state, and coastal compartments discussed in text. Arrows indicate location of photographs

Fig. 9.6 Field aspects of Campartment I. (**a**): Active sea cliffs fronted by reef build ups; (**b**): Active cliffs fronted by vegetated talus deposits; (**c**): Pluvial erosion of the Barreiras Formation providing sediments to the coastline; (**d**): Mega-salient formed by the presence of offshore reefs; (**e**): Active cliffs at Cabo Branco, an iconic landmark of the Paraíba state (See Fig. 9.5 for location of photographs)

Compartment II extends for 40 km from Cabo Branco to the settlement of Lucena and contains the Paraíba River mouth. Quaternary deposits reach their greatest extent and active cliffs are absent (Fig. 9.8a), with paleo-cliffs completely covered with vegetation (Fig. 9.8b). Fringing and beachrock reefs outcrop during low tides, protecting the shoreline, decreasing wave energy and favoring the development of mega-salients, such as in Lucena, Cabedelo and Ponta do Hotel Tambaú (Fig. 9.8a, c). The coral-algal reefs present in this compartment, especially in front of Cabedelo, probably grew on beachrock reefs, resulting in their nearly straight character. The development of salients has helped retain sediments in this compartment. These sediments are finer resulting (Fig. 9.7) in lower gradient beaches. The beaches in this

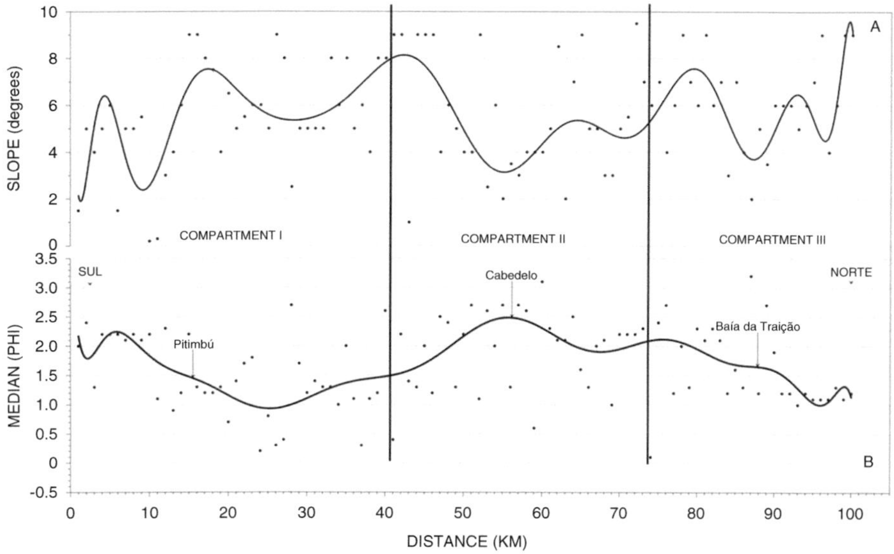

Fig. 9.7 Longshore variation of Median (D50) and beach slope at the State of Paraíba. Sample positions were projected to a north-south straight line. Therefore total distance (100 km) in the graph is not the same as the actual shoreline length

area are usually completely protected from waves during the low tide, thus presenting characteristics of sheltered beaches (Fig. 9.8d). Between the reef formations and the shoreline is a lagoon, which has transversal bars and seagrass coverage, especially in the northern area of this compartment (Fig. 9.8d).

Compartment III extends for 55 km from Lucena to the Rio Grande do Norte border. This compartment has a straighter coast and coral-algal reefs are absent, except for the long beachrock reef that blocks the entrance of the Traição Bay, as mentioned before (Fig. 9.2e). Cliffs are close to the shoreline and are active in some sectors (Fig. 9.9a). The Barreiras Formation is covered by aeolian deposits, which increase in size northwards with the dunes stabilized by vegetation (Fig. 9.9b). Moreover, these dunes represent the main heavy mineral extraction site of the entire Brazilian coastal zone (Mataraca). Salients have formed in areas protected by beachrock reefs, such as in the Traição Bay (Fig. 9.9c). The beaches in this region have a greater exposure to wave action combined with an increase in grain size and slope (Fig. 9.5) and are typically reflective to low-tide terrace in character (Fig. 9.9d).

9.5.1 Shoreline Behavior

Based on field observations during overflights and through the comparison of vertical aerial photographs from different dates, Neves (2003) and Neves et al. (2006), evaluated the behavior of the Paraíba coastline, and classified it into four categories (Fig. 9.10):

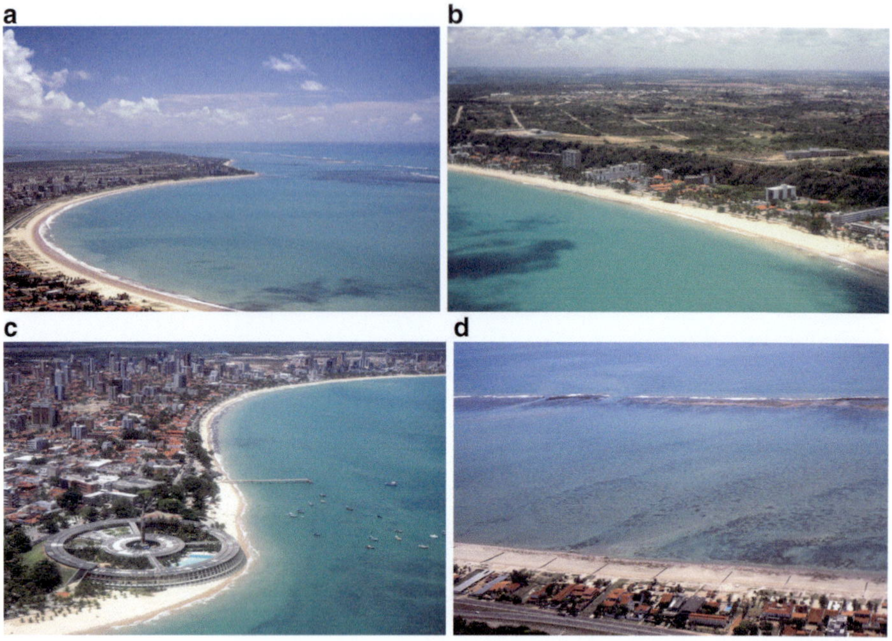

Fig. 9.8 Field aspects of Campartment II. (**a**): Mega-salients formed in lee of offshore reefs have resulted in extensive progradation of the shoreline in the vicinity of the Paraíba river mouth; (**b**): Inactive sea cliff left behind by shoreline progradation; (**c**): Mega–salient and hotel at Tambaú beach. The hotel was built in 1971 and since then has maintained its position relative to the shoreline. Thus, the present shoreline configuration does not reflect erosional retreat as the photograph might suggested at first sight; (**d**): a shallow lagoon separating the beach from the offshore reefs. Note the presence of shore-transverse sand waves on the lagoon floor (See Fig. 9.5 for location of photographs)

Eroding Coast includes all sectors of the coastline showing features indicative of erosional retreat, such as active non-vegetated cliffs of the Barreiras Formation, micro-cliffs on sandy terraces with collapsed vegetation, damage to cultural features, etc. (Fig. 9.11a). This category covers approximately 42 % of the Paraíba coastline, and is concentrated in Compartments I and III.

Prograding Coast includes the sectors of the coastline that are protected by fringing and bachrock reefs that induce the formation of sandy shoreline salients. This category also includes small indentations in the coastline, which favor sediment trapping. In these sectors, sandy terraces and well-developed foredunes are commonly found (Fig. 9.11b). Thirty-three percent of the Paraíba coastline is prograding, with a higher concentration in Compartment II, characterized by a tendency of sediment accumulation during the Holocene.

Equilibrium Coast this category includes the sectors of the Barreiras Formation cliffs that are separated from the direct action of waves by talus deposits and vegetated sandy terraces located at the base of these cliffs (Fig. 9.11c). In these sectors,

Fig. 9.9 Field aspects of Campartment III. (**a**): Active sea cliffs fronted by vegetated talus; (**b**): Blowouts along the shoreline and on top of the Barreiras Formation; (**c**): Beachrock reef at Traição bay; (**d**): Exposed low-tide terrace beach during low tide (See Fig. 9.5 for location of photographs)

the coastline does not experience significant oscillations, even considering seasonal variations in the beach prism. Twenty-one percent of the coast is classified in this category. These stable sectors of the coastline are more commonly found in Compartment I, and are associated with indentations in the coastline between small promontories supported by outcrops of carbonate formations of the Paraíba Basin.

Coastline Stabilized by Engineering Works includes all stretches where various types of coastal protection structures (e.g. walls, breakwaters, gabions, rip-raps) have been built (Fig. 9.11d). This category includes 4 % of the coast and is predominantly found in the central compartment, which is also the most densely populated. In these sectors, the recreational quality of the beaches is very low.

9.6 Beach Safety

Information on beach safety is very scarce for the study area. Observations of local newspapers indicate that the most dangerous beaches in Paraíba are located in Compartment I, where reflective beaches predominate (Fig. 9.12). Of these beaches, the Barra do Gramame beach (Fig. 9.13a) is the most dangerous, according to the

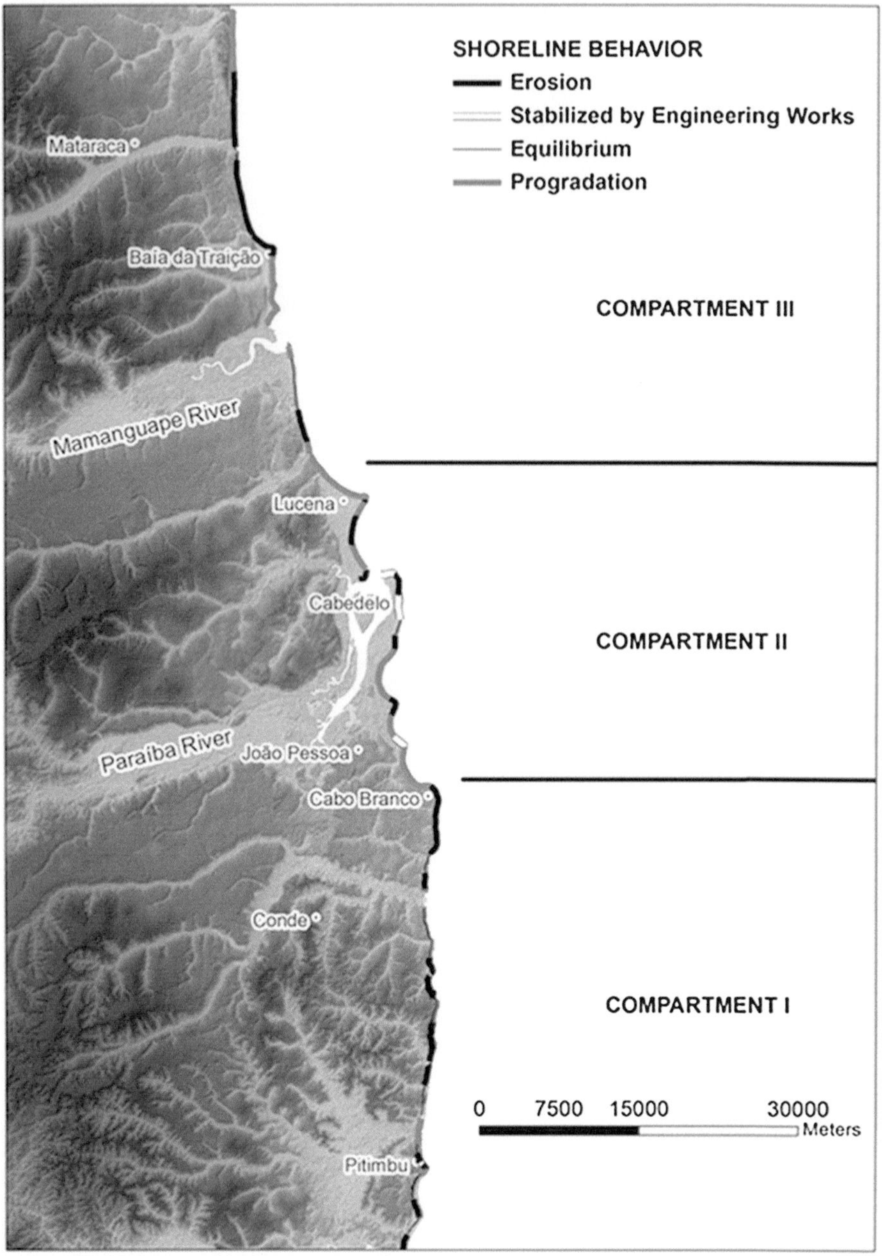

Fig. 9.10 Shoreline behavior at Paraíba State

Fig. 9.11 Examples of categories of shoreline behavior adopted in this work. (**a**): Erosion; (**b**): Prograding foredune ridges; (**c**): Equilibrium; (**d**): Stabilized by engineering works

Fire Department of Paraíba, because of the currents present at the mouth of the Gramame river. According to the Fire Department, drownings result when beachgoers attempt to cross the river mouth during high tide. In the other two compartments there are no references to dangerous beaches, except for the Ponta do Lucena beach (Fig. 9.13b). This beach is located on the left margin of the Paraíba River, where there are perpendicular sand banks to the beach indicating the action of strong tidal currents.

Due to their predominantly reflective characteristic, there would be no other apparent reason for the Paraíba beaches to be hazardous for swimming. However, the concentration of the majority of incidents in Compartment I seems to result from the combination of two factors: (i) higher concentration of coastal population in this sector, and (2) the steep slope of these beaches, with deep water right off the beach face, in combination with exposure to higher waves, as previously mentioned.

In Compartment II, which has the highest population density in the state, the presence of reefs, most of which are exposed at low tide, shelter the beaches making them safer for beachgoers. In Compartment III, although there are no reported incidents, probably due to low population density, swimming risk would be similar or even higher to that observed in Compartment I because of the lack of protection afforded by the presence of offshore reefs, resulting in higher energy reflective and low-tide terrace beaches (Fig. 9.9d).

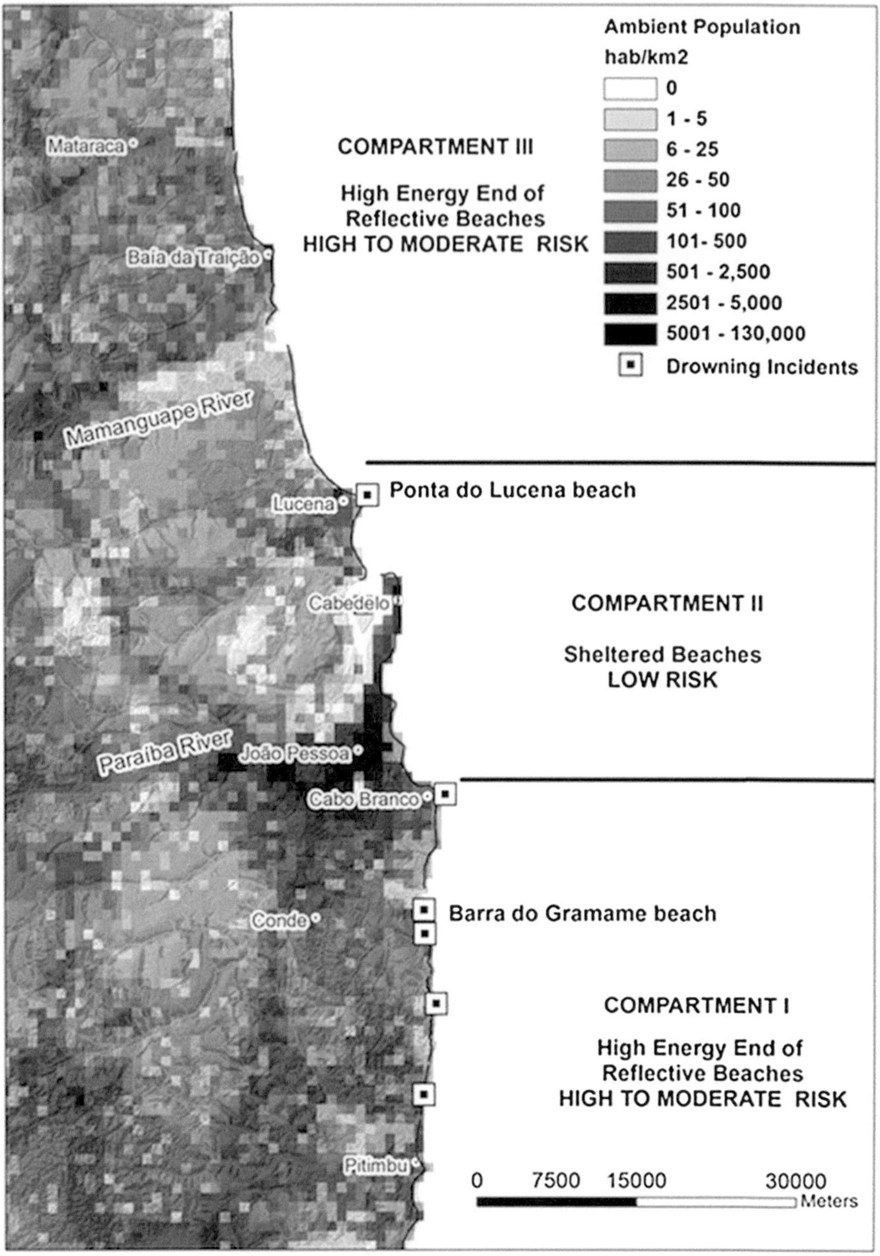

Fig. 9.12 Distribution of ambient population along the coastal zone and risk for beach goers in each coastal compartment. Also shown is the location of beaches where most incidents are reported. Ponta do Lucena and Barra do Gramame beaches are where most drownings take place. The Ambient Population is from the Landscan 2010 dataset (The LandScan 2010™ High Resolution global Population Data Set copyrighted by UT-Battelle, LLC, operator of Oak Ridge National Laboratory under Contract No. DE-AC05-00OR22725 with the United States Department of Energy)

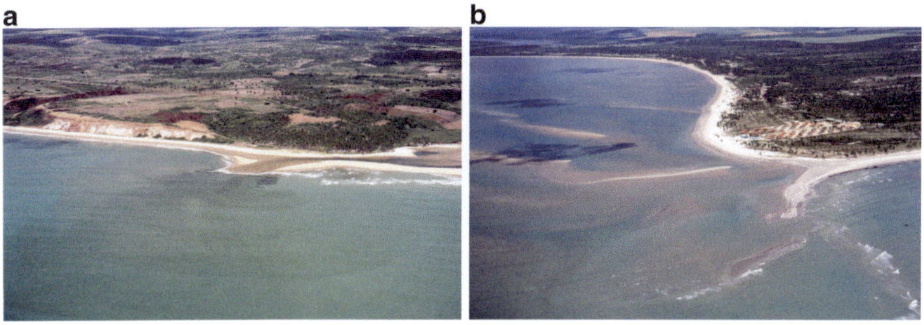

Fig. 9.13 The two most dangerous beaches of the Paraíba state. (**a**): Barra do Gramame beach where drownings usually happen when beach goers try to cross the river mouth during high tide; (**b**): Ponta do Lucena beach where drownings appear to be associated with strong tidal currents at the mouth of the Paraíba river

9.7 Paraíba Beaches and Geological Heritage

Overall, the Paraíba beaches are similar to the Alagoas beaches. Despite their relatively short coasts, the control played by geological heritage in the modal morphodynamic stages of beaches is evident. The main source of sediments for the beaches is the erosion of the Barreiras Formation cliffs, resulting in predominantly medium- to coarse-grained sediments with different degrees of wave exposure, afforded by the biogenic and beachrock reefs. The lower breaker waves result in a dominance of reflective beaches, with low-tide terrace beaches in the most exposed sectors, somewhat modified by tides (higher RTR values). In Compartment II, where Holocene accumulation of sediments predominated, sediments are fine-grained, with a more protected shoreline due to fringing reefs, resulting in low-energy sheltered beaches, with shallow "lagoons" separating the shoreline from the reefs.

In Compartment I, the long-term tendency for coastal retreat results in a steeper shoreface profile. The presence of reefs does not significantly reduce wave energy and the beaches although reflective to low-tide terrace, are exposed to higher waves. In Compartment III, with the exception of the Traição Bay sector that is protected by the beachrock, there are no reef formations close to the shore. Although these beaches are exposed to higher levels of energy, they maintain their reflective characteristic, illustrating the control of sand size in contributing to modal beach state.

A more subtle, though still important, control in the determination of the characteristics of the beaches of Paraíba is the structure of the Paraíba Sedimentary Basin itself (Fig. 9.14). Compartments I and II are located in the Paraíba Basin, where the thickest section of carbonate rocks is found (Maria Farinha and Gramame formations). The greatest thickness of these rocks is observed in the João Pessoa Graben. The northen limit of this graben coincides approximately with the limits between Compartments II and III. In turn, Compartment III is found upon the Mamanguape High, a high of the crystalline basement upon which the thickness of the sedimentary rocks of the Paraíba Basin is minimum.

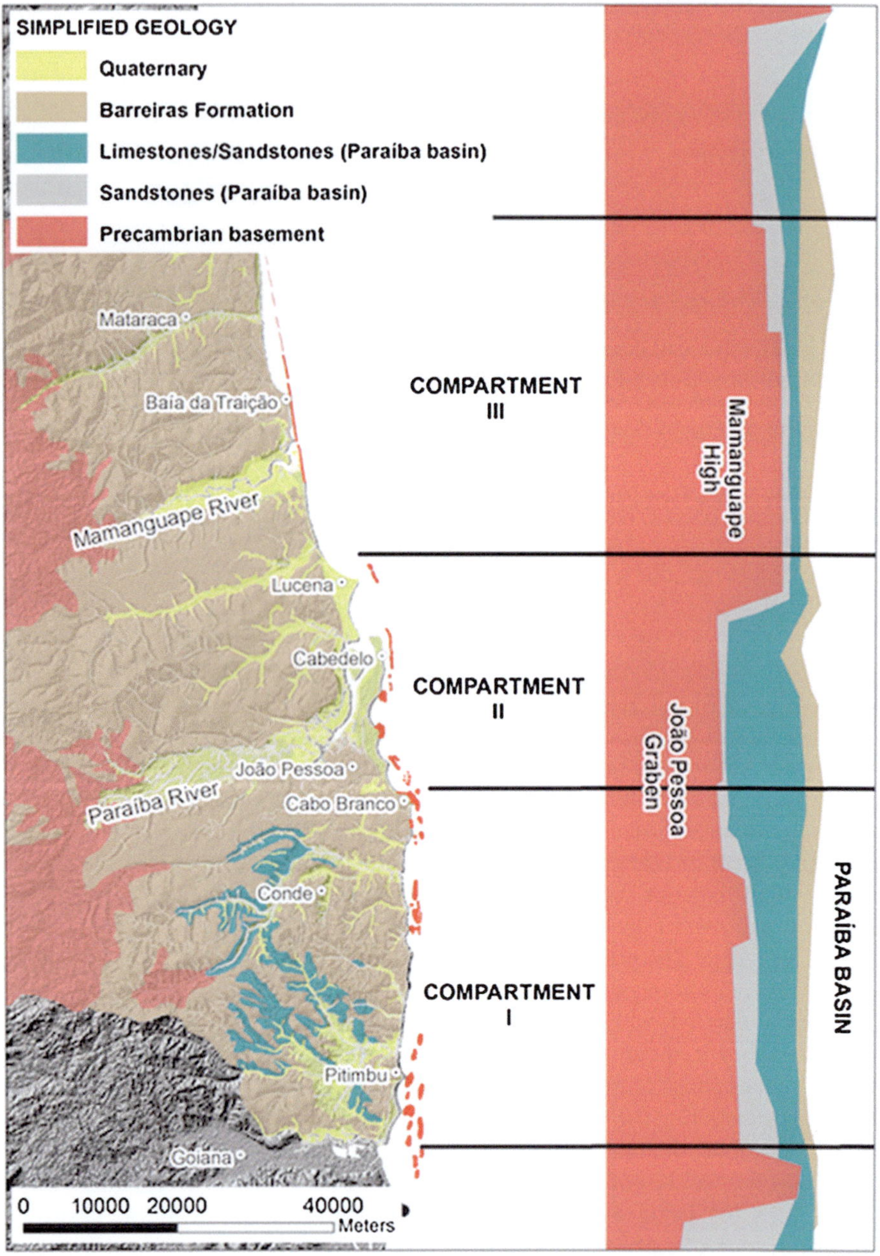

Fig. 9.14 Geological framework of the Paraíba coastal zone and its apparent control on the physiography of the coastal compartments. See text for discussion (Modified from Barbosa and Lima Filho 2005)

Thus, in this starved coast, where the coastline presents a long-term erosive tendency, the presence of carbonate rocks in Compartments I and II, at the base of the Barreiras Formation, favored the development of consolidated substrates, left behind by coastline retreat, upon which reefs developed, reducing wave energy along the coast. In Compartment I, this aspect combined with coarse beach sands reinforces the reflective/sheltered characteristic of these beaches. In turn, in Compartment II, the long-term tendency of sediment accumulation might have been favored by the presence of the João Pessoa Graben. The more depositional character of this compartment results in finer beach sand which, associated with the reef protection has produced sheltered low gradient beaches. Finally, Compartment III, located upon a high of the Paraíba Basin basement, does not have conditions favorable for development of consolidated substrates near the shoreline. The coarser beach sands contribute to the modal reflective state of this sector, despite experiencing higher wave energy.

Acknowledgments This study is a contribution of inctAmbTropic—Brazilian National Institute of Science and Technology for Tropical Marine Environments, CNPq/FAPESB Grants: 565054/2010-4 and 8936/2011. ACSP Bittencourt and JML Dominguez thank CNPq for a Research Fellowship.

References

Almeida FFM, Hasui Y, Neves BBB, Fuck RA (1981) Brazilian structural provinces: an introduction. Earth Sci Rev 17:1–29

Amarante OC, Zack J, Brower M, Sá AL (2001) Atlas do Potencial Eólico Brasileiro. Ministério de Minas e Energia, Brasília. 45p

Barbosa CMBM (1983) Geoquimica dos sedimentos carbonáticos da plataforma continental do Estado da Paraiba. Dissertação de Mestrado, Univ. Federal de Pernambuco, 100p

Barbosa JA, Lima Filho M (2005) Os domínios da bacia da Paraíba. 3 Cong Bras P&D Petróleo e Gás, Rio de Janeiro CD-ROM

Barbosa JA, Neumann VH, Lima Filho M, Souza EM, Moraes MA (2007) Estratigrafia da faixa costeira Recife-Natal (Bacia da Paraíba e Plataforma de Natal, NE Brasil). Estud Geol 17(2):3–30

Beurlen K (1967a) Estratigrafia da faixa sedimentar costeira Recife-João Pessoa. Bol Geol 16(1):43–53

Beurlen K (1967b) Paleontologia da faixa sedimentar costeira Recife-João Pessoa. Bol Geol 16(1):73–79

Bittencourt ACSP, Martin L, Dominguez JML, Silva IR, Sousa DL (2002) A significant longshore transport divergence zone at the Northeastern Brazilian coast: implications on coastal Quaternary evolution. Anais Acad Bras Ciênc 74(3):505–518

Bittencourt ACSP, Dominguez JML, Martin L, Silva IR, Sousa DL (2005) Longshore transport on the northeastern Brazilian coast and implications to the location of large scale accumulative and erosive zones: an overview. Mar Geol 219:219–234

Brito Neves BB, Santos EJ, Van Schmus WR (2000) Tectonic history of the Borborema Province, northeastern Brazil. In: Cordani UG, Milani EJ, Thomaz Filho A, Campos DA (eds) Tectonic Evolution of South America. 31 International geological congress, Rio de Janeiro, pp 151–182

Kousky VE (1979) Frontal influences on northeast Brazil. Mon Weather Rev 107:1142–1153
Mabesoone JM, Alheiros MM (1991) Nota Explicativa dos Mapas. In: Revisão Geológica da faixa sedimentary costeira e Pernambuco, Paraíba e parte do Rio Grande do Norte. Estudos e Pesquisas, Série B 10:21–31
Masselink G, Short AD (1993) The effect of tide range on beach morphodynamics and morphology: a conceptual model. J Coast Res 9:785–800
Neves SM (2003) Erosão Costeira no Estado da Paraíba. Thesis, Universidade Federal da Bahia, Brazil
Neves SM, Dominguez JML, Bittencourt ACSP (2006) Paraíba. In: Muehe D (ed) Erosão e Progradação no litoral brasileiro. Ministério do Meio Ambiente, Brazil, pp 173–178
Pianca C, Manzzini PLF, Siegle E (2010) Brazilian offshore wave climate based on NWW3 reanalysis. Braz J Oceanogr 58:53–70
Rossetti DF, Bezerra FHR, Góes AM, Valeriano MM, Andrades-Filho CO, Mittani JCR, Tatumi SH, Brito-Neves BB (2011) Late Quaternary sedimentation in the Paraíba Basin, Northeastern Brazil: landform, sea level and tectonics in Eastern South America passive margin. Palaeoogr Palaeoclimatol Palaeoecol 300:191–204
Rossetti DF, Bezerra FHR, Dominguez JML (2013) Late Oligocene–Miocene transgressions along the equatorial and eastern margins of Brazil. Earth Sci Rev 123:87–112
Salles FJP, Bentes FCM, Santos JA (2000) Catálogo de Estações Maregráficas. Fundação de Estudos do Mar, Rio de Janeiro
Xavier RA, Dornellas PC, Maciel JS, Bú JC (2012) Caracterização do regime fluvial da bacia hidrográfica do rio Paraíba – PB. Rev Tamoios 08(2):15–28

Chapter 10
Tropical Sandy Beaches of Pernambuco State

Pedro de Souza Pereira, Tereza Cristina Medeiros de Araújo, and Valdir do Amaral Vaz Manso

Abstract Pernambuco state is located in the Northeast region of Brazil and has a 187 km long coastline containing a range of coastal environments including beaches, estuaries, reefs and headlands all with varying degrees of preservation and occupation. The coast of which 156 km (83 %), is sandy beaches, can be divided into three physiographic sectors. Sand deposits are limited to a narrow strip of the Holocene beach area, deposited between the low-water line and the Holocene terraces. They are predominately quartz, with some beaches containing heavy minerals. The shoreface and the inner shelf has a highly complex underwater topography owing to the presence of beachrock, coral and algae reefs structures at several different depths, as well as paleochannels and other features. The coast has three well defined zonation in terms of wave energy: (1) Goiana to Paulista city, has a high wave convergence zone offshore with a high dissipation/diffraction rates across the inner shelf due to its wide reefs and shallow shelf resulting in low waves at the shore; (2) Paulista to Cabo city, an intermediate wave convergence zone marked by few divergence zones at low tide; (3) Ipojuca to São José da Coroa Grande city there are fewer convergence zones but the inner shelf has numerous discontinuous reefs inducing refraction and diffraction patterns. This pattern is controls the shoreline morphology and beach morphodynamic behavior. Based on this the beaches can be divided into three sectors: with the reef protected northern beaches are tide-modified; the exposed southern beaches are wave-dominated; and the middle sector beaches where the beachrock reefs controls the relation between wave and tide are geomorphologically controlled.

P.d.S. Pereira (✉) • T.C.M. de Araújo
Geological Oceanography Laboratory, Department of Oceanography,
Center of Geoscience and Technology, Federal University of Pernambuco,
Av. Arquitetura, s/n, Cidade Universitária, Recife, PE 50740-550, Brazil
e-mail: pedro.pereira@pq.cnpq.br; tcma@ufpe.br

V.d.A.V. Manso
Marine Geophysical and Geology Laboratory, Department of Geology,
Center of Geoscience and Technology, Federal University of Pernambuco,
Av. Arquitetura, s/n, Cidade Universitária, Recife, PE 50740-550, Brazil
e-mail: vazmanso@uol.com.br

A.D. Short, A.H. da F. Klein (eds.), *Brazilian Beach Systems*, Coastal Research Library 17, DOI 10.1007/978-3-319-30394-9_10

10.1 Introduction

10.1.1 Geology

Pernambuco state is located in the Northeast region of Brazil and has a 187 km long coastline containing a range of coastal environments including beaches, estuaries, reefs and headlands all with varying degrees of preservation and occupation (Fig. 10.1). The coast of which 156 km (83 %), is sandy beaches, can be divided into three physiographic sectors: the northern sector between Goiana and Olinda cities is characterized by a extensive coral and algae reef development and linear sandy beaches; the middle sector between Paulista and Cabo de Santo Agostinho cities has linear beachrock reefs; and the southern sector between Cabo de Santo Agostinho and São José da Coroa Grande city, has a irregular coastline with several embayed beaches, where several well developed coral and algae reefs can be found (Coutinho et al. 1994).

The coastal plain was developed during the Quaternary and is of considerable geological and geomorphological complexity, since it represents a transitional between marine and continental environment, where several different geomorphological features can be found. It is dominated by meso-Cenozoic sediments of which Quaternary sediments are the most important deposits. Its development is closely related to sea-level fluctuations, sediment supply and the dominant coastal processes that control the morphology and the sediment distribution. It has a young transgressive character with a predominance of mangrove-lined estuaries.

Another striking feature is the absence of well-developed dune system along the whole littoral. This characteristic is due to the lack of favorable conditions for sand accumulation such as large beaches and the predominance of lower velocity wind most of time and the low reef-attenuated breaker wave climate, which reduced wave height and onshore sediment transport.

The beachrock, coral and algal reefs play a decisive role in the wave energy modification, distribution of sediments and consequent changes in coastal morphology. The reef lines protect the coast from the direct wave attack as waves are attenuated as they propagate across the reef crests. However, at the reef edges wave diffraction and refraction occurs and can cause occasional erosion, besides the formation of a more crenulate coastline with alternating small embayed beaches and protruding sandy forelands.

10.1.2 Climate

The climate of the equatorial region of Brazil is determined by the Intertropical Convergence Zone (ITCZ), giving rise to northeasterly and southeasterly trade winds, which converge over this region. The ITCZ migrates meridionally during the year, reaching farthest north in August-September and coming closest to the Equator in March-April. It exerts significant control over the regions rainfall and wind regimes (Nimer 1979; McGregor and Nieuwolt 1998).

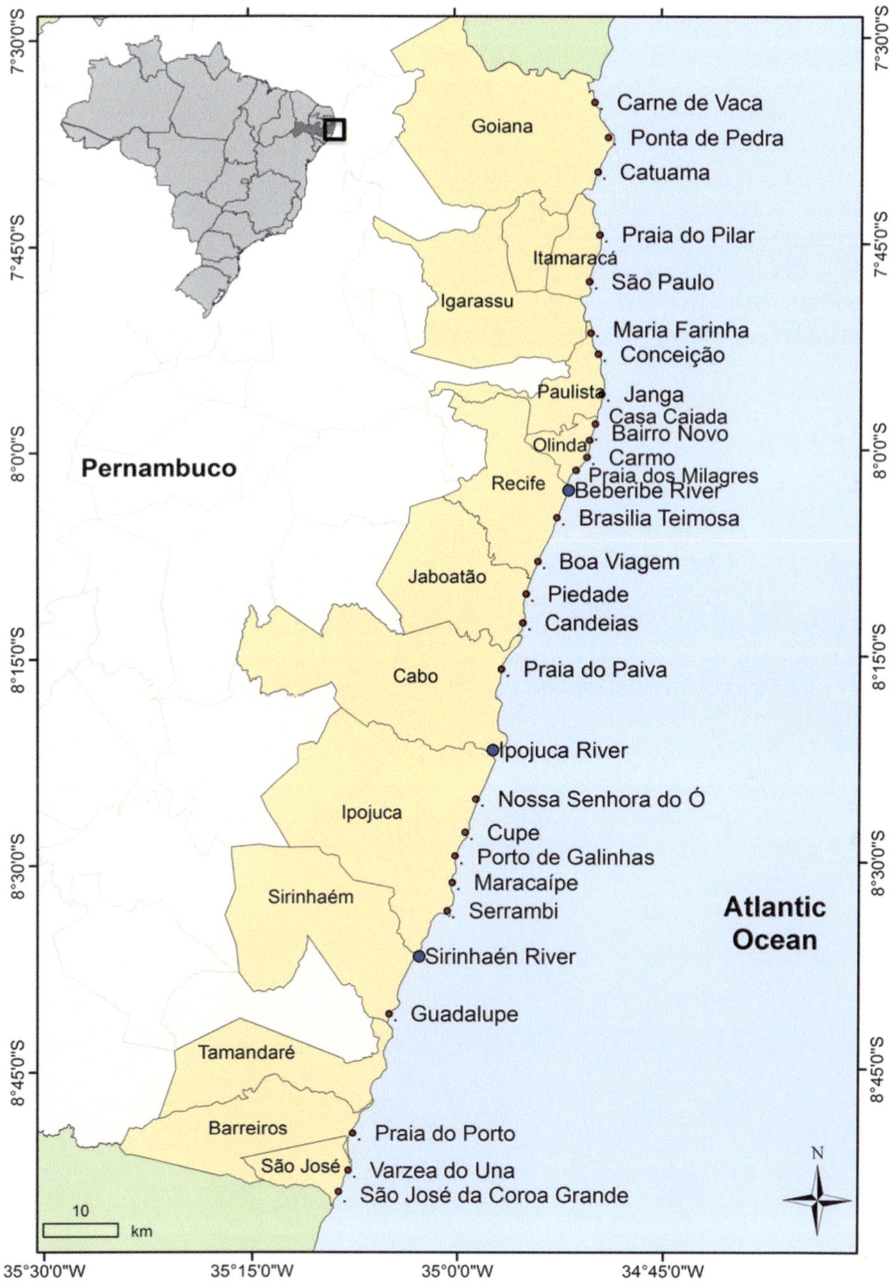

Fig. 10.1 Pernambuco state location highlighted in the Brazil inset and its coastal cities and major beaches

The Pernambuco coastal zone has a Tropical Atlantic climate (Aw Koppen classification) with an annual average temperature of 27 °C and annual rainfall around 2000 mm, with distinct dry and wet seasons. This regime is characterized by a bimodal pattern with a dry season (or summer) that starts in October and ends in March; and a rainy season (winter) that starts in April and ends in September. The wind regime is governed by the general atmospheric pressure distribution pattern of the South Atlantic Ocean (Nimer 1979). During the dry season, the main winds are from south-southeast with a high occurrence of east-east-northeast winds, while during the rainy season, the southeast to south-southeast winds predominates. The mean wind velocities vary from 3.1 to 4.7 m s^{-1} with highest velocities between April and September (Cavalcanti and Kempf 1970).

10.1.3 Waves and Tides

Wave climate may be defined as the set of prevailing wave conditions that occur in a particular oceanic or coastal region over a defined time interval (Inman and Masters 1994). Therefore, based on the available dataset, the incident waves are predominately from southeast (Fig. 10.2) and generated mainly by trade winds, particularly during the rainy season. Less energetic waves from the south-southeast, generated by higher latitude cold fronts, are also noticeably present, (Pianca et al. 2010).

The mean wave height is between 1.5 and 2 m and the mean peak period between 8-and 9 s. The highest and the longest waves observed arrive from south-southeast with a height of 4.8 m and period of 14.3 s, while the lowest wave arrive from southeast with a height of 0.21 m and the shortest wave is from south with a peak period of 4.7 s. At the shore however breaker waves height is often attenuated by the extensive reef systems.

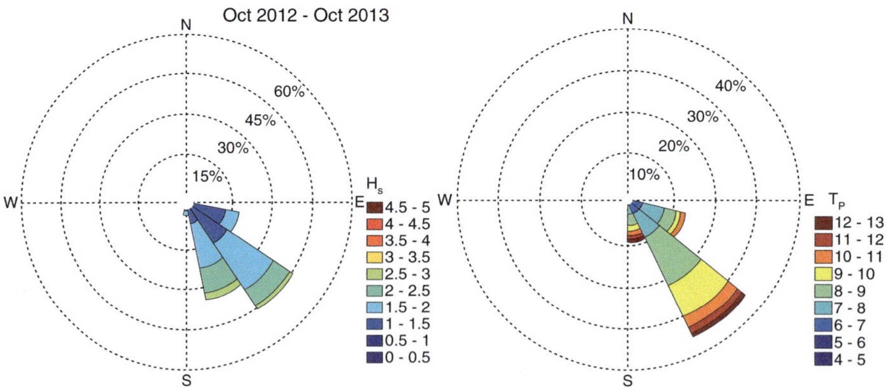

Fig. 10.2 Wave climate offshore of Pernambuco coast: significant wave height rose (*left* figure); peak period rose (*right* figure). Data collect by the National Buoy Program (PNBOIA – 69,154, Longitude: −34.55° Latitude: −8.15°)

The tides are basically controlled by astronomic sources with a very small influence of non-periodic sources that can generate low storm surges. At the Recife harbor, the tides are semidiurnal with a F=0.09 and a period of 12.42 h, with two high and low tides per lunar day, with a small diurnal inequality.

Regular tide forecasting is provided by the Directory of Hydrography and Navigation from Brazilian Navy, with the tides at Recife having a mean height of 1.67 m, with the spring tides reaching 2.7 m and classified as meso-tidal.

10.1.4 Coastal Sediments

Sand deposits are limited to a narrow strip of the Holocene beach area, deposited between the low-water line and the Holocene terraces. They are predominately quartz, with some beaches containing heavy minerals. Due to the prevailing weather conditions at the coast and low reef-attenuated breaker wave heights the beaches are practically devoid of dunes as mentioned before.

Mud deposits associated with estuarine mangroves are present along much of the Pernambuco estuarine coast. The sandy fraction is mainly composed by quartz, mica and small contents of heavy minerals. The clay fraction is composed mainly of poorly crystallized kaolinite, montmorillonite and illite. In general, the coastal sediments have high percentage of biodetritus.

10.1.5 Shoreface Features and Bathymetry

The Pernambuco shoreface and the inner shelf has a highly complex underwater topography owing to the presence of beachrock, coral and algae reefs structures at several different depths, as well as paleochannels and other features (Fig. 10.3). The three types of reefs are spatially distributed based on their age and vary considerably in height and width as well as their longshore continuity. These differences create a complex scenario along the whole shelf and coast that significantly affects the inshore wave climate and the hydrodynamic processes. Thus the impact of these structures is essential to the understanding of the coastal processes.

According to Dominguez et al. (1990) the reefs of Pernambuco coast, that are called in this work coastal reefs or just reefs, are basically of three types: beachrock, coral and algae reefs. Beachrocks are lithified beach sediments from the intertidal zone that are consolidated by the calcium carbonate precipitation. Normally they are shore parallel narrow and straight (Fig. 10.4). The coral and algae reefs are generally biological reef that have grown on top of the beachrock as pointed by Dominguez et al. (1990). Those authors were the only ones to characterize the reef system along the whole coast of Pernambuco, and were not able to significantly differentiate the coral and algae reef in detail as is shown in Fig. 10.3. Due to biological growth on top of the beachrock, this type of reef varies considerably in width and length and are not as straight as the beachrock reefs.

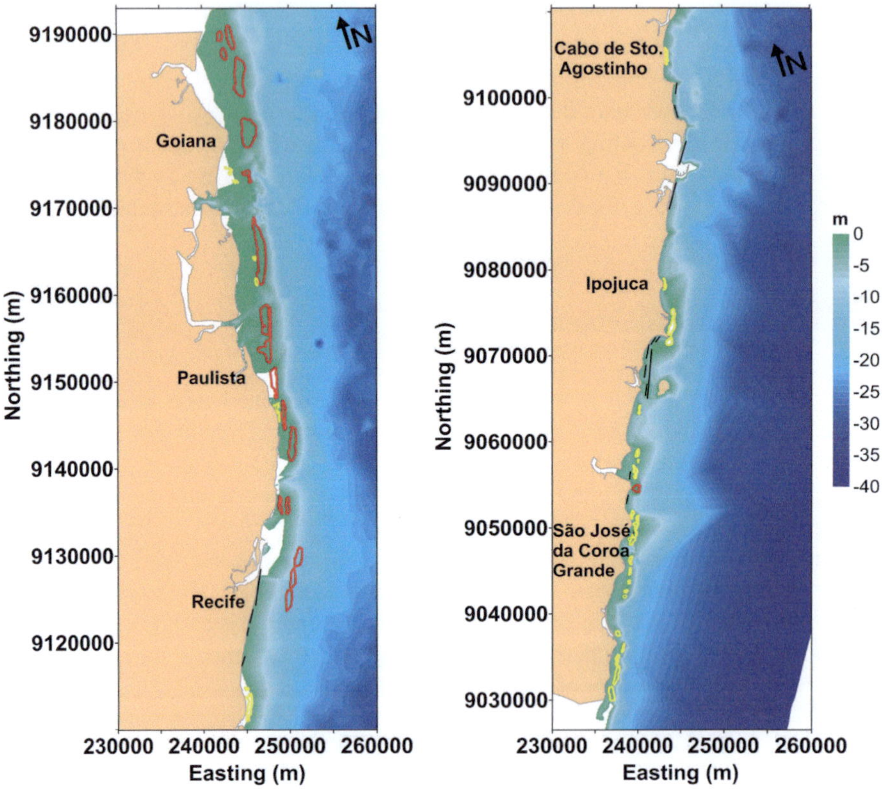

Fig. 10.3 Bathymetry along Pernambuco state coastline highlighting the reefs present along the coast. The *red lines* are coral and algae reefs with depth around 2 m and the *yellow lines* are coral and algae reefs at depths less then 1 m. The *black lines* are beachrock reefs parallels to the coast (Dominguez et al. 1990). The coast was split in half for better illustration

The reef lines significantly modifies wave energy arriving at the coast, not only by substantially reducing energy due to wave breaking over the reefs, but also due to the energy redistribution as the waves propagates over and around the reefs, modifying the spectral energy distribution as well as producing high frequency waves (or secondary waves). Figure 10.4 shows in detail the reef system in front of Recife that consist of three beachrock reef lines, one close to the shore and the other two on the shoreface.

Based on the morphological properties and sediment characteristics, Coutinho (1979) had divided Pernambuco shelf in to three parts: inner shelf, mid shelf and outer shelf.

The inner shelf is limited to the 20 m bathymetric isoline. It has low relief with some irregularities caused by reef lines, channels and sandy features. This shallow part of the inner shelf is covered by terrigenous sand with little gravel and mud. There is also a high quantity of biogenic sediments that are strongly reworked by the wave action.

The mid shelf is located between 20 and 40 m bathymetric isoline, and has more irregular relief, and is covered by coarse biogenic material, which can contains up

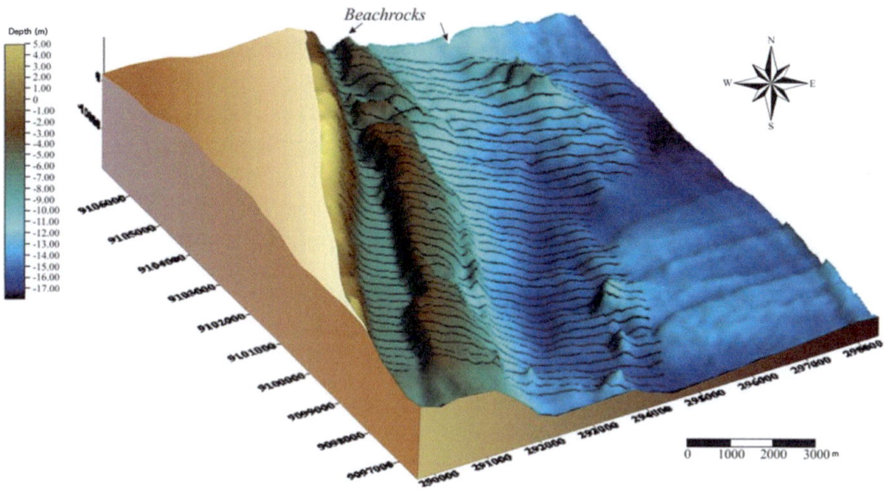

Fig. 10.4 Inner shelf detailed bathymetry along Recife city. Note the presence of two reef lines along the coast (Figure source: Gregório 2009). The *black dots* are the navigation routes

to 90 % calcium carbonate. The sediment color indicates that living algae is abundant. Also, this sediment does not show any signs of wave reworking.

The outer shelf begins beyond 40 m depth and is covered by biodetritus sand, algae gravel and gray mud. The *Halimedaceae* algae tend to be more abundant and the calcium carbonate sediment content can be higher then 75 %. The carbonate aggregation is very bioturbated, especially the sand made from algae reefs.

The shelf has a mean width of 33 km widening to 40 km in the northernmost sector, and narrowing to 30 km between Ipojuca and Sirinhaém rivers (Araújo et al. 2004). The narrowness is attributed to its structural configuration as well as to its low sedimentation rates. The shelf break ranges from 50 to 65 m with a mean depth of 60 m. The inner shelf has the steepest slope, which after 30 m becomes gentler across the mid and lower shelf.

10.1.6 Previous Beach Studies

The first studies carried out along the Pernambuco coast were reviews by Pontes and Araújo (2006). The first study was by Ottmann et al. (1959) who investigated Piedade Beach, a beach protected by two lines of beachrock reef in the metropolitan region of Recife. They classified the reefs as a beachrock, which is formed by sediment from a former shoreline. According to the authors such feature controls the beach behavior since it acts as a barrier for the incident wave energy, either reducing or in places totally dissipating the waves. Thirty-six years passed until the next work was published by Silva (1995) who studied the northern beaches of Pontas de Pedra. They observed changes in the profile caused by the low incoming wave energy, especially during spring tides.

Pontes and Araújo (2006) point out that most of the surveyed beaches where located in the central state area, the metropolitan area, which in turn is close to most of the local research centers. The authors also said that different approaches where carried out by the researches, however, the two-dimensional beach topographic surveys where the main methodology used to understand physical processes and to monitor and understand erosion issues.

Since the work of Pontes and Araújo (2006) there have been studies by Gregório and Araújo (2008), Macedo et al. (2012), Viegas et al. (2012), and Mallmann et al. (2014). In 2009 the MAI project (FINEP/UFPE 2009) was an integrated monitoring project to study hydrodynamic and sedimentological processes along the beaches of the metropolitan area. The beaches surveyed showed a variety of morphologies and behaviors. The principal conclusion was that several beaches along this length of Pernambuco coast tend to be eroded during the spring tides, specially the equinoctial. The combination of equinoctial tide with the incidence of southeast swells erodes the beach profile, especially at the beach face, berm and backshore, with the erosion even reaching the dunes.

Viegas et al. (2012) were the first to study the beaches at the northern part of the state where the beaches have a low breaker wave height and are modified by the tide. At those beaches sandbars are absent and the most striking feature is a broad low tide terrace.

Mallmann et al. (2014) where the first to investigate three-dimensional beach topography along a 45 km section of coast. They discuss the presence and absence of megarips, cusps and rhythmic bars by analyzing those features in satellite images, which in turn where used to assess the energy regime and whether the beach behavior is controlled by waves, tide or ancient morphology.

Currently the perspective is that more beaches will be studied, specifically with video and high precision GPS for longer period of time since without time series of beach change it is impossible to differentiate erosion and accretion episodes caused by tides, climate, oceanographic cycles or even anthropogenic actions (Muehe 2003). This information will provide a better understanding of the natural system and provide information for management policies of natural areas as well where structures have been built along the coast.

10.2 Pernambuco Beach Systems

The coast of Pernambuco has three well defined zonation in terms of wave energy: (1) Goiana to Paulista city, has a high wave convergence zone offshore with a high dissipation/diffraction rates across the inner shelf due to its wide reefs and shallow shelf resulting in low waves at the shore; (2) Paulista to Cabo city, an intermediate wave convergence zone marked by few divergence zones at low tide; (3) Ipojuca to São José da Coroa Grande city there are fewer convergence zones but the inner shelf has numerous discontinuous reefs inducing refraction and diffraction patterns. This pattern is controls the shoreline morphology and beach morphodynamic behavior as will be shown in the following sections.

10.2.1 Waves, Tide and Reef Interaction

The coastal reefs play a major role in modulating incident wave energy at the beach and consequently the beach morphodynamics. According to previous studies (e.g. Nelson 1994; Costa 2010) the water depth over the beachrock, algae and coral reefs determine the amount of wave attenuation across the reef and consequently the level of wave energy remaining to transport sand inside the reef and to break at the shore.

At low tide, when a large portion of the reef can emerged, waves break across the crest resulting in a limited amount of wave energy inside the reef. On the other hand, at high tide, especially during the spring tide, the level may be 2–3 m above the reef surface, allowing substantially more wave energy to propagate across the reef. As a result, tidal modulation of wave height becomes an important process with higher waves at high tide and low waves at low tide. This reef-tidal modulation can be found along the whole coast of Pernambuco where the reefs are present, as occurs on the beaches of Boa Viagem and Candeias (Fig. 10.5).

The degree of wave modulation depends on the geometrical characteristics of the reefs particularly the width and length of the reef platform, the average depth and especially roughness (Costa 2010). These factors are directly related to the geological history of the location and distribution of reef-building species in case of coral and algae reefs. In Fig. 10.5 it is possible to see the significant correlation between the tide and the breaker wave height. According to Costa (2010), the more substantial barrier reef at Candeias (Fig. 10.5 lower panel) experiences greater wave dissipation then the fringing reef in front of Boa Viagem (Fig. 10.5 upper panel). As a result, the tide modulation changes the degree of beach exposure to the incoming wave energy, thereby changing the rates of erosion/deposition and hence morphological variations.

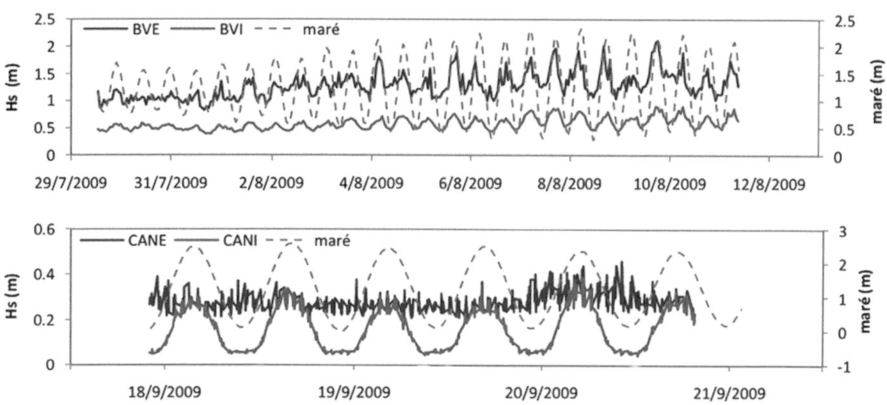

Fig. 10.5 Wave (*solid line*) and tide level (*dashed line*) time series for Boa Viagem beach (*upper panel*) and Candeias beach (*lower panel*). The wave series were measured seaward of the reef (*light gray*) and landward of the reef (*dark gray*) (Adapted from Costa (2010). For the beaches location see Fig. 10.1)

It is in this context, that the incident wave height at the beach is directly related to the reef presence, whether on the inner shelf and surf zone or at the backshore. The beaches can be classified using dimensionless fall velocity ($\Omega = H_b/W_s\ T$) (Wright and Short 1984) and relative tide range ($RTR = TR/H_b$) (Masselink and Short 1993) where H_b is the wave breaking height, W_s is sediment fall velocity, T is the mean wave period and TR the spring tide range. In beaches where waves break over the reef the incident wave height to be used at the Ω and at RTR for beach differentiation will substantially lowered by the reef. Thus, in addition to the forcing process (tide and wave) the beach morphodynamics is controlled by the presence of the reef and their modulation of wave energy.

Besides the wave height modulation effect caused by the reef another very important wave modification process is the change in incident direction caused by refraction and diffraction, which in turn creates longshore gradients in the angle of wave approach and wave energy, thus generating convergence and divergence zones of sediment erosion and/or deposition. This longshore variability of the incident wave energy and direction is responsible for the formation of many coastal features along Pernambuco state coast such as salient, tombolos and spits and even erosional hot spots.

Figure 10.6 shows results from a wave propagation model (WAPO; Silva 2013). According to Mariño-Tapia et al. (2010), the WAPO model solves in detail the reef influence over the waves better then other classical models. The results are for the main incident wave conditions ($H_s = 2.0$ m; $T_p = 8$ s; and Dir = 135°) during low and high tide. From the figure it is possible to see several convergence zones marked by light and dark red, especially during high tide, and several divergence zones, marked by light green and blue. From the simultaneous analysis of this figure together with Fig. 10.2, one can state that there is a significant increase in wave height cause by refraction and focusing over the reefs shown in Fig. 10.2. On the other hand, the reef shadow zones are low energy zones marked by wave diffraction. Also, Fig. 10.6 illustrates the tidal effect on wave attenuation, with the incoming waves significantly smaller at low tide compared to high tide, especially close to the shore where at low tide the divergence zones becomes more evident.

10.2.2 Beach Types and States

Beach morphology depends not just on wave action or tide, but the interaction between these processes (Davis and Hayes 1984). However, with the geographical variations in wave height and tidal range, the effect of one may become more important than the other.

In this context, Short (1991) attempted to develop a conceptual model that includes the effect of the tide on beach morphodynamics where it was possible to distinguish three beach types on a macrotidal environment based on wave height. However, only in the subsequent studies of Masselink and Short (1993), Masselink (1993) and Short (2006) that it had become possible to better differentiate beach types using the RTR.

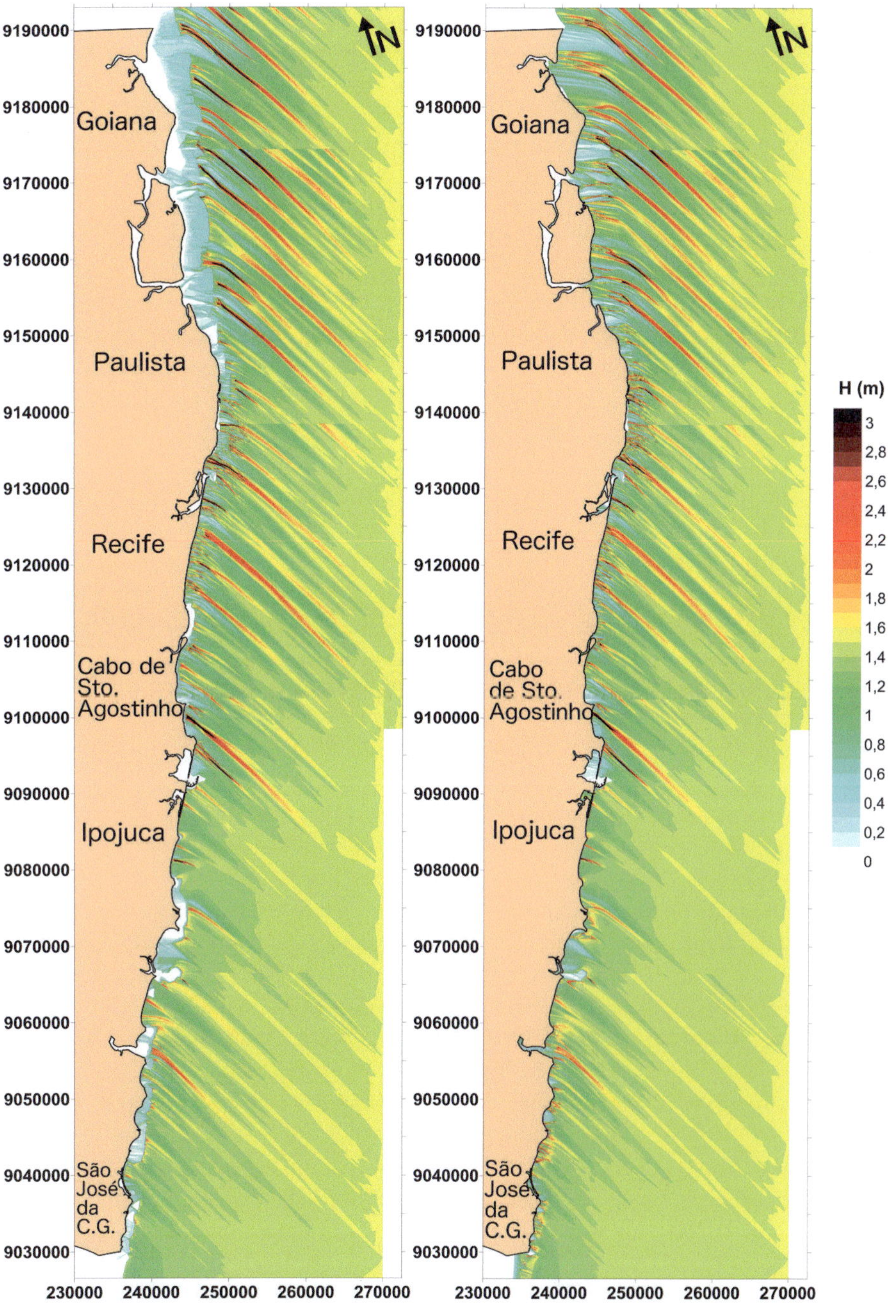

Fig. 10.6 WAPO wave model results for the whole coast of Pernambuco at low tide (*left* panel) and at high tide (*right* panel). For this simulation run the mean wave condition was used, $H_s = 2.0$ m, $T_p = 8.0$ s and Dir = 135°, using the wave data shown at Fig. 10.2

According to Davies and Hayes (1984), tides have a passive role in the sediment transport and consequently on the morphological changes. Their primary role is to change the sea level by alternately submerging and exposing, a large portion of the beach profile and the inner surf zone, whose main effect would be not only retard the transport rates but also the morphological changes. However, the tide effects do more than just delay the rate of beach profile changes, it also shifts the wave processes zones (shoaling, breaking, surf and swash zone), so that, during high tide, the surf zone decreases its width while the shoaling, and the swash motions becomes more intense. However at low tide, the shoaling zone increases as the swash importance decreases. The cross-shore displacement of wave energy zones during the tidal cycle results in a flatter slope with a lower and smoothed profile.

This section of Brazilian coast is exposed to moderate waves generated by the dominant southeasterly trade winds and a mesotidal regime. It is in this context that the balance between the tidal and wave forces, in an environmental bordered by reefs, is crucial for the beach state differentiation and understanding of the predominant coastal processes. In addition, the geological heritage and its evolution during the Quaternary have generated shoreface heterogeneities, which in turn modulate the wave climate along the coast.

On beaches that are free from the reef influence or from a low gradient shoreface and are fully exposed to the wave climate, their behavior is controlled predominately by the incident waves. Some beaches in the central state area such as Praia do Paiva (Fig. 10.7) and southern beaches as Cupe, Maracaípe and Várzea do Una, are all intermediate according to Wright and Short (1984) and commonly show three-dimensional surf zone features as rhythmic and transverse bars as well as rip currents.

Mallmann et al. (2014) discuss the beach morphodynamic behavior at Ipojuca based on satellite images. Figure 10.8 shows three satellite images of Cupe beach, which considerable show temporal variation in beach state especially the number of rip channels, a typical intermediate beaches feature. Their results also show a diversity of beach states in neighboring beaches due to the presence and absence of reefs, which drastically affects the incident wave height. The authors classified the beaches into three types, wave-dominated, tide-modified and geomorphologically-controlled.

In places where the reefs are located at the shoreface and in the surf zone the beaches are increasingly modified by the tide action. This situation is observed at Boa Viagem, Piedade, Porto de Galinha, Tamandaré and São José da Coroa Grande beaches, which have a low gradient beach profile and no surf zone or three-dimensional features. During winter months however the higher waves forms cusps on the high tide beach face. At these beaches the profile is characterized by a reflective high tide beach plus low tide terrace (R+LTT) (Masselink and Short 1993; Short 2006), (Fig. 10.9).

In the northern part of the state there are several reef lines, which dissipate most wave energy and have shallow and flat shorefaces in their lee. Figure 10.10 shows the results of SWAN wave modeling for the northernmost beach, Carne de Vaca, where substantial wave dissipation occurs crossing the two reef lines, at 2 and at 3 km. The model input parameter was a wave from southeast with a H_s of 2.0 m,

Fig. 10.7 Oblique photography of Paiva Beach captured at low tide showing the presence of rip current channels cutting the surf zone. On this day the beach was classified according Short (2006) to be Reflective + Transverse Bar and Rip (Photo courtesy of Roberto L. Barcellos)

which is reduced to 0.2 m a more than 90 % reduction in wave energy. This pattern also occurs on the beaches of Itamaracá Island, as well as Catuãma and Pontas de Pedra.

Figure 10.11 shows a set of beach profiles surveyed at Carne de Vaca during low tide. It clearly shows the steep and narrow high tide beach face and the very gentle intertidal tide profile with a relatively long and shallow terrace (Viegas et al. 2012). This information together with a break in the profile slope at the beach face indicates a tide-modified (R + LTT) beach (Masselink and Short 1993; Short 2006). Also it has a mean RTR = 6.25 indicating a tide-modified beach state. However, it

Fig. 10.8 Satellite images from Cupe Beach showing different morphodynamic stages classified according to Wright and Short (1984): (**a**) Rhythmic Bar and Beach (RBB) on 17/09/2007; (**b**) Low Tide Terrace (LTT) on 11/03/2010; (**c**) Transverse Bar and Rip (TBR) on 05/11/2010 (Images source: Google Earth)

Fig. 10.8 (continued)

Fig. 10.8 (continued)

Fig. 10.9 Oblique photography of Boa Viagem beach captured at low tide where three reef lines are visible in the surf zone and the low tide terrace is totally exposed (*wet dark brown sand*) as well as the absence of features such beach cusps. At this day the beach was classified as Low Tide Terrace (Photo courtesy of Roberto L. Barcellos)

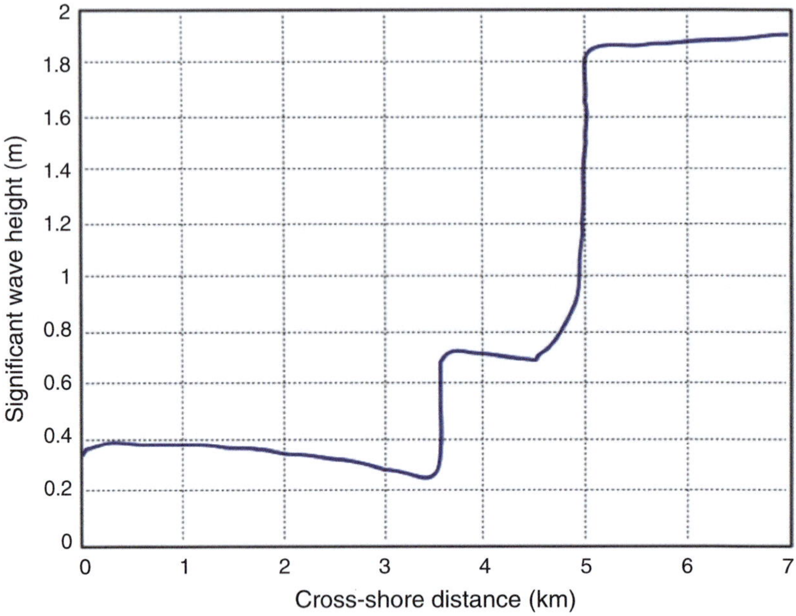

Fig. 10.10 Simulation of a 2 m wave height from SE across the reef system located in front of Carne de Vaca beach. Note the damping in wave height at a cross-shore distance of 2 and 3.5 km caused by wave breaking over the reef. The shore is at 0 km

needs to be mentioned that this is the modal state since the measured range of RTR for this beach it was between 3 and 30, meaning that in some circumstance it can be even tide-dominated.

Table 10.1 contains the RTR calculated for certain beaches along the Pernambuco coast where hydrodynamic data and topographic surveys were carried out at different times periods. From the table it is possible to see that most of the beaches are classified as reflective beach during high tide as in the model proposed by Short (2006). However, during low tide the main stages found were Low Tide Terrace and Transverse Bar and Rip. From the table it is clearly shown that there is a differentiation in terms of H_b, RTR and beach stage among the selected cases. The primary cause of such differentiation are the reefs as previously discussed. The beaches of the northern sector have lower H_b values and higher RTR values (from 4.84 to 10.37) while the southern sector has higher H_b values and lower RTR values (from 1.77 to 3.01), the middle sector has values H_b and RTR between both sectors.

According to the RTR shown in Table 10.1, the northern beaches are clearly tide-modified beaches and the southern surveyed beaches wave-dominated. However, the middle sectors beaches show how the presence of reef affects the dominant energy. Most of the middle sector beaches shown in Table 10.1 have a

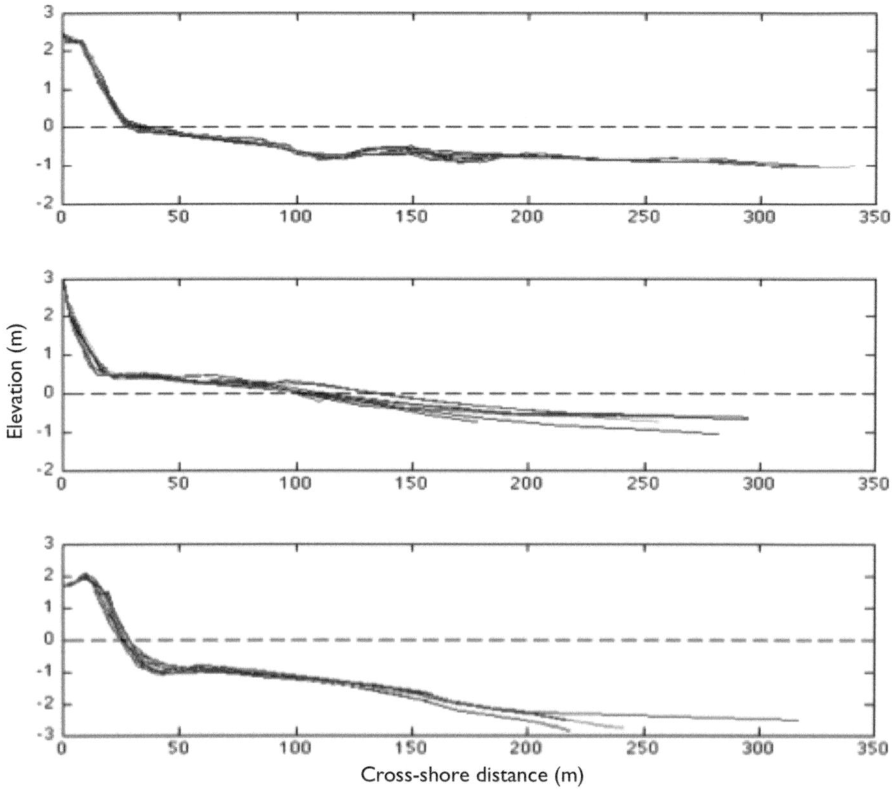

Fig. 10.11 Set of tide-modified beach profiles surveyed between October of 2010 and September of 2011 at Carne de Vaca beach. The profiles are 1 km apart. The *top* one is the northernmost and the *bottom* one is the southernmost

beachrock reef right in front of them, or at the surf zone, acting almost as a "sand bar" or at the beach face (Fig. 10.3). So far this type of longshore differentiation agrees with the physiographic sectors proposed by Coutinho et al. (1994).

It needs to be highlighted that the wave data shown in Table 10.1 were not collected on the shadow of fringing reefs. This fact is important since the shadow zone caused by this type of reef significantly affects the incoming wave height creating diffraction zone, resulting in a higher RTR zone. As a result of this diffraction and dissipation process on the lee side of the reef, tombolos and salients are generated due to gradient on longshore sediment transport, forming several crenulated bays as the bays of Tamadaré and the salient of São José da Coroa Grande (Fig. 10.12a, b). This crenulate shoreline shape also behaves differentially in terms of beach morphodynamics. Mallman et al. (2014) have classified the beaches on the lee side of the Ipojuca's reef as low tide terrace.

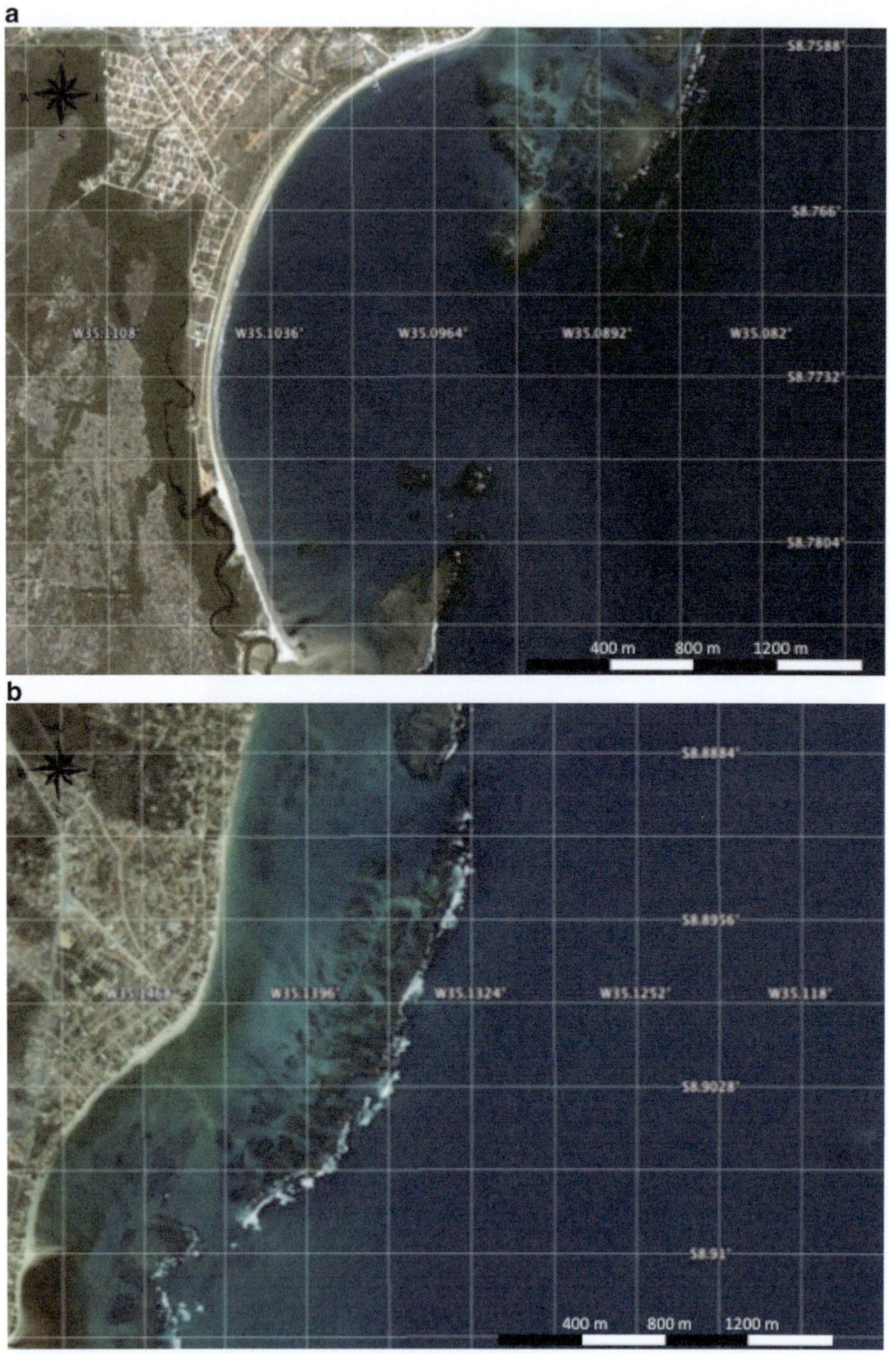

Fig. 10.12 Tamandaré bay (**a**) and the salient where the city of São José da Coroa Grande is located. Both geomorphological features are a result of the diffraction and refraction process caused by the reef presence

Table 10.1 Relative tide range and beach stage calculated based on the average conditions of breaking height.

	City	Beach	TR[a]	H_b[b]	RTR[c] = TR/H_b	Beach state[d]	References
Northern sector	Goiana	Carne de Vaca	2.8	0.58	4.84	R+LTT	Viegas et al. (2012)
	Goiana	Carne de Vaca	2.8	0.45	6.25	R+LTT	Viegas et al. (2012)
	Itamaracá	São Paulo	2.8	0.40	7.00	R+LTT	Martins (1997)
	Itamaracá	Pilar	2.8	0.40	7.00	R+LTT	Martins (1997)
	Paulista	Nossa Senhora do Ó	2.8	0.40	7.00	R+LTT	FINEP/UFPE (2009)
	Paulista	Conceição	2.8	0.27	10.37	R+LTT	FINEP/UFPE (2009)
	Paulista	Maria Farinha	2.8	0.29	9.66	R+LTT	FINEP/UFPE (2009)
Middle sector	Recife	Boa Viagem	2.8	0.88	3.18	R+LTT+BL	Duarte (2002)
	Recife	Boa Viagem	2.8	0.62	4.52	R+LTT+BL	Duarte (2002)
	Recife	Boa Viagem	2.8	0.90	3.11	R+LTT+BL	Duarte (2002)
	Jaboatão	Piedade	2.8	0.61	4.59	R+LTT	FINEP/UFPE (2009)
	Jaboatão	Piedade	2.8	0.60	4.67	R+LTT+BL	FINEP/UFPE (2009)
	Jaboatão	Piedade	2.8	0.76	3.68	R+LTT+BL	FINEP/UFPE (2009)
	Cabo	Paiva	2.8	0.98	2.86	R+TBR	Madruga Filho (2004)
	Cabo	Paiva	2.8	0.95	2.95	R+TBR	Madruga Filho (2004)
	Cabo	Itapuama	2.8	0.90	3.11	R+TBR	Madruga Filho (2004)
Southern sector	Ipojuca	Maracaípe	2.8	1.58	1.77	R+TBR	Macedo et al. 2012
	Barreiros	Praia do Porto	2.8	0.72	3.89	R+TBR	Regina et al. (2013)
	Barreiros	Várzea do Uná	2.8	0.93	3.01	R+TBR	Regina et al. (2013)

[a]*TR* tide range, [b]*Hb* mean breaking height, [c]*RTR* mean relative tide range
[d]Beach stages: *R+LTT* Reflective+Low Tide Terrace, *R+LTT+BL* Reflective+Low Tide Terrace+Beachrock Line, *R+TBR* Reflective+Transverse Bar and RIP

10.2.3 Beach Hazard and Safety

Because of the high diversity of beach morphology and geological control, the Pernambuco beaches have several different types of hazards that vary along the coast. At present there is a lack of scientific research on beach hazards and safety, however, several physical permanent and non-permanent hazards exist along the coast in addition to biological hazards.

For this section of Brazilian coast the main hazards are the reefs, not just because of it hardness but also due to a variety of species that occurs over it like coral, algaes, sea urchins, etc., where the beach users can injure themself. The reef also change the local bathymetry and can lead to a variation on alongshore depths that may not be clear for the beach user.

Another hazardous feature that are rip currents, which are strong currents that originate near the shore and extending through the surf zone and flow seaward (Komar 1998; Brander and Short 2000). According Brander (1999), they are an integral component of coastal circulation in several beach states and represent an important mechanism for the transport of water and sediment coast of the ocean.

A recent study by Maia et al. (2014) have shown that beaches along some cities of the Recife Metropolitan region have permanent topographically controlled rip currents located in gaps in the lines of beachrock. A good example of such type of rip is provided by an Argus video station composed by five cameras installed at the northern part of Boa Viagem beach. At this site using a time exposure image (Fig. 10.13) it is possible to see the waves breaking over the beachrock reefs as well as at the shoreline. Along the most seaward wave breaker line there is a clear gap at X = −180 m and Y = 200 m which cause the water return. According to personal communications of several lifeguards that work in this area drownings commonly occur as water flows seaward through these gaps.

On open sand beaches, not protected by beachrock or algae and coral reefs, such as Cupe and Paiva, beach rip currents are a striking feature as shown in Figs. 10.7 and 10.8. Field surveys as well as video images have shown that at these two places the average rip spacing is close to 100 m. Once again personal lifeguards communications have also indicated that this two beaches are the most dangerous beaches along the coast of Pernambuco due to these dynamic beach rips.

On tide-modified beaches Pontas de Pedra and Itamaracá Island, the main hazard are the relatively fast change of depth. Due to their gentle slope, these beaches are very attractive for beach user that do not have swimming skills and appear very safe not just because of the gentle slope but also because of the low wave conditions.

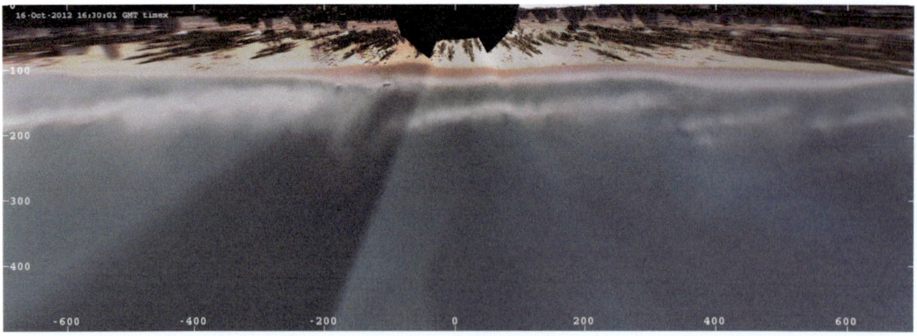

Fig. 10.13 Rectified time exposure image of Boa Viagem beach showing a permanent rip current located at X = −180 m (longshore distance) and Y = 200 m (cross-shore distance). The *white* pattern over the image is showing the wave dissipation over the beachrock and at the shoreline

However, during low tide the beach users walk out across the tidal flats and stay over few hours in the water and as the tide rises they get caught due to rapid change in depth, resulting in drownings. This processes is also known as tidal cut-off.

However the most feared hazard along the Pernambuco coast are the sharks. According to Hazin et al. (2008) between the years of 1992 and 2006 over 47 persons where injured by sharks along the Recife Metropolitan region of which 17 resulted in to death. The researchers says that the majority of the victims where surfers and young males, with the attacks usually occurring in shallow water, close to the shore at beaches with a narrow channel bordered by an adjacent reef. Most of the attacks were caused by bull sharks (*Carcharhinus leucas*; Hazin et al. (2008)) but some incidents may also be attributed to tiger sharks (*Galeocerdo cuvier*; Gadig and Sazima (2003) and Hazin et al. (2008)). Hazin et al. (2008) found that Boa Viagem and Piedade beaches have the highest numbers of shark attacks, however, beaches like Paiva and Candeias are also on the list. Recent discoveries have suggesting a relation between permanent rip currents and the attacks.

Despite the hazardous aspect already presented here morphological features like inlets and rocks also needs attention due to many threats that they can expose the beach users. Normally the inlets have a bimodal cross-shore current that will have high velocities depending on the tide and river discharge. Those current can easily carry a person out of their safety zone. Also, due to changes in the current the bathymetry nearby the inlets may change very quickly, exposing the users to a new threat. Rocks are a hazards owing to their shape and hardness, which can cause several injuries when the beach users try to climb over it or even dive from it.

10.3 Anthropogenic Effects Over the Coast

10.3.1 Coastal Erosion Along the Recife Metropolitan Area

The oldest reference to coastal erosion along Pernambuco coast dates from 1914 and was written by Ferraz who stated that the erosion of Olinda city was occurring due to changes in the littoral drift caused by the extension of Recife harbor. This was followed by mangrove deforestation and their landfill, as occurred at the Beberibe River month (located at the division between the cities of Recife and Olinda), which resulted in accelerated shoreline erosion at Olinda.

In 1953 a study of the erosion situation along the Olinda coast was undertaken by the *Laboratoire Dauphinois d'Hydraulique Neyrpic* (Grenoble, France) which in turn recommended the construction of two semi-submerged breakwaters and three short groins which where deployed at Milagres and Carmo beaches. However, the problem was not satisfactorily solved and resulted in problems developing at the neighboring northern beaches (Bairro Novo, Casa Caiada and Janga).

The growth Recife has extended the urban area along its rivers and coast. As a consequence the beaches of Candeias, Piedade and Boa Viagem, which were stable, began to show signs of erosion due to the unplanned urbanization occurring along the foreshore. This development prevented the natural exchange of sediment from subaerial beach to the subtidal zone and also longshore sand transport. As a result some emergency hard engineering solutions were constructed to protect the infrastructures behind the beach, most of the time without the proper knowledge of the coastal processes and without any technical support. This did not solve the problem but rather had magnified and transferred it to neighboring downdrift areas.

In terms of the whole coast, the erosion problem is occurring along one-third of Pernambuco's beaches with several different factors contributing to the problem (Manso et al. 2006). On some beaches the erosion is a direct product of anthropogenic modifications caused by construction too close to the beaches (impermeability of beach ridges) and even the backshore as occurred in Boa Viagem beach (Fig. 10.14), as well as the beaches of Olinda and Paulista cities.

Other beaches have also shown signs of erosion resulting from modification of the sediment supply. In some cases this has been caused by natural factors such as divergence of the longshore current forming cell circulation patterns as is the case of the most embayed beaches, formed in lee of reefs, with examples being the beaches of Serrambi, Tamandaré and Guadalupe.

Fig. 10.14 The urban evolution along Boa Viagem beach during the last three decades. *Upper panel*: In 1989 the beach system was well preserved with the presence of a wide backshore and dune vegetation zone; *middle panel*: in 1995 with the dune vegetation and part of the backshore occupied by a sidewalk, road and a revetment; *lower panel*: the beach in 2002 without backshore and a revetment

Due to the natural characteristics of the area as well as the historical occupation of the littoral, several processes can be cause erosion along the Pernambuco coast especially in the metropolitan region. Natural processes such as wave refraction, beachrock fracturing, reduction of the sediment supply and changes in the wave climate are some examples of processes that may be responsible for erosion at certain coastal locations. In terms of the historical occupation, changes in the river courses, removal of riparian vegetation, mangrove landfill, building of dams, deforestation of coastal vegetation and occupation of the beach are some examples of anthropogenic factors that may be causing erosion. The anthropogenic interference is the most significant as its enhance the processes and does not allow the natural processes to establish an equilibrium state. However, it is not an easy task to establish the actual contribution of each of these factors due to the lack of long term monitoring programs.

10.3.2 Hard and Soft Engineering Solutions

In order to solve or reduce the coastal erosion problem in the Recife Metropolitan Area different types of structures have been deployed over time, all aiming to protect public and/or private properties, and in most cases constructed on an emergency basis. However, many of these structures have been shown to be partially ineffective as a means of protection. Figure 10.15 shows examples of such structures.

According to the MAI project (FINEP/UFPE 2009) there are 20 km of coastal structures along the Recife Metropolitan coastline, with 4390 m in Jaboatão Guararapes city; 3440 m in Recife; 7610 m in Olinda and 4650 m in the Paulista city (FINEP/UFPE 2009), the structures are:

- Jaboatão Guararapes: 3260 m revetment and seawalls (74 %), 530 m of groins and jetties (12 %) and a 600 m breakwater (14 %). In this city the structures have been built to protect the land, by armoring "shoreline" at the expense of the beach, with a consequent adverse impact on local tourist visitation;
- In Recife, the main coastal structure is a 2100 m long revetment at Boa Viagem beach and another one of 1340 m on the beach at Brasília Teimosa. This type of structure also has the objective of protecting the land and not the beach;
- In Olinda, there are 1700 m of revetment and seawalls (22 %), 250 m of groins and jetties (3 %) and 5660 m of breakwaters (75 %). These structures in the form of riprap, breakwaters and groins provide good protection for infrastructure close to the coastline, however, they cause significant changes in the rates of the longshore transport, with negative downdrift effects;
- In Paulista, 1850 m revetments and seawalls (40 %), 280 m of groins and jetties (6 %) and 2520 m of breakwaters (54 %) have been constructed.

However, even with all the coastal engineering structures already deployed along the coastline, the erosion problem remains since the causes have not been addressed. The structures implemented along the Recife Metropolitan coast aim to protect the

Fig. 10.15 Hard engineering structures deployed at Recife Metropolitan region in the last four decades. (**a**) groin system deployed at Bairro Novo beach (Olinda); (**b**) breakwater Milagres beach (Olinda); (**c**) revetment along the coast of Boa Viagem (Recife); and (**d**) detached breakwaters at Candeias beach after their segmentation and beach nourishment (Images are courtesy of CPRH)

land by armoring the coastline at the expense of the beaches. So although they show positive effects in protecting the land, they often they have negative effects on the beach and sediment dynamics.

The central coast of Recife Metropolitan Region has a serious erosion problem, given that erosion is present along most of the coast, has intensified, and is expected to worsen in the future. As an answer to that the state government had started a new project in which most of the beaches of the area will be nourished and several hard engineering structures will be covered by sand or will be reduced in size so they can not interfere as much with the longshore transport as most structures have in the past. At present only the beaches of Jaboatão dos Guararapes city have been nourished.

10.4 Summary and Conclusions

The coast of Pernambuco has a range of coastal environments including beaches, estuaries, reefs and headlands with varying degrees of preservation and occupation. The coast is dominated by fine to medium quartz sediment in a mesotidal environment receiving southeast waves. The shelf can be divided in to three parts in which the most prominent feature are the shore parallel beachrock and coral and algae reefs which have complex underwater relief and are present at several different depths.

This beachrock and reefs significantly affect the inshore wave climate and the coastal processes. In particular the variation of water depth over the reefs modifies wave height and direction thereby affects beach morphodynamics and shoreline plan form. As a result, depending on the resulting relative tide range, the beaches can be classified as wave-dominated (Wright and Short 1984) or the tide-modified (Masselink and Short 1993), with the impact of the reefs locally changing the beach state from wave-dominated to tide-modified.

Based on this the beaches can be divided into three sectors: the reef protected northern beaches are tide-modified; the exposed southern beaches are wave-dominated; and the middle sector beaches where the beachrock reefs controls the relation between wave and tide are geomorphologically-controlled.

The combination of the diversity of beach environments and varying levels of wave energy have resulted in Pernambuco having a high diversity of beach hazards along the coast, with the main hazards being the reefs, variation in water depth, permanent and non-permanent rip currents, inlets, tidal currents, rocks and biological hazards such as coral and sharks.

Owing to it high occupation density, Pernambuco coast has an erosion problem largely caused by the unorganized urbanization processes that even occupies the backshore zone. This urbanization processes has prevented the natural sediment processes of shifting sediment from subaerial beach profile to the submerged zone and also alongshore. In addition changes in the river courses, removal of riparian vegetation, mangrove landfill, building of dams, deforestation of coastal vegetation, inappropriate coastal structures defenses, also contribute to the erosion problem along the Pernambuco coast.

Acknowledgements This study is a contribution of inctAmbTropic—Brazilian National Institute of Science and Technology for Tropical Marine Environments.

References

Araújo TCM, Seoane JCS, Coutinho PN (2004) Geomorfologia da plataforma continental de Pernambuco. In: Leça E, Leitão SN, Costa MF (eds) Oceanografia um cenário tropical. Universidade Federal de Pernambuco – Departamento de Oceanografia, Recife, pp 39–57

Brander RW (1999) Sediment transport in low-energy rip current systems. J Coast Res 15(3):839–849

Brander RW, Short AD (2000) Morphodynamics of a large-scale rip current system. Muriwai Beach. New Zealand. Mar Geol 165:27–39

Cavalcanti LB, Kempf M (1970) Estudo da Plataforma Continental na área do Recife (Brasil). Trabalhos Oceanográficos/UFPE. 9/11:149–158

Costa MBSF (2010) Influência da estrutura recifal na transformação das ondas do litoral de Recife e Jaboatão dos Guararapes/PE-Brasil. Dissertação de Mestrado, Universidade Federal de Pernambuco

Coutinho PN, Farias CC (1979) Contribuição à origem dos recifes do nordeste. In: Atas do IX Simpósio de Geologia do Nordeste, Natal

Coutinho PN, Manso VAV, Lima Filho MF, Lima ATO, Medeiros AB, Pedrosa FJA, Martins MHA, Chaves NS, Sampaio AS, Lira ARA, Menezes MCB, Brito MF (1994) Coastal quaternary of Pernambuco. Brazil. In: Abstracts 14th international sediment congress, Recife, pp 21–26

Davies RA, Hayes MO (1984) What is a wave-dominated coast? Mar Geol 60:313–329

Dominguez JML, Bittencourt ACSP, Leão ZMAN, Azevedo AEG (1990) Geologia do Quaternário Costeiro do Estado de Pernambuco. Rev Bras Geosci 20:208–215

Duarte RX (2002) Caracterização Morfo-Sedimentológica e Evolução de Curto e Médio Prazo das Praias do Pina. Boa Viagem e Piedade. Recife/Jaboatão Dos Guararapes – PE. Dissertação de Mestrado, UFPE

FINEP/UFPE (2009) Integrated environmental monitoring – MAI-PE. Final report. FINEP, Recife

Gadig OBF, Sazima IA (2003) Non-fatal attacks by the tiger shark, *Galeocerdo cuvier*, on the northeast coast of Brazil (Condrichthyes, Carcharhinidae). Arq Ciênc Mar 36:119–122

Gregório MN (2009) Evolução da linha de costa e caracterização da plataforma continental interna adjacente à cidade do Recife – PE. Tese de Doutorado. Univ Fed Pernambuco

Gregório MN, Araújo TCM (2008) Interannual morphological changes of Boa Viagem Beach – Northeast Coast of Brazil. Boletim Paranaense de Geociências 62:61–84

Hazin FHV, Burgess GH, Carvalho FC (2008) A shark attack outbreak off Recife. Pernambuco. Brazil: 1992–2006. Bull Mar Sci 82(2):199–212

Inman DL, Masters PM (1994) Status of research on the nearshore. Shore Beach 62:11–20

Komar PD (1998) Beach processes and sedimentation. Prentice Hall, Englewood Cliffs, p 544

Macedo RJA, Manso VAV, Pereira NS, França LG (2012) Transporte de Sedimentos e Variação da Linha de Costa em Curto Prazo na Praia de Maracaípe (PE) Brasil. Revista de Gestão Costeira Integrada 12:343–355

Madruga Filho JD (2004) Aspectos Geoambientais entre as praias do Paiva e Gaibu. Município do Cabo de Santo Agostinho (Litoral Sul de Pernambuco). Tese de Doutorado, UFPE

Maia JCB, Pereira PS, Lessa RPT (2014) Variação espaço- temporal das correntes de retorno em municípios da região metropolitana do Recife. Q Environ Geosci 5(2):166–176

Mallman DB, Pereira PS, Façanha PRSMN, Santos FMM (2014) Classificação morfodinâmica das praias arenosas do município de Ipojuca através da análise semântica de imagens verticais. Pesquisas em Geociências 41(2):169–189

Manso VAV, Coutinho PN, Guerra NC, Soares Jr CFA (2006) Erosão e Progradação do Litoral Brasileiro (Pernambuco). In: Dieter Muehe (ed) Erosão e Progradação do Litoral Brasileiro, vol 1. Brasília, pp 79–196

Mariño-Tapia I, Silva-Casarín R, Enriquez-Ortiz C, Mendoza-Baldwin E, Mancera EE, Ruiz-Rentería F (2010) Wave transformation and wave-driven circulation on natural reefs under extreme hurricane conditions. Coast Eng Proc, p 9

Martins MHA (1997) Caracterização Morfológica e Vulnerabilidade do Litoral da Ilha de Itamaracá-PE. Dissertação de Mestrado, UFPE

Masselink G (1993) Simulating the effects of tides on beach morphodynamics. J Coast Res SI 15:180–197

Masselink G, Short AD (1993) The effect of tide range on beach morphodynamics and morphology: a conceptual beach model. J Coast Res 9(3):785–800

McGregor GR, Nieuwolt S (1998) Tropical climatology, 2nd edn. Wiley, New York

Muehe D (2003) Beach morphodynamic research in Brazil: evolution and applicability. J Coast Res 35:32–42

Nimer E (1979) Climatologia do Brasil. In: Série Recursos Naturais e Meio Ambiente, vol 4. IBGE, Rio de Janeiro, p 422

Nelson RC (1994) Depth limited design wave heights in very flat regions. Coast Eng 23:43–59

Ottmann F, Nóbrega R, Coutinho PN, Oliveira SPB (1959) Estudo topográfico e sedimentológico de um perfil da praia de Piedade – Recife – Pernambuco, vol 1. Trabalhos do Instituto de Biologia Marinha, Recife, pp 19–37

Pianca C, Mazzini PL, Siegle E (2010) Brazilian offshore wave climate based on NWW3 reanalysis. Brazil J Oceanogr 58(1):53–70

Pontes PM, Araújo TCM (2006) Monitoramento morfológico das praias do Estado de Pernambuco-Brasil: Análise temporal e espacial. Trop Oceanol 34:1–11

Regina E, Manso VAV, Pereira PS (2013) Caracterização morfodinâmica das paias de Barreiros. litoral sul de Pernambuco. In: XIV Congresso da Associação Brasileira de Estudos do Quaternário, Natal

Silva AMC (1995) Aspectos erosivos na praia de Pontas de Pedra – PE. Monografia. Universidade Federal Rural de Pernambuco

Silva R (2013) Manual del usuario y documento de referencia del programa MWAPO3. Instituto de Ingeniería de la Universidad Nacional Autónoma de México. p 28

Short AD (1991) Macro-meso tidal beach morphodynamics – an overview. J Coast Res 7(2):417–436

Short AD (2006) Australian beach systems – nature and distribution. J Coast Res 22:11–27

Viegas VS, Lacerda CH, Pereira PS (2012) Morfodinâmica das praias adjacentes a foz do Rio Goiana (Pernambuco-Brasil). In: Anais do Congresso Brasileiro de Oceanografia. Rio de Janeiro, 13–16 Novembro 2012

Wright LD, Short AD (1984) Morphodynamics variability of surf zones and beaches: a synthesis. Mar Geol 56:93–118

Chapter 11
The Sandy Beaches of the States of Sergipe-Alagoas

José Maria Landim Dominguez, Abílio Carlos da Silva Pinto Bittencourt, Adeylan Nascimento Santos, and Lucas do Nascimento

Abstract The beaches of Sergipe and Alagoas illustrate the role of geological inheritance and the effect of sediment grain size in controlling beach morphodynamics. Overall, under the same wave energy levels, beaches located on prograding sections of the coast have finer sediments and are dissipative, whereas on coastal sections starved of sediments the sand is coarser and the beaches are more reflective. This results from the fact that distant sourced sands, typical of large rivers, such as the São Francisco, are typically fine. These sands are transported southwards feeding the Sergipe shoreline, where dissipative beaches dominate.

Moreover locally sourced cliff sands, as is the case of the Alagoas shoreline, tends to be coarser, because they result from the physical breakdown, by wave action, of older coarser geologic units (e.g. Barreiras Formation). The erosive remnants and marine terraces of these older units, provide foundations for fringing reefs, which further protects the shoreline from wave energy. As a result reflective beaches are characterist of the Alagoas state.

11.1 Introduction

The coastline of Sergipe (SE) and Alagoas (AL) extends for approximately 370 km (Bittencourt et al. 1983; Barbosa et al. 1986; Araújo et al. 2006) (Fig. 11.1). In this chapter, the sandy beaches of these two states are combined, not only to avoid the artificialness imposed by political boundaries, but also to illustrate the role played by the São Francisco river, located at the border of the two states, in the

J.M.L. Dominguez (✉) • A.C.S.P. Bittencourt
Institute of Geosciences, UFBA, Campus Ondina, Salvador, BA, Brazil
e-mail: landim@ufba.br; abilio@pq.cnpq.br

A.N. Santos
School of Engineering and Information Technology, UNIFACS, Salvador, Bahia, Brazil
e-mail: adeylan.santos@pro.unifacs.br

L. do Nascimento
Prefeitura Municipal de Camaçari, Camaçari, Bahia, Brazil
e-mail: lucasnascimento.lec@gmail.com

A.D. Short, A.H. da F. Klein (eds.), *Brazilian Beach Systems*, Coastal Research Library 17, DOI 10.1007/978-3-319-30394-9_11

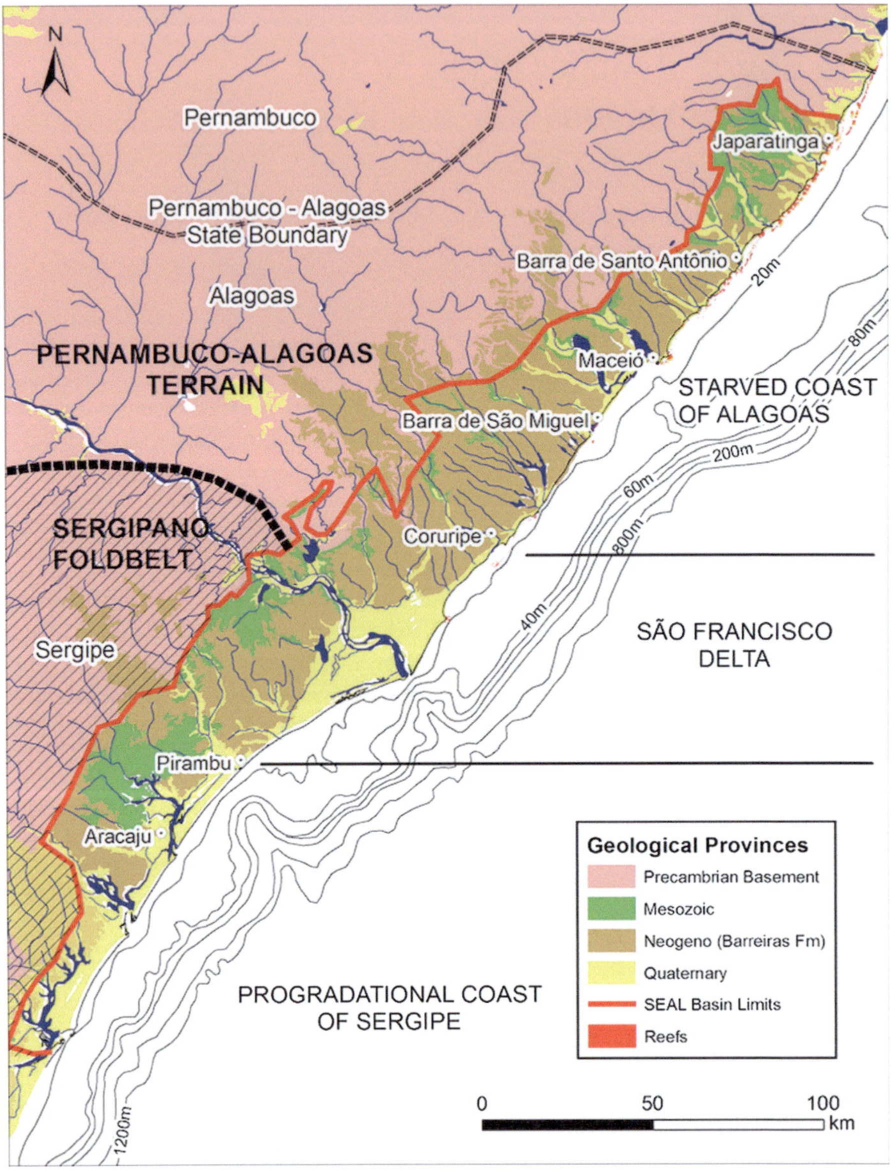

Fig. 11.1 Simplified geological map of the Sergipe-Alagoas coastal zone. Also shown are the boundaries of the three compartments used herein to describe major modal beach morphodymanic states

morphodynamics of the sandy beaches. The São Francisco is one of the largest rivers in Brazil and its fluvial input and sediment dispersion patterns along the coastline, have significantly influenced the character of sandy beaches in the area.

The characterization of the beaches in Sergipe-Alagoas was performed following the same approach used for Bahia, including the compilation of data gathered mainly over the past 15 years, most of which were either unpublished or only available in gray literature with very restricted circulation and availability. Textural data were collected from the beach face over different periods (Barbosa 1997; Oliveira 2003; Rodrigues 2008; Santos 2010), at 1 km intervals, accompanied by measurements of beach-face slope and type and number of breaking waves in the surf zone at the date/time of sampling. Samples were also collected from the lower course of the São Francisco river, also at 1 km intervals, during the summer of 2012. All sediment samples underwent dry sieving grain size analysis. No treatment was performed prior to the grain size analysis to eliminate bioclastic components and plant fragments.

By integrating this database with information on coastal geology, wave transformation modelling and sediment dispersion along the shoreline, we can attain a better comprehension of the spatial variability of the modal beach states in Sergipe and Alagoas and their main controlling factors.

11.2 Coastal Geology and Geomorphology

The geologic framework of the coastal zone of Sergipe-Alagoas was established during the separation of the South American and African continents that began in the Early Cretaceous (145 Ma). As a result of the continent separation and Atlantic Oean formation, the following pre-Quaternary geologic units are now present in the coastal zone (Fig. 11.1):

(i) Metasedimentary rocks, which compose the Sergipana Fold Belt of Neoproterozoic age, formed during the Brasiliano/Pan-African cycle (~600–560 Ma) (Almeida 1977; Oliveira 2012),
(ii) Rocks from the granitic-migmatitic basement intruded by Neoproterozoic plutons with batholitic dimensions, denominated the Pernambuco-Alagoas terrain (Silva Filho et al. 1996, 1997, 2002; Lyra de Brito et al. 2009).
(iii) Mesozoic sedimentary rocks deposited in the Sergipe-Alagoas sedimentary basin formed during the separation of the South American and African continents (Castro Jr 1988; Mohriak et al. 2000). The Sergipe sub-basin is characterized by a relatively thin terrestrial sedimentary cover and by a series of depocenters in the shelf region and deep-water province (Ponte 1969; Ojeda 1982). On the other hand, the Alagoas sub-basin is characterized by a thick syn-rift sedimentary sequence (Neocomian–Barremian) on the terrestrial and shelf portion, and a relatively thin sedimentary sequence deposited during the thermal phase of the basin evolution (Albian–Cenomanian until the present),
(iv) Detrital deposits of marine-transitional origin (Barreiras Formation), deposited during the Middle-Lower Miocene transgression, which flooded the continental margin of Brazil (Rossetti et al. 2013).

The oldest units lie inland and have a hilly relief, while the Miocene deposits of the Barreiras Formation form tablelands that dominate the geomorphology of the coastal zone (Fig. 11.2a, b) and outcrop over stretches of the coastline along the

a

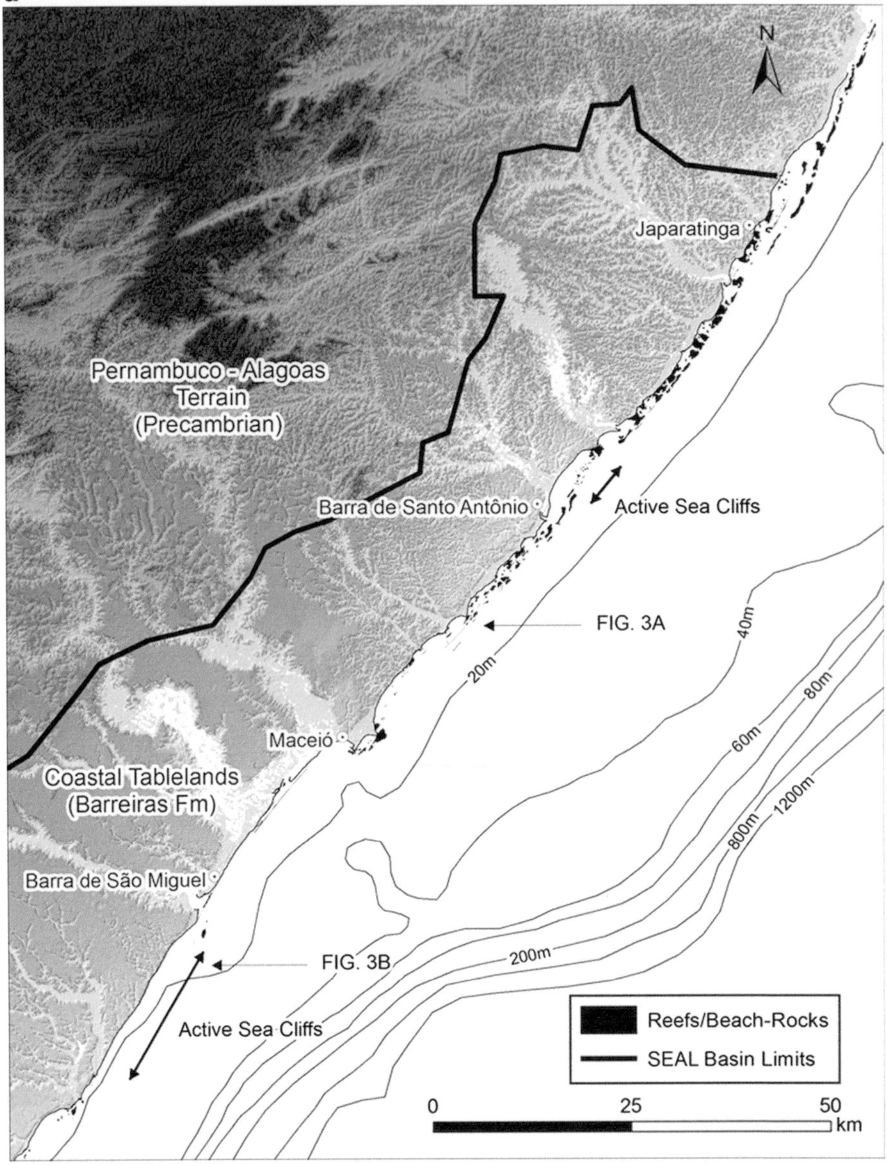

Fig. 11.2 Contrasting coastal zone geomorphologies north and south of the São Francisco delta. (**a**) Alagoas state north of the delta. Note the absence of Quaternary plains with the Barreiras formation very close to the shoreline and exhibiting stretchs of active sea cliffs. Also numerous coral and beachrock reefs bordering the shoreline particularly north of Maceió city. (**b**) Sergipe state and the São Francisco delta (Note extensive progradation of the shoreline, forming wide beach-ridge deposits. Fossil sea cliffs carved into the Barreiras formation are located several kilometers inland. Coral reefs are absent). Arrows indicate location of images shown in Fig. 11.3

b

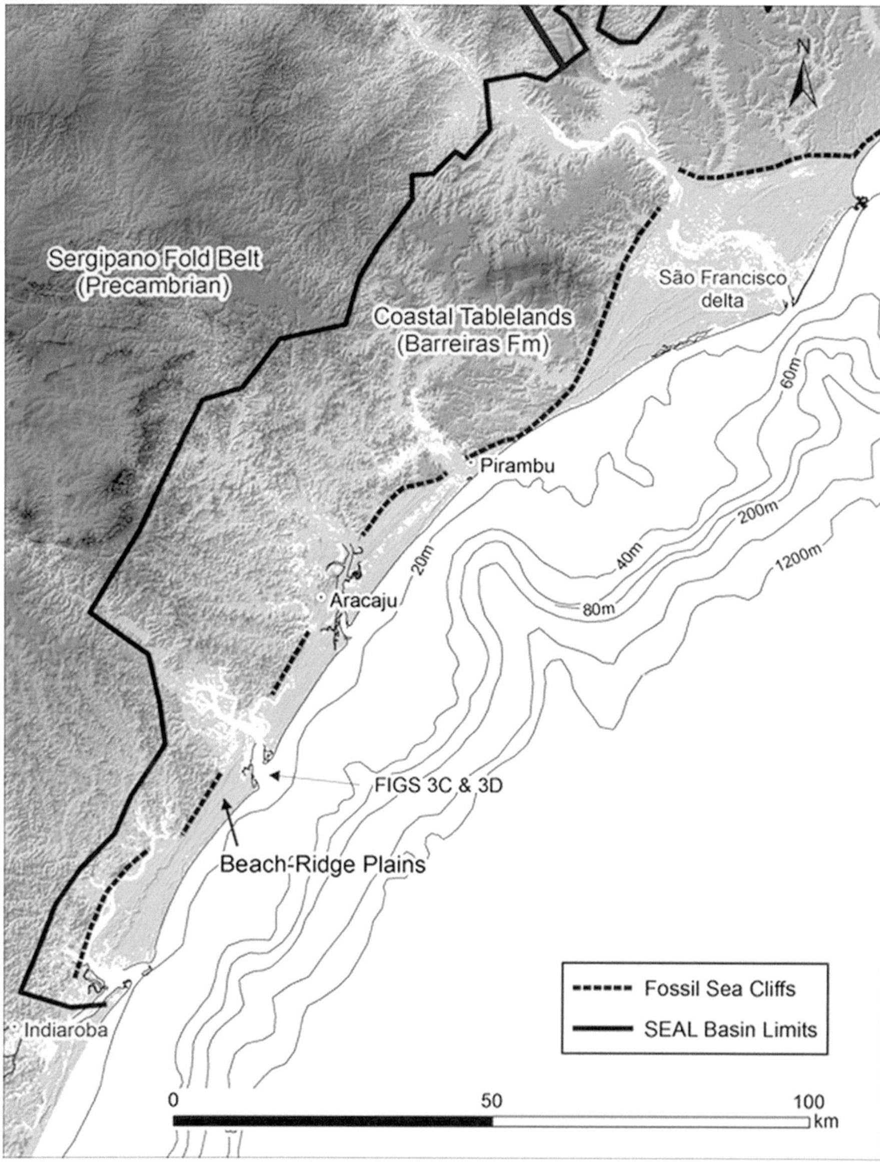

Fig. 11.2 (continued)

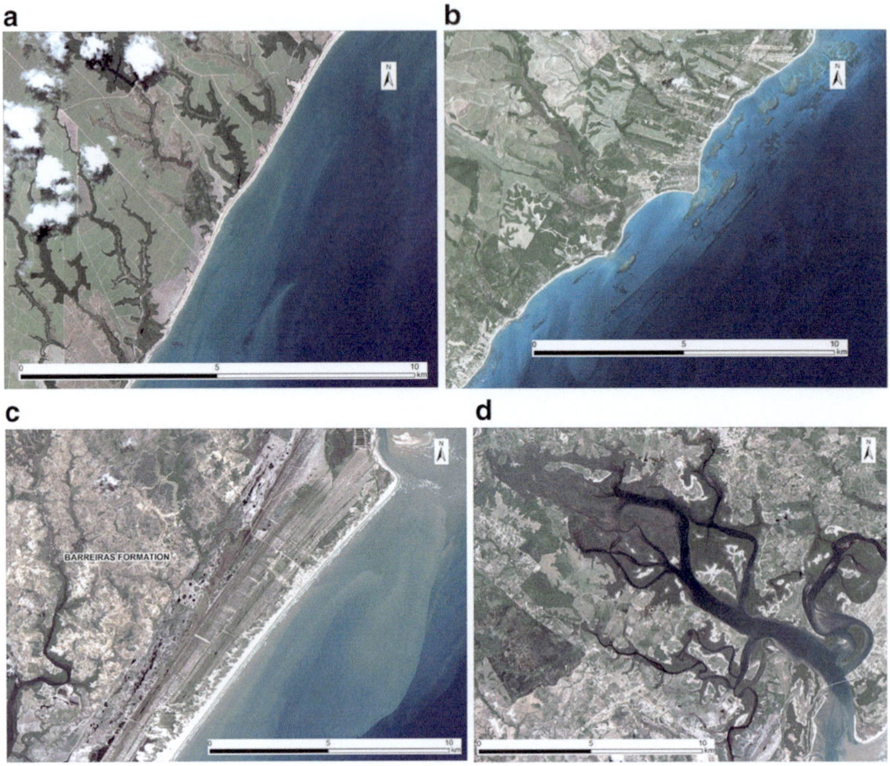

Fig. 11.3 Morphological aspects of the shoreline. Alagoas state: (**a**) active sea cliffs, a major source of coarse sediments to the shoreline. (**b**) numerous coral and shore parallel beachrock reefs border and shelter the shoreline. Sergipe state: (**c**) wide beach-foredune ridge plains bordering the shoreline, which is devoid of coral and beachrock reefs and active sea-cliffs. (**d**) Large estuarine areas are associated with every major river emptying into the Sergipe coast, provide a sediment-sink and prevent delivery of coarse sediment to the shoreline. See Fig. 11.2 for image location (Source: Google earth)

state of Alagoas, forming active cliffs (Fig. 11.3a). In the state of Sergipe, fossil cliffs are located several kilometers inland from the coastline (Figs. 11.2b and 11.3b).

Mesozoic sedimentary rocks outcrop mainly in the river valleys in Sergipe. However, in Alagoas they can reach the shoreline locally, where active cliffs are formed, which are capped by the sediments from the Barreiras Formation (Fig. 11.1).

The geomorphology of the coastline/coastal zone presents two very contrasting types of landscape, limited by the delta of the São Francisco river. To the northeast of the delta (Alagoas), the coastline presents several stretches with cliffs, beachrock and reefs located at variable distances offshore. The latter developed both on beachrock and marine-cut terraces sculpted in Mesozoic sedimentary rocks (Figs. 11.2a and 11.3c). To the southwest of the delta (Sergipe), the shoreline is dominated by broad beach-foredune ridge plains, which reach up to 9 km in width (Figs. 11.2b and 11.3b) and are bordered by large estuaries (Figs. 11.2b and 11.3d). Beachrock and coral reefs are absent in this section.

The continental shelf adjacent to the study area increases in width from 15 km at its southwestern extremity to 35 km at its northeastern extremity (Fig. 11.1). Shelf sedimentation characteristics are also contrasting between the northeastern and southwestern areas of the São Francisco River delta. The Sergipe inner continental shelf extends out to a depth of 20 m and has essentially siliciclastic sedimentation (Guimarães 2010), while the Alagoas shelf is dominated by more bioclastic sediments (Fontes et al. 2011; Assis 2015).

11.3 Sediment Supply

There are three main sources of sediments that supply the beaches of Sergipe-Alagoas (Fig. 11.4):

(i) Fluvial sediments carried mainly by the São Francisco River, one of the largest rivers in South America, with a length of 2863 km and drainage basin of 634,000 km^2 (Souza et al. 2003; Knoppers et al. 2006; Medeiros et al. 2007). Until the 1950s, the mean flow and material transported in suspension at the river mouth were 3010 $m^3\ s^{-1}$ and 69×10^5 t $year^{-1}$, respectively (Medeiros et al. 2007). After the construction of large dams, the values of mean flow and material transported in suspension decreased, respectively, to 1760 $m^3\ s^{-1}$ and 2.28×10^5 t $year^{-1}$ (Medeiros et al. 2007). The inputs of the small rivers that

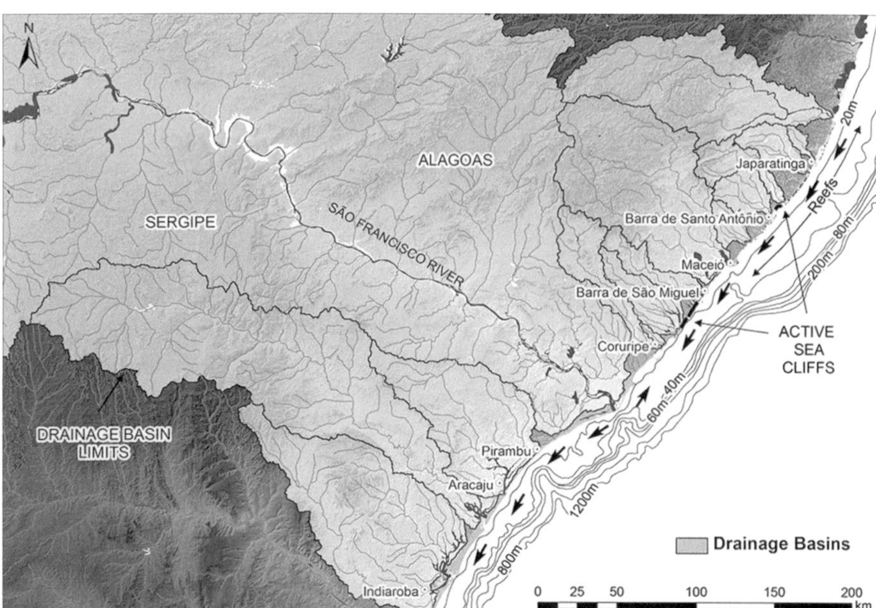

Fig. 11.4 Major drainage basins emptying along the Sergipe-Alagoas states. Also indicated are: sections of the coastline with active sea cliffs and direction of sediment transport along the coast

flow into the states of Sergipe (discharges range from 2.24 $m^3 s^{-1}$ to 14 $m^3 s^{-1}$) and Alagoas (discharges between 5.98 $m^3 s^{-1}$ and 25.8 $m^3 s^{-1}$) are added to these values (SEMARH 2015a, b). Unfortunately, there is no consistent long-term data series for the sediment load of these rivers;
(ii) Siliciclastic sediments eroded from the active cliffs of the Barreiras Formation and from the Mesozoic sedimentary rocks that outcrop along the coastline of the state of Alagoas;
(iii) Bioclastic sediments produced in the biogenic reefs that border the shoreline, particularly along the northeastern half of Alagoas, where some beaches are composed mainly by bioclastic fragments produced on the reef top.

In the submerged portion of the region, as mentioned previously, siliciclastic sediments dominate across a narrow shore-parallel area that extends offshore out to depths of 10–15 m. Offshore from this area, there are only bioclastic sands and gravels, composed mainly by coralline algae fragments (Nascimento 2011). Apparently, there are no important sources of siliciclastic sands on the continental shelf that could nourish the shoreline during the drop in relative sea-level (3–4 m) that affected the coast of Brazil during the last 5700 years (Dominguez and Bittencourt 1996).

11.4 Coastal Processes: Waves, Tides, Winds and Precipitation

Waves from east-southeast with periods of 6 and 8 s and heights ranging from 1 to 2 m dominate the coast (Pianca et al. 2010). These waves are generated by the trade winds, during the autumn and winter, while waves from the south-southeast associated with higher latitude cold fronts also reach the region (Pianca et al. 2010). Thus, due to the general orientation of the coastline, the transport of sandy sediments is predominantly from the northeast to the southwest (Oliveira 2003, Santos 2010, Rodrigues 2014) (Figs. 11.4 and 11.5).

The study area has a semidiurnal micro-meso-tidal regime, with maximum range of 2.38 m within some estuaries (Fig. 11.5) (Salles et al. 2000).

The climate is hot and humid and strongly influenced by the moist air masses from the Atlantic Ocean. Coastal rainfall varies between 1250 and 1900 mm per year (Fig. 11.5), with the rainy season concentrated during April, May and June. Total precipitation lower than 1300 mm per year occur only in the São Francisco river delta (Nimer 1989). Also in this region, there are as many as four to five consecutive dry months, favoring dune development. Winds are stronger in the region of the mouth of the São Francisco river and decrease in speed to the northeast and southwest, with a more significant decrease to the northeast of Maceió (Fig. 11.5).

The dominant direction of coastal currents on the continental shelf is northeast to southwest, which, associated with the transport promoted by the action of waves, results in a dominant southwestwards transport of sediments from the São Francisco River. This has a direct impact on the characteristics of the coastal zone and beach morphodynamics as is discussed in Sect. 11.5.

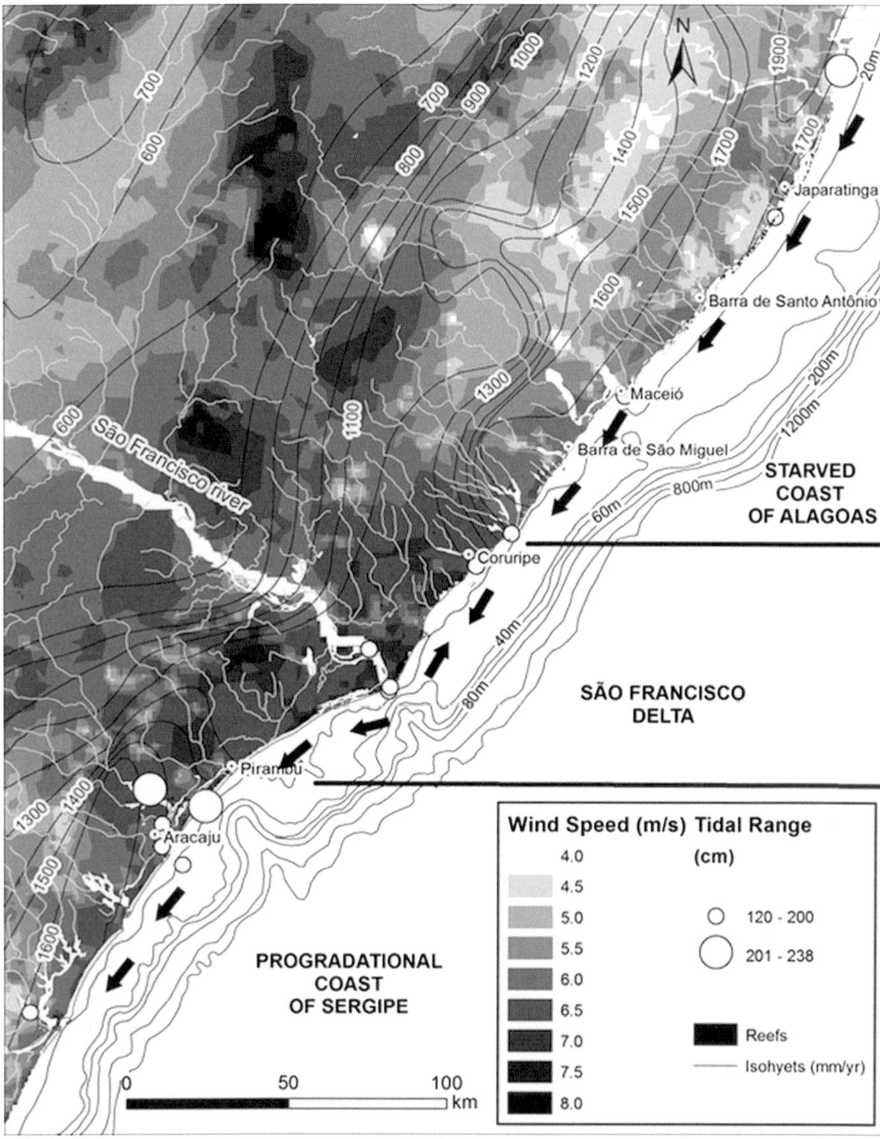

Fig. 11.5 Spatial distribution of annual precipitation (isohyets in mm year^{-1}), tidal range and wind speeds (measured 50 m above earth surface). Also shown are the limits of the major coastal compartments and areas of reef build-ups. Wind speed data from Amarante et al. (2001). Direction of sediment transport along the shoreline indicated by thick arrows

11.5 Coastal Compartments

The Sergipe-Alagoas coast can be divided into three compartments based on physiographic characteristics and the predominant beach state (Fig. 11.1):

11.5.1 The Starved Coast of Alagoas

The Starved Coast of Alagoas extends for 200 km from the northern border with Pernambuco to Coruripe (Fig. 11.6) – this compartment is characterized by a reduced input of sediments to the coastline. As a consequence, the coast has active cliffs and is dotted with beachrock reefs, some of which are located up to 3.8 km offshore, and by fringing coral reefs (Maida and Ferreira 1997; Castro and Pires 2001; Leão et al 2003; Rudorff and Gherardi 2008), particularly northeastward of Maceió. Many of the coral reefs developed over ancient beachrock reefs and marine-cut terraces carved in the Cretaceous rocks of the Alagoas basin. This compartment has two distinct sub-compartments:

Sub-Compartment I Between Maceió and São José da Coroa Grande the beaches are protected from wave action by extensive fringing coral reef located just off the coast, usually forming a lagoon during low tide. These reef formations produce salients and tombolos connected by crenulated embayments (Fig. 11.3b). Most of the beaches are sheltered (Jackson et al. 2002) and protected from waves during low tide. The beach face is usually concave (Fig. 11.7a), with a steeper reflective upper portion that is only exposed to wave action during high tides, (Fig. 11.7b) and has a higher concentration of coarse grains (Fig. 11.7c). The lower portion of the profile, on the other hand, presents a low slope and is completely protected from wave action during low tides forming a tide-modified reflective plus low-tide terrace beach (Fig. 11.7a). Although the D_{50} value of the beach sediments is predominantly in the range of fine sand, an expressive tail of medium to coarse sand is also present (Fig. 11.8). Since most of human occupation occurred near the shoreline, walls frequently represent the internal beach face limit over many stretches. In two coastal stretches, Mesozoic sedimentary rocks from the Alagoas basin outcrop on the coastline in cliffs and are fronted by marine-cut terraces (Figs. 11.6 and 11.7d).

Sub-Compartment II Between Coruripe and Maceió (Fig. 11.6) the coast is more exposed to waves due to the near absence of fringing reefs, with the exception of the city of Maceió and isolated occurrences of beachrock (Fig. 11.9a), some of which are located a few kilometers offshore. This compartment includes a stretch of approximately 19 km in length of active cliffs carved in the Barreiras Formation, which act as a source of beach sediments (Figs 11.9b, c). As a consequence, beaches along this compartment have a reflective character (Fig. 11.9d), especially in front of the cliffs.

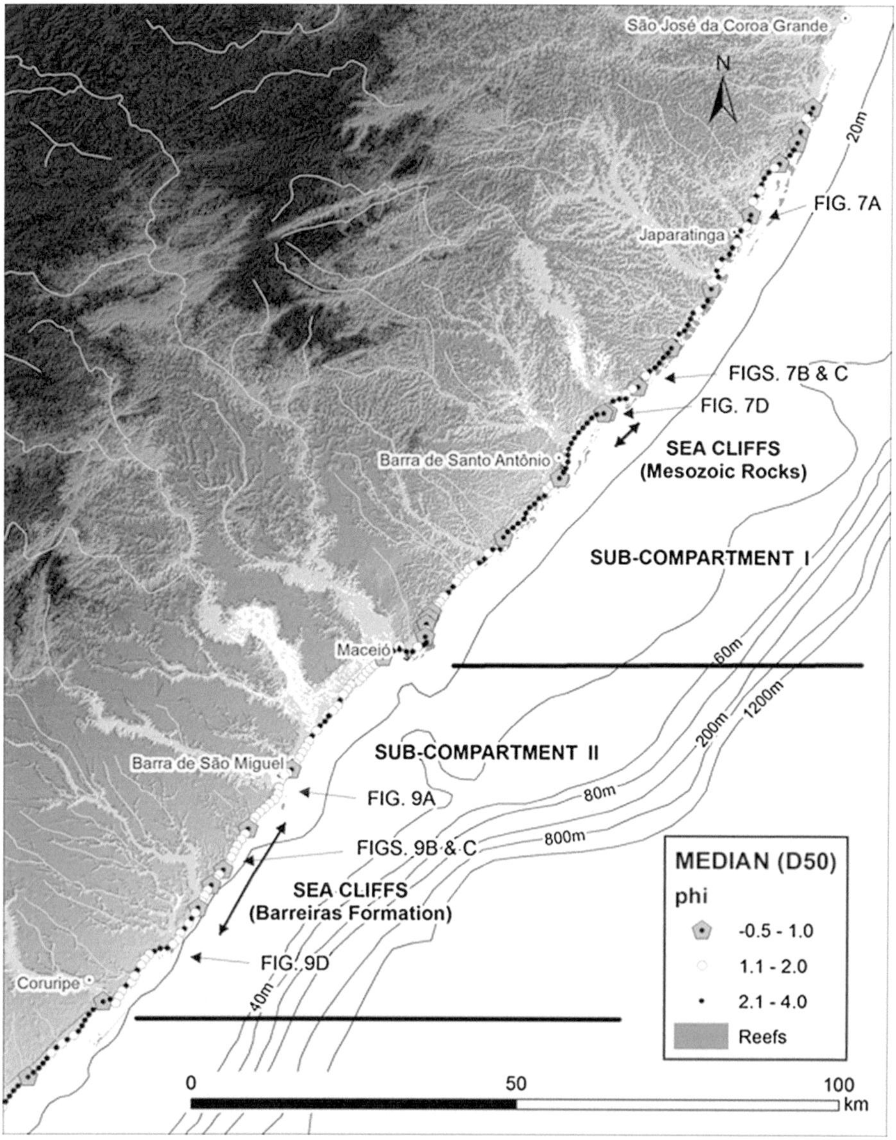

Fig. 11.6 Distribution of median grain size (D_{50}) of beach face sediments along the Starved Coast of Alagoas compartment and its two sub-compartiments. The two stretches of the shoreline where active sea-cliffs are indicated. Arrows indicate location of photos shown in Figs. 11.7 and 11.9

Fig. 11.7 Field aspects of beaches in the sub-compartment I, of the Sediment Starved Coast of Alagoas. Beaches are sheltered showing a concave geometry (**a**), with steeper slopes characterizing the upper beach-face which exhibit reflective characteristics and a major percentage of coarse sediments (**b** and **c**). On two stretches of the shoreline, mesozoic sedimentarary rocks outcrop forming active sea cliffs (**d**) (see Fig. 11.6 for location of photographs)

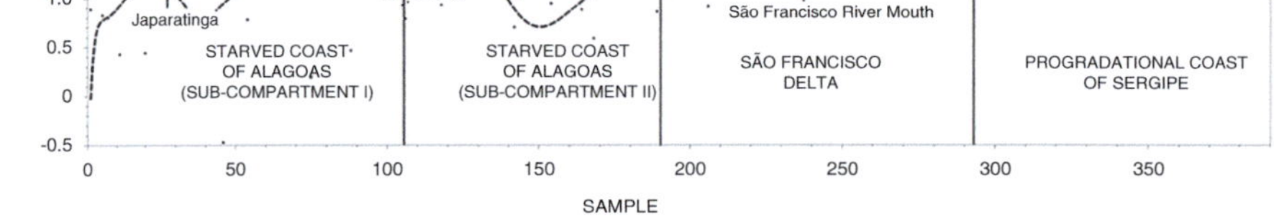

Fig. 11.8 Alongshore variation of the D_{85}, D_{50} and D_{15} values on the Alagoas and Sergipe coast

Fig. 11.9 Field aspects of beaches in the sub-compartment II of the Sediment Starved Coast of Alagoas. (**a**) beachrock reefs are still present but less numerous; (**b**) sand is coarser and the beach face steeper; (**c**) a long section of active sea-cliffs, carved into the Barreiras formation; and (**d**) the dominant morphodynamic state is reflective but exposed to higher energy levels than in sub-compartment I (See Fig. 11.6 for location of photographs)

11.5.2 São Francisco River Delta

São Francisco river Delta compartment corresponds to the wave-dominated delta of the São Francisco river, which occupies the center of the study area and extends, for the purposes of the present study, for 120 km from Coruripe to Pirambu (Fig. 11.10). A clear gradient is observed in this compartment from medium to coarse sands in the vicinity of the river mouth, which then laterally change to fine sand to both the northeast and southwest, following the sediment dispersion pattern induced, respectively, by the effective northeast longshore transport north of the river mouth (Santos 2010), and to the southwest (Barbosa 1997, Oliveira 2003) south of the river mouth. It is noteworthy that due to the tendency of most fluvial sediments being transported to the southwest, the section of coast with medium sand is also longer in this direction (Fig. 11.10). The beaches in this compartment go from intermediate to dissipative, following the pattern of change in beach sediment grain size and wave energy which is slightly higher north of the river mouth (Fig. 11.11a–d). The greater aridity and longer dry season (3–4 consecutive months) in this compartment, associated with greater wind velocity, favors the development of dune fields on both sides of the river mouth (Fig. 11.11a; Dominguez et al. 1992; Dominguez 1996; Barbosa 1997).

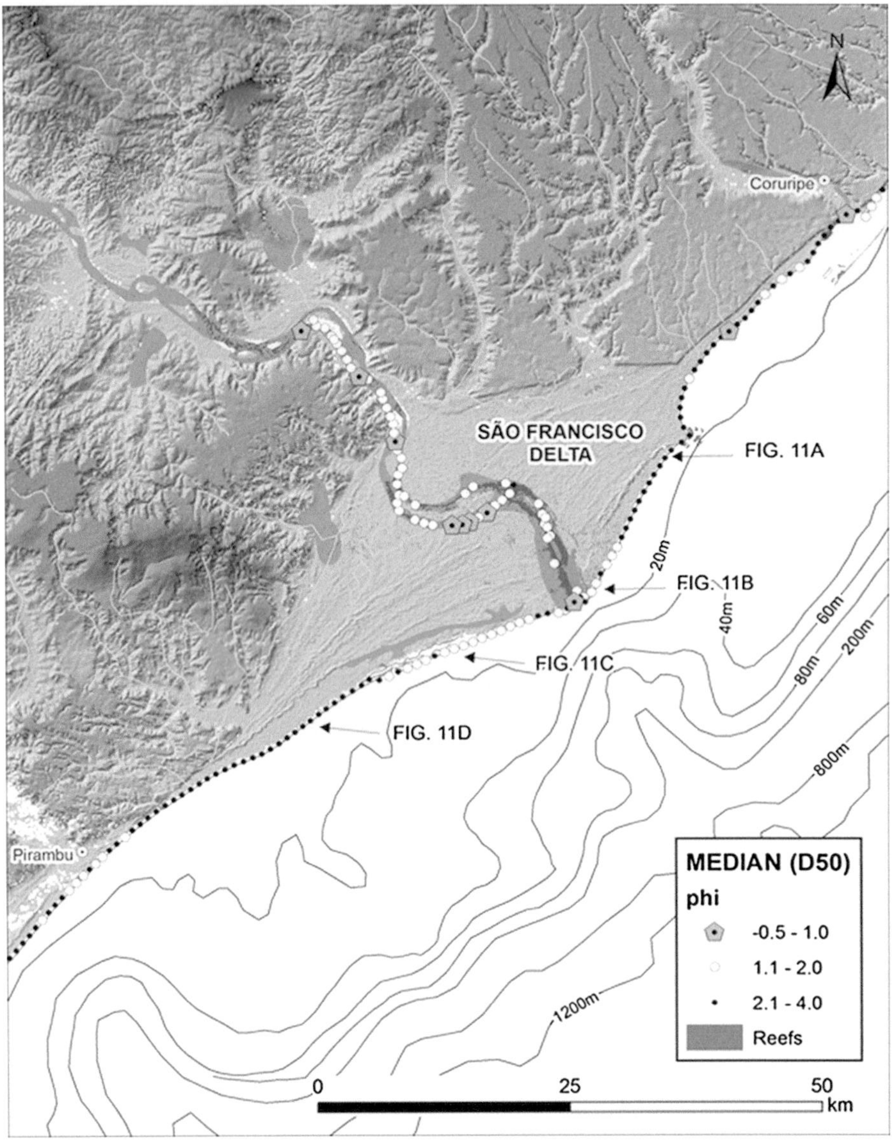

Fig. 11.10 Values of median grain size (D_{50}) along the São Francisco Delta compartment and its lower river course. Arrows indicate location of photos shown in Fig. 11.11

a b c d

Fig. 11.11 Field aspects of sandy beaches along the São Francisco Delta compartment. (**a**) high-energy dissipative beaches are dominant north of the river mouth. Intermediate beaches are common near the São Francisco river mouth where sands are coarser. (**b**) north of the river mouth, transverse bar and rip beach. (**c**) south of the river mouth, longshore bar–trough beach. (**d**) in the downdrift direction, away from the river mouth, as the sands get finer and finer, dissipative beaches dominate (See Fig. 11.10 for location of photographs)

11.5.3 Progradational Coast of Sergipe

This compartment extends from Pirambu south to the Real River (Fig. 11.12). The coastal zone has progradational characteristics, with wide beach deposits (Suhayda et al 1977) associated with marine isotopic stages 5 and 1 bordering the shoreline (Figs. 11.2b and 11.3c). Small blow-outs border the shoreline as well. In this 96 km long compartment the shoreline forms three broad beach arcs with predominantly fine sediment on the beach face. The beaches are typically dissipative, resulting from the combination of greater exposure to wave energy (absence of reef constructions and beachrock) and fine sediments (Suhayda et al. 1977; Fig. 11.13a–d). The dominance of fine sand in this compartment may have resulted from the combination of the contributions from the São Francisco River and the other three rivers that flow into this section, which have large estuarine areas (Figs. 11.2b and 11.3d) that act as filters retaining coarser sediments.

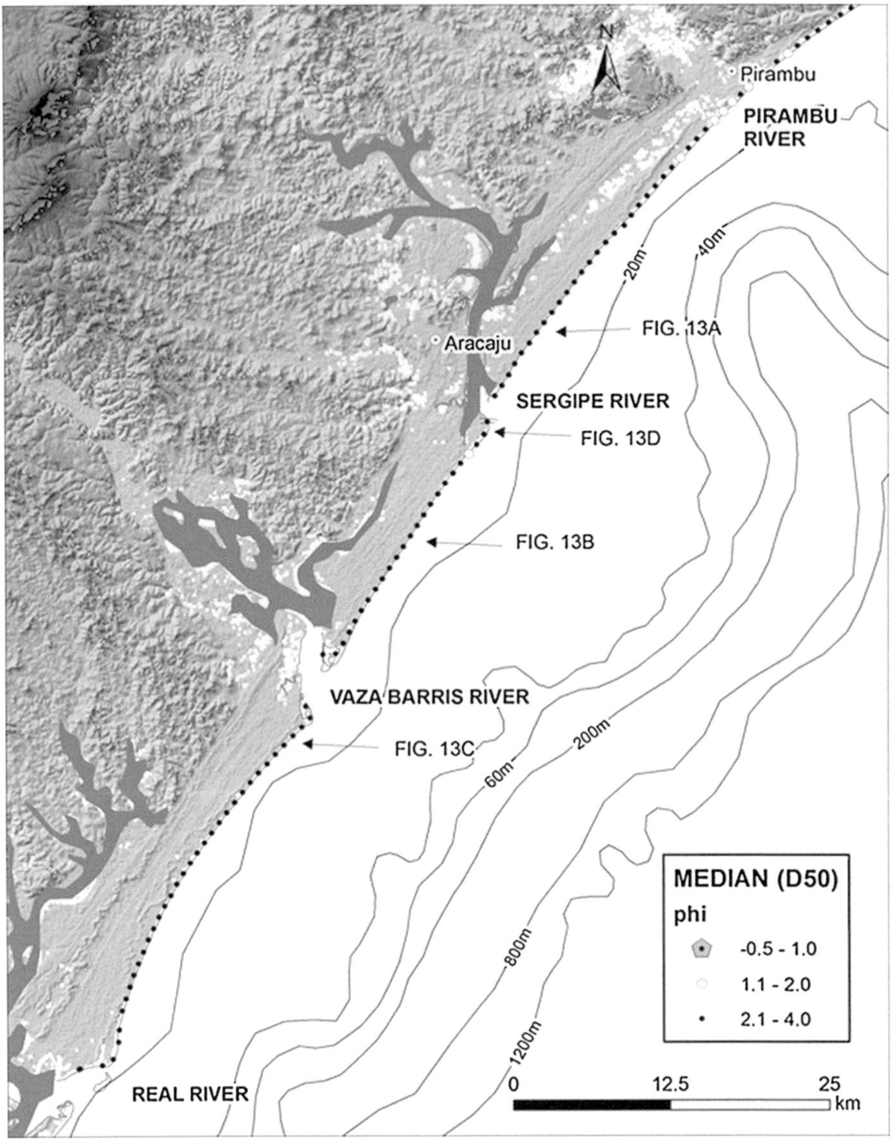

Fig. 11.12 Values of median grain size (D_{50}) along the Progradational Coast of Sergipe. Arrows indicate location of photos shown in Fig. 11.13

11.6 Shoreline Behavior

Based on the evaluations of Oliveira (2003), Araújo et al (2006), Bittencourt et al. (2006), Santos (2010) and Rodrigues (2014), the shoreline behavior of the study area can be summarized as follows (Fig. 11.14):

Fig. 11.13 Field aspects of the Progradational Coast of Sergipe beaches. **(a):** dissipative beach south of the São Francisco river. **(b):** dissipative beach south of Aracaju city. **(c)** and **(d):** dissipative urban beaches at Aracaju city. (See Fig. 11.12 for location of photographs)

(i) The Starved Coast of Alagoas Compartment – due to the reduced sediment supply, this compartment has the most severe coastal erosion in the study area, particularly in sub-compartment I, where coastline occupation density is greater (Santos 2010). Although sheltered beaches dominate this sub-compartment, it is also the one that has the most severe cases of erosion.

Land use practices, with construction too close to the shoreline, exacerbates the problem (Fig. 11.15a, b), resulting in the proliferation of rigid shoreline protection structures that degrade the recreational quality of the beaches. Santos (2010), evaluating the behavior of the shoreline in Alagoas based on field surveys and comparison of aerial photographs from the period 1955 to 1995, concluded that 58 % of the shoreline was subjected to erosive processes, 21 % was in balance and 21 % in progradation, with this latter percentage

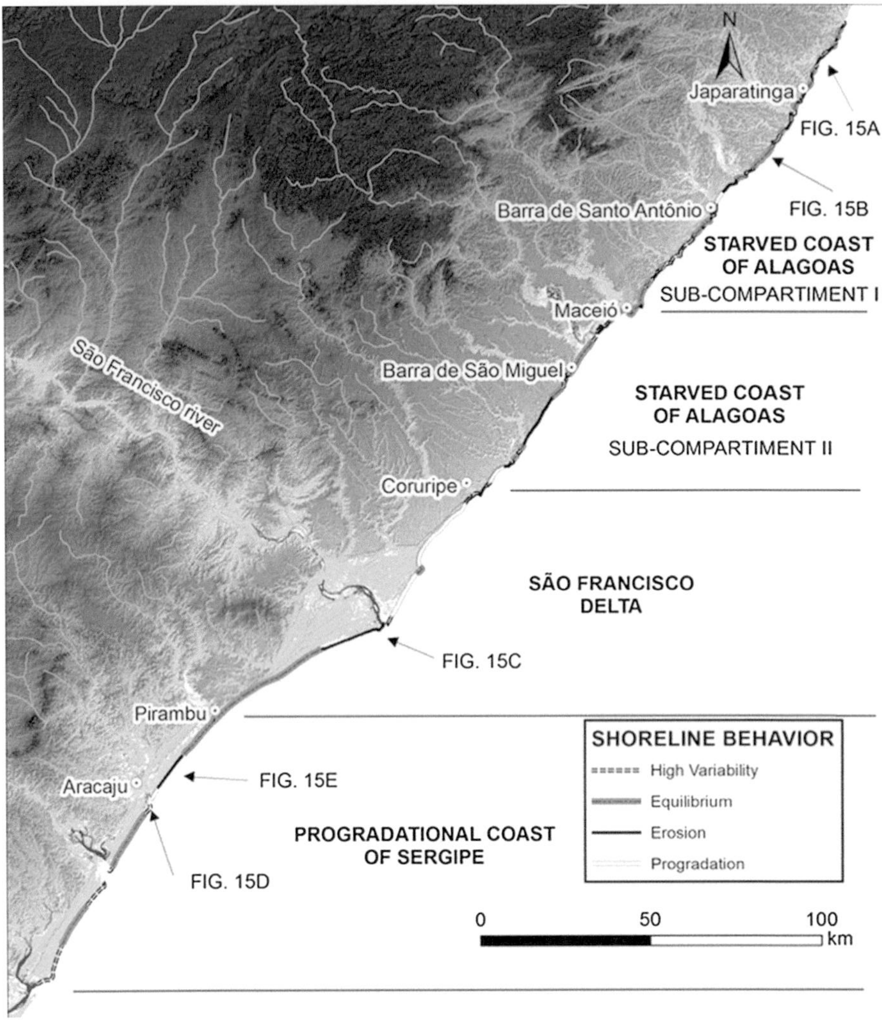

Fig. 11.14 Coastline beahaviour along the Sergipe and Alagoas coast. Arrows indicate location of photos shown in Fig. 11.15

associated with the delta of the São Francisco River. Even sparsely populated sections, such as sub-compartment II, showed evidence of continued erosive shoreline retreat, in the form of active cliffs;

(ii) São Francisco River Delta Compartment – although this coastal compartment has been progradational in the long term, erosive retreat has also occurred associated with the lateral migration of the mouths of small rivers such as at Pirambu. Likewise, the mouth of the São Francisco River has been affected by severe erosion, which resulted in the destruction of the Cabeço village

Fig. 11.15 Coastal erosion in Alagoas and Sergipe states. Sediment Starved Coast of Alagoas – (**a**) and (**b**) inappropriate occupation of the shoreline with no provision for setbacks results in man-made erosion leading to widespread construction of sea walls. São Francisco Delta compartment – (**c**): shoreline retreat at the river mouth resulted in destruction of the Cabeço coastal village, originally built besides the lighthouse. Progradational Coast of Sergipe compartment. (**d**): Shoreline erosion is very common in association with lateral displacement of tidal channel talweg as illustrated at the Sergipe river mouth. (**e**): the construction of the offshore port of Sergipe caused development of a coastal salient starving downdrift beaches of sediments and producing shoreline retreat south of structure (photo taken facing south) (See Fig. 11.14 for location of photographs)

(Fig. 11.15c). In this case, erosion has been attributed to the retention of sediments and changes in flow regimes associated with the construction of large dams (Bittencourt et al. 2007). Finally, another section that has undergone erosion is the Pontal do Peba beach, located 25 km northeast of the São Francisco River mouth.

(iii) The Progradational Coast of Sergipe Compartment – in this generally progradational compartment, coastal erosion results from processes of lateral migration of fluvial channels, similar to what occurs in ebb tidal deltas (Fig. 11.15d) (Oertel 1977; Fitzgerald 1984; Bittencourt et al. 2001), or are associated with the effects of sediment retention in the salient formed after the construction of an offshore port (Inácio Barbosa Terminal) (Fig. 11.15e).

11.7 Beach Safety

No consistent statistics were found regarding drownings along the coasts of Sergipe and Alagoas. Only newspaper reports and some fragmented data from lifeguard services were found for both states. The available information is, therefore, anecdotal, though coherent with the beach morphodynamic states that were previously described (Fig. 11.16). Thus, the safest beaches for swimming are those situated in sub-compartment I of the Starved Coast of Alagoas, due to the low wave energy resulting from the protection delivered by the fringing reefs. Sub-compartment II is dominated by reflective beaches, which would supposedly make these beaches safer. However, the reflective character of these beaches seems to be more related to the grain size of the sands than by the sheltered character of the beaches themselves, which is reduced over this stretch, given the small number of fringing and beachrock reef. The more exposed character makes these beaches less safe for swimming, particularly during the high tides. In the other two compartments (Delta of the São Francisco River and Progradational Coast of Sergipe), bathing risk is high due to the dissipative character of the beaches and more intense winds.

11.8 Sediment Input, Texture of the Beach Sand and Morphodynamics

The beaches of Sergipe-Alagoas illustrate the role played by sediment supply and sediment size to beach state. Sergipe and Alagoas are separated by the São Francisco River, which represents the greatest point source of sediment to the region. While in Sergipe a general trend of long term progradation of the coastline predominated during marine isotopic stages 5 and 1, as shown by the wide strandline deposits that accumulated during these two high-sea periods, in Alagoas erosion has dominated during the same period, leading to the very limited or no beach deposits. This asymmetry observed on the coastline is also displayed in the shelf sediment cover.

The longterm erosional coastal sections exhibit active cliffs and marine-cut terraces, particularly if the geological substrate undergoing erosion is made up of sedimentary rocks. In the regions tropical climate these terraces serve as substrate for sessile organisms, which can eventually develop into reef constructions, which, in turn, protect the shoreline from wave action, generating sheltered beaches. The

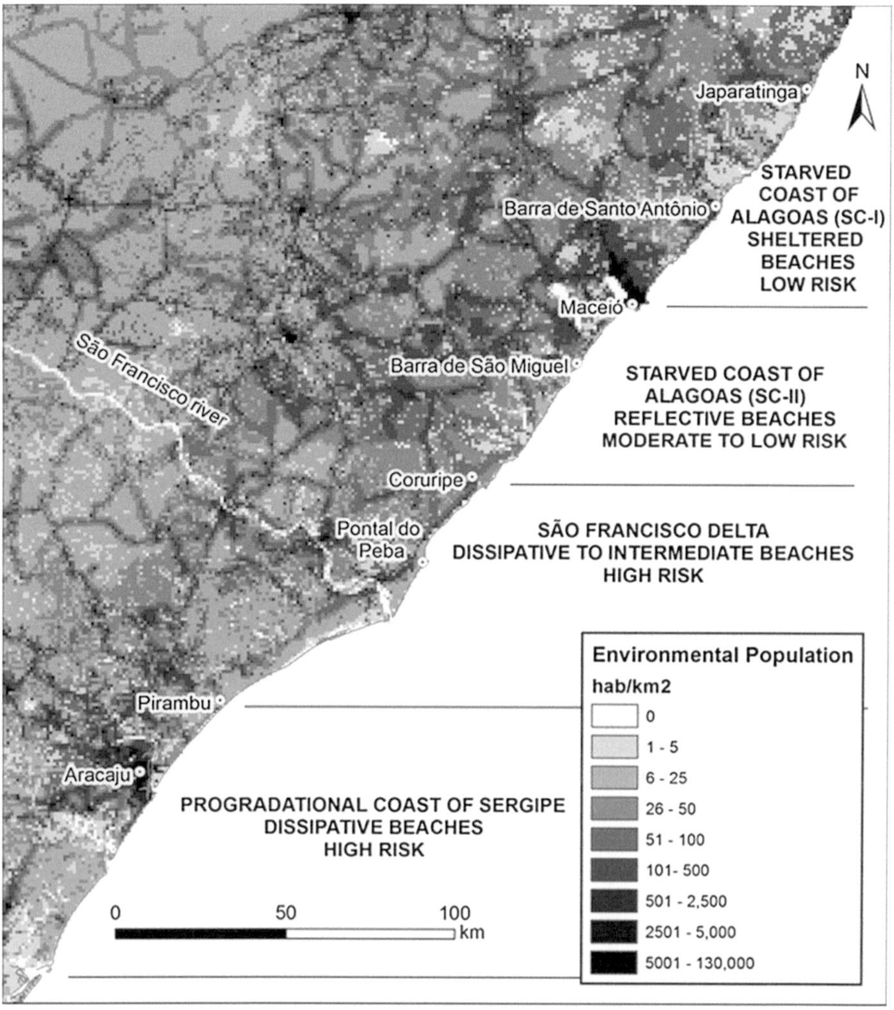

Fig. 11.16 Beach risk as related to modal morphodynamic beach state and environmental population. The Ambient Population is from the Landscan 2010 dataset (The LandScan 2010™ High Resolution global Population Data Set copyrighted by UT-Battelle, LLC, operator of Oak Ridge National Laboratory under Contract No. DE-AC05-00OR22725 with the United States Department of Energy)

tropical climate also favors the cementation of the beach prism, forming beachrock reefs that are later exhumed by the retreat of the shoreline, thus also contributing towards shoreline protection and serving as a substrate for the development of reef constructions. The main implication for the type of beach present in the area is the predominance of either reflective beaches, due to local sources of coarse sands, originated from the erosion of cliffs, or sheltered tide-modified beaches, which during the high tides may exhibit reflective behavior.

In the *Starved Coast of Alagoas*, the reduced input of fluvial sediments in association with outcropping rocks of variable levels of resistance to erosion, produced two sectors with distinct beach morphodynamic characteristics. In Sub-Compartment I, Mesozoic rocks from the Alagoas basin outcrop near the coastline forming active cliffs. The tendency of long term erosion has produced several marine-cut terraces that have served as substrate for the development of reef constructions. As a consequence, this section is characterized by the presence of several salients that separate small bays, in which sheltered tide-modified beaches exhibit a concave profile, and are exposed to wave energy only during high tides. The upper portion of the beach-face is steeper and concentrates coarser sediments, while the low tide terrace has fine sands, characteristically enricheded in bioclasts. In this sub-compartment, the sediment collected from the mid-point of the beach-face presents a very characteristic tail of coarse sediments, with great heterogeneity and greater dispersion in median values and sorting (Fig. 11.8). This grain size distribution results from a combination of mixed sediment sources (siliciclasts contributed by small rivers, cliff erosion and bioclasts that are produced locally) with lower energy levels incapable of fully sorting these sediments. Coarse sand reflective beaches predominate in Sub-Compartment II, where there is a large section of coast with active cliffs carved in the Barreiras Formation, with few structures that are able to reduce wave energy, with the exception of some beachrock reefs.

In *the São Francisco River Delta Compartment*, the river sediment are transported to either side of the river mouth, with the greater volume, transported southwestwards. The beaches in the vicinity of the river mouth, where medium sands predominate, are intermediate, either longshore bar-trough or transverse bar, and distally as the sands fine and become typically dissipative beaches with the development of dune fields, possibly as a result of reduced precipitation and increase of wind speed (Barbosa 1997). Grain size of the beach sediment varies regularly and is very well sorted (Fig. 11.8).

In the *Progradational Coast of Sergipe* the beaches present very uniform characteristics, with fine to very fine sands with typically dissipative beaches occurring along the whole compartment. The median values and the values of D_{15} and D_{85} are practically constant. Beach sands exhibit the best sorting in the whole Sergipe-Alagoas shoreline (Fig. 11.8). The good sorting may result from the combination of the natural filtering of river sediments at existing wide estuarine areas and the distal contributions from the São Francisco River. Winds over this coastal section also have higher velocities compared to those along the Alagoas coast, which may result in higher waves.

11.9 Summary

Many studies have highlighted the role of geological inheritance (Short 2010) and the effect of sediment grain size (McNinch 2004) in controlling beach morphodynamics. The beaches of Sergipe and Alagoas illustrate not only these aspects, but also the importance of sediment supply, in this case exemplified by the São Francisco River, whose inputs regionally influence the highly contrasting morphodynamic states of the beaches of these two states.

Overall, under the same wave energy levels, beaches located on prograding sections of the coast have finer sediments and are dissipative, whereas on coastal sections starved of sediments, in which long term shoreline erosion is the rule, the sand is coarser and the beaches are more reflective. This results from the fact that distant sourced sands, typical of large deltas, such as the São Francisco, displays a longshore fining (Frings 2008) in this case supplying the southern Sergipe coast with fine sand. Moreover locally sourced cliff sands tends to be coarser, as in the case of starved coasts, because they result from the physical breakdown, by wave action, of older coarser geologic units (e.g. Barreiras Formation). The erosive remnants and marine terraces of these older units, provide foundations for fringing reefs, which further protects the shoreline from wave energy.

Thus, overall, sediment supply and grain size exerted an important control in the longshore variation of beach morphodynamic in Sergipe and Alagoas.

Acknowledgments This study is a contribution of inctAmbTropic—Brazilian National Institute of Science and Technology for Tropical Marine Environments, CNPq/FAPESB Grants: 565054/2010-4 and 8936/2011. ACSP Bittencourt and JML Dominguez thank CNPq for a Research Fellowship.

References

Almeida FFM (1977) O Cráton do São Francisco. Rev Bras Geosci 7(4):349–364

Amarante OC, Zack J, Brower M, Sá AL (2001) Atlas do Potencial Eólico Brasileiro. Ministério de Minas e Energia, Brasília

Araújo TCM, Santos RCAL, Seoane JCS, Manso VAV (2006) Alagoas. In: Erosão e Progradação do Litoral Brasileiro. Ministério do Meio Ambiente, Brasília, pp 198–212

Assis HMB (coord) (2015) Carta Batimétrica da Plataforma Continental Rasa do Estado de Alagoas. Recife. CPRM, escala 1:250.000. Folhas de Referência:SC.25-Y-A, SC.25-V-C, SC.25-V.A, SC.24-Z-B e SC.24-X-D

Barbosa LM (1997) Campos de Dunas Costeiras Associadas à Desembocadura do Rio São Francisco (SE/AL): origem e controles ambientais. Dissertation, Universidade Federal da Bahia, Brazil

Barbosa LM, Bittencourt ACSP, Dominguez JML, Martin L (1986) The quaternary coastal deposits of the state of Alagoas: influence of the relative sea-level changes. In: Quaternary of South America and Antarctica Peninsula, vol 4. A.A. Balkema, Rotterdam, pp 269–290

Bittencourt ACSP, Martin L, Dominguez JML, Ferreira YA (1983) Evolução paleogeográfica quaternária da costa do Estado de Sergipe e da costa sul do Estado de Alagoas. Rev Bras Geosci 13:93–97

Bittencourt ACSP, Lessa GC, Dominguez JML, Martin L, Vilas-Boas GS, Farias FF (2001) High and low frequency erosive and constructive cycles in estuarine beaches: an example from garcez point, Bahia/Brazil. Anais Acad Bras Ciências 73:599–610

Bittencourt ACSP, Oliveira MB, Dominguez JML (2006) Sergipe In: Erosão e Progradação do Litoral Brasileiro. Ministério do Meio Ambiente, Brasília, Brazil, pp 213–218

Bittencourt ACSP, Dominguez JML, Fontes LCS, Sousa DL, Silva IR, Silva FR (2007) Wave refraction, river damming, and episodes of severe shoreline erosion: the São Francisco river mouth, Northeastern Brazil. J Coast Res 34:930–938
Castro Jr ACM (1988) Structural evolution of the sergipe-alagoas basin.Brazil. Dissertation, Rice University, USA
Castro CB, Pires DO (2001) Brazilian coral reefs: what we already know and what is still missing. Bull Mar Sci 69:357–371
Dominguez JML (1996) The São Francisco Strandplain – a paradigm for wave-dominated deltas? In: De Batist M, Jacobs P (eds) Geology of siliciclastic shelf seas. Geological Society, London, special publication 117, pp 217–231
Dominguez JML, Bittencourt ACSP, Martin L (1992) Controls on quaternary coastal evolution of the east-northeast coast of Brazil: roles of sea-level history, trade winds and climate. Sediment Geol 80:213–232
Dominguez JML, Bittencourt ACSP (1996) Regional assessment of long term trends of coastal erosion in northeastern Brazil. Anais Acad Bras Ciências 68:355–371
Fitzgerald DM (1984) Interactions between the ebb- tidal delta and landward shoreline: price inlet, south caroline. J Sediment Petrol 4:1303–1318
Fontes LCS, Suffredini M, Mendonça JB, Queiroz EV, Silva DS, Santos JR (2011) Sedimentos biodetríticos da plataforma continental sul de Alagoas. XIII Congresso da Associação Brasileira de Estudos do Quaternário. In: Proceedings. Buzios Rio de Janeiro, Brasil. CD-ROM
Frings RM (2008) Downstream fining in large sand-bed rivers. Earth Sci Rev 87:9–60
Guimarães CRP (2010) Composição dos Sedimentos Superficiais e da Fauna Bêntica na Plataforma Continental de Sergipe. Dissertation, Universidade Federal da Bahia, Brazil
Jackson NL, Nordstrom KF, Eliot I, Masselink G (2002) 'Low energy' sandy beaches in marine and estuarine environments: a review. Geomorphology 48:147–162
Knoppers B, Medeiros PRP, Souza WFL, Jennerjahn T (2006) The São Francisco estuary, Brazil. Handbook of environmental chemistry, part H. Springer, Heidelberg, pp 51–70
Leão ZMAN, Kikuchi RKP, Testa V (2003) Corals and coral reefs of Brazil. In: Cortés J (ed) Latin American coral reefs. Elsevier, Amsterdam, pp 9–52
Lyra de Brito MF, Silva Filho AFS, Guimarães IP (2009) Caracterização geoquímica e isotópica do batólito Serra do Catu e sua evolução da interface dos domínios Sergipano e Pernambuco-Alagoas, Província Borborema. Rev Bras Geosci 39:324–337
Maida M, Ferreira B (1997) Coral reefs of Brazil: an overview. In: Proceedings of the 8th international coral reef symposium, Panama 1, pp 263–274
McNinch JE (2004) Geologic control in the nearshore: shore-oblique sandbars and shoreline erosional hotspots, Mid-Atlantic Bight, USA. Mar Geol 211:121–141
Medeiros PRP, Knoppers BA, Santos RC Jr, Souza WFL (2007) Aporte Fluvial e Dispersão de Matéria Particulada em Suspensão na Zona Costeira do Rio São Francisco (SE/AL). Geochim Bras 21:212–231
Mohriak WU, Mello MR, Bassetto M, Vieira IS, Koutsoukos EAM (2000) Crustal architecture, sedimentation, and petroleum systems in the Sergipe–Alagoas Basin, northeastern Brazil. In: Mello MR, Katz BJ (eds) Petroleum systems of south Atlantic margins: AAPG memoir, vol 73. American Association of Petroleum Geologists, Tulsa, Oklahoma, pp 273–300
Nascimento AA (2011) Sedimentação Holocênica na Plataforma Continental de Sergipe, Nordeste do Brasil. Thesis. Universidade Federal da Bahia, Brazil
Nimer E (1989) Climatologia do Brasil. IBGE, Rio de Janeiro
Oertel GF (1977) Geomorphic cycles in ebb-tidal deltas and related patterns of shore erosion and accretion. J Sediment Petrol 47:121–1131
Ojeda HAO (1982) Structural framework, stratigraphy, and evolution of Brazilian marginal basins. AAPG Bull 66:732–749
Oliveira MB (2003) Caracterização Integrada da Linha de Costa do Estado de Sergipe – Brasil. Thesis. Universidade Federal da Bahia, Brazil

Oliveira EP (2012) Capítulo XI – Faixa de Dobramento Sergipana. In: Barbosa JSF (ed) Geologia da Bahia: pesquisa e atualização. Companhia Baiana de Pesquisa Mineral, Salvador, pp 179–198

Pianca C, Manzzini PLF, Siegle E (2010) Brazilian offshore wave climate based on NWW3 reanalysis. Braz J Oceangr 58:53–70

Ponte FC (1969) Estudo morfo-estrutural da bacia Alagoas-Sergipe. Bol Téc Petrobrás 12:439–474

Rodrigues TK (2008) Análise das Mudanças da Linha de Costa das Principais Desembocaduras do Estado de Sergipe com Ênfase no Rio Sergipe. Thesis. Universidade Federal da Bahia, Brazil

Rodrigues TK (2014) Impacto das mudanças climáticas na zona costeira do município de Aracaju-SE: estudo de caso utilizando o SMC (Sistema de Modelagem Costeira). Dissertation, Universidade Federal da Bahia, Brazil

Rossetti DF, Bezerra FHR, Dominguez JML (2013) Late Oligocene–Miocene transgressions along the equatorial and eastern margins of Brazil. Earth Sci Rev 123:87–112

Rudorff FM, Gherardi DFC (2008) Coral reef detection using SAR/RADARSAT-1 images at Costa dos Corais, PE/AL, Brazil. Braz J Oceangr 56:85–96

Salles FJP, Bentes FCM, Santos JA (2000) Catálogo de Estações Maregráficas. Fundação de Estudos do Mar, Rio de Janeiro

Santos AN (2010) A Dinâmica Costeira e seus Efeitos sobre a Ocupação Urbana na Orla Marítima do Estado de Alagoas. Dissertation, Universidade Federal da Bahia, Brazil

SEMARH (2015a) http://www.semarh.al.gov.br. Accessed 15 Mar 2015

SEMARH (2015b) http://www.semarh.se.gov.br. Accessed 15 Mar 2015

Short AD (2010) Role of geological inheritance in Australian beach morphodynamics. Coast Eng 57:92–97

Silva Filho AF, Guimarães IP, Kozuch M (1996) Mineral chemistry and tectonic significance of Neoproterozoic ultrapotassic plutonic rocks from Cachoeirinha Salgueiro fold belt, northeastern Brasil. Int Geol Rev 38:649–664

Silva Filho AF, Guimarães IP, Brito MFL, Pimentel MM (1997) Geochemical signatures of the main neoproterozoic late tectonic granitoids from the Proterozoic Sergipano Belt, Brazil: significance for the brasiliano orogeny. Int Geol Rev 33:639–653

Silva Filho AF, Guimarães IP, Van Schmus WR (2002) Crustal evolution of the pernambuco-Alagoas complex, borborema province, NE Brazil: Nd isotopic data from neoproterozoic ganitoids. Gondwana Res 5:409–422

Souza WFL, Knoppers B, Balzer W, Leipe T (2003) Fluxos de água e sedimentos na costa leste do Brasil: relações entre a tipologia e as presses antrópicas. Geochim Bras 17:057–074

Suhayda JN, Hsu SA, Roberts HH, Short AD (1977) Documentation and analysis of coastal processes, northeast coast of Brazil. Coastal Studies Unit Technical Report 238, 98 pp

Chapter 12
Beaches in the State of Bahia: The Importance of Geologic Setting

José Maria Landim Dominguez, Abílio Carlos da Silva Pinto Bittencourt, Lucas do Nascimento, and Adeylan Nascimento Santos

Abstract The state of Bahia with approximately 1000 km of shoreline, a great heterogeneity of exposed rock types and subjected to varying degrees of sediment supply allowed us to investigate the controls exerted by these factors in determining the beach types.

The modal morphodynamic states of the beaches of Bahia is dominantly controlled by sediment grain size, which in its turn results from the long term history of the coastal zone. Locally sourced sediments, eroded from the coastal tablelands (Barreiras Formation), are predominantly coarse grained favoring reflective beaches (Sediment Starved Southern Coast). Distally sourced sediments, as in the case of large rivers are predominantly fine to very fine sands, resulting in a dominance of dissipative and high-energy intermediate beaches (Deltaic Coast of the Jequitinhonha River). Shoreline stretches nourished by small rivers are characterized by medium size sands and a dominance of intermediate high energy beaches (Northern Littoral Coast Compartment). Finally, the stretch of the coast fronted by Mesozoic Rifts, characterized by a great heterogeneity of sedimentary rock types and small sediment supply, have a very irregular shoreline, bordered by fringing reefs. In this section low energy sheltered beaches dominate.

J.M.L. Dominguez (✉) • A.C.S.P. Bittencourt
Institute of Geosciences, UFBA, Campus Ondina, Salvador, BA, Brazil
e-mail: landim@ufba.br; abilio@pq.cnpq.br

L. do Nascimento
Prefeitura Municipal de Camaçari, Camaçari, Bahia, Brazil
e-mail: lucasnascimento.lec@gmail.com

A.N. Santos
School of Engineering and Information Technology, UNIFACS, Salvador, Bahia, Brazil
e-mail: adeylan.santos@pro.unifacs.br

A.D. Short, A.H. da F. Klein (eds.), *Brazilian Beach Systems*, Coastal Research Library 17, DOI 10.1007/978-3-319-30394-9_12

12.1 Introduction

Bahia has the longest open coastline in Brazil (approximately 1000 km) (Fig. 12.1) and its coastal landforms include cliffs and rivers, a wave-dominated delta, beach-ridge plains, coral reefs and large bays. The evolution of these landforms during the Cenozoic has affected certain key features of sandy beaches along the Bahia coast. This chapter summarizes the characteristics of these beaches based primarily on data collected over the previous 15 years, most of which are unpublished or were reported only in dissertations, theses and monographs. These data were collected during overflights and field surveys along the coastline. During the field surveys, samples of beach-face sediment were collected at 1-km intervals, the beach face gradient was measured, and the type and number of breakers in the surf zone were recorded simultaneously. This database enabled quantitative assessment of the spatial variation of these parameters along the Bahia coastline, which enabled us to gain an understanding of the spatial variability of the modal morphodynamic states of Bahia's beaches and their main controlling factors when combined with geologic data and modeling of wave transformation (refraction) and sediment characteristics along the coast. This is the main focus of this chapter.

12.2 Geology and Geomorphology of the Coastal Zone

Bahia's coastline configuration was established and controlled by the separation of the South American and African Plates beginning in the Early Cretaceous (145 Ma) (Silva et al. 2007). This separation largely produced the general coastline physiography. The interactions between this physiography and other factors, including variations in sea level and climate, the supply of sediments, and the evolution of marginal sedimentary basins shaped the final coastal configuration and produced the variety of coastal features that are now present.

The following pre-Quaternary geologic units are exposed along the coast (Fig. 12.1):

(i) Metamorphic rocks (primarily high grade) that compose the São Francisco Craton of Archean–Paleoproterozoic age (Alkmim et al. 2001; Barbosa and Sabaté 2004) and their marginal fold belts of Neoproterozoic age (Almeida 1977; Cruz et al. 2012; Oliveira 2012).
(ii) Mesozoic sediments deposited in basins formed during the split of the South American and African Plates (Recôncavo, Camamu and Almada basins: Mohriak et al. 2008).
(iii) Detrital material of marine–transitional origin (Barreiras Formation) deposited during an Early–Middle Miocene transgression that flooded the continental edge of Brazil (Rossetti et al. 2013); these deposits underlie the coastal plains, and form coastal tablelands.

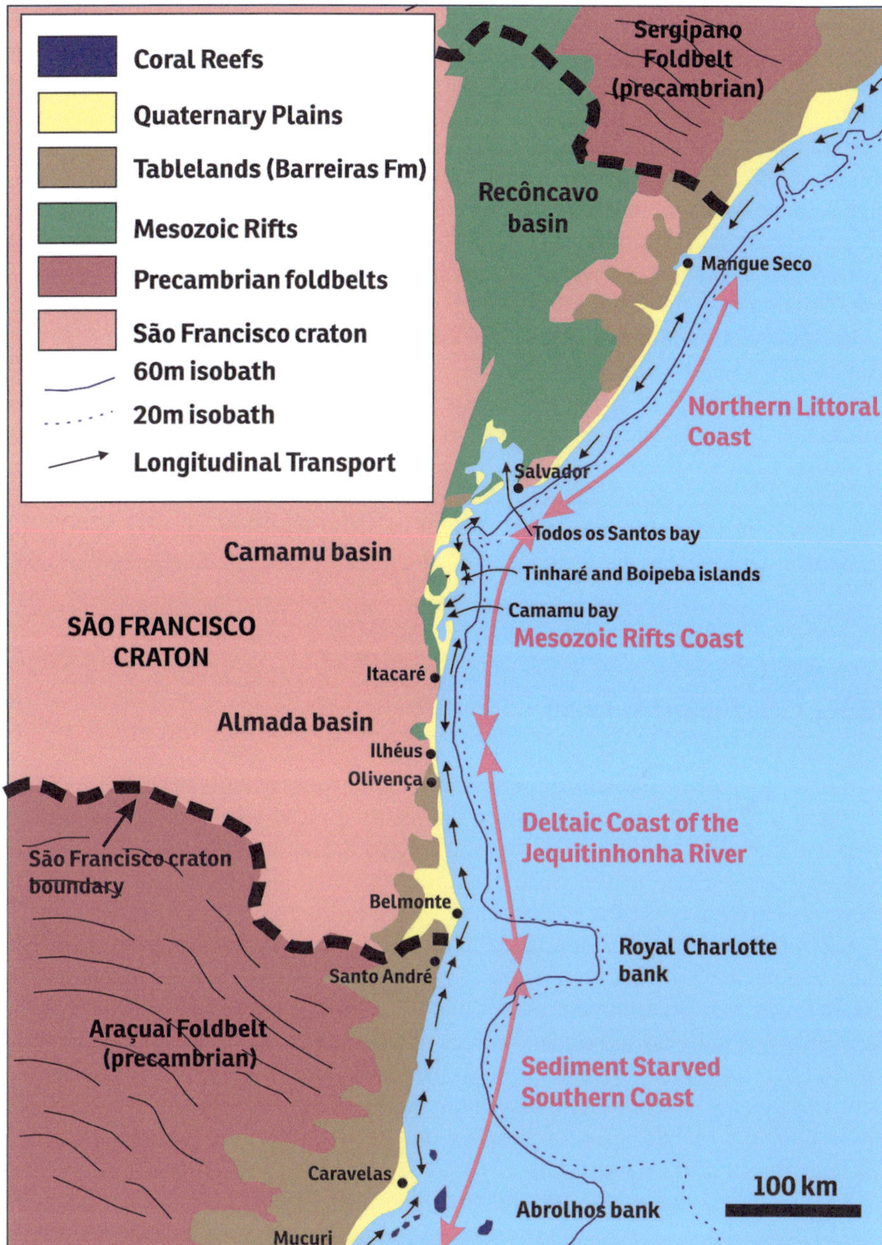

Fig. 12.1 Geological framework of the Bahia coastal zone and its major coastal compartments. Bahia's coast extends from Mangue Seco to Mucuri

The northern-central coastal zone faces the São Francisco Craton and is generally narrow. The Recôncavo, Camamu and Almada sedimentary basins are located in this section. Differential erosion between rocks of the older cratonic basement (more resistant) and the Mesozoic sedimentary rocks (less resistant) produced a topographic lowering of the region dominated by the latter rocks (Dominguez and Bittencourt 2009; Dominguez 2015). These low-lying areas were flooded during high sea-level stands, including the current highstand, resulting in bays and estuaries and the complete filling of the Almada basin. The most-rugged stretch of coast corresponds to the area where these bays have not yet been filled (Todos os Santos and Camamu bays and associated estuaries) (Fig. 12.1).

In southern Bahia, the portion of the coastal region beyond the limit of the São Francisco Craton, widens and is primarily underlain by the widespread Barreiras Formation, which forms active sea cliffs along the shoreline. This wide coastal zone is associated with a wide continental shelf that has been affected by volcanic activity dating from the beginning of the Tertiary (40–60 Ma) (Mohriak 2006; Sobreira and França 2006), which favored the development of the Royal Charlotte and Abrolhos Banks (200 km maximum width; Fig. 12.1). The largest coral reefs in the western Atlantic Ocean are located on the shallowest portions of these banks (Leão et al. 2003).

12.2.1 Sediment Sources

Two main sources of siliciclastic sediments supply the coastal beaches of the State of Bahia (Fig. 12.2): these are fluvial sediments transported by rivers of various sizes and material eroded from active cliffs of the Barreiras Formation. The latter is typically coarser, due to the close proximity to the source area and because the Barreiras Formation itself is composed of coarse sediments.

Siliciclastic sediments dominate a narrow strip fringing the coastline to maximum depths of 10–15 m (Dominguez et al. 2012, 2013). Offshore of this belt, bioclastic sands mainly composed of coralline algae fragments, are present. Therefore, there are apparently no significant sources of siliciclastic sands along the continental shelf that may have supplied the coast. However, locally, particularly adjacent to bank reefs, a few beaches may consist dominantly of bioclastic fragments produced on reef tops, as in the case of beaches on the islands of Tinharé and Boipeba (Rebouças et al. 2011) (Fig. 12.1).

The Jequitinhonha River transports a significant amount of sediment to the coastal zone (27×10^6 t year^{-1}) (Bernal 2009) and has built, along with the Pardo River a typical wave-dominated delta. The other rivers emptying along the coast have lower discharges, as is the case with the Contas (100 m^3 s^{-1}), Paraguaçu (83 m^3 s^{-1}) and Itapicuru (27 m^3 s^{-1}) rivers. The contribution of fluvial deposits is even less along the southern coast of Bahia. The Mucuri River is the largest river flowing to this stretch of coastl (liquid discharge of 74 m^3 s^{-1}; solid discharge of 1.2×10^6 t year^{-1}) (Queiroz 2003).

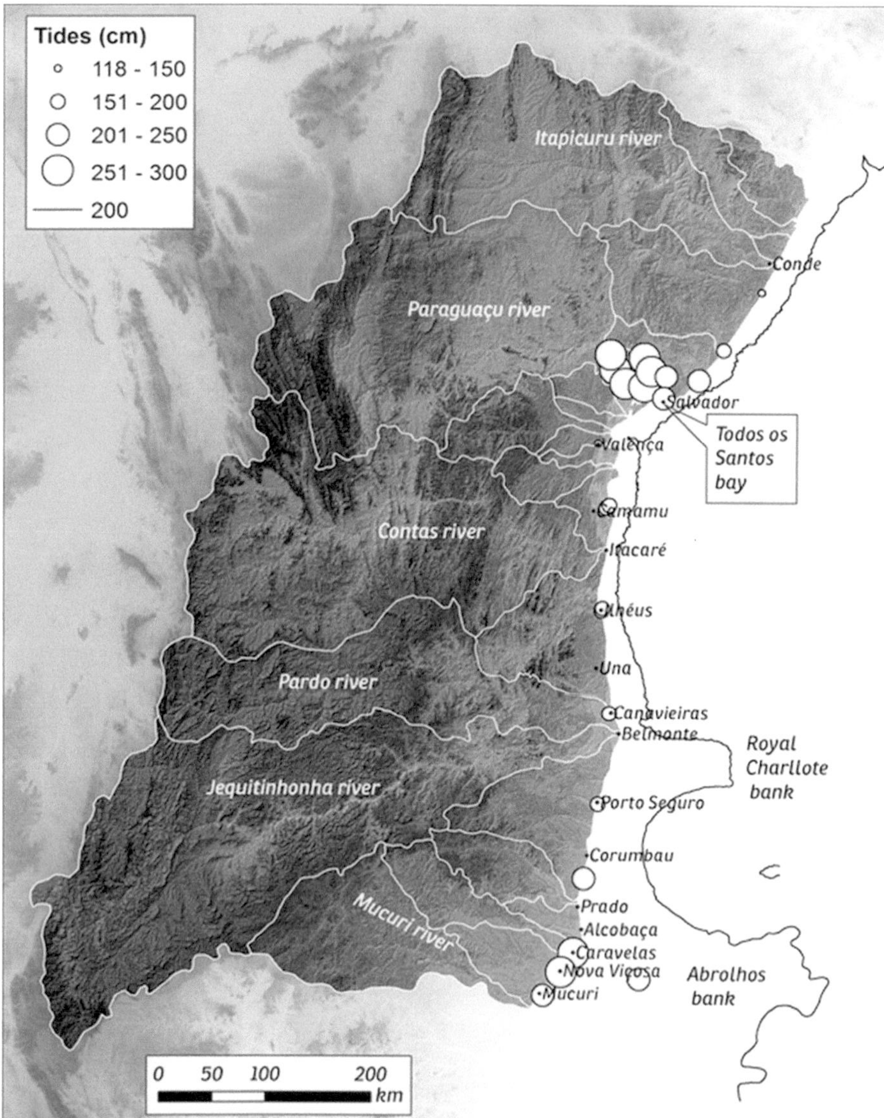

Fig. 12.2 Digital elevation model of drainage basins emptying in the study area. Also shown is the variation in tide range along the shoreline

The 3–4 m drop in relative sea level that affected the coast of Brazil during the last 5700 years (Angulo and Lessa 1997; Martin et al. 2003) may have also contributed sediment to coastline progradation through a mechanism that acted in accordance with the inverse of the Bruun rule (Bruun 1962).

12.3 Coastal Processes: Waves, Tides, Winds and Precipitation

The highest average annual rainfall in the coastal area of Bahia usually occurs in the section between Salvador and Ilhéus (1800–2600 mm $year^{-1}$) (Fig. 12.2), with rainfall decreasing southward toward Caravelas (1400 mm $year^{-1}$) and northward toward Mangue Seco (1400 mm $year^{-1}$). The tide ranges from 1.18 to 2.71 m (Fig. 12.2), with the largest tides observed within Todos os Santos Bay and in the area of Caravelas (Sales et al. 2000; Cirano and Lessa 2007). Mean wave heights and periods range from 1 to 3 m and from 6 to 10 s, respectively (Pianca et al. 2010). The wave propagation directions vary during the year depending on changes in the wind regime. East–northeast waves predominate in the spring and summer, southern waves are more common during the autumn and winter, due to changes in the direction of trade winds and a greater frequency of cold fronts.

Figure 12.1 shows the dispersion of sandy sediments along the coastline by wave action as modeled by Bittencourt et al. (2000) and Dominguez et al. (2009). A significant effect of the seasonal variation in the wave directional frequency is a rotational circulation on beaches enclosed between rocky headlands, as reported by Farias et al. (1985) for Armação Beach in the city of Salvador (Fig. 12.2).

The combination of seasonal changes in the wave direction and extensive coral reefs positioned offshore has produced local convergences in the sediments transport in southern Bahia (Fig. 12.1). This convergence has resulted in a tombolo at Corumbau and a mega-salient at Caravelas, where a broad progradation of the shoreline occurs in an otherwise sediment starved region (Fig. 12.1). The coral reefs also protect long sections of coast from wave action (Fig. 12.3), which has affected beach morphodynamics, including modifications imposed by tidal action, by altering the relative tidal range (RTR) (Masselink and Hegge 1995).

12.4 Coastal Compartments and Beach Types

Bahia's coastal zone can be subdivided into four compartments based on their physiographic features (Dominguez et al. 2012). These compartments, described below, are designated the Northern Littoral Coast, the Mesozoic Rift Coast, the Deltaic Coast of the Jequitinhonha river, and the Sediment Starved Southern Coast (Fig. 12.1). The compartments exhibit various beach morphodynamic stages given their different geologic heritages in addition to their very different physiographic characteristics.

The Northern Littoral Coast Compartment Extends from Salvador to the border with the state of Sergipe (Mangue Seco) (Fig. 12.4). This compartment is characterized by a nearly linear northeast–southwest-aligned coast, and the presence of small rivers. The crystalline basement is exposed in numerous locations along the coastline, particularly in the southern portion. The supply of sediments, although small

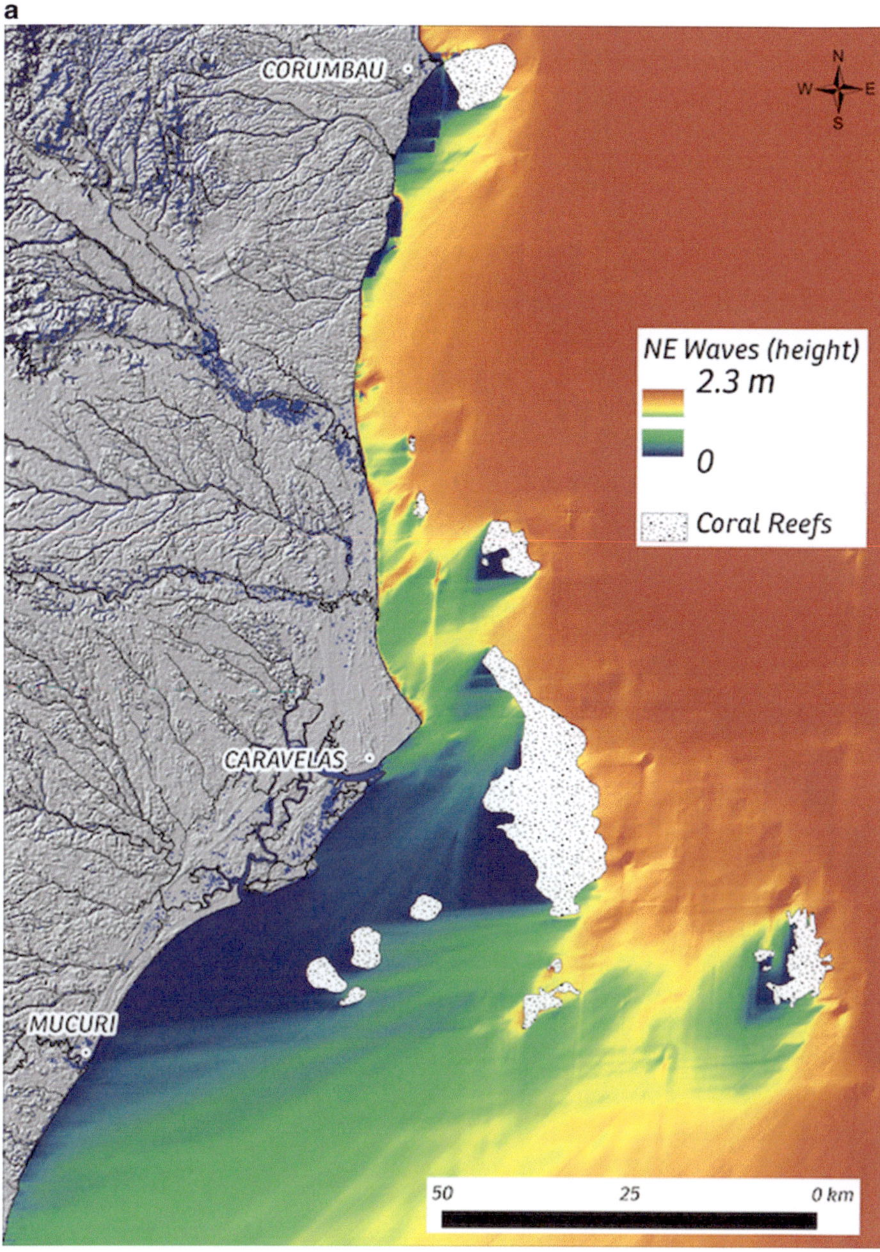

Fig. 12.3 Effects of coral reefs on wave refraction and attenuation along the "Sediment Starved Southern Coast" of Bahia (see Fig.12.1 for location). (**a**) NE waves. (**b**) SSE waves

b

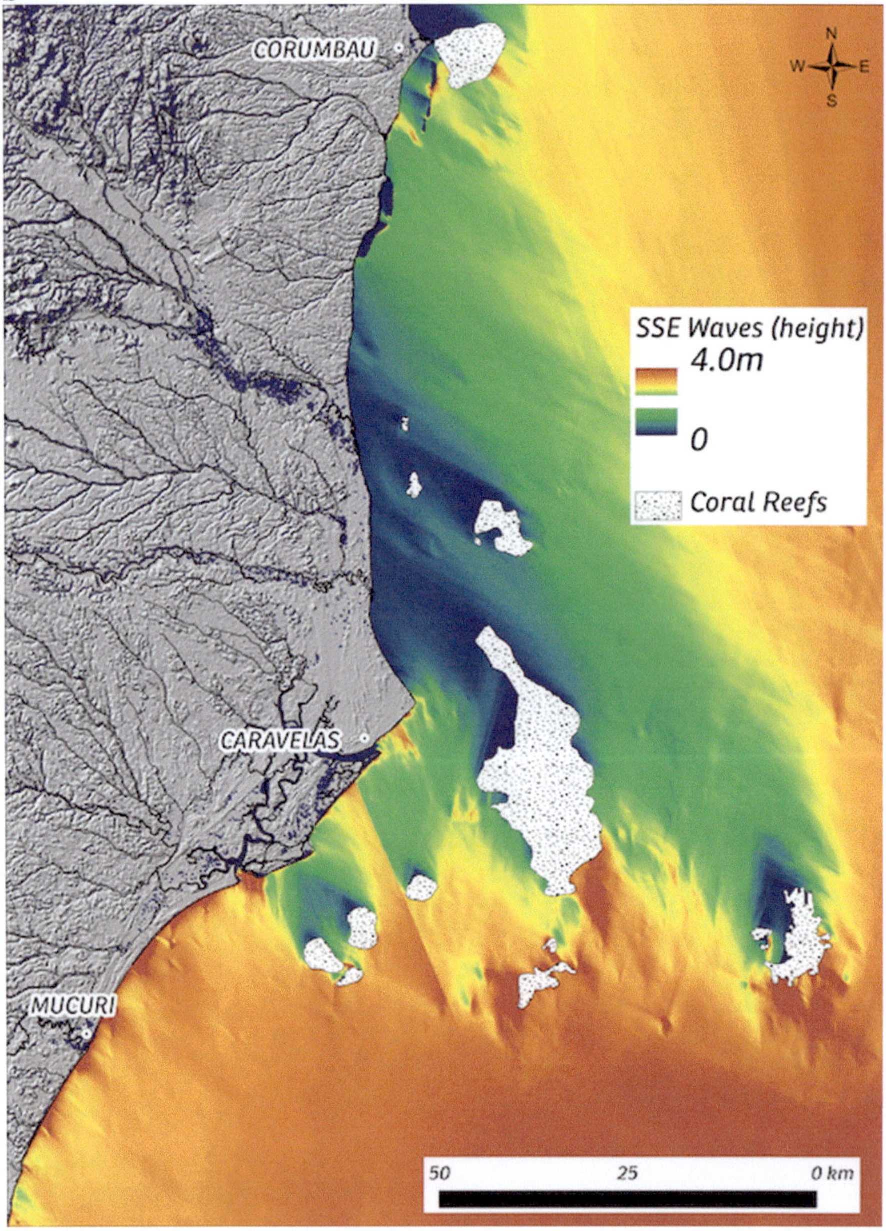

Fig. 12.3 (continued)

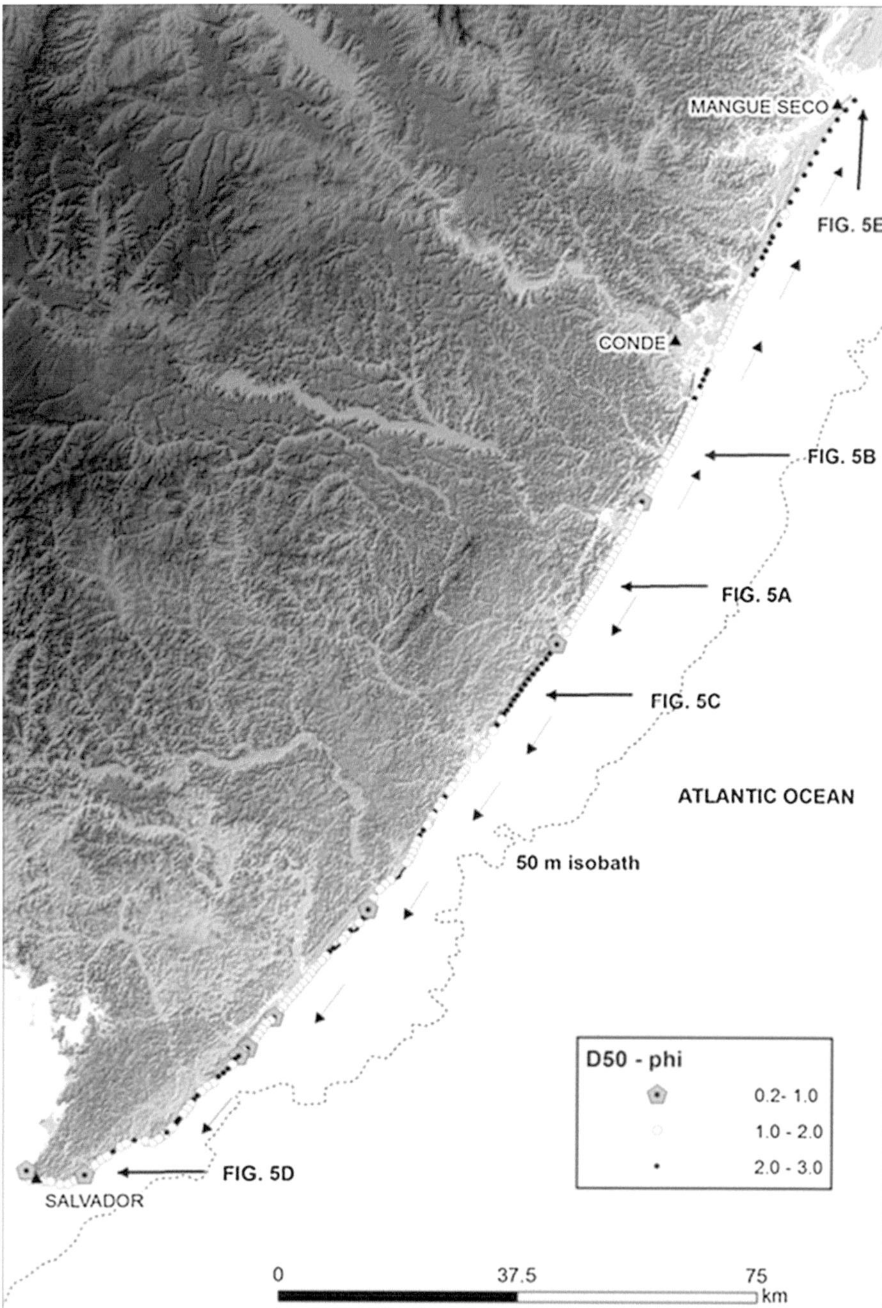

Fig. 12.4 Variation of D_{50} values of beach face sediments along the Northern Littoral Coast. Arrows indicate location of photos shown in Fig. 12.5. Small arrows show sediment dispersal along the shoreline

Fig. 12.5 Photographs of beaches present in the Northern Littoral Coast compartment. (**a**) a foredune ridge a few meters high is very common in this compartment; (**b**) example of transverse bar and rip intermediate beach; (**c**) example of longshore bar-trough intermediate beach; (**d**) the beaches at Salvador city are limited by small rocky promontories, which favors seasonal beach rotation, such as at Armação beach shown in photo; and (**e**) dissipative beach near the locality of Mangue Seco (see Fig. 12.4 for location of photographs)

(Genz et al. 2003), has enabled the accumulation of beach deposits associated with marine isotope stages 5e and 1, and possibly with older isotope stages including stages 9 or 11 (Dominguez et al. 2009; Dominguez and Bittencourt 2012). This is the only compartment on the Bahia coast where coastal dunes are present. These dunes are predominantly of the blowout type and are present on beach deposits of isotope stages 5e and 9/11. On deposits associated with the current highstand only foredune ridges are present. In most places, the coast is bordered by a foredune approximately 5–6 m high, most likely resulting from a coast which has remained essentially stationary or experienced minor erosional retreat (Fig. 12.5a). This compartment is also characterized by the presence of beachrock reefs and exposures of crystalline basement, particularly in the southernmost stretch, in the city of Salvador. Medium-grained sands predominate along the beach face, which, combined with moderate to high wave energy (particularly in the winter), results in a dominance of intermediate beaches at various morphodynamic stages (longshore bar-trough and transverse bar and rip states; Fig. 12.5b and c). Coarse sands are present near the river mouths. A gradual decrease in particle size is generally observed in the transport direction. The presence of numerous crystalline basement exposures in the southern portion of this section limits the longshore continuity of the beaches and creates conditions for their seasonal rotation in response to changes in wave directional frequency, as happens at Armação Beach (Farias et al. 1985; Fig. 12.5d). Conversely, the beach sediment at the north end of this compartment become increasingly finer, which is accompanied by a change in the morphodynamic behavior of the beach to more dissipative, with numerous blow-outs in the backshore (Fig. 12.5e).

The Mesozoic Rifts Coast Extends from the mouth of Todos os Santos Bay to Itacaré and is characterized by exposures of Mesozoic sedimentary rocks of the Recôncavo, Camamu and Almada basins (Figs. 12.1 and 12.6). The progressive lowering of relative sea level since the Middle Miocene (Miller et al. 2005) and the humid climate in this region favored differential erosion between the sedimentary rocks (less resistant) and crystalline basement rocks (more resistant), which caused a topographic lowering, particularly during the Quaternary low sea-level stands (Dominguez and Bittencourt 2009). The resulting low-lying areas were flooded during high sea-level stands, such as the current stand, which created bays, including Camamu and Todos os Santos and their islands, tidal channels and indented coastlines (Dominguez et al. 2009; Dominguez and Bittencourt 2012; Dominguez 2015). Erosional terraces carved into sedimentary rocks of the Mesozoic basins, some of which were colonized by corals, forming fringing reefs, are common in this area (Leão et al. 2003). The tops of these reefs act as a source of bioclastic sediment for the beaches, which is evident in the composition and particle sizes (Rebouças et al. 2011).

The Mesozoic Rifts compartment with its indented outline and the presence of nearby fringing coral reefs exhibits the highest variability in beach particle size and wave energy. The sediment sources, except for the Contas River, are also local because regional rivers flow into the local bays and large estuaries, where the thickest deposits have accumulated. Other sediment sources include current and past erosion of Jurassic aeolian sandstones, which are exposed, for example, at Morro de São Paulo. Other rocks exposed along the coastline are limestones of Cretaceous age (Algodões Formation). Therefore, most sources of sand available to this stretch of coastline are fine to medium grained, except along the sections where carbonate detritus from eroding reef supplies coarse sand and bioclastic gravel, as on the islands of Tinharé and Boipeba. The beaches in this section include high-energy intermediate types with transverse and longshore bars on the Maraú Peninsula (Fig. 12.7a), reflective to low-energy intermediate beaches enclosed between bordering fringing coral reefs (Fig. 12.7b), sheltered beaches behind coral reefs, several of which exhibit a variety of features reported by Jackson et al. (2002) and Masselink et al. (2006) (Fig. 12.7c), and beaches fronted by reef flats, which only receive waves at high tide

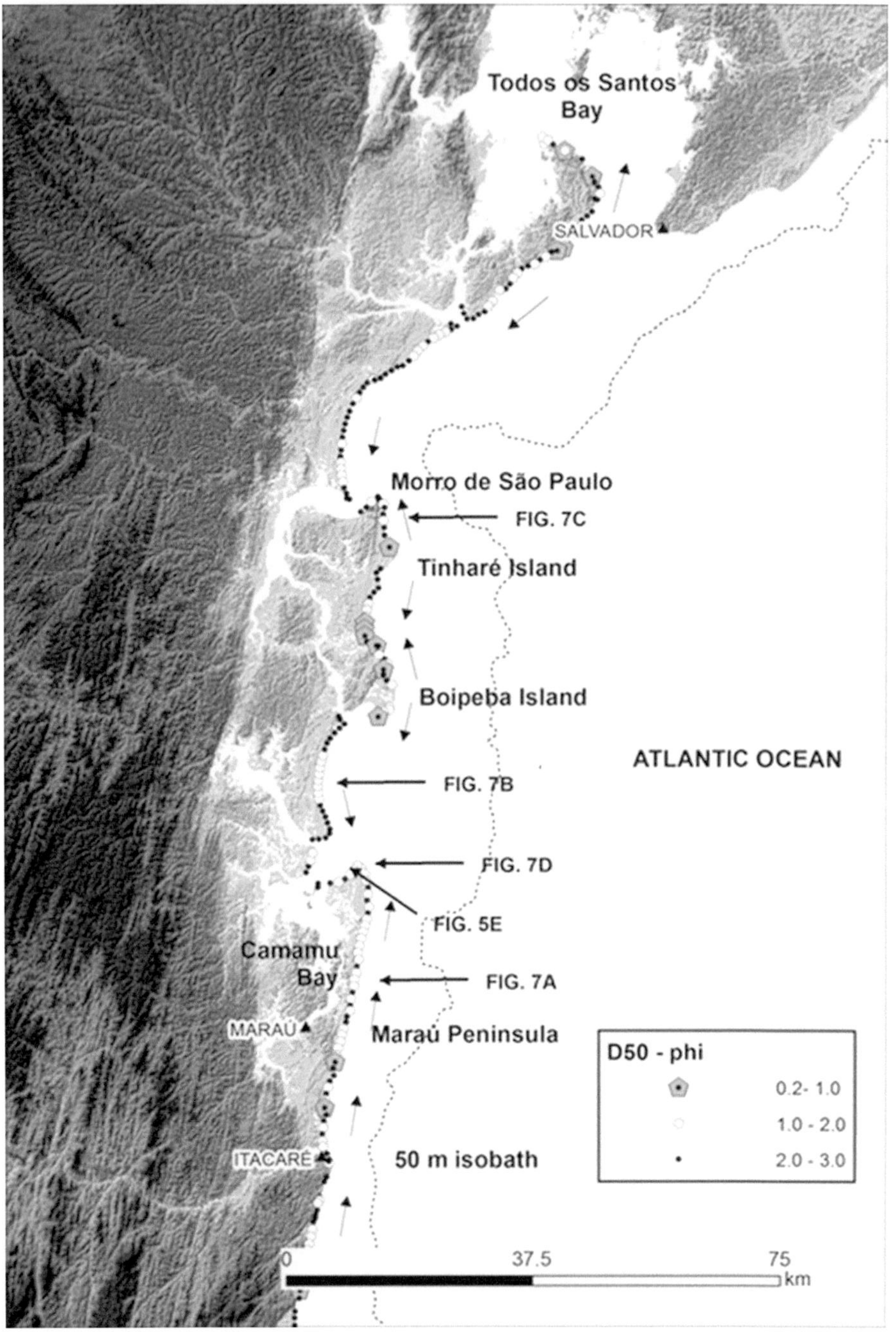

Fig. 12.6 Variation of D_{50} values of beach face sediments along the Mesozoic Rifts Coast compartment. Arrows indicate location of photos shown in Fig. 12.7. Small arrows show sediment dispersal along the shoreline

Fig. 12.7 Photographs of beaches present in the Mesozoic Rifts Coast compartment. (**a**) example of transverse bar and rip intermediate beach at the Maraú peninsula. Camamu bay seen in background; (**b**) reflective beaches are common in embayments between fringing reefs; (**c**) sheltered beach behind fringing reefs at Tinharé island; (**d**) beach and intertidal reef flats at the Maraú Peninsula; and (**e**) sheltered beach at the interior of Camamu bay (see Fig. 12.6 for location of photographs)

(Fig. 12.7d). Low-energy (reflective) beaches with numerous intertidal bars are present in the interior of Camamu and Todos os Santos Bays (Fig. 12.7e).

The Deltaic Coast of the Jequitinhonha River The Jequitinhonha is the main river flowing through Bahia and, has a typical wave-dominated delta (Dominguez 1983, 1987). The sediment delivered by this river is transported predominantly northward by the longshore drift. The coastal deposits generally consist of fine sands, except for coarse sand in the immediate vicinity of the mouth of the Jequitinhonha River (Fig. 12.8), and displays a clear trend of decreasing grain size with distance from the river mouth (Gasparini et al. 2004; Frings 2008) and entrainment in longshore transport (Fig. 12.1). Because of the lack of coral reefs along the coastline, the wave energy is higher which combines with the fine sand to promote the development of beaches with more-dissipative to intermediate character (Figs. 12.9a, b, d), except in the immediate vicinity of the river mouth, where the coarse sands favor beaches with more-reflective characteristics (Fig. 12.9c).

The Sediment Starved Southern Coast Extends from the village of Santo André to the border with the state of Espírito Santo (Ponta dos Lençóis) (Fig. 12.10). This compartment contains the largest area of coastal tablelands of the Barreiras Formation (Fig. 12.1). River basins that drain into this section are small, which explains why this section receives virtually no input of fluvial siliciclastic sediments. The lack of sediments results in the presence of numerous active cliffs carved into the coastal plains. The erosion of these cliffs by waves, rainfall and landslides is the main source of coastal sediments, which produced the Caravelas mega-salient and the Corumbau tombolo behind the Abrolhos and Itacolomis reef complexes, respectively (Fig. 12.3). The presence of these coral reefs, induce wave refraction, and convergence of sediment transport, resulting in the accumulation of the sandy plains (Andrade et al. 2003; Bittencourt et al. 2008; Dominguez et al. 2009). The reduced availability of sediment in this compartment promoted the development of coral reefs along sections where suitable substrates were available, including erosional terraces in the lateritic horizons of the Barreiras Formation, on volcanic rocks and on karstic surfaces carved in continental shelf carbonates in the region of Abrolhos (Bittencourt et al. 2008). The sheltering from waves by coral reefs; the greater energy dissipation due to the wider continental shelf; the increase in tidal action; and the supply of coarse sands derived from the Barreiras Formation resulted in a dominance of reflective beaches (Fig. 12.11a, b, c and d), sometimes with well-developed low-tide terraces (Fig. 12.11e).

12.4.1 Shoreline Behavior

The sandy beaches of Bahia were assigned to four categories, i.e. eroding, equilibrium, prograding and highly variable (Fig. 12.12), in terms of their shoreline behavior between 1960 and 2000 (Dominguez et al. 2006).

(i) Eroding shorelines were characterized by evidence of continued erosional retreat, e.g. vegetation with exposed roots, cliffs and threatened property, based on field observations; vertical aerial photographs, satellite images; and interviews with residents. The most severe evidence of erosion in Bahia includes the following: sediment retention by engineering works associated with port facilities (e.g. the Port of Ilhéus); sediment retention on unstable capes (e.g., the Caravelas Plain); negative long-term sediment budget (e.g. the Sediment

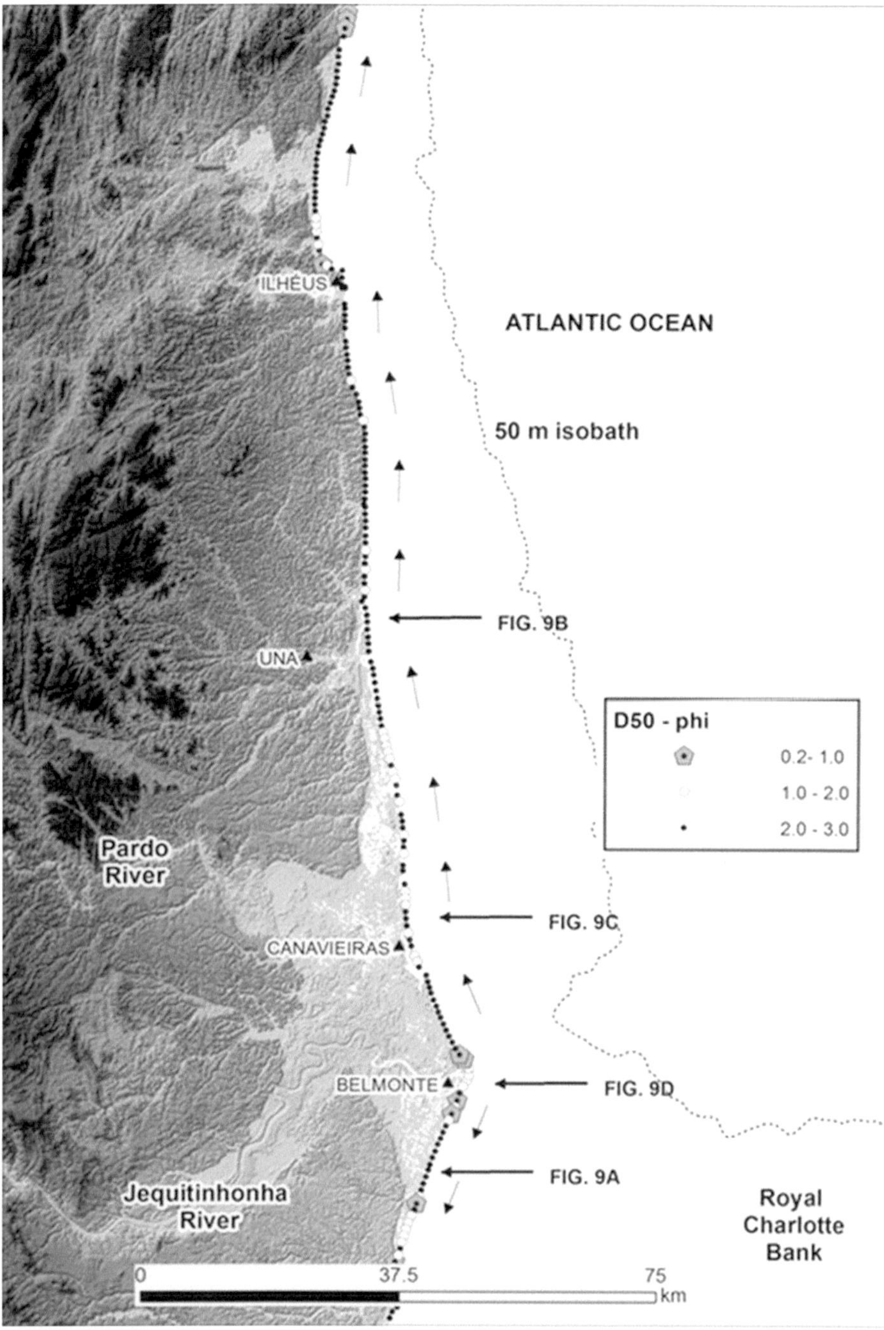

Fig. 12.8 Variation of D_{50} values of beach face sediments along the Deltaic Coast of the Jequitinhonha river compartment. Arrows indicate location of photos shown in Fig. 12.9. Small arrows show sediment dispersal along the shoreline

Fig. 12.9 Photographs of beaches along the Deltaic coast of the Jequitinhonha river compartment. (**a** and **b**) dissipative and high energy intermediate beaches area dominant in this compartment; (**c**) reflective beach at the Jequitinhonha river mouth and (**d**) example of rhythmic bar and beach intermediate beach (see Fig. 12.8 for location of photographs)

Starved Southern Coast); reductions in solid and liquid stream discharges resulting from natural processes or human intervention; and longshore migration of small river mouths.

(ii) Equilibrium coastlines display no significant alteration of the coastline position during the four-decade study period, although seasonal variations may be observed. This category includes coastal sections characterized by long segments of straight coastlines (e.g. Northern Littoral Coast, Maraú Peninsula) and the presence of wide embayed beaches (e.g. coastal plains of Guaibim, Pratigi, and northern portion of the Caravelas Plain).

(iii) Prograding coastlines that displayed significant progradation during the last four decades. The sections with the greatest coastline progradation in the study area are located immediately north and south of the Jequitinhonha River mouth, where the shoreline prograded as much as 500 m.

(iv) Highly variable coastlines occur where the coastline position displays high spatial and temporal variability associated with the transport and deposition of sediments. In these sections, progradation and erosion have alternated. The processes associated with this type of behavior include the sedimentary dynamics of small river mouths and the entrances of bays and estuaries, owing to changes in mouth bars (ebb deltas) and adjacent coast.

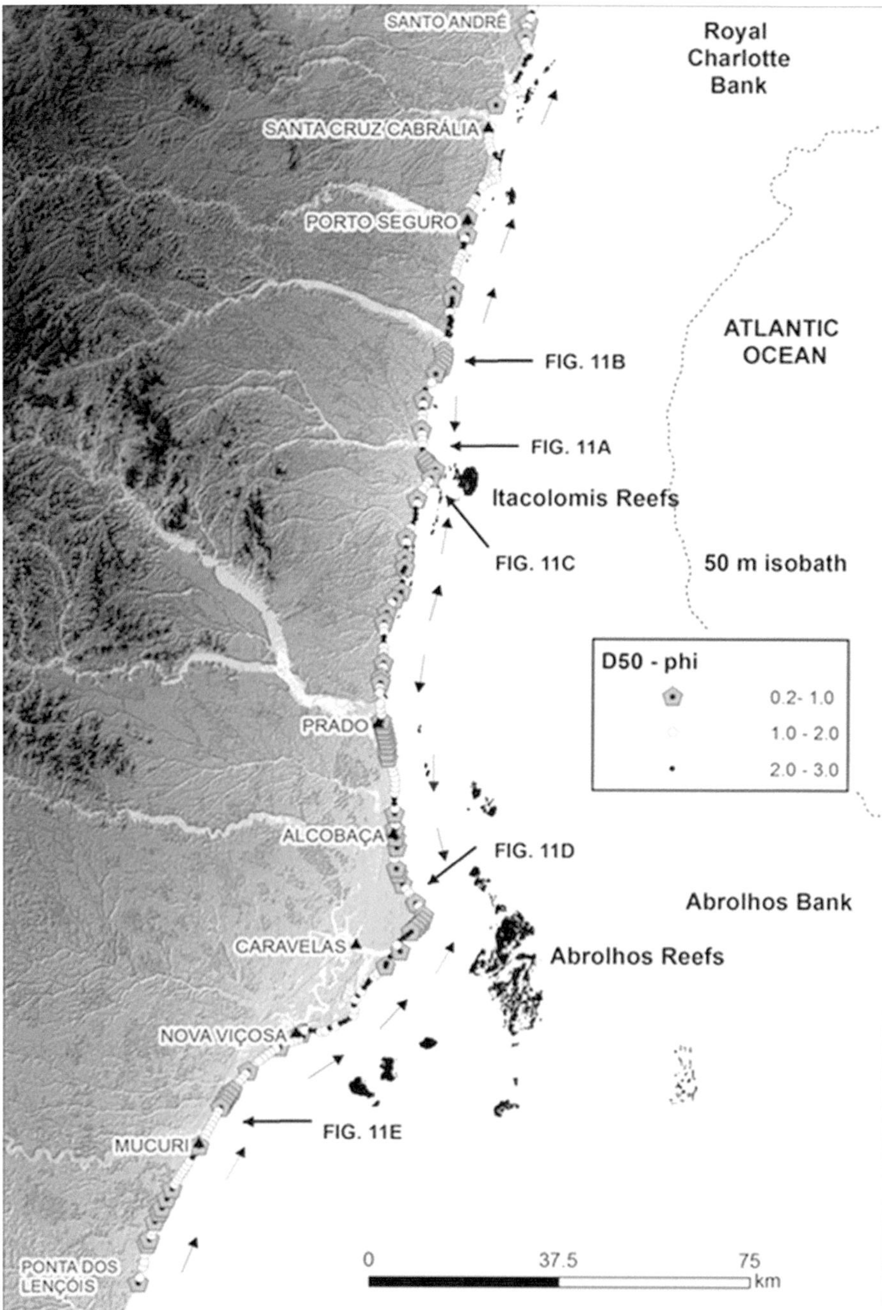

Fig. 12.10 Variation of D_{50} values of beach face sediments along the Sediment Starved Southern Coast Compartment. Arrows indicate location of photos shown in Fig. 12.11. Small arrows show sediment dispersal along the shoreline

Fig. 12.11 Photographs of beaches present in the Sediment Starved Southern coast. (**a**) reflective beach at Corumbau sheltered by the Itacolomis reefs; (**b**) active cliffs on the Barreiras Formation and associated reflective beaches. (**c**) reflective beach at Corumbau; (**d**) reflective beach at Caravelas strandplain; and (**e**) reflective beach with well developed low-tide terrace (see Fig. 12.10 for location of photographs)

Quantitatively, the shoreline behavior in the study area indicates that eroding sections (26 %) may be explained by processes typically associated with dispersion and accumulation of sediments along the coastline, river mouth dynamics, human intervention and long-term trends of negative sediment budgets. Sixty percent of the coast is categorized as balanced, at least during the four-decade study period, with the most significant cases of progradation (6 %) associated with the small number of local river mouths.

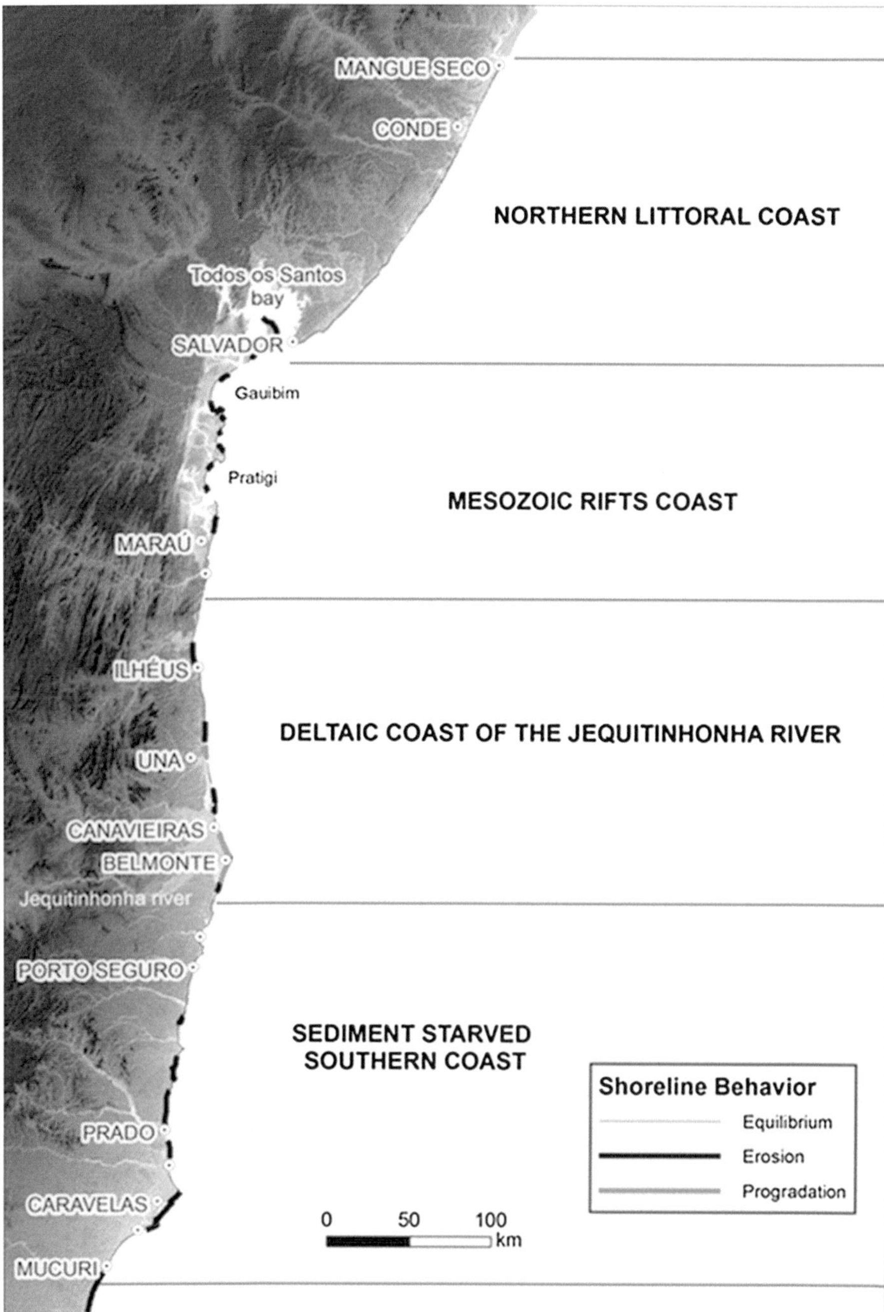

Fig. 12.12 Shoreline behavior of Bahia state coastline between 1960 and 2000

12.5 Beach Safety: Case Study of Salvador Beaches

Studies of beach safety in Bahia are scarce. Carvalho (2002), performed a detailed analysis of Salvador's urban beaches and found that spring is the season with the highest percentage of accidents (44.8 %) involving beachgoers. This timing of accidents apparently results from the following combination of factors: (i) a high numbers of beachgoers due to improved bathing conditions following the winter and the occurrence of important national holidays (Independence Day, Children's Day and All Souls' Day); (ii) an average incident wave height in the spring of approximately 1.13 m versus 1.08 m in the summer; (iii) steady trade winds producing stronger coastal currents than in the summer; (iv) a beach profile characterized by transverse and longshore bars that are inherited from the most-energetic winter events; and (v) astronomical tide height that is approximately 8 % greater than that during the summer. The last three factors enhance the rip currents, which leads to a higher number of accidents involving beachgoers when combined with the increased public use of the local beaches. The risk increases with more bathers exposed to high-velocity rip currents along the beach.

These findings from the Salvador beaches may be extrapolated to other beaches in Bahia and used to develop a risk framework, as shown in Fig. 12.13. The beaches in the Mesozoic Rift compartment, excluding the Maraú Peninsula, pose the lowest risk to beachgoers, given the dominance of reflective and sheltered beaches. The beaches of the Sediment Starved Southern Coast compartment, although predominantly reflective, display one characteristic that adversely affects beach safety: active cliffs subject to collapse. The beaches in the Deltaic Coast of the Jequitinhonha River and the Northern Littoral Coast compartments North Shore sections have the highest risk, given their dominance of high-energy intermediate (longshore and transverse bar) and dissipative morphodynamic states. The southern portion of the Northern Littoral compartment (metropolitan area of Salvador city) exhibits the highest values of ambient population (average number of people over a 24-h period; LandScan dataset: http://web.ornl.gov/sci/landscan/), and, therefore, beach safety is a key concern there. Although beaches in the Deltaic Coast of the Jequitinhonha River compartment also pose a high risk to beachgoers, this risk is of lesser importance, due to the low population, except at Ilhéus, which is an important tourist destination and where accidents involving beachgoers are common.

12.6 Geologic Inheritance and Texture/Composition of Beach Sands

Several studies have highlighted the role of the geologic inheritance in controlling beach morphodynamics (McNinch 2004; Jackson et al. 2002; Short 2010). These authors primarily emphasize issues including the degree of embayment of beaches, the presence of rips and megarips, beach rotation and glacial heritage. The effect of

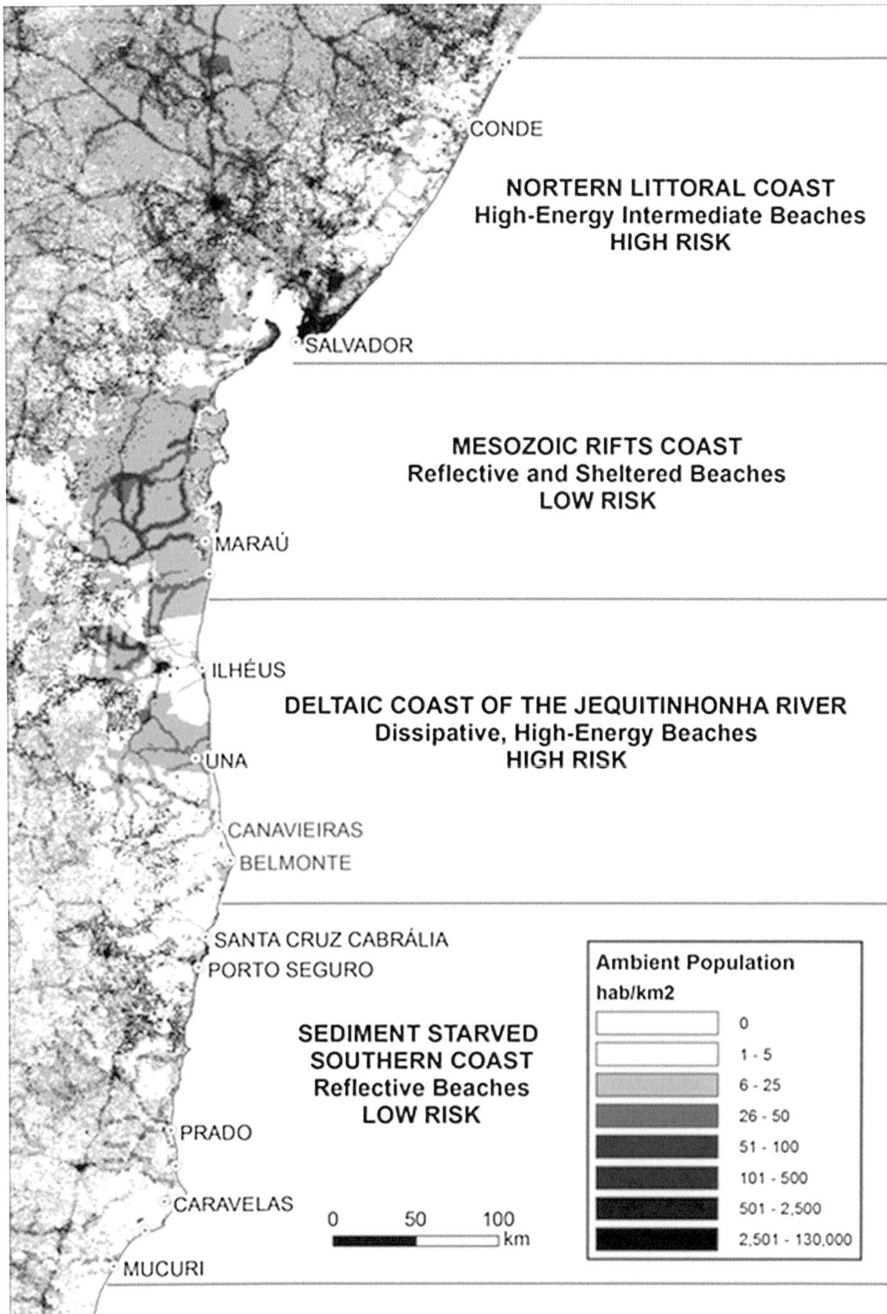

Fig. 12.13 Distribution of ambient population in the Bahia coastal zone and level of beach safety. The portions of the coastal zone exhibiting greatest population concentration are the state capital (Salvador) and Todos os Santos bay, followed by the Porto Seguro region. Otherwise the coastal zone is scarcely populated. The Ambient Population is from the Landscan 2010 dataset (The LandScan 2010™ High Resolution global Population Data Set copyrighted by UT-Battelle, LLC, operator of Oak Ridge National Laboratory under Contract No. DE-AC05-00OR22725 with the United States Department of Energy)

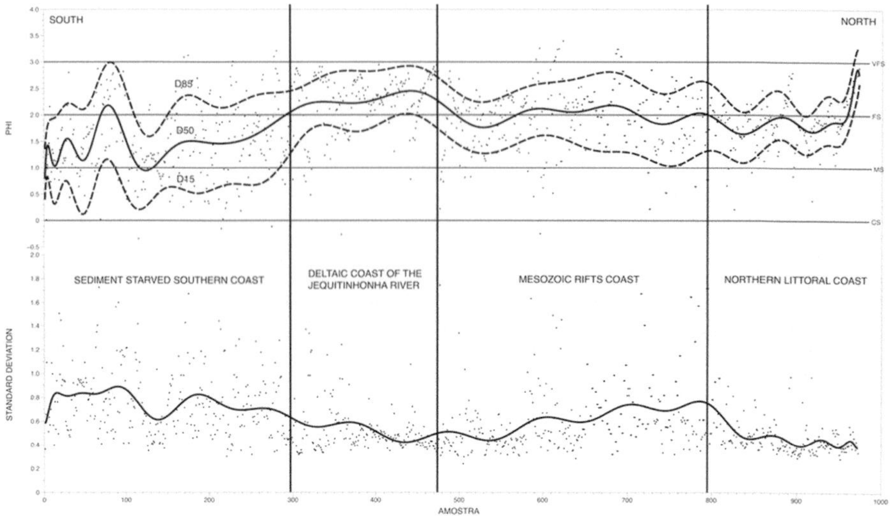

Fig. 12.14 (**a**) variation in the values of D_{50}, D_{15} and D_{85} of beach face sands along the shoreline of Bahia; and (**b**) variation of degree of sorting (Folk's formulation) of beach sands along the shoreline of Bahia

geologic heritage is primarily analyzed in the present study in regard to factors controlling the beach particle size, which in turm exerts a major control on the morphodynamic state of Bahia's beaches.

Based on the foregoing findings, a conceptual model was developed for relating the geological heritage to beach morphodynamics. The main control exerted by the geology is on the sediment particle size available to this coastline. The large river drainage basins tend to supply finer sediments (Gasparini et al. 2004; Frings 2008), as occurs in the Deltaic Coast of the Jequitinhonha river compartment (Figs. 12.8 and 12.14). The combination of high wave energy and fine sediments produce a dominance of high-energy dissipative and intermediate beaches (transverse/longshore bar). Conversely, small and medium rivers tend to transport coarser sediments to the coastline, which occurs in the Northern Littoral Coast Compartment (Figs. 12.4 and 12.14). Therefore, high-energy intermediate beaches (transverse and longshore bars) predominate in this stretch. Furthermore, foredunes a few meters tall are associated with these beaches because the supply of sediment is just sufficient to maintain the coastline in a position of balance or low retreat. A trend of the sediments fining in the direction of drift was also observed in both sections (Fig. 12.14). The unimpeded transport of sediments without obstructions driven by constant erosion and transport by waves sorts beach sediments as indicated by the longshore fining in part of the southern and the two central compartments (Fig. 12.14).

Where fluvial contributions are absent, the main source of beach sediment is erosion of older sediments or rocks exposed along the coastline. The main source of sediments in the Sediment Starved Southern Coast Compartment is the erosion of active cliffs in the Barreiras Formation, which consists primarily of coarse sediment (Figs. 12.10 and 12.14a). These coarser, partly gravelly sands in combination with the reduction in energy provided by the offshore coral reefs promote the development of reflective beaches with steep gradients. This reduction in wave height and a slight increase in the tide height in this section increase the RTR, resulting in tide-modified beaches typified by a reflective high tide beach and the common presence of low-tide terraces. Reduced reworking by waves and restricted transport of sediments results in the most poorly sorted beach sand among the Bahia beaches (Fig. 12.14b).

The Mesozoic Rifts Coast Compartment is also characterized by a reduced supply of sediments because all the rivers there except the Contas are small and flow into bays and estuaries, where the coarser sediments settle out. Consequently, the sediments supplied to the coastline are typically fine sands and silts. The eroded sediments from the few rock outcrops in the region, consisting mostly of eolian fine-grained sandstones of Mesozoic age (Tinharé Island), are added to these fluvial sediments (Figs. 12.6 and 12.14a). Other minor exposed rocks consist of limestone and crystalline basement. The fine sands are ultimately trapped in the ebb-tide deltas at the mouths of the bays and estuaries. The erosion of Mesozoic sedimentary rocks resulting from wave action creates erosional terraces that are colonized by coral reefs. These reefs lead to decreased wave energy and localized production of bioclastic sediments of gravel and coarse sand sizes supplying the coastline (Figs. 12.6 and 12.14a). Therefore, a large number of reflective enclosed beaches, typically fetch limited and sheltered, and beaches fronted by reef/rock flats, in which sediment is mobilized by waves only during high tide or high-energy events, are formed. Thus, the beaches form only thin sandy prisms resting on older rocks or coral reefs. Reduced wave energy and impeded transport of sediments also lead to more poorly sorted beach sediments in this compartment (Fig. 12.14b).

12.7 Summary

The state of Bahia with approximately 1000 km of shoreline, a great heterogeneity of exposed rock types and subjected to varying degrees of sediment supply allowed us to investigate the controls exerted by these factors in determining the beach types.

The modal morphodynamic states of the beaches of Bahia is dominantly controlled by sediment grain size, which in its turn results from the long term history of the coastal zone. Locally sourced sediments, eroded from the coastal tablelands (Barreiras Formation), are predominantly coarse-grained favoring reflective beaches (Sediment Starved Southern Coast). Distally sourced sediments, as in the case of large rivers are predominantly fine to very fine sands, resulting in a dominance of dissipative and high-energy intermediate beaches (Deltaic Coast of the Jequitinhonha River). Shoreline stretches nourished by small rivers are characterized by medium size sands and a dominance of intermediate high-energy beaches (Northern Littoral Coast Compartment). Finally, the stretch of the coast fronted by Mesozoic Rifts, characterized by a great heterogeneity of sedimentary rock types and small sediment supply, have a very irregular shoreline, bordered by fringing reefs. In this section low-energy sheltered beaches dominate

Acknowledgments This study is a contribution of inctAmbTropic—Brazilian National Institute of Science and Technology for Tropical Marine Environments, CNPq/FAPESB Grants: 565054/2010-4 and 8936/2011. ACSP Bittencourt and JML Dominguez thank CNPq for a Research Fellowship.

References

Alkmim FF, Marshak S, Fonseca MA (2001) Assembling west Gondwana in the neoproterozoic: clues from the São Francisco craton region, Brazil. Geol 29:319–322

Almeida FFM (1977) O cráton do São Francisco. Rev Bras Geosci 7(4):349–364

Andrade ACS, Dominguez JML, Martin L, Bittencourt ACSP (2003) Quaternary evolution of the caravelas strandplain – southern Bahia state – Brazil. Anais Acad Bras Ciênc 75:1–26

Angulo RJ, Lessa GC (1997) The Brazilian sea-level curves: a critical review with emphasis on the curves from the paranaguá and cananéia regions. Mar Geol 140:141–166

Barbosa JSF, Sabaté P (2004) Archean and paleoproterozoic crust of the São Francisco craton, Bahia, Brazil: geodynamic features. Precambrian Res 133:1–27

Bernal JMS (2009) Contribuição do Aporte Fluvial de Sedimentos para a Construção da Planície Deltáica do Rio Jequitinhonha – BA. Thesis, Universidade Federal da Bahia, Brazil

Bittencourt ACSP, Dominguez JML, Martin L, Silva IR (2000) Patterns of sediment dispersion coastwise the State of Bahia – Brazil. Anais Acad Bras Ciênc 72:272–287

Bittencourt ACSP, Leão ZMAN, Kikuchi RKP, Dominguez JML (2008) Deficit of sand in a sediment transport model favors coral reef development in Brazil. Anais Acad Bras Ciênc 80:205–214

Bruun P (1962) Sea level rise as a cause of shore erosion. J Waterways and Harbors Div 88:117–130

Carvalho MP (2002) Fatores meteorológicos, oceanográficos, morfodinâmicos, geológicos e urbanos relacionados à incidência de afogamentos nas praias da Costa Atlântica de Salvador. Thesis, Universidade Federal da Bahia, Brazil

Cirano M, Lessa GC (2007) Oceanographic characteristics of baía de todos os Santos, brazil. Rev Bras Geof 25(4):363–387

Cruz S, Alkmim F, Pedreira AJ, Teixeira LR, Soares ACP, Gomes LCC, Souza JS, Leal ABM (2012) Capitulo X – Orógeno Araçuaí. In: Barbosa JSF (org) Geologia da Bahia: pesquisa e atualização. Companhia Baiana de Pesquisa Mineral, Salvador, Bahia, Brazil, pp 131–177

Dominguez JML (1983) Evolução quaternária da planície costeira associada à foz do Rio Jequitinhonha (BA): influência das variações do nível relativo do mar e da deriva litorânea de sedimentos. Thesis, Universidade Federal da Bahia, Brazil

Dominguez JML (1987) Quaternary sea level changes and the depositional architecture of beach – ridge strandplains along the coast of Brazil. University of Miami, Coral Gables

Dominguez JML (2015) The todos os Santos Bay—an ephemeral high stand feature incised into an aborted cretaceous rift. In: Vieira BC et al (eds) Landscapes and landforms of Brazil, World geomorphological landscapes. Springer, Cham, pp 55–63

Dominguez JML, Bittencourt ACSP (2009) Geologia – cap II. In: Hatge V, Andrade JB (eds) Baía de todos os Santos – aspectos oceanográficos. EDUFBA, Salvador, pp 25–66

Dominguez JML, Bittencourt ACSP (2012) Zona Costeira – Capítulo XVII. In: Barbosa JSF (ed) Geologia da Bahia: pesquisa e atualização. Companhia Baiana de Pesquisa Mineral, Salvador, pp 395–425

Dominguez JML, Bittencourt ACSP, Santos AN, Andrade ACS, Lavenere-Wanderley AMO, Silva IR, Queiroz IG, Freitas LB, Nascimento L, Silva RP (2006) Bahia. In: Muehe D (ed) Erosão e Progradação no litoral brasileiro. Ministério do Meio Ambiente, Brazi, pp 219–225

Dominguez JML, Andrade ACS, Almeida AB, Bittencourt ACSP (2009) The Holocene barrier strandplains of the state of Bahia. In: Dillenburg SR, Hesp PA (eds) Geology and geomorphology of Holocene coastal barriers of Brazil. Elsevier, Amsterdam, pp 253–288

Dominguez JML, Nunes AS, Rebouças RC, Silva RP, Freire AFM, Poggio CA (2012) Plataforma Continental – Capitulo XVIII. In: Barbosa JSF (ed) Geologia da Bahia: pesquisa e atualização. Companhia Baiana de Pesquisa Mineral, Salvador, pp 427–496

Dominguez JML, Silva RP, Nunes AS, Freire AFM (2013) The narrow, shallow, low-accommodation shelf of central Brazil: sedimentology, evolution, and human uses. Geomorphology 203:46–59

Farias FF, Bittencourt ACSP, Zanini A, Dominguez JML (1985) Variações temporais e espaciais na dinâmica de sedimentação da Praia de armação – Salvador, Bahia. Rev Bras Geociên 15:48–54

Frings RM (2008) Downstream fining in large sand-bed rivers. Earth Sci Rev 87:9–60

Gasparini NM, Tucker GE, Bras RL (2004) Network-scale dynamics of grain-size sorting: implications for downstream fining, stream-orofile concavity, and drainage basin morphology. Earth Surf Process Landf 29:401–421

Genz F, Lessa GC, Cirano M, Lima GMP (2003). As tendências de longo prazo das vazões fluviais no litoral do Estado da Bahia. IX Congresso sobre Planejamento e Gestão da Zona Costeira dos Países de Expressão Portuguesa. In: Proceedings. Recife, Pernambuco, Brazil. CD-ROM

Jackson NL, Nordstrom KF, Eliot I, Masselink G (2002) 'Low energy' sandy beaches in marine and estuarine environments: a review. Geomorphology 48:147–162

Leão ZMAN, Kikuchi RKP, Testa V (2003) Corals and coral reefs of Brazil. In: Cortés J (ed) Latin American coral reefs. Elsevier, Amsterdam, pp 9–52

Martin L, Dominguez JML, Bittencourt ACSP (2003) Fluctuating Holocene sea levels in eastern and southeastern Brazil: evidence from multiple fossil and geometric indicators. J Coast Res 19:101–124

Masselink G, Hegge B (1995) Morphodynamics of meso- and macrotidal beaches: examples from central Queensland, Australia. Mar Geol 129:1–23

Masselink G, Kroon A, Davidson-Arnott RGD (2006) Morphodynamics of intertidal bars in wave-dominated coastal settings – a review. Geomorphology 73:33–49

McNinch JE (2004) Geologic control in the nearshore: shore-oblique sandbars and shoreline erosional hotspots, Mid-Atlantic Bight, USA. Mar Geol 211:121–141

Miller KG, Kominz MA, Browning JV, Wright JD, Mountain GS, Katz ME, Sugarman PJ, Cramer BS, Christie-Blick N, Pekar SF (2005) The phanerozoic record of global sea-level change. Science 310:1293–1298

Mohriak WU (2006) Interpretação geológica e geofísica da bacia do espírito Santo e da região de abrolhos: petrografia, datação radiométrica e visualização sísmica das rochas vulcânicas. Bol Geoc Petrobras 14:133–142

Mohriak W, Nemcok M, Enciso G (2008) South Atlantic divergent margin evolution: rift-border uplift and salt tectonics in the basins of SE Brazil. In: Pankhurst RJ, Trouw RAJ, Brito Neves BB, De Wit MJ (eds) West gondwana: pre-Cenozoic correlations across the south Atlantic region, vol 294, Geological society, London, spec publ., pp 365–398

Oliveira EP (2012) Capítulo XI - Faixa de Dobramento Sergipana. In: Barbosa JSF (ed) Geologia da Bahia: pesquisa e atualização. Companhia Baiana de Pesquisa Minera, Salvador/Bahia, pp 179–198

Pianca C, Mazzini PL, Siegle E (2010) Brazilian offshore wave climate based on NWW3 reanalysis. Braz J Oceanogr 58:3–70

Queiroz IG (2003) Produção de sedimento e alterações no regime hidrossedimentológico da bacia hidrográfica do rio Mucuri – repercussão na zona costeira. Thesis, Universidade Federal da Bahia

Rebouças RC, Dominguez JML, Bittencourt ACPS (2011) Provenance, transport and composition of dendê coast beach sands. Braz J Oceanorg 59:339–347

Rossetti DF, Bezerra FHR, Dominguez JML (2013) Late Oligocene–Miocene transgressions along the equatorial and eastern margins of brazil. Earth Sci Rev 123:87–112

Sales FJP, Bentes FCM, Santos JA (2000) Catálogo de estações maregráficas. Fundação de estudos do Mar. Rio de Janeiro, Brazil

Short AD (2010) Role of geological inheritance in Australian beach morphodynamics. Coas Eng 57:92–97

Silva OB, Caixeta JM, Milhomem PS, Kosin MD (2007) Bacia do recôncavo. Bol Geoc Petrobras 15:423–431

Sobreira JFF, França RL (2006) Um modelo tectono-magmático alternativo para a região do complexo vulcânico de Abrolhos. Bol Geoc Petrobras, Rio de Janeiro 14:143–147

Chapter 13
The Beaches of Espírito Santo

Jacqueline Albino, Nery Contti Neto, and Tiago Castro Alves Oliveira

Abstract Espírito Santo coastline covers more than 400 km with a high diversity beaches influenced by inland geology and inner continental shelf. Three main sectors were characterized: beaches associated with well-developed coastal plains; beaches preceded by shore platforms and limited by sedimentary cliffs; and embayed beaches limited by cristallyne outcrops. These distinct sectors trigger distinct morphodynamical processes as well as vulnerability to erosion and/or flooding.

13.1 Introduction

Espírito Santo is located in southeastern Brazil, between Rio de Janeiro and Bahia states. Its coast is 411 km in length and faces southeast into the South Atlantic ocean (Fig. 13.1). Continental shelf morphology and sedimentology, shoreline orientation, coastal geology and regional oceanographic characteristics and processes all influence the wave energy regime and coastal geomorphology, including the sediment composition and state of the beaches.

The tidal amplitude increases slightly from south to north (IH-Cantabria 2013). The south-central portions of the state have micro-tides (average tidal range ~1.0 m), while meso-tidal conditions occur inside the Abrolhos continental shelf (Fig. 13.3), in the north (average tidal range ~ 2.0 m).

The offshore wave climate is very uniform along the coast, with only variations in the continental shelf inducing some differences in breaker wave height. Based on wave hindcast data at 30 m water depth waves average 1.0–1.5 m along central and

J. Albino (✉) • N. Contti Neto
Department of Oceanography and Ecology, Universidade Federal do Espírito Santo, Fernando Ferrari, 514 – Campus Universitário – Goiabeiras, 29075-910 Vitória, ES, Brazil
e-mail: jacqueline.albino@ufes.br; nerycn@gmail.com

T.C.A. Oliveira
Environmental Engineering Department Universidade Federal do Espírito Santo, Av. Fernando Ferrari, 514 – Campus Universitário – Goiabeiras, 29075-910 Vitória, ES, Brazil
e-mail: tiago.c.oliveira@ufes.br

A.D. Short, A.H. da F. Klein (eds.), *Brazilian Beach Systems*, Coastal Research Library 17, DOI 10.1007/978-3-319-30394-9_13

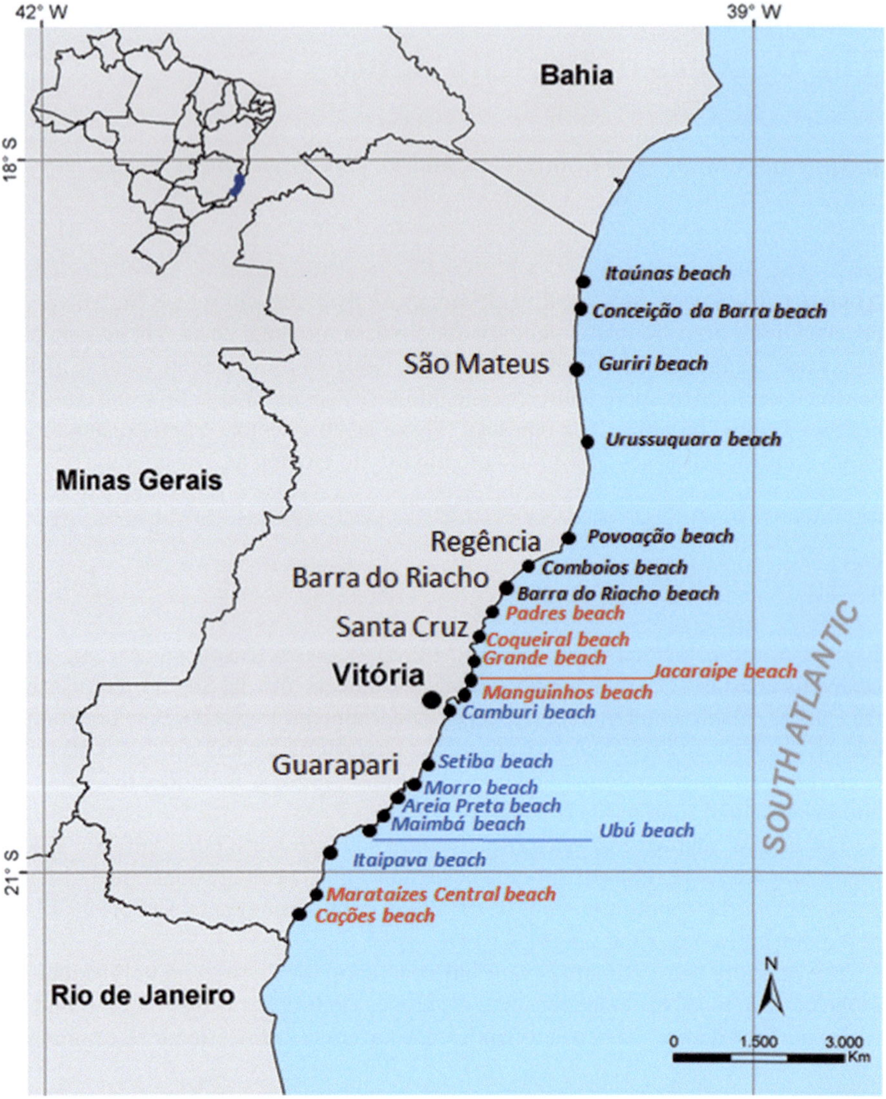

Fig. 13.1 Map of Espírito Santo showing the locations mentioned in text. In *black*: exposed siliciclastic beaches associated with sandy terraces. In *red*: siliciclastic and bioclastic sand beaches adjacent to sedimentary cliffs and marine abrasion terraces. In *blue*: embayed and pocket sandy beaches associated with rocky indented coastlines

south coasts while along the northern coast they average from 0.5 to 1.0 m, with waves coming mostly from east (Reguero et al. 2012).

Oceanic sandy beaches occupy 71 % of the coast. Three beach types are presented in Fig. 13.1: embayed and pocket sandy beaches associated with the rocky indented coastlines (25 % of the coast); siliciclastic and bioclastic sand beaches adjacent to sedimentary cliffs and marine abrasion terraces (22 %); and exposed siliciclastic beaches associated with sandy terraces (24 %). The remainder of the coast consist of tidal flats, estuaries and artificial coasts resulting from nourishment and hard structures, such as seawalls, breakwater and ports, mainly in the metropolitan region of Victoria.

The embayed beaches are subjected to rotational and oscillatory processes, and are represented by beaches in dynamic equilibrium since they are constrained within small coastal cells without significant fluvial input. The abrasion terrace beaches have their morphodynamics and sedimentology directly subordinated to the spatial distribution of the terraces, which also change the pattern of nearshore waves. Greater morphological mobility and more efficient longshore sediment transport are observed on the more exposed beaches. These beaches can be found in deltaic plains such as the Doce River, where they receive riverine inputs (Dominguez et al. 1981).

Flooding and erosion problems occur where the beaches are narrow and lack volume. In addition, in areas of intense urbanization beach sands have been lost particularly during periods of high wave energy.

This chapter reviews the ocean beaches of Espírito Santo. The geographical distribution, geomorphology and mineralogical composition of coastal geological units, as well as the morphology and sedimentary characteristics of the inner continental shelf, are presented in Sect. 13.2. The contribution of marine and continental sediments and composition of sand beaches are presented in Sect. 13.3. Section 13.4 describes the distribution of waves and tides along the coast. The interaction between the geological control and oceanographic processes in coastal systems, and the determination of geomorphological provinces are discussed in Sect. 13.5; while beach morphodynamic processes for each system are presented in Sect. 13.6. Finally, the use and occupation of the beaches and their vulnerability to erosion and flooding are discussed in the last Sect. 13.7.

13.2 Geology and Geomorphology

13.2.1 The Coast

Three geological-geomorphological units are recognized along Espírito Santo coast: the hills and outcrops composed of Precambrian crystalline rocks with an uneven relief; the Neogene plateau of Barreiras Formation whose surface slopes slightly seaward; and the Quaternary fluvial and coastal plain (Martin et al. 1996; Albino et al. 2006) (Fig. 13.2). The heterogeneous spatial distribution of these units generates a highly diverse coastline.

Sedimentary Coastal Plains The Quaternary sedimentary plains are limited along Espírito Santo coastline as their geological evolution is associated with sea level fluctuations and availability of fluvial sediments. The most extensive are located in the vicinity of Doce River mouth and in the drowned valleys of the São Mateus, Piraquê-Açu, Reis Magos, Jucu, Itapemirim and Itabapoana rivers (Fig. 13.3). The relative fall in sea level which followed the last transgressive maximum of 5100 years

Fig. 13.2 Coastal geomorphology features along Espírito Santo: (**a**) Coastal plain in the vicinity of the Doce River, Urussuquara beach; (**b**) crystalline promontories, Guarapari; and (**c**) sedimentary cliffs of Barreiras Formation, Cações Beach, Marataizes

BP, resulted in the formation of sandy marine terraces composed of siliciclastic sands (Martin et al. 1996), progradation of the shoreline and filling of estuaries by sandy-muddy sediments, and the development of extensive mangrove areas.

The Tablelands The sedimentary deposits of Barreiras Formation extend along the entire coastline narrowing towards the south. In the north, near the border with Bahia, they are approximately 80 km wide. South of the deltaic plain of Doce River down to Vitória, the tablelands narrow down to 13 km finally ending close to the beach, in vivid red cliffs. Abrasion terraces are associated with active cliffs due to the processes of waves undermining and exposure of lateritic structures. Active cliffs alternating with rocky outcrops can also be found in the extreme south of the state.

Morais et al. (2006) described the Barreiras Formation facies in the onshore portion of Espírito Santo basin, between Vitória and the northern border of Rio de Janeiro state. It is composed predominantly of layered beds of sandy sediments, with extensive to sub-tabular lenses of gravel interspersed with muddy sediments. The sediments are usually white-grayish, strongly mottled reddish-purple due to the presence of iron oxide/hydroxide. The facies indicates sediment deposition associated with suspension and gravity flows, associated with an interlaced distal fluvial environment, together with some by tractive processes. In contrast, the estuarine deposits found in the north and northeast of Brazil are suggestive of transition sedimentary environments during higher sea level (Rossetti et al. 2013).

Monazite, zircon, tourmaline, rutile, sillimanite and opaque, predominantly ilmenite, occur in Barreiras Formation. This assemblage reflects high-grade metamorphic source (Force 1980), while the abundance of ultra-stable minerals (zircon, tourmaline, rutile) indicates sediment maturity. Based on this, Nascimento et al. (2011) state that those minerals could have been transported from the inland of the state during the deposition of the Formation.

The Rocky Outcrops The crystalline outcrops along the south-central coast are represented by hills, mountains, headlands and islands. The poly-metamorphic character displayed by these rocks suggests the action of metamorphic events related to several orogenic cycles, among them, the Transamazonic and Brasiliano, as well as the anorogenic phenomena associated with major tectonic fracture (Cordani 1973).

The rocks consist of granites together with migmatite, charnockite and kinzigitic gneisses, as well as charnockite cores (Coutinho 1974; Machado Filho et al. 1983; Brasil 1987). The migmatite gneisses of the western portion are well oriented with uniform gneissification and sometimes sharp banding, characteristic of ribboned and/or migmatitic gneisses. The mineralogical composition varies between granite and granodiorite, with associated quartz-microcline-plagioclase (oligoclase), biotite and minor amounts of hornblende (Cordani 1973). The kinzigitic gneisses of the eastern band exhibit mineralogical associations of quartz-plagioclase-potassium feldspar-biotite-garnet-cordierite-sillimanite (Coutinho 1974).

The outcrops are responsible for the high content of heavy minerals in the nearby sandy beaches, particularly those bordered by rocky outcrops and Tablelands deposits. Sillimanite, zircon, andaluzite, rutile and monazite are widespread and found in high concentrations.

13.2.2 The Continental Shelf

Espírito Santo continental shelf has three distinct provinces (I, II and III in Fig. 13.3). The first, Bahia Sul-Espírito Santo extends from Belmonte (Bahia) in the north to near the Doce River mouth (Espírito Santo). The physiography of the continental shelf in this province is associated with the southern portion of the Abrolhos Bank, whose origin is related to volcanic activities during the Eocene (52 and 4 Ma. Mohriak 2005). The average width in this portion is 230 km (França 1979).

The second province, along the central coast (II, Fig. 13.3), is marked by the narrowing of the continental shelf (shelf break varies between 60 and 80 m depth) to an average width of 50 km; which then widens southward. The inner shelf is characterized by an irregular bottom morphology, associated with the submerged abrasion terraces formed by lateritic concretions, which favors the growth of carbonate organisms such as coralline algae, bryozoans, mollusks and benthic foraminifera (Albino and Suguio 2011). These organisms are responsible for 75 % of the sedimentary composition of the continental shelf (Kowsmann and Costa 1979).

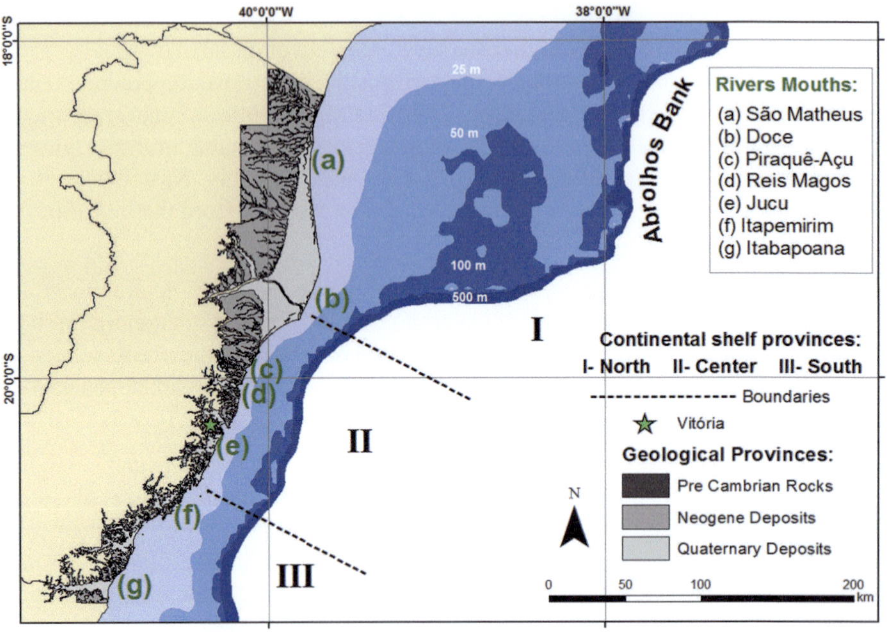

Fig. 13.3 Geological and Continental Shelf provinces along Espírito Santo (I, II and III). Major river mouths are identified by letter (*a*) to (*h*)

The third province occupies the southern part of the state and has a typical irregular inner shelf morphology and average width of 77 km (III Fig. 13.3).

13.3 Coastal Sediments Supply and Characteristics

Coastal sediments sources include fluvial sand and mud, marine carbonate and a mix of sediment eroded from the exposed Barreiras Formation. The influence of fluvial deposits is localized to the vicinity of river mouths, and particularly the Doce and São Mateus rivers in the north, and Jucu and Itabapoana rivers in the south (Fig. 13.3).

Inner shelf sediments adjacent to Doce River mouth are mainly sandy and muddy sand siliciclastics, supplied by the river (Albino and Suguio 2010). The offshore dispersion of Doce River fluvial sediments extend about 15 km off the coast, where the fluvial sediments are gradually replaced by biosiliciclastic sands that dominate the bottom surface of outer continental shelf. South of Doce River mouth the shoreface contains predominantly medium to coarse siliciclastic sands. Bioclastic and mixed sands have been observed only on the southern continental shelf (Albino and Suguio 2010). At Doce River mouth and along the beaches of the coastal plain sands are predominantly medium/average siliciclastic with unstable heavy minerals reinforcing the fluvial influence (Fig. 13.4a).

Small riverine input from other watersheds draining into the coast is neglected because fluvial sediments are deposited in the drowned valleys and river estuaries on the tablelands of Barreiras Formation and between the crystalline headlands (Albino et al. 2006). Locally, lithoclastic terrigenous sand or mud deposits still occur in the shoreface and continental shelf where riverine sediment inputs are observed (Kowsmann and Costa 1979; Albino and Suguio 2010).

Up to 80 % of $CaCO_3$ can be found on sands in the central-northern shore, as carbonate production along the continental shelf is intense, making it the main component on sandy beaches in the region (Fig. 13.4b). They contain carbonate detritus and shells of different shapes and sizes. The main components are coralline algae and mollusks especially in high-energy beaches due to increased resistance to abrasion and fragmentation when deposited with harder siliciclastic sands. As carbonate sediments are easily broken, the closer the source area, the higher the diversity and the bigger the carbonate sediment (Albino and Suguio 2010).

The crystalline headlands and active cliffs of Barreiras Formation are sources of past and current sediments to beaches of center-south of Vitoria. In the coastal city of Guarapari, 50 km south of Vitoria, where the crystalline outcrops are interspersed with the active cliffs of Barreiras Formation, the contribution of heavy mineral from

Fig. 13.4 (**a**) Heavy minerals at Doce River mouth; (**b**) commonly found bioclasts on cliffed and terrace beaches, Santa Cruz beach; and (**c**) high heavy mineral concentration along rocky coastal beaches, Meaipe beach

both sources developed heavy minerals placers, including grenades, sillimanites (kinzigitic gneiss), monazite, rutile and zircon (Fig. 13.4c). The beach of Areia Preta ("black sand", in portuguese) is internationally known for its high concentration of monazite and ilmenites. The high concentration of heavy minerals in this beach ranks it among the most radioactive beaches with Radio (Ra) along the Brazilian coast and Thorium (To), worldwide (Veiga et al. 2006).

13.4 Wind, Waves and Tides

The wind regime in Espírito Santo is seasonal and bimodal northerly and southerly. East and northeast winds occur during summer generated by the South Atlantic Anticyclone, a semi-permanent high pressure system. Stronger southeast and south winds are more common during winter and fall and are generated by polar fronts moving northward from southern Brazil. As the wind approaches the coast, local topography becomes more influential and may locally affect wind direction and velocity (Amarante et al. 2009).

Information about wave climate in Espírito Santo waters is very scarce and is mainly based on occasional short-term observations. Due to this lack of information, wave hindcasting has been used to reconstruct the wave climate. Figure 13.5 presents the directional histograms of Espírito Santo waves for one offshore point (Off1) beyond the shelf-break (about 1000 m water depth) and three onshore points (On1, On2 and On3) at about 30 m water depth. This analysis used wave data between 1968 and 2008 from the Global Ocean Wave (GOW) model (Reguero et al. 2012). The GOW model was calibrated and validated globally using instrumental measurements of 21 buoys and altimetery data extracted from satellite images (Reguero et al. 2012).

Because of the seasonal wind regime offshore wave climate is also bimodal and seasonal, with northeast and east swell being the most frequent with wave heights between 0.5 and 1.5 m, whilst waves from southeast quadrant are less frequent but higher, with waves commonly reaching 2 m (Stech and Lorenzzetti 1992; Pianca et al. 2010).

Along the northernmost sector of Espírito Santo, where the continental shelf is both wide and relatively shallow due to Abrolhos bank, swells are predominately easterly and reduced in height- 40 % of waves with less than 1 m, as indicated by On1. Further south, at On2 and On3, higher waves arrive more frequently from the northeast, averaging between 1 and 1.5 m high (Fig. 13.5).

Tides are micro-tidal along the south-central coast ranging the maximum tidal range from 0.8 to 1.9 m increasing to meso-tidal in the north, reaching 2.0–2.3 m adjacent to the Abrolhos continental shelf (IH-Cantabria 2013).

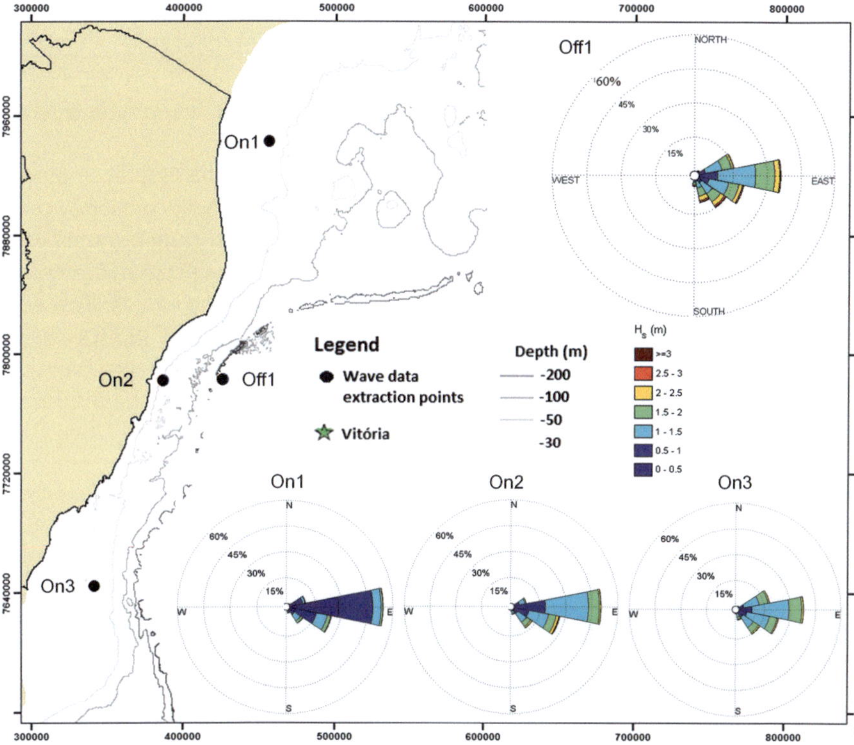

Fig. 13.5 Model results of incoming wave climate

13.5 Coastal Provinces and Geomorphology

Espírito Santo's coast can be divided into three geomorphic provinces: Coastal plains; cliffs and shore platforms; and coasts with rocky outcrops. The Quaternary coastal plain varies in width along the state: it is wide along the fluvial-marine plain associated with the deltaic plain of Doce River and smaller river valleys (Martin et al. 1996, 1997) which act as a sink for both fluvial and marine sediments. On the other hand, shore platforms associated with the Barreiras Formation, are responsible for the narrowing of the Quaternary deposits, and even narrower in Precambrian hills and crystalline headlands along the coast of rocky outcrops, where local sediment sources are composed of marine and fluvial input, as well as material from coastal erosion.

13.5.1 Coastal Plains

Espírito Santo coast is characterized by the proximity of crystalline outcrops to the coast, which limits the extent of most of the coastal plains. However, two extensive plains have developed: the Doce River deltaic plain extends along the northern coast for approximately 160 km, with a maximum width of 40 km near the main valley, where the shoreline has experienced considerable progradation (Fig. 13.3). The Itabapoana River plain, located in the south, is 21 km long and 5 km wide and is covered by well preserved Restinga vegetation, a low height halophytic tree and shrub species developed under poor sandy substrates.

According to Dominguez et al. (1981), the dimensions of the Holocene progradation cannot be explained solely by the accumulation of sediments presently transported by Doce River. These authors suggest that a sea level lowering between 4 and 5 m, during the last 5000 years, was a contributing factor in the progradation of the coastal plain. In addition the Abrolhos Bank and its relatively wide, shallow sedimentary platform have protected the coast from the direct action of the waves, and contributed to the progradation of the coast over the past 2500 years.

This age corresponds to the period when Doce River completed the infilling of the coastal lagoons and supplied fluvial load to the coast. Only when Doce River began to flow directly into the ocean, did the fluvial-marine processes became more intense, enabling the development of broad marine terraces. The relative changes in sea level and the successive changes in hydrodynamic conditions due to paleoclimatic changes, associated with hydraulic jet effect, were responsible for the asymmetry and disagreements between the sandy terraces (Martin et al. 1993, 1996).

Doce River acts as the main sediment source to the adjacent coast and inner continental shelf. According to Albino and Suguio (2010), the fine to coarse sands of the bedload are deposited in the vicinity of the mouth and the adjacent inner continental shelf. The coarse to very coarse sands originate from the reworking of sediments from eroded beach ridges. The high degree of exposure to the incident waves causes high mobility of adjacent beaches, with erosion of about 70 m of the river mouth beach observed during occasional river flooding (Albino et al. 2006). The river, however, also acts as a hydraulic jetty and effectively blocks the longshore transport of sand.

Large mangrove areas are located in the Piraquê- Jucu, Itabapoana, Itapemirim and São Mateus estuaries and the urban mangrove of Vitória, amounting a total of about 70 km^2 (Schaeffer-Novelli et al. 1990). Not with standing being under legal protection, mangrove areas are threatened by industrial and domestic pollution, and the Vitória Bay mangroves, e.g, have lost 40 % of its original area mainly due to embankments and dredging for civil construction and coastal engineering works.

13.5.2 Cliffs and Rock Platforms

Between Riacho River and Vitória (Fig. 13.3) the Quaternary coastal plain and beaches are limited in development, usually restricted to the base of the Barreiras Formation sea cliffs and the rock platforms (formerly known as wave-cut terraces). Due to the limited accommodation space between the platforms and cliffs, deposits are limited and small coastal plains only develop where river valleys are present and act as sediment traps.

Fig. 13.6 Development of cliffs and platforms associated with marine abrasion processes

The cliffs and rock platforms are associated with the marine abrasion process, as illustrated in Fig. 13.6. The Neogene Barreiras Formation, in southeastern Brazil, consists of continental deposits that were deposited on the inner continental shelf as alluvial fans or braided rivers when the sea level was more than 100 m below present (Morais 2007). Rock platforms were carved into this formation during subsequent episodes of relatively high Quaternary sea level and today form the seabed along the shoreface and inner continental shelf of the central coast, where they are encrusted with carbonate-secreting organisms. The death assemblages of these organisms are reworked by the waves and tidal currents onto adjacent beaches, which are composed of coralline algae, mollusk shells, bryozoans, echinoid and some foraminifera.

The highest carbonate diversity is observed either in deeper or calmer coastal waters, where wave action cannot reach the seafloor and sediment fragmentation is scarcer. As a consequence, carbonate in those places is associated with granules and coarser sands. Approaching the strandline, more resistant mollusk fragments and coralline algae become dominant, in addition to siliciclastic grains. Albino and Suguio (2010) found that the less resistant bryozoans and equinoderms fragments, found in medium to fine sands, are very scarce in shallow waters.

On a regional scale, the beaches are stable in meso-term according to sediment availability and morphological accommodation capacity between the beach and inner shelf. The beaches are fronted by a low gradient shoreface where biogenic encrustation is present. The carbonate detritus makes up more than 50% of the beach sand and the sediment budget is positive. The tendency of the coast is toward stability as sediments become available, and the rock platforms act to retain them.

13.5.3 Rocky Headlands and Embayed Beaches

The crystalline rocks and Neogene sedimentary deposits alternate along the central-southern coastline forming a series of headlands and embayed beaches. The coastal plains are discontinuous and beaches have variable exposure to the incident waves. The length and width of beaches and extent of coastal terraces is controlled by the

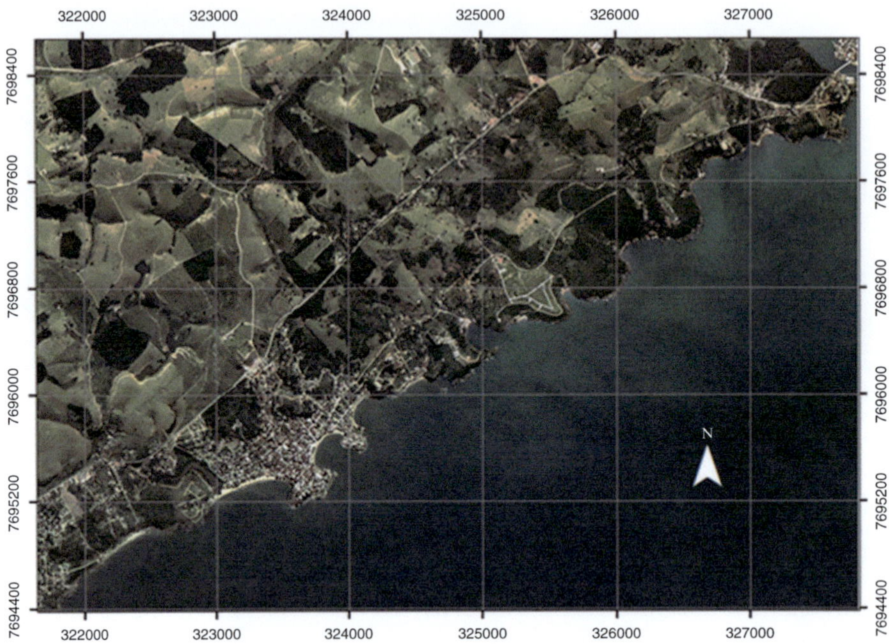

Fig. 13.7 Embayed beaches along the rocky outcrop shoreline, in southern Espírito Santo. The degree of exposition to waves controls sediment transport (Source: Geobases 2010)

headlands and contribution of terrigenous fluvial input. The coastal terraces are rarely reached by waves, except during severe storm events. Therefore, these sheltered beaches are in good condition and protect the coast from erosion. They also present high diversity of sandbanks and associated fauna and have limited urban development.

Beach morphodynamic processes along this indented coastline are dependent on the degree of wave exposure and embaymentization (Albino et al. 2006), as shown in Fig. 13.7. Wave refraction and diffraction is prominent, and features such as tombolos, spits and bars are common. Coastal sediment cells, limited laterally by rocky headlands, consist predominantly of siliciclastic coarse to medium sand, with sediment transport associated with swell direction and restricted to within the embayments, where reversals in sediment transport and beach rotation may occur.

13.6 The Beach Systems

13.6.1 Doce River Coastal Plain

Regional Beach and Sediment Characteristics The wide beaches adjacent to Doce River are directly influenced by the evolution of the coastal plain, where river sediments are continuously supplied to the coast. At the river mouth the sediments are poorly sorted angular siliciclastic grains which suggests they are supplied by the river. Southward of the Doce River mouth towards Barra do Riacho the sediments coarsen and increase in roundness (Albino and Suguio 2010). To the north the beach exhibits less reworked siliciclastic grains, while the sand becomes coarser and rounder.

The second major influence on Doce River plain evolution is the action of longshore sediment transport and the flat morphology (Dominguez et al. 1992; Martin and Suguio 1992). Bittencourt et al. (2007) suggest that the current wave conditions, as well as the morphology of the continental shelf, seem to have remained with the same characteristics from the late Quaternary to present time. Consequently, seasonal fluctuation in wave direction result in reversals in longshore transport and contribute to the high mobility of the beaches.

Beach Morphology and Mobility Near river mouth constant river input has led to shoreline progradation while the exposure to waves and fluctuation in loads and river discharges maintains high beach mobility (Fig. 13.8). Alternating shoreline construction and erosion during the evolution of Doce River beach-ridge plain have been identified by Dominguez et al. (1983) and Bittencourt et al. (2007), a process that continues to the present (Albino and Suguio 2010; Oliveira et al. 2015).

Beach Types Beaches associated with Doce River coastal plain are moderate to steep beaches composed of coarse to medium siliciclastic sand and range from intermediate to reflective (Fig. 13.9). The high mobility of profiles near the mouth is reflected in the seasonal variation in beach type. The presence of dissipative bars near the river mouth is probably due to the incorporation of fine fluvial sands that reduce the nearshore gradient. Longitudinal bars are favored by surfers, while beaches with moderate slope are preferred for sport fishing.

Further south, the Comboios and Barra do Riacho beaches do not receive river sediments and have coarser sand and steeper reflective beaches. Along the northern coastal plain intermediate beaches occur in association with northward longshore sediment transport and finer sand lower gradient beaches, as seen in Fig. 13.10.

As a result of river input, which causes ongoing progradation and low level of urban development there is no evidence of erosion or sediment imbalance along the coastal plain (Albino et al. 2006). An exception is in the vicinity of river mouths and/or sandy spits, where occupation can be denser, and shoreline mobility higher.

Dissipative beaches occupy the northernmost portion of Doce River plain and are also a favorable place for dune formation. As there is abundant sediment supply and the easterly wind is persistent and strong enough to transport the available sand, it is possible to find foredunes all along this section, the largest reaching 30 m high in the touristic village of Itaúnas (close to São Mateus River). In the last 40 years, is has been migrating an average of up to 5 m yr^{-1} both southward and northward, due to northeast and southeast winds, respectively.

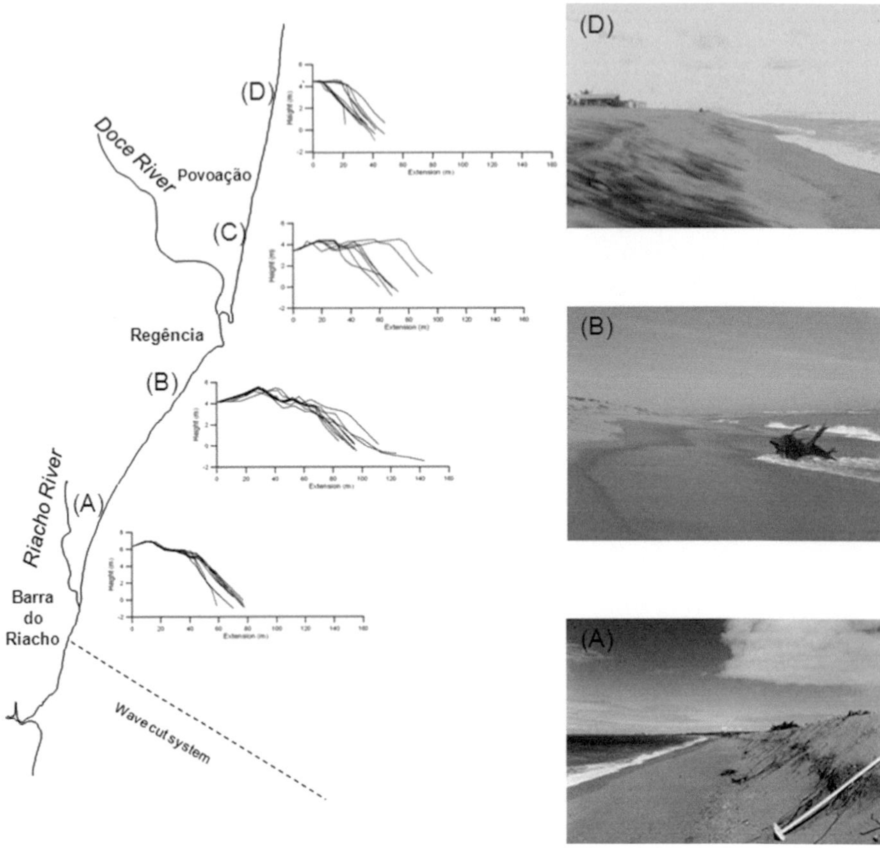

Fig. 13.8 Seasonal beach profile changes on Doce River terraces. Results represents monthly field data from January 1996 to January 1997

13.6.2 Cliff and Rock Platform Systems

Regional Beach Characteristics This section of the coast has variable beach morphology dependent on the rock platform distribution, together with variable breaker wave height and grain size. As a result three beach types can be identified: (a) exposed beaches: straight or slightly curved beaches exposed to east and southeast waves; (b) embayed and pocket beaches: laterally limited by promontories of the Barreiras Formation or/and preceded by small or large platforms; (c) estuarine beaches: located at the mouth of estuaries, laterally limited by sedimentary promontories. The beaches there are also narrow as they are squeezed between the rock platforms and the cliffs, while the rock platform controls the mobility and morphodynamic processes. Moreover, rock platforms form beaches with mixed sand composition including bioclastic fragments. Based on Short (2006) these beaches can be classified as reflective plus low tide rock flats.

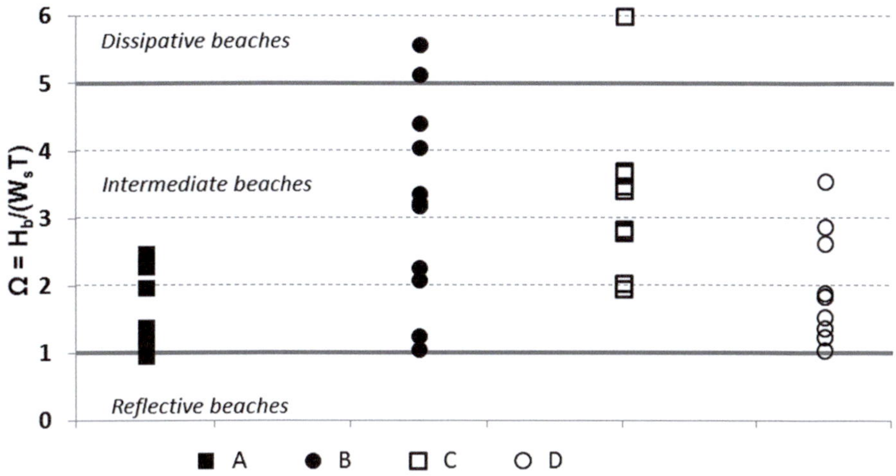

Fig. 13.9 Beach types along the Doce River coastal plain beaches. For location of (*A*), (*B*), (*C*) and (*D*), please refer to Fig. 13.8

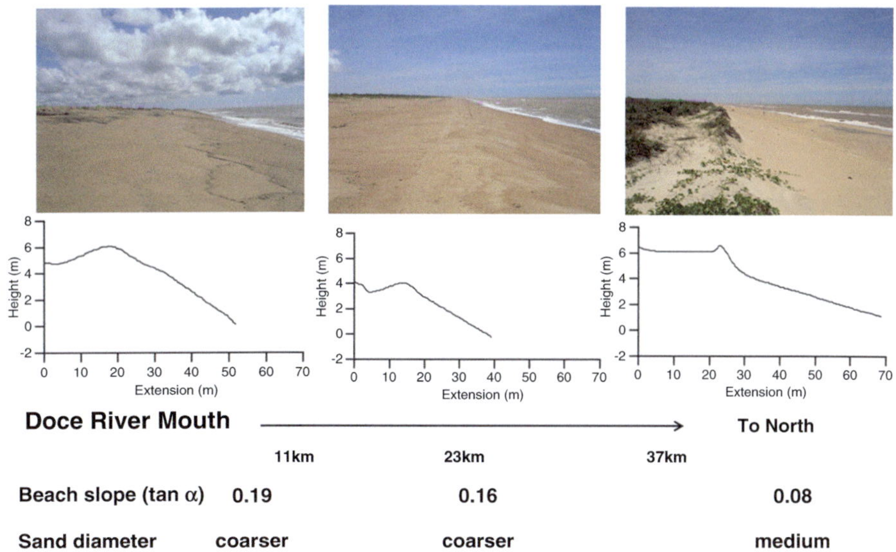

Fig. 13.10 Average beach slope and sand diameter along northern plain of Doce River, from Urussuquara beach to Guriri beach, please see Fig. 13.1

Waves are largely dissipated cross the rough shallow rock platforms (Albino et al. 2006). Visual observation suggests that since the rock platform beaches retained most of their sediment even under severe meteorological and oceanographic conditions the waves are largely dissipated across the platforms with an average breaker height of only 1 m (Fig. 13.11).

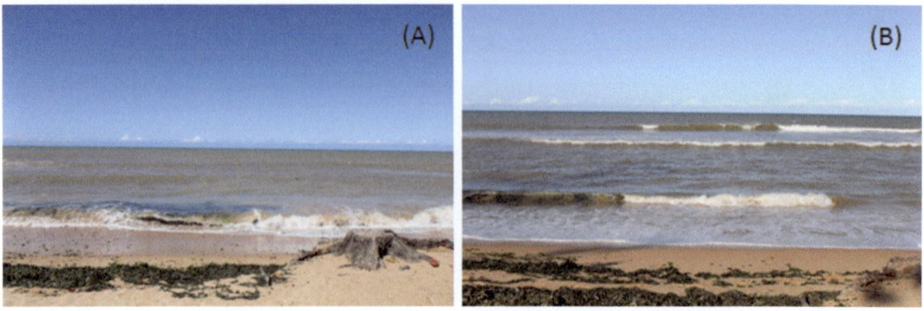

Fig. 13.11 The influence of tidal level on breaking waves (**a**) surging waves during high tide; and (**b**) spilling waves at low tide dissipated by the platform at Manguinhos beach

Masselink and Short (1993) pointed out that tidal level causes the height and types of breakers to vary. During low tide, when the rock platform becomes exposed, dissipation is stronger and breakers spill across the platform. During high tide, waves are higher and cross the platform to break as surging and collapsing breakers.

Sediment Characteristics Along the rock platforms, the coastal bioclastic fragments have different degrees of resistance to disintegration. Bioclastic fragments (bryozoan, equinoderm, foraminifera and worm tubes) are more susceptible to disintegration producing finer grains which constitute 10–40 % of medium-to-fine sands, while the more resistant coralline algae and mollusk shelles result in coarser grains and are the dominant bioclastic fragments (Fig. 13.12). Therefore, mixed sand beaches will tend towards coarseness with increased siliciclastic, coralline algae and shell fragments.

Tanner (1995) observed that bioclastic fragments tend to become finer when associated with hard quartz grains, as a consequence of the enhanced abrasion of the bioclastic fragments. Moreover, they tend to be coarser near the source (Oehmig and Michels 1993). However, irrespective of composition abundance, it remains true that hydrodynamic effects are responsible for the final composition of a sedimentary deposit, according to grain physical properties, such as shape, size, density and angularity.

Beach Morphology and Mobility The cliffs and rock platforms favour the development of narrow stable beaches. These beaches are protected from direct incident wave attack and loss of material seaward. Therefore, the rock platforms play an import role on sediment retention in the beach profile while maintaining a balanced sediment cell.

The rock platforms particularly influence the beach and face slope variability. On beaches where terraces or outcrops are closer to the emerging portion, they are more protected and tend to be composed of coarse to medium sands with steeper slopes. On exposed beaches or when they are fronted by a discontinuous wave cut terrace the beach face slope is more variable. This variability decreases on beaches where the terraces are more continuous (Fig. 13.13).

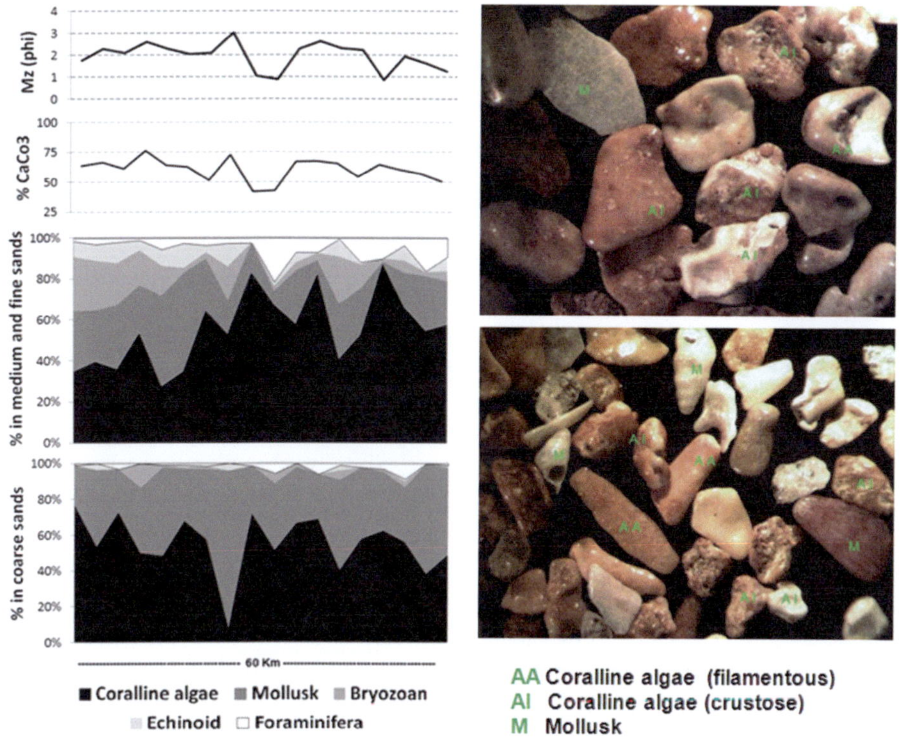

Fig. 13.12 Principal components and grain size distribution of bioclastic debris in beaches located between Vitoria and Barra do Riacho

The distribution and morphology of the rock platforms controls the beach volume. The exchange of materials between the upper and lower parts of the beach (berm and beach face, respectively) dominate the changes in beach profile, with limited exchange seaward of the terraces. The beach accommodation space is reduced where the terraces are closer to the shoreline, so the sand tends to deposit more vertically. This reflects on the berm height. By contrast, beaches with wider terraces and greater accommodation space have wider and more flattened profiles (Fig. 13.13).

Beach Types Breaker wave process depends on the distance between the mean water level and the rock platform whose uneven distribution results in complex waves propagation, as suggested by Bray and Hooke (1997). Along this coast the rock platforms restrict the observed morphodynamic state to intermediate-reflective at high tide and dissipative at low tide. These observations are similar to the low-tide terrace state suggested by Wright and Short (1984) and the reflective plus low tide rock flats of Short (2006). Fully dissipative characteristics are found on the more exposed low gradient beaches composed of fine sand.

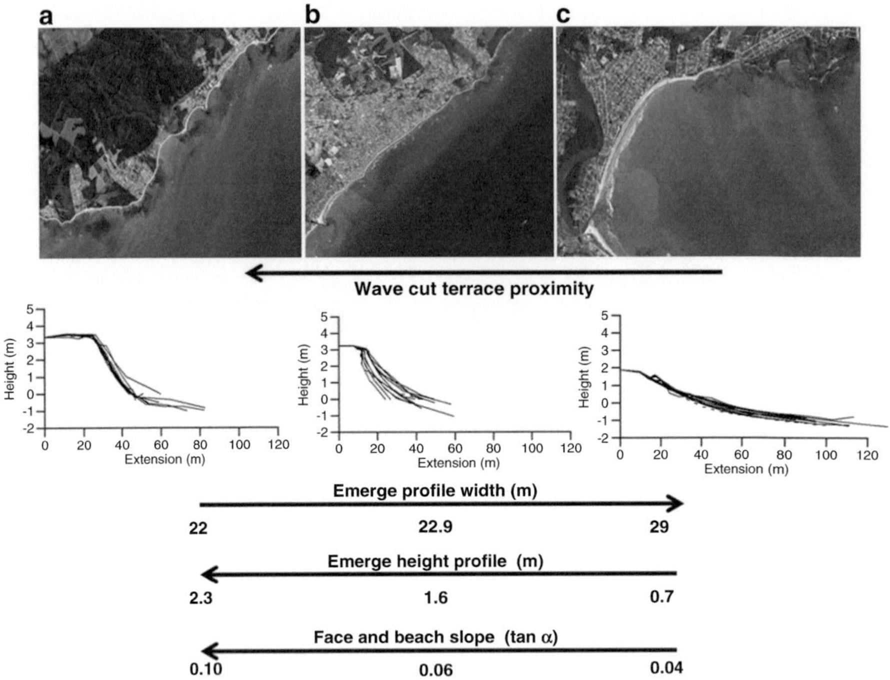

Fig. 13.13 Mobility and morphological characteristics of rock platform beach profiles: (**a**) Padres beach; (**b**) Jacaraipe beach; and (**c**) Grande beach. Results represents monthly field data from January 1996 to January 1997 (Images source: Google earth)

The predicted beach states obtained by applying the dimensionless fall velocity (Gourlay 1968) showed a large variation along the platform beaches (Fig. 13.14), due to both temporal and spatial variations in wave height and sediment fall velocity. Variation of the grain size composition controls the fall velocity while breaker wave height is controlled by platform width and morphology. Shallow rock platforms result in a low Ω is low with small variations and usually reflective beaches. Deeper rock platforms have a more variable Ω and beach ranging between dissipative and intermediate.

13.6.3 Embayed Beaches Systems

Regional Beach Characteristics More than 20 embayed beaches are located between Jucu and Itapemirim rivers. By definition, embayed beaches are located between bedrock headlands, each one representing one sediment cell and working as a closed system with limited or no longshore sediment transport (e.g. Albino et al. 2016). As a consequence sediment transport is restricted to within the cell, as the rocky headlands act as a sediment trap. They can also induce beach rotation. It

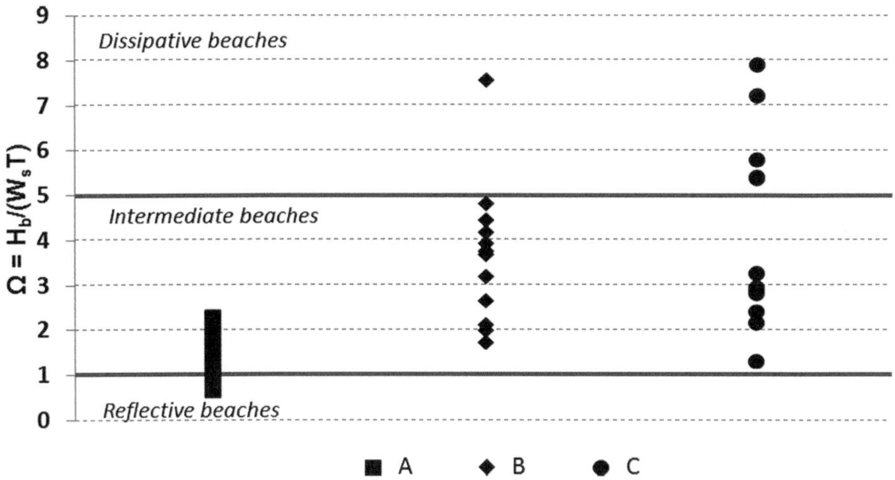

Fig. 13.14 Beach state of rock platform beaches. For location of (*A*), (*B*) and (*C*), please refer to Fig. 13.3

is possible for onshore/offshore exchange to occur, and during high storm events sediment may be transported around those headlands. However, as the headlands generally extend beyond the breaker zone, regular swells cannot transport the sediment load.

Sediment Characteristics The second characteristic of embayed beaches is that the sediment source is closed and restricted to the same beach. As the sources in this sector varies from fluvial, crystalline outcrops and the erosion the Barreiras Formation, sediment is also very diverse and its transport is a function not only of wave and current energy, but also sediment characteristics. Also due to the presence of both Barreiras Formation and crystalline outcrops surrounding and limiting the size of the beaches, beaches like Ubú, Maimbá, Morro and Areia Preta, have very high heavy mineral content (Fig. 13.15).

In a process known as hydraulic equivalence, quartz and feldspar of a given size may be transported altogether with a smaller heavy mineral grain. In contrast, for a given wave energy condition, lighter minerals may be withdrawn from the subaerial beach, but not the heavy minerals, that will, over time, increase the concentration of the latter, developing placer deposit. As a result, heavy minerals concentration can act as a good erosion proxy.

Although heavy minerals have been used as an indicator of sediment transport since Rittenhouse (1943) and Trask (1952), the relationship between size, density and shape of each mineral and direction of transport can be very difficult, and may be restricted to a small area around the heavy mineral source. In addition, on long beaches, the high number of sources make this approach even harder.

Fig. 13.15 Heavy mineral presence on the foreshore of Ubu beach

Beach Mobility and Type The third common characteristic of the beaches in this sector is that sediment mobility is related to the degree of exposure to incident waves. In Fig. 13.16, black dots represent erosion spots, while the colored shoreline relates to wave exposure. It is possible to note that most of the eroding beaches are oriented to north-northeast quadrant. As the maximum strength of the longshore current occurs when waves reach the coast with an angle of 45°, the more energetic east-southeast and southeast waves transport sediment on these beaches. In contrast, beach exposed to west-northwest and north-northwest and to the lower northeast waves, are experiencing accretion.

Fruta and Baleia beaches are separated by a headland and have opposite beach types (Fig. 13.17). Fruta is located south of the promontory and is dissipative year round with very fine sand and a very low gradient and while protected from northeast waves is eroding. Baleia Beach, on the northern side, is a steep reflective beach composed of coarse sand and exposed to incident waves and is accretionary (Albino, 1996). It is also contains beach sandstones that limit its size and sediment loss. The headland that divides the Baleia and Fruta beaches represents a physical barrier to sediment exchanges. Sediment is however transported south from Ponta da Fruta to Sol beach and deposited in Setiba, the southernmost beach in the arc. Figure 13.18 shows the beach types.

Klumb-Oliveira and Albino (2014) found along Praia do Morro beach sediment is eroded from the beach during winter and deposited in the bars and rips, while in summer the bars and rips move shoreward and rebuild the beach. Beach morphology varies from dissipative in erosional areas and under storm conditions to transverse bar and rip and low-tide terrace in depositional areas and under fine weather conditions (Fig. 13.19).

In the same study, a 40 years analysis of aerial images concluded that the interdecadal variation of beach width and morphology is greater over the same decade that in a longer period. As an oscillating beach, seasonal variations may occur, and shorter time scale studies may present inconclusive results. However, if one takes into account a larger time scale, the beach can be considered stable (Klumb-Oliveira and Albino 2014).

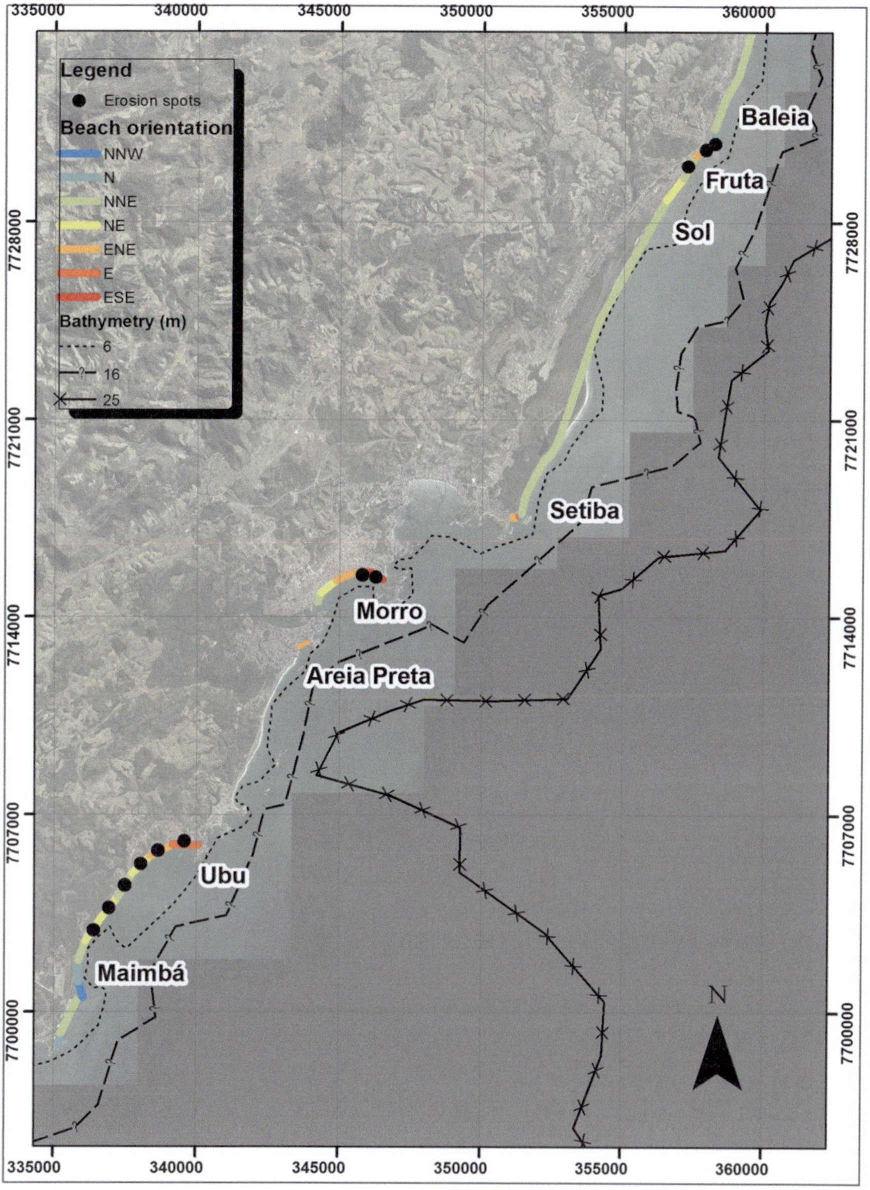

Fig. 13.16 Degree of exposure of eight different beaches along the rocky coast of Espírito Santo

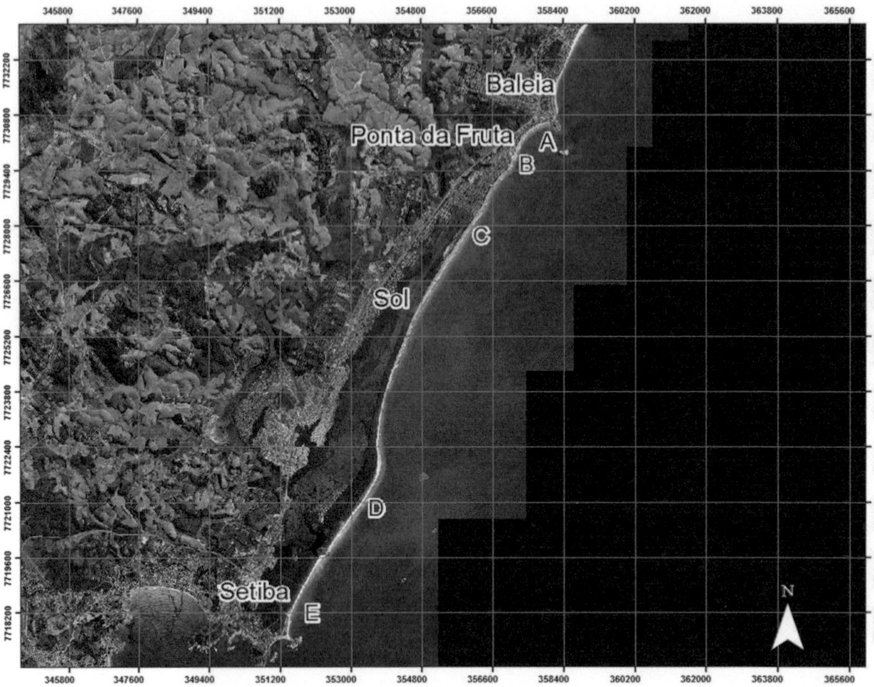

Fig. 13.17 Headland dividing Baleia and Ponta da Fruta beaches. The figure also shows Sol and Setiba beaches; the letters *A* to *E* marks samples location

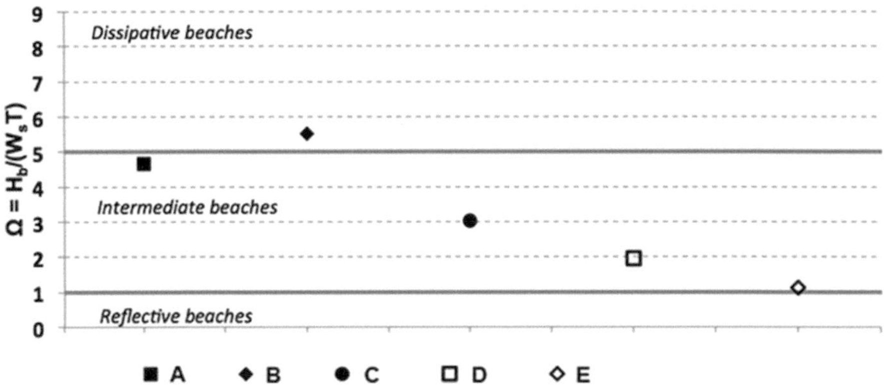

Fig. 13.18 Beach types. Station A represents Baleia beach (in the northernmost part of the arch) B to C are Sol beach and E Setiba beach, in the south (see Fig. 13.7 for locations)

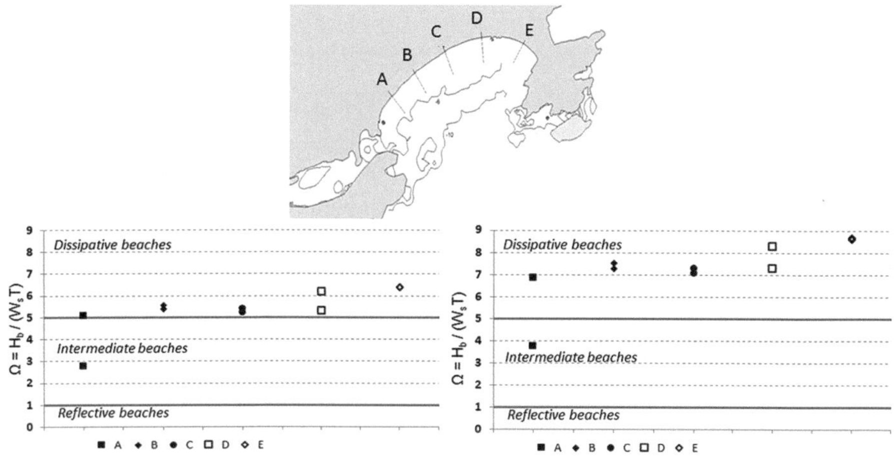

Fig. 13.19 Morro beach types under fine weather (*left*) and storm conditions (*right*). Location of Stations *A* to *E* are shown on the map above

13.7 Beach Use and Abuse

A total of 49 % of the state's population lives in 13 coastal cities, out of 78 municipalities in the state (IBGE 2010). Oil production, port and touristic activities have generated an expansion of coastal urban centers, which have had severe morphological impacts of the shoreline.

Reserves of light and heavy oil, as well as gas, are located on the mainland, continental shelf and ultra-deep water. The state stands out in regard to its production and reserves of oil and natural gas. This industry has influenced various sectors of the economy such as housing and service in the cities where the extractive fields and support facilities are located (IBGE 2010). The municipalities associated with the coastal plain of the Doce River and the southern plain around the river Itabapoana have strong ties with the oil industry.

In Espírito Santo, most harbours, port terminals and industrial activities are concentrated around the capital, Vitória, and Guarapari where the embayed-estuarine geomorphology favors the port installation. The highway infrastructure and intense urbanization of the surroundings coast are the main factors effecting the coast. Nicolodi and Peterman (2010) attributed medium to very high vulnerability to this region due to relatively high dense occupation compared to the mean density of the state.

Camburi Beach, located in Espírito Santo Bay, is the largest and most important beach of Vitória. It is an important area of leisure and recreation for the inhabitants of Espírito Santo capital. Its waterfront is lined with hotels, restaurants and kiosks and it is very important to the state's tourist economy. In addition, since the construction of Tubarão port complex in the 1960s (Fig. 13.20), this bay is also an important industrial site.

Wave propagation studies showed that the wave height distribution along the beach was substantially altered by the construction of the breakwater and port access channel. This caused diffraction of waves around Ponta do Tubarão and refraction by the dredged channel, generating a convergence of east-northeast waves orthogonal into the southwest portion of the beach (Melo and Gonzales 1995). As a result of the new equilibrium shape, the southern beach retreated, while to the north-east, the longshore sand transport is deposited sediments upstream of the groin (Fig. 13.20).

A variable beach morphodynamics is observed along the 6 km long Camburi beach. In the southwestern portion (A in Figs. 13.20 and 13.21), the beach is wide, reflective, with coarse sand. After the adjustment period, longshore transport were responsible for sediment erosion and beach type changed to an intermediate low tide terrace (Fig. 13.21a). Towards the northeast sediment is being deposited and rips and cusps are common (Fig. 13.21b) while mobility is high. At the northern end of the beach, there is little mobility of the profiles because of the diffraction processes (Fig. 13.21c) with deposition of fine sand and a low gradient beach face.

Urban pressure on the dynamic range of the beach is responsible for other morphodynamic imbalances. Along the south-central sectors, Albino et al. (2006)

Fig. 13.20 Camburi beach, in the state capital Vitória, with urban, recreational and industrial activities (Source: Geobases 2010)

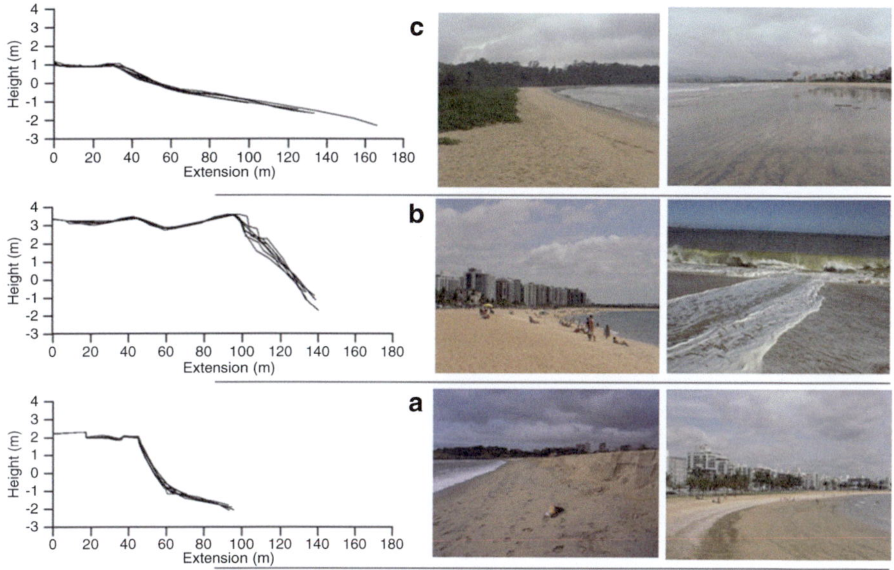

Fig. 13.21 Mobility and morphological characteristics of Camburi beach profiles

observed beach erosion and destruction of urban structures. The intense occupation of the beach foredunes and ridges increases erosion vulnerability due to reduced sediment availability. Roads, parking places and kiosks are frequently threatened by erosion. The Rodovia do Sol (ES 010), an important touristic route which extends along the coast is also at risk in some locations (Fig. 13.22).

Many attempts have been made to protect the coast along the state, with Conceição da Barra (mouth of São Mateus River) and Marataízes (southwards Itapemirim River) the most eye-catching. After years of unsuccessful shoreline protection developments, newly designed breakwaters were constructed in both places, in order to reduce offshore sediment loss, eliminate rip currents due to the structures and improve the aesthetic appearance (Elfrink et al. 2006).

13.8 Summary and Conclusions

While having a relatively small coastline comparing to other states, Espírito Santo has a high degree of beach diversity. The continental shelf width, bimodal waves and wind regimes, different tidal amplitudes and shoreline orientation due to geological control, affects beach type, morphology and mobility.

Sandy beaches in the state can be divided into three different sectors (Table 13.1). On the northern coastal plains under influence of Doce River, longshore

Fig. 13.22 (**a**) Maimbá; (**b**) Jacaraípe and; (**c**) Itaipava beaches: narrow beaches located in front of urban structures in the hazard zone along the south central coast of the state of Espírito Santo

transport dominates and affects grain size and beach type that ranges from reflective to dissipative states. Sediment has been supplied by the river since the Quaternary.

The central coast of Espírito Santo is dominated by Tertiary sedimentary cliffs and abrasion terraces, which restrict the width of the coastal plains. Erosion of the cliffs constantly supplies sediment to the beaches, with terrigenous sediments like quartz and heavy minerals being mixed with carbonate detritus from the nearshore. The abrasion terraces controls the low-tide terrace beach type and sediment transport is dependent on the shape and density of the grains.

In the south crystalline rock headlands alternate with pocked beaches, and three common characteristics can be found: little sediment transport, restricted to embayed sediment cells; sediment sources close to the beach; and sediment transport a function of the degree of exposure to incident waves. Interannual variability may be greater than interdecadal, thus studies must focus on longer time scales.

As an important economic spot in Brazil, Espírito Santo's coast have been subjected in recent decades to anthropogenic impact, though port construction, urban occupation and uncontrolled touristic development. As a consequence beach sediment transport and morphology have been changed in specific places: erosion has

Table 13.1 Main characteristics of the beaches found in each sector

	Doce River coastal plain	Cliffs and platform beaches	Embayed beaches system
Coastal province and geomorphology	Sedimentary coastal plain	Tablelands	Rocky outcrops
Sediment source and characteristics	Siliciclastic fluvial sands	Mixed sands: siliciclastic from Barreiras and bioclastic sand from platform biogenic encrustation	Siliciclastic sand from the outcrops and Barreiras weathering with bioclastic contribution
Wave exposure	Exposed	Protected by the shore platform	Exposed, semi-exposed and protected according to the embaymentisation
Tide signal	Mainly wave-dominated ambient due to Doce River, but estuaries can be found	Influences wave breaking: during low tide waves are dissipated by the exposed shore platform; during high tide, waves break as surging and collapsing	Wave-dominated ambient
Beach types and coastal plain/beach extent	Wide coastal plain with Reflective and intermediate beaches	Dependent on the spatial distribution of the shore platform. Narrow coastal plain	Related to embaymentisation and temporal variation due to beach oscillation. Very narrow coastal plain
Use vulnerability	Occupation of frontal dunes	Flooding	Erosion and inundation due the dense occupation

been exacerbated by engineering projects where the beach is exposed to more energetic waves and longshore transport is interrupted. Efforts are now being been made to better understand beach morphodynamics and to minimize coastal hazards.

References

Albino J (1996) Morphodynamics and Coastal Processes on Baleia, Fruta and Sol beaches, State of Espirito Santo. Braz An Acad Bras Ciênc 68(3):425–432

Albino J, Suguio K (2010) Sedimentation processes and beach morphodynamics active at the Doce River mouth, Espírito Santo State. Braz An Acad Bras Ciênc 82(4):1031–1044

Albino J, Suguio K (2011) The influence of sediment grain size and composition on the morphodynamic state of mixed siliciclastic and bioclastic sand beaches in Espírito Santo state, Brazil. Rev Bras Geomorfologia 12(2):81–92

Albino J, Girardi G, Nascimento KA (2006) Espírito Santo. In: Muehe D (ed) Progradação e Erosão do Litoral Brasileiro. Ministério de Meio Ambiente (MMA), Brasília, pp 227–263

Albino J, Jimenez JA, Oliveira TCA (2016) Planform and mobility in the Meaípe-Maimbá embayed beach on the South East coast of Brazil. Geomorphology 253:110–122. doi:10.1016/j.geomorph.2015.09.024

Amarante OAC, Silva FJL, Andrade PEP, Parecy E (2009) Atlas eólico: EspíritoSanto. Agência de Serviços Públicos de Energia do Estado do Espírito Santo (ASPE), Vitória

Bittencourt ACSP, Dominguez JML, Martin L, Silva IR, De Medeiros KOP (2007) Past and current sediment dispersion pattern estimates through numerical modeling of wave climate: an example of the Holocene delta of the Doce River, Espírito Santo. Braz An Acad Bras Cienc 79(2):333–341

Brasil MME (1987) Projeto RADAM BRASIL 1987 Geologia – vol 34, Folha Rio Doce

Bray MJ, Hooke JMH (1997) Prediction of soft-cliff retreat with accelerating sea-level rise. J Coast Res 13(2):453–467

Cordani U (1973) Evolução geológica pré-cambriana da faixa costeira do Brasil entre Salvador e Vitória. Tese de livre Docência no Departamento de Mineralogia e Petrologia do Instituto de Geociências da USP, São Paulo, p 98

Coutinho JMV (1974) Os minerais pesados de areia na foz do rio Doce. In: SBG (ed) Proceedings of the XXVIII Brazilian geological congress, vol 5, Porto Alegre, pp 61–77

Dominguez JML, Bittencourt ACSP, Martin L (1981) Esquema evolutivo da sedimentação quaternária nas feições deltaicas dos rios São Franscisco (SE/AL), Jequitinhonha (BA), Doce (ES) e Paraíba do Sul (RJ). Rev Bras Geoc 11(4):227–237

Dominguez JML, Bittencourt ACSP, Martin L (1983) O papel da deriva litorânea de sedimentos arenosos na construção das planícies costeiras associadas a desembocaduras dos rios São Franscisco (SE/AL), Jequitinhonha (BA), Doce (ES) e Paraíba do Sul (RJ). Rev Bras Geoc 13(4):93–105

Dominguez JML, Bittencourt ACSP, Martin L (1992) Controls on quaternary coastal evolution of the east-northeastern coast of Brazil: roles of sea-level history, trade winds and climate. Sediment Geol 80:213–232

Elfrink B, Acetta D, Mangor K (2006) Innovative shoreline protection scheme for the city of Marataízes, Espirito Santo, Brazil. Paper presented at the International Conference on Coastal Engineering (ICCE), Clearwater, Florida, E.U.A

Filho L, Ribeiro MW, Gonzales SR (1983) Geologia. In: Projeto RADAMBRASIL: Levantamentos de recursos naturais, folha SE 23/24. IBGE, Rio de Janeiro/Vitória 32:27–247

Force ER (1980) Geology of titanium-mineral deposits. Special Paper of the Geol Soc Am, Tucson (AZ, USA) 1(259):73–76

França AM (1979) Geomorfologia da margem continental leste brasileira e das áreas oceânicas adjacentes. Projeto REMAC Série n. 7, 89–127p, In: PALMA, J.J.C

GEOBASES Sistema Integrado de Bases Geoespaciais do Estado do Espírito Santo (2010) Governo do Estado do Espirito Santo. http://www.geobases.es.gov.br/

Gourlay MR (1968) Beach and dune erosion tests. Delft Hydraulics Laboratory, Report No.M935/M936

Instituto Brasileiro de Geografia e Estatística – IBGE (2010) http://www.ibge.gov.br/cidadesat/default.php. Accessed 15 Feb 2014

IH-Cantabria (2013) SMC-Brasil, Documento temático de níveis. Universidade de Cantábria. 140 pp (In Portugues)

Klumb-Oliveira AL, Albino J (2014) Mobilidade e Erosão na Enseada da Praia do Morro em diferentes escalas de tempo. Rev Bras Geomorfologia 15:103–117

Kowsmann RO, Costa MPA (1979) Sedimentação quaternária da margem continental brasileira e das áreas oceânicas adjacentes. Projeto Remac n. 8, p 55

Martin L, Suguio K (1992) Variation of coastal dynamics during the last 7,000 years recorded in beach-ridge plains associated with the central Brazilian coast. Palaeogeogr Palaeoclimatol Palaeoecol 99:119–140

Martin L, Suguio K, Flexor JM (1993) As flutuações de nível do mar durante o Quaternário superior e a evolução geológica de "deltas" brasileiros. Boletim IG-USP. Publ Especial 15:186

Martin L, Suguio K, Flexor JM, Archanjo JD (1996) Coastal quaternary formations of the southern part of the State of Espírito Santo (Brazil). An Acad Bras Cienc 68(3):389–404

Martin L, Suguio K, Dominguez JML, Flexor JM (1997) Geologia do Quaternário Costeiro do Litoral Norte do Rio de Janeiro e do Espírito Santo. CPRM (Serviço Geológico do Brasil) e FAPESP (Fundação de Amparo à Pesquisa de São Paulo), São Paulo, p 112

Masselink G, Short AD (1993) The effect of tide range on beach morphodynamics and morphology: a conceptual beach model. J Coast Res 9(3):785–800

Melo E, Gonzales JA (1995) Coastal erosion at Camburi Beach (Vitória, Brazil) and its possible relation to port works. In: Proceedings of the COPEDEC IV international conference on coastal and port engineering in development countries, vol I, September, Rio de Janeiro, RJ, pp 397–411

Mohriak WU (2005) Interpretação geológica e geofísica da Bacia do Espírito Santo e da região de Abrolhos: petrografia, datações radiométricas e visualização sísmica das rochas vulcânicas. Bol Geoc Petrobrás 14:133–142

Morais RMO (2007) Sistemas fluviais terciários na área emersa da Bacia do Espírito Santo (Formações Rio Doce e Barreiras). Thesis, Univ Rio de Janeiro

Morais RMO, Mello CL, Costa FO, Santos PF (2006) Fácies sedimentares e ambientes deposicionais associados aos depósitos da Formação Barreiras no Estado do Rio de Janeiro, vol 6. Geologia USP/Série Científica, São Paulo, pp 19–30

Nascimento Jr DR, Aguiar VAP, Giannini PCF (2011) Minerais Pesados das areias praiais de Guarapari (ES): distribuição, proveniência e fatores de risco à saúde. In: Abstracts of the XIII ABEQUA Congress, Búzios, 9–14 October 2011

Nicolodi JL, Petermann RM (2010) Climate changes and vulnerability of the Brazilian coastal zone in its environmental, social, and technological aspects. J Integr Coast Zone Mang 10(2):151–177

Oehmig R, Michels KH (1993) The isolation of settling-velocity fractions of sand-size material: new possibilities for the settling tube technique. J Sediment Res 64:687–689

Oliveira TCA, Albino J, Venancio ID (2015) Littoral longshore sediment transport along the deltaic plain of Rio Doce. Quat Environ Geosci 6(1):20–25. doi:10.5380/abequa.v6i1.36738

Pianca C, Mazzini PLF, Siegle E (2010) Brazilian offshore wave climate based on NWW3 reanalysis. Braz J Oceanogr 58(1):53–70

Reguero BG, Menéndez M, Méndez FJ, Minguez R, Losada IJ (2012) A Global Ocean Wave (GOW) calibrated reanalysis from 1948 onwards. Coast Eng 65:38–55

Rittenhouse G (1943) Transportation and deposition of heavy minerals. Bull Geol Soc Am 54:1725–1750

Rossetti DF, Bezerra FH, Dominguez JML (2013) Late Oligocene–Miocene transgressions along the equatorial and eastern margins of Brazil. Earth-Sci Rev 123:87–112

Schaeffer-Novelli Y, Cintrón-Molero G, Adaime RR, Camargo TM (1990) Variability of mangrove ecosystems along the Brazilian coast. Estuaries 13(2):204–218

Short AD (2006) Australian beach systems-nature and distribution. J Coast Res 22(1):11–27

Stech JL, Lorenzzetti JA (1992) Response of the South Brazil Bight to the passage of Wintertime cold fronts. J Geophys Res 97(C6):9507–9520

Tanner WF (1995) Carbonates. In: Environment clastic granulometry. Florida Geological Survey

Trask PD (1952) Sources of beach sand at Santa Barbara, California, as indicated by mineral grain studies. Beach Erosion Board Tech. Memo No. 28, U.S. Army Corps of Engineers

Veiga R, Sanches N, Anjos RM, Macario K, Bastos J, Iguatemy M, Aguiar JG, Santos AMA, Mosquera B, Carvalho C, BaptistaFilho M, Umisedo NK (2006) Measurement of natural radioactivity in Brazilian beach sands. Radiat Meas 41(2):189–196

Wright LD, Short AD (1984) Morphodynamic variability of surf zones and beaches: a synthesis. Mar Geol 56:93–118

Chapter 14
The Beaches of Rio de Janeiro

Dieter Muehe and Flavia Moraes Lins-de-Barros

Abstract This chapter presents the beaches of the state of Rio de Janeiro within a geomorphological perspective that includes the characterization of sediment types, exposure to waves, geology and relief and associated processes, especially the beach morphodynamic, their sedimentary budget and erosive tendencies. The study took as its foundation the knowledge of the coast acquired during the research carried out over several decades by the Marine Geography Laboratory of the Federal University of Rio de Janeiro, including the monitoring of the beaches and extensive field work. In addition to this knowledge, access to the publications of other research groups of this and other universities was also of crucial importance for a more complete overview. Thus the chapter presents the diversity of beach environments to be found in Rio de Janeiro state on the basis of a regionalization into two macro-compartments separated by Cape Frio where the shoreline undergoes an abrupt change of direction. This change is responsible for a considerable difference in the exposure of the beaches to southerly waves, resulting in distinct dynamic processes within these two compartments. Each of the compartments has then been divided into smaller regions, in the attempt to group segments with similar geomorphological characteristics. Thus, in the eastward-facing northern compartment which extends from the northern limit of the state to Cape Frio, there are two sub-regions. The first includes the beaches along the Paraiba do Sul coastal plain and the second encompasses the coastline from Macaé to Arraial do Cabo including the coastal plains between Rio das Ostras and Cape Frio. In this latter sub-region a second division was highlighted to address specifically the beaches of the Armação dos Búzios (Cape Búzios) and Arraial do Cabo (Cape Frio), which stand out from the rest of that set. Then in the second macro-compartment where the coast is exposed to the south, there are four sub-regions of the double barriers and the lagoon system between Arraial do Cabo and Maricá; the sheltered beaches of Guanabara bay; the exposed beaches of the metropolitan region; and the beaches in the vicinity of the Serra do Mar coastal range.

D. Muehe (✉)
Graduate Program in Geography, Universidade Federal do Espirito Santo, Vitória, Brazil
e-mail: dieter.muehe@gmail.com

F.M. Lins-de-Barros
Departamento de Geografia/Programa de Pós Graduação em Geografia, Universidade Federal do Rio de Janeiro, Rio de Janeiro, Brazil
e-mail: flaviamlb@gmail.com

A.D. Short, A.H. da F. Klein (eds.), *Brazilian Beach Systems*, Coastal Research Library 17, DOI 10.1007/978-3-319-30394-9_14

14.1 Introduction

The coast of the State of Rio de Janeiro extends for roughly 653 km from the Barra do Itabapoana, which borders Espírito Santo in the north, to the locality of Trindade in the south where it borders São Paulo state. Based on the coastal orientation there are two macro-compartments separated by Cape Frio, here called the 'east' and 'south' coasts.

The east coast trends in a general northeast-southwest direction modified by the delta-shaped configuration of the coastal plain of the Paraíba do Sul river and by the rocky headlands of Cape Buzios and Cape Frio, with their embayed beaches (Fig. 14.1); while the south coast extends west of Cape Frio. It trends in an east-west direction, interrupted only by the Guanabara Bay and by rocky promontories separating the various beach arcs. The second, south-facing compartment, is exposed directly to the southerly storm waves associated with the frequent passage of higher latitude cold fronts and the accompanying southerly swell generated by South Atlantic higher latitude storms dissociated from local wind systems (Muehe and Valentini 1998). Melo (1993) quotes as an example of extreme cases for Rio de Janeiro, with swell periods arriving from the southern quadrant with periods of 10 to 16 s and significant heights of up to 4 m. The east coast, despite also being subject to the alternation of good-weather and storm waves, is more protected from the storm wave direct action by its orientation, as well as the presence of embayments to the north of Cape Frio and Cape Buzios, together with differing wave exposure on the north and south flanks of the Paraíba do Sul coastal plain. In general waves tend to have shorter periods and lower heights than on the south coast. Tide range for the entire coast is micro-tidal (<2 m), with spring tide range around 1 m and neap tide range around 0.5 m.

Sediment transport is driven by wave regimes of opposing directions, and consequently subject to a bimodal regime of longshore transport (Muehe 2013). On the south coast the residual longshore sediment transport, resulting from the action of storm waves from the south and lower waves from the eastern quadrant, has a tendency to equilibrium while on the east coast the residual transport is to the north.

14.2 Geology and Geomorphology

The coastal geology of Rio de Janeiro state is mainly characterized by (1) a crystalline basement of pre-Cambrian age; (2) a continental coast formed during the Mesozoic period; (3) basalt dykes on rocky headlands also of Mesozoic age; (4) a fault-system of Cenozoic age; and (5) a Quaternary coastal plains. In addition there are 1.8 Ba old rocks, which form the so-called *Terreno de Cabo Frio*. These rocks represent the Congo Craton, most of which is located on the African continent (Ferrari 2012).

The rocky basement is associated with the Mantiqueira Province, one of the structural provinces described by Almeida et al. (1981). It extends north from Uruguay to the south of Bahia and is thought to be the most complex structural

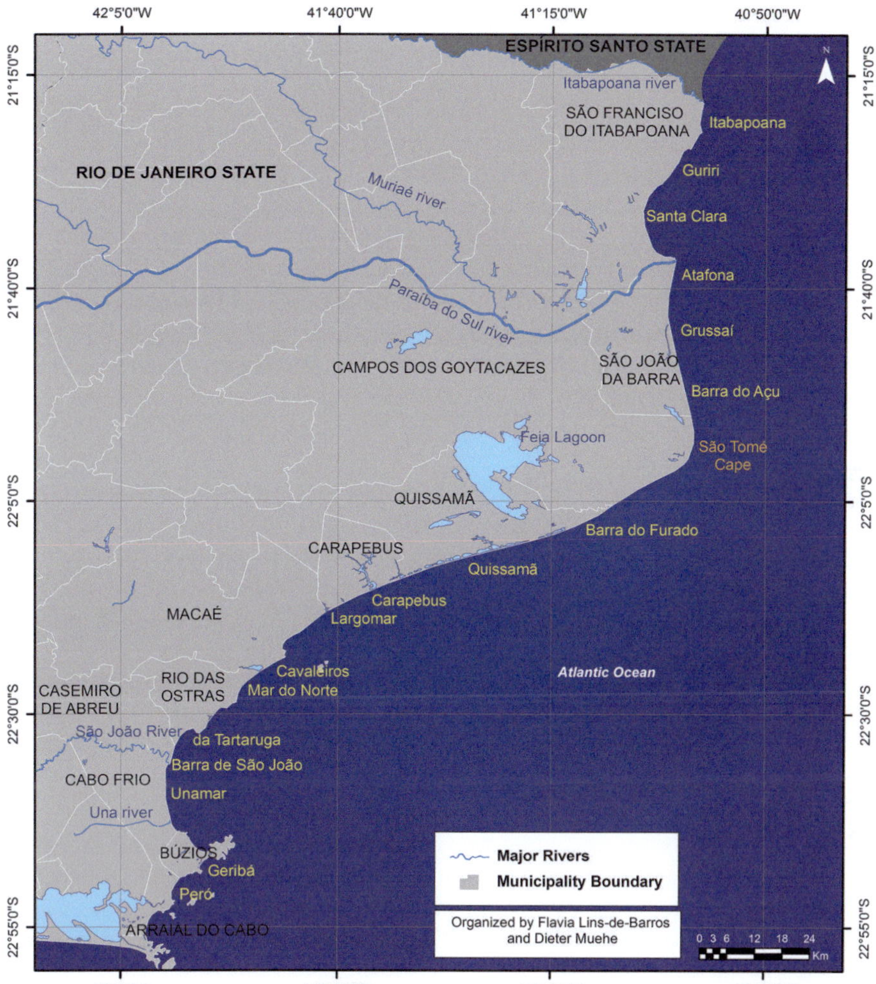

Fig. 14.1 Location of the east-facing coast dominated by the Paraiba do Sul delta-like coastal plain

province affected by the neoproterozoic/Cambrian Orogenic Cycle (Brasiliano Orogeny) in South America. The rocks possess distinct, complex lithological associations, represented principally by gneissic rocks on a granite base. The orogenic rocks, which resulted from the Brasilian Cycle, were initially denominated the Brasília Belt, the Ribeira Belt and the Araçuí Belt. The state of Rio de Janeiro is situated in the centre of the Ribeira Belt where the Brasiliano I Orogenesis (> 600 Ma) in the eastern portion of the state, the Rio Doce Orogenesis (ca 560 Ma) in the coastal portion (Figueiredo and Campos Neto 1993) and the Buzios Orogenesis (ca 520 Ma), described by Schmitt et al. (1999). The most characteristic rock of Rio de Janeiro city is the lenticular or augen gneiss, which forms the mountain peaks of

Pedra da Gávea, Corcovado, Pico da Tijuca, Andaraí and others. These Precambrian rocks have been fractured due to tectonic movements, which occurred between 130 and 70 Ma associated to the extensional tectonic process, which led to the breaking up of the Gondwana supercontinent in the Cretaceous period.

This extensional tectonic process is associated with a series of fractures related to the rifting and opening of the South Atlantic Ocean. In the Cenozoic period this process, labeled the "South-Atlantic Event" (Schobbenhaus et al. 1984) affected the whole of the east coast of the South-American Platform. Ancient zones of weakness were reactivated, in what was called the "Wealdenian Reactivation" (Almeida 1967) and their resulting structures controlled in part the formation of the coastal basins of Campos and Santos, which are separated by the structural high of Cape Frio. There was also an extensive alkaline lava flow about 130 Ma, which gave rise to basalt dikes. These can be found along the coast of Rio de Janeiro, especially in the city of Rio de Janeiro, in Armação dos Búzios, Cabo Frio, Arraial do Cabo and also in the islands of the bays of Angra dos Reis and Parati (Ferrari 2012). This rifting was also responsible for the generation of the fractures with northeast-southwest orientations: (associated with the reactivations of the pre-existing structures of the Ribeira Belt); northwest-southeast (zones of accommodation of regional efforts associated with the faults in the transfer of the rift); and east-west (related to the oceanic fracture zones), according to Stanton et al. (2010). The latter is associated to the oceanic fracture zone of Rio de Janeiro, and responsible for the sharp change in the direction of the shoreline from Cape Frio to the Marambaia barrier (Muehe and Valentini 1998), which exposes this coast to the southerly storm waves.

The most recent fault system, developed during the last phase of the opening of the Atlantic Ocean, gave rise to a set of horsts and grabens which include the mountain ranges of Serra da Mantiqueira, the Serra do Mar and the coastal range interspersed with depressions, including the Guanabara graben which covers the whole of the low-lying area which extends from Sepetiba Bay, in the west, to the locality of Barra de São João, in the east (Ferrari 2012). The infilling of the various valleys with sediments gave rise to the small sedimentary basins such as those of Resende, Macacu and Barra de São João.

Finally, are the Quaternary coastal plains whose origin is related to the deposition of continental sediments, which have been reworked by coastal processes including Pleistocene sea-level oscillations. The penultimate marine transgression in Brazil occurred about 120,000 year BP, when the climate was similar to the present (Flexor et al. 1984; Bittencourt et al. 1982), and a sea level about 7–8 m higher than present. Between 120,000 and 7000 year BP, sea level remained below the present level and during that period various depositional features were formed giving rise to marine terraces as well as fluvial erosional features such as deep valleys, which are well preserved on the continental shelf (Bittencourt et al. 1982). The last marine transgression started at about 17,000 years BP with the sea level reaching about 5 m above present between 6500 and 7000 years BP decreasing afterward with two short periods of rapid rise between 3800–3600 and 2700–2500 years BP. The double beach barriers with associated lagoons are result of these sea level fluctuations with the inner barrier related to the Pleistocene transgression and the outer to the Holocene transgression (Muehe and Valentini 1998).

14.3 Climate

The climate of Rio de Janeiro state is tropical, with rain throughout the year with maximum rainfall in summer together with highest temperatures. The climate of the coast is strongly influenced by the Atlantic Tropical High Pressure Mass with its northeast trade winds and by the Atlantic Polar Mass. The former brings hot and humid conditions and is responsible for the intense heat, which predominates in summer months. The latter is cold and strongest between May and September accompanied by an increase in the intensity and frequency of the cold fronts, responsible for the sharp drop in temperature that can last for up to a week.

Beyond the regional influence of these air masses, the coast of Rio de Janeiro state is locally influenced by the relief of the Serra do Mar mountain range resulting in orographic rainfall, especially in the south of the state due to its proximity to the coast. The heavy rainfall often generates mass movement of sediment from mountain slopes to the beaches, and presents a risk to the population, which occupy the steep slopes. In the metropolitan region, the so-called summer rains are responsible for recurring catastrophic episodes of landslides and flooding. In Cabo Frio and Arraial do Cabo the strong northeast wind, associated to a narrow continental shelf, causes the upwelling of cold water, the reason the cape's name, and also makes the climate much drier along this coastal zone. The low rainfall associated with this phenomenon, together with the presence of fine sand and the northeast trade winds, has favored the formation of extensive dune fields.

14.4 Rio de Janeiro Beach Systems

The Rio de Janeiro coastal system consists of the two major east and south coast compartments and several secondary compartments, which are listed in Table 14.1.

14.4.1 The East Coast

From the Itabapoana River at the northern border between Rio de Janeiro and Espírito Santo the coastline extends south-eastward to the town of Arraial do Cabo Cabo Frio at Cabo Frio (Fig. 14.1). Immediately south of the border is the small Itabapoana river coastal plain which is dominated by 20 km of eroding sedimentary cliffs, fronted by narrow beaches, which tend to be submerged during high tide (Fig. 14.2). Sediment eroded from the cliffs is negligible but was an important source under transgressive sea level episodes. The morphology then changes abruptly to the wide delta-like regressive beach ridge plain of the Paraíba do Sul River, with Cape São Tomé at its most prominent extremity. The coastal plain has about 150 km of coastline and a near continuous reflective beach, locally interrupted by the Paraiba do Sul estuary and by a low swampy region, indicating the former position of the

Table 14.1 Ocean and bay beach length of the Rio de Janeiro coast

Location		Geographical limits	Dominant features	Length (km)	Rocky coast (km)	Beaches (km)	Beaches (%)
Open ocean beaches	**East coast**	Itabapoana – Macaé	Paraíba do Sul Coastal Plain	178	0	178	100
		Macaé – Búzios	Rio das Ostras and Cape Búzios coastal plain	63	7	55	87
		Búzios – Arraial do Cabo	Pocket beaches and beach-dune systems	83	56	27	33
		Overall length		**324**	**63**	**260**	**80**
	South coast	Arraial do Cabo – Itaipuaçu	Barrier beaches of the Lake Region	104	3,7	100	96
		Itaipuaçu – Marambaia	Ocean beaches os Rio de Janeiro and Niterói	122	39	83	68
		Marambaia – Trindade	Rocky coast with pocket beaches	103	86	17	17
		Overall length		**329**	**129**	**200**	**61**
Guanabara Bay urban beaches		East side (Niterói)	Beaches of Icaraí and São Franciso	19	12	7	37
		West side (Rio de Janeiro)	Beaches of Flamengo and Botafogo	16	13	3	19
		Overall length		35	25	10	56
Southern Bay Beaches		Sepetiba and Ilha Grande	Rocky coastline and bay beaches	**500**	**379**	**121**	**24**

See Figs. 14.1 and 14.20 for locations

Paraíba do Sul estuary before its displacement to its present position on the north flank of the delta. As the result of this channel switching a Holocene beach ridge plain was deposited on the north flank of the delta separated from the Pleistocene plain on the southern flank (Fig. 14.3). The beach-ridge plain is fronted by a transgressive Holocene barrier with a narrow lagoon located between the barrier and the Pleistocene coastal plain (Fig. 14.4). The main source of the barrier sediments is the continental shelf, as attested by the high degree of rounded and polished sands, with a relatively low contribution of river sands (Martin et al. 1984). Access to beaches is difficult because of an absence of roads and use remains restricted to residents from the few local urban areas. South of of Macaé (Fig. 14.5), the coast consist of outcrops of the rocky basement which forms major promontories at Macaé, Cape

Fig. 14.2 Sedimentary cliffs of the Barreiras Formation. Beaches almost disappear during high tide (Photo: D. Muehe)

Búzios and Cape Frio; marine terraces between Rio das Ostras and Búzios (Praia Rasa) segmented by the estuary of the São João River, in the city of Barra de São João; and by coastal terraces fronted by foredunes as on the beaches of Tucuns, Peró and on the beach arc between the city of Cabo Frio and Arraial do Cabo.

14.4.1.1 Beaches of the Paraiba do Sul Coastal Plain

While the Paraiba do Sul delta shoreline appears homogeneous there are significant differences on either side of São Tome cape. In northern Holocene terrace of beach ridges, the beaches are attached to the terrace, as in Farol de São Tomé, where fishing boats are pushed offshore and pulled back by a tractor (Fig. 14.6). In contrast the southern sequence of Pleistocene beach ridges, is fronted by a narrow Holocene transgressive beach-barrier including small shallow lagoons, with barrier segments either eroding and overwashing or with large and well vegetated foredunes. There is also the low-lying area of the previous river valley fronted by a low barrier, as in Barra do Furado.

Based on beach profiles, beach mobility, (i. e. the standard deviation of the variability of the beach width), and grain size (Bastos and Silva 2000), the coast between Atafona and Cabiunas (Fig. 14.3) has a southward increase in grain size, with medium-sized sand (0.44 mm) at Atafona grading to very coarse sand (1.15 mm) at Cape São Tomé and Cabiunas. Accordingly the beach face slope is lowest in Atafona (around 5°) and becomes steep (8°) afterward. The generally low beach mobility

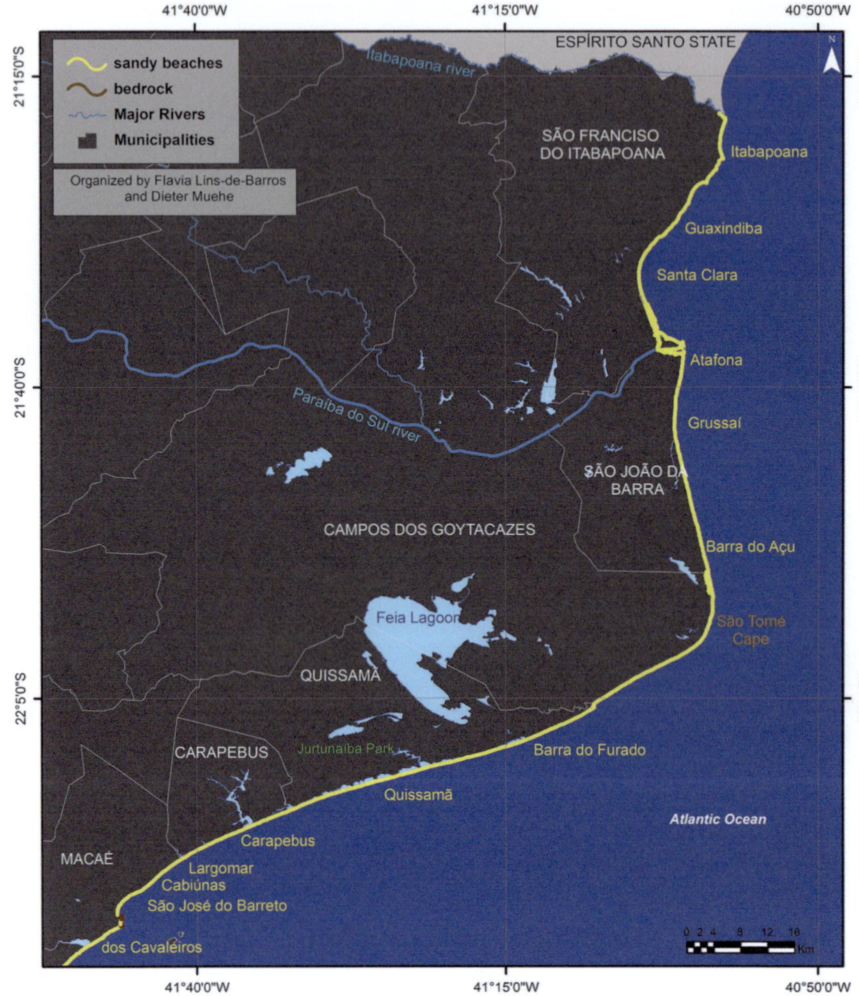

Fig. 14.3 The coastline from Itabapoana to Macaé with the wide beach ridge plain of the Paraíba do Sul River

also decreases from 19 m in Atafona to 4 m in Cabiunas, while the morphodynamic beach state is predominantly reflective, excepting at Atafona with an Ω value of 2.6 (Low Tide Terrace). The reflective beach state (Ω=<1) with low beach mobility (0.31 m) and coarse grain size also occurs a little to the south of Cabiunas at Lagomar (Fig. 14.3; Muehe 1998). Monthly profiles surveyed over 2 years and another set of 19 profiles at different time intervals are shown in Fig. 14.7. Similar characteristics have also been found by Machado (2010), in front of the Jurubatiba National Park, consisting of a segment between the Furado channel and Lagomar.

A common feature of all these profiles is the sharp break in slope between the steep beach face 7–9° and the low gradient of the shoreface (1.5–3°) a result of the

Fig. 14.4 Jurubatiba restinga, near Carapebus. Pleistocene beach ridge plain with Holocene beach barrier in front and associated lagoons (Photo: Flavia Lins-de-Barros)

Fig. 14.5 The city of Macaé (Photo: Flavia Lins-de-Barros)

Fig. 14.6 Fishing boats on the beach in Farol de São Tomé (Photo: Dieter Muehe)

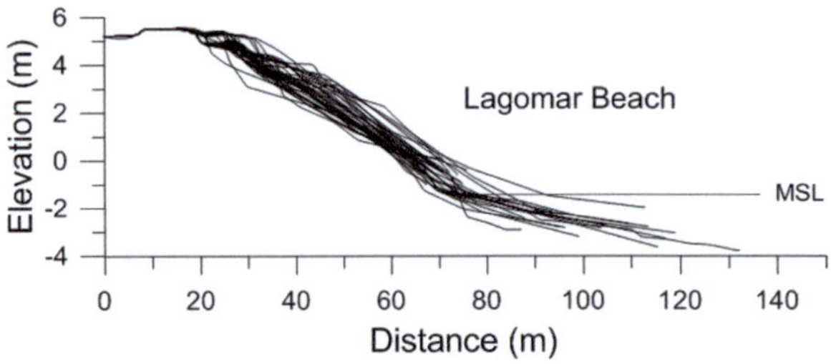

Fig. 14.7 Lagomar beach. A steep, reflective very coarse-grained beach face with fine shoreface sediments

decrease in shoreface sediment grain size due to the deposition of fine sediments from the Paraiba do Sul river. The sediment budget tends to be negative, dependant essentially on longshore sediment transport to maintain equilibrium. Interruption of this transport, as occurred at the Barra do Furado jetties (Fig. 14.3), has resulted in substantial progradation of the shoreline south of the jetties and substantial erosion along almost 10 km of the beach-barrier on the downdrift side of the jetties (Muehe et al. 2006).

Most of the eroding beaches are inaccessible except by off-road vehicles. Therefore concern first focuses on places near urbanized areas undergoing severe erosion as in Atafona at the Paraiba do Sul river outlet (Fig. 14.8). Here a 4 km long coastal segment, has receded about 240 m in 45 years (from 1954 to 2000; Ribeiro 2007). Tens of houses have been destroyed both by erosion and by foredune migration driven by the strong northeasterly winds.

Fig. 14.8 Atafona beach, at the Paraíba do Sul river outlet, showing severe erosion of the coastline (Photo: Dieter Muehe)

14.4.1.2 Rio das Ostras to Cape Búzios Coastal Plain

The coast between Rio das Ostras and Cape Buzios (Fig. 14.9) is about 30 km in length and dominated by a fluvio-marine coastal plain. It consists of an east-facing embayment with an almost continuous beach terminating at the tombolo of Cape Búzios. Two small interruptions occur, one at the São João river outlet at the town of Barra de São João, and the second one in the form of a small promontory near the southern end of the embayment. The former, approximately in the middle of the beach arc, strongly influences the morphology of the beach due to the input of fine river sediments, which are transported towards the southern end of the embayment. The finer sediments combined with wave height reduction due to the diffraction of waves around the Cape, has resulted in the beach state changing from reflective in the north to dissipative in the south.

Comparison of aerial orthophotographs for 1976 and 2000 (Lins-de-Barros 2010) revealed two segments experiencing coastal erosion at Virgem Beach and on the northern part of Abricó Beach (see location at Fig. 14.9), with the rest of the coast being stable during the period analysed. On Virgem Beach the retreat of the back beach averaged 8 m, with a maximum of 12 m. The same rate was found at Abricó Beach by Muehe et al. (2011) by comparing ortho-photographs and by monitoring beach profiles from 1999 to 2011 (Figs. 14.10 and 14.11).

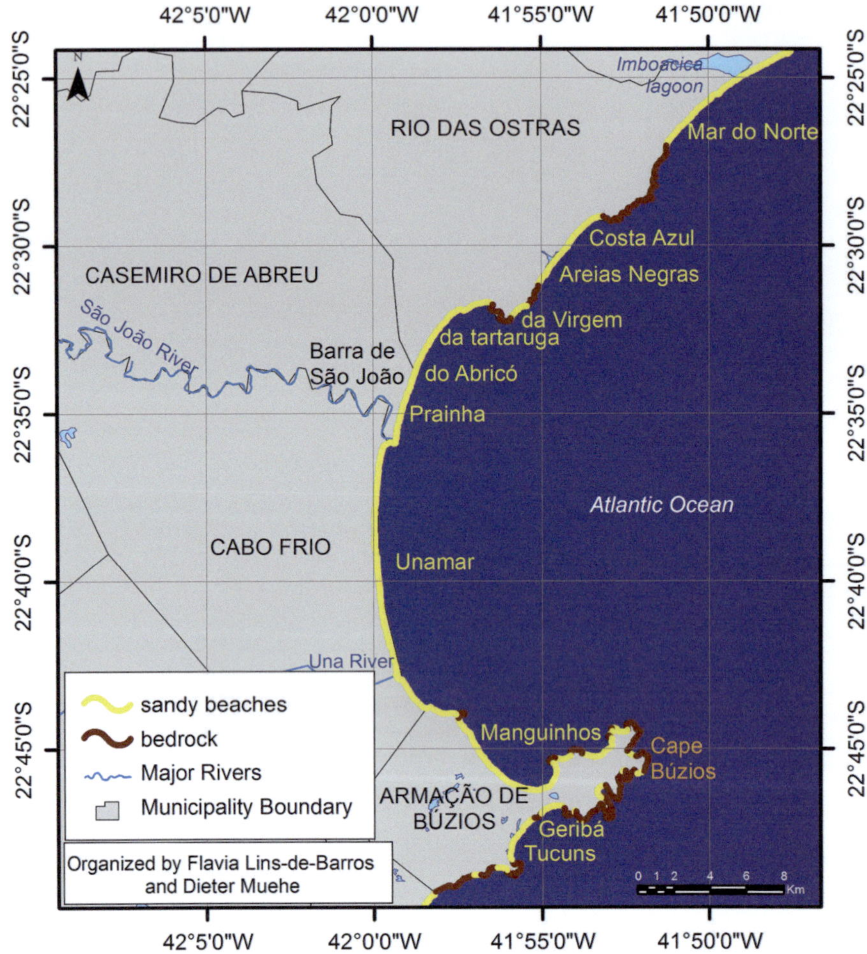

Fig. 14.9 The coastal segment between Rio das Ostras and Cape Búzios

14.4.1.3 The Pocket Beaches and Beach-Dune Systems from Cape Búzios to Arraial do Cabo (Cape Frio)

Cape Búzios and Arraial do Cabo are characterized by the alternation of basement rocky outcrops and Quaternary coastal plains. From Cape Búzios to Arraial three distinct morphological compartments occur comprising the beaches of Tucuns, Peró and Foguete in Cabo Frio (Fig. 14.12) backed by dune fields driven by the north-easterly trade winds and orientated obliquely to the shoreline (Muehe et al. 2010). Foredune-beach profiles measured at each beach arc show the intermediate beach types with well-developed foredunes in each of the locations (Fig. 14.13).

Fig. 14.10 Erosion of the back beach at Abricó Beach (Photo: Flavia Lins-de-Barros)

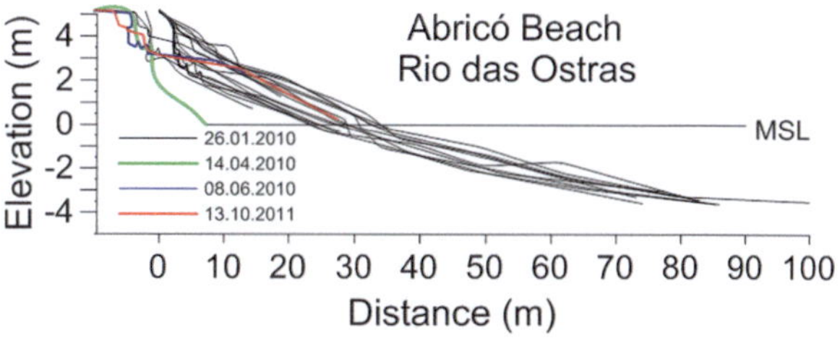

Fig. 14.11 Beach profile at Abricó Beach, Rio das Ostras, between 1999 and 2011 (Source: Muehe et al. (2011))

Due to the indented coastline both Búzios and Arraial do Cabo favour development of embayed or pocket beaches (Fig. 14.14). In Búzios the varied exposure to storm and fair weather conditions has resulted in a diversified morphodynamic behaviour of the beaches located around the cape, as shown by Bulhões (2011) and Bulhões et al. (2013) who identified a north, and a south sector. While the former is protected against the southern storms, the latter is largely exposed to the high-energy swell and storm waves from the south. Examples of these environments are the low energy, morphologically stable reflective Canto Beach and the erosional intermediate Brava Beach of the north sector and the high energy, storm exposed, episodically erosive intermediate Geribá Beach of the south sector (Fig. 14.15; Bulhões et al. 2013).

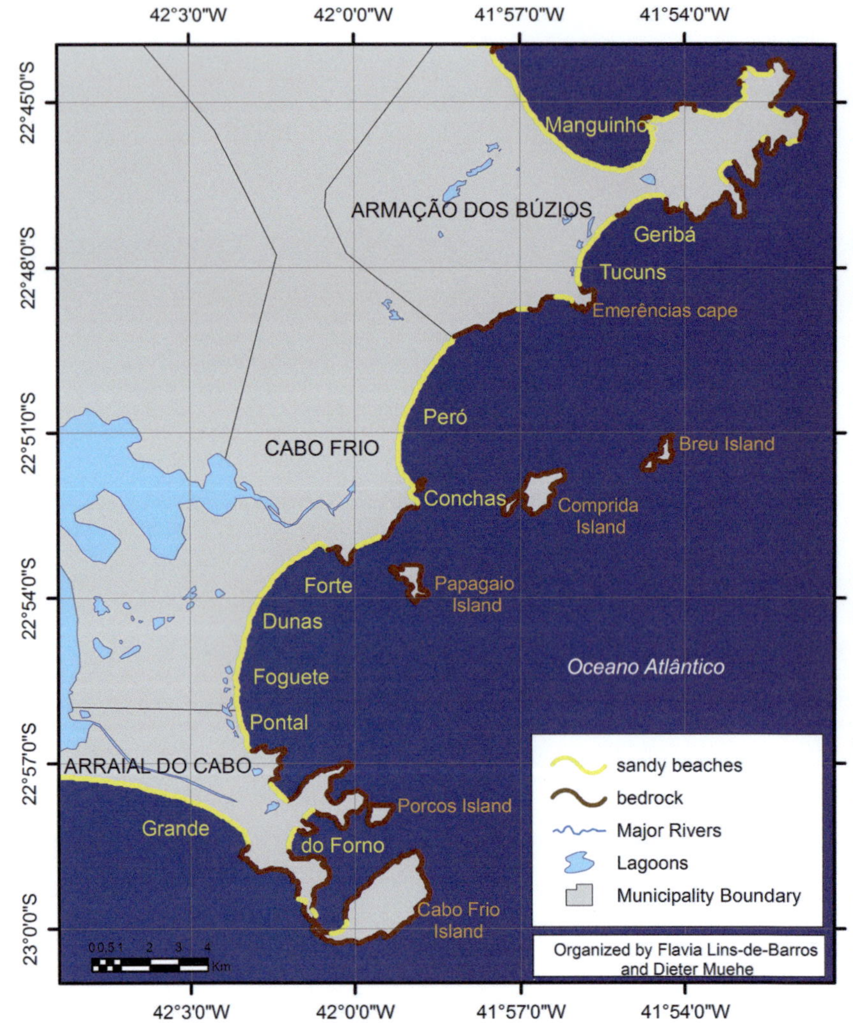

Fig. 14.12 Location of the different beach arcs with beach-shoreface profiles

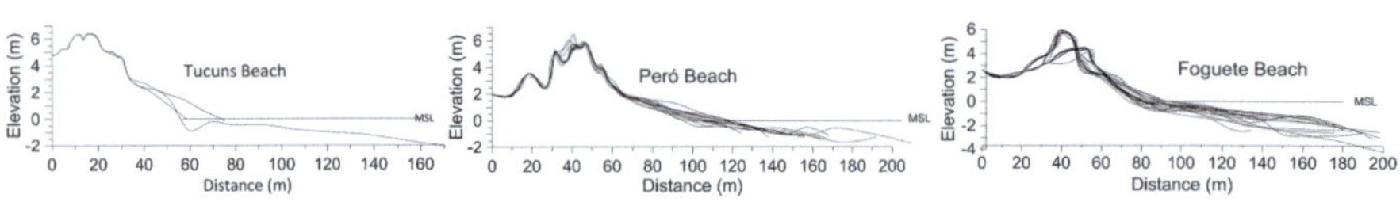

Fig. 14.13 Beach and foredune profiles of the beaches of Tucuns, Peró and Foguete (See location in Fig. 14.12)

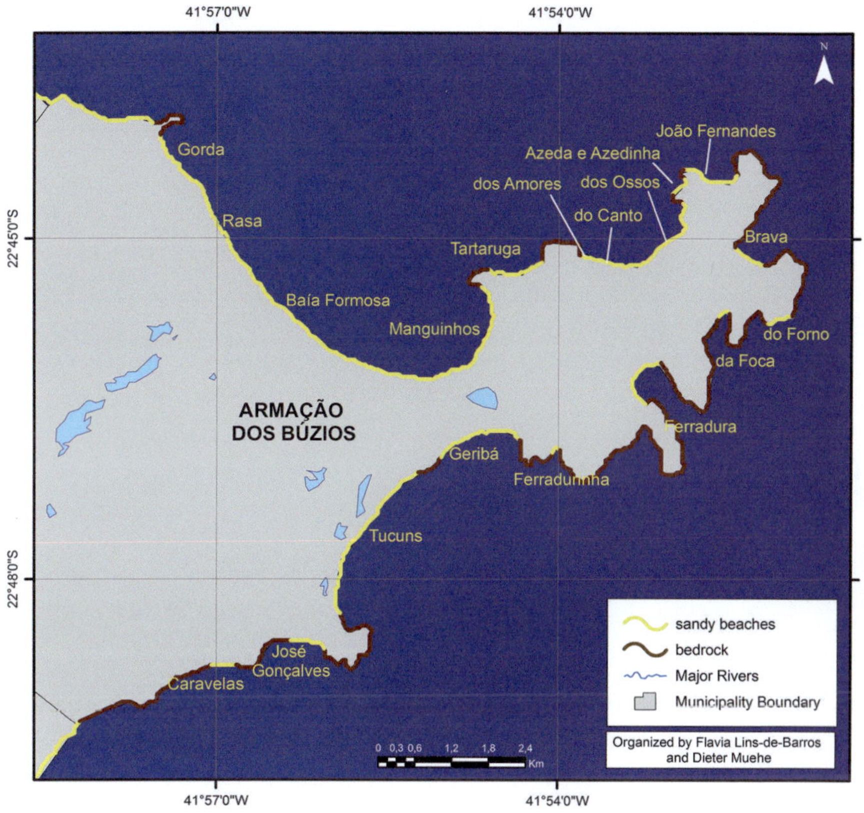

Fig. 14.14 The indented coastline of Cape Búzios with the north and south sectors exposed, respectively, to moderate and storm wave environments

Lins de Barros (2010) compared aerial photographs around Búzios between 1976 and 2000, commented on the shoreline stability of several beaches. However, the extremely rapid growth of the city and its encroachment on some beaches and dunes represent a risk during storms, especially on more exposed beaches such as Geribá.

Three embayed beaches composed of very fine quartz sand are located around Arraial do Cabo (Fig. 14.16). Prainha Beach, exposed to the northeast, is an intermediate with a usually Low Tide Terrace beach, while Forno and Anjos Beach are exposed to the southeast but protected from storm waves by the Cabo Frio Island which forms the Cabo Frio (cold cape), the name due to the upwelling of cold water in response to the strong and persistent northeasterly wind. Due to their relatively deep water and protected site on the northeastern sector of Anjos Beach the embayment was chosen for the location of a small port, a marina for fishing boats and an anchorage for small fishing and pleasure boats (Fig. 14.17). Beach profiles

Fig. 14.15 Geribá Beach, an exposed beach with foredunes and restinga vegetation (Photo: Dieter Muehe)

measured on Anjos Beach, at monthly intervals in 2001 and at irregular intervals in 2005 and 2006, show a reflective beach state in the northeastern segment of the beach with occasional intermediate state in the southwestern extremity where wave energy is higher (Fig. 14.18).

On Cape Frio sediment transport is northwards resulting in a decrease in grain size from medium to fine and very fine sand, which favours the development of higher foredunes in the north sector. A good relation between the above-mentioned decrease in grain size and beach-dune morophology was reported by Fernandez et al. (2006) on the beach arc between Cabo Frio and Arraial do Cabo where an initial reflective beach changed gradually toward the north from a Low Tide Terrace to a Longshore Bar-Trough to a Dissipative domain with an associated increase in the foredune height and the development of a considerable dune field at the distal end of the sediment transport cell (Fig. 14.19).

14.4.2 The South Coast

The south coast extends from Arraial do Cabo (Cabo Frio) in the east, down to the border with São Paulo, in the west. The most striking features of this coastal compartment are the long, straight often double barriers separated by interbarrier

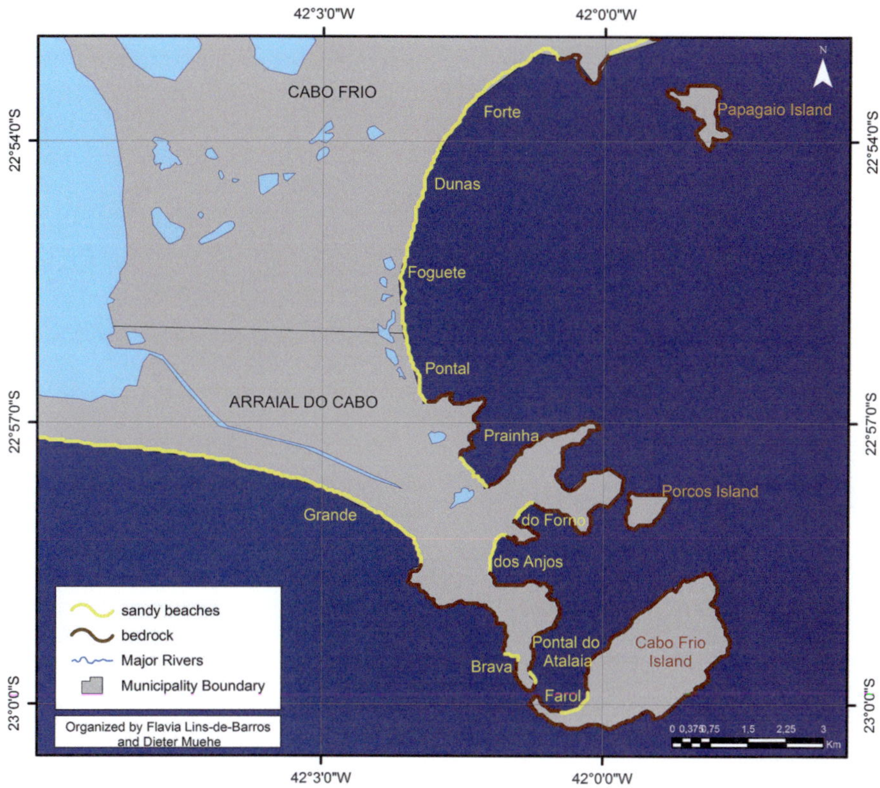

Fig. 14.16 Location of Arraial do Cabo and Cabo Frio

lagoons (Fig. 14.20). The origin and age of these barriers has been intensely debated since the 1940s by various authors such as Lamego (1940, 1945), Ponçano et al. (1979), Suguio and Martin (1981), Ireland (1989), Martin and Suguio (1989), Muehe and Corrêa (1989) and Turcq et al. (1999). The accepted theory at the present time is that the barriers migrated by landward translation to their present position, the inner barrier being related to the penultimate Pleistocene marine transgression and the outer barrier being associated with the Holocene transgression.

The east-west orientation of the coastline, unlike the generally northeast-southwest oriented eastern Brazilian coast, exposes the beaches to the high southerly waves and strong winds associated with cold fronts. The beaches are mainly composed of very mature quartz sands, a result of an almost total absence of input of terrigenous sediments due to the blocking of coastal drainage by the beach barriers. The beaches are long with only occasional interruptions by rocky promontories. Longshore sediment transport tends to move in both directions along the beaches, alternately driven by the southeast and southwest waves with a long-term trend to equilibrium with no net transport.

Fig. 14.17 Anjos Beach with marina and anchorage (Photo: Flavia Lins-de-Barros)

Fig. 14.18 Anjos Beach in Arraial do Cabo is a reflective very fine sand beach with well developed foredune

14.4.2.1 Barrier Beaches of the Lake Region from Arraial do Cabo to Itaipuaçu

The landscape of the Lake region is characterized by a series of small lakes and back barrier lagoons, whose presence led to the designation of the area. Araruama, the largest lagoon, (Fig. 14.20), is situated behind an inner Pleistocene barrier, while the smaller ones are found in the interbarrier depression between the inner and outer barriers, and completely isolated from any fluvial contribution.

Topographic cross profiles measured at various locations show the double barriers (Fig. 14.21). Note the different shapes of the landward profile of the Pleistocene

Fig. 14.19 Aerial photo of the beach arc between Cabo Frio and Arraial do Cabo with indication of the direction of both longshore and aeolian sediment transport (Photo: Ernesto Galiotto)

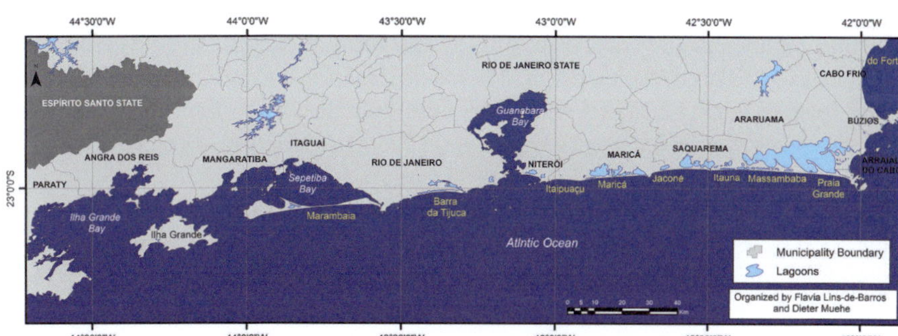

Fig. 14.20 Location of beaches from Cape Frio to Ilha Grande Bay

barrier. While in Vilatur (near Saquarema) the profile slopes continuously backwards, on Praia Seca there is a sharp erosional rupture due to the retrogradation of the barrier owing to the expansion of the Araruama lagoon. In Brejo do Espinho the dune field drops towards the lagoon while in Gaivotas (on the eastern half of Massambaba Beach) only the Holocene barrier is present, possibly due to erosion of the older barrier (Muehe 2006).

Muehe and Corrêa (1989) observed occasional overwashing of the outer barrier by storm waves, as well as a backshore erosional scarp. Further, the frontal barrier is narrow compared to the older one (from about 50 m up to 400 m in width) and

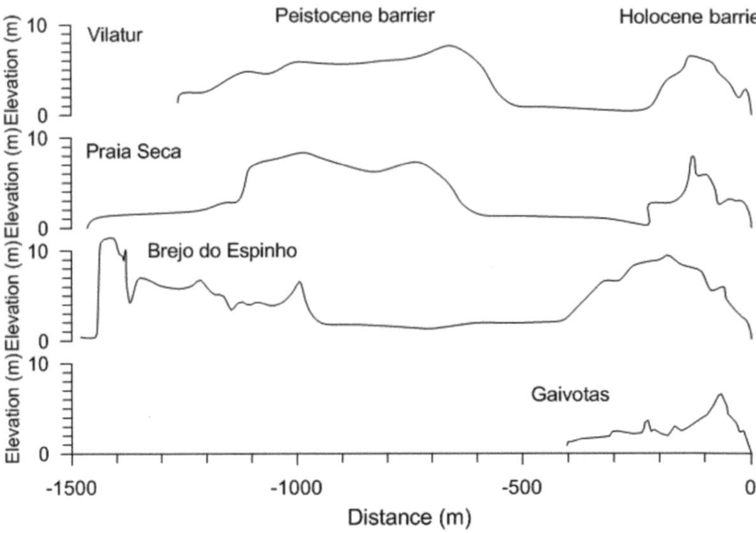

Fig. 14.21 Topographic cross profiles showing the double and single barriers with the depression filled by lagoons. The Araruama Lagoon is behind the Pleistocene barrier while the ocean beach is in front of the Holocene barrier for location see Fig. 14.22 (Source: Muehe 2006)

low with heights varying from 5 to 10 m, reinforcing their transgressive character and reflecting the high wave-energy. The occurrence of overwash deposits on this barrier, formed during a higher sea-level stand than at present one, indicates what might happen if there is a sea-level rise or an increase in storm wave frequency or subsidence.

The grain size varies from very coarse grained (1.0 mm) reflective beach of Itaipuaçu in the west to the fine grained (0.18–0.25 mm) intermediate beach of Massambaba in the east. This pattern is more a result of the foreshore and inner continental shelf grain size distribution than to longshore sediment transport, as shown by Muehe and Corrêa (1989) in a study of the Massambaba beach. In spite of the great variation in the beach width and consequently sediment volume, together with occasional destruction of houses and a road on top of the barrier, the beaches have shown a remarkable resilience in returning to their previous position (Muehe 2011).

Long-term sequential beach profiles have been surveyed on Massambaba beach at Monte Alto and Vilatur, near Saquarema, (Fig. 14.22). In spite of a 10–12 m retreat of the backshore scarp during a severe storm in 2001, the beach has always returned to their previous position (Muehe 2011). Considering the lack of continental sediment input, the source of the recovery must be found on the inner continental shelf. This is in accordance with the findings of Denny et al. (2013) and Schwab et al. (2013) for the coast of South Carolina and Fire Island, New York, respectively.

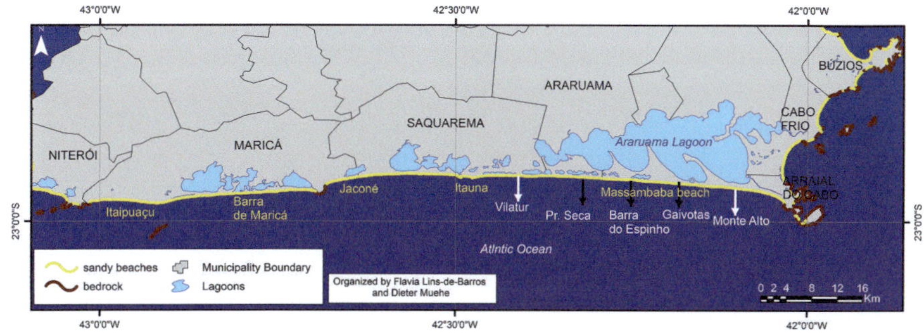

Fig. 14.22 Location of topographical cross profiles *(black arrows)* and long term beach profiles (*white arrows*) between Vilatur and Monte Alto near Arraial do Cabo (*black arrows*)

Fig. 14.23 Guanabara Bay entrance view from Cristo Redentor (Photo: Flavia M. Lins-de-Barros)

14.4.2.2 Guanabara Bay Sheltered Beaches

Guanabara Bay (Fig. 14.23) is the second largest bay on the Brazilian coast. It has an average depth of approximately 3 m in its most inland section, 8 m in its central part and 17 m in the entrance channel. The main sandy beaches are found near the entrance to the bay, with Flamengo, Botafogo and Urca, on the western side of the entrance and Icaraí, São Francisco and Jurujuba on the eastern side, in Niterói

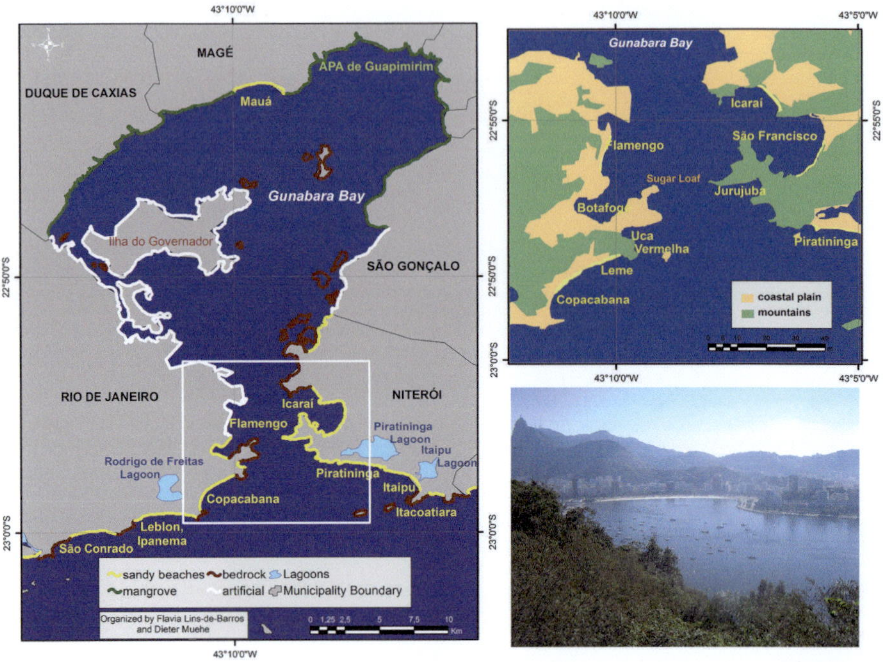

Fig. 14.24 Guanabara Bay and Botafogo cove at the western side of the bay entrance (enlarged) (Photo Flavia Lins-de-Barros)

(Fig. 14.24). These beaches are partially protected from storm waves though waves from the south can reach the beaches of Niterói, and exceptional storm waves from the southeast can reach the beaches of Botafogo and Flamengo.

The interior shoreline of Guanabara Bay has been modified by landfills including the area now occupied by the center of the city and its port. Artificial coastlines are also found at Niterói and São Gonçalo on the eastern side the bay. At the back of the bay, in the municipalities of Duque de Caxias and Magé, there is a predominance of mangrove swamps, which have been highly impacted by the pollution coming from the rivers, which drain the Baixada Fluminense where approximately three million inhabitants live. The only remnant of relatively well-preserved mangrove forest is to be found in the legally protected Environmental Area of Guapimirim.

14.4.2.3 Niterói and Rio de Janeiro Ocean Beaches

The cities of Rio de Janeiro and Niterói contain several beaches located within their densely populated urban nucleus. The urbanized oceanic margin of the municipality of Rio de Janeiro include Leme-Copacabana, Leblon-Ipanema, Vidigal, Barra da Tijuca-Recreio dos Bandeirantes, Prainha and Grumari (Fig. 14.25), these last two without any urban occupation. In Niterói, the urbanized coast covers the beaches of Piratininga, Itaipu and Itacoatiara (Fig. 14.26). These beaches have varying degrees

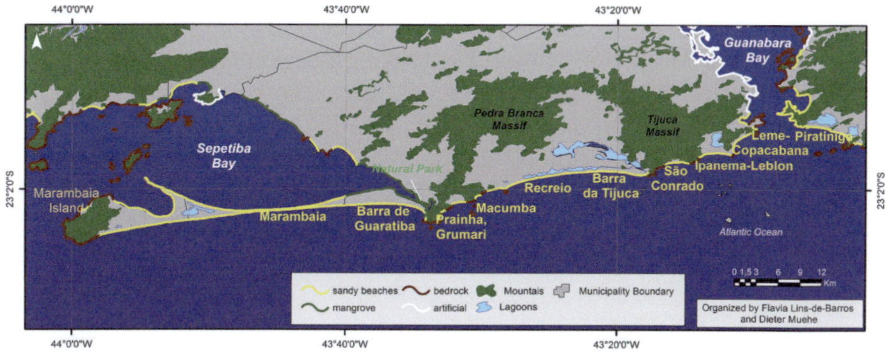

Fig. 14.25 Location of the ocean beaches of Niterói and Rio de Janeiro

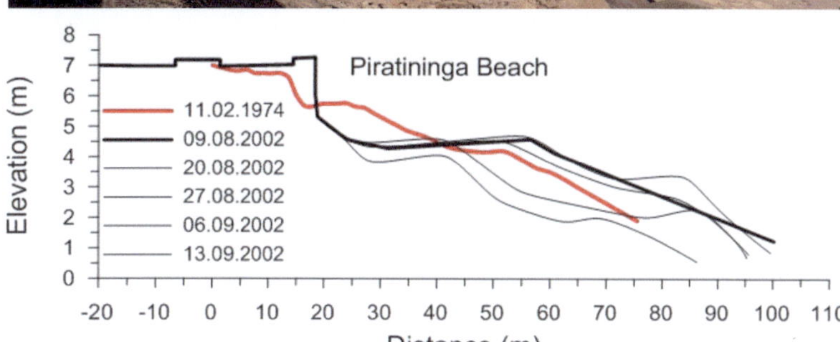

Fig. 14.26 Piratininga Beach after one of the several destructions of the wall. The beach still maintains its position and sediment volume (Photo: Dieter Muehe)

Fig. 14.27 Copacabana beach (Photo: Flavia Lins-de-Barros)

of exposure, with those lying near the entrance to Guanabara Bay, such as Piratininga (Niterói), Leme and Copacanana (Rio de Janeiro) being more sheltered.

Because they are located in heavily modified urban areas, the beaches of this sector have undergone many interventions including landfill, dredging and the building of seawalls. These modifications have changed the original condition of the beach morphodynamics especially of Copacabana Beach. In Niteroi, a seawall constructed on Piratininga Beach, in order to protect against erosion, has been destroyed and rebuild several times. Nevertheless, despite the presence of the wall the beach profile maintains its position and sediment volume of about 300 m^3/m (Fig. 14.26).

Copacabana Beach (Fig. 14.27) underwent substantial sand nourishment in the 1960s to increase the number of traffic lanes. This has been considered successful because the beach has, since then, remained stable. Sand was dredged from the inner continental shelf, near to the entrance of Guanabara Bay, and from the Botafogo inlet, in the interior of the Bay (Vera-Cruz 1972). The textural characteristics, such as rounding of the sand grains, were not identical with those of the original beach, which reduced the quality of the nourished sand. In addition, berm encroachment over the shoreface and increase in grain size contributed to an increase in shoreface gradient, which altered the beach morphodynamics. Prior to the enlargement of the beach, a former geology student who became later a distinguished marine geologist, surveyed the beach using the well-known Emery method. He identified the alternation of storm and fair weather in a work that became the first continuous (1 year) beach monitoring in Brazil (Kowsmann 1970).

To the west, Ipanema-Leblon beach is strongly influenced by alternation of easterly fair-weather and southerly storm waves and related reversal of the longshore sand transport. The two beaches are separated by the Jardim de Alah channel, which connects the Rodrigo de Freitas Lagoon to the sea (Fig. 14.28). Constant dredging is neces-

Fig. 14.28 The Ipanema-Leblon Beach in the foreground, and Copacabana Beach to the right. At the center of the image the Rodrigo de Freitas Lagoon (Source: Orthofoto 2008)

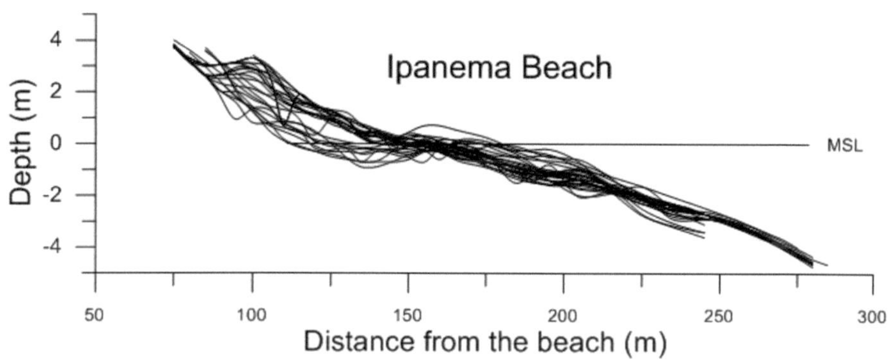

Fig. 14.29 Daily beach and shoreface profiles measured along the Ipanema pier from 7 to 18 November, 1977 (Source: Muehe and Dobereiner 1977)

sary to keep it open. During the building of a sewage system a pier constructed across the surf zone in Ipanema, was used as a platform for a daily monitoring of the bathymetry (Muehe and Dobereiner 1977). The profiles show the high variability of the surf zone morphology typical of an intermediate Bar-Trough beach state (Fig. 14.29).

The most extensive beach arc in the city of Rio de Janeiro and also the most exposed to the strong southerly waves begins at Barra da Tijuca Beach (Fig. 14.30), and includes the Recreio dos Bandeirantes Beach, and extends as far as Macumba Beach, with a total length of about 21 km. The beach is bordered in the east by the Joatinga Channel, which connects the Barra da Tijuca lagoon to the sea and to the west by the Sernambetiba Channel. This coastal segment belongs to the area known as Baixada de Jacarepaguá and similar to the segment between Maricá and Cape Frio it is composed of double barrier-lagoon system.

Grumari, Prainha, Funda, Praia do Meio, Perigoso, and Inferno are environmentally protected beaches with no human construction on them. They are enclosed by

Fig. 14.30 Barra da Tijuca Beach and lagoon. After the construction of a sewage system the pier depicted in the photo was removed (Photo: Flavia Lins-de-Barros)

the Guaratiba Precambrian mountain range, which also separates them from Marambaia Beach (Fig. 14.31). At the eastern end of the Marambaia Beach lies the Praia da Guaratiba, with a tidal channel, Barra de Guaratiba, connecting to Sepetiba Bay. Behind this beach there is an extensive mangrove swamp, which is part of the biological and archaeological reserve of Guaratiba.

The Marambaia beach barrier, known as *restinga da Marambaia*, is 42 km long and in its middle segment narrows to a single barrier located in front of Sepetiba Bay. Because it is a military area it is the best-preserved beach in Rio de Janeiro, as access to the public is restricted. A large dune field increases the height of the barrier but blowouts allowed localized overwash. Furthermore, the erosion on the lagoon side of the barrier has raised concern as to the possibility of a breach during a storm event with the consequent penetration of waves into the bay endangering the port located on the opposite shore.

14.4.2.4 Southern Beaches Influenced by the Serra do Mar Mountain Range

The southern coast of the State of Rio de Janeiro is strongly influenced by the proximity of the Serra do Mar mountain range, which has led to the formation of embayed and pocket beaches separated by rocky headlands and promontories as

Fig. 14.31 Marambaia beach and mangrove area of the Natural Park of Guaratiba (Photo: Flavia Lins-de-Barros)

well as the formation of various islands, the largest of which, the Ilha Grande, is situated in the municipality of Angra dos Reis (Fig. 14.32). There is also a narrow fluvial-marine plain developed along this coast with extensive mangrove forests and swampy areas of low-lying alluvial land, as in the back of the Parati, Ribeira and Mangaratiba bays. Sandy beaches are found, such as those at the mouth of the Mambucaba river (Figs. 14.32 and 14.33) or on Sul Beach, on the oceanic side of Ilha Grande.

14.4.2.5 Embayed Beaches from Paraty to Angra dos Reis

The extreme south of the state of Rio de Janeiro, on its border with São Paulo, has an extremely indented coastline which has, as its most notable features, the Juatinga point, an impressive peninsula mountain massif, with an altitude of more than 1000 m and which shades the Paraty parts of the Angra dos Reis coasts, and the Saco de Mamanguá (Fig. 14.34), a deep inlet in the rocky basement erroneously regarded as being a fjord. Numerous small embayed beaches occupy parts of the generally rocky shore.

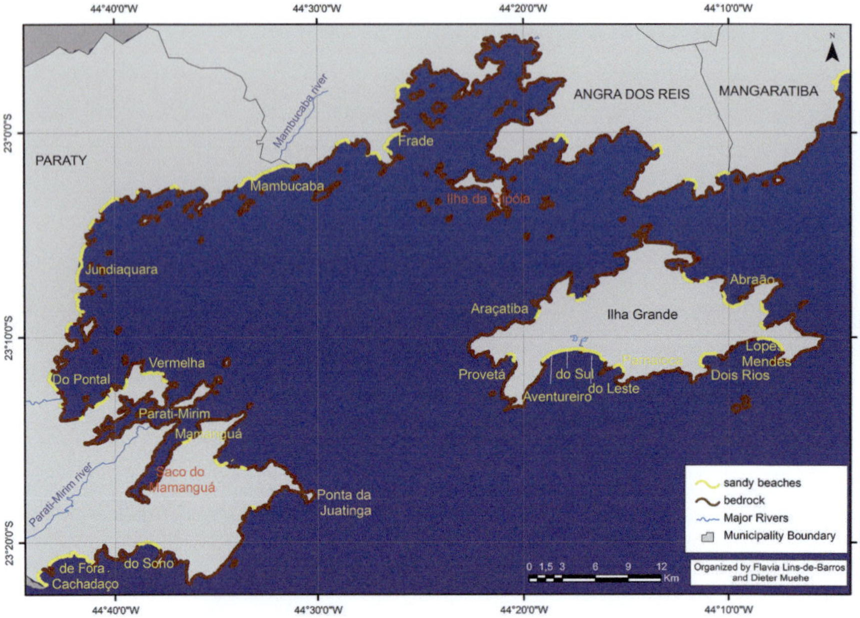

Fig. 14.32 Location of Parati and Angra dos Reis beaches

Fig. 14.33 Batanguera Beach, near Mambucaba Beach (Photo: Flavia Lins-de-Barros)

Fig. 14.34 A small and sheltered Beach inside the Mamanguá cove (Photo: Flavia Lins-de-Barros)

14.4.2.6 Ilha Grande

Ilha Grande is the largest island on the coast of Angra dos Reis. It has an area of 193 km^2 with a mountainous relief whose highest peaks are the Pico da Pedra D'Água (1031 m) and the Pico do Papagaio (982 m). More than 80 % of the land area of the island belongs to the Parque Estadual da Ilha Grande, an environment conservation unit created in 1981 in which no kind of occupation is allowed. The island consists of numerous peninsulas and inlets, and more than 20 embayed beaches.

In geologic and geomorphologic terms, the island is part of the same unit that gave rise to the Serra do Mar, the Coastal Massifs and the Guanabara Bay. It is composed predominantly of metamorphic rocks. Quaternary coastal plains are poorly developed, with the most extensive being located behind the exposed Leste and South Beach where remnants of a former lagoon system are found.

Other smaller Quaternary plains, such as at the beaches of Parnaioca, Dois Rios and Lopes Mendes (Fig. 14.35), have developed in the southeastern sector of the island. In the northern sector the beaches are embayed small and sheltered. The Vila do Abraão, the most populated settlement on the island, is situated in this sector.

Fig. 14.35 Lopes Mendes Beach on the south-facing side of Ilha Grande (Photo: Flavia Lins-de-Barros)

14.5 Summary and Conclusions

During his stay in Brazil in the middle of the past century, Andre Cailleux, a well-known French geologist, was surprised by the high proportion of well-rounded and polished quartz grains (*emmoussées luisants*) in the sand of the beaches of Rio de Janeiro. While a proportion of about 30 % of this kind of sand usually typifies beach sediment, the proportion he found was up to 80 % and more. These findings have been extensively confirmed by sediment analysis of beaches and inner shelf along the coast from Niterói to Arraial do Cabo (Muehe 1979). The reason for this high proportion is the almost absolute lack of continental sediment input to the shelf, as all the sand is trapped by the beach barrier-lagoon systems of the coast. Even in the east sector, above Cape Frio, sediments delivered by rivers are predominantly fine sands and silt bearing little relation to the beach sediments, which are basically reworked relict shelf sands. Nevertheless, in spite of the lack of a continental source of sediment, the beaches have shown a remarkable resilience in their ability to return to their previous position independently of back beach erosion. It seems therefore that the inner continental shelf constitutes a vital factor in the maintenance of the beaches. As beaches offer one of the most favored leisure options, as also a basic attraction for domestic and foreign tourism in view of the high quality of the beaches, there is concern as to how the coast will cope with the change in the climatic-oceanographic environment. Thus preliminary investigations into the potential source of borrowed material for the beach restoration carried out in Rio de Janeiro (Oliveira and Muehe 2013; Medeiros et al. 2014) have indicated that there are compatible offshore sediments to replenish the different urban beaches of the state.

Considering the evolution of coastal research the first systematic contributions dated back to the middle of the twentieth century with the publications of Alberto Ribeiro Lamego about the beach barrier-lagoon systems of Rio de Janeiro (Lamego 1940, 1945). The author interpreted the evolution of the barriers (*restingas*) as being spits whose eastward growth gradually closed the coastal embayments given rise to the lagoons. Only at the end of the 1970s and during the 1980s more detailed investigations became available (Muehe and Corrêa 1989; Muehe 1982, 1984; Ireland 1989 and Turcq et al. 1999) rejecting the hypothesis of lateral progradation in favour of the development of transgressive beach barriers formed by landward translation in response to a rising sea level. In the 1980s the first studies on the formation of the Paraíba do Sul coastal plain were published by Dias and Gorini (1980), Dominguez et al. (1981), Martin et al. (1984) and Dominguez (1989) Nowadays, in Brazil, the term *restinga* is used in relation to any morphologic deposit of marine sands.

Since then the number of authors, research projects and themes studying the beaches of the state of Rio de Janeiro have increased enormously. However, there are still some gaps. Outstanding are the need for chronological analysis, and geophysical investigations for the better understanding of the evolutionary processes of the coastal barrier. Surprisingly is an almost absence of basic studies of most of the urban beaches of Rio de Janeiro city. There is also a need for a coherent topographic and bathymetric monitoring program of the beaches and inner shelf together with oceanographic measurements. It also becomes increasingly necessary that investigations of the beaches should create an interface with ecological and social studies in order to achieve more adequate forms of management and preservation of these environments and their ecosystems.

References

Almeida FFM (1967) Origem e evolução da plataforma brasileira. DNPM/DGM, Rio de Janeiro, (Boletim 241) pp 1–36, apud. Silva LC (2011) Meso/Neoproterozóic. In: Silva LC,77 Cunha HCS (eds) Geologia do Estado do Rio de Janeiro. Texto explicativo do mapa geológico do Estado do Rio de Janeiro, chapter 2, 2nd edn. CPRM, 2001. Brasília. Available at http://www.cprm.gov.br/publique/cgi/cgilua.exe/sys/start.htm?infoid=612

Almeida FFM, Hasui Y, Brito Neves BB et al (1981). Brazilian structural provinces: an introduction. Earth Sci Rev 17:1–29, apud. Silva et al (2011) Meso/Neoproterozóic. In: Silva LC, Cunha HCS (eds) Geologia do Estado do Rio de Janeiro. Texto explicativo do mapa geológico do Estado do Rio de Janeiro, chapter 5, 2nd edn. CPRM, 2001. Brasília. Available at http://www.cprm.gov.br/publique/cgi/cgilua.exe/sys/start.htm?infoid=612

Bastos AC, Silva CG (2000) Caracterização morfodinâmica do litoral norte fluminense, RJ, Brasil. Rev Bras Oceanogr 48(1):41–60

Bittencourt ACSP, Dominguez JML, Flexor J-M et al (1982) Evolution du littoral brésilien au Quaternaire: influence des variations du niveau de la mer. In: Abstarcts of te Latin America regional conference. Comissiono n the coastal environment, IGU, Rio de Janeiro

Bulhões E (2011) Exposição de praias, clima de ondas e processos litorâneos em praias de enseada: exemplos e aplicações em Armação dos Búzios, Rio de Janeiro, Brasil. Doctor thesis, Universidade Federal Fluminense, p 230

Bulhões E, Fernandez GB, Rocha TB (2013) Morphodynamics of embayed beaches. Case study in Cape Buzios, Rio de Janeiro, Brazil. In: Conle DC, Masselink G, Russell PE, O'Hare TJ (eds) Proceedings of the 12th international coastal symposium (Plymouth, England). J Coast Res Special Issue 65:1739–1744

Denny JF, Schwab WC, Barnhardt WA et al (2013) Holocene sediment distribution on the inner continental shelf of northeastern South Carolina: implication for the general sediment budget and long-term shoreline response. Cont Shelf Res 56:56–70, http://dx.doi.org/10.1016/j.csr.2013.02.004

Dias GTM, Gorini, MA (1980) Evolution du complexe deltaique du fleuve Paraíba do Sul, Brésil. In: Abstracts of the 26 Congres Geologique International, Paris, p 458

Dominguez JML (1989) Ontogeny of a strandplain – evolving concepts on the evolution of the Doce River beach-ridge plain (east coast of Brazil). In: Extended abstracts of the international symposium on global changes in South America during the Quaternary: past-present-future. ABEQUA/INQUA, vol 1, São Paulo, pp 235–240

Dominguez JML, Bittencourt ACSP, Martin L (1981) Esquema evolutivo da sedimentação quaternária nas feições deltaicas dos rios São Francisco (Se/Al), Jequitinhonha, (Ba), Doce (ES) e Paraíba do Sul (RJ). Rev Bras Geoc 11(4):227–237

Fernandez GB, Pereira TG, Muehe D, Rocha TB (2006) Aplicação de critérios morfodinâmicos na diferenciação de setores ao longo do arco praial entre Cabo Frio e Arraial do Cabo – RJ. VI Simpósio Nacional de Geomorfologia (SINAGEO), Goiânia, 6–10 September 2006

Ferrari A (2012) Geologia. In: Serra M (ed) Guia de História Natural do Rio de Janeiro. Cidade Viva/Instituto Cultural da Cidade Viva, Rio de Janeiro, p 281

Figueiredo MCH, Campos Neto MC (1993) Geochemistry of the Rio Doce magmatic arc, southeastern Brazil. An Acad bras Ciênc 65(Supl 1):63–81, apud. Silva et al (2011) Meso/Neoproterozóic. In: Silva LC, Cunha HCS (eds) Geologia do Estado do Rio de Janeiro. Texto explicativo do mapa geológico do Estado do Rio de Janeiro, chapter 5, 2nd edn, CPRM, 2001. Brasília: Available at http://www.cprm.gov.br/publique/cgi/cgilua.exe/sys/start.htm?infoid=612

Flexor J-M, Martin L, Suguio K et al (1984) Gênese dos cordões litorâneos da parte central da costa brasileira. In: Lacerda LD, Araújo DSD, Cerqueira R, Turcq BJ (eds) Restingas: origem, estrutura, processos. Universidade Federal Fluminense, Niterói, pp 35–45

Ireland S (1989) The Holocene sedimentary history of the coastal lagoons of Rio de Janeiro State, Brazil. In: Tooley MJ, Shennan I (eds) Sea-level changes, The institute of Britsh geographers special publications series 20. Blackwell, Oxford, pp 25–66

Kowsmann RO (1970) Variações de curto e longo prazo de um perfil de praia em Copacabana, Rio de Janeiro. Instituto de Pesquisas da Marinha 39:1–15. Rio de Janeiro

Lamego AR (1940) Restingas na costa do Brasil. Departamento Nacional da Produção Mineral, Divisão de Geologia e Mineralogia, Rio de Janeiro, p 63

Lamego AR (1945) Ciclo evolutivo das lagunas fluminenses. Departamento Nacional da Produção Mineral, Divisão de Geologia e Mineralogia, Rio de Janeiro, Bol 118, p 48

Lins-de-Barros F (2010) Contribuição metodológica para análise local da vulnerabilidade costeira e riscos associados: estudo de caso da Região dos Lagos, Rio de Janeiro. Doctor thesis, Universidade Federal do Rio de Janeiro, p 297

Machado GMV (2010) Análise morfossedimentar da praia, antepraia e plataforma continental interna da linha de costa do Parque Nacional de Jurubatiba – Rio de Janeiro. Quat Environ Geosci 2(1):1–17. Available at http://ojs.c3sl.ufpr.br/ojs2/index.php/abequa/article/view/13816/11703

Martin L, Suguio K (1989) Excursion route along the Brazilian coast between Santos (State of São Paulo) and Campos (North of State of Rio de Janeiro). International symposium on global changes in South America during the quaternary. Special publication n 2, 136p
Martin L, Suguio, K, Dominguez et al (1984) Evolução da planície costeira do Rio Paraíba do Sul (RJ) durante o Quaternário: influência das flutuações do nível do mar In: Annals of the 33 Congresso Brasileiro de Geologia, Rio de Janeiro 1:84–97
Medeiros N, Dias MS, Ayres Neto A et al (2014) Mapeamento acústico de areias submersas para recuperação de praias do Rio de Janeiro, Brasil. Rev Gestão Costeira Integr 14(1):149–158. doi:10.5894/rgci429
Melo E (1993) The sea sentinel project: watching waves in Brazil. Paper presented at the Proceedings of the VIII symposium on coastal and ocean management – coastal zone 93, New Orleans, 19–23 July 1993
Muehe D (1979) Sedimentology and topography of a high energy coastal environment between Rio de Janeiro and Cabo Frio – Brazil. An Acad Bras Cênci 51(3):473–481
Muehe D (1982) Evidence of landward translation of beach barriers east of Guanabara bay. In: Latin American regional conference, Rio de Janeiro. Abstracts. International Geographical Union
Muehe D (1984) Evidências de recuo dos cordões litorâneos em direção ao continente no litoral do Rio de Janeiro. In: Lacerda LD, Araújo DSD et al (eds) Restingas: origem, estrutura, processos. Annals of the Simpósio sobre restingas brasileiras, Universidade Fderal Fluminense, Niterói, pp 75–80
Muehe D (1998) Estado morfodinâmico praial no instante da observação: uma alternativa de identificação. Rev Bras Oceanogr 46(2):157–169
Muehe D (2006) Gênese da morfologia do fundo da Lagoa de Araruama e cordões litorâneos associados. Abstracts of the Simpósio Nacional de Geomorfologia, Goiânia, 2006
Muehe D (2011) Erosão costeira: tendência ou eventos extremos? O litoral entre Rio de Janeiro e Cabo Frio, Brasil. Rev Gestão Costeira Integr 11(3):315–325. doi:10.5894/rgci282
Muehe D (2013) Erosão costeira, mudanças do clima e vulnerabilidade. In: Guerra AJT, Jorge MCO (eds) Processos erosivos e recuperação de áreas degradadas. Oficina de Textos, São Paulo, pp 160–189
Muehe D, Corrêa CHT (1989) Dinâmica de praia e transporte de sedimentos ao longo da restinga da Massambaba. Rev Bras Geoc 10(3):387–392
Muehe D, Dobereiner C (1977) Dinâmica do fundo marinho ao longo do pier de Ipanema, Rio de Janeiro. An Acad Bras Ciênc 49(2):281–286
Muehe D, Valentini E (1998) O litoral do Estado do Rio de Janeiro. Fundação de Estudos do Mar (FEMAR), Rio de Janeiro p 99
Muehe D, Lima CF, Barros FM (2006) Rio de Janeiro. In: Muehe D (ed) Erosão e progradação do litoral brasileiro, vol 1. Minstério do Meio Ambiente, Brasília, pp 265–296
Muehe D, Belligotti FM, Lins-de-Barros FM, Oliveira JF, Maia LFPG (2010) Potential vulnerability to climate change of the beach-dune system of the Peró coastal plain, Rio de Janeiro, Brazil. Pan-Am J Aquat Sci 5:267–276
Muehe D, Fernandez GB, Bulhões EMR et al (2011) Avaliação da vulnerabilidade física da orla costeira em nível local, tomando como exemplo o arco praial enre Rio das Ostras e o Cabo Búzios/RJ. Rev Bras Geomorfol 12(2):45–58
Oliveira JF, Muehe D (2013) Identificação de áreas de sedimentos compatíveis na plataforma continental interna para recuperação de praias entre as cidades de Niterói e Macaé – Rio de Janeiro, Brasil. Rev Gestão Costeira Integr 13(1):89–99. doi:10.5894/rgci362
Ponçano WL, Fúlfaro VJ, Gimenez AF (1979) Sobre a origem da baía de Sepetiba e das restingas da Marambaia, RJ. In: Annals of the Simpósio Internacional de Geologia, Rio Claro, 2:291–304
Ribeiro GP (2007) Aplicação da dinâmica do campo de dunas em Atafona, São João da Barra (RJ), como requisito para interpretação do processo de erosão costeira. Programa de Pós-Graduação em Geologia do Quaternário, Museu Nacional, Universidade Federal do Rio de Janeiro, Rio de Janeiro p139

Schmitt, RS, Trouw RAJ, Van Schmus WR (1999) The characterization of a Cambrian (~520 Ma) tectonometamorphic event in the coastal domain of the Ribeira Belt (SE Brazil), using U/Pb in syntectonic veins. In: Actas of the SEGEMAR, simpósio sudamericano de geologia isotópica, 2, Cordoba, pp 363–366, apud. Silva et al (2011) Meso/Neoproterozóic. In: Silva LC, Cunha HCS (eds) Geologia do Estado do Rio de Janeiro. Texto explicativo do mapa geológico do Estado do Rio de Janeiro, chapter 3, 2nd edn, CPRM, 2001. Brasília: Available at http://www.cprm.gov.br/publique/cgi/cgilua.exe/sys/start.htm?infoid=612

Schobbenhaus C, Campos DA, Derze GR et al (1984) Geologia do Brasil. Texto Explicativo do Mapa Geológico do Brasil e da área oceânica adjacente incluindo depósitos minerais, escala 1:2.500.000, DNPM-MME, Brasília, p 501, apud. Silva LC (2011) Meso/Neoproterozóic. In: Silva LC, Cunha HCS (eds) Geologia do Estado do Rio de Janeiro. Texto explicativo do mapa geológico do Estado do Rio de Janeiro, chapter 2, 2nd edn, CPRM, 2001. Brasília: Available at http://www.cprm.gov.br/publique/cgi/cgilua.exe/sys/start.htm?infoid=612

Schwab WC, Baldwin WE, Hapke CJ et al (2013) Geologic evidence on onshore sediment transport from the inner continental shelf: Fire Island, New York. J Coast Res 29(3):526–544, http://dx.doi.org/10.2112/JCOASTRES-D-12-00160.1

Stanton N, Schmitt R, Galdeano A et al (2010) Crustal structure of the Southeastern Brazilian margin, Campos Basin, from aeromagnetic data: new kinematic constraints. Tectonophysics 490:15–27

Suguio K, Martin L (1981) Progress in research on Quaternary sea level changes and coastal evolution in Brazil. Columbia, University of South Carolina. In: Proceedings of the variations in sea level in the last 15.000 years, magnitude and causes Congress. Columbia, pp 1–17

Turcq B, Martin L, Flexor JM, Suguio K, Pierre C, & Tasayco-Ortega L (1999) Origin and evolution of the Quaternary coastal plain between Guaratiba and Cabo Frio, State of Rio de Janeiro, Brazil. In: Knoppers BA, Bidone ED, Abrão JJ (eds), Environmental Geochemistry of Coastal Lagoon Systems of Rio de Janeiro, Brazil, vol. 6. Niterói: UFF/FINEP, pp 25–46

Vera-Cruz D (1972) Artificial nourishment of Copacabana Beach. In: Proceeding of the thirteenth coastal engineering conference, 2:1451–1463. ASCE, Vancouver, Canada, 10–14 July 1972

Chapter 15
The Beaches of the State of São Paulo

Michel Michaelovitch de Mahiques, Eduardo Siegle, Javier Alcántara-Carrió, Filipe Galiforni Silva, Paulo Henrique Gomes de Oliveira Sousa, and Cristina Celia Martins

Abstract The geological heritage of the coastal province of Sao Paulo state, characterized by the presence of the Serra do Mar mountain chain, combined with the hydrodynamics and resulting shoreline orientation and wave exposition defines six compartments with different kind of beaches. Thus, the relative position of the Serra do Mar mountain chain mainly controls the type of beach morphodynamics along the coast of the state of São Paulo. It is closer to the littoral in the North, leading to the dominance of reflective beaches, and further in the South, where dissipative beaches, associated with Late Quaternary coastal plains prevail. In the northern sector, wave energy is lower due to the presence of São Sebastião Island and a very irregular shoreline. In contrast, the southern sector is characterized by a large coastal plain and beaches exposed to most of the incident waves. Two compartments have been defined in the northern sector and three more in the southern one, with a transitional compartment among them. Moreover, the area as a whole is intensely occupied, either by an industrial zone (i.e. the Santos harbour area) or by tourism activities.

15.1 Introduction

15.1.1 Geological and Geomorphological Framework

The São Paulo coast occupies the centre of the São Paulo Bight, which corresponds to the arc-shaped part of the southern Brazilian margin (Zembruscki 1979). Its geomorphology is strongly controlled by the presence and the relative position of

M.M. de Mahiques (✉) • E. Siegle • J. Alcántara-Carrió
F.G. Silva • P.H.G.d.O. Sousa
Oceanographic Institute of the University of São Paulo (IOUSP),
Praça do Oceanográfico 191, 05508-120 São Paulo, SP, Brazil
e-mail: mahiques@usp.br; esiegle@usp.br; javier.alcantara@usp.br;
filipe.galiforni.silva@gmail.com; sousaph@gmail.com

C.C. Martins
Fundação de Estudos e Pesquisas Aquáticas, Fundação de Estudos e Pesquisas Aquáticas,
Avenida Afranio Peixoto, 412, Butantã, 05507-000 Sao Paulo, SP, Brazil
e-mail: cristina@fundespa.org.br

A.D. Short, A.H. da F. Klein (eds.), *Brazilian Beach Systems*, Coastal Research Library 17, DOI 10.1007/978-3-319-30394-9_15

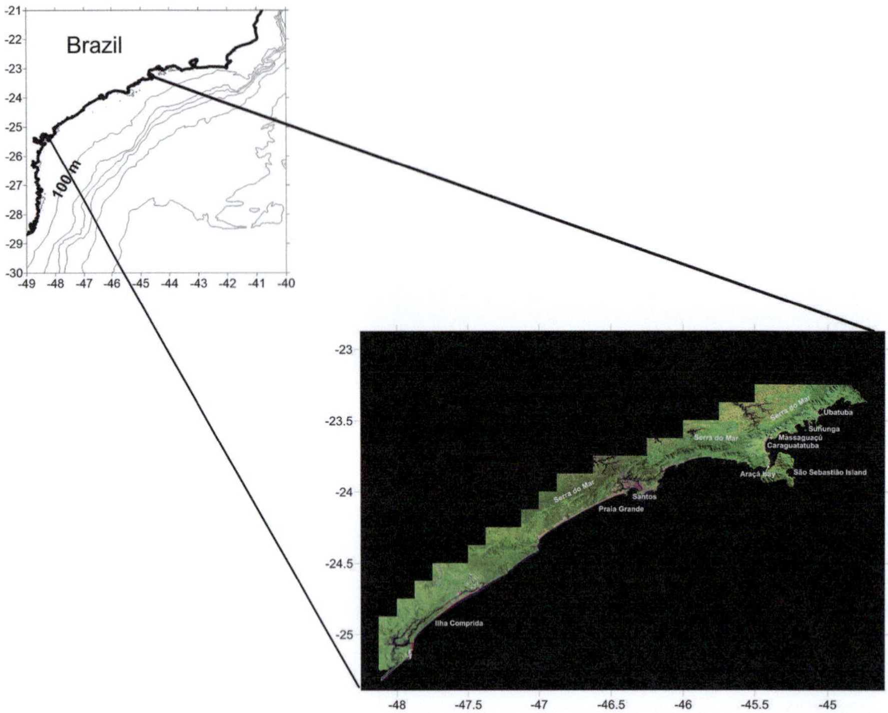

Fig. 15.1 Location and mosaic of satellite images of the São Paulo coast, showing the beach compartments discussed in this chapter

the Serra do Mar crystalline massif. This coastal range is associated with the development of the Santos Basin since the Juro-Cretaceous (146 Ma) during the rifting and breakup of the Gondwana supercontinent. The progressive uplift of the Serra do Mar Mountain chain since the late Cretaceous led to the re-direction of the drainage systems westwards to the Paraná Basin, with only small rivers run directly to the coast, leading to a restricted input of fluvial sediments to the coastline.

Moreover, the close approach of the mountains to the sea, in the northernmost portion of the region, results in a highly irregular coastline with islands and numerous headland embayed beaches, which typifies this sector. In contrast, in the southern bight, the greater distance of the mountain chain from the sea favours the development of longer coastal plains, interrupted by estuarine channels. The relative position of the Serra do Mar is also reflecting the coastal geology, with lower grade metamorphic rocks prevailing in the south, and granites, gneisses and migmatites being present in the north. Therefore, the São Paulo coast is divided into two sectors, with a general boundary close to the city of Santos, corresponding to the transition between the Coastal Plain and Estuaries Coast to the south, and the northern Scarped Coast (Litoral das Escarpas Norte) (Besnard 1950a, b; Muehe 2012) (Fig. 15.1).

The adjacent ocean floor also reflects the relative position of the Serra do Mar mountain chain. In the south, the isobaths run parallel to the coast, with a more complex morphology in the north, where the presence of islands and submerged rock outcrops are more common.

15.1.2 Holocene Sea-Level Changes and the Formation of Coastal Plains

Although the first references to Holocene palaeo-sea-levels in Brazil were made nearly a century ago (Hartt 1870; Branner 1904), more systematic studies started in the mid-1960s. Since then, more than 100 publications focusing on Holocene sea-level history have been published. During the 1970s and 1980s, based on thousands of radiocarbon dates, Holocene relative sea-level change curves were determined for areas between 5°S and 34°S (Angulo et al. 2006 and references therein).

The sea-level curves for the eastern Brazilian coast for the middle and late Holocene were originally defined by various reconstructions of sea-levels, both in space and over time, obtained from a database of more than 700 radiocarbon dates (Suguio et al. 1985; Angulo and Suguio 1995). Based on these data, paleo-sea-level trends have been determined for several sectors of the Brazilian coast.

A mid-Holocene sea-level highstand, followed by a drop towards the present time, is evident in these studies, although the Holocene relative sea-level history still has many controversial aspects. The main debates concerning the paleo-sea-level for the last 7000 years include the uncertainty in the elevation of the maximum highstand and the presence or absence of high-frequency sea-level oscillations (Martin et al. 2003; Angulo et al. 2006).

One model proposes two or three periods of sea-level lower than the present after 5600 year BP (Suguio et al. 1985; Angulo and Suguio 1995), and a second model, based entirely on vermetidae tubes, assumes that the present sea-level constitutes the lowest level of the last ca. 7000 years (Angulo and Lessa 1997; Angulo et al. 1999). According to these latter authors, the great majority of the indicators used to infer the secondary sea-level oscillations in previous studies derive from shell middens, which can be unreliable paleo-sea-level indicators. The latter sea-level model has since been corroborated by Ybert et al. (2003).

Martin et al. (2003) produced the first reservoir-corrected, astronomically calibrated sea-level change curve for a sector of the Brazilian coast. The results obtained by these authors suggest the occurrence of three main events of submergence of the coast (7800–5600, 3700–3500 and 2300–2100 year BP) interspersed among periods of emergence. Controversially, Angulo et al. (2006) suggest that a progressive decline in the sea-level occurred since the mid-Holocene maximum, based on a large data set.

The relationship between Holocene sea level changes and coastal plains evolution on the coast of the State of São Paulo was reported in several papers, including the work of Suguio and Martin (1978). Nevertheless, this work deals with a more regional evolutionary model. On the other hand, detailed work on evolution of coastal Holocene deposits have been done on Comprida Island, a "false" 70 km-long barrier island, since its deposits are essentially of regressive origin (Nascimento et al. 2008; Giannini et al. 2009; Guedes et al. 2011).

15.1.3 Regional Climate

The São Paulo coast is located near the tropical and subtropical boundary and its climate is classified as Tropical Atlantic. Changes in the wind and rainfall regimes in the region have been attributed to variations of the anticyclone associated with the South Atlantic Subtropical High (SAS), which is responsible for the dominance of northeast winds both in spring and summer (Bastos and Ferreira 2000), and the northward penetration of polar cold fronts. The cold front systems are active throughout the year, and have a strong influence on temperature and rainfall regimes (Nobre and Shukla 1996).

The rainfall regime in the southern Brazilian region follows a seasonal pattern with regional precipitation during the winter and early spring (May to September) mainly due to the extratropical circulation regime as a result of migratory cyclones along the subtropical Atlantic coast (Vera et al. 2002). Conversely, summer precipitation (September to April) is associated with the activity of the South American Summer Monsoon (SASM). During this time, the SASM is associated with the South Atlantic Convergence Zone (SACZ), which is responsible for the intensity and location of the summer precipitation. The SACZ exhibits significant variations in terms of intensity and geographical extensions at different time scales (Paegle and Mo 2002). The monsoon starts to decay when the convection changes northward to the equator, following the decrease of solar heating in the South American subtropical zone.

Although the SASM is weaker and has a shorter life span than the Asian monsoon system, it is still responsible for more than 80 % of the annual mean precipitation in its activity centre (23°S) and for 50 % of the total summer precipitation in the coast southward of 25°S (Cruz et al. 2009).

Other climatic changes have been attributed to many factors, such as variations in the intensity of El Niño-Southern Oscillation (ENSO) (Martin et al. 1993); latitudinal shifts of the ITCZ (Jaeschke et al. 2007); and latitudinal changes in the Meridional Overturning Circulation (Cruz et al. 2009).

The weather dynamics of the southwest Atlantic results mainly from the advance of the Migratory Polar Anticyclone (MPA), brought about by the displacement and weakening of the Tropical Atlantic Anticyclone (TAA), which predominates in the region during most of the year. During these events, the winds alternate from the northeast and east to the southeast and southwest, and they are preceded by low-pressure systems. The displacement of these systems, followed by MPAs, causes the highest energy events along the southern and southeastern Atlantic coasts (Monteiro 1969; Sant'Anna Neto 1990).

There are two preferential zones for cyclogenesis in southern South America (Innocentini and Caetano Neto 1996). The first is located to the east of the Andean Cordillera, around 43°S, and the second over Uruguay (33°S). During the autumn and winter months especially, the cyclones that develop in this latter zone are displaced eastwards over the South Atlantic, where they develop into storms.

15.1.4 Sediment Sources

Due to the orientation of the Serra do Mar mountain range and its location close to the coast the drainage basins that run to the coast are very limited in size and volume of water. The most important basin that runs directly to the coast is the Ribeira River, with an area of approximately 25,000 km^2. The outflow of the river in its lower course varies from approximately 300 to more than 1200 $m^3.s^{-1}$, with the variation in flow influenced strongly by the subtropical humid climate. Other river basins present outflows which barely exceed 50 $m^3.s^{-1}$.

On the southern sector the coastal sediments are mainly derived from the reworking of the Late Pleistocene and Mid-Holocene transgressive fine sands, which presently cover the coastal plains. Coarser sediments, present on the northern sector reflective beaches are essentially originated by the wave erosion of the adjacent rock outcrops or by local drainages.

Biogenic sediments are scarce, corresponding solely to patches of shells fragments, mainly located close to rock outcrops, where hydrodynamic conditions do not allow the deposition of siliciclastic sediments.

15.1.5 Human Disturbance

The coast of the State of São Paulo is probably the most affected by anthropogenic activities from the whole Brazilian coastline. Thus, the presence of the biggest harbour of Latin America, associated with a decadal occupation related to tourism and nautical activities and, more recently, due to a flourishing oil and gas exploration on the adjacent margin, has led to considerable modification of the coast. These activities result in severe impacts to the beach environments.

15.2 The São Paulo Beach Systems

15.2.1 Coastal Hydrodynamics

The coastal processes acting on the São Paulo beaches are closely linked to the variations in the incoming waves caused by the South Atlantic Tropical Anticyclone (SATA) and MPA. The SATA acts as a semi-fixed system, moving north–south (between the 15°N and 5°S) with maximum atmospheric pressure between 1020 and 1023 mb and high temperatures. This system is responsible for the northeasterly winds with constant frequency throughout the year with maximum intensities between the 10° and 40°S. It generates low energy northeast and east waves, compared to the MPA waves (Tessler and Cazzoli y Goya 2005; Pianca et al. 2010).

The MPA is formed by cold air masses in sub polar latitudes (60°S). It has high atmospheric pressure (1014–1036 mb) and is characterized as being evolving systems in search of thermal equilibrium. Thereby, they change position meridionally, preceded by cold fronts (Fonzar 1994). The frontal systems periodicities vary seasonally and generate energetic waves from the southern and southeastern quadrants. Waves generated by trade winds reach the region during most of the year, while the most energetic waves are dependent on the frontal systems and their frequency each month.

Systematic data about waves off the southeastern Brazilian coast are rare, the only existing records being associated with major engineering projects in the coastal region. Despite this, the main wave systems have a close relationship with the climatic rhythm, as determined by the prevalence of either the TAA or the MPA, which, generally speaking, give rise to waves from the east-northeast quarter or from the southeast and northeast quarters, respectively. Most wave heights are from 0.5 to 2.0 m (here treated as moderate-energy waves). The lower come from the east and southeast, and those above 2 m (moderate- to high-energy waves) usually come from the south and southwest. This dynamic behaviour, arising from the advance of the MPAs towards the tropical regions, together with the general orientation of the coastline, causes the change in direction of the waves and wave-induced coastal currents. The latter present a shift to the southwest under the influence of tropical systems and a shift to the northeast in response to the advance of polar systems (Ponçano et al. 1999).

According to Pianca et al. (2010) southerly waves prevail most of the year, except during spring, when easterly waves prevail. Highest waves, arriving from southeast, can reach up to 6 m, during fall. A seasonal wave climate analysis has been presented by Silva et al. (2014) for the southern São Paulo continental shelf and is represented in the directional histograms of Fig. 15.2.

The beach responses to the varying wave conditions are related to the distinct geological characteristics of each sector of the São Paulo coast, which definine its degree of exposure and orientation. Thus, the southern sector, between Ilha Comprida and Praia Grande, is characterized by a large coastal plain and beaches exposed to most of the incident waves (from northeast to south) with coastline orientation of about 45–60°. In general terms, the dominant longshore drift is to the northeast, with southwest inversions during periods of low frequency frontal systems. These variable patterns generate longshore drift cells with divergence and convergence centers throughout the systems (Tessler 1988; Souza 1997; Nascimento Jr 2006; Sousa et al. 2013; Silva et al. 2014). These convergence and divergence areas and variations in wave height along the coast, are responsible for erosive and depositional processes along this sector of the coast.

The whole São Paulo coast has a microtidal regime, with maximum tide amplitude of 1.3 m. There are no references, in the area, about the effect of the tide regime on beach morphodynamics.

15.2.2 Beach Compartment, Types and Morphodynamics

There are 318 ocean beaches on the São Paulo coast. There were almost no systematic beach studies until the 1970s and even after, mainly from academic studies (M.Sc. and D.Sc. thesis), most of them never published in scientific journals. Souza

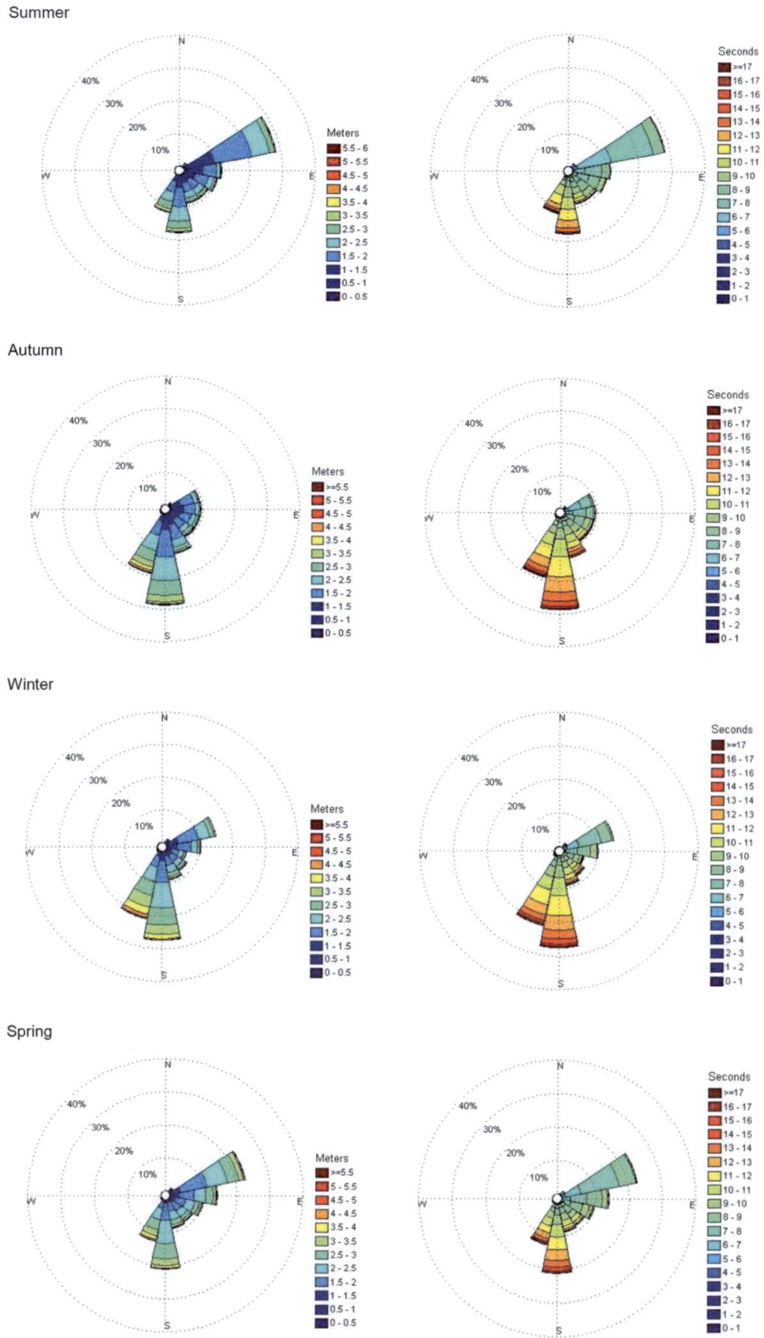

Fig. 15.2 Directional wave histograms for each season on the southern São Paulo continental shelf. *Left*: Significant wave height (m); *Right*: wave period (s) (Silva et al. 2014)

Fig. 15.3 Panoramic view of Sununga Beach, a typical enclosed-reflective beach on the Northern Scarps Coast

(2012) presented a partial review of the works on beach morphodynamics that have been conducted in the area. Thus, based in geomorphological characteristics, here they are grouped into six coastal compartments.

15.2.2.1 Bays and Islands of the Northern Scarps Coast

This compartment is characterized by the occurrence of a set of islands and semi-enclosed bays, these later with areas not larger than 10 km^2 each (except for Caraguatatuba Bay) with varying shapes and orientation (Souza 1990; Mahiques et al. 1998; Rodrigues et al. 2002). The islands and bays contain both small pocket beaches (Martins et al. 2010) (Fig. 15.3), and longer embayed beaches up to a few kilometres in length. Their morphodynamics varies from essentially reflective to intermediate with well-defined cusps (Sousa et al. 2013). The texture and composition of the beach sediments are heterogeneous and their distribution is strongly influenced by the competence of the small drainage basins as well as their degree of exposure to the dominant waves.

For the northern beaches, between São Sebastião and Ubatuba, the proximity of the coastal range (Serra do Mar) results in small bays and headland-embayed beaches with variable orientation. This causes incident waves to approach the coastline at different angles, resulting in locally specific sediment transport patterns. Therefore, each beach has specific morphodynamic behaviour, making it important study each beach separately. Studies such as those by Barros (1996), Hurtado et al. (2004),

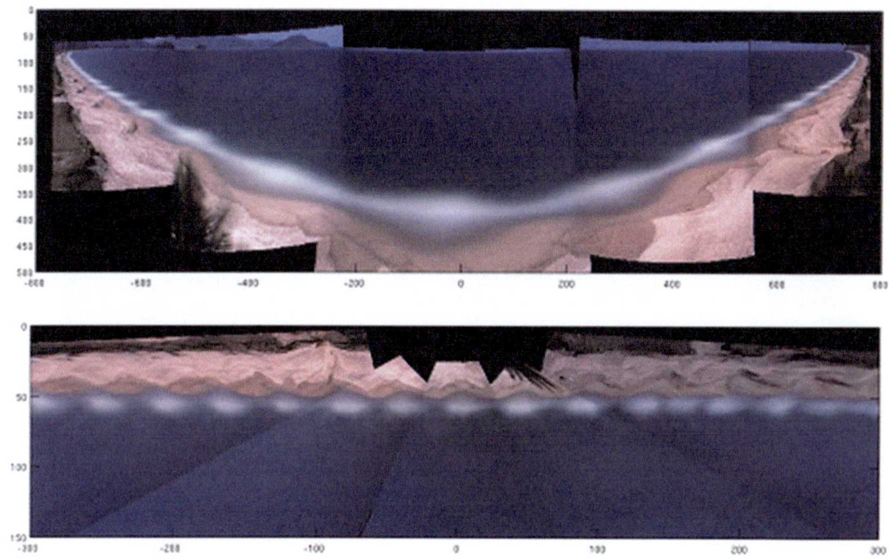

Fig. 15.4 Topography of Massaguaçú beach seen through time-exposure Argus video images

Nuber (2008), Rogacheski (2010), Martins et al. (2010), Mascagni (2012), Cardoso (2013), Ribeiro et al. (2013) and Sousa et al. (2013) are examples of studies conducted on specific beaches in the region. The region has beaches in both static and dynamic equilibrium (Silveira et al. 2010) as well as eroding beaches, with Massaguaçú Beach (Caraguatatuba) the most erosive and most studied of the beaches.

The Massaguaçú embayment contains a curved south to southeast facing beach bordered by headlands. It is composed of medium to coarse sand and is sheltered from the southern waves by the Ilhabela Island. It is a reflective beach with well-developed rhythmic beach cusps along its central portion. Recently, Massaguaçú Beach has been studied through an Argus video system (Holman 1994) in order to assess temporal evolution of the cusps and its interactions with swash processes (Santos et al. 2013; Pianca 2014). Figure 15.4 shows an example of merged Argus images for the Massaguaçú beach. In its centre-northern portion the beach experienced intense erosion that reached the backing road (Rogacheski 2010; Nuber 2008; Sousa et al. 2013; Ribeiro et al. 2013). Longshore sediment transport gradients and higher wave energy in this portion of the beach are the most likely cause for this erosional hot spot (Sousa 2013), while a higher frequency and increased intensity of storms may have aggravated this process (Rogacheski 2010). Based on a temporal evolution of coastal vulnerability through coastal state indicators, Ribeiro et al. (2013), concluded that occupation may not be the main contributor to the erosion, despite it having a relatively high occupation rate.

The northern coast is also characterized by the large variation in beach type (Tessler et al. 2006; Silveira et al. 2010), due to their variable orientation and morphodynamics, together with hydrodynamic forcings at the different time scales. Sununga Beach (Fig. 15.3), for example, is a reflective pocket beach exposed to the

Fig. 15.5 Northern sector of Caraguatatuba Bay

south to southwest and experiences short term beach rotation process (at daily scale), mainly associated with moderate to high energy waves related to frontal systems (Martins et al. 2010). These authors also concluded that the alternating erosive/depositional processes at the beach extremities are symmetric in duration, although different in magnitude, with a time lapse of 1–2 days. In the same bay, the intermediate to dissipative Lázaro Beach is more protected from the incoming southern waves. Therefore, unlike Sununga Beach, its morphology is not associated to frontal systems (Cardoso 2013).

In this sector, the Caraguatatuba Bay beaches represent an exception which is largely explained by São Sebastião Island and a local structural control, which shelter the beaches and which together with fine sand maintains a series of low-gradient beaches (Figs. 15.5 and 15.6).

15.2.2.2 São Sebastião Channel and Ilhabela Island

The São Sebastião Channel can be considered a particular feature within the Northern Scarps Coast. It is a deep channel, parallel to the coastline, isolated from the ocean by Ilhabela Island, whose landscape is supported by three alkaline eocenozoic volcanoes. The channel is 24 km long, between 2 and 6 km wide with depths ranging from 20 to 45 m from the mouth to its narrowest portion. The sandy beaches located both on the island and continent side of the channel represent, in fact, local accumulations of sediments in the channel's indentations. More effective wave action is only observed at the southern entrance of the channel.

Fig. 15.6 Central part of the Caraguatatuba Beach, a dissipative beach, which is an exception on the Northern Scarps Coast

It is also adjacent to the São Sebastião channel that we find the only mangroves and tidal flats on the open coast: in the Araçá bay. Araçá is a relatively small bay protected from the incident swell waves by the Ilhabela Island. Recent studies have investigated its hydrodynamic conditions. About two-thirds of the bay is exposed at low tide and land reclamation for harbour activities has reduced the size of the bay (Dottori and Siegle 2014).

15.2.2.3 Bertioga Coastal Plains

This compartment marks the beginning of the transition between the Northern Scarps Coast and the Coastal Plains and Estuaries Coast. Nevertheless, Muehe (2012) includes it in the northern sector. It forms a set of coastal plains, separated by headlands of the crystalline basement. Each of these plains may contain one or two larger drainage systems, which are responsible for the input of fine sand to the coastline (Martins 2000). The beach systems adjacent to these plains can show a highly variable morphology, from intermediate-reflective to intermediate-dissipative, separated by headlands that result in longshore transport segmentation between the different sectors. It is a region characterized by beaches in dynamic equilibrium (Silveira et al. 2011), with longshore transport strongly related to seasonal wave incidence (Souza 1997). Bertioga beach, for example, is characterized by rapid morphodynamic response to varying forces and dominant cross-shore transport and longshore drift to the northeast or southwest according to incident waves (Martins 2000). The beaches to the west in this region usually experience southwestward longshore transport, while the beaches to the east (eg Boracéia) experience longshore transport mainly to northeast, with divergence/convergence cells found in both (Souza 1997).

15.2.2.4 The Santos Area

The Santos area is the most populous on the coast of São Paulo. Also, it has the largest harbour of Latin America as well as one of the largest industrial complexes in Brazil.

The crescentic Santos coastal plain is 40 km long and 15 km at its widest point, and bounded by Serra de Mongaguá to the south and the rocky part of Santo Amaro Island to the north. In the central and northeastern parts of the plain it is drained by a network of estuarine channels that surround São Vicente and Santo Amaro islands. According to Suguio and Martin (1978), the genesis of this plain is related to the last two transgressive-regressive events, during the Isotope Stage 5e (Cananéia Transgression) and the Holocene Climatic Maximum (Santos Transgression).

Most of the beaches in Santos Bay are located on its northern innermost side, bordering the cities of Santos and São Vicente. They all exhibit an intense degree of urbanization, being separated one from the other by seven artificial channels that have been built since the beginning of the twentieth century as a way to reduce the flooding of the cities and which now act as sources of particulate matter (natural and artificial) to the coastline.

Ponta da Praia Beach, located at the easternmost sector of this beach arc, is undergoing a severe erosion, which has been intensified following the deepening and widening of the dredged channel of Santos harbour (Italiani 2014) (Fig. 15.7).

The Santos and São Vicente area beaches are located inside the Santos Bay, oriented southward and therefore exposed to the southerly waves. Farinnacio et al. (2009) suggest there are two longshore drift cells, diverging in approximately the middle of the bay. One cell to the east, acting between the central portion of the bay and its eastern end, close to the harbour channel; and another between the middle and São Vicente, to the west. The authors also concluded that anthropogenic actions have substantially influenced the area, generating erosional and depositional sectors along the bay, mainly due to the interruption caused by the Santos submarine outfall and by the artificial tombolo that connected the Porchat Island to the western end of the bay about 1950. Secondly, the set of drainage channels that reach the beach also changed the local sediment budget.

15.2.2.5 Itanhaém-Peruíbe Coastal Plains and Adjacent Beaches

The Itanhaém-Peruíbe coastal plains, together with the Bertioga one, are the least studied of the São Paulo coast (Giannini 1989; Cazzoli y Goya and Tessler 2000). This sector includes the northernmost part of the Coastal Plains and Estuaries Coast and is characterized by the progressive retreat of the Serra do Mar mountain chain as well as by the development of coastal plains with areas of tens of square kilometres. It also marks the beginning of dissipative beaches backed by dune fields.

Fig. 15.7 Accelerated erosion of the Ponta da Praia Beach, in Santos Bay

Fig. 15.8 Southern sector of the Ilha Comprida beach, a 70 km long dissipative beach along the southernmost coast of São Paulo

15.2.2.6 The Cananeia-Iguape Compartment

The Cananeia-Iguape compartment is the southernmost in São Paulo. The most conspicuous feature of the area is the Comprida Island, a 70-km sand barrier which encloses the Cananéia-Iguape lagoonal complex. Detailed work on the sedimentary morphology and evolution of this island was undertaken by Giannini et al. (2009). The Comprida Island contains two main erosive hot spots: one in the central-southern area and another at the northern portion of the island (Suguio 2003). These areas are related to gradients in the longshore drift and higher wave power. The northern sector, with higher occupation rates has been experiencing high erosion rates with coastline retreat of up to 10 m during a 1-year surveyed period (Silva et al. 2014). The island can be divided in two main sectors: the first 16 km of beach from south to north, containing intermediate to dissipative beaches; and the remaining sector dominated by dissipative beaches (Nascimento Jr, 2006) (Fig. 15.8). Seasonal varying wave climate generates well defined beach profile differences (Karniol-Marquez and Mahiques 2010), although the island's beaches respond differently to the same incident wave pattern (Silva et al. 2014).

15.3 Anthropogenic Impacts on the São Paulo Coastal Zone

Responses to inadequate land management and use of soil in coastal environments are manifest in severe catchment erosion (Xue et al. 2009), flooding (Jongman et al. 2012) and high damage losses (Hinkel et al. 2014). Furthermore, it is important to point out that, generally, anthropogenic influences, especially occupation, intensify coastal erosion (Sousa et al. 2011).

Table 15.1 Population density of some coastal cities of São Paulo

	City	Population density (hab/km^2)
Cities around Santos harbour	Santos	1494.26
	São Vicente	2247.88
	Guarujá	2026.80
	Praia Grande	1781.87
Cities around São Sebastião harbour (North Coast)	Caraguatatuba	207.88
	São Sebastião	185.00
	Ilhabela	81.13
South Coast	Ilha Comprida	47.01
	Cananéia	9.86
	Iguape	14.58

Data are based on the information available at http://www.ibge.gov.br

Anthropogenic activity along the São Paulo coast is more relevant along the centre-north coast of the state, being most dominant in the Santos lowlands, where several economic activities and urbanization are strongly linked to Brazil's largest harbour (Table 15.1). Santos has had severe environmental impacts related to pollution (Torres et al. 2009) and coastal erosion (Italiani 2014), much of it related to port installation and activities. The cities of Santos, São Vicente, Guarujá and Praia Grande have high population densities with values from 1494 hab/km^2 in Santos up to 2247 hab/km^2 in São Vicente. It is all related to the development of industries, companies and services related to harbour activities. The smaller São Sebastião harbour did not promote a boom of occupation on the north coast of the state and consequently, cities such as São Sebastião, Ilhabela have lower population densities when compared to the cities around Santos. Finally, there are no harbours on the south coast, and the cities of Ilha Comprida, Cananéia and Iguape (Table 15.1) show the lowest population density in the region.

Some studies involving coastal erosion risk and vulnerability have been developed for the São Paulo coast. Souza and Suguio (2003) built a map of coastal erosion risk for the state's coastline. They concluded that approximately 42 % of the state's sandy beaches present high to very high erosion risk, reaching up to 66 % of the state's coastline. However, the authors suggest that these risks are mainly associated with natural rather than anthropogenic activities.

Several beaches along the area have relatively high occupation rates. Massaguaçú beach demonstrates the anthropogenic impacts that affect life and routine of local population (Fig. 15.9). To better understand such impacts at that site, Santos et al. (2013) and Sousa et al. (2013) analysed the evolution of occupation process and linked them to local coastal processes. Massaguaçú is a 7.5 km long embayed beach located in the northeast of Caraguatatuba beach and on the northern side of São Sebastião Island. Across from Cocanha Beach, in the northeast portion of Massaguaçú, there are Cocanha Islands and the larger Tamanduá Island. Massaguaçú is experiencing coastal erosion in the centre of the bay and accretion at each end (Nuber 2008).

Most of the impacts on Massaguaçú beach occur in the area of up to 500 m from the water line. Figure 15.9 displays the evolution of occupation in the area from

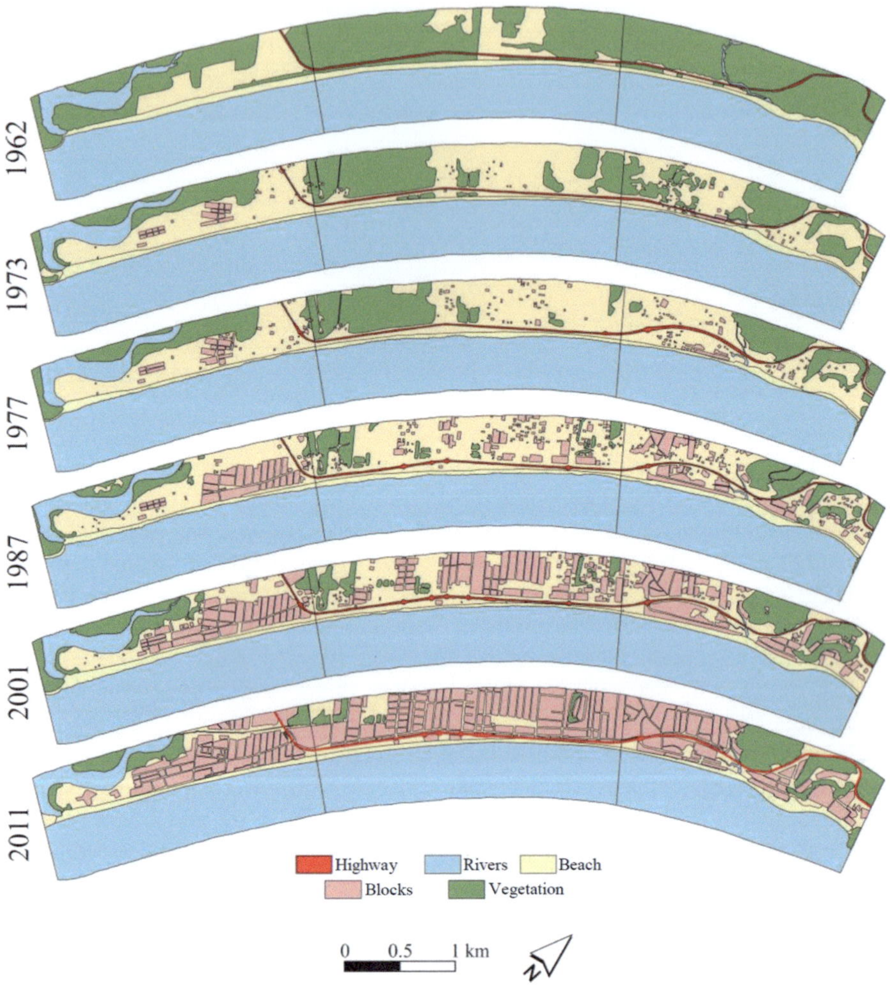

Fig. 15.9 Evolution of occupation in Massaguaçú from 1962 to 2011 (Adapted from Ribeiro et al. 2013)

1962 to 2011. There was little occupation and considerable vegetation along the coast in 1962, however by 1973 development has commenced. From 1973 to 1977, it spread rapidly and vegetation started being cleared. From 1977 to 1987, vegetation was mainly restricted to the Serra-do-Mar highlands, except in the central area and the occupation process continued to intensify. Between 1987 and 2011, most of the coast was occupied and 10 years later, in 2011, the occupation became massive in Massaguaçú. Areas without occupation are due to the steep slopes of the Serra-do-Mar and the headlands that border the beach and act as natural barriers for this process, contributing to the preservation of vegetation in these areas.

Figure 15.9 also shows the change on the Bracuí river course in the northern part of the beach, mainly between 1962 and 1977 and the presence of Rio-Santos

Fig. 15.10 Structures build in 2005 and 2006 (*above* – Photo: E. Siegle) and a larger seawall build in 2012–2013 (*below* – photo: P.H.G.O. Sousa)

highway that was built before 1962. The change in the course of Bracuí River contributed to the increasing number of flood events caused by elevation of fluvial water level. As river does not have competence to break through beach sediments, local authorities open the channel artificially to release water accumulated in the backbarrier and prevent flooding.

In the central part of Massaguaçú beach the highway was built parallel and close to the beach. Thus, it affects sediment exchanges between emerged and submerged zones, especially during storm surges, leading to a negative sediment budget. The central area of Massaguaçú is experiencing severe coastal erosion and the highway has been damaged (Fig. 15.10a). To protect the road, several attempts were made with the construction of small structures along the central part of the beach until a larger seawall was built in 2013 (Fig. 15.10b).

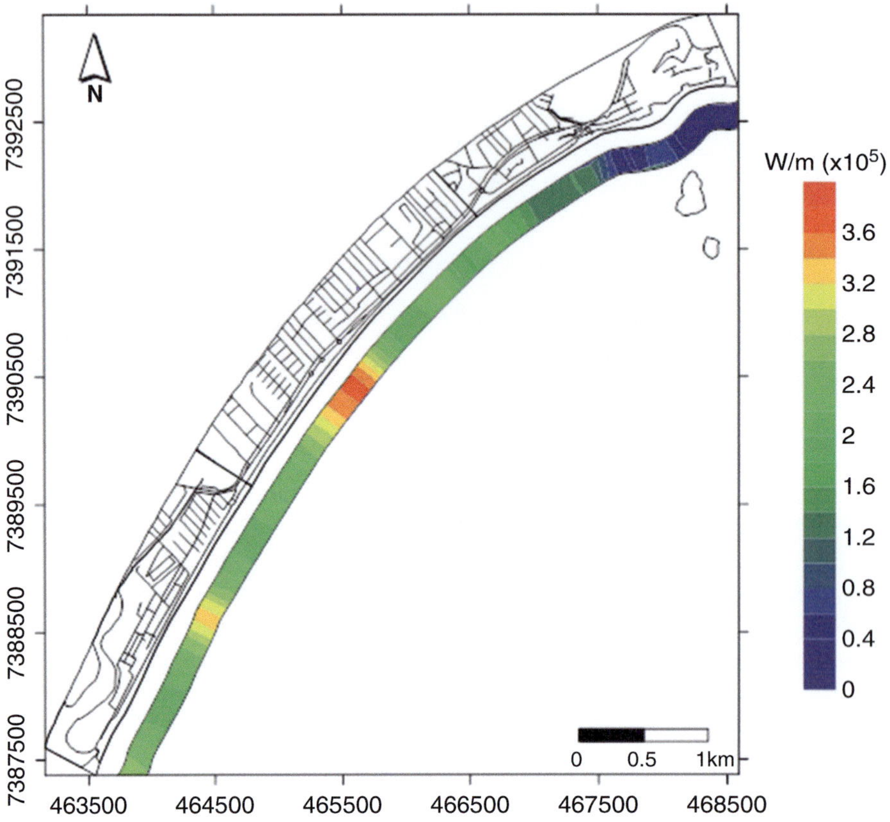

Fig. 15.11 Distribution of wave power along Massaguaçú coast

Fig. 15.12 Destruction of buildings constructed directly on the berm of Ilha Comprida Beach

Through the analysis of the local wave climate, Sousa (2013) observed that July was when the highest occurrence of energetic wave events reach the São Paulo coast, with the most frequent and energetic waves in Massaguaçú arrive from south (H=2 m, T=10 s) and east (H=1.5 m, T=8 s). Wave climate was used to calculate wave power after Holthuijsen (2007) and the results indicate higher wave power coinciding with the area of erosion, while areas with lower wave power occur in the south and in the northern shadow zone, caused by islands (Fig. 15.11).

Beach erosion is not exclusive to highly populated areas but it is almost always associated with human activity. Figure 15.12 shows an example of building destruction in Ilha Comprida, the less populated sector of the coast of the state, but in which constructions have been built on the beach berm.

References

Angulo RJ, Lessa G (1997) The Brazilian sea-level curves: a critical review with emphasis on the curves from Paranaguá and Cananéia regions. Mar Geol 140:141–166

Angulo RJ, Suguio K (1995) Re-evaluation of the maxima of the Holocene sea-level curve for the State of Paraná, Brazil. Palaeogeogr Palaeoclimatol Palaeoecol 112:385–393

Angulo RJ, Giannini PCF, Suguio K, Pessenda LCR (1999) The relative sea-level changes in the last 5500 years southern Brazil (Laguna-Imbituba region, Santa Catarina State) based on vermetid 14C ages. Mar Geol 159:327–339

Angulo RJ, Lessa GC, Souza MC (2006) A critical review of mid to late-Holocene sea-level fluctuations on the eastern Brazilian coastline. Quat Sci Rev 25:86–506

Barros MO (1996) Dinâmica e evolução dos ambientes praiais da Fazenda, do Puruba e de Itamambuca, litoral norte do Estado de São Paulo. Dissertation, University of São Paulo

Bastos CC, Ferreira NJ (2000) Análise Climatológica Da Alta Subtropical Do Atlântico Sul. In: Proceedings of the 11th Brazilian meteorological congress, Sociedade Brasileira de Meterologia, Rio de Janeiro, 2000

Besnard W (1950a) Considerações gerais em torno da região lagunar de Cananéia-Iguape, I. Bol Inst Paulista Oceanogr 1:9–26

Besnard W (1950b) Considerações gerais em torno da região lagunar de Cananéia-Iguape, II. Bol Inst Paulista Oceanogr 2:3–28

Branner JC (1904) The stone reefs of Brazil, their geological and geographical relations. Bull Mus Compar Zool 44:1–7

Cardoso DG (2013) Variabilidade praial nas praias do Lázaro, Domingas Dias e Sununga, litoral norte de São Paulo à passagem dos sistemas frontais (Ubatuba, SP). Dissertation, University of São Paulo

Cazzoli y Goya S, Tessler MG (2000) Variações morfológicas espaço temporais entre as praias de Cibratel e Itanhaém-Suarão, Estado de São Paulo. Rev Bras Oceanogr 48:151–166

Cruz FW, Vuille M, Burns SJ et al (2009) Orbitally driven east–west antiphasing of South American precipitation. Nat Geosci 2:210–214

Dottori M, Siegle E (2014) Hydrodynamic forcings at the entrance of an intertidal flat: Araçá Bay, Brazil. In: Abstracts of the 17th PECS – Physics of Estuaries and Coastal Seas, Porto de Galinhas, Brazil, 2014. http://www.pecs2014-brazil.org/Xabs/xabs_189_Dottori_Siegle.pdf

Farinnacio A, Cazzoli y Goya S, Tessler MG (2009) Variações da linha de costa nas baías de Santos e São Vicente. Quat Environ Geosci 1:42–48

Fonzar BC (1994) A circulação atmosférica na América do Sul: os grandes sistemas planetários e subsistemas regionais que atingem o continente (localização e trajetórias). Cad Geoc, Rio de Janeiro 11:11–33

Giannini PCF (1989) Evolução quaternária na planície costeira de Peruíbe Itanhaém (SP) segundo resultados granulométricos: um confronto de métodos de tratamento estatístico. Bol Par Geoc 38:1–50

Giannini PCF, Guedes CCF, Nascimento DR Jr et al (2009) Sedimentology and morphological evolution of the Ilha Comprida barrier system, Southern São Paulo Coast. In: Dillenburg S, Hesp P (eds) Geology and geomorphology of Holocene coastal barriers of Brazil, vol 107, Lecture notes in earth sciences. Springer, Berlin, pp 177–224

Guedes CCF, Giannini PCF, Sawakuchi AO et al (2011) Determination of controls on Holocene barrier progradation through application of OSL dating: the Ilha Comprida Barrier example, Southeastern Brazil. Mar Geol 285:11–16

Hartt CF (1870) Geology and physical geography of Brazil. Fields, Osgood, Boston, 625p

Hinkel J, Lincke D, Vafeidis AT et al (2014) Coastal flood damage and adaptation costs under 21st century sea-level rise. PNAS. doi:10.1073/pnas.1222469111

Holman RA (1994) The ARGUS Program. Bulletin of the Coastal Imaging Lab. Oregon State University, Corvallis, OR 97331

Holthuijsen LH (2007) Waves in oceanic and coastal waters. Cambridge University Press, Cambridge, 404p

Hurtado SN, Cazzoli Y, Goya SV (2004) Evolution and dynamic of Itamambuca beach and comparative evaluation of morphodynamic studies of beaches. J Coast Res SI39:568–571

Innocentini V, Caetano Neto ES (1996) A case study of the 9 August 1998 South Atlantic storm: numerical simulations of the wave activity. Weath Forecast 11:78–88

Italiani DM (2014) Resposta morfodinâmica à alimentação artificial da Ponta da Praia, Santos, SP. Dissertation, University of São Paulo

Jaeschke A, Rühlemann C, Arz H, Heil G, Lohmann G (2007) Coupling of millennial-scale changes in sea surface temperature and precipitation off northeastern Brazil with high-latitude climate shifts during the last glacial period. Paleoceanography. doi:10.1029/2006PA001391

Jongman B, Ward PJ, Aerts JGJH (2012) Global exposure to river and coastal flooding: long term trends and changes. Glob Environ Chang 22:823–835

Karniol-Marquez M, Mahiques MM (2010) Variações morfológicas no prisma praial da Ilha Comprida (Sudeste do Brasil) – Subsídios para uma Gestão Costeira Sustentável. Gest Cost Integr 10:361–379

Mahiques MM, Tessler MG, Furtado VV (1998) Characterization of energy gradient in enclosed bays of Ubatuba region, South-Eastern Brazil. Estuar Coast Shelf Sci 47:431–446

Martin L, Fournier M, Mourguiart P et al (1993) Southern Oscillation signal in South American palaeoclimatic data of the last 7000 years. Quat Res 39:338–346

Martin L, Dominguez JML, Bittencourt ACSP (2003) Fluctuating Holocene sea-levels in eastern and southeastern Brazil: evidence from multiple fossil and geometric indicators. J Coast Res 19:101–124

Martins CC (2000) Variações morfológicas e sedimentares de curto período em perfis praiais, Praia da Bertioga/SP. Dissertation, University of São Paulo

Martins CC, Dias JMA, Mahiques MM (2010) Daily morphological changes determined by high-energy events on an embayed beach: a qualitative model. Earth Surf Process Landf 35:487–495

Mascagni ML (2012) Dinâmica morfo-sedimentar da praia de Boiçucanga, SP, Brasil. Dissertation, University of São Paulo

Monteiro CAF (1969) A frente polar atlântica e as chuvas de inverno na fachada sul-oriental do Brasil: contribuição metodológica à análise rítmica de tempo no Brasil. PhD Thesis, University of São Paulo

Muehe D (2012) O litoral brasileiro e sua compartimentação. In: Cunha SB, Guerra AJT (eds) Geomorfologia do Brasil, 8th edn. Bertrand Brasil, Rio de Janeiro, pp 273–349

Nascimento DR Jr (2006) Morfologia e Sedimentologia ao Longo do sistema Praia-Duna Frontal de Ilha Comprida, SP. Dissertation, University of São Paulo

Nascimento DR Jr, Giannini PCF, Tanaka APB, Guedes CCF (2008) Mudanças morfológicas da extremidade NE da Ilha Comprida (SP) nos últimos dois séculos. Rev Inst Geoc – USP 8:25–39

Nobre P, Shukla J (1996) Variations of sea surface temperature, wind stress, and rainfall over the tropical Atlantic and South America. J Clim 9:2464–2479

Nuber E (2008) Evolução morfológica e sedimentológica do arco praial de Massaguaçu, litoral norte do Estado de São Paulo. Dissertation, University of São Paulo

Paegle JN, Mo KC (2002) Linkages between summer rainfall variability over South America and sea surface temperature anomalies. J Clim 15:1389–1407

Pianca C (2014) Space-time variability of coastal morphology: results from video remote sensing. Dissertation, University of São Paulo

Pianca C, Mazzini PLF, Siegle E (2010) Brazilian offshore wave climate based on NWW3 reanalysis. Braz J Oceanogr 58:53–70

Ponçano WL, Tessler MG, Freitas CGL, Mahiques MM (1999) Tendências regionais de transporte de sedimentos arenosos ao longo das praias paulistas. Rev Univ Guarulhos, Geociências IV:102–115

Ribeiro JS, Sousa PHGO, Vieira DR, Siegle E (2013) Evolução da vulnerabilidade à erosão costeira na Praia de Massaguaçú (SP), Brasil. Gest Cost Integr 13:253–265

Rodrigues M, Mahiques MM, Tessler MG (2002) Sedimentação atual nas enseadas de Ubatumirim e Picinguaba, região norte de Ubatuba, Estado de São Paulo, Brasil. Rev Bras Oceanogr 50:27–45

Rogacheski CEA (2010) Dinâmica sedimentar e a caracterização de zonas de erosão acenuada (ZEA) ao longo do arco praial de Massaguaçu, SP. Dissertation, University of São Paulo

Sant'Anna Neto JL (1990) Ritmo Climático e a gênese das chuvas na zona costeira paulista. Dissertation, University of São Paulo

Santos HH, Siegle E, Sousa PHGO (2013) Beach cusps: spatial distribution and time evolution at Massaguaçú beach (SP), Brazil. In: AGU Spring Meeting. http://adsabs.harvard.edu/abs/2013AGUSMOS31B..07D

Silva FG, Sousa PHGO, Siegle E (2014) Wave climate analysis and its relation with erosion processes along the beach system of Ilha Comprida (SP), Brazil. In: Proceedings of the 17th Physics of Estuaries and Coastal Seas (PECS) conference. Porto de Galinhas, Brazil. http://www.pecs2014-brazil.org/Xabs/xabs_167_Silva_etal.pdf

Silveira LF, Klein AHF, Tessler MG (2010) Headland-bay beach planform stability of Santa Catarina state and of the northern coast of São Paulo state. Braz J Oceanogr 58(2):101–122

Silveira LF, Klein AHF, Tessler MG (2011) Classificação morfodinâmica das praias do Estado de Santa Catarina e do litoral norte do Estado de São Paulo utilizando sensoriamento remote. Braz J Aquat Sci Technol 15:13–28

Sousa PHGO, Siegle E, Tessler MG (2011) Environmental and anthropogenic indicators for coastal risk assessment at Massaguaçú Beach (SP) Brazil. J Coast Res SI64:319–323

Sousa PHGO, Siegle E, Tessler MG (2013) Vulnerability assessment of Massaguaçú Beach (SE Brazil). Ocean Coast Manag 77:24–30

Souza CRG (1990) Considerações sobre os processos sedimentares quaternários e atuais na região de Caraguatatuba, litoral norte do Estado de São Paulo. Dissertation, University of São Paulo

Souza CRG (1997) As células de deriva litorânea e a erosão nas praias do estado de São Paulo. PhD Thesis, University of São Paulo

Souza CRG (2012) Praias arenosas oceânicas do estado de São Paulo (Brasil): síntese dos conhecimentos sobre morfodinâmica, sedimentologia, transporte costeiro e erosão costeira. Rev Dep Geogr – USP, Special Volume 30 years, 307–371

Souza CRG, Suguio K (2003) The coastal erosion risk zoning and the State of São Paulo plan for coastal management. J Coast Res 35:530–547

Suguio K (2003) Tópicos de geociências para o desenvolvimento sustentável: as regiões litorâneas. Rev Inst Geoc USP 1:1–40

Suguio K, Martin L (1978) Quaternary marine formation of the states of São Paulo and Southern Rio de Janeiro. In: International symposium on coastal evolution in the quaternary. Brasilian National Working Group for the IGCP, São Paulo. Special Publication, 55p
Suguio K, Martin L, Bittencourt ACSP et al (1985) Flutuações do nível relativo do mar durante o quaternário superior ao longo do litoral brasileiro e suas implicações na sedimentação costeira. Rev Bras Geoc 15:273–286
Tessler MG (1988) Dinâmica sedimentar quaternária no litoral sul paulista. PhD Thesis, University of São Paulo
Tessler MG, Cazzoli y Goya (2005) Processos costeiros condicionantes do litoral brasileiro. Rev Dep Geogr – USP 17:11–23
Tessler MG, Cazzoli y Goya SC, Yoshikawa PS, Hurtado SN (2006) São Paulo. In: Muehe D (ed) Erosão e Progradação do Litoral Brasileiro. MMA, Brasília, pp 297–346
Torres JR, Abessa DMS, Santos FC, Maranhão LA, Davanso MB, Nascimento MRL, Mozeto A (2009) Effects of dredging operations on sediment quality: contaminant mobilization in dredged sediments from the Porto of Santos, SP, Brazil. J Soils Sedim 9:420–432
Vera CS, Vigliarolo PK, Berbery EH (2002) Cold season synoptic scale waves over subtropical South America. Mon Weather Rev 130:684–699
Xue Z, Feng A, Yin P, Xia D (2009) Coastal erosion induced by human activities: a Northwest Bohai sea case study. J Coast Res 25:723–733
Ybert JP, Bissa WM, Catharino ELM, Kutner M (2003) Environmental and sea-level variations on the southeastern Brazilian coast during the Late Holocene with comments on prehistoric human occupation. Palaeogeogr Palaeoclimatol Palaeoecol 189:11–24
Zembruscki SG (1979) Geomorfologia da margem continental sul brasileira e das bacias oceânicas adjacentes. In: Chaves HAF (ed) Geomorfologia da margem continental brasileira e das bacias oceânicas adjacentes, vol 7. Série Projeto REMAC, PETROBRAS-CENPES-DINTEP, Rio de Janeiro, pp 129–177

Chapter 16
The State of Paraná Beaches

Rodolfo José Angulo, Carlos Alberto Borzone, Mauricio Almeida Noernberg, Clécio José Lopes de Quadros, Maria Cristina de Souza, and Leonardo Cruz da Rosa

Abstract Paraná beaches are composed primarily of mature polycyclic fine quartz sand and are characterized by wave-dominated beaches along the open coast and tidal-modified beaches inside the estuaries. At the estuary mouths large ebb-tidal-deltas influence beach sand transport and budget and cause shifts in the shoreline of hundreds of meters during periods of a few (<10) years. In contrast, the open ocean coast, away from the influence of ebb-tidal deltas, has remained stable (<10 m shift) over the last 5–6 decades. Erosion problems on the Paraná coast are related to natural coastline shift induced by ebb-tidal delta dynamics as well as human destruction of foredune ridges and constructions over the beach and the beach dynamic fringe. At present time, several erosion problems remain without correct environmental and long- term solutions.

16.1 Introduction

16.1.1 Regional Setting

The State of Paraná beaches extend for 126 km between Barra do Ararapira (25°18′S) and Barra do Saí (25°58′S) (Angulo and Araújo 1996) and faces east into the Atlantic Ocean (Fig. 16.1). This coastal sector is characterized by an up to 55 km wide Quaternary coastal plain and large estuarine complexes, backed by the

R.J. Angulo (✉) • C.J.L. de Quadros • M.C. de Souza
Departamento de Geologia, Laboratório de Estudos Costeiros – Lecost,
Universidade Federal do Paraná, Setor de Ciências da Terra,
Centro Politécnico, Av. Cel. Francisco H. dos Santos 100, Caixa Postal 19001, 81531-970
Curitiba, Paraná, Brazil
e-mail: fitoangulo@gmail.com; cleciodequadros@yahoo.com.br;
cristinasouza2527@gmail.com

C.A. Borzone
Laboratório de Ecologia de Praias, Centro de Estudos do Mar,
Universidade Federal do Paraná, Setor de Ciências da Terra,
Av. Beira-mar, s/n Caixa Postal 61, 83255-976 Pontal do Sul,
Pontal do Paraná, Paraná, Brazil
e-mail: capborza@ufpr.br

A.D. Short, A.H. da F. Klein (eds.), *Brazilian Beach Systems*, Coastal Research Library 17, DOI 10.1007/978-3-319-30394-9_16

mainland plateaus (800–1000 m high) and Serra do Mar mountain ranges that reach more than 1500 m in altitude (Fig. 16.1).

The Paraná coastal region contains two main geologic domains: the Achaean to Mesozoic crystalline rocks and the Cenozoic sedimentary deposits. The crystalline rocks are mainly Achaean to Cambrian granulites, gneisses, granites and migmatites, and Mesozoic diabase and diorites that occur as a dike complex related to the opening of the South Atlantic Ocean (Mineropar 1989). The Cenozoic sedimentary deposits are dominated by continental Miocene to Quaternary alluvial fans, Quaternary fluvial, colluvial and talus deposits and at the coast by Quaternary coastal deposits including present and paleo-estuarine, deltaic, tidal-deltas, tidal-flat, aeolian, beach, shoreface, and inner-shelf sediments (Angulo 2004).

During the mid- to late-Holocene sea-level lowering (~3 m) and the positive sedimentary budget initiated coastal progradation, that increases in width from south to north. At Barra Velha (25° 39′S), the Holocene coastal plain is only few hundred meters wide increasing northward to reach 5 km at Paranaguá (25° 34′S) (Fig. 16.2). At the same time, estuarine complexes were infilled since mid- to late Holocene owing to sedimentation coupled with the fall in sea-level (Angulo et al. 2009a).

16.1.2 Drainage

The drainage system of the Paraná coastal zone is dominated by watersheds of Paranaguá (~3.9×10^3 km^2) and Guaratuba (~1.9 10^3 km^2) estuarine complexes (Angulo 1992), and by several smaller watersheds such as Saí-Guaçu (~133 km^2, Noernberg et al. 1997) and Matinhos (33 km^2, Milani and Canali 2000) (Fig. 16.3). On the northern Superagüi coastal plain small rivers (~156 km^2, Noernberg et al. 1997) flow into the Mar do Ararapira estuary, which in the 1960s was connected to the Baía dos Pinheiros estuary by the construction of an artificial channel (Fig. 16.3).

The catchment areas of Paranaguá and Guaratuba basins are characterized by high-gradient incised mountainous rivers valleys flowing onto a low-gradient floodplain with rivers, estuarine head-deltas and estuaries occupying the coastal plain.

The main estuarine complexes are the Paranaguá (with a water body of ~552 km^2 and mangroves and salt marshes covering ~287 km^2) (Noernberg et al. 2006) and

M.A. Noernberg
Centro de Estudos do Mar, Laboratório de Oceanografia Costeira e Geoprocessamento, Universidade Federal do Paraná, Setor de Ciências da Terra, Av. Beira-mar, s/n Caixa Postal 61, 83255-976 Pontal do Sul, Pontal do Paraná, Paraná, Brazil
e-mail: m.noernberg@ufpr.br

L.C. da Rosa
Departamento de Engenharia de Pesca e Aquicultura, Laboratório de Ecologia Bentônica, Universidade Federal de Sergipe, Centro de Ciências Agrárias Aplicadas, Av. Marechal Rondon, s/n, 49100-000 Jardim Rosa Elze, São Cristóvão, Sergipe, Brazil
e-mail: leonardo.rosa@rocketmail.com

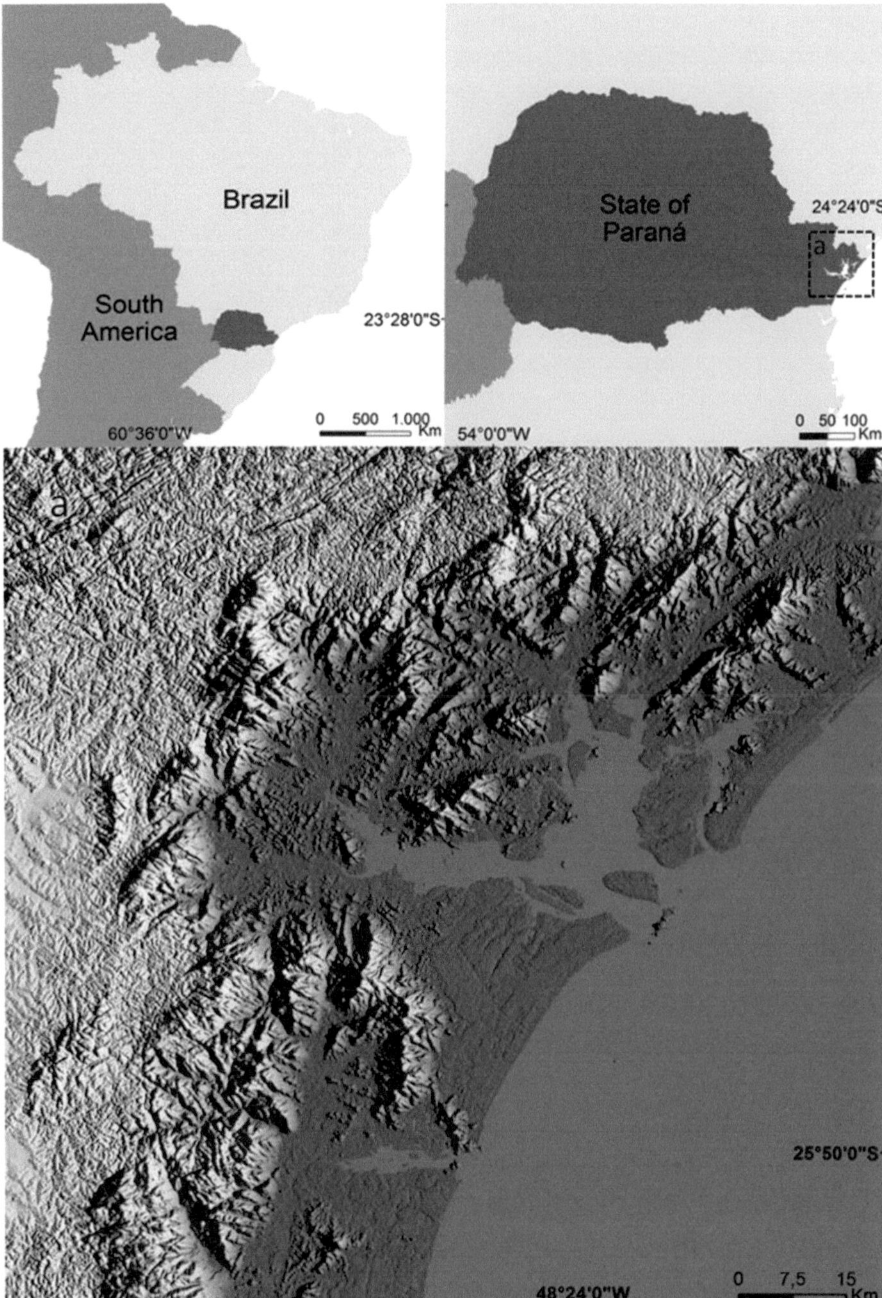

Fig. 16.1 Location and (*a*) relief of the State of Paraná coastal zone

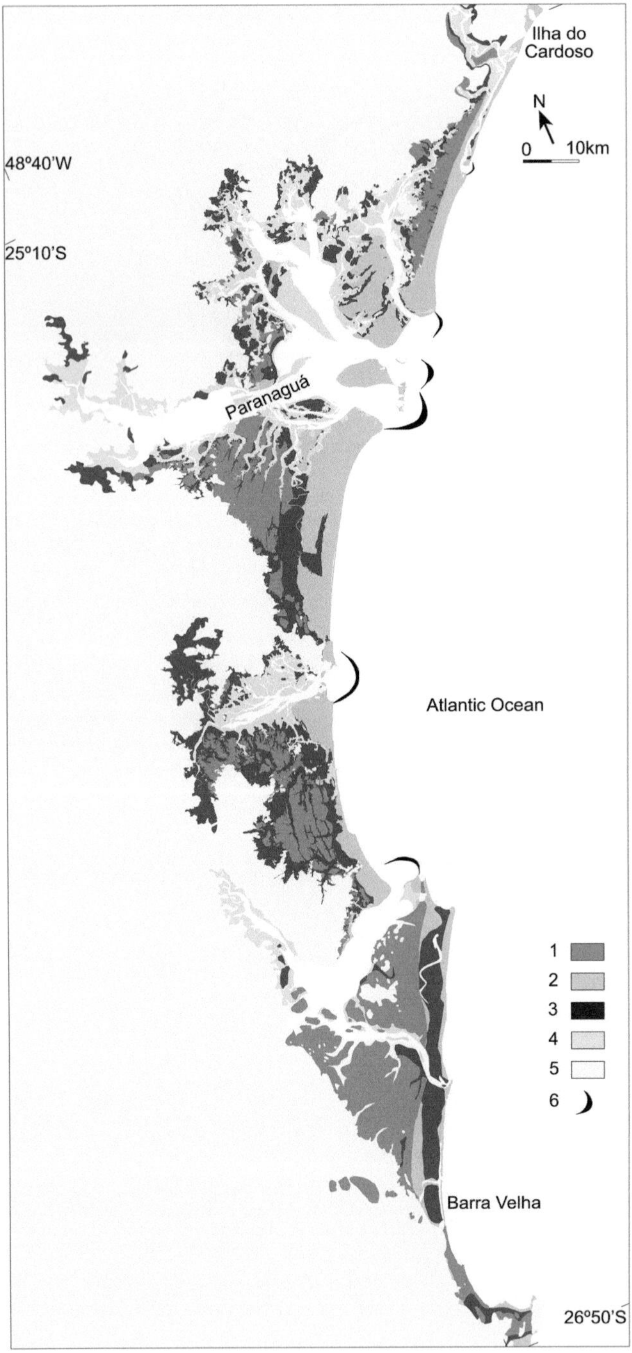

Fig. 16.2 Quaternary geology from Barra Velha to Ilha do Cardoso coastal plains. (*1*) Pleistocene barriers, (*2*) Holocene barriers, (*3*) paleo-estuarine plains, (*4*) present tidal-flats, (*5*) other units (*6*) ebb-tidal deltas (After Angulo et al. 2009a)

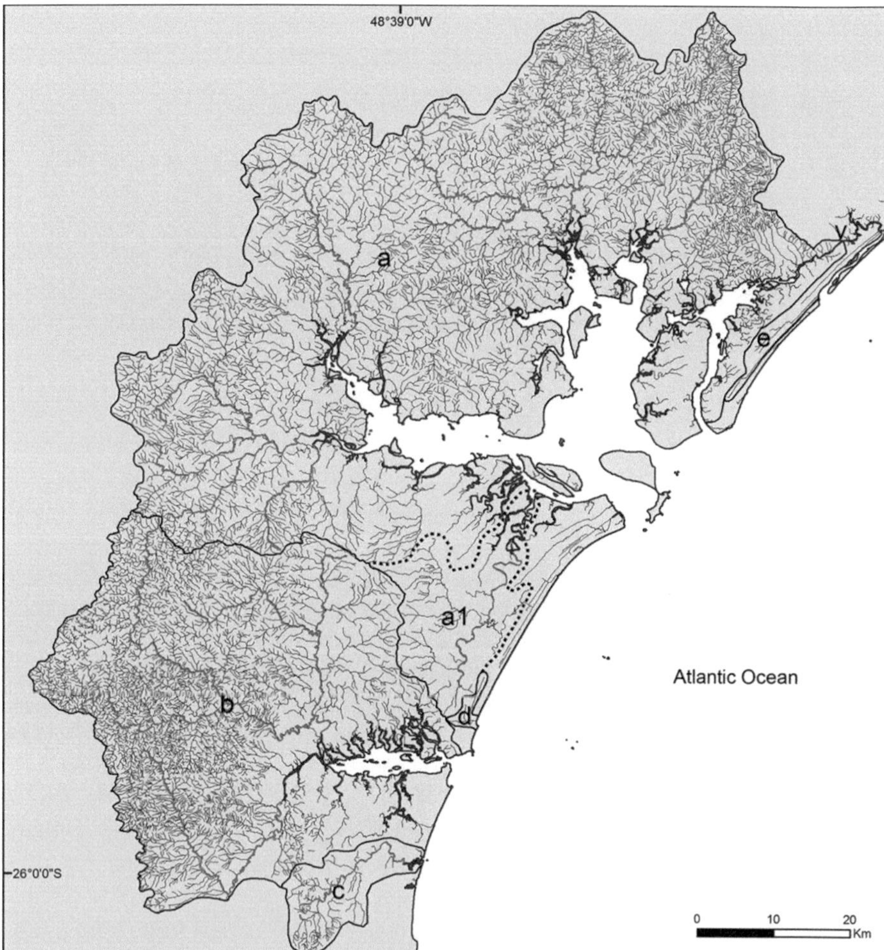

Fig. 16.3 Watersheds of State of Paraná coastal zone: (*a*) Paranaguá; (*a1*) Guaraguaçu; (*b*) Guaratuba; (*c*) Saí-Guaçu; (*d*) Matinhos; and (*e*) Ararapira. (*v*) Varadouro artificial channel

Guaratuba (~50 km^2 water and ~81 km^2 mangroves-salt marshes) (Noernberg et al. 1997). Despite the catchment area of Paranaguá being twice the size of the Guaratuba, its drainage density is 1.1 rivers/km^2 whereas in Guaratuba is 1.9 rivers/km^2 (Noernberg et al. 2006). The larger drainage density results in Guaratuba Bay having pronounced stratification in the water column (Noernberg et al. 2006). Both systems are relatively shallow with tidal flats representing 24.6 % of the area in Paranaguá and 24.0 % in Guaratuba, and greatest depths reaching up to 30 m and 27 m respectively at the estuarine mouths (Noernberg et al. 2006).

The two other Paraná estuaries are Mar do Ararapira, located at the northern state border (~6.5 km^2 water body and ~2.2 km^2 of mangroves and salt marshes on the Paraná side) and the Saí-Guaçu, located on the southern state border (~0.35 km^2 water body and ~3 km^2 of mangroves and salt marshes).

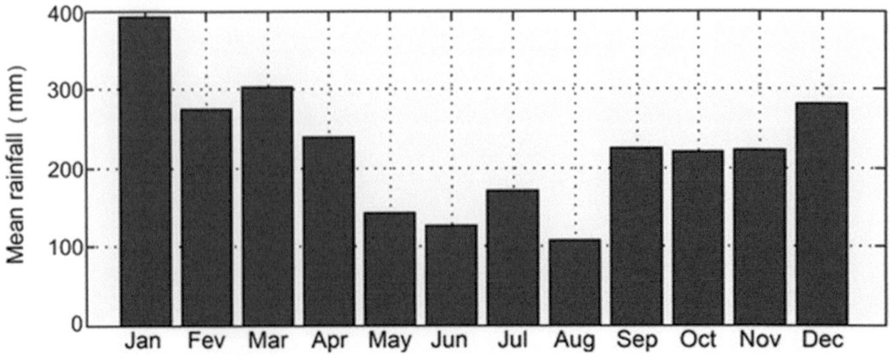

Fig. 16.4 Mean monthly rainfall between 1996 and 2010 at Pontal do Sul (After Dourado and Fomin 2015)

16.1.3 Climate

The local climate is controlled by the displacement of the semi-permanent anti-cyclone gyre in the South Atlantic and by the passage of cold polar masses in winter. These cold fronts are the main atmospheric disturbances bringing winds from south to southwest and frequently causing storm surges. The Serra do Mar Mountains act as a barrier against cold fronts and causes a concentration of stationary fronts in the bay region (Lana et al. 2000), generating orographic rainfall. Because of this phenomenon, the Paraná coast has the highest rainfall in southern Brazil (Labraga et al. 2000), with a mean annual rainfall of 2363 mm during the 1997–2003 period (Vanhoni and Mendonça 2008). The rainfall is seasonal with a summer maximum associated with the movement of the Tropical Atlantic air mass over the area (Vanhoni and Mendonça 2008) (Fig. 16.4).

Prevailing winds are from east, southeast and south 56 % of the time, with northeast winds less than 10 %. These directions also represent the strongest winds (≥7.4 m s^{-1}) (Dourado and Fomin 2015, Fig. 16.5), which during storm surges arrive from southeast and may reach up to 25 m s^{-1} (Cazal et al. 2011). The storm surges are always associated with extra-tropical cyclones acting over the coast (Diniz and Kousky 2004). The cyclones form off south/southeast Brazilian coast move eastward or southwestward with velocities about 9 m s^{-1} and travel distances of up to 2700 km with a mean life span of 3 days (Reboita 2008).

16.1.4 Wave and Tides

The tide along the Paraná coast is semidiurnal with diurnal inequalities and nonlinear co-oscillation interactions. The spring tide range is about 1.5 m and the tidal wave propagates into the estuaries as a progressive wave with range increasing to 2.2 m in the inner of the Paranaguá estuarine complex, transitioning from

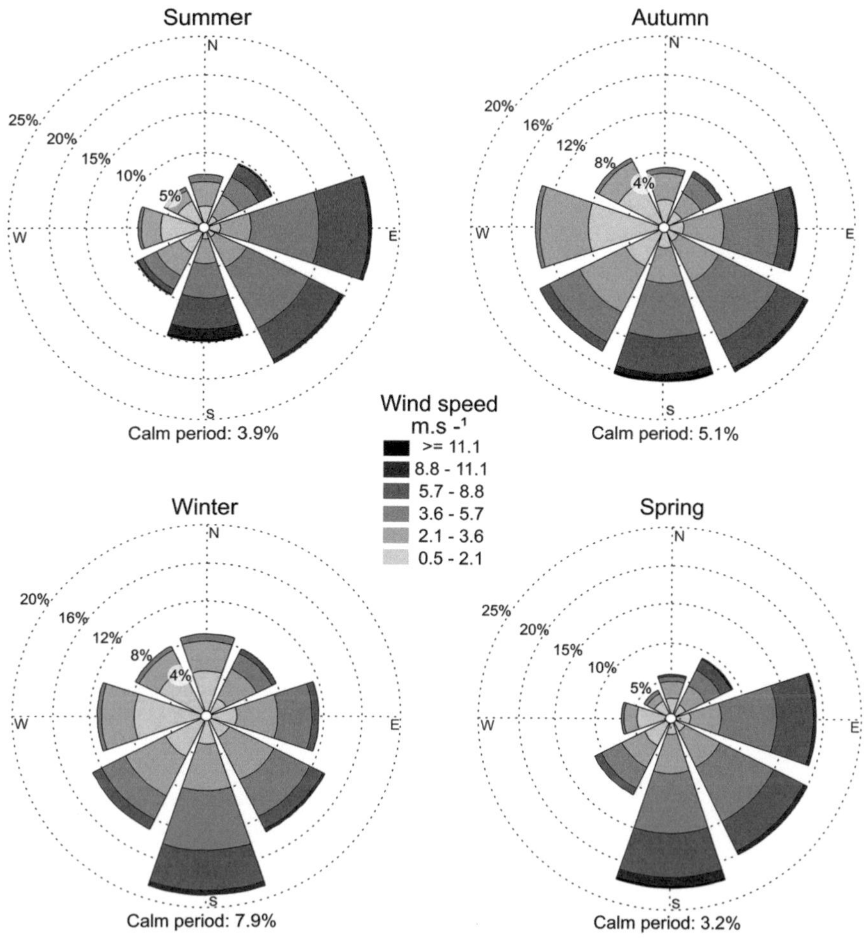

Fig. 16.5 Wind speed and direction seasonal variability between 1996 and 2010, at Pontal do Sul (After Dourado and Fomin 2015)

micro-tidal to meso-tidal conditions (Marone and Jamiyanaa 1997). The main tidal constituents for a point offshore and other one at Paranaguá bay inlet are shown in Table 16.1. During storm surges the water levels can reach to 80 cm above astronomical tides (Marone and Camargo 1994).

Beach erosion occurs when the storm surges coincide with high swell. The frequency of swells (mean peak wave period $T_p \geq 10$s) at the coast is about 36 % with the largest swells arriving in autumn, while winds waves arrive 63 % of the time (Nemes 2011). The predominant wave direction (measured from July 2009 to June 2010) is south-southeast (28 %) followed by southeast (25 %), south (21 %) and east-southeast (16 %) (Nemes 2011).

The mean significant wave height (H_s) at the depth of 18 m is 1.6 m with a maximum significant wave height of 4.8 m, the T_p is 8.4 s and maximum of 17.8 s

Table 16.1 Main tidal constituents for two points, one 30 km offshore – 30 m depth (D30) and in the mouth of the Paranaguá bay inlet (IL), (for location see Fig. 16.1), with their amplitudes (h in cm) and phases (φ in degrees) (after Marone et al. 2015)

	D30		IL	
	φ	h	φ	h
Constituent	(°)	(cm)	(°)	(cm)
M2	68	33	82	37
S2	72	21	82	25
O1	72	11	77	11
M4	140	11	180	8
K1	131	7	130	6
K2	57	7	62	9
M3	193	7	227	7
N2	138	6	159	7
MS4	230	5	257	3
MN4	93	4	132	3
MO3	41	3	65	3
MK3	74	3	78	3
M2	68	33	82	37
S2	72	21	82	25
O1	72	11	77	11
M4	140	11	180	8
K1	131	7	130	6
K2	57	7	62	9

Notes: *M2* Principal lunar semidiurnal constituent, *S2* Principal solar semidiurnal constituent, *O1* Lunar diurnal constituent, *M4* Shallow water overtides of principal lunar constituent, *K1* Lunar diurnal constituent, *K2* Lunisolar semidiurnal constituent, *M3* Lunar terdiurnal constituent, *N2* Larger lunar elliptic semidiurnal constituent, *MS4* Shallow water quarter diurnal constituent, *MN4* Shallow water quarter diurnal constituent, *MO3* Shallow water lunar third-diurnal constituent, *MK3* Shallow water terdiurnal

(Fig. 16.6). The highest energy at the coast is associated with winds from south-southeast, a rise in sea level (storm surge) and large waves, all of which also intensifies the coastal currents.

In shallow waters (~10 m), off the north sector of Pontal do Sul-Matinhos arc (25° 38′ 30″ S and 48°24′ 32″ W, ~2 km from the coast), the longitudinal currents are dominated by the low frequency oscillation, followed by the high frequency (tides) with predominately northward flow reinforced by the longitudinal wind component with a maximum bottom coastal current of 54.2 cm s^{-1} (Noernberg and Alberti 2014). Tides signal are however present in both the shore normal and long-shore components currents.

In the navigation channel of the southern Paranaguá inlet the transverse tidal currents are more energetic than longshore currents. The spring ebb currents average is 60 cm s^{-1} (Fig. 16.7). However during high-energy events the longshore currents are intensified reaching 70 cm s^{-1} and capable of transporting bottom sediments (Noernberg et al. 2007).

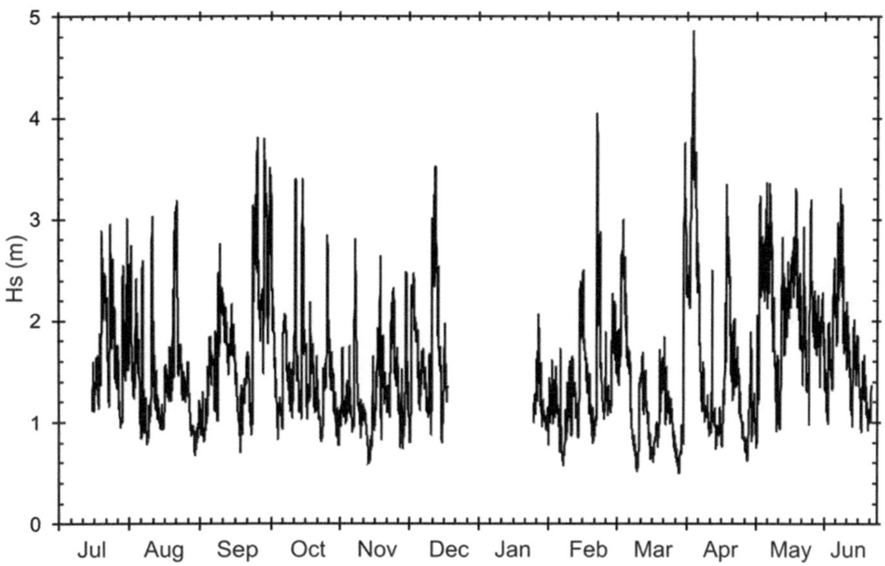

Fig. 16.6 Significant wave height (H_s) variability at the depth of 18 m between July 2009 and June 2010

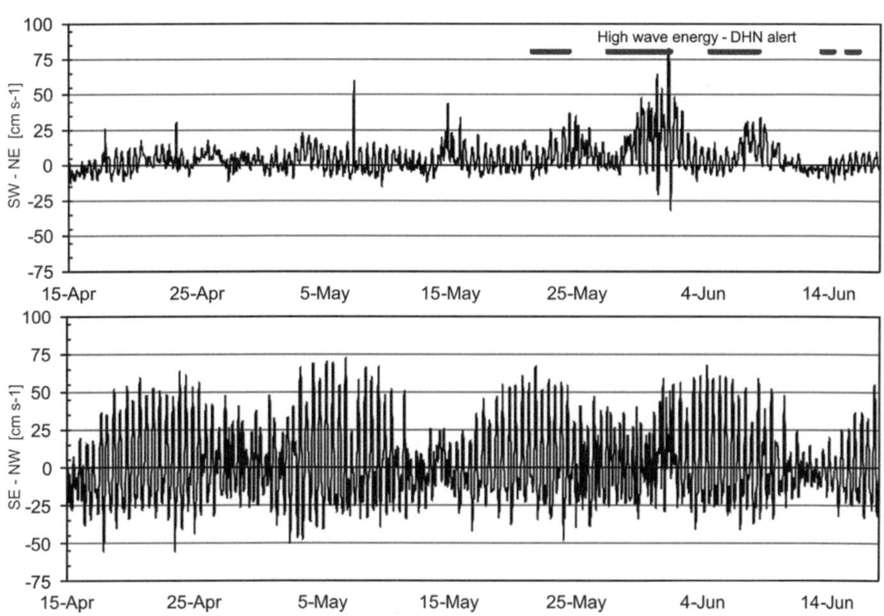

Fig. 16.7 Bottom currents in access channels of Paranaguá harbor – southern hemisphere autumn 2009. Transverse component (SE-NW) with positive values to ocean (NW – ebb currents) and longitudinal component (NE-SW) with positive values to NE. Periods of high wave energy is showed at the top from Brazilian Navy Hydrography and Navigation Service (DHN) (Source: Noernberg et al. 1997)

16.1.5 Coastal Sediments

16.1.5.1 Texture and Composition

The Paraná beaches are mainly composed of well-sorted medium to very-fine quartzose sand, with variable rates of heavy mineral and bioclastic carbonate (Bigarella et al. 1966, 1969, 1970/1971; Paranhos Filho 1996; Angulo et al. 1996; Mihály 1997; Giannini et al. 2004). The heavy mineral assemblage consists of zircon, tourmaline and hornblende (>15 % each) in addition to rutile, staurolite, pistacite/epidote, apatite, kyanite, sillimanite, garnet, titanite, oxy-hornblende, hypersthene, monazite and cassiterite (Angulo et al. 1996; Mihály 1997; Giannini et al. 2004; Freitas et al. 2014). The bioclastic grains are predominately mollusc shell fragments.

16.1.5.2 Source of Sediments

The source of Paraná beach sediments can be analyzed at different space and temporal scales. At regional scale (hundred of km) and Quaternary time (~2.6 Ma) the sediment source need to be evaluated in relation to sea-level changes. The ultimate sources of coastal sediments are the deeply weathered Achaean to Phanerozoic rocks whose sediments have been transported to the coast by rivers throughout the Quaternary. The sediments are deposited in deltas, estuaries, beaches, aeolian dunes and on the shelf. However, as Quaternary sea-level has been lower (tens to more than 100 meters) than present 94 % of the time, most of these deposits are located on the continental shelf. Numerous transgressive/regressive cycles during the Quaternary have reworked these sediments several times, resulting in mature polycyclic beach sediments at the coast. Evidences of this origin are (a) the highly mature texture – rounded, well sorted medium to fine to very fine quartzose beach sand; (b) a high concentration of heavy mineral on some Holocene paleobeach deposits that can reach 30–40 cm in thickness and (c) the contrast between the coarse angular crystal grains of the crystalline rock mineral and the fine rounded grains of the beach sand.

At present, the deposition of coarse fluvial sediments –sand and gravel – occurs mainly at the estuary head-deltas (Fig. 16.8). While the middle part of the estuaries is coincident with the maximum turbid zone (MTZ) with fine muddy bottom sediments (Fig. 16.9) suggesting that no sand is transported through the estuary to the open coast, because the fluvial sand cannot be transported through the low energy estuary zone (MTZ). However, within these low-energy zones are channels where fluvial and/or tidal currents may be able to transport sand size sediments through the inner part to the outer part of the estuaries.

At present time fluvial sand could also reach the beach system from the rivers that drain crystalline rocks terrains and flows into the outer part of the estuary, such as the Guaraguaçu River (330 km^2 catchment area, Fig. 16.3), where the sediments

Fig. 16.8 Estuarine head-delta of Cubatão River at Guaratuba estuarine complex in 1994

are transported by ebb tidal currents; or where rivers flows directly to the sea, such as the Saí-Guaçu and Matinhos rivers.

Large quantities of fine sediment and organic matter are presently transported by rivers and estuaries to the open sea, as can be seen at satellite images (Fig. 16.10) and deposited at the present (Veiga et al. 2006) and Holocene shoreface and inner shelf (Souza et al. 2012).

It can be concluded that Paraná beaches are mainly composed of mature polycyclic fine quartz sand derived mainly from the continental shelf and transported onshore by waves and currents (Souza et al. 2012). At the coast however the main longshore transport is wave-driven northerly currents, with rates on the order of 10^4–10^5 m^3 per year (Sayão 1989; Lessa et al. 2000; Lamour 2000; Lamour et al. 2006). Part of the sand could also be derived from erosion of Holocene and Pleistocene barriers and possibly by contemporary fluvial sediments supply.

16.2 Paraná Beach System

16.2.1 Previous Work on Paraná Beaches

Pioneer work on Paraná beaches was performed by J.J. Bigarella and collaborators during the middle of the twentieth century (Bigarella 1946, 1965, 1973; Bigarella et al. 1957, 1961, 1966, 1969, 1970/1971; Bigarella and Sanches 1966). This work focused on the beach mineralogy, texture and sedimentary structures. The 1980s

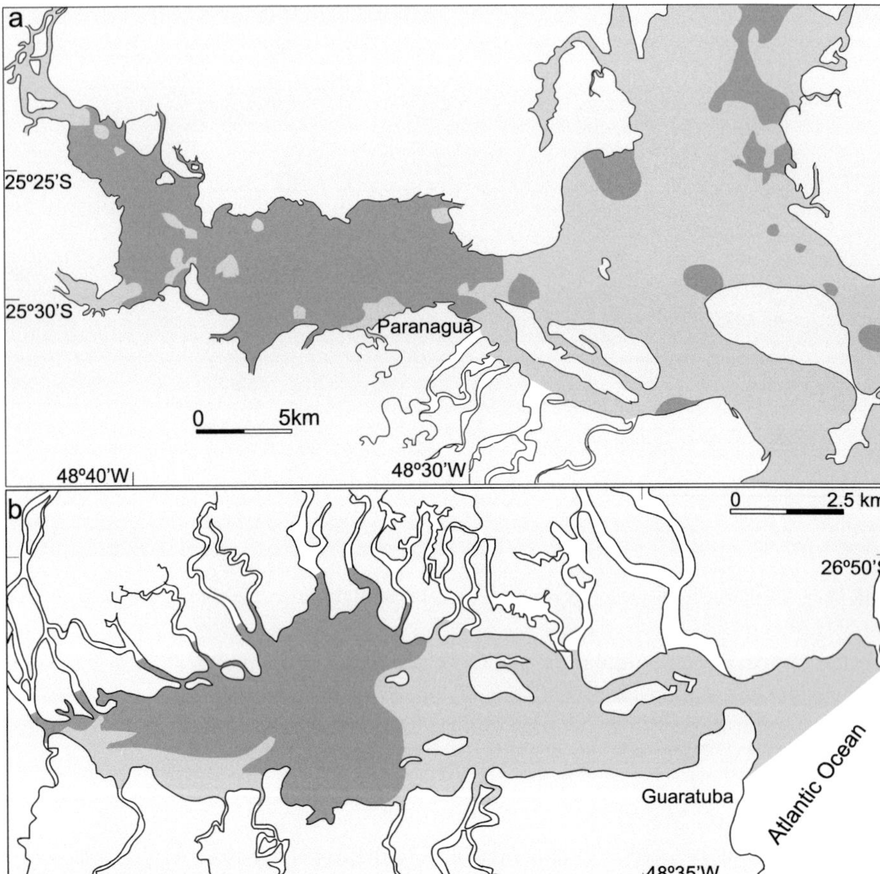

Fig. 16.9 Surface sedimentary facies of Paranaguá (**a**) and Guaratuba (**b**) estuarine complex. Sandy sediments (*medium gray*) and sandy-mud and muddy-sand sediments (*dark gray*) (After Angulo et al. 2009a). The sandy sediments areas at the estuarine head correspond to head-estuarine deltas and the muddy sediment areas correspond to MTZ

saw the beginning of published works concerning interaction of coastal dynamics and land use (Angulo 1984, 1993a, b; Angulo et al. 2006a) and a regional scale beach morphodynamic classification was proposed (Angulo 1993a; Angulo and Araújo 1996). In the 1990s research began on Paraná beach morphodynamics using the Australian approach, as well as its relation to benthic fauna (Borzone et al. 1996, 1998, 1999, 2003, 2007; Souza and Borzone 1996; Borzone and Souza 1997; Soares et al. 1997; Barros et al. 2001; Gandara-Martins 2007; Rosa and Borzone 2008). There was also work on beach texture and mineralogy (Angulo et al. 1996), erosion (Tessler and Mahiques 1993; Paranhos Filho et al. 1994; Paranhos Filho 1996; Mihály 1997; Araújo 2001; Mihály and Angulo 2002; Giannini et al. 2004; Angulo and Souza 2005; Angulo et al. 2006b, 2009b), beach volumetric variations

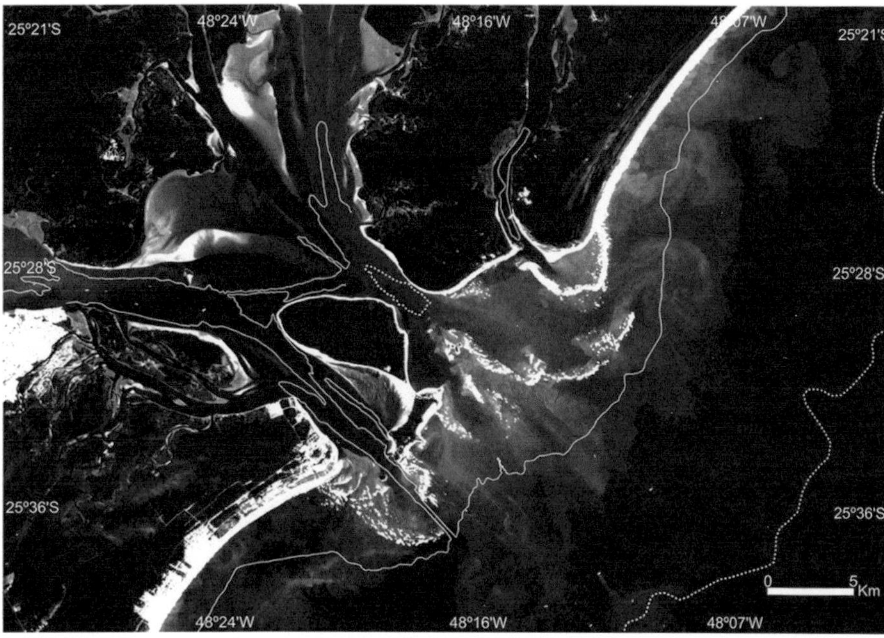

Fig. 16.10 Breaking waves over the ebb-tidal deltas shoals and estuarine plumes at the Paranaguá estuarine complex inlets and isobaths of 10 m (*solid white line*) and 20 m (*dashed white line*)

(Angulo and Soares 1994; Quadros 2002; Bessa Jr and Angulo 2003; Souza and Angulo 2003; Quadros et al. 2002, 2007, 2016), interaction between coastal dynamics and land (Angulo 1984, 1993b; Angulo and Araújo 1996; Angulo et al. 2006b, 2009b; Bessa Jr 2003; Bessa Jr and Angulo 2003; Paranhos Filho 1996; Paranhos Filho et al. 1994; Pierri et al. 2006; Sampaio 2006) and beach hazards and risk (Angelotti and Noernberg 2010).

16.2.2 Regional Characterization

The Paraná coast faces southeast and consists of several mainland beach arcs and two islands – Ilha do Mel and Ilha das Peças, and some headlands and rocky islands. There are also three large estuarine complexes: Guaratuba, Paranaguá and Canal do Superagüi-Baía dos Pinheiros; and two smaller estuaries, which define the north – Mar do Ararapira – and south – Rio Saí-Guaçu – state borders (Fig. 16.11). Large ebb-tidal deltas (named *barras* in Portuguese), are associated with the estuarine inlets and have considerable influence on wave approach and adjacent shoreline dynamics. Some beaches occurs inside the estuaries, however most of their shore is characterized by tidal-flats covered mainly by mangroves totaling ~377 km^2 (Fig. 16.12).

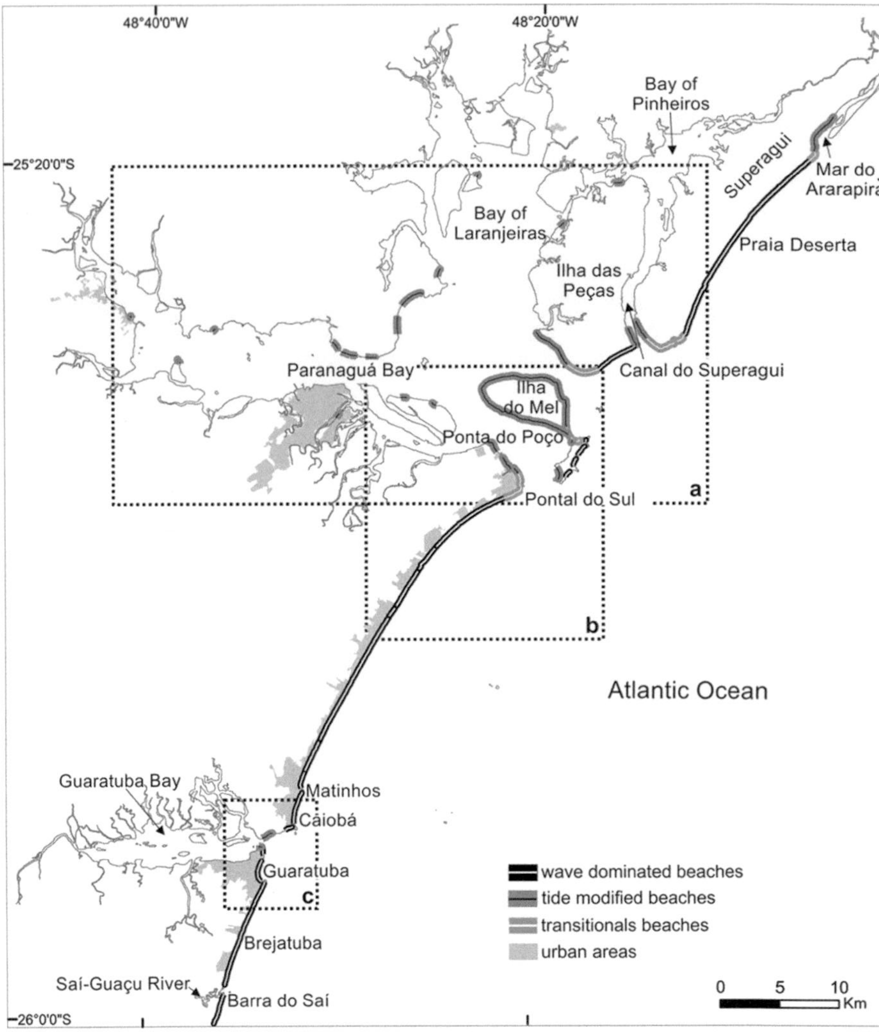

Fig. 16.11 Location of wave-dominated, tide-modified and transitional beaches of State of Paraná

Paraná beaches are characterized as both wave-dominated and tidal-modified beach types. The wave-dominated beaches are located on the open coast and on exposed sections and range from dissipative to intermediate. Along sectors were shoals of the ebb-tidal delta dissipate part of the wave energy the beaches transform from dissipative to reflective. The estuarine beaches occurred inside the estuaries where fetch is large enough to allow sufficient wave generation to form tide-modified beaches.

The open-coast beaches are all wave-dominated and occur within five beach arcs or embayments: from south to north: Brejatuba (10.1 km long), Guaratuba (2.3 km),

Fig. 16.12 Estuarine coast with large tidal-flats covered by mangroves at Guaratuba bay in 1994

Caiobá-Matinhos (3.0 km), Matinhos-Pontal do Sul (34.5 km) and Praia Deserta (21.3 km) (Fig. 16.11). Moreover at the ends of some of these arcs the beaches become increasingly influenced by inlet and ebb tide shoal dynamics. The open-coast beaches have long-term (decadal) stability and the shoreline has not shifted more than 10 m over the past five decades (Angulo 1993a; Angulo and Soares 1994). In contrast the inlet beaches associated with the Saí-Guaçu, Guaratuba, Paranaguá, Canal do Superagüi and Ararapira inlets (Fig. 16.11) are dominated by waves, wave-generated currents, and ebb- and flood-tidal currents (Fig. 16.10). As a consequence, sections of the shorelines of these beaches have moved more than 1 km during the last five decades (Angulo 1993a; Angulo and Araújo 1996), including the shoreline at Ilha das Peças that has moved 600 m in 3 years (Angulo 1993a; Angulo and Araújo 1996 Fig. 16.13).

16.2.3 Beach Morphodynamics

16.2.3.1 Ocean Beaches

Morphodynamic studies on Paraná wave-dominated ocean beaches have demonstrated the importance of the abundant fine sand to morphodynamic behavior. Extensive low gradient beach profiles dominate most of the coast (Table 16.2). In the north, Deserta beach has a high dimensionless fall velocity (Ω), and a relative homogeneous dissipative profile along its 21.3 km length (Souza and Borzone 1996; Gandara-Martins 2007). The Paranaguá inlet dynamics influences adjacent beaches,

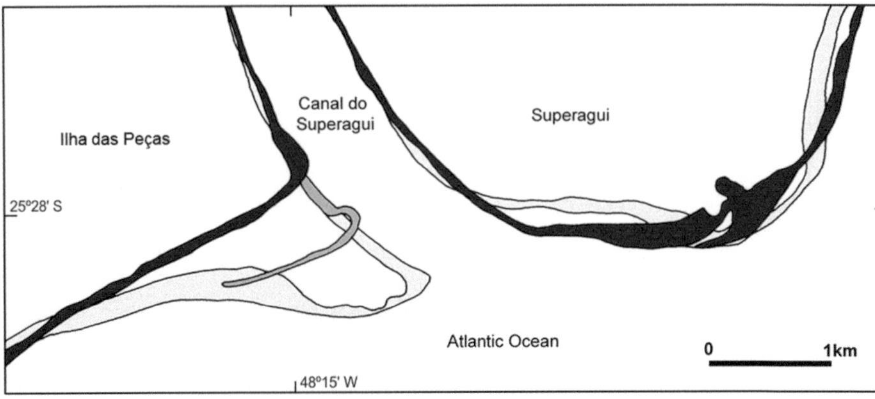

Fig. 16.13 Coastal changes at the Canal do Superagüi inlet between 1992 and 1980 (After Angulo and Araújo 1996). Subaerial beach in 1952 (*light gray*), 1955 (*medium gray*) and 1980 (*dark gray*)

like the northern portion of Ilha das Peças and the northern portion of the Matinhos-Pontal do Sul beach-arc (Atami beach). The beaches here contain double or multi-bar systems with more intermediate values of Ω (Borzone et al. 1998).

At Ilha do Mel, three oceanic beaches are located between the eastern headlands and exhibit dissipative morphodynamics with extensive (>100 m wide) and very low gradient profiles (1.14°) (Borzone et al. 1996). Recent substantial shoreline progradation resulting from an alteration to the nearshore profile changed the modal state to a more intermediate condition on one of these beaches (Praia de Fora Sul).

In middle section of the Matinhos-Pontal do Sul beach-arc, an increase in grain size results in a shift to an intermediate state with a persistent bar-through system (Marisol to Praia de Leste) and rhythmic bar and beach (Monções to Caravelas) in the middle of the arc (Borzone et al. 1998; Quadros et al. 2016, Fig. 16.11 and Table 16.2). Under higher swell conditions the beaches of this arc shift to more dissipative conditions (Quadros et al. 2016). At Caiobá-Matinhos beach-arc, this condition persists, with the exception of Caiobá beach that has finer sediment and more dissipative conditions (Barros et al. 2001).

Northern Guaratuba Inlet is under some tidal influence (RTR>3), and has the only strictly reflective beach (Praia Mansa) along the Paraná coast (Barros et al. 2001; Borzone et al. 2007). Southern Guaratuba inlet, which has a gradual increase in grain size from north to south, produces intermediate beaches, with predominately rhythmic and transverse bar beach systems (Barros et al. 2001).

16.2.3.2 Estuarine Beaches

The estuarine beaches occur mainly inside Paranaguá estuarine complex, around its inlets and along the northern border of Antonina-Paranaguá axis as well as along western border of Baía das Laranjeiras. These beaches vary in length from less than

Table 16.2 Morphodynamic parameters of the wave-dominated oceanic Paraná beaches (for location see Fig. 16.11)

	Beach face				Wave			
	Width	Slope	Mean Ø	Sorting	H_b	Period		
Beach name	(m)	(°)	(Φ)	(Φ)	(m)	(s)	Ω	RTR
Deserta[a]	83	1.5	2.65	0.40	1.15	9.3	6.5	1.48
Peças Norte[a]	84	1.4	2.71	0.33	1.07	12.0	5.2	1.59
Ilha do Mel[b]	103	1.3	2.79	0.25	1.17	10.8	5.9	1.45
Atami[b]	105	1.1	2.88	0.34	0.69	7.0	3.8	2.46
Atami[c]	100	0.5	2.53	0.65	0.75	7.7	4.2	2.27
Marisol[c]	113	1.1	2.46	0.65	0.90	7.9	5.1	1.89
Ipanema[c]	107	1.4	2.47	0.70	0.90	8.4	5.1	1.89
Guarapari[c]	71	1.5	2.43	0.59	0.85	8.5	4.8	2.0
Santa Terezinha[c]	53	2.0	2.43	0.52	9.00	8.8	5.1	1.89
Praia de leste[c]	81	2.2	2.41	0.66	8.00	8.5	4.5	2.13
Praia de leste[b]	75	1.9	2.50	0.40	0.75	7.0	2.9	2.26
Praia de leste[d]	65	1.4	2.48	0.45	0.80	8.7	3.8	2.12
Monções	55	3.3	2.36	0.87	0.80	8.6	3.9	2.13
Caravelas	39	4.1	2.36	0.65	0.90	8.4	4.4	1.9
Caiobá[d]	75	1.3	2.84	0.29	1.25	10.0	7.3	1.36
Mansa[d]	60	2.9	1.78	0.83	0.53	9.0	1.6	3.21
Guaratuba[d]	80	1.8	2.30	0.62	0.87	9.7	3.5	1.95
Brejatuba[d]	50	1.9	2.23	0.59	1.15	10.3	3.9	1.48

Sources: [a]Gandara-Martins (2007); [b]Borzone et al. (1996); [c]Quadros et al. (2016); [d]Borzone et al. (1998); *RTR* Relative tidal range

100 m to kilometers (5.4 km). Although refracted ocean waves may be present, the dominant process of sediment reworking is by locally generated wind waves in a fetch-limited environment, which are characterized by low height (<0.15 m) and short period (<4 s). These low-energy sand beaches have large relative tidal ranges and consequently a higher RTR (~10) and are tide-modified grading to tide-dominated (Borzone et al. 2003; Rosa and Borzone 2008) (Table 16.3).

The continuous ship traffic of Paranaguá Harbor produced substantial boat wakes that also contribute to estuarine beach formation and maintenance.

The estuarine beach morphology is characterized by a short and steep beach face with a broad and flat low-tide terrace terminating at a prominent break in slope (Nordstrom 1992; Rosa and Borzone 2008). However, when close to channels, the strong ebb and flood currents may prevent low-tide terrace formation (Rosa and Borzone 2008). The inlet influence also produces beaches with more complex morphology, where convex beach faces and low-tide terrace with bars and troughs can be observed. On the other hand, beaches located in the inner estuarine have straight beach face and low-tide terrace profiles (Rosa and Borzone 2008).

Table 16.3 Morphodynamic parameters of the tide-modified estuarine Paraná beaches (for location see Fig. 16.11)

	Beach face				Tidal flat				Wave				
	Width	Slope	Mean Ø	Sorting	Width	Slope	Mean Ø		Hb	Period			
Place	(m)	(°)	(Φ)	(Φ)	(m)	(°)	(mm)	Sorting	(m)	(s)	TR	Ω	RTR
Brasília	24	4.2	2.45	0.49	>25	0.7	0.15	0.35	0.15	3.5	1.92	1.9	13
Ilha das Cobras	18	6.9	1.04	0.56	–	–	–	–	0.23	2.8	2.07	1.1	9
Piaçaguera	12	4.8	1.95	0.44	70	0.8	0.24	0.45	0.08	2.2	2.09	1.0	26
Europinha	11	5.4	0.41	1.09	>33	0.8	0.41	1.30	0.08	2.5	2.42	0.2	30
Ponta da Pita	14	6.6	0.39	1.28	>15	0.6	0.45	1.55	0.05	2.0	2.74	0.2	55
Ilha das Gamelas	16	6.4	1.02	0.99	121	0.2	0.19	0.75	0.08	3.6	2.10	0.3	26
Ponta do Pasto	16	7.4	1.72	0.55	>20	1.2	0.22	0.65	0.12	2.1	2.07	1.3	17
Ponta da Cruz	15	6.5	1.76	0.86	>12	1.3	0.26	1.10	0.20	3.1	2.07	1.5	10
Techint	21	4.2	2.19	0.70	145	0.2	0.19	0.55	0.30	3.7	1.92	2.8	6
Ponta do Poço	19	5.9	2.65	0.39	–	–	–	–	0.13	3.9	1.92	1.7	15
Coroazinha	22	4.3	2.30	0.43	18	0.6	0.20	0.55	0.13	3.2	1.92	1.5	15
Vila das Peças	36	4.2	2.53	0.42	–	–	–	–	0.10	2.3	1.74	2.1	17
Peças Cemitério	31	3.1	2.60	0.40	102	0.4	0.15	0.40	0.13	3.7	1.74	1.7	14
Superagui Vila	45	2.9	2.46	0.50	–	–	–	–	0.12	3.5	1.75	1.5	14
Superagui Peças	20	4.4	2.55	0.40	24	0.8	0.17	0.50	0.07	2.0	1.76	1.7	24

Notes: *TR* Tidal range, *RTR* Relative tidal range.

The sedimentary composition of these beaches shows sharp shifts along the estuary gradient. The tide-modified beaches located close to inlets have both a beach face and low-tide terrace composed by well-sorted fine sands while the inner more tide-dominated beaches have beach faces composed by poorly-sorted coarser sands while the low-tide terrace to tidal flats become muddy (Rosa and Borzone 2008). These granulometric differences between beach face and low tide terrace denotes a lack of sedimentary exchange between them, as is typical of tide-dominated beaches (Short 2006).

16.2.3.3 Seasonal Variation

Seasonal variations in morphodynamics states and sediment characteristic were studied at two oceanic beaches: Atami beach (Soares et al. 1997; Quadros et al. 2007) and Caravelas beach (Quadros et al. 2007) (Figs. 16.11 and 16.14). The first beach had Ω values varying from 2.6 to 12.4, with values greater than 5 most of the year. Intermediate values occurred during spring and summer, when the subtidal beach exhibits a strong profile variation as a consequence of bar formation, together with a decrease in sorting. Dissipative conditions occurred during winter, producing smoother profiles and maximum sorting (Soares et al. 1997) (Fig. 16.14). Caravelas beach, with a narrow profile and low Ω values, showed the strongest total profile variation associated with high energy events (cold fronts). Sediment exchange between the subaerial to the subtidal portion of the beach was evident at both beaches (Quadros et al. 2007) (Fig. 16.14).

In the Matinhos–Pontal do Sul beach-arc the subaerial beach sand volume increased during summer season when sea waves ($T_p < 10$ s) dominated and decreased during fall and spring when swell ($T_p > 10$ s) predominated (Quadros et al. 2016). Quadros et al. (2016) also recorded the impact of a high wave event between 25 May and 07 June of 2000 at two locations on this beach-arc (Atami and Caravelas). They found erosion of the subaerial beach portion (6.9 m^3m^{-1} and 69.2 m^3m^{-1}) and deposition of the subtidal beach (109.7 m^3m^{-1} and 66.6 m m^{-1}) resulted in a positive budget of 102.8 m^3m^{-1} and negative of 2.6 m^3m^{-1} respectively.

16.2.4 Beach-Dune Interactions

The regressive Quaternary coastal plain of Paraná consists of foredune ridges and paleo-foredune ridges (Angulo 1993c). The paleo-foredune ridges are conspicuous across the Holocene coastal plain but are very rare on the Pleistocene terrains. The Holocene ridges are 1–8 m high, 20–250 m wide and extend several kilometers longshore (Fig. 16.15) and are covered by dense vegetation of bushes and trees. Paleo-soils suggest a long period of stabilization (Bigarella et al. 1970/1971). Unfortunately, most of these ridges have now been obliterated by urbanization.

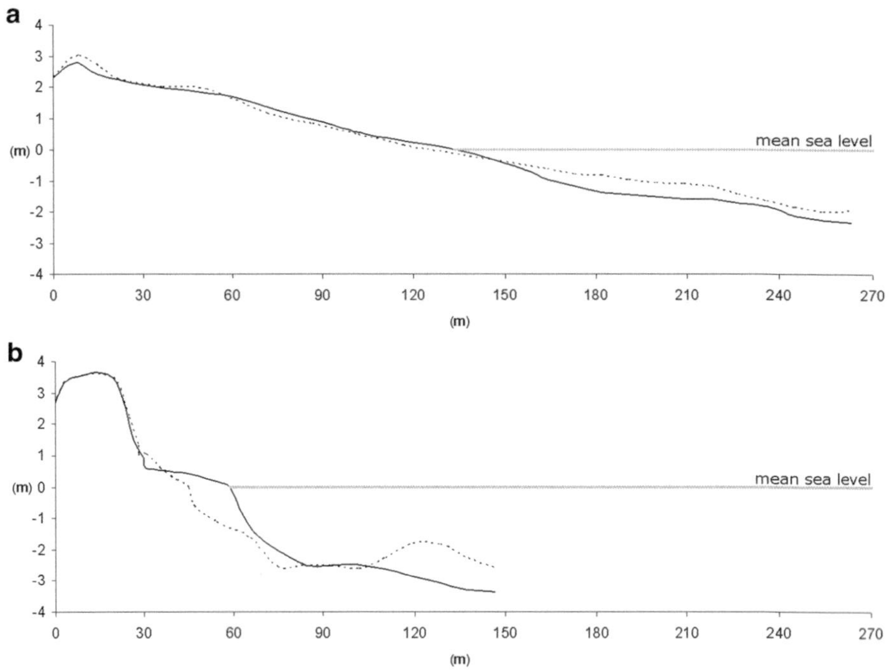

Fig. 16.14 Atami (**a**) and Caravelas (**b**) beach profiles before (May 25th, *solid line*) and after (June 7th, *dashed line*) a storm event in May 2000 (After Quadros et al. 2007) (For location see Fig. 16.11)

On the open-sea coast there are usually one or more foredune ridges formed during the past few decades. However no detailed research into foredune evolution has been undertaken along the Paraná coast. Paleo-blowouts on the paleo-ridges suggest that the south and southeast winds were most effective in constructing the foredunes (Fig. 16.15), which is in agreement with the strongest current wind direction (Angulo 1993c) (Fig. 16.5). The most effective conditions to accrete foredunes - strong winds and low rain – seems to occur during the spring season (Angulo 1993c). During storms, waves can reach and total or partially erode the foredune ridge (Fig. 16.16), which then recover during fair wave windy periods.

16.2.5 Longshore Sand Transport and Sediment Cells

At a regional scale the Paraná coast is part of a coastal sediment system that extends at least from Barra Velha (25°39′S) to Ilha do Cardoso (25°09′S) (Lessa et al. 2000). Sand is transported by wave-driven longshore currents both north and south by the two opposing south-southeast and east-northeast wave systems. At the estuarine inlets tidal currents interrupt the longshore currents and sand is transported on and

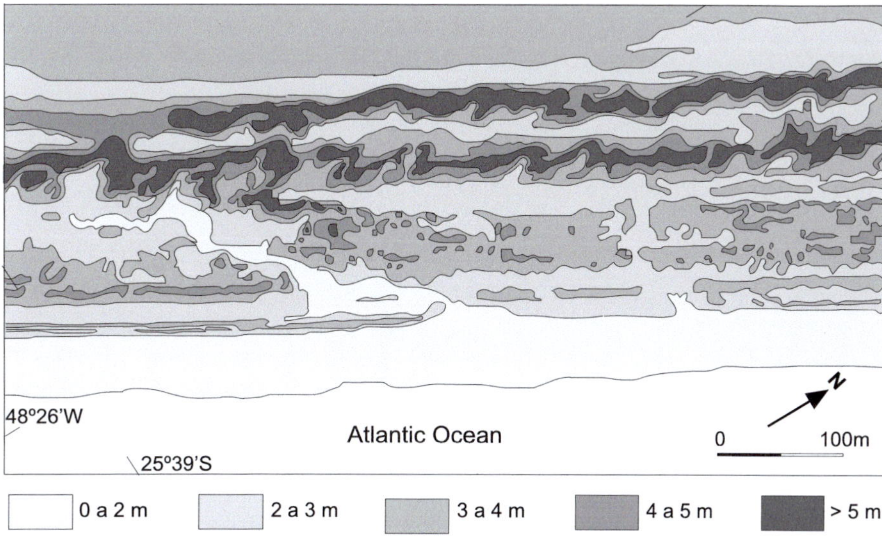

Fig. 16.15 Paleo-foredune ridges at the northern part of the Matinhos – Pontal do Sul beach arc (After Angulo 1993c)

Fig. 16.16 Foredune ridges partially eroded after a storm at Praia Deserta beach-arc at 11 March 1987

off-shore by wave and tidal currents, with tidal currents segregated into ebb- and flood-dominated tidal channels, similar to the Hayes (1975) model. They also have large shallow ebb-tidal deltas resulting from large tidal prisms and interaction of wave generated and ebb-tidal currents. Sand volumes of these deltas range form 10^8 m^3 (southern inlet of Paranaguá, Lamour et al. 2006) to 2×10^5 m^3 (Ararapira inlet, Angulo et al. 2006a, Fig. 16.17). The frontal lobes of the ebb-tidal deltas correspond to shoal areas <10 m depth where wave generated currents can transport the sand from one side to the other side of the inlet. On the ebb-tidal deltas the flood tidal

Fig. 16.17 Ararapira ebb-tidal delta at July 1994

currents transport sand through the flood-tidal channels to inside the main ebb-tidal channel, while ebb-tidal currents transport the sand to the frontal lobe, where the sand can then be transported by wave generated currents. As a consequence, the tidal delta can be considered as part of a single sand-integrated, transport system extending from Barra Velha to Ilha do Cardoso (Fig. 16.2). However successive dredging and deepening of the access channels to the harbors of Paranaguá and São Francisco since the middle of the twentieth century, has caused a decrease in downdrift sand tidal-delta volume and severe coastal erosion along downdrift beaches (Angulo et al. 2006a).

16.2.5.1 Headland Bypassing

At Morro do Cristo sand moves around the point and emerges on Guaratuba beach as migratory sand spit (Fig. 16.18). The spit is part of the headland sand bypassing system that transfers sand from Brejatuba beach, around the headland and onto Guaratuba beach. The arrival and migration of the spit causes tens of meters of change in the shoreline of the beach on a scale of years to decades. Headland sand bypassing also occurs commonly in Santa Catarina and follows the model proposed by Short and Masselink (1999).

16.2.5.2 Beach Stability

As previously mentioned, Paraná beaches possess varying degrees of stability. Most of the open-sea beaches are presently in a state of dynamic equilibrium, with the shoreline changing less than 10 m over a period of 4–5 decades (Angulo 1993b; Angulo and Soares 1994). The sand budget is maintained by a sand reservoir contained in the foredunes. The sand eroded from the beach and foredune during

Fig. 16.18 View of the sand spits at the south part of the Guaratuba beach-arc in (**a**) 1994; and (**b**) 2014 when the spit has migrated along the beach impounding a lagoon. It is part of the northward headland sand bypassing around the headland (For location see Fig. 16.11)

storms returns to the beach during lower energy wave conditions and gradually accumulates in the upper berm and foredune. During storms the morphodynamics state shifts to dissipative returning to an intermediate state during fair weather waves conditions (Quadros 2002). In addition a rotation along the log-spiral Guaratuba Beach was observed, induced by alternating periods of east-northeast and south-southeast wave dominance (Bessa and Angulo 2003).

The beaches influenced by inlet dynamics are much more unstable, with annual or interanual changes in the ebb-tidal deltas inducing meters to hundred of meter changes at the shoreline, as recorded at Guaratuba, Paranaguá, Canal do Superagüi and Ararapira inlets (Angulo 1993b). At Prainha Beach the shift of the main-ebb-tidal-channel of the Guaratuba ebb-tidal delta caused the shoreline to shift 200 m in 26 years (Fig. 16.19, Angulo and Araújo 1996). At Superagüi and Ilha das Peças

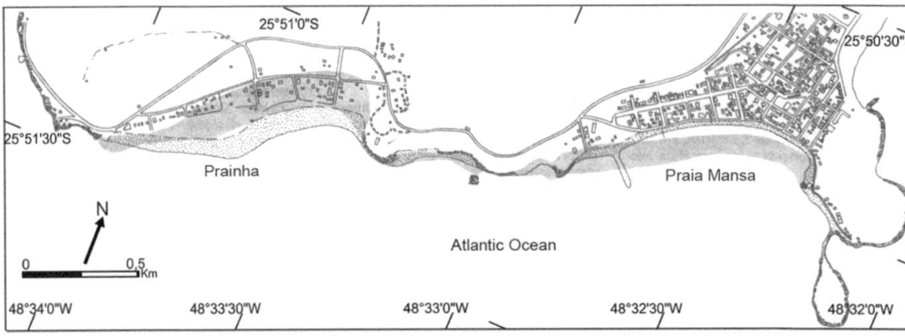

Fig. 16.19 Changes of coastline and beach morphology at Praia Mansa and Prainha induced by main-ebb-tidal-channel shift (After Angulo and Araújo 1996). Subaerial beach in 1953 (*gray area*) and 1980 (*dotted area*)

beach morphology changed in response to changes in the morphology of the Canal do Superagüi ebb-tidal delta (Fig. 16.13, Angulo and Araújo 1996).

At Mar do Ararapira and Ilha do Mel two centennial to millennial scale changes in the coastal configuration were identified, changes that could happen again during the next decades to centuries. Mar do Ararapira Inlet migrates southward promoting severe erosion at the southern margin and extensive progradation at the northern coast (Tessler and Mahiques 1993; Miháły and Angulo 2002; Angulo et al. 2009b). Simultaneously, inside the Mar do Ararapira estuary tidal currents eroded the estuarine concave margins and at one place the erosion might cut the spit, which isolates the estuary from the sea (Fig. 16.20). This process could occur during the present decade and cause opening of a new inlet 6 km north of the present one. In turn, new inlet formation would promote: (a) closure of the present inlet; (b) collapse of the present ebb-tidal delta; and (c) coastal erosion and progradation of the adjacent shoreline on the scale of hundreds of meters (Angulo et al. 2009b; Fig. 16.21).

Ilha do Mel is located at between the two inlets – north and southeast – of Paranaguá estuarine complex (Figs. 16.1 and 16.11) and due to inlet processes is highly unstable. In the long term – millennial – longshore transport promoted the northward migration of the southeast inlet (Lessa et al. 2000), with the ongoing migration prevented by Ilha do Mel crystalline rocky hills. However, the tidal current have now begun to erode the island's coastal plain located landward from the hills (Araújo 2001). During the 1990s erosion progressed to where the narrow eroded isthmus was overwashed by storm waves (Giannini et al. 2004; Fig. 16.22). This process will probably continue until an inlet opens. After such an opening, the littoral transport could promote: (a) closure of the present southeast inlet; (b) linking of the south part of the Ilha do Mel to the Pontal do Sul coastal plain; (c) total erosion of the northwest part of the island; or (d) formation of only a single inlet. However the dredging of the access channel to the Paranaguá harbour seems to have interrupted the littoral drift and may avoid those dramatic changes to the Ilha do Mel coast.

Along the island's western coast (Fig. 16.23) and along the Saco do Limoeiro coast beach erosion alternates with beach progradation over periods of years. Those changes are attributed to oblique sand bars migration (Fig. 16.24).

Fig. 16.20 Narrow part of the Ararapira spit in December 2008 where the opening of a new inlet could occur during the current decade (After Angulo et al. 2009b)

Fig. 16.21 Expected coastal changes after Ararapira new inlet opening (After Angulo et al. 2009b). Present (*solid line*) and expected (*dashed line*) coastlines and present (*p*) and expected (*e*) inlets (*i*) and ebb-tidal deltas (*d*). Areas expected to be eroded (*light gray*), accreted (*dark gray*) or silting (*medium gray*) after new inlet opening

16.3 Beach Use and Abuse

16.3.1 Beach Development and Management

The Paraná coastal zone was settled between 7000 and 1000 years ago by a prehistoric population, called sambaquieiros (Gaspar 1996), whose occupation left almost 300 shell-middens along the coastal zone (Bigarella 1950/1951a, b; Parellada and Gottardi Neto 1994). Later, the region was occupied by Carijós (Santos 1850) a linguistic group related to Tupi-Guarani (Rodrigues 1985). At the beginning of the sixteenth century Portuguese arrived in the region and settled at Ilha da Cotinga on the inner part of Baía de Paranaguá (Bigarella 1991). Until the 1920s the occupation

Fig. 16.22 Ilha do Mel isthmus where over-washed by storm waves in 1996

Fig. 16.23 Cliff (*c*), beach (*b*) and mangroves (*m*) at the west coast of Ilha do Mel in 2003

was concentrated inland and on the estuarine coasts, with the beaches unoccupied except for a few (<10) small fishing settlements (Andriguetto-Filho 2002). The Paraná beaches were use as a transport route between the ports of Paranaguá and Guaratuba. Part of the trip was made over the beach between Pontal do Sul and Matinhos by foot, horse and wagons, and more recently by cars and buses (Bigarella 1991). At Paraná the use of beaches for recreation began in the 1920s (Bigarella 1991). However, until the 1950s there were few coastal settlements until the release of many housing lots in the 1960s and 1970s (Pierri et al. 2006; Sampaio 2006). During the 1980s several laws protected environmental areas, including the *APA de Guaraqueçaba* (Environmental Protected Area of Guaraqueçaba) instituted in 1985 and the *Parque Nacional do Superagüi* (National Park of Superagüi) instituted in 1989, which covers the entire northern Paraná coastal zone. At present, there is a marked contrast in the occupation and use of the Paraná beaches; three different sectors can be recognized: the north sector, the Ilha do Mel, and central-south sector.

Fig. 16.24 Rhythmic enlargement (*e*) of subaerial beach in 1980 (**a**) generated by sand waves (**b**) at Saco do Limoeiro at Ilha do Mel

16.3.1.1 The Northern Sector

The north sector of the Paraná coast contains the Superagüi and the islands of the Peças coast (Fig. 16.11), where beaches remain in natural conditions with little occupation (Fig. 16.25). Only along the estuarine beaches do small fisherman villages exist, which are gradually shifting from fishing to tourism (Fig. 16.26).

16.3.1.2 The Ilha do Mel Sector

The Ilha do Mel is located between the two Paranaguá estuarine complex inlets (Figs. 16.1 and 16.11) and after Iguaçu falls – is the second most popular tourist destination in Paraná. Island occupation is restricted and no cars or motorcycles are

Fig. 16.25 Praia Deserta beach-arc in 1994

allowed. Most of the island (~90 %) is protected by environmental laws and only three main sectors are occupied – Encantadas, Nova Brasília and Fortaleza (Fig. 16.11). Several partly successful attempts have been made to regulate island occupation, however conflicts remain as to the island's future.

In addition, there are several problems related to the island's dynamic beaches. In the Fortaleza-Farol das Conchas beach-arc large changes in coastal morphology have occurred since the 1980s and have accelerated towards the end of twentieth century. In 1980 the beach-arc had a typical log-spiral shape with a wider subaerial beach next to the Farol das Conchas headland (Fig. 16.27). During the 1990s, a large sand spit grew in lee of the headland and isolated a small bay were mangrove developed (Fig. 16.28). By 2002 a 0.32 km^2 coastal plain emerged and the former concave coastline shifted to convex (Fig. 16.29). This change was generated by sand supply from the east bypassing Farol das Conchas headland (Giannini et al. 2004). The sand accumulated in the shoaling wave and surf zones and changed the direction of wave approach. As a consequence, intense erosion is occurring along the central and northern sectors of the beach-arc and is threatening to destroy some residences (Fig. 16.30). Simultaneously intense progradation occurred at the southern and eastern sectors where coastline has prograded more than 400 m (Fig. 16.29). By the end of the twentieth century, erosion was also severe along the island's isthmus and it was overwashed by storm waves (Fig. 16.22). At that time, it was expected that a new inlet would form and split the island, but the growing sand spit inhibited the erosion at the isthmus and northward of the beach-arc. During 2000–2010 substantial progradation (~350 m) occurred at Encantadas de Fora beach (Fig. 16.31). The causes of these coastal changes are not fully understood. On the one hand, the island is strongly influenced by shifts of the ebb-tidal delta shoals, which could induced coastline shift of hundred of meters in few (<10) years (Angulo 1993a).

Fig. 16.26 (**a**) Vila Fátima and (**b**) Tibicanga fisherman villages in 2004

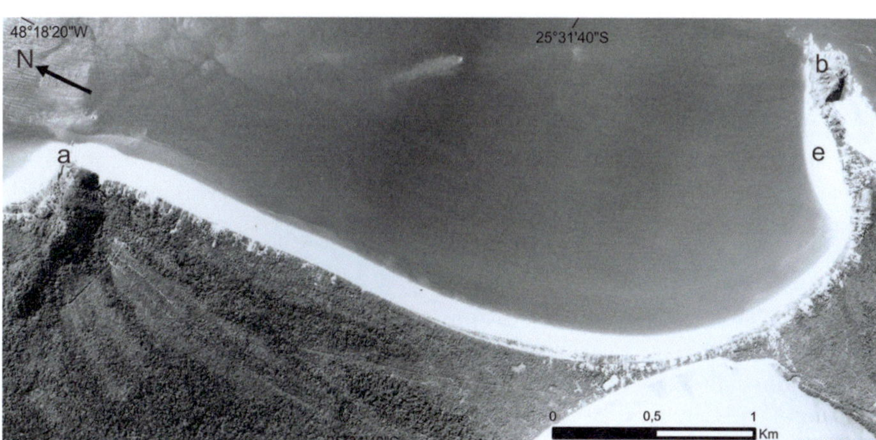

Fig. 16.27 Fortaleza-Farol das Conchas (*a–b*) beach-arc in 1980, with a slight subaerial beach enlargement (*e*) at the south part of the beach-arc

Fig. 16.28 Coastal plain emerged between 1980 and 1994 at the south part of Fortaleza-Farol das Conchas beach-arc. Coastline in 1980 (*dashed line*)

Fig. 16.29 Coastal plain emerged since 1980 in 2006 at the south part of the Fortaleza – Farol das Conchas beach-arc. Coastline in 1980 (*dashed line*)

Conversely, irregular disposal of sand dredging from the navigation channel of Paranaguá harbor probably induced intense progradation at the eastern part at Morro do Farol das Conchas – Forte beach arc and Encantadas de Fora beach (Angulo et al. 2006b).

16.3.1.3 The Central-Southern Sector

The central-south sector of Paraná coast extends from Ponta do Poço to Barra do Saí (Fig. 16.11) and, in contrast to the northern sector, is almost entirely occupied by continuous-linear-littoral-occupation (Deschamps and Kleinke 2000). This sector includes Guaratuba, Matinhos and Pontal do Paraná municipalities, which in the 1990 had high rates (5–11 %) of population growth (Deschamps and Kleinke 2000).

Fig. 16.30 Beach erosion at the north part of the Fortaleza – Farol das Conchas beach-arc in 2004

Fig. 16.31 (**a**) Encantadas de Fora beach in 2004; and (**b**) after substantial progradation in 2014. Note progradation in front of building

Fig. 16.32 Beira Mar Avenue built over the dynamic beach fringe at the southern part of the Matinhos – Pontal do Sul beach arc in 1994

By 2000, 70 % of the residences (~44,000) of those municipalities were second-residence and occupied occasionally, whereas only 30 % (~19,000) were occupied by permanent residents (Deschamps and Kleinke 2000).

The Paraná coastal urban regulation was precarious until the 1980s when State laws developed constructive regulations to prevent environmental and urban degradation (Sampaio 2006). However, there were several cases, where the state regulations were not respected. A classic case was the road (Beira Mar Avenue) built along the coast between Matinhos and Praia de Leste, which in some sectors were built on the foredune ridges and the dynamic beach fringe. After a few years the storm waves attacked the works and so began a classic auto-cyclic process of beach erosion – straightness and lowering – that has progressed northward (Fig. 16.32). Stone seawalls were built several times to protect the road, but the beach erosion has accelerated and the subaerial beach has been almost completely eroded along some sectors (Fig. 16.33). Until 2000 there was irregular squatter occupation of the foreshore just beyond the southern end of the road. The settlement was partially destroyed during a storm (Fig. 16.34), after which it was completely removed by authorities. As a consequence the subaerial beach and the foredune ridges have recovered in this section (Fig. 16.35), which is a good example of how the removal of hazardous occupation can help restore the beach.

The beach to the north of Matinhos headland presents a typical log-spiral form (Fig. 16.36). In the southern spiral occupation has advanced over the beach and altered the natural shoreline form causing a chronic beach erosion problem (Fig. 16.36). This beach sector could be easily restored by removing the few (<10) constructed building.

The Caiobá-Matinhos beach-arc has experienced erosion problems since the 1970s after the construction of Atlantic Avenue, which was built on the foredune ridge and on the beach in some regions (Angulo 1993a; Figs. 16.37 and 16.38). Since then, the subaerial beach has been eroded and infrastructure destroyed several

Fig. 16.33 Stone walls built to protect the road at the southern part of the Matinhos – Pontal do Sul beach-arc in the 1990s

Fig. 16.34 Damage caused by a high-energy wave event associated with both spring tides and storm surge in May 2001

times resulting in the construction of erosion control works, including seawalls, gabion-walls, and gabion-groins (Fig. 16.39). At the present time, erosion continues along the northern part of the beach-arc (Fig. 16.40), but in the central and southern part of the arc there has been substantial progradation, beach recovery, and development of several foredune ridges (Fig. 16.41). These changes were induced by the shifting of the Guaratuba ebb-tidal delta lobe, which generated a convergent wave refraction zone, causing accretion in the wave-shadow zone in lee of the shoal. At the same time, delta-margin erosion occurred immediately north of the shoal (Fig. 16.42).

Fig. 16.35 Beach and foredune naturally recovered after buildings were removed in 2004. The place is the same as in Fig. 16.34. Note the street lamp for comparison

Fig. 16.36 Vertical aerial photography (1997) of the Matinhos beach near to the headland showing the occupation of the beach (*arrows*) has altered the natural shoreline form, which has caused a chronic beach erosion problem

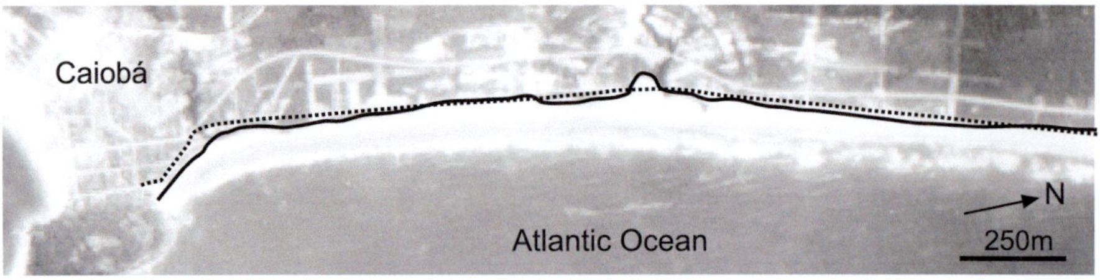

Fig. 16.37 Coastline in 1953 and (*solid line*) Atlantic Avenue (*doted line*) built partially over the beach and the dynamic beach fringe at Caiobá-Matinhos beach-arc

Fig. 16.38 Erosion problem at the central part of the Caiobá-Matinhos beach-arc in 1980. Note the earlier rock-wall that was built on the beach and destroyed by waves

Fig. 16.39 Gabions sea-wall and groins damaged after a storm in 1994 at the north part of the Matinhos-Caiobá beach-arc

Close to the Guaratuba Inlet there are three beaches: – Mansa, Prainha and Caieiras – which experience periods of progradation and erosion caused by shifting of the ebb-tidal delta shoals and channels (Figs. 16.11 and 16.42). At Mansa in the 1970s intense beach erosion occurred during storm events. This erosion was related, in part, to changes of the Guaratuba ebb-tidal delta shoals that decreased wave protection and allow higher wave action to reach this usually low-energy reflective beach. The erosion progressed until it endangered infrastructure. A study performed by the Civil Engineering National Laboratory of Portugal (LNEC 1977) proposes beach nourishment and construction of a terminal groin at the downdrift (west) end of the beach to avoid sand loses. The groin was built but the nourishment was not possible owing to technological problems. Rocks and sediments were deposited as an improvised erosion control works. In 1978, a gabion seawall and small groins

Fig. 16.40 Erosion problem at the north part of the Caiobá-Matinhos beach-arc in 2014

Fig. 16.41 Substantial beach and foredune progradation occurred at the southern end of the Caiobá-Matinhos arc as a result of the convergent wave zone related to the frontal lobe of the Guaratuba ebb tidal-delta in 2014 (See Fig. 16.42)

were built along the beach (Fig. 16.43) that induced apparent beach recovery resulting in a 40 m wide backshore (Fig. 16.44). Some people claim that the beach recovery is related to gabion seawall and groins, but others suggest that recovery is related to increases of wave protection as a consequence of changes on the ebb-tidal delta shoals and the rock groin that prevents longshore beach sand losses (Fig. 16.44).

At Prainha and Caieiras beaches periods of erosion and progradation are related to the shift of the main ebb channel and marginal linear bars of the ebb-tidal delta (Angulo 1993b, Angulo et al. 2006b). At Prainha after a period of progradation the

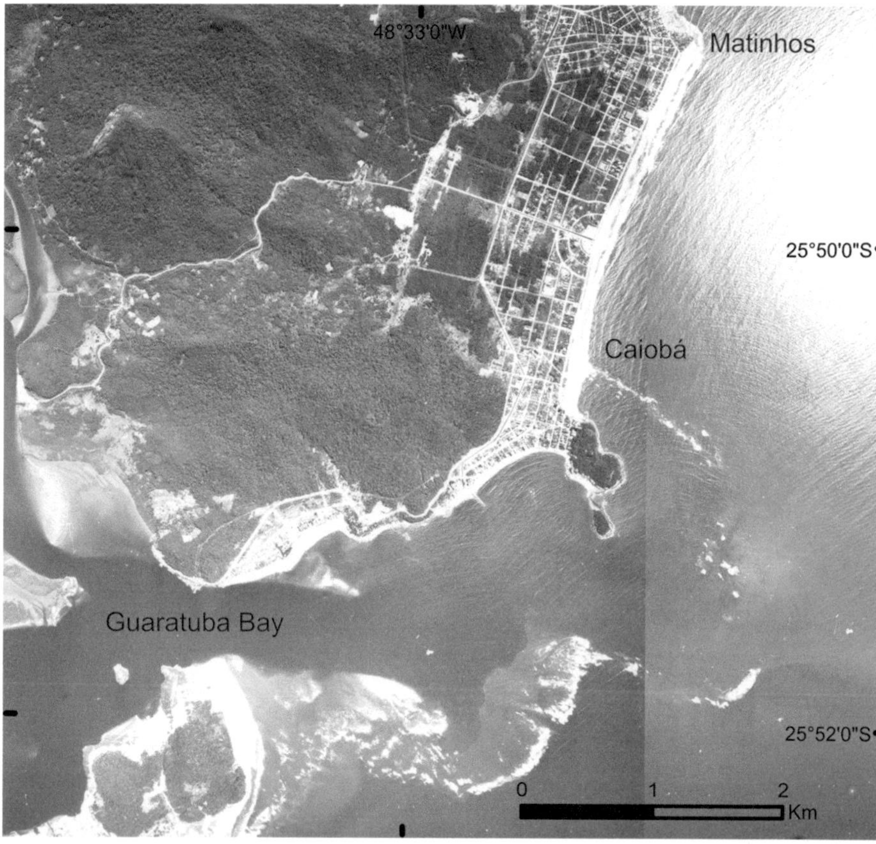

Fig. 16.42 *Vertical* aerial photograph (1980) showing the Guaratuba ebb-tidal delta. Note that the ebb-tidal delta frontal lobe reaches the south part of the Caiobá- Matinhos beach-arc and modifies wave refraction generating a convergent wave refraction zone and beach progradation as shown in Fig. 16.41

new emerged area was occupied with residences (Fig. 16.19), then after a new period of erosion, rock walls were built to protect the residences. This process of occupation of new emerged area was common on the Paraná coast. At Pontal do Sul, significant progradation occurred between 1953 and 1980 and was subsequently entirely occupied. A new area emerged between 1980 and 2000, however attempts to occupy the new land were prevented by environmental laws.

A different case of beach erosion occurs between Pontal do Sul and Ponta do Poço in the southeast Paranaguá inlet (Fig. 16.45). In the 1960s a network of drainage channels were excavated at the coastal plain. One these channels interrupted the littoral transport, causing downdrift erosion (Fig. 16.45).

Some foredunes are also impacts by urban runoff and people traffic. Between Pontal do Sul and Barra do Saí surface drainage of pluvial waters cuts and erodes

Fig. 16.43 Gabions sea-wall and groins built in the 1980s at Praia Mansa

Fig. 16.44 Praia Mansa beach in 1994 were the rock groin built in the 1970s traps sand transported by westward longshore currents

the foredune ridge thereby diminishing beach erosion protection (Fig. 16.46). Also, on those beaches and at Ilha do Mel pedestrians tracks reduce the vegetation and induce foredune breaching and erosion (Fig. 16.47).

16.3.2 Beach Hazards and Risk

Studies about beaches hazards and safety in Paraná coast are almost nonexistent. There is only one detailed study dealing with this subject on Paraná coast; however it is restricted to the beaches of Pontal do Paraná (Angelotti and Noernberg 2010). The report analyzed information about infrastructure, permanent and temporary risks, bathers' density, lifeguards towers, and rescue statistics. In general, most

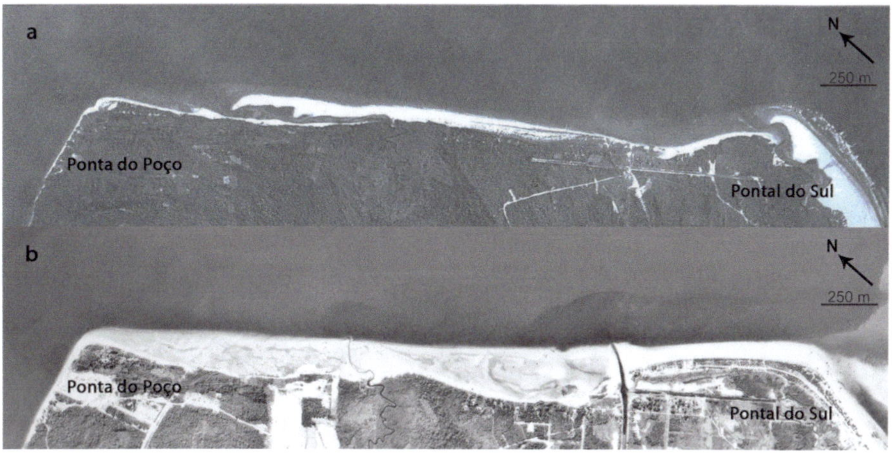

Fig. 16.45 Aerial photographs of the coastal sector between Pontal do Sul and Ponta do Poço at the southeastern Paranaguá inlet, in 1954 (**a**) and 1980 (**b**). Note the spits oriented toward the estuary indicating the predominant drift direction. After the opening of the drainage channel the littoral drift was interrupted causing intense costal erosion down drift and the change of coastal morphology from beach to sand flat (For location see Fig. 16.11)

bathers ignore the presence of beach hazards like: surf zone currents; tides which continually change water depth and current velocities; wind-driven currents; and rip currents (Angelotti and Noernberg 2010). Consequently there are a hundreds of rescues during the summer season, which averaged 386 rescues per year between 1998 and 2004 (Angelotti and Noernberg 2010).

The hazards associated with sea bathing are exacerbated by human perception of the beach environment. The statistics shows that only 54 % of them bathe in patrolled places, 74 % do not ask lifeguards about sea conditions, and 17 % do not observe the sea conditions before they go swimming. Additionally, only 50 % of the bathers have swimming skills, nevertheless 44 % of the bathers go to depths with the water level at the neck level or above (Angelotti and Noernberg 2010).

Cases of injuries caused by the cnidarian *Physalia physalis* are registered during summer at the health centers. However, the occurrence of this cnidarian at the beaches seems to be related with particular oceanographic and meteorological conditions, notwithstanding more detailed studies are necessary to elucidate this occurrence and all the aspects related to beach hazards and safety on the Paraná coast.

16.4 Summary and Conclusions

Paraná beaches are composed primarily of mature polycyclic fine quartz sand and are characterized by wave-dominated beaches along the open coast and tidal-modified beaches inside the estuaries. At the estuary mouths large ebb-tidal-deltas

Fig. 16.46 Local beach and foredune erosion caused by drainage pluvial waters pipes at the central stable part of the Pontal do Sul-Matinhos beach-arc in 1995

Fig. 16.47 Foredune erosion induced by trails at Ilha do Mel in 2003

influence beach sand transport and budget and cause shifts in the shoreline of hundred's of meters during periods of a few (<10) years. In contrast, the open ocean coast, away from the influence of ebb-tidal deltas, has remained stable (<10 m shift) over the last 5–6 decades. Beaches are backed by low (<5 m) foredune ridges that constituted a buffer to beach erosion during high energy waves events. Intermediate to dissipative morphodynamic stages dominates the open coast beaches, whereas estuarine beaches usually have a high tide reflective beach fronted by wide sand to mud tidal flats. At the inlets, complex transition between these morphodynamics conditions occurs, with significant annual variations in the beach morphology.

At present time, the use of Paraná coast presents a major contrast. The northern – Superagüi and Ilha das Peças – coast is protected by several environmental laws and has very low occupation. The Ilha do Mel coast located between two inlets has intense tourist use, but limited constructions and no motorized vehicles. Erosion problems at those coastal sectors are related to natural coastline shifts induced by inlet and ebb-tidal delta dynamics and also probably to occasional dredging sand disposal. In contrast, the southern coastal sector is densely occupied with erosion problems related to natural coastline shift induced by ebb-tidal delta dynamics as well as human destruction of foredune ridges and constructions over the beach and the beach dynamic fringe. At present time, several erosion problems remain without correct environmental and long- term solutions.

Acknowledgments To Andrew Short for the friendship, the stimulating discussions, the willingness to participate in the field trip, the suggestions on the manuscript including comments on headland bypassing and, last but not least, to correcting the English.

References

Andriguetto-Filho JM (2002) Sistemas técnicos de pesca no litoral do Paraná: caracterização e tipificação. In: Raynaut C, Zanoni M, Lana PC, Floriani D, Ferreira ADD, Andriguetto-Filho JM (eds) Desenvolvimento e Meio Ambiente – em busca da interdisciplinaridade. Editora da UFPR, Curitiba, pp 213–233

Angelotti R, Noernberg MA (2010) Risk analysis of sea bathing in Pontal do Paraná county, PR – season 2003/2004. Braz J Aquat Sci Technol 14(1):65–75, http://dx.doi.org/10.14210/bjast.v14n1.p65-75

Angulo RJ (1984) As praias do Paraná: problemas decorrentes de uma ocupação inadequada. Análise conjuntural IPARDES 6(1):1–4. Reprint in 2000 in Revista Paranaense de Desenvolvimento 99:97–103

Angulo RJ (1992) Geologia da planície costeira do Estado do Paraná. PhD, thesis, Instituto de Geociências, Universidade de São Paulo, São Paulo, p 334

Angulo RJ (1993a) A ocupação urbana do litoral paranaense e as variações da linha de costa. Bol Paranaen Geoc 41:73–91

Angulo RJ (1993b) Variações na configuração da linha de costa no Paraná nas últimas quatro décadas. Bol Paranaen Geoc 41:52–72

Angulo RJ (1993c) Morfologia e gênese das dunas frontais do litoral do Estado do Paraná. Rev Bras Geoc 23(1):68–80

Angulo RJ (2004) Mapa do Cenozóico do litoral do Estado do Paraná. Bol Paranaen Geoc 55:25–42, (CD-ROM with map) http://dx.doi.org/10.5380/geo.v55i0.4281
Angulo RJ, Araújo AD (1996) Classificação da costa paranaense com base na sua dinâmica, como subsídio à ocupação da orla litorânea. Bol Paranaen Geoc 44:7–17
Angulo RJ, Soares CR (1994) Beach profile variations between 1981 and 1993 in southern coast of Paraná State. Bol Paranaen Geoc 42:203–215
Angulo RJ, Souza MC (2005) Geologia e geomorfologia. In: Marques MCN, Britez RM (eds) História natural e conservação da Ilha do Mel. Editora UFPR, Curitiba, pp 19–34
Angulo RJ, Giannini PCF, Kogut JS, Prazeres Filho HJ, Souza MC (1996) Variação de características sedimentológicas através de uma sucessão de cordões litorâneos como função da idade deposicional, no Holoceno da Ilha do Mel, estado do Paraná. Bol Paranaen Geoc 44:77–86
Angulo RJ, Souza MC, Lamour MR (2006a) Coastal erosion problems induced by dredging activities in navigation channels of Paranaguá and São Francisco harbor, southern Brazil. J Coast Res SI 39:1801–1803
Angulo RJ, Soares CR, Marone E, Souza MC, Odreski LLR, Noernberg MA (2006b) Paraná. In: Muehe D (ed) Erosão e progradação do litoral brasileiro. Ministério do Meio Ambiente, Brasília, pp 347–400
Angulo RJ, Lessa GC, Souza MC (2009a) The Holocene barrier systems of Paranaguá and northern Santa Catarina coasts, Southern Brazil. In: Dillenburg SR, Hesp PA (eds) Geology and geomorphology of Holocene coastal barriers of Brazil, Lecture notes in earth sciences, 107. Springer, Berlin, pp 135–176
Angulo RJ, Souza MC, Muller ME (2009b) Previsão e consequências da abertura de uma nova barra no Mar do Ararapira, Paraná-São Paulo, Brasil. Quat Environ Geosci 01(2):67–75, http://dx.doi.org/10.5380/abequa.v1i2.14577
Araújo AD (2001) Dinâmica sedimentar e evolução paleogeográfica do Saco do Limoeiro na Ilha do Mel, e sua relação com o canal de acesso ao porto de Paranaguá. MSc, dissertation, Universidade Federal do Paraná, Curitiba, p 73
Barros FCR Jr, Borzone CA, Rosso S (2001) The Macroinfauna of six beaches near Guaratuba Bay, Southern Brazil. Braz Arch Biol Technol 44(4):351–364
Bessa O Jr (2003) Interferência entre a ocupação urbana e a dinâmica natural no litoral sul do Paraná. PhD, thesis, Universidade Federal do Paraná, Curitiba, p 156
Bessa O Jr, Angulo RJ (2003) Volumetric variations on the beaches of the south seashore of Paraná State in Brazil. J Coast Res SI 35:209–215
Bigarella JJ (1946) Contribuição ao estudo da planície litorânea do Estado do Paraná. Arq Biol Tecnol 1:75–111
Bigarella JJ (1950/1951a) Contribuição do estudo dos sambaquis no Estado do Paraná, 1 Regiões adjacentes às baías de Paranaguá e Antonina. Arq Biol Tecnol 5–6(17):231–292
Bigarella JJ (1950/1951b) Contribuição ao estudo dos sambaquis no Estado do Paraná, I Regiões adjacentes à baía de Guaratuba. Arq Biol Tecnol 5–6(18):293–321
Bigarella JJ (1965) Nota explicativa acompanhando a folha geológica de Paranaguá. Bol Universidade Paraná Geol 13:1–15
Bigarella JJ (1973) Morphoscopic and mineralogic differences between water-laid and wind-laid deposits. Bol Paranaen Geoc 31:15–111
Bigarella JJ (1991) Matinho: Homem e Terra – Reminiscências… Prefeitura Municipal de Matinhos, Associação de Defesa e Educação Ambiental – ADEA, p 212
Bigarella JJ, Sanches J (1966) Contribuição ao estudo dos sedimentos praiais recentes. V – Praia suspensa do saco do Tambarutaca, Município de Paranaguá-PR. Bol Paranaen Geogr 18–20:151–175
Bigarella JJ, Doubek A, Salamuni R (1957) Planta geológica provisória da baía de Guaratuba. Bol Universidade Paraná Geol 8:1–6
Bigarella JJ, Salamuni R, Marques Filho PL (1961) Método para avaliação do nível oceânico à época da formação dos terraços de construção marinha. Bol Paranaen Geogr 4–5:11–115

Bigarella JJ, Freire SS, Salamuni R, Viana R (1966) Contribuição ao estudo dos sedimentos praias recentes, II Praias de Matinhos e Caiobá. Bol Universidade Federal Paraná Geog Física 6:1:109
Bigarella JJ, Alessi AH, Becker RD, Duarte GK (1969) Textural characteristics of the coastal dune, sand ridge and beach sediments. Bol Paranaen Geoc 27:15–80
Bigarella JJ, Duarte MG, Becker RD (1970/1971) Structural characteristics of the dune, foredune, interdune, beach, beach-dune ridge and sand ridge deposits. Bol Paranaen Geoc 28–29:9–72
Borzone CA, Souza JRB (1997) Estrutura da macrofauna bentônica no supra, meso e infralitoral de uma praia arenosa do sul do Brasil. Oecologia Brasiliensis 3:197–212
Borzone CA, Souza JRB, Soares AG (1996) Morphodynamic influence on the structure of inter and subtidal macrofaunal communities of subtropical sandy beaches. Rev Chil Hist Nat 69:565–577
Borzone CA, Tavares YAG, Barros FCR Jr (1998) Beach morphodynamics and distribution of Mellita quinquiesperforata (Leske, 1778) on sandy beaches of southern Brazil. In: Mooi R, Telford M (eds) Proceedings of the 9th international echinoderm conference, San Francisco, California, USA. Balkema, Rotterdam, pp 581–586
Borzone CA, Gutsei KC, Soares CR (1999) Sedimentos e macrofauna bentônica da zona de águas rasas da praia de Matinhos, Paraná, Brasil. Rev Atlântica 21:43–58
Borzone CA, Melo SG, Rezende KV, Vale R, Krul R (2003) Macrobenthic intertidal communities from wave to tide dominated beach environments: a case study in two Brazilian beaches. J Coast Res SI 35:472–480
Borzone CA, Lorenzi L, Calil P (2007) Cross-shore mysidacea distribution in two sandy beaches with contrasting morphodynamics. Rev Bras Zool 24(4):943–949
Cazal HGSV, Fomin IM, Oliveira BA, Sutil UA, Oliveira E, Dourado MS (2011) Caracterização Sazonal da Precipitação e do Vento no Litoral Paranaense. In: 14th Congresso Latino-Americano de Ciências do Mar, Balneário Camboriú, Brazil, CD-ROM
Deschamps MV, Kleinke MLU (2000) Os fluxos migratórios e as mudanças socioespaciais na ocupação contínua litorânea do Paraná. Rev Paranaen Desenvolvimento 99:45–49
Diniz FA, Kousky EV (2004) Ciclone no Atlântico Sul – análise sinóptica e observação. In: 13th Anais Congresso Brasileiro de Meteorologia, Fortaleza, Brazil CD-ROM
Dourado MS, Fomin IM,(2015) Estado médio e variabilidade da atmosfera do litoral paranaense. Rev Bras Meteorologia (accepted)
Freitas TG, Souza MC, Angulo RJ (2014) Variação decadal da granulação e da mineralogia ao longo das praias desde as baías de São Francisco até a de Paranaguá (in prep)
Gandara-Martins AL (2007) A macrofauna bentônica das praias arenosas expostas do Parque Nacional de Superagui, PR. MSc, dissertation, Universidade Federal do Paraná, Curitiba, p 79
Gaspar MD (1996) Análise das datações radiocarbônicas dos sítios de pescadores, coletores e caçadores. Bol Museu Paranaen Emílio Goeldi 8:81–91
Giannini PCF, Angulo RJ, Souza MC, Kogut JS, Delai MS (2004) A erosão da costa leste da Ilha do Mel, baía de Paranaguá, estado do Paraná: modelo baseado na distribuição espacial de formas deposicionais e propriedades sedimentológicas. Rev Bras Geoc 34(2):231–242
Hayes MO (1975) Morphology of sand accumulations in estuaries. In: Cronin LE (ed) Estuarine research, geology and engineering, vol 2. Acad Press, New York, pp 3–22
Labraga JC, Frumento O, López M (2000) The atmospheric water vapor cycle in South America and the tropospheric circulation. J Clim 13:1899–1915
Lamour MR (2000) Dinâmica sedimentar do canal da Galheta via de acesso ao porto de Paranaguá, PR. MSc, dissertation, Universidade Federal do Paraná, Curitiba, p 100
Lamour MR, Angulo RJ, Soares CR (2006) Bathymetric evolution of critical silting sectors on Galheta channel, access way to Paranaguá port, Paraná state – Brazil. J Coast Res 23:49–58
Lana PC, Marone E, Lopes RM, Machado EC (2000) The subtropical estuarine complex of Paranaguá Bay, Brazil. In: Seeliger U, Lacerda LD, Kjerfve B (eds) Coastal marine ecosystems of Latin America. Springer Verlag, Berlin, pp 131–145

Lessa GC, Angulo RJ, Giannini PCF, Araújo AD (2000) Stratigraphy and Holocene evolution of a regressive barrier in south Brazil. Mar Geol 165:87–108. doi:10.1016/S0025-3227(99)00130-9

LNEC – Laboratório Nacional de Engenharia Civil (1977) Combate à erosão na praia de Caiobá, Paraná. Technical report (unpublished), p18

Marone E, Camargo R (1994) Marés meteorológicas no litoral do estado do Paraná: o evento de 18 de agosto de 1993. Nerítica 8(1–2):73–85

Marone E, Jamiyanaa D (1997) Tidal characteristics and a variable boundary numerical model for the M2 tide for the Estuarine Complex of the Bay of Paranaguá, PR, Brazil. Nerítica 11(1–2):95–107

Marone E, Noernberg MA, Netto Jr JPB, Etri T, Mayerle R (2015) Estuarine dynamics of an interlinked complex system: Paranaguá Bay, Brazil. (in prep)

Mihály P (1997) Dinâmica sedimentar do litoral norte paranaense e extremo sul paulista. MSc, dissertation, Universidade Federal do Paraná, Curitiba, p 104

Mihály P, Angulo RJ (2002) Dinâmica da desembocadura do corpo lagunar do Ararapira. Rev Bras Geoc 32(2):217–222

Milani JR, Canali NE (2000) O sistema hidrográfico do rio Matinhos: uma análise morfométrica. Rev Ra'ega 4:139–152

Mineropar – Minerais do Paraná SA (1989) Mapa geológico do Estado do Paraná. Departamento Nacional da Produção Mineral (DNPM), Brasília, 1map

Nemes DD (2011) Caracterização das ondas de superfície na plataforma interna do estado do Paraná. MSc, dissertation, Universidade Federal do Paraná, Curitiba, p 134

Noernberg M, Alberti AL (2014) Variabilidade oceanográfica na plataforma continental interna do Paraná: condição de primavera. Rev Bras Geof (in press)

Noernberg MA, Lautert LFC, Araújo AD, Odresky LL (1997) Base de dados digital do litoral paranaense em sistema de informações geográficas. Nerítica 11:191–195

Noernberg MA, Lautert LFC, Araújo AD, Marone E, Angelotti R, Netto JPB Jr, Krug LA (2006) Remote sensing and GIS integration for modeling the Paranaguá estuarine complex – Brazil. J Coast Res SI 39:1627–1631

Noernberg MA, Marone E, Angulo RJ (2007) Coastal currents and sediment transport in Paranaguá estuary complex navigation channel. Bol Paranaen Geoc 60–61:45–54

Nordstrom KF (1992) Estuarine beaches: an introduction to the physical and human factors affecting use and management of beaches in estuaries, lagoons, bays, and fjords. Elsevier Science Publishers Ltd, London/New York, p 235

Paranhos Filho AC (1996) O processo erosivo e as variações morfodinâmicas praias na Ilha do Mel (Baía de Paranaguá-PR). MSc, dissertation, Universidade Federal do Paraná, Curitiba, p 239

Paranhos Filho AC, Soares CR, Angulo RJ (1994) Nota sobre a erosão na Ilha do Mel. Bol Paranaen Geoc 42:153–159

Parellada CI, Gottardi Neto A (1994) Inventário de sambaquis do litoral do Paraná. Bol Paranaen Geoc 42:121–152

Pierri N, Angulo RJ, Souza MC, Kim MK (2006) A ocupação e o uso do solo no litoral paranaense: condicionantes, conflitos e tendências. Desenvolvimento e Meio Ambiente 13:137–167, http://dx.doi.org/10.5380/dma.v13i0.9849

Quadros CJL (2002) Variações morfológicas e volumétricas associadas à incidência de sistemas frontais em duas praias arenosas do litoral paranaense. MSc, dissertation, Universidade Federal do Paraná, Curitiba, p 57

Quadros CJL, Marone E, Prata VP Jr, Araújo AD (2002) Sistema para perfilagem de fundo na zona de arrebentação utilizando sensor de pressão. Pesq Geoc 28:173–178

Quadros CJL, Marone E, Angulo RJ, Martins GJ, Netto JPB Jr (2007) Morfodinâmica sedimentar associada à incidência de sistemas frontais em duas praias do litoral paranaense. Bol Paranaen Geoc 60–61:65–74, http://dx.doi.org/10.5380/geo.v60i0.11086

Quadros CJL, Angulo RJ, Souza MC, Marone E (2016) Subaerial beach volume variation and morphodynamic of two oceanic wave-dominated beach-arcs at the Paraná coast, southern Brazil (in prep.)

Reboita MS (2008) Ciclones Extratropicais sobre o Atlântico Sul: Simulação Climática e Experimentos de Sensibilidade. PhD, thesis, Universidade de São Paulo, p 316

Rodrigues AD (1985) Relações internas na família lingüística Tupi-Guarani. Rev Antropologia 27–28:33–53

Rosa LC, Borzone CA (2008) Uma abordagem morfodinâmica na caracterização física das praias estuarinas da baia de Paranaguá, sul do Brasil. Rev Bras Geoc 38(2):237–245

Sampaio R (2006) Ocupação das orlas das praias paranaenses pelo uso balneário. Desenvolvimento e Meio Ambiente 13:169–186, http://dx.doi.org/10.5380/dma.v13i0.9850

Santos AV (1850) Memória histórica da cidade de Paranaguá e seu município. Publicação da Secção de História do Museu Paranaense 1951, vol 1, p 347

Sayão OJ (1989) Littoral drift along some beaches in Brazil. In: Magoon OT, Converse H, Miner D, Tobin LT, Clark D (eds) Proceedings of the 6th symposium on coastal and ocean management, Charleston, USA, vol 4. American Society of Civil Engineers, New York, pp 3638–3746

Short AD (2006) Australian beach systems – nature and distribution. J Coast Res 22:11–27, http://dx.doi.org/10.2112/05A-0002.1

Short AD, Masselink G (1999) Embayed and structurally controlled beaches. In: Short AD (ed) Beach and shoreface morphodynamics. Wiley, Chichester, pp 230–250

Soares CR, Borzone CA, Souza JRB (1997) Variações morfológica e sedimentológicas ao longo de um ciclo anual numa praia arenosa no sul do Brasil. Oecologia Brasiliensis 3:245–258

Souza MC, Angulo RJ (2003) Decadal and inter-annual variations of shoreline and beach volumes in Itapoá (Santa Catarina, Brazil). J Coast Res SI 35:202–208

Souza JRB, Borzone CA (1996) Distribuição de Callianassídeos (Crustacea: Decapoda: Thalassinidea) em praia do litoral paranaense, com especial referência a Callichirus major (Say, 1818). Arq Biol Tecnol 39(3):553–565

Souza MC, Angulo RJ, Assine ML, Castro DL (2012) Sequence of facies at a Holocene storm-dominated regressive barrier at Praia de Leste, southern Brazil. Mar Geol 291–294:49–62. doi:10.1016/j.margeo.2011.10.009

Tessler MG, Mahiques MM (1993) Utilization of coastal geomorphic features as indicators of longshore transport: examples of the southern coastal region of the State of São Paulo, Brasil. J Coast Res 9:823–830

Vanhoni F, Mendonça F (2008) O clima do litoral do estado do Paraná. Rev Bras Climatalogia 3–4:49–63

Veiga FA, Angulo RJ, Sá F, Odreski LLR, Lamour MR, Disaró ST (2006) Origin of mud deposits in a wave dominated shallow inner continental shelf of the State of Paraná coast, southern Brazil. J Coast Res SI 39:262–265

Chapter 17
Santa Catarina Beach Systems

Antonio Henrique da F. Klein, Andrew D. Short, and Jarbas Bonetti

Abstract The Santa Catarina coast is located in southern Brazil between 26° and 29.3°S. The coast trends south then southwest for about 430 km, and contains 922 km of open coast and bay shoreline, with 246 sandy beaches occupying 60 % of the shore. Based on geology and orientation the coast is divided into five coastal provinces, each of which contains a different suite of beach systems. The provinces and their beaches range for the highly embayed Penha to Baia Norte province which is dominated by short low energy reflective beaches, to the exposed high energy Cabo Santa Marta to Torres province which has long near continuous multi-barred dissipative beaches. This chapter reviews the provinces and their beach systems including spatial and temporal controls, as well as beach erosion, hazards and safety.

17.1 Introduction

The Santa Catarina coast is located in southern Brazil between 26° and 29.3° S. The coast trends south then southwest for about 430 km, and contains 922 km of open coast and bay shoreline (Fig. 17.1). The state has a narrow coastal plain, usually less than 20 km in width, rising inland to the Serra do Mar and Geral escarpments, which trend north-south through the state paralleling the Atlantic coast and dividing the state between a larger elevated plateau region to the west and the eastern coastal plain (Santa Catarina-GAPLAN 1986; DNPM 1986; Scheibe 1986; Bortoluzzi 1987). The climate is humid sub-tropical, which maintains numerous streams that rise in the wooded slopes of the scarps with most rivers flowing west while shorter streams flow east to the coast (Nimer 1989). The Serra do Mar coastal rain forests, which once covered the state, including the coast, has now been largely cleared for agriculture and development. According to the IBGE (2010) there were 6.2 million

A.H. da F. Klein (✉) • J. Bonetti
Departamento de Geociências, Laboratório de Oceanografia Costeira,
Universidade Federal de Santa Catarina, Florianópolis, SC 88040-900, Brazil
e-mail: antonio.klein@ufsc.br; jarbas.bonetti@ufsc.br

A.D. Short
School of Geosciences, University of Sydney, Sydney, NSW 2006, Australia
e-mail: andrew.short@sydney.edu.au

A.D. Short, A.H. da F. Klein (eds.), *Brazilian Beach Systems*, Coastal Research
Library 17, DOI 10.1007/978-3-319-30394-9_17

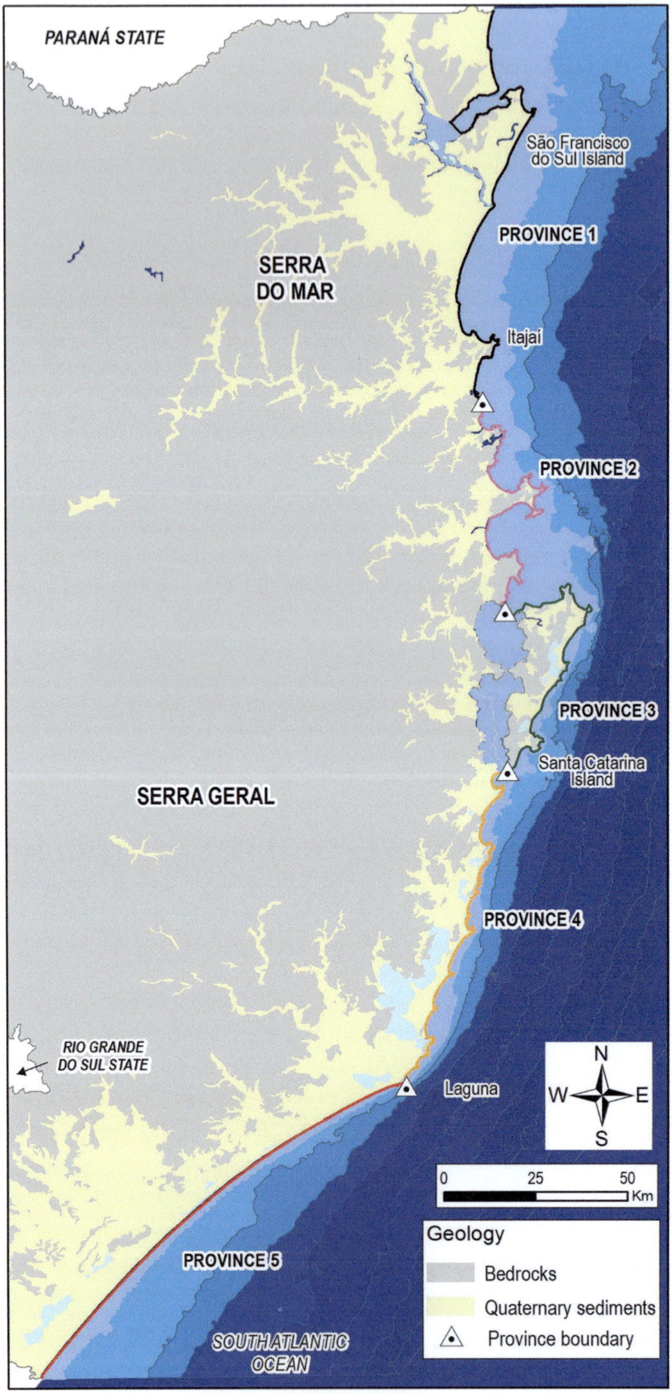

Fig. 17.1 Map of Santa Catarina with simplified geomorphology and location of five coastal provinces

people residing in the state, with 38 % living in the 36 coastal municipalities, representing about 258 inhabitants per square kilometer.

The coast includes numerous beaches, islands, bays, inlets, salt marshes, mangroves, estuaries and lagoons. The abundance of coastal environments combined with a mild subtropical climate has resulted in use of the coast traditionally for fishing and transport, and more recently for housing, recreation, bathing, surfing, and general tourism (Polette and Raucci 2003; Klein et al. 2006). The intense use of sections of the coast for urban and tourist development started in the 1970s and has resulted in extensive areas of habitat loss including sectors of shoreline degradation and armoring (Klein et al. 2006, 2009). It is the aim of this chapter to review the nature and dynamics of Santa Catarina's beach systems, as well as the hazards to which the coast, its beaches and beach users are exposed.

17.1.1 Geology

The Santa Catarina coast contains three geological zones (Scheibe 1986; DNPM 1986; Bortoluzzi 1987). In the north Cenozoic sediments extend south to Penha, with only occasional headlands resulting in longer beach systems (Horn Filho et al. 1994). In the centre is a belt of Proterozoic granitoids and Archean migmatites that outcrops at the coast between Itajaí and Cape Santa Marta (Scheibe 1986; Bortoluzzi 1987; DNPM 1986). The bedrock dominates this section of coast and plays a significant role in size, orientation and morphodynamics of the beach systems, with generally smaller embayed beaches of variable length, orientation and sediments ranging from fine to very coarse (Klein 2004; FitzGerald et al. 2007; Hesp et al. 2009; Klein et al. 2010; McBride et al. 2013). South of Santa Marta fine uniform Quaternary sediments dominate all the way to the border at Torres, with the Archean granulities lying inland but not exposed at the coast, however outcrops of Parana Basin occur on the coast and on the inner shelf (Scheibe 1986; Villwock 1987; Bortoluzzi 1987; Caruso Jr 1993).

17.1.2 Climate

Santa Catarina has a humid sub-tropical climate (Koppen: Cfa) dominated by the sub-tropical South Atlantic high pressure system, which is centered between 18 and 35°S and maintains an easterly flow of humid air onto the coast (Santa Catarina – GAPLAN 1986; Nobre et al. 1986; Nimer 1989). Mean temperature ranges from 23 °C in summer to 14 °C in winter, while coastal precipitation ranges from 1250 to 1400 mm (Santa Catarina – GAPLAN 1986). In summer, occasional rainfall exceeding 100 mm d^{-1}, results in coastal flash flooding with catastrophic consequences to the beaches, especially in urban areas with low soil infiltration capacity.

The state straddles climate boundary between the more tropical north and the increasingly more temperate south (Santa Catarina – GAPLAN 1986; Nimer 1989). During winter the warm tropical flow is punctuated by cold fronts associated with sub-polar lows, which bring cooler weather, strong southerly winds and frontal rain (Nobre et al. 1986; Klein 1997), occurring in 6 and 11 day cycles (Stech and

Lorenzetti 1992) with an 8 day average (Rodrigues et al. 2004). The tropical-temperate boundary is represented by the southern limit of mangroves on Santo Antonio Lagoon (Municipality of Laguna, Soares et al. 2012), with temperate salt marsh dominating the coastal wetlands to the south (Schaeffer-Novelli et al. 1990; Soares et al. 2012).

17.1.3 Hurricane Catarina

The development of a Hurricane Category 2 (Saffir-Simpson scale) called *Catarina* in the western South Atlantic Ocean basin, in late March 2004, has the distinction of being the first documented tropical cyclone in the South Atlantic (Pezza and Simmonds 2005; McTaggart-Cowan et al. 2006). Although other systems have been observed to undergo brief periods of tropical development (eastern South Atlantic in April 1991, western South Atlantic in January 2004: McAdie and Rappaport 1991), and more recently Anita tropical storm in 2010, none has threatened the coast of Catarina. At least three people died as a consequence of *Catarina* and an estimated damage of US$ 350,000 was reported (Marcelino et al. 2014). However, no erosion or inundation of infrastructure located close to the beach were record in southern Brazil (Tabajara 2004; Marcelino et al. 2014).

17.1.4 Ocean Surface Temperature

The general circulation in the southwest Atlantic Ocean is characterized by the flow of the warm tropical Brazil Current (average temperature 24 °C, salinity < $35^{0}/_{00}$) toward the south, while the sub-Antarctic Malvinas current flows north (Legeckis and Gordon 1982; Olson et al. 1998). In the warmer season a thermocline forms due to the penetration of the South Atlantic Central Water (ACAS, temperature < 20 °C) which causes a stratification of the water column. In winter the ACAS recedes onto the slope. The general pattern has more homogeneous conditions during the summer and more heterogeneous during the winter. However, there is the outcrop of ACAS near the island of Santa Catarina in the hot season (Pereira et al. 2009) and seasonal east coast upwelling occurs at Cape Santa Marta during northerly winds.

17.1.5 Drainage

The humid climate maintains numerous rivers with headwaters on the Geral Plateau. Most drain west to the Rio Uruguay, with only 37 % draining to the Atlantic coast. There is only one moderate size coastal river, the Itajai-Acu, together with several minor drainage systems. Only the Itajaí (fine sand), Itapocú (fine to medium sand) and Tijucas (coarse sand and mud) rivers are actively contributing sediments to the coastal environment. The Cubatão and Tubarão rivers deposit sand as they flow into an intra-bay delta (Herrmann 1999) and an intra-lagoonal delta (Nascimento Jr

2010) system respectively. The remaining streams are still infilling their estuaries. Much of the Quaternary sand south of Cape Santa Marta, is believed to be ultimately derived from the La Plata River.

17.1.6 Waves and Tides

The Santa Catarina coast faces east into the South Atlantic receives waves from four wave directions (Araujo et al. 2003, Fig. 17.2).

- The South Atlantic high generates east-northeast seas, with period between 3 and 8 s and height between 0.5 and 1.5 m, which arrive 10 % of the time.
- Longer period easterly waves generated by the sub-tropical high arrive about 50 % of the time with heights between 1 and 1.5 m and periods of 6–11 s.
- Long period (7–16 s) south-southeast swell arrives 25 % of the time with heights between 1 and 2 m.
- Low pressure cold fronts generate higher seas between 1 and 3.5 m, with 4–8 s periods 10 % of the time.

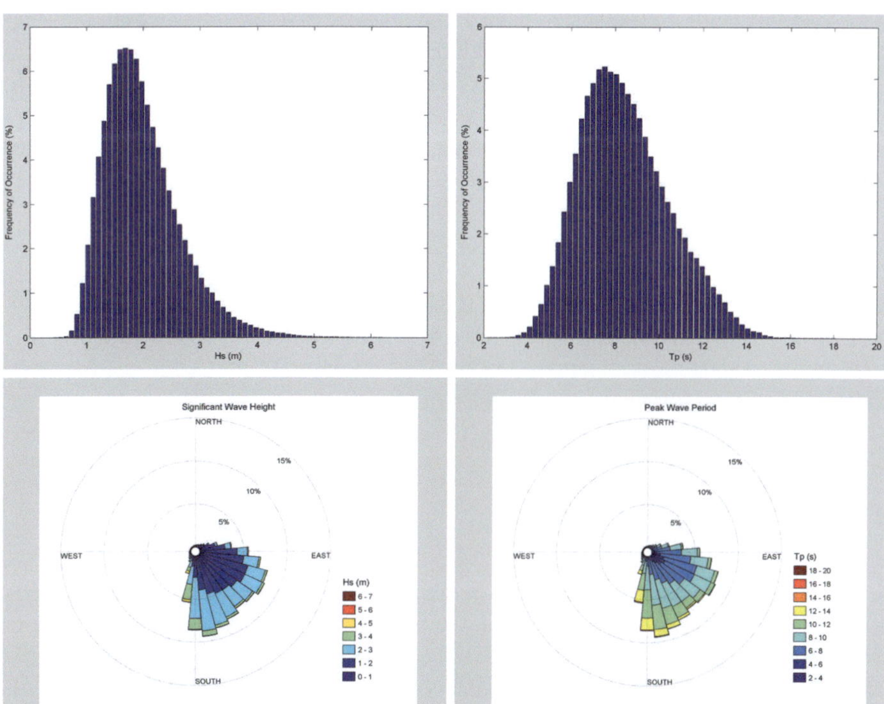

Fig. 17.2 Santa Catarina waves: *Upper* – 60 year plot of significant wave height (*left*) and peak period (*right*) (SMC data base). *Lower* – directional significant wave height (*left*) and peak period (*right*). Point in front of SC Island was validated by Gomes da Silva et al. (2015). Location coordinates: 46°45′W, 27°42′S

Alves (1996), Alves and Melo (2001), Araujo et al. (2003), Klein (2004), Miot da Silva (2006), Hesp et al. (2009), Signorin (2010) and Pianca et al. (2010) summarized the wave climate off Santa Catarina as having a dominant south swell with a period of 12 s and mean height between 1 and 1.5 m, increasing to 2 m in winter, while locally generated seas arrive from the east to northeast with a 7 s period and mean height of 1–1.25 m, particularly during the autumn and winter. Melo et al. (2006) reported an extreme wave with H_{max} of 13 m in front of Santa Catarina Island.

Other important studies include: the probability of occurrence of extreme waves in SC, based on a 1 year record (Straioto 2009); a study of wave generation zone in the Atlantic with a reanalysis of data from Araujo et al. (2003), that are in agreement with the AECID report (Miranda 2013); and a validation of wave parameters from the past 60 years based on Downscaled Ocean Waves (DOW) reanalysis database (SMCBrasil) (Gomes da Silva 2014).

The wave attenuation (refraction and diffraction) along the embayed sections of the SC coast are very important and early studies of wave refraction process along the SC coast were developed by Alves (1996) and Alves and Melo (2001). Oliveira (2013) and Ribeiro (2014) reported that in southern corner of Itapocorói Bay wave height decreases almost 78 % (e.g.: from 3.5 m to 0.5 m wave height) and change the direction by 120° (from S-SE to N-NW).

Tides along the SC coast are micro-tidal ranging from 1.05 m in the north to 0.46 m in the south. Storm surges have a frequency of 5–7 days throughout the year, with greater intensity occurring between the seasons (summer to fall and winter to spring) (Stech and Lorenzetti 1992; Trucollo 1998). The persistent moderate to occasionally high waves and micro-tides maintain a wave-dominated coast along the entire open coast with a relative tide range (RTR) less than 2 (Klein and Menezes 2001; Klein 2004; Oliveira et al. 2014):

$$RTR = TR / H_b \tag{17.1}$$

where TR = mean spring tide range (m) and H_b = breaker wave height (m) (Masselink and Short 1993). Howewer, lower waves within sheltered bays result in some tide-modified and tide-dominated sections (RTR > 3, Hb < 0.5 m) (Klein and Menezes 2001; Klein 2004; Oliveira et al. 2014).

17.1.7 Coastal Sediments

Coastal sediment are derived from three sources. In the south the ultimate source is the La Plata River located 1000 km to the south of SC, the second largest river in South America with a sediment load of 91×10^6 t $year^{-1}$ (Sytiski et al. 2005). Waves have moved its sand northward along the coast of Uruguay and Rio Grande do Sul and southern Santa Catarina throughout the Quaternary. Siegle and Asp (2007) and later Hesp et al. (2009) suggested sediment trapping is taking place south of Cape Santa Marta. In Santa Catarina this sand is predominately uniform very fine to fine-grained quartz and composes the extensive beaches, barriers and dunes along the southern SC coast (Martins and Eichler 1969; Martins 1970; Giannini 1993).

Sand from smaller rivers and streams deposited on the shelf at lower sea levels has also been supplied throughout the state with sand moving onshore during the Holocene sea level transgression (Abreu de Castilhos 1995; Souza and Corrêa 2006; Hesp et al. 2009; Hein et al. 2013; Porpilho et al. 2015). This sand ranges from fine to coarse sand with areas of considerable local variation (Fig. 17.3).

Holocene fluvial sediment supply to the coast is restricted to three river systems, the Itajai-Acu river, which has supplied fine sand to the Navegantes barrier system (Hein et al. 2014); and the Tijucas river which has supplied mud to the regressive Tijucas barrier (Caruso Jr and Araújo 1997; Klein and Menezes 2001; Buynevich et al. 2005; Asp et al. 2005; FitzGerald et al. 2007, 2008). Elsewhere the smaller rivers, bays and estuaries are acting as sediment sinks and not supplying fluvial sand to the coast. However, there are intra-bay delta (Herrmann 1999) and intra-lagoonal delta (Nascimento Jr 2010) systems as a result of sand input from Cubatão and Tubarão rivers respectively (see Fig. 17.1). The net longshore sediment transport in Santa Catarina is to north-northeast. This is indicated by the net northerly sediment transport (Siegle and Asp 2007; CPE 2009); tidal inlets migrating northward at an average of 100 m year^{-1} (Piérri 2005; Klein et al. 2006; Cassiano and Siegle 2010; Vieira da Silva 2009; Bhering 2012); and headland bay beach orientation (Klein et al. 2003b; Klein 2004; Silveira et al. 2011). Lessa et al. (2000) suggested that the source of sediment to Parana coast is from north of Santa Catarina, where the littoral cell starts (at Piçarras beach). However, there is also some local alongshore reversal in the sediment transport, mainly in the protected area of headland bay beaches (Miot da Silva 2006; Mazzer et al. 2008). In this case the inlet migration inside of these bays is to the south (Siegle et al. 1998; Camargo 2009; Araujo 2008; Bhering 2012).

17.1.8 Coastal Provinces and Geomorphology

The geomorphology of the coast has been described by Silveira (1964), Suguio and Tessler (1984), Suguio and Martin (1987), Bortoluzzi (1987), Vilwock (1994), Diehl and Horn Filho (1996), Muehe (1998), Horn Filho (2003), Klein (2004), Fitzgerald et al. (2006), Fitzgerald et al. (2007) and Hesp et al. (2009). Based on both Quaternary deposits and bedrock geology Klein (2004) and FitzGerald et al. (2007, 2008) divided the coast into four major segments (see Fig. 17.1):

1. The northern coast between Baia de Guaratuba (Parana) and Penha has sediment abundant strand plains consisting of regressive foredune ridge barriers and estuaries.
2. Between Penha and Papagaio Island, including Santa Catarina Island, is a bedrock-dominated coast with re-entrants and bays containing both regressive and transgressive barrier systems.

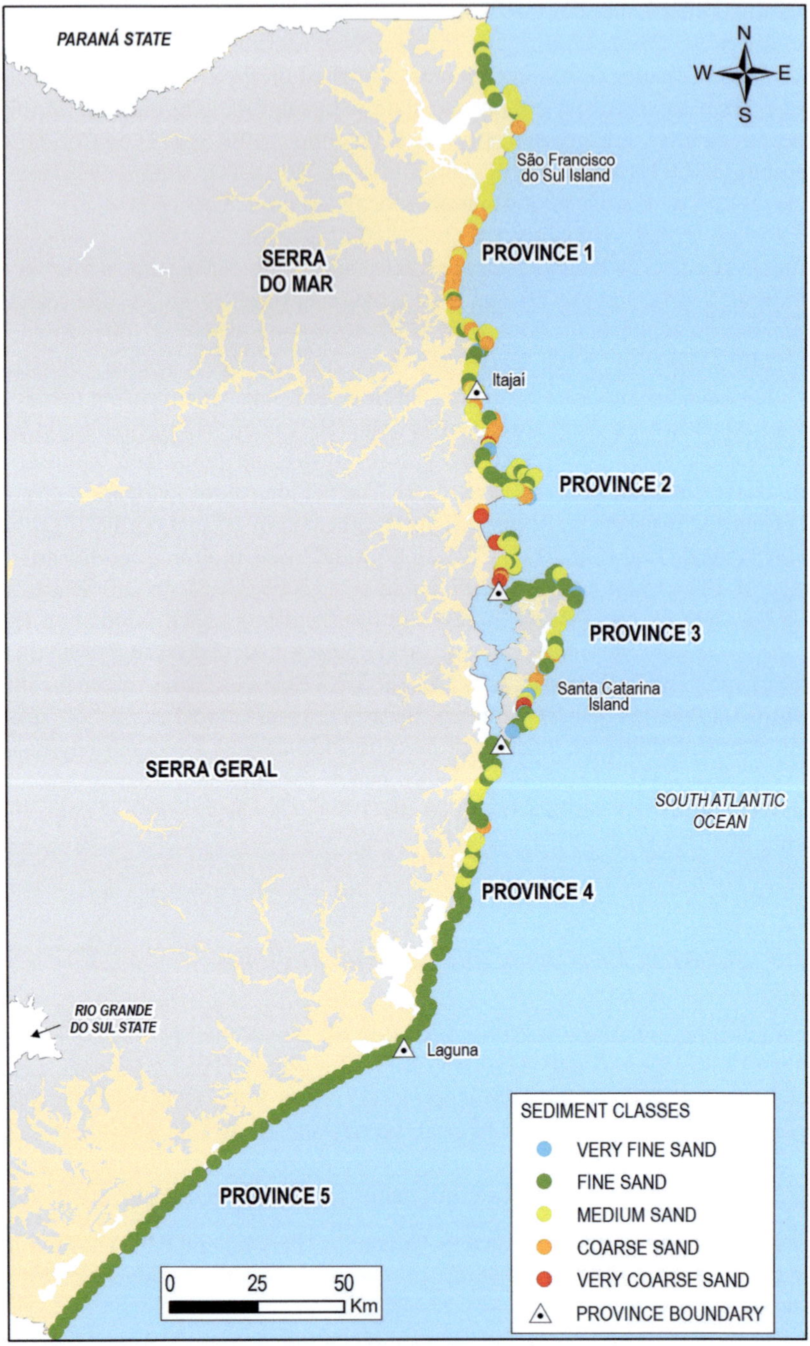

Fig. 17.3 Map of distribution of median beach sand size along the coast of Santa Catarina

Table 17.1 Santa Catarina coastal provinces

FitzGerald et al. (2007, 2008)	This chapter	Beach characteristics
1. Barra do Sai to Penha	**1. Barra do Sai to Penha**	Embayed intermediate
2. Penha to Papagaios Island	**2. Penha to Baia Norte**	Short, embayed, reflective
	3. Santa Catarina Island	Variable, embayed reflective-intermediate
	North and South Bays*[a] *(Florianópolis Bays)	*(short, tide-modified & tide-dominated)*
3. Papagaios Island to Cabo Santa Marta	**4. Papagaios Island to Cabo Santa Marta**	Variable, embayed intermediate
4. Cabo Santa Marta to Torres	**5. Cabo Santa Marta to Torres**	Long multi-bar dissipative

[a]North & South bays contain numerous low energy beaches but are not considered in this chapter

3. South of Papagaio Island to Cape Santa Marta the bedrock headlands decrease in size and increase in spacing resulting in longer transgressive barrier systems, with active dune fields.
4. The southernmost sector between Cape Santa Marta and the border at Torres is a long straight southeast facing coast dominated by Pleistocene and Holocene transgressive barriers and numerous active dune fields.

In this chapter on beach systems FitzGerald's and Klein's regions are used, with the division of the second province, based on beach characteristics, into two provinces and an additional sub-province, containing the sheltered North and South Bays (Table 17.1). The beach characteristics of the five open coast provinces are presented in Sect. 17.2.1.

For each one of these provinces a segmentation of Santa Catarina coastline was performed using five different coastal types: headland, inlet, infrastructure, beach and coastal vegetation. Examples of the five types of coast are shown in Fig. 17.4. Headlands were related to the rocky sectors, in most cases as features projected into the sea; discontinuities of the coastline at river mouths, lagoons and bays were classified as inlets; the infrastructure class represents the landfills located directly in contact with the sea, in addition to coastal engineering structures such as groins, seawalls and other man-made structures; beaches were identified as elongated sandy deposits located between the sea and the dune or other natural or anthropic feature in the backshore; and coastal vegetation was represented, in most cases, by mangroves and salt marshes (*Spartina* sp.).

Table 17.2 shows the distributions of the five classes along 921.6 km of measured coastlines (it must be noted that in this research part of Babitonga Bay and both South and North bays were included). Beaches were the most common feature (57.8 %), followed by headlands (27.6 %) and coastal vegetation (7.7 %). The other classes: inlet and infrastructure together were identified in less than 65 km (7.1 %) of the state's coast.

Fig. 17.4 Examples of the five defined Santa Catarina coastal type classes (Photos by J. Bonetti)

Table 17.2 Total distribution of the coastal types in Santa Catarina State

Coastal type	Length (km)	Percentage
Headland	254.2	27.6
Inlet	15.9	1.7
Infrastructure	47.8	5.2.
Beach (open and sheltered)	532.5	57.8
Coastal vegetation	71.2	7.7
Total	921.6	100.0

17.2 Santa Catarina Beach Systems

Beaches are the dominant coastal landfrom in SC occupying almost 60 % of the coast, followed by rocky coast and headlands (28 %), which induce many embayed beaches. The following section examines the nature and extent of the beach systems in each of the five coastal provinces coast (Table 17.3).

17.2.1 Regional Variation in Beach Systems

The coast of Santa Catarina is bedrock-controlled coast exposed to persistent moderate east through southerly waves and micro-tides resulting in a predominately wave-dominated coast and numerous embayed beaches on the open shore. The regional variation in beach systems and their state is primarily controlled by

Table 17.3 Characteristics of open coast provinces and beaches

Province:	1	2	3	4	5	**Total**/Mean
Coast length (km)	190	155	104	127	120	**697**
Beach length (km)	120	80	63	86	120	**464**
Beach number	62	111	30	37	5	**246**
Mean beach length (km)	2.1	0.6	2.1	2.3	24.0	1.91
(σ)	(4.2)	(1.0)	(2.7)	(2.6)	(26)	
RTR	1.2	1.9	1.1	0.9	0.7	1.1
(σ)	(0.5)	(1.2)	(0.2)	(0.3)	(0.2)	
Mean grain size (mm)	0.3	0.32	0.30	0.19	0.21	0.28
(σ)	(0.14)	(0.19)	(0.14)	(0.03)	(0.005)	(0.15)
Mean beach state[a]	5.1	6.0	5.1	4.0	1.9	5.3
(σ)	(1.8)	(1.5)	(1.1)	(1.7)	(1.2)	(1.8)
Rip number[b]	223	77	174	123	73	**670**
Mean rip spacing (m)	206	150	205	160	422	
Topographic rip number	12	10	11	27	5	**65**
Embaymentization	0.8	0.6	0.7	0.6	0.9	
(σ)	(0.2)	(0.3)	(0.1)	(0.2)	(0.1)	
Orientation (degrees)	116	152	123	100	152	
(σ)	(91)	(104)	(84)	(34)	(14)	
Beach gradient (degrees)	5	5	4.1	3	1.8	
(σ)	(2)	(2)	(1.5)	(2)	(0.2)	

[a]*1* dissipative, *2* longshore bar & trough, *3* rhythmic bar & beach, *4* transverse bar & rip, *5* low tide terrace, *6* reflective
[b]Rip data based on Google Earth imagery

variation in breaker wave height and grain size. Along the micro-tidal Santa Catarina coast the deepwater wave height (H_o) averages 1.0–2.0 m (Fig. 17.2). At the shore however the major control on breaker wave height (H_b) is the regional orientation, and locally the considerable geology and bedrock control, resulting in H_b ranging from 1.5 m on fully exposed beaches to ~0 m in very sheltered bays.

Based on the proposed subdivision of the coast (Table 17.1) there is one semi-sheltered province (2), three moderately to well-exposed provinces (1, 3 and 4) and the very exposed southern province (5). Each of these provinces is discussed based on beach provincial data and related characteristics presented in Table 17.3.

17.2.1.1 Province 1 – Guaratuba-Farol de Cabeçudas

The province begins at Guaratuba (Barra do Sai) and extends south as seven long exposed near continuous beaches which occupy the first 40 km of coast down to the northern shore of Babitonga Bay at São Francisco do Sul (Fig. 17.5). São Francisco Island has 50 km of open coast shore containing 11 beaches. Most are moderately

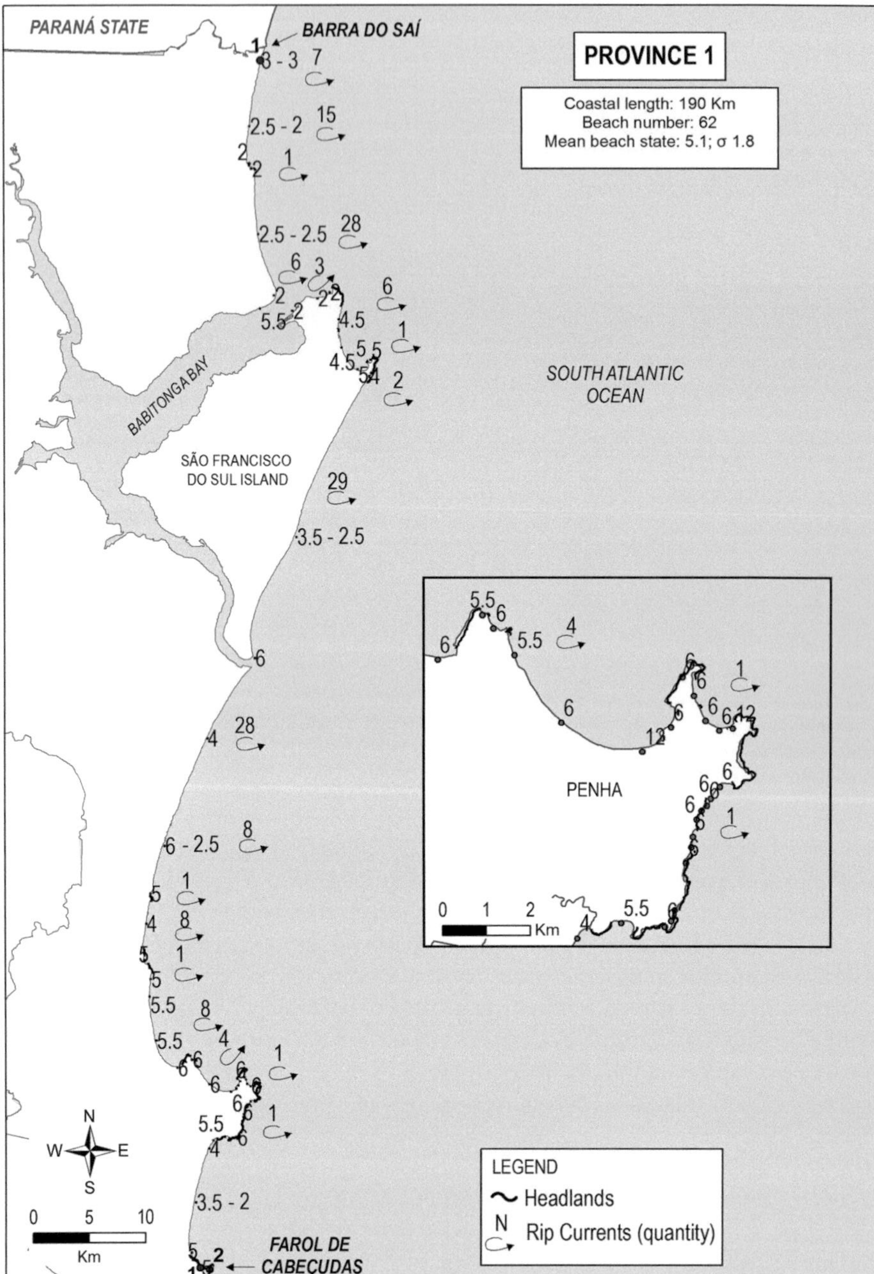

Fig. 17.5 Santa Catarina province 1 showing beach states (1 to 6) and rip currents (number above curved arrow). *1* dissipative, *2* longshore bar & trough, *3* rhythmic bar & beach, *4* transverse bar & rip, *5* low tide terrace, *6* reflective

sheltered and cluttered around Point Jose Dias and Ubatuba, with the 26 km long Grande beach occupying most of the eastern shore of the island where it is well exposed with a continuous intermediate beach system. South of the island the beaches run near continuously for 40 km to the rocky Porto Belo region, followed by the longer Navegantes beach and finally the rocky Farol de Cabeçudas. The longer beaches are rip-dominated beaches, with shorter, coarser grained and more sheltered LTT to R beaches along the rocky sections.

In summary the 190 km long province is an east-facing shore containing 62 beaches, which occupy 106 km (70 %) of the coast, including several moderately long well-exposed beaches which are aligned by the dominant waves to face southeast (orientation =131°). The beaches have a mean length of 2 km, are moderately embayed (embaymentization = 0.8), have predominately fine-medium sand (mean = 0.3 mm, σ = 0.14 mm) with a moderate slope (mean = 5°). Their exposure results in energetic intermediate beach states (RBB-TBR), with half the beaches and 52 km of coast consisting of higher energy double bar systems. Most of the beaches systems are rip-dominated with a total of 223 rips observed in Google Earth images with an average spacing of 206 m, together with 12 topographic rips.

17.2.1.2 Province 2 – Farol de Cabeçudas –Currais

This province begins at the rocky protruding Farol de Cabeçudas and apart from Balneario Camboriu and Itapema and Tijucas bays is a predominately rocky shore with numerous small both sheltered and exposed reflective beaches (Fig. 17.6). Even in the four bays waves are generally low with LTT to R conditions in Balneario Camboriu and Itapema; and tide-dominated B + MF in Tijucas. The province has a total of 77 rips with an average spacing of 150 m, together with 10 topographic rips. Of the 117 beaches 88 are reflective, another 20 LTT, with the remainder split between the northern intermediate beaches and the tide-dominated Tijucas beaches.

This crenulate bedrock-dominated province contains half the states beaches (111), however with an average length of only 0.6 km they represent only 18 % (80 km) by length, and occupy only 51 % of the province. This province contains many small embayed and/or sheltered reflective beaches (77 %), with embaymentization = 0.6, of variable orientation (mean orientation = 142°, σ = 102°) composed of fine to medium sand (mean = 0.32 mm, σ = 0.19 mm) and a moderate (mean = 5°) to steep (>8°) slope. It also contains Santa Catarina's only open coast tide-dominated beaches at Tijucas.

17.2.1.3 Province 3 – Santa Catarina Island

Santa Catarina Island has two shores, the sheltered western shore and the exposed north and east-facing coast, which is discussed here. The 25 km long northern shore has 15 sheltered reflective beaches, which occupy 17 km of the shore, the remainder being granite headlands (Fig. 17.7). The 79 km long east coast contains 15 generally

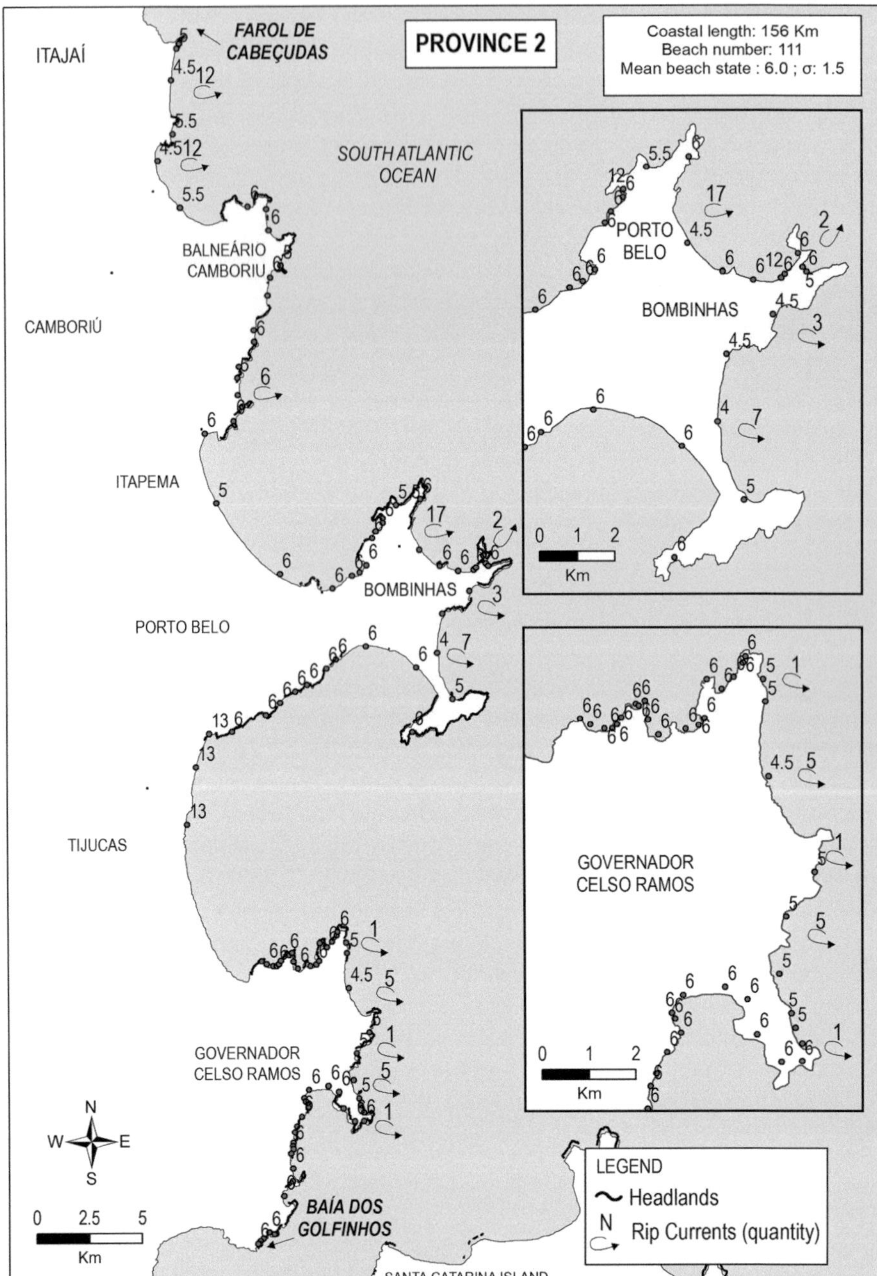

Fig. 17.6 Santa Catarina province 2 showing beach states (1 to 6) and rip currents (number above curved arrow). *1* dissipative, *2* longshore bar & trough, *3* rhythmic bar & beach, *4* transverse bar & rip, *5* low tide terrace, *6* reflective, *13* beach + mud flats

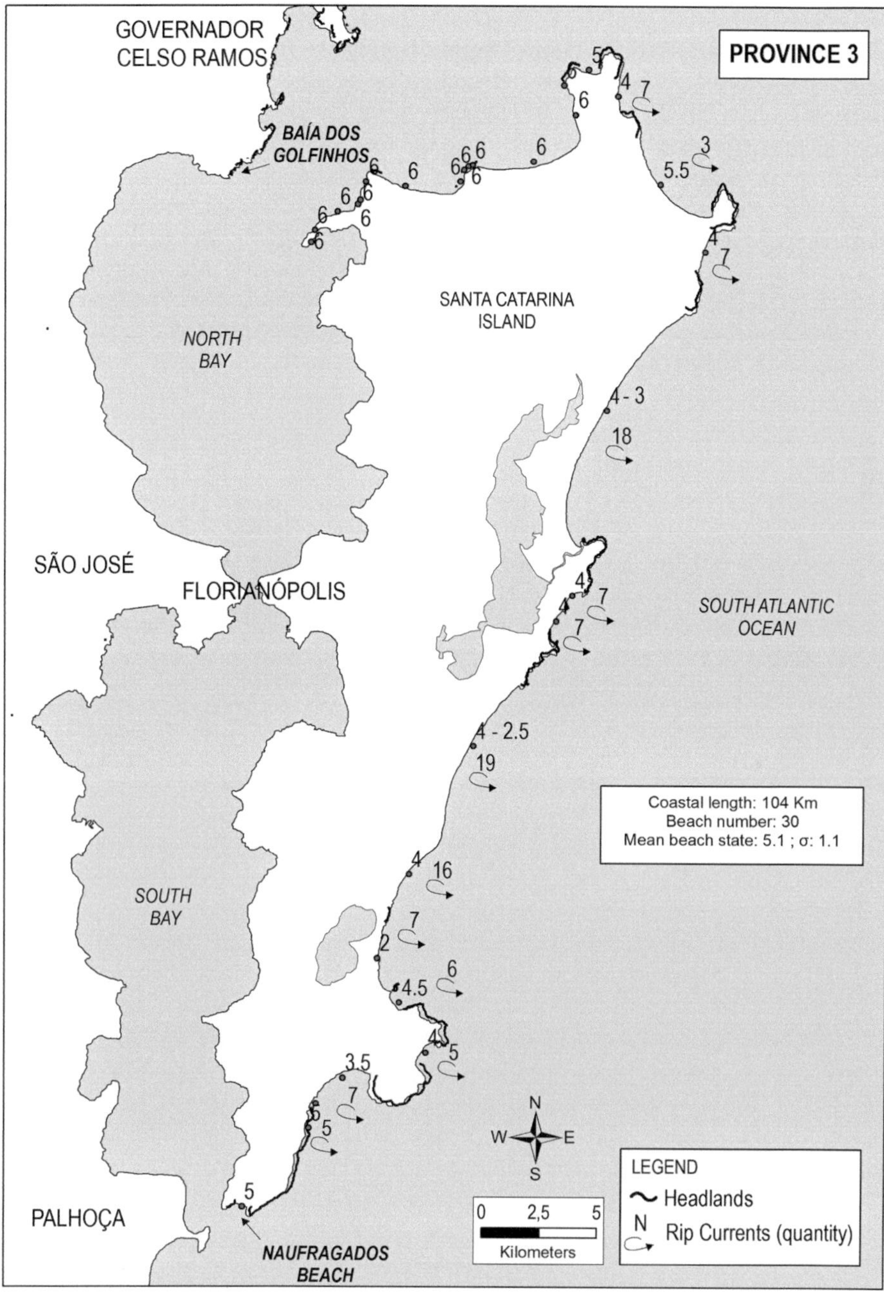

Fig. 17.7 Santa Catarina province 3 including North and South bays, showing beach states (1 to 6) and rip currents (number above curved arrow). *1* dissipative, *2* longshore bar & trough, *3* rhythmic bar & beach, *4* transverse bar & rip, *5* low tide terrace, *6* reflective

east-southeast-facing embayed beaches, (embaymentization=0.7, orientation =123°), each separated by prominent granite headlands. In total the beaches occupy 61 % of the coast. The beaches are composed of fine to medium quartz sand (mean=0.3 mm, σ=0.14 mm), which generates a lower slope (mean=4°). Within each embayment are predominately rip-dominated TBR-RBB beach systems, including two longer double bar systems of Mocambique and Joaquina, both popular surfing beaches. These two beaches are also backed by the most extensive transgressive dune systems on the island. One hundred and seventy-four rips were observed with an average spacing of 205 m, together with 11 topographic rips. In total the 30 beaches occupy 63 km (61 %) of coast. They average 2.1 km in length with the lower energy LTT dominating, but also with 20 km of higher energy intermediate double bar system (LBT-RBB).

17.2.1.4 North and South Bays

North and South bays, which separate Santa Catarina Island from the mainland, have shoreline lengths of 62 and 98 km respectively. Their shorelines include the sheltered western shore of Santa Caterina Island as well as the sheltered mainland shore. They contain 119 and 273 beaches respectively. Most of the beaches are small, short and tide-modified to tide-dominated. The eastern shores of both bays have extensive mangrove forest.

17.2.1.5 Province 4 – Point Papagaio – Cape Santa Marta

South of Santa Catarina Island the coast is straighter, faces east and extends for 127 km south to Cape Santa Marta (Fig. 17.8). It contains 38 beaches (mean length=2.2 km, embaymentization=0.6, orientation=100°), located in about 20 headland-bound embayments composed of fine quartz sand (mean=0.19 mm, σ=0.03 mm), the finer sand lowering the mean slope to 3°. They occupy 86 km of coast (68 %) with the more energetic TBR beach state dominating. These result in 123 rips being present with an average spacing of 160 m. There are also 27 topographic rips against the many headlands. In addition there are 57 km of double bar beaches of which 24 km reach a fully dissipative triple-bar system. This is also the beginning of the major south-trending transgressive dune systems, with about eight major dune systems backing the higher energy beaches.

17.2.1.6 Province 5 – Cape Santa Marta – Torres Inlet

The southern province commences on the south side of Cape Santa Marta where there are two exposed embayed beaches, followed by a basically near continuous 120 km long 400–500 m wide triple-bar fully dissipative beach system (mean beach state=1) (Fig. 17.9) divided by the Camacho, Rincão, Araranguá and Mampituba

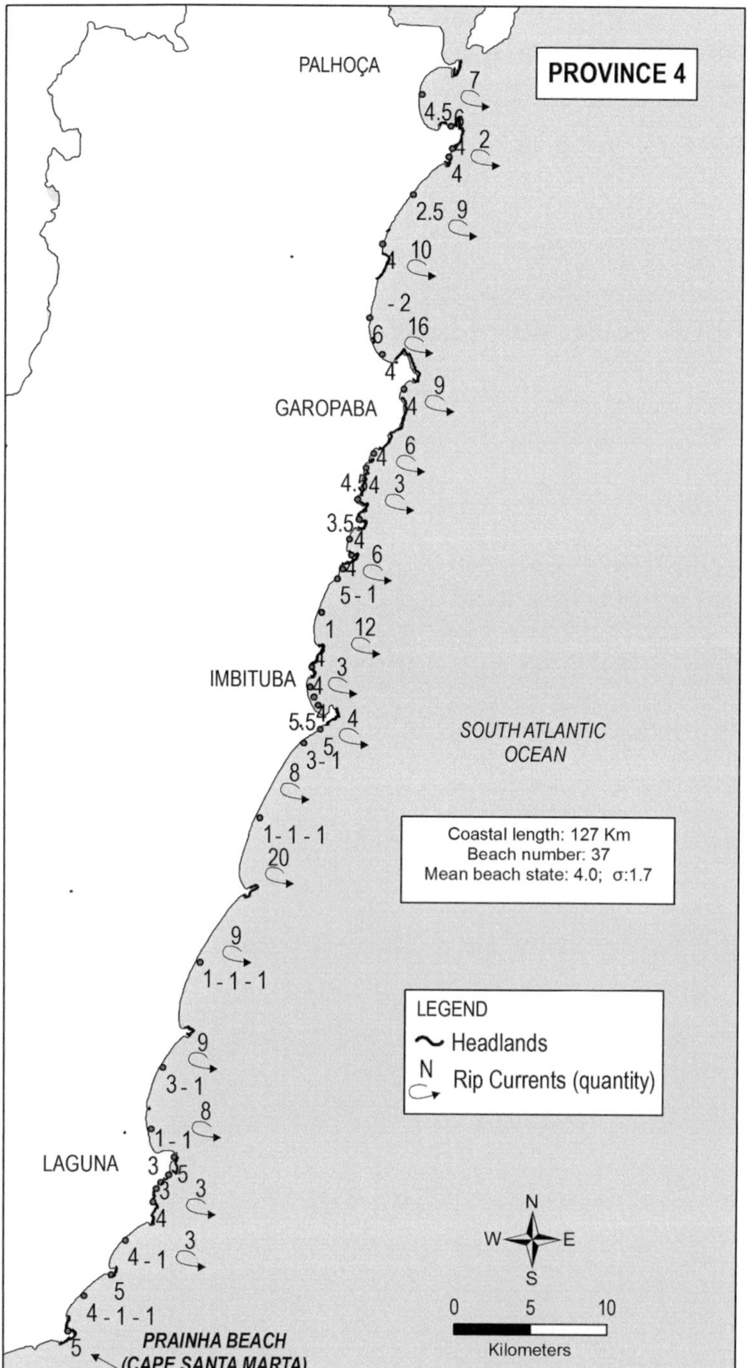

Fig. 17.8 Santa Catarina province 4, showing beach states (1 to 6) and rip currents (number above curved arrow). *1* dissipative, *2* longshore bar & trough, *3* rhythmic bar & beach, *4* transverse bar & rip, *5* low tide terrace, *6* reflective

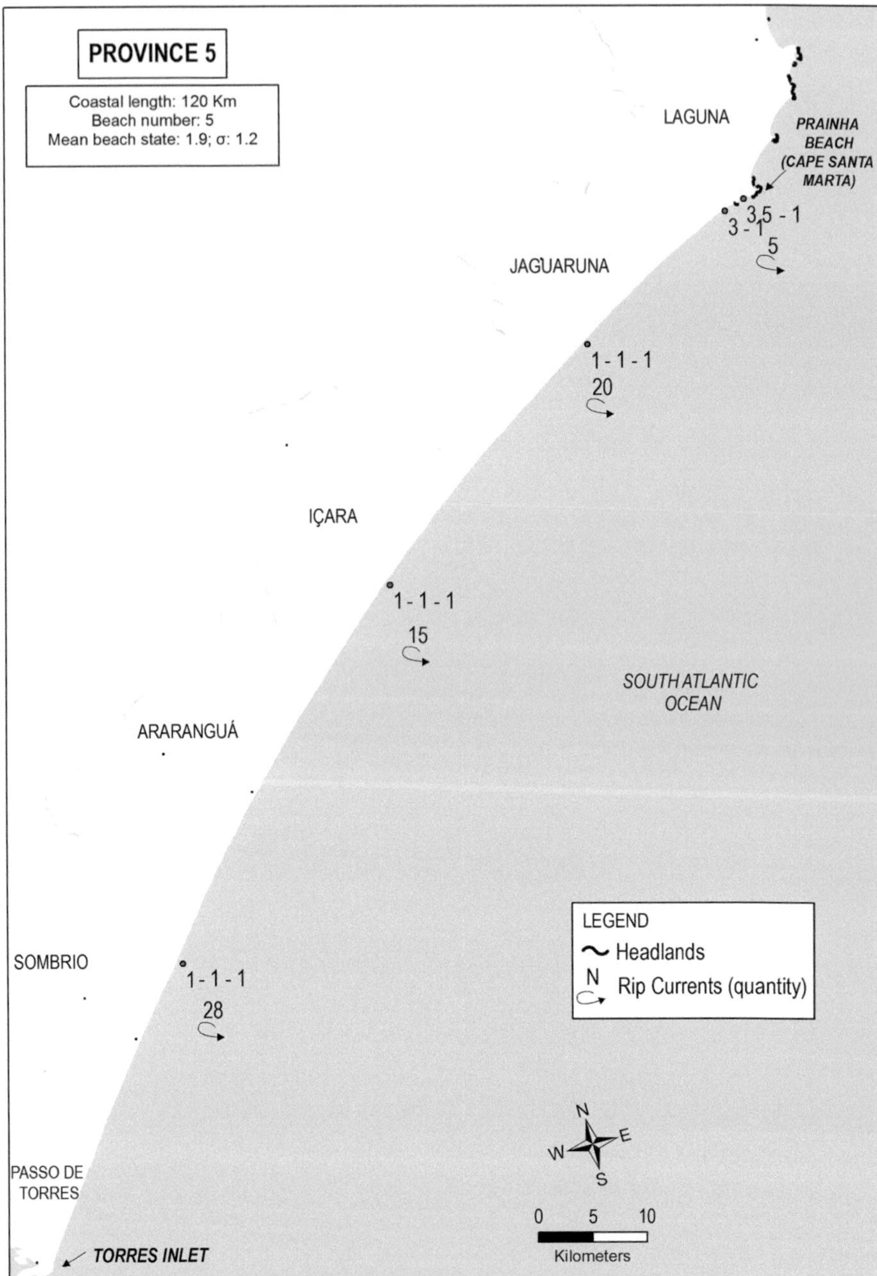

Fig. 17.9 Santa Catarina province 5, showing beach states (1 to 6) and rip currents (number above curved arrow). *1* dissipative, *2* longshore bar & trough, *3* rhythmic bar & beach, *4* transverse bar & rip, *5* low tide terrace, *6* reflective

inlets into three long beaches composed of very uniform fine quartz sand (mean = 0.21 mm, σ = 0.005 mm), with wide low gradient beaches (mean slope = 1.8°). The five long beaches that make up this province have an average length of 24 km, with an embaymentization of only 0.9 and a southeast orientation (152°), which exposes them to the full force of the dominant southerly waves. They occupy 98.6 % of the coast. The beaches are backed by continuous massive south-trending transgressive dunes. This section was also observed to have 73 rips with a mean spacing of 422 m double that of the other provinces, but only four topographic rips owing to the few headlands.

In provinces 4 and 5 the slope between the shoreline and wave base (~20 m deep) decreases from between 1:200 and 1:20 (0.3–3°) in the north to 1:500 (0.12°) in the south, as fine sand, long dissipative beaches and massive sandy barriers dominate province 5 (Hesp et al. 2009).

In total the SC coast contains 246 open coast beaches totaling 464 km in length (See Table 17.3), with an average R-LTT beach state (mean = 5.3, σ = 1.8). Based on beach number the typical beach is short, low energy and reflective, however based on beach length exposed multi-bar beaches occupy 212 km (46 %) of the shore, including 62 km of fully dissipative beaches. In addition North and South bays contain another 392 small tide-modified and tide-dominated beaches spread along 160 km of sheltered shoreline.

Besides the regional variation in beach state, many beaches contain longshore variations in wave height and/or sand size that results in intra-beach modification in beach state. Klein and Menezes (2001), Klein (2004) and later Miot da Silva et al. (2012) and Oliveira et al. (2012, 2014) found that along the embayed beaches the beach state was dependent both on the longshore variation in H_b, as well as being modulated by changes in sediment size, with the beach state ranging from a single R/LTT in the southern corner of embayments to at times a two or three bar RBB-D in the more exposed north.

17.2.2 Beach Type/State, Number and Length

Santa Catarina has 246 beach systems along its open coast, together with 119 beaches in Baia do Norte and 273 beaches in Baia do Sul, a state total of 631. This section is concerned with the open coast beaches which are predominately wave-dominated, with just eight tide-dominated systems (Table 17.4). The wave-dominated contain the full range of beach states from dissipative to reflective (Fig. 17.10). The nature and distribution of the beach states is discussed below. Table 17.4 summarises the range of beach states along the SC open coast. There are two ways of reading this table, using the number of beaches and their states, or the length that each beach state covers.

Based on beach number the most common beach type is the lower energy *reflective*, which account for 122 (51 %) of the beaches (Table 17.4). These however average only 0.75 km in length (σ = 1.5 km) and represent only 20 % (95 km) of the

Table 17.4 Santa Catarina open coast beach state by number and length

Beach state[a]	Beach state[a]	Number	Total (km)	Mean (km)	σ (km)
1	D	7	143.9	20.6	20.0
2	LBT	8	21.3	2.7	2.4
3	RBB	10	39.8	4.0	3.0
4	TBR	39	83.8	2.2	2.5
5	LTT	47	71.1	1.5	4.1
6	R	127	95.3	0.75	1.5
Sub-total		**238**	**455.2**	**1.91**	**5.4**
11	R+SF	1	2.3	2.3	–
12	R+TSF	4	1.4	0.4	–
13	R+TMF	3	5.0	1.7	–
Sub-total		**8**	**8.7**	**1.1**	–
Total		**246**	**464**	**1.9**	–

[a]Wave-dominated: 1=*D* dissipative, 2=*LBT* longshore bar & trough, 3=*RBB* rhythmic bar & beach, 4=*TBR* transverse bar & rip, 5=*LTT* low tide terrace, 6=*R* reflective
Tide-dominated: 11=*B+SF* beach+sand flats, 12=*B+TSF* beach+tidal sand flats, 13=*B+TMF* beach+tidal mud flats (See Short 2006). On multi-bar systems based on beach state of inner bar

state's beaches by length. They tend to occur in the central bedrock-controlled province 2 and usually are located in sheltered embayments of varying size and orientation (Fig. 17.6). Mean embaymentisation in 0.64, the smallest of the five provinces, and mean orientation in 134° (σ = 96°). They occur in both low wave sheltered locations as well as some more exposed locations composed of medium to coarse beach sediments, such as Costa Brava (Taquarinhas, Taquaras, Estaleiro, Estaleirinho, Flamingo) (Fig. 17.10f). The open coast reflective beaches usually have a moderate to steep beach faces (mean=4.7°), with a 2.5–3.0 m high berm, which decrease to less than 1.5 m on lower energy reflective beaches. During high wave events the beach faces is eroded and deposited as an attached bar or terrace and the backshore-foredune may have scarp up to 3 m high (Klein and Menezes 2001).

Rip-dominated intermediate beaches occur on 104 (43 %) of the beaches and occupy 216 km or 47 % of the sandy coast. The intermediate are dominated by the lower energy LTT and TBR both in terms of number of beaches and beach length. *Low tide terrace* (LTT) beaches account for 40 beaches (17 %) and with and average length of 1.5 km occupy 71 km (16 %) of the sandy coast. This beach state tends to occur in moderately sheltered embayments, as well as some more sheltered locations with very fine sand, so fine it precludes the development of a reflective high tide beach (Fig. 17.10e). These systems consist of a 20–200 m wide dissipative low tide terrace, which may or may not have a steeper high tide beach (Oliveira et al. 2012, 2014). They are what Klein and Menezes (2001), Klein et al. (2010) classified as 'sheltered dissipative'.

The rip-dominated *transverse bar and rip* (TBR) is one of the more hazardous beach systems for beach users. In SC there are 39 (16 %) TBR beaches, which have an average length of 2.2 km and occupy 84 km (18 %) of the sandy coast. Their typical rip spacing is 190 m (σ=110 m). They occur along longer reasonably well-exposed

Fig. 17.10 Examples of Santa Catarina beach types; (**a**) Dissipative beach at Ibiraquera, Province 5 (numbers indicate 3 bars); (**b**) Longshore bar and trough at Moçambique beach, Province 3; (**c**) Rhythmic bar and beach at Campo Bom, Province 5; (**d**) embayed transverse bars and rips at Galheta, Province 3; (**e**) low tide terrace at Do Luz, Province 4; (**f**) reflective beach and cusps at Estalerio, Province 2; (**g**) beach and mud flats at Tijucas, Province 2 (Photos by: RS Ribeiro, M Muler, AD Short)

embayments composed of fine to medium sand, particularly along provinces 3 and 4. They are highly visible with their associated rhythmic shoreline topography and alternating shallow bars and deeper rip channels (Fig. 17.10d).

The three *higher energy beach states* (RBB, LDT and D) represent the higher energy beaches along the SC coast and all occur on generally longer well-exposed east and southeast-facing beaches in provinces 4 and 5 (Fig. 17.10a, b, c). All three tend to receive similar levels of wave energy, the major difference being variations in sand size, with more medium sand resulting in RBB and LBT beaches and the finer sand, particularly in the south (provine 5) producing the multi-bar fully dissipative beaches (Fig. 17.10a).

There are a total of 25 (10 %) of these beach systems, however because of their greater lengths they occupy 205 km (45 %) of the sandy coast, particularly along the longer embayed and more exposed northern (e.g. Praia Grande – province 1) and southern (eg. Morro dos Conventos – provinces 4 and 5). The higher energy intermediate have shore parallel troughs separating the bar from the beach, with rip currents crossing the bar on average every 200 m in province 1–4 and 400 m in province 5. Dissipative inner bars only occur on seven beaches (3 %), but these average 20.6 km in length with a total length of 144 km (31 %). The dissipative beaches usually have two to three shore parallel bars whose spacing increases offshore, as is discussed in the Sect. 17.2.3.

In addition there are eight tide-dominated beaches located on the 'open' coast, all deep within coastal embayments. They have a mean embaymention ratio of 0.5 and an average length of 1.1 km. They are exposed to very low waves, and even though tides remain micro the RTR>>3 (Table 17.4; Fig. 17.10g). Most of the beaches in North and South bays are tide-modified to tide-dominated the latter containing tidal flats.

17.2.3 Bar Number, State and Spacing

The above relationships show a clear correlation between beach state and beach length. This relationship is also borne out in Table 17.5 which summarizes the occurrence of beaches with no bar, one, two and three bars along the coast, while Table 17.6 summaries the beach states associated with each bar and the average bar spacing on two and three bar systems. In relation to beach states, the no bar beaches are all Reflective, which is also indicated by their short average length of just 0.8 km, a product of their usually deeply embayed and sheltered location. The one bar beaches tend to be longer (mean length = 1.2 km) intermediate beaches, ranging from LTT to RBB with a mean beach state of 4.4 (TBR/LTT). The two bar beaches occur on still longer beaches (mean length = 7.6 km) and have inner bar beach states ranging from R to RBB (mean beach state 3.6 (TBR/RBB)), while the outer bar ranges from RBB to D (mean beach state 1.8). The three bar beaches are associated with the longest (mean length = 14 km) most exposed beaches composed of uniform

Table 17.5 Santa Catarina bar number by beaches and length

Number bars[a]	Beach type	Number beaches	Total length (km)	Mean length (km)
Tidal flats	Tide-modified	8	9	1.1
0 bar	Reflective	127	95	0.75
1 bar	Intermediate	80	95	1.2
2 bar	Inter-Diss	25	182	7.3
3 bar	Dissipative	6	83	13.8
Total		**246**	**464**	**1.9**

[a] 0 bar = reflective, 1 bar = intermediate, 2 & 3 bars = dissipative

Table 17.6 (a) Beach state[a] (means & σ) associated with each bar on 0, 1, 2 and 3 bar systems; and (b) mean bar location (distance offshore for province 5)

a. Number bars	0 bar	1 bar	bar 2	bar 3
0 bar	6	–	–	–
1 bar	–	4.4 (0.9)	–	–
2 bars	–	3.6 (1.5)	1.8 (0.7)	–
3 bars	–	1.5	1	1
b.	Distance offshore (m)			
2 bars	–	65	168	
3 bars	–	36–81	110–204	238–324

[a] *1* D, *2* BT, *3* RBB, *4* TBR, *5* LTT, *6* R

fine quartz sand (mean = 0.2 mm) and all have dissipative outer bar/s, with possibly of a LBT inner bar, particularly during prolonged periods of lower waves.

The shoreward decrease in beach state within multi-bar systems is to be expected, as wave breaking on the outer bar/s successively lowers wave height shoreward resulting in lower inner bar wave height and subsequently lower energy beach states (see Short and Aagaard 1993 and Short 1992).

Table 17.6 shows that overall the bars have a mean offshore location (distance from the shoreline) of 65 m, 168 m and 280 m, the exponential increase in spacing in agreement with a standing wave mode of formation. When the inner bar is attached the three bars have mean crest locations of 36 m, then 110 m and 238 m. When the inner bar is detached the whole system moves offshore with bar crest locations at 81 m, 204 m and 324 m.

On a section of coast with essentially identical wave-tide conditions, the shift between two to three bars is a product of overall sediment size and beach gradient, with finer sand and lower surf zone gradients favoring three bars (Short and Aagaard 1993). Whether the inner bar is attached or detached will be a product of antecedent wave conditions, with long periods of lower waves favoring a shift to an attached inner bar, while detached bars should follow storm and high wave periods (Short and Aagaard 1993).

17.2.4 Controls on Beach State

Wave-dominated beaches occur in areas where the RTR<~3. Santa Catarina has a spring tide range of 1.2 m in the north decreasing to 0.46 m in the south. This implies the beaches will be wave-dominated when H_b begins to exceed ~0.4 m which is the case along most of the open coast. With an average open coast deepwater wave height of 1.5 m all exposed beaches will have RTR of 0.9 (RTR = 1.2/1.3 = 0.9), the RTR rising as wave refraction-diffraction and attenuation lower wave height into embayed beaches and bays, until a lowered wave height of ~0.3 m will result in an RTR of ~4 (RTR = 1.2/0.3 = 4). Tide-dominated beaches occur in a handful of sheltered deeply embayed beaches, where wave height ~0.1 m resulting in an RTR of ~12 (RTR – 1.2/0.1 = 12). As Table 17.3 shows on the open coast the mean RTR reached a minimum of 0.7 ± 0.2 in the exposed province 5 and a maximum of 1.9 ± 1.2 is the more sheltered province 2, as is to be expected.

Another major control on beach state is sediment size, expressed through sediment fall velocity (W_s) in the dimensionless fall velocity (Gourlay 1968)

$$\Omega = H_b / TW_s \tag{17.2}$$

where $\Omega < 1$ produces reflective beaches, 1–5 intermediate and >6 dissipative (see Chap. 1). As T remains relatively uniform along the coast it is changes in H_b and W_s that control beach state. At one extreme dissipative beaches require both high waves and fine to very fine sand, while coarse sand and coarser will always result in reflective beaches, with intermediate beaches favoring fine to medium sand. In Santa Caratina all the fully dissipative beaches in province 4 and 5 are composed with fine sand (mean = 0.2 mm) as well as being fully exposed to higher waves. At the other extreme the reflective beaches tend to occur in province 1–3 where not only is sand coarser (mean = 0.3–0.48 mm) but also geological controls reduce breaker wave height through refraction and attenuation within embayments (Klein and Menezes 2001), with attenuation as much as 78 % recorded by Ribeiro (2014) in Itapocorói Bay.

These two parameters H_b and W_s, together with T, which remains unchanged during shoaling, and tide range which is micro thoughout, control the range of beach types and states along the Santa Catarina coast. As mentioned above deepwater wave height (H_o), which is affected by changes in shoreline geometry and wave shoaling, refraction, diffraction and attenuation will control H_b, while local and regional changes in sediment texture will control W_s. As a consequence spatial changes in H_b and W_s result in the considerable variation in beach type and state along the coast, in both space and time. This is illustrated in Fig. 17.11, which plots the location of beaches in each of the five provinces based on their Ω and RTR. While tide range (~0.8 m) remains essentially a constant, the RTR ranges from 0.5 to 9 owing to the range of H_b from ~0.1 to 1.6 m, though most beaches plot RTR<3 indicating they are wave-dominated. Likewise Ω ranges from near zero to 11 based on both H_b and W_s (very fine to coarse sand), indicting the full range of

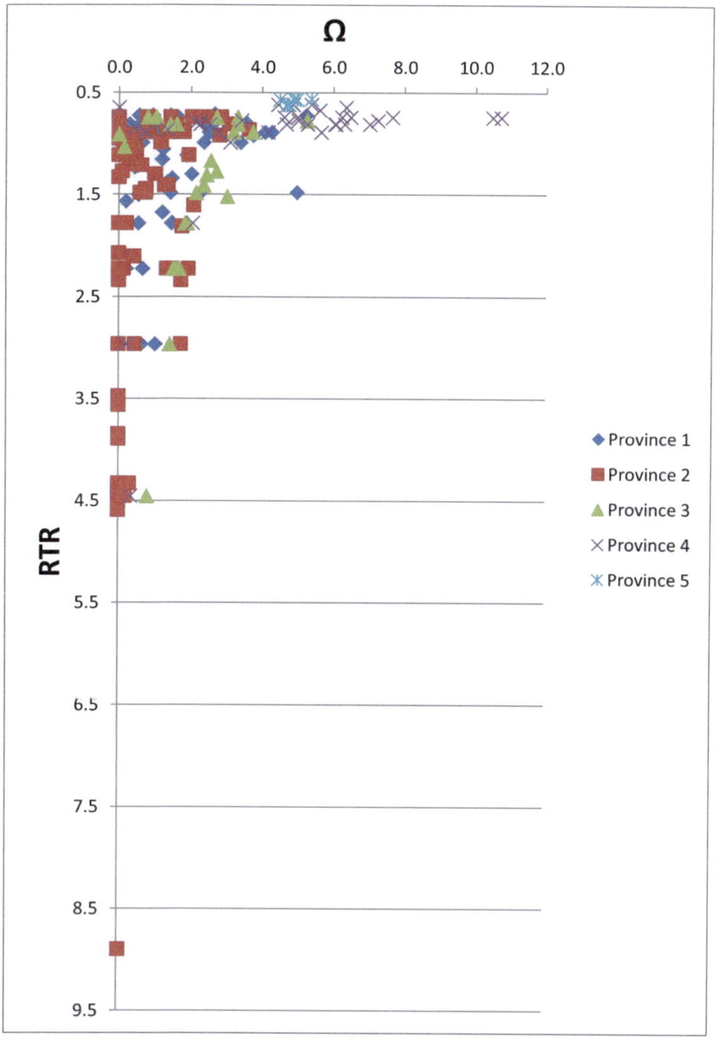

Fig. 17.11 Plot of Santa Catarina beaches for each province based on Ω and RTR

wave-dominated beach states (reflective-intermediate-dissipative). As expected the highest energy beaches occur in the most exposed provinces 4 and 5, with province 4 having the highest Ω, owing to finer sand, rather than higher waves. All the dissipative beaches in province 5 however cluster around an Ω of 5, while in province 4 Ω extends from 1 to 11. The lowest energy beaches occur in the bedrock-dominated and heavily embayed and sheltered province 2, with some beaches having an RTR > 3, including the tide-dominated Tucias B + MF reaching an RTR of 9. The northern province 1 has a mixture of both high energy (e.g. Grande Ω = 5) and lower energy beaches; while province 3, Santa Catarina island, also has both the lower energy northern shore beaches, with an $\Omega < 1$ and RTR up to 4.5 and more exposed eastern shore beaches when $\Omega = 5$.

17.2.5 Temporal Variation in Beach State

The data presented in Table 17.3 is based on limited temporal observations of most of the beach systems using Google Earth and field observations. It is expected that while the data should represent the average state of these beaches, there will be temporal variation in beach states in response to changing wave conditions, particularly on the intermediate beaches and the inner bar/s on the 2 and 3 bar systems, with least change occurring on the sheltered fully reflective beaches.

17.2.6 Longshore Sand Transport and Sediment Cells

An essential component of all beach systems is their sediment characteristics and sand budget. The sediment size, texture, sorting, and mineralogy will determine together with waves and tides, the beach type, slope, color and nature, with only beach type discussed above. The geological substrate and boundaries will determine the beach length and influence wave refraction and diffraction, which in turn affect spatial and temporal variation in beach type and state. The long term stability of the beach is however affected by its sediment budget which may be positive leading to beach accretion, negative leading to erosion or stable and balanced leading to a stable shoreline. Knowledge of the budget is therefore critical to understanding the Quaternary/Holocene evolution of each beach system, and its contemporary behavior and management. The Quaternary/Holocene budget can be gauged from the size of the system and its nature, whether regressive, transgressive or stable, including the size of sediment sinks, as detailed for the Brazilian coast in Dillenburg and Hesp (2009). The contemporary budget is more difficult to determine, as it involves knowledge of decadal scale shoreline behavior and on longer coast rates of longshore sand transport including on some headland sand bypassing and overpassing. At this stage in our knowledge of the Santa Catarina coast we can at best provide qualitative indications of budget, with modeling capable of generating some rates for longer beaches. For an overview of Santa Catarina's Holocene barrier evolution see FitGerald et al. (2007, 2008), Hesp et al. (2009) and Hein et al. (2012, 2013).

The entire coast is dominated by southerly waves, which produces northerly longshore sand transport. Aeolian transport is also to the north on Santa Catarina Island and northwards (province 1–3), while northerly winds and southerly aeolian transport dominates province 4 and 5 down into Rio Grande do Sul. This being the case this brief overview of the transport will begin in the south and work north.

Table 17.7 Estimated rates of longshore sand transport in province 5

Rate (1000 m^3 $year^{-1}$)	Location	References
147–389	Province 5	Pitombeira (1975)
1379	Province 5	Lima et al. (2001)
218–574	Province 5	Fontoura (2004)
98	Province 5	Alfredini (1999)
49–55	Province 5	CPE (2009)
???	Province 5	Motta et al. (2010)
20–200	Prainha	Abreu (2011)
200–500	Grande	Abreu (2011)
Range: 20–1379		

Province 5 The 115 km long section of continuous beaches between Torres Inlet and Cape Santa Matra faces southeast into the prevailing southerly waves, which generate a net northerly transport of very uniform fine quartz sand (mean = 0.2 mm). This transport commences in Uruguay and continues along the beaches of Rio Grande do Sul and the southern Santa Catarina beaches from Torres to at least Cape Santa Marta and probably beyond. The estimates of north longshore sand transport however vary by up to two-orders of magnitude (Table 17.7). The most detailed study by CPE (2009) based on 12 years of wave hindcasting at Morro dos Conventos Beach (Ararangua inlet) suggest a net northerly movement of between 45,000 and 55,000 m^3 $year^{-1}$, substantially less than the more than 1 M m^3 $year^{-1}$ suggested by Lima et al. 2001. Clearly more investigations are required to derive a more accurate rate. The CPE (2009) modeling also indicates that reversals in transport occurs both seasonally and annually the reversal driven by changes in wave height and direction.

The net northerly transport is also evident in morphological features such as downdrift offset and erosion at jetties, as at Torres (Fig. 17.12a); the northerly deflection river mouth spits, as at Araranguá (migrating at ~100 m $year^{-1}$ – Vieira da Silva et al. 2011; Bhering 2012) (Fig. 17.12b) and in particular in the northerly distribution of increasingly fine to very fine sand.

On the other hand, Siegle and Asp (2007) estimated the non-dimensional potential intensity of longshore drift (per unit area). They suggest that near Cape Santa Marta there is a reduction of sediment transport with a reversal. This is evident in morphological features such as longshore stability of Camacho Inlet near by Cape Santa Marta (Vieira da Silva 2009; Vieira da Silva et al. 2011; Fig. 17.12d) and in southerly migration of Urussange inlet (Bhering 2012; Fig. 17.12c).

What is interesting on this coast is that the aeolian transport is southward (30 m $year^{-1}$ in barchan dune system at Praia Grande do Sul – Laguna- Jaguaruna – Giannini 1993; Giannini et al. 2005), resulting in some degree of recycling of the wave transported material, however there have been no calculations of the rate of southerly aeolian transport. It is however expected to be at least an order of magnitude less than the northerly wave-driven transport.

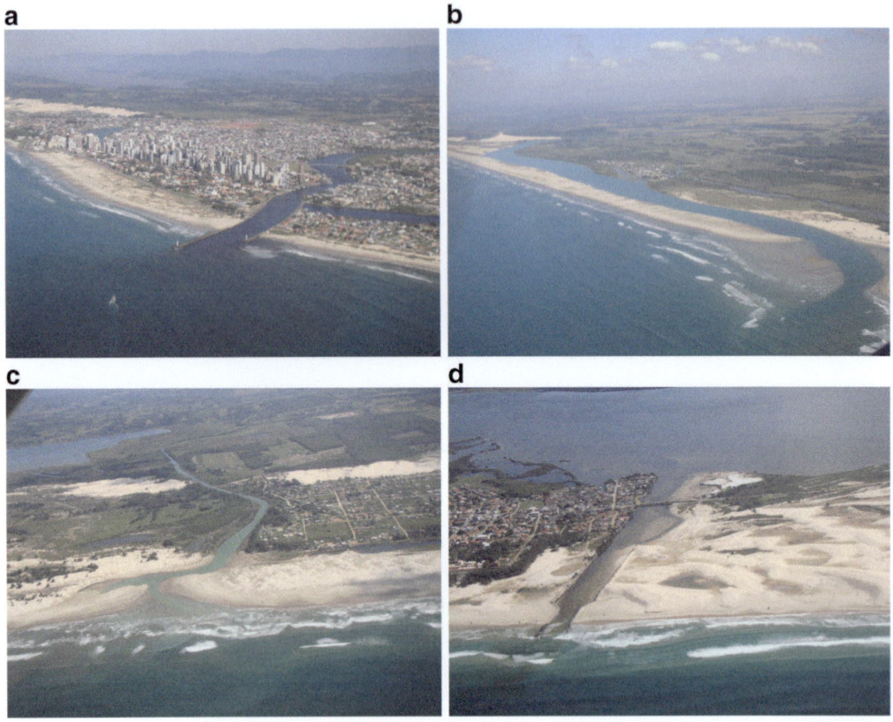

Fig. 17.12 Province 5 inlets: (**a**) Torres Inlet; (**b**) Ararangiá inlet; (**c**) Urussanga inlet; (**d**) Camacho inlet (Photos by R S. Ribeiro, M Muler, A D Short)

Province 4 At Cape Santa Marta the coast trends north for 91 km to the mouth of South Bay, with about 20 headlands producing a series of embayed beaches. The change in shoreline orientation combined with the headlands interrupt and reduce the rates of longshore transport. Headland bypassing is likely to occur around many of the headlands, however wind-driven overpassing is in the reverse direction to the south. It has been proposed that much of the sand for the extensive barriers, particularly at Pinheira and Caieiras may well be of shelf origin, and not a downdrift sink (Hein et al. 2012).

Province 3 Santa Catarina Island consists of 30 embayed beaches. Some are separate sediment cells, including Pantano do Sul, while others appear to be part of interrupted but linked northerly sediment transport cells. The cells are interrupted by headlands, islands and changes in coastal orientation and where the interruption is permanent the cell remains closed as at Pantano do Sul. However Mazzer (2007) suggest a bypassing between Pantano do Sul and Solidão Beach and no bypassing between Matadeiro and Armação beach based on numerical simulation. The Pantano do Sul bypassing is corroborated by Souza and Corrêa (2006) who found large nearhore megaripples migrating in a northerly direction in front of north Pantano do Sul beach. At Armação beach Abreu de Castilhos et al. (1997, 1998) estimated

Fig. 17.13 Examples of province 3 sediment transport: (**a**) Headland overpassing dune at Ingleses beach; (**b**) Ponta das Canas spit; (**c**) the large terminal Daniella spit (Photos by R S Ribeiro, M Muler, A D Short)

longshore transport to south of 1.36×10^{-4} m^{3}h^{-1} and to north of 6.24×10^{-4} m^{3}h^{-1} with is a net transport to northeast direction. This beach is also an erosional hot spot (south) (Dalbosco 2013).

Where the cell is open sand can the transferred to the next northerly cell via subaqueous headland bypassing, and on some beaches via aeolian headland overpassing. These include Armacao-Campeche-Joaquina system; and the northern Mocambique-Santinho-Ingleses-Brava-Bom Jesus-Ponto Grossa system. This latter appears to have headland bypassing (Porpilho et al. 2015; Vieira da Silva et al. 2016a, b), as well as overpassing on the order of 10,000 m^{3}year^{-1} (Boeyinga et al. 2010; Fig. 17.13a), leading to the formation of active recurved spits at Ponta das Canas (Nunes 2002), Jureré and Forte (Vieira da Silva et al. 2016a, b) (Fig. 17.13b), as well as uniform moderately sorted medium sand and well sorted fine sand, with an average diameter ranging from 0.20 mm between Santinho to Daniela (Vieira da Silva et al. 2016a, b) and finally the Holocene development of the substantial spit at Daniela (Diehl 1997; Diehl et al. 1998, Fig. 17.13c). This spit appears to be the terminus for fine to medium quartz sand transport (mean = 0.2–0.3 mm) around the northern part of the island, it is more than 1 km^2 in size and may contain several million cubic meters of sand. There is also the possibility it is linked to the strong tidal flow throught the bay entrance and could be contributing sand to the tidal delta (Vieira da Silva et al. 2016).

Province 2 This central bedrock-dominated section of coast is the least likely to have longshore transport, and appears to contain numerous small embayed and pocket closed sediment cells, with little or no exchange between cells. This is evident by changes in sediment type between adjoining embayments (mean 0.1–0.6 mm) (Klein and Menezes 2001; Klein 2004), the development of the Tijucas open coast mud environment, and the lack of feature indicative of intra-embayment transport, including spits and transgressive dunes.

Menezes (2008) obtained a sediment budget to the Balneário Camboriú bay and suggested that the bay is in budget, but as with other embayed beach in SC has different cells along the coast. He suggested that no sediment is exported, in agreement with Klein (2004). In this very dense urban area Cruz (2004) and Menezes (2008) analyised the amount of sediment carried from the beach by the human body during the summer time. They found an average of 20 g per person, that in 20 years represent a shoreline change of 0.17 m, using Bruun's Rule.

Province 1 The northern cell contains longer well-exposed wave-dominated beaches which are experiencing northerly transport. There is morphological evidence of headland bypassing in the recurved spits between Ponta Jose Dias and Capri; the orientation of the headland-bay beaches (Klein et al. 2003b; Klein 2004; Silveira et al. 2012); and the northerly deflection of Lagoa da Barra Velha and the Rio Acarai and Sai Guacu river mouths. Barra Velha tidal inlet has average northerly migration of almost 100 m year^{-1} (Piérri 2005; Klein et al. 2006; Cassiano and Siegle 2010; Bhering 2012). There are erosional hot spot areas at Piçarras and Barra Velha beaches and Menezes (2008) found an erosional hot spot at Gravatá beach, at the northern end of the long Navegantes system, where the sediment is transported in an offshore direction near the north headland.

To the north of São Francisco Island, Abreu (2011) used sand fluorescent tracers to establish an annual alongshore sediment buget in the study area. The calculation of littoral drift indicate a net alongshore sediment transport towards the north of between 200,000 and 550,000 m^{3}year^{-1} on exposed beaches (Grande beach) and between 20,000 and 200,000 m^{3}year^{-1} on semi-exposed beaches (Prainha beach). Silveira et al. (2012) proposed a sediment budget box model for this area and suggested an inlet bypassing around of 100,000–150,000 m^{3}year^{-1}, by the ebb tial delta of São Francisco entrance. In the north, Itapoa area there is a complex cell, where the behavior is a function of ebb shoal delta as well. The evolution of this area is presented by Souza (1999) and Silveira et al. (2012). Finally, Lessa et al. (2000) suggested that the source of sediment to Parana Coast is from north of Santa Catarina, with the littoral cell starting at Penha-Piçarras beach.

In summary, there is net northerly wave-driven sand transport along the SC coast, with greatest rates expected along the long exposed beaches in province 1, and least to zero in the deeply embayed beaches of province 2. There is however considerable divergence in the estimates of transport rates, which on the open coast vary by up to two orders of magnitude (Table 17.7). More research is required in this area to obtain more accurate rates. In addition aeolian headland overpassing and subaqueous headland bypassing is occurring on many headlands and is the subject of current research (De Camargo et al. 2016; Vieira da Silva et al. 2016; Pinto et al. 2015; Porpilho et al. 2015), which will hopefully provide greater insight into the mechanisms of the longshore sand transport.

17.3 Beach Hazards and Risk

17.3.1 Coastal Erosion and Inundation

Two distinctive erosional processes can be identified along SC coast and may be associated with either the gradual change in shoreline position or to specific high waves events. Although many authors have identified sectors of eroding coast (see Sect. 17.1.9) the process has not always been verified. In some places, like Camacho and Itapocu inlets, jetties were placed in order to stabilize the inlet. In others, like Barra da Lagoa beach, severe storms caused heavy erosion followed by beach recovery.

Klein et al. (2006) presented a synthesis of the nature and magnitude of erosive processes observed along provinces 1 and 2. The authors identified local hot spots (Barra do Sul, Barra Velha, Piçarras, Gravatá/Navegantes and Barra Sul/Balneário Camboriú) and detailed the measures that have been undertaken for coastal protection. A similar study was performed by Horn Filho (2006) for Florianópolis. The beaches recognized as being under the highest degree of risk were: Ingleses, Canasvieiras, Barra da Lagoa, Armação, Pântano do Sul and Naufragados. Both studies associated the coastal degradation, with coastal expansion and inadequate management plans.

Although erosional trends related to a negative sediment budget can be recognized in some of these spots (as in Piçarras, for example), in many of them the erosion problems seem to be more related to episodic storm events and surges. As previously mentioned, SC is regularly exposed to the influence of storms associated to polar fronts and extra-tropical cyclones. Some of these are strong enough and/or develop so close to the coast to develop storm surges, resulting in coastal flooding and beach erosion (Bonetti et al. 2012). Rudorff et al. (2014), based on reports from the state's Civil Defense, reported a concentration of these events in the months of May to September and an average of 3.28 registered occurrences per year of infrastructure damages between 1997 and 2010.

From 2000 to 2010 storm surges affected 13 municipalities, which declared states of emergency, and one which declared a state of public calamity (Rudorff et al. 2014). As expected, the problems related to this phenomenon concentrate in areas where the urbanization is placed close to the coastline, in many cases replacing original foredunes. Figure 17.14 shows the effects of mentioned 2010 storms over

Fig. 17.14 Different effects of coastal storms in non-developed (**a**, **b**) and developed (**c**, **d**) sectors of the coast of Florianópolis (Photos by J. Bonetti)

Barra da Lagoa / Moçambique beach system. As seen in Fig. 17.14a, b, severe erosion of the foredune resulted in no permanent damage since the area is a park with no human occupation, while Fig. 17.14c, d show that the same events resulted in severe damage to the existing assets located on the beach at the southern end of beach.

The ten most affected municipalities, according to Rudorff et al. (2014), correlate to the ones with higher levels of urbanization and armouring close to the coast: Florianópolis (average frequency of 5–6 episodes/year); Balneário Barra do Sul, Barra Velha, Itajaí, Navegantes, Balneário Camboriú, Bombinhas, Içara (3–4) and Garopaba and Itapema (1–2).

17.3.2 Beach Hazards and Safety

The beaches of Santa Catarina vary considerably in beach type and state, though they are dominated in terms of beach length by the higher energy dissipative (31 %) and rip-dominated intermediate beaches (46.5 %) (see Table 17.3). Based on Google

Earth there are usually 670 beach rips along the coast together with 64 permanent topographic rips. These average about 200 m in spacing for province 1–4, doubling to 420 m for province 5, in keeping with the global observations of Short and Brander (1999) for east coast swell environments. While the less energetic and least hazardous reflective beaches are the most common they make up only 20 % of the beaches by length. As a consequence SC has numerous hazardous beaches characterized by variable surf zone morphology (bars, channels and trough), variable surf zone currents (onshore, longshore and offshore), including strong narrow rip currents, together with breaking waves. All of these present a physical hazard to beach users (Klein et al. 2005). The result of the interaction of beach users with these hazards has been a level of risk involving accidents leading to rescues and first aid and in some cases drowning. In order to mitigate this risk lifeguards are maintained on most popular beaches particularly during the summer vacation periods. However even with lifeguards rescues and drowning are at an unacceptably high rate, in common with much of the Brazilian coast (Szpilman 2000).

In order to address this problem Klein et al. (2005) commenced a study of beach safety management along the SC coast. Using results of data collected between 1995 and 2003, they found four main factors play a significant role in the beach accidents: (1) landscape defines the use of coastline and the number of beach users with embayed beaches being more popular; (2) rip currents are the main natural hazards for bathers (responsible for ~ 82 % of accidents – see Figs. 17.5, 17.6, 17.7, 17.8, and 17.9); (3) the number of people on the beach contributes to beach accidents. As the number of people increase, bathers tend to move further into the surf zones, causing more accidents to occur; and (4) social factors include hazard signage perception and type of beach users. The majority of accidents occur on beaches where physical hazards are low with low wave (H_b ~1.0 m) and weak rip currents (<0.5 m s^{-1}), conditions that are particularly prevalent during the summer time. They are also usually well-signed with red flags on the beach and yellow flags on the lifeguards towers. This implies that the number and type of beach users plays a significant role in the high level of risk experienced on these beaches.

The analysis of hazards (their nature and magnitude) associated with sea bathing in SC demonstrates the importance of obtaining data of all types, not only environmental as proposed by Short and Hogan (1994), but also social, economic and cultural as suggested by Hoefel and Klein (1998), Klein et al. (2003a, 2005), and Mocellin (2006). Such information serves as a basis for the development of more effective public safety programs. The ~15,000 accident records in the project's database represent about 30 % of total accidents on SC beaches. Several initiatives are under way to obtain a higher rate of return on rescue records and hence to improve safety measures (Klein et al. 2005). The results of hazard analysis associated with bathing serve as a tool for the Fire Departament (legally responsible for the lifeguarding at Brazilian beaches) to assess its response capabilities and to implement change that contibuite to both accident prevention and effective operational procedures (mitigation).

Most importantly a number of recommendatiuons based on the project, were proposed by Klein et al. (2003a) and Mocellin (2006) including a Risk model for SC coast (Klein et al. 2005). These include: (1) wider publicity about the meaning of the warning flags used along the SC coast; (2) increasing the size of flags to make them more visible; (3) instructing lifeguards to undertake more preventative actions, such as distributing educational folders, talking to beach users and using equipment such as megaphones and jet skis to warn bathers away from dangerous spots; (4) increasing the number of lifeguards by hiring civilian lifeguards to work with local authorities where the recommended number is based upon the average ratio of rescues per lifegurds and number of people on the beach; (5) involving lifeguards in higher education and research on issues related to beach hazards (e.g. Mocellin 2006); (6) establishing a civil lifeguards association (Civil Fire Department); (7) expanding the campaign "Water that reaches navel is a sign of dangers" by the Fire Departament of the Santa Catarina, especially on the Santa Catarina Island; and (8) supporting the "Dolphin Project" which has the aim of transmitting notions of beach hazards, citizenship and environmental education to children and adolescents between 9 and 14 years old.

At the end of project the methodology, database and procedures were totally accepted by the Fire Department, indicating the success of the beach management project developed by the UNIVALI/CTTMAr. The most concrete and pleasing result of the project has been the decrease in the number of fatal accidents by ~80 % along the 13 patrolled beaches of the central northern Santa Catarina coast, approximately 100 km of coastline, between the 1995/1996 and 2001/2002 summer seasons.

17.4 Summary and Conclusions

The 922 km long Santa Catarina coast has a subtropical climate and based on the regional geology its beach systems can be divided into five open coast provinces. Tides are micro and deepwater waves average 1.5 m arriving from the east through south. Most of the open coast faces east to southeast and is composed of fine to medium sand resulting in predominately wave-dominated beaches, with tide-modified and tide-dominated beaches occurring in a few very sheltered embayments and within the larger North and South bays. The full range of wave-dominated beaches occurs along the coast, with the northern province 1 dominated by moderate length embayed higher energy intermediate beaches; the bedrock dominated province 2 by short embayed and pocket reflective beaches; Santa Catarina Island (province 3) by a mix of north-facing reflective and more exposed rip-dominated embayed intermediate; province 4 is another embayed east-facing shore dominated by more exposed intermediate beaches; while the southern province 5, south of Cape Santa Marta faces southeast into the highest waves which combine with fine sand to maintain long fully dissipative multi-bar beaches backed by transgressive dunes. Northerly longshore sand transport is uninterrupted in province 5, while in province 1, 3 and 4 headlands, bays and inlets interrupt the flow resulting in sediment sinks and some headland bypassing and overpassing, as well as discrete sediment cells, which also exist in the more embayed province 2, but within local closed sediment cells. The sheltered North and South bays dominated by small tide-modified and tide-dominated pocket beaches.

All the moderate to higher energy beaches are rip-dominated with up to 700 beach rips occurring along the coast together with more than 60 topographic rips. These combine with breaking waves and variable surf zone topography to present a moderate to high level of hazard to the beach users. During the high usage summer and vacation periods levels of beach risk are high and rescues, first aid and drowning common. A project commenced in 1995 is addressing this unacceptably high level of risk and has had success in reducing the number of rescues and drowning.

There has been considerable urban, second house and tourist development of the coast since the 1970s in places leading to developments in the transgressive dunes and on the foredunes and beach. Some beachfront locations have been damaged by waves resulting in the presence of seawalls and groins, and beach nourishment has already been performed in some places. In addition there are pronounced natural erosion hot spot areas with negative sediment budget in province 1 (Itapoa, Barra Velha, Piçarras): province 2 (Gravatá); and province 3 (Armação beach).

Acknowledgments The authors would like to acknowledge CNPq – *Science Without Borders Program* which provided funding for a 3 years (2013–2015) Visiting Professor Grant at UFSC – Graduate Geographer Program and Graduate Oceanographer Program, project no. 400302/2012-8 (Coastal Hazards and Headland Sediment Bypassing Assessment in Different Coastal Environments), part of which resulted in the compilation and editing of this book. This work also had the financial support of the following programs: Programa de Formação em Recursos Humanos em Geologia da Petrobras (PFRH-PB 240); Fundo Clima Ministério do Meio Ambiente, project MMA-Riscos; and SMC Project (MMA) and Bolsa de Produtividade em Pesquisa – Nível 2 [Proc. No. 303550/2012-0]). At UFSC, we thank José Mauricio de Camargo and Charline Dalinghaus for assisting with figures and tables.

References

Abreu JJ (2011) Transporte sedimentar longitudinal e morfodinâmica praial: exemplo do litoral norte de Santa Catarina. 2011. Unpublished PhD. thesis, Programa de Pós-graduação em Geografia, Universidade Federal de Santa Catarina, Florianópolis, SC, Brasil, 484p

Abreu de Castilhos JJ (1995) Estudo evolutivo, sedimentológico e morfodinâmico da praia da Armação, Ilha de Santa Catarina. Unpublished MSc. dissertation, Programa de Pós-Graduação em Geografia, Universidade Federeal de Santa Catarina, Florianópolis, SC, Brasil, 134p

Abreu de Castilhos JJ, Howa H, Resseguier A (1997) Application de la technique de tracagefluorescent dans I evaluation du transport cotier a la plage d'Armação, ile de Santa Catarina, Bresil. In: Sierra De Ledo B, Klingebiel A, Soriano Sierra E (eds) Atas, Colóquio Franco-brasileiro – Manejo costeiro da Ilha de Santa Catarina. UFSC, Florianópolis, pp 95–106

Abreu de Castilhos J, Howa H, Resseguier A (1998) Quantificação de transporte sedimentar na praia da Armação, Ilha de Santa Catarina, SC. Geosul (UFSC) 14:273–278

Alfredini P (1999) The longhsore trasnport evaluation used as a tool for the brazilian coastline litoral processes knowledge. Anais do VII Congresso Brasileiro da Associação Brasileira de Estudos do Quaternário (ABEQUA), 13p

Alves JHGM (1996) Refração do espectro de ondas oceânicas em águas rasas: aplicações à região costeira de São Francisco do Sul, SC. Florianópolis. Unpublished MSc. dissertation, Programa de Pós-Graduação em Engenharia Sanitária e Ambiental, Universiade Federal de Santa Catarina, Florianópolis, SC, Brasil, 89p

Alves JHGM, Melo E (2001) Measurements and modeling of wind waves at northern coast of Santa Catarina, Brazil. Rev Bras Oceanogr 49(1/2):13–28

Araujo RS (2008) Morfologia do Perfil Praial, Sedimentologia e Evolução Histórica da Linha de Costa das Praias da Enseada do Itapocorói – Santa Catarina. Unpublished MSc. dissertation, Mestrado em Tecnologia e Gestão Ambiental – Estratégias, Instrumentos e Tecnologias para Gestão Ambiental, Universidade do Vale do Itajaí, Itajaí, SC, Brasil, 145p

Araujo CES, Franco D, Melo E, Pimenta F (2003) Wave regime characteristics of the southern Brazilian coast. In: Proceedings of the sixth international conference on coastal and port engineering in developing countries, COPEDEC VI, Colombo, Sri Lanka, Paper no. 097; p 15 (published in CD, no pages)

Asp NE, Buynevich I, Siegle E, Fitzgerald D, Klein AHF, Cleary W, Angulo R (2005) Coastal geomorphology of Tijucas, SC-Brazil: preliminary Holocene evolution model. In: 10 Congresso Associac¸˜ao Brasileira de Estudos do Quaternário, Proceedings... Guarapari, Brazil: ABEQUA, pp 6–12

Bhering ÁFP (2012) Análise da estabilidade das desembocaduras ao longo do litoral do estado de Santa Catarina, Brasil. Unpublished BSc. report, Oceanografia, Universidade Federal de Santa Catarina, Florianópolis, SC, Brasil, 58p

Bonetti J, Klein AHF, Muler M, De Luca CB, Silva GV, Toldo EE Jr, González M (2012) Spatial and numerical methodologies on coastal erosion and flooding risk asserssment. In: Finkl C (ed) Coastal hazards. Chapter 16. Coastal research library series. Springer, Dordrecht, pp 423–442

Bortoluzzi CA (1987) Esboço geomorfológico de Santa Catarina. In: Silva LC da, Bortoluzzi CA (eds) Texto explicativo para o mapa geológico do estado de Santa Catarina – E: 1:500.00. Florianópolis: Série Textos Básicos de Geologia e Recursos Minerais de Santa Catarina. N1. Series: Mapas e Cartas Sinteses, 3, Seção Geologia, 3. DNPM/SC. pp 133–167

Boeyinga J, Dusseljee DW, Pool AD, Schoutens P, Verduin F, van Zwicht BNM, Klein AHF (2010) The effects of a bypass dunefield on the stability of a headland bay beach: a case study. Coast Eng 57:152–159

Buynevich IV, Asp NE, Klein AHF, Siegle E, FitzGerald D, Cleary W, Angulo R (2005) Mud in the surf: nature at work in a Brazilian Bay. EOS Transactions. Am Geophys Union, USA 86(33):301–304

Camargo JM (2009) Descrição geomorfológica e análise da estabilidade quanto a posição das desembocaduras do rio Piçarras e rio Iriri na enseada do Itapocorói – SC. Unpublished BSc. report, Oceanografia, Universidade do Vale do Itajaí, Itajaí, SC, Brasil, 72p

Caruso F Jr (1993) Mapa Gelógico da Ilha de Santa Cataria – Escala 1:100.000. Texto Explicativo e Mapa. Notas Técnicas, Porto Alegre 6:1–28

Caruso Jr F, Araújo SA (1997) A planície de cheniers da Baía de Tijucas, litoral de Santa Catarina. In: Proceedings of the 10th Semana Nacional de Oceanografia (Itajaí, Santa Catarina, Brazil), pp 40–43

Cassiano GF, Siegle E (2010) Migração lateral da desembocadura do Rio Itapocú, SC, Brasil: evolução morfológica e condicionantes físicas. Rev Bras Geofísica 28(4):537–549

CPE (Coastal Planning and Engineering do Brasil) (2009) Modelagem Numérica Como Subsídio aos Estudos à Fixação da Barra do Rio Araranguá. Relatório Técnico, 89p

Cruz CR (2004) Avaliação da quantidade de areia retirada pelas pessoas das praias e sua contribuição no balanço sedimentar litorâneo. Unpublished BSc. report, Oceanografia, Universidade do Vale do Itajaí, Itajaí, SC, Brasil

Departamento Nacional Da Produção Mineral – DNPM (1986) Mapa geológico do Estado de Santa Catarina. E = 1:500.000. Departamento Nacional Da Produção Mineral – DNPM, Florianópolis

Dalbosco ALP (2013) Análise histórica da linha de costa e dos padrões hidrodinâmicos e de transporte de sedimentos da praia da Armação, ilha de Santa Catarina, SC. Unpublished MSc. Dissertation, Programa de Pós-Graduação em Engenharia Sanitária em Ambiental, Universidade Federal de Santa Catarina, Florianópolis, SC, Brasil, 189p

De Camargo JM, Klein, AHF, Short AD (2016) Spit evolution as a result of headland bypassing, Santa Catarina, Brazil. 14th International Coastal Symposium, Sydney, Abstracts, p. 143

Diehl FL (1997) Aspectos geoevolutivos morfodinâmicos e ambientais do pontal da Daniela, ilha de Santa Catarina (SC). Unpublished MSc. Dissertation, Programa de Pós-Graduação em Geografia, Universidade Federal de Santa Catarina. Florianópolis, SC, Brasil, 131p

Diehl FL, Horn Filho NO (1996) Compartimentação geológico-geomorfológica da zona litorânea e planície costeira do Estado de Santa Catarina. Notas Técnicas 9:39–50

Diehl LF, Klein AHF, Calliari LJ, Horn Filho NO (1998) Coastal evolution of the Daniela spit, Santa Catarina Island, southern Brazil. J Coast Res SI (26):156–164

Dillenburg SR, Hesp P (2009) Geology and geomorphology of holocene costal barriers of Brazil. Springer, Berlin, 380p

Fitzgerald DM, Cleary WJ, Buynevich IV, Hein C, Klein AHF, Asp NE, Angulo RJ (2006) Variability of strandplain development in Santa Catarina, Brazil. Quaternary land-ocean interactions: driving mechanisms and coastal responses, IGCP-495, Santa Catarina, Brazil

FitzGerald DM, Cleary WJ, Buynevich IV, Hein CJ, Klein AHF, Asp N, Angulo R (2007) Strandplain evolution along the southern coast of Santa Catarina, Brazil. J Coast Res SI (50): 152–156

FitzGerald DM, Clearly WJ, Buynevich IV, Hein CJ, Klein AHF, Asp N, Angulo R (2008) Diverse styles of coastal evolution along Santa Catarina, Brazil. XX American Quaternary Association (AMQUA), Quaternary ice sheet-ocean interactions and landscape responses, Pennsylvania State University, PA, pp 15–16

Fontoura JAS (2004) Quantificação do transporte longitudinal de sedimentos não coesivos na zona de surfe da costa do Rio Grande do Sul (aplicação à praia do Cassino e adjacências dos molhes leste e oeste). Unpublished PhD. Thesis, Programa de Pós-Graduação em Geociências, Universidade Federal do Rio Grande do Sul, RS, Brasil, 298p

Giannini PCF (1993) Sistemas deposicionais no Quaternário costeiro entre Jaguaruna e Imbituba, SC. Unpublished PhD. thesis, Programa de Pós-Graduação em Geologia Sedimentar, Instituto de Geociências, Universidade de São Paulo, São Paulo, SP, Brasil, 2 v., 2 mapas, 439 p

Giannini PCF, Assine ML, Barbosa L, Barreto AMF, Carvalho AM, Claudino-Sales V, Maia LP, Martinho CT, Peulvast JP, Sawakuchi AO, Tomazelli LJ (2005) Dunas e paleodunas eólicas costeiras e interiores. In: Souza CRG, Suguio K, Oliveira PE, Oliveira AM (ed) Quaternário do Brasil (capítulo 11). Abequa, ANP, FAPESP, IG, UnG. Ribeirão Preto, SP, Holos Editora, pp 235–257

Gomes da Silva P (2014) Exposição à inundação costeira nas praias dos Ingleses, Moçambique e Barra da Lagoa, Florianópolis, SC. Unpublished MSc. dissertation, Programa de Pós-Graduação em Geografia, Universidade Federal de Santa Catarina, Florianópolis, SC, Brasil, 237p

Gomes da Silva P, Klein AHF, González M, Gutierrez O, Espejo A (2015) Performance assessment of the database downscaled ocean waves (DOW) on Santa Catarina coast, South Brazil. Anais da Academia Brasileira de Ciências [online]

Gourlay MR (1968) Beach and dune erosion tests. Delft Hydraulics Laboratory, Report No.M935/M936

Hein CJ, FitzGerald DM, Cleary WJ, de Menezes JT, Klein AHF, Albernaz MB (2012) Diverse sedimentological signatures of the mid-Holocene highstand in Brazil. Eos Transactions, AGU, Fall Meeting Suppl., Abstract EP33B-0898

Hein CJ, FitzGerald DM, Cleary WJ, Albernaz MB, Menezes JT, Klein AHF (2013) Evidence for a transgressive barrier within a regressive strandplain system: implications for complex coastal response to environmental change. Sedimentology 59:1–34

Hein CJ, FitzGerald DM, Menezes JT, Cleary WJ, Klein AHF, Albernaz MB (2014) Coastal response to late-stage transgression and sea-level highstand. Geol Soc Am Bull 1:B30836.1
Herrmann MLP (1999) Evolução do baixo curso do rio Cubatão, SC. Período 1957 a 1985. Geosul 14(27):128–134
Hesp PA, Giannini PCF, Martinho CT, Miot da Silva G, Asp NE (2009) The Holocene barrier systems of the Santa Catarina coast, southern Brazil. In: Dillenburg SR, Hesp PA (eds) Geology and geomorphology of Holocene coastal barriers of Brazil. Springer, Berlin, pp 93–134
Hoefel FG, Klein AHF (1998) Environmental and social decision factors of beach safety in the central northern coast of Santa Catarina, Brazil. Notas tec FACIMAR 2:155–166
Horn Filho NO (2003) Setorização da Província Costeira de Santa Catarina em base aos aspectos geológicos, geomorfológicos e geográficos. Geosul 18(35):71–98
Horn Filho NO (2006) Ilha de Santa Catarina. In: Muehe D et al (eds) Erosão e progradação do litoral brasileiro. Ministério do Meio Ambiente, Brasília, MMA, pp 413–436
Horn Filho NO, Strenzel GMR, Borges SF, Santos CR dos, Camargo G (1994) Aspectos geológicos do Quaternário costeiro de um segmento do litoral norte catarinense. In: 38 Congresso Brasileiro de Geologia. Boletim de Resumos Expandidos… Balneário Camboriú/SC, SBG, pp 389–390
IBGE – Instituto Brasileiro de Geografia e Estatística (2010) Sinópse do Censo Demográfico 2010 para Santa Catarina. População residente, total, urbana total e urbana na sede municipal, em números absolutos e relativos, com indicação da área total e densidade demográfica, segundo as Unidades da Federação e os municípios – 2010. Disponível em: http://www.censo2010.ibge.gov.br/sinopse/index.php?uf=42&dados=8. Accessed in Dec. 2014
Klein AHF (1997) Um método indireto para a determinação do estágio morfodinâmico de praias oceânicas arenosas. In: VI Congresso da Associação Brasileira de Estudos do Quaternário. Anais… ABEQUA: Curitiba/PR, pp 76–78
Klein AHF (2004) Morphodynamics of headland-bay beaches: examples from the coast of Santa Catarina State, Brazil. Unpublished PhD thesis. Faculdade de Ciências do Mar e do Ambiente. Universidade do Algarves, Faro, Algarve, Portugal, 198p
Klein AHF, Menezes JT (2001) Beach morphodynamics and profile sequence for a headland bay coast. J Coast Res 17(4):812–835
Klein AHF, Santana GG, Diehl FL, Menezes JT (2003a) Analysis of hazards associeted with sea bathing: results of five year work oceanic beaches of Santa Catarina State, Southern Brazil. J Coast Res SI (35):107–116
Klein AHF, Benedet L, Hsu JRC (2003b) Stability of headland bay beaches in Santa Catarina: a case study. In: Klein AHF et al (eds) Proceedings of the Brazilian symposium on sandy beaches: morphodynamics, ecology, uses, hazards and management. J Coast Res SI (35):151–166
Klein AHF, Mocellim O, Menezes JT, Berribilli M, Vintém G, Dafferner G, Diehl FL, Sperb RM, Santana GG (2005) Beach safety management on the coast of Santa Catarina, Brazil. Z Geomorphol 141:47–58
Klein AHF, Menezes, JT, Diehl FL, Abreu JGN, Polette M, Sperb RM, Sperb RC (2006) Litoral Centro Norte. In: Dieter Muehe. (Org.). Erosão e Progradação do Litoral Brasileiro,1st edn. Ministério do Meio Ambiente/Secretaria de Qualidade Ambiental nos Assentamentos Humanos, Brasilia, pp 402–412
Klein AHF, Araújo RS, Polette M, Sperb RM, Freitas D, Sprovieri FC, Pinto FT (2009) Ameliorative strategies at Balneário Piçarras Beach – Brazil. In: Williams AT, Micallef A (eds) Beach management guidelines: principles and practice. Earthscan Publisher, London, pp 241–255
Klein AHF, Ferreira Ó, Dias JMA, Tessler MG, Silveira LF, Benedet L, Menezes JT, Abreu JGN (2010) Morphodynamics of structurally controlled headland-bay beaches in southeastern Brazil: a review. Coast Eng 57:98–111
Legeckis R, Gordon AL (1982) Satellite observations of the Brazil and Falkland currents – 1975 to 1976 and 1978. Deep-Sea Res 29:375–401

Lessa GC, Angulo RJ, Giannini PCF, Araújo AD (2000) Stratigraphy and Holocene evolution of a regressive barrier in south Brazil. Mar Geol 165:87–108

Lima SF, Almeida LESB, Toldo EE Jr (2001) Estimativa da capacidade do transporte longitudinal de sedimentos a partir de dados de ondas para a costa do Rio Grande do Sul. Pesquisas em Geociências 28:99–107

Marcelino EV, Rudorff FM, Goerl RF, Marcelino IPVO (2014) Furacão Catarina. In: Herrmann MLP (ed) Atlas de Desastres Naturais do Estado de Santa Catarina: período de 1980 a 2010, 2nd edn. IHGSC/Cadernos Geográficos, Florianópolis, pp 162–172

Martins LR (1970) Sedimentos praiais de Santa Catarina – II: Laguna Imbituba. Notas e Estudos. Escola de Geologia, Universidade Federal do Rio Grande do Sul 2(1):1–33

Martins LR, Eichler BB (1969) Propriedades texturais dos sedimentos litorâneos de Santa Catarina. 11: areias de praia e dunas, trecho Laguna-Imbituba. Notas e Estudos. Escola de Geologia, Universidade Federal do Rio Grande do Sul 2(1):27–44

Masselink G, Short AD (1993) The effect of tide range on beach morphodynamics and morphology: a conceptual model. J Coast Res 9:785–800

Mazzer AM (2007) Proposta metodológica para análise de vulnerabilidade da orla marítima à erosão costeira: aplicação em praias arenosas da costa sudeste da Ilha de Santa Catarina (Florianópolis, Santa Catarina – Brasil). Unpublished PhD. thesis, Programa de Pós-Graduação em Geociências, Universidae Federal do Rio Grande do Sul, 168p

Mazzer AM, Dillenburg SR, Souza CRG (2008) Proposta metodológica para análise de vulnerabilidade à erosão costeira no sudeste da ilha de Santa Catarina, Brasil. Rev Bras Geosci 38:2–17

McAdie CJ, Rappaport EN (1991) Diagnostic report of the National Hurricane Center. NOAA Tech. Rep. 4 (1), National Hurricane Center, Coral Gables, FL, 45p

McBride RA, Anderson J, Buynevich IV, Cleary WJ, Fenster MS, FitzGerald DM, Harris MS, Hein CJ, Klein AHF, Liu B, Menezes JT, Pejrup M, Riggs SR, Short AD, Stone GW, Wallace DJ, Wang P (2013) Morphodynamics of barrier systems:a synthesis. In: Shroder J (Editor in Chief), Sherman DJ (ed) Treatise on Geomorphology (Coastal and submarine geomorphology), 1st edn. Academic Press, San Diego, 10:174–244

McTaggart-Cowan R, Bosart LF, Davis CA, Atallah EH, Gyakum JR, Emanuel KA (2006) Analysis of Hurricane Catarina (2004). Mon Weather Rev 34:3002–3053

Melo FE, Straioto KMGT, Franco D, Romeu MA (2006) Distribuição estatística de alturas de ondas individuais em Santa Catarina: resultados preliminares. In: II Semminário e Workshop em Engenharia Oceânica, novembro de 2006, FURG, Rio Grande, pp 14, publicado em CD sem paginação, disponível em: http://www.semengo.furg.br/2006/19.pdf

Menezes JT (2008) Balanço de sedimentos arenosos da enseada de Balneário Camboriú-SC. Unpublished PhD. thesis, Programa de Pós-Graduação em Geociências, Universidade Federal do Rio Grande do Sul, Porto Alegre, RS, Brasil, 167p

Miot da Silva G (2006) Orientação da linha de costa e dinâmica dos sistemas praia e duna: praia de Moçambique, Florianópolis, SC. Unpublished PhD. thesis, Universidade Federal do Rio Grande do Sul, Porto Alegre, RS, Brasil, 293p

Miot da Silva G, Mousavi SMS, Jose F (2012) Wave-driven sediment transport and beach-dune dynamics in a headland bay beach. Mar Geol 323–325:29–46

Miranda JR (2013) Identificação das zonas de geração de ondulações longiquas incidentes na costa de Santa Catarina entre 2002 e 2005. Unpublished BSc. report, Oceanografia, Universidade Federal de Santa Catarina, Florianópolis, SC, Brasil 191p

Mocellin O (2006) Determinação do nível de risco público ao banho de mar das praias arenosas do litoral centro norte de Santa Catarina. Unpublished MSc. dissertation, Programa de Pós-graduação em Ciência e Tecnologia Ambiental, Universidade do Vale do Itajaí, Itajaí, SC, Brasil, 162p

Motta, LM, Toldo Jr EE, Almeida LESB; Nunes, JC (2010) Balanço de sedimentos arenosos no litoral médio do Rio Grande do Sul. In: 22 Salão de iniciação científica, Porto Alegre, RS

Muehe D (1998) O litoral brasileiro e sua compartimentação. In: Guerra A, Cunha SB (eds) Geomorfologia do Brasil. Bertrand Brasil, São Paulo, pp 273–349
Nascimento Jr DR (2010) Evolução sedimentary holocênica do delta do rio Tubarão, estado de Santa Catarina. Unpublished PhD. thesis, Programa de Pós-Graduação em Geoquímica e Geotectônica, Universidade de Sâo Paulo. São Paulo, SP, Brasil, 277p
Nimer E (1989) Climatologia da Região Sul. Rev Bras Geogr 34(1):142–154
Nobre CA, Cavalcanti IFA, Gan MA, Nobre PA, Kayano MT, Rao VB, Bonatti JP, Satyamurti P, Uvo CB, Cohen JC (1986) Aspectos da climatologia dinâmica do Brasil. Climanálise N Especial, 65p
Nunes MG (2002) Estudo morfossedimentar do sistema praial-lagunar de Ponta das Canas, Ilha de Santa Catarina, SC. Unpublished MSc. dissertation, Programa de Pós-Graduação em Geografia, Universidade Federal de Santa Catarina, Florianópolis, SC, Brasil,133p
Oliveira JG (2013) Modelagem numérica das respostas do perfil praial a eventos extremos em praias de enseada: estudo de caso da Enseada de Itapocorói, Santa Catarina. Unpublished BSc. report, Oceanografia, Universidade Federal de Santa Catarina, Florianópolis, SC, Brasil, 162p
Oliveira UR, Oliveira UR, Parizotto BADM, Horn Filho NO, Bonetti J (2012) Setorização morfossedimentar das praias arenosas da Ilha de Santa Catarina (SC) utilizando técnicas de análise estatística multivariada. Gravel 10:1–11
Oliveira UR, Barletta RC, Peixoto JRV, Horn Filho NO (2014) Distribuição espacial das características morfodinâmicas das praias arenosas da costa oceânica da Ilha de Santa Catarina, SC, Brasil. Pesquisas em Geociências 41(2):89–103
Olson DB, Podest AG, Evans RH, Brown OB (1998) Temporal variations in the separation of Brazil and Malvinas currents. Deep-Sea Res 35(12):1971–1990
Pereira MD, Schettini CAF, Yukioma C (2009) Caracterização de feições oceanográficas na plataforma de Santa Catarina através de imagens orbitais. Braz J Geophys 27(1):81–93
Pezza AB, Simmonds I (2005) The first south Atlantic hurricane: unprecedented blocking, low shear and climate change. Geophys Res Lett 32:L15712. doi:10.1029/2005GL023390
Pianca C, Mazzini PLF, Siegle E (2010) Brazilian offshore wave climate based on NWW3 reanalysis. Braz J Oceanogr 58(1):53–70
Piérri GCS (2005) Variabilidade da desembocadura do rio Itapocú: análise da evolução morfológica em escalas de médio e curto termo. Unpublished BSc. report, Oceanografia, Universidade do Vale do Itajaí, Itajaí, SC, Brasil, 83p
Pinto MW, Meireles R, Cooper A, Klein AHF (2015) Santinho/Ingleses transgressive dunefiled system-Santa Catarina Island (Brasil): temporal variability in vegetation, manmade structures and dune migration. Coastal Sediments 2015, San Diego
Pitombeira ES (1975) Estimativa do volume anual de transporte litorâneo na costa do Rio Grande do Sul. Unpublished MSc. Dissertation, Instituto de Pesquisas Hidráulicas (IPH), Universidade Federal do Rio Grande do Sul (UFRGS), Centro de Hidrologia Aplicada, Porto Alegre, Brasil, 98p
Polette M, Raucci GD (2003) Methodological proposal for carrying capacity analysis in Sandy beaches: a case study at central beach of Balneário Camboriú (Santa Catarina, Brazil). J. Coast Res SI (35):94–106
Porpilho D, Klein AHF, De Camargo RSV, Prado MFV, Short AD, Da Silva GV, Toldo EE Jr (2015) Bedform classification in front of Santinho Headland, Santinho beach – Santa Catarina Island. Coastal Sed 2015, San Diego
Ribeiro PJC (2014) Transformações de ondas em praias de enseada: estudo de caso na Enseada do Itapocorói – SC. Unpublished MSc. dissertation, Programa de Pós-Graduação em Geografia, Universidade Federal de Santa Catarina, Florianópolis, SC, Brasil, 140p
Rodrigues MLG, Franco D, Sugahara S (2004) Climatologia de frentes frias no litoral de Santa Catarina. Rev Bras Geofísica 22(2):135–151
Rudorff FM, Bonetti J, Moreno DA, Oliveira CAF, Murara PG (2014) Maré de Tempestade. In: de Paula Herrmann ML (ed) Atlas de desastres naturais do estado de Santa Catarina: período de

1980 A 2010 2ª Edição atualizada e revisada, 2nd edn. IHGSC/Cadernos Geográficos, Florianópolis, pp 151–154

Santa Catarina – GAPLAN (1986) Atlas de Santa Catarina. Santa Catarina – GAPLAN, Florianópolis. 173p

Schaeffer-Novelli Y, Cintrón-Molero G, Adaime RR, Camargo TM (1990) Variability of mangrove ecossystems along the brazilian coast. Estuaries 13(2):204–218

Scheibe LFA (1986) A geologia de Santa Catarina – Sinópse provisória. Geosul 1:7–38

Short AD (1992) Beach systems of the central Netherlands coast: processes and morphology and structural impacts in a storm-driven multi-bar system. Mar Geol 107:103–137

Short AD (2006) Australian beach systems; nature and distribution. J Coast Res 27:11–27

Short AD, Aagaard T (1993) Single and multi-bar beach change models. J Coast Res SI 15:141–157

Short AD, Brander R (1999) Regional variation in rip density. J Coast Res 15:813–822

Short AD, Hogan CL (1994) Rips and beach hazards, their impact on public safety and implications for coastal management. J Coast Res SI 12:197–209

Signorin M (2010) Análise e comparação do clima de ondas do litoral Norte, Centro e Sul de Santa Catarina utilizando modelagem numérica. Unpublished BSc. report, Oceanografia, Universidade do Vale do Itajaí, Itajaí, SC, Brasil, 82p

Siegle E, Asp NE (2007) Wave refraction and longshore transport patterns along the southern Santa Catarina coast. Braz J Oceanogr 55:109–120

Siegle E, Klein AHF, Toldo Jr EE, Abreu JGN (1998) Morphology and stability of the Camboriú River inlet, Santa Catarina – Brazil. J Coast Res SI (26):94–100

Silveira JD (1964) Morfologia do litoral. In: Azevedo A (ed) Brasil: a terra e o homem, vol 1. Cia. Editora Nacional, São Paulo, pp 253–305

Silveira LF, Klein AHF, Tessler MG (2011) Morphodynamical classification of beaches from Santa Catarina State and north coast of São Paulo State using remote sensing. Braz J Aquat Sci Technol 15(2):13–28

Silveira LF, Benedet L, Signorin M, Bonanata R (2012) Evaluation of the relationships between navigation channel dredging and erosion of adjacent beaches in southern Brazil. ICCE2012. DOI: http://dx.doi.org/10.9753/icce.v33.sediment.106

Soares MLG, Estrada GCD, Fernandez V, Tognella MMP (2012) Southern limit of the Western South Atlantic mangroves: assessment of the potential effects of global warming from a biogeographical perspective. Estuar Coast Shelf Sci 101:44–53

Souza MC (1999) Mapeamento da planície costeira e morfologia e dinâmica das praias do Município de Itapoá, Estado de Santa Catarina: subsídios à ocupação. Unpublished MSc dissertation, Curso de Pós-Graduação em Geologia Ambiental, Departamento de Geologia, Universidade Federal do Paraná, Curitiba, 196 p

Souza JAG, Corrêa ICS (2006) Feições sedimentares presentes na plataforma continental interna da enseada do Pântano do Sul, ilha de Santa Catarina, SC – Resultados preliminares. In: Semana acadêmica dos alunos de pós-graduação em geociências, 1., 2006, Porto Alegre. Resumos... Porto Alegre: UFRGS, 2006, pp 129–132

Szpilman D (2000) Afogamento. Rev Bras Medicina Esportiva, Rio de Janeiro 6(4):131–144

Stech JL, Lorenzetti JA (1992) The response of the South Brazil Bight to the passage of wintertime cold fronts. J Geophys Res 97(66):9507–9520

Straioto KMGT (2009) Caracterização de ondas extremas no litoral de Santa Catarina. 2009. Unpublished MSc. dissertation, Programa de Pós-Doutorado em Engenharia Sanitária e Ambiental, Universidade Federal de Santa Catarina, Florianópolis, SC, Brasil, 89p

Suguio K, Tessler MG (1984) Planícies de cordões litorâneos do Brasil: origem e nomenclatura. In: de Lacerda LD et al (eds) Restingas: origem estruturas e processos. CEUFF, Niterói, pp 195–216

Suguio K, Martin L (1987) Classificação de costas e evolução geológica das planícies litorâneas quaternárias do sudeste e sul do Brasil. In: ACIESP (ed) Simpósio de Ecossistemas da Costa

Sul e Sudeste Brasileira. Anais, vol 1. Academia de Ciências do Estado de São Paulo, São Paulo, pp 1–28

Sytiski JPM, Vörösmarty CJ, Kettner AJ, Green J (2005) Impact of humans on the flux of terrestrial sediment to the global coastal ocean. Science 308(5720):376–380. doi:10.1126/science.1109454

Tabajara LL (2004) Resultante geológica causada pela passagem do Ciclone Catarina no litoral Norte do Rio Grande do Sul. Gravel 2:125–130

Trucollo EC (1998) Maré meteorológica e forçantes atmosféricas locais em São Francisco do Sul – SC. Unpublished MSc. dissertation, Programa de Pós-Graduação em Engenharia Sanitária e Ambiental, Universidade Federal de Santa Catarina, Florianópolis, SC, Brasil, 100p

Vieira da Silva G (2009) Análise da estabilidade quanto à posição das desembocaduras do rio Araranguá, da Barra do Camacho e do Rio da Madre, litoral sul e centro do estado de Santa Catarina. Unpublished BSc. report, Oceanografia, Universidade do Vale do Itajaí, Itajaí, SC, Brasil, 66p

Vieira da Silva G, de Carvalho JLB, Klein AHF (2011) Camacho inlet Southern Brazil instability causes and user conflicts. Ocean Coast Manag 54:569–576

Vieira da Silva G, Muler M, Prado MFV, Short AD, Toldo Jr EE, Klein AHF (2016a) Shoreline changes analysis and insights into sediment transport path – Example of Santa Catarina Island north shore, Brazil. J Coast Res SI. http://dx.doi.org/10.2112/JCOASTRES-D-15-00164.1

Vieira da Silva G, Toldo EE Jr, Klein AHF, Short AD, Woodroffe CD (2016b) Headland sand bypassing – quantification of net sediment transport in embayed beaches, Santa Catarina Island North Shore, Southern Brazil. Mar Geol 379:13–27. http://dx.doi.org/10.1016/j.margo.2016.05.008

Villwock JA (1987) Processos costeiros e a formação de praias arenosas e campos de dunas ao longo da costa sul e sudeste brasileira. In: Academia de Ciências do Estado de São Paulo (org) Simpósio sobre Ecossistemas da Costa Sul e Sudeste Brasileira: Síntese dos conhecimentos. Anais, vol 2. Academia de Ciências do Estado de São Paulo, Cananéia, pp 380–398

Vilwock JA (1994) A Costa Brasileira: geologia e Evolução, vol 7. Notas Técnicas CECO/IG/UFRGS, Porto Alegre, pp 38–49

Chapter 18
Ocean Beaches of Rio Grande do Sul

Lauro Júlio Calliari and Elírio Ernestino Toldo Jr.

Abstract Beaches along the Rio Grande do Sul (RS) coast are fully exposed with the exception of the extreme northern end where a rocky headland composed of sandstone, basalt and volcano-clastic sequences provide a small degree of protection to coastal dynamics. The beaches generally consist of fine quartz sand with a low gradient swash zone usually containing low-relief beach cusps. Beaches range between intermediate and dissipative, with beach state controlled in part by grain-size variations through the contribution of shell fragments, gravel and coarse quartz sand. This wave-dominated coast is episodically exposed to storm waves, which favors the formation and maintenance of two to three shore parallel sand bars. Most of the beaches possess distinct hazard to bathers associated to their beach state and hazard rating. The longshore bar and trough (LBT) and rhythmic bar and beach (RBB) states are very common and the troughs are usually deeper than the height of the bathers. The combination of higher beach population during summer seasons and rip-dominated intermediate beaches result in the most hazardous beaches along the northern littoral and in the extreme south. Other beach hazards are related to the presence of washouts and flooding, as well as, to episodic deposition of fluid mud at the surfzone and beach as a result of anthropogenic influence. Approximately 80 % of the coastline is undeveloped and the main impacts are associated to car traffic and litter represented by solid waste and plastic beads. Foredunes along the littoral are affected by changes in coastal orientation in relation to the prevailing northeast winds and beach morphodynamics and range from well-developed to non-existent.

L.J. Calliari (✉)
Laboratório de Oceanografia Geológica, Universidade Federal do Rio Grande, Rio Grande, RS, Brazil
e-mail: lcalliari@log.furg.br

E.E. Toldo Jr.
Centro de Estudos de Geologia Costeira e Oceânica, Universidade Federal do Rio Grande do Sul, Porto Alegre, RS, Brazil
e-mail: toldo@ufrgs.br

A.D. Short, A.H. da F. Klein (eds.), *Brazilian Beach Systems*, Coastal Research Library 17, DOI 10.1007/978-3-319-30394-9_18

18.1 Introduction

Beaches from Rio Grande do Sul (RS) are located on the large sedimentary surface of the Pelotas basin, and far from the Precambrian shield (Fig. 18.1). The coast is a 615 km long Holocene sandy barrier with a uniform northeast-southwest orientation and a slight sinuosity, gentle regional slope, few inlets and poorly developed ebb deltas. The surf zone is wide, shallow and dissipative with multiple bars. The coastal plain consists of unconsolidated deposits, which do not receive contribution of modern sands, as most fluvial bedload is retained in lagoons and other coastal plain environments. The surf-zone sediment transport is dominated by waves and the regional coastal processes are primarily controlled by the wave energy flux parallel to the beach. The littoral is exposed to small semi-diurnal microtides. However, the

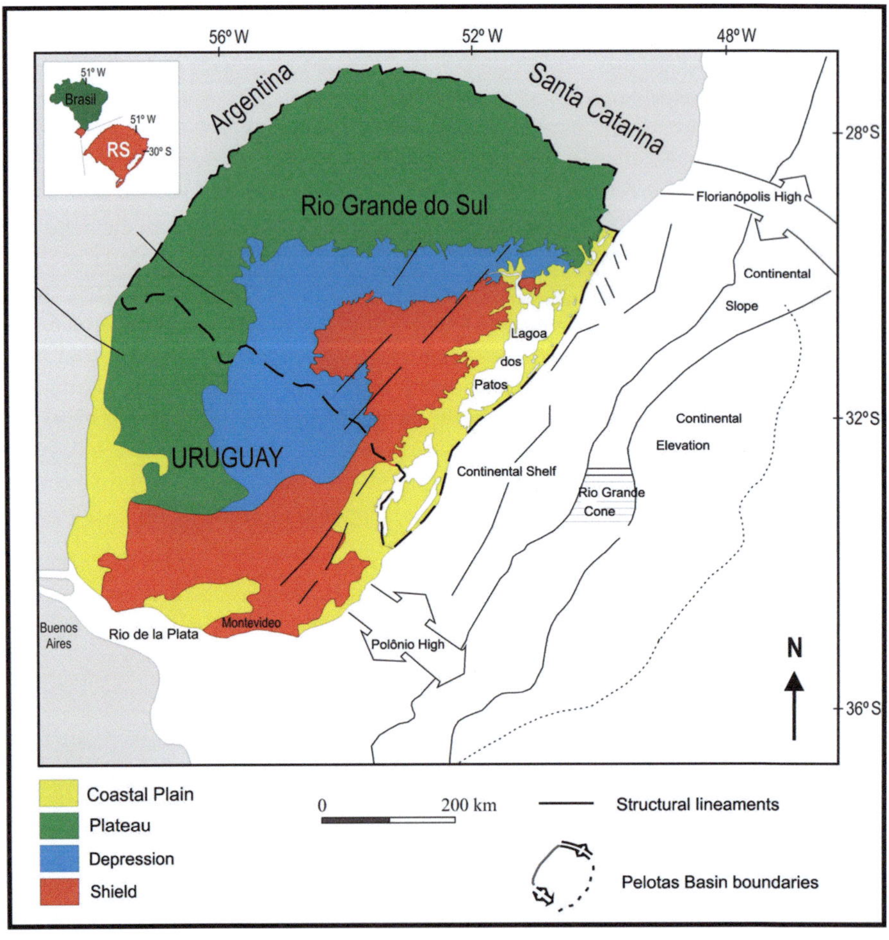

Fig. 18.1 Physiography and main geological structures of the Precambrian shield and the Pelotas basin (Modified from Carraro et al. 1974; Corrêa 1987)

meteorological tide can exceed by several times the astronomical tide. This chapter reviews the geology, sediment supply and coastal processes that formed and maintain the beach systems.

18.1.1 Geology

The Pelotas basin is the southernmost of the marginal basins that comprises the Brazilian continental margin. The Florianópolis structural high marks the northern limit while the Cabo Polônio structural high extends south into Uruguay, with the highest topographic areas forming the western boundary of the basin (Fig. 18.1). The continental shelf is located in the Pelotas basin, which had its origin during the Cretaceous and received the clastic sediments derived from the dissection of the adjacent uplands (Corrêa et al. 2007).

The southern side of Pelotas is limited by the basement while the northern part by Paleozoic and Mesozoic sediments of the Parana basin. This northern part rises to elevations up to 1500 m, while the southern relief is smoother. The coastal plain is the sub-aerial portion of the Pelotas basin and has an area of 33,000 km^2, which consists of a low-lying surface containing numerous lagoons and coastal lakes (Fig. 18.1).

The RS coastal plain, contains four barrier-lagoon–type depositional systems deposited during high stands in Quaternary glacio-eustatic sea-level cycles (Tomazelli et al. 2000; Dillenburg et al. 2000). These deposits have been classified, from the oldest to the most recent, as barriers I, II, III and IV and consist of lagoonal, aeolian, beach and marine facies aligned in a northeast-southwest orientation. The development of each of these depositional systems was initiated about 400 ka, and are associated with the last four transgressive-regressive events. The oldest three barriers are Pleistocene, and the most recent, barrier IV, along with its ocean beaches is Holocene.

The adjacent continental shelf has a width between 150 and 180 km, reaching maximum depths of approximately 100–140 m, with slopes ranging from 0.5 to 1.5 m km^{-1} (Corrêa 1996).

18.1.2 Drainage and Sediment Supply

The RS hydrographic network is dominated by the Guaíba, Camaquã, Velhaco, São Lourenço and Pelotas rivers, with the Jacuí, Taquari, Sinos, Gravataí and Caí rivers tributaries of the Guaíba River. The São Gonçalo channel, which flows into the northern end of Lagoa dos Patos estuary, receives waters from the Jaguarão, Taquari and Cebolati rivers. These rivers form the southeast hydrographic basin, which includes the 180,000 km^2 surface of the Lagoa dos Patos and Lagoa Mirim (Fig. 18.2).

The combination of a higher relief hinterland, with an extensive drainage network, and a humid subtropical climate, as well as the hydraulic behavior of the Patos-Guaiba system, result in a predominance of clastic sediments being deposited in the lagoon system (Toldo et al. 2000, 2006b).

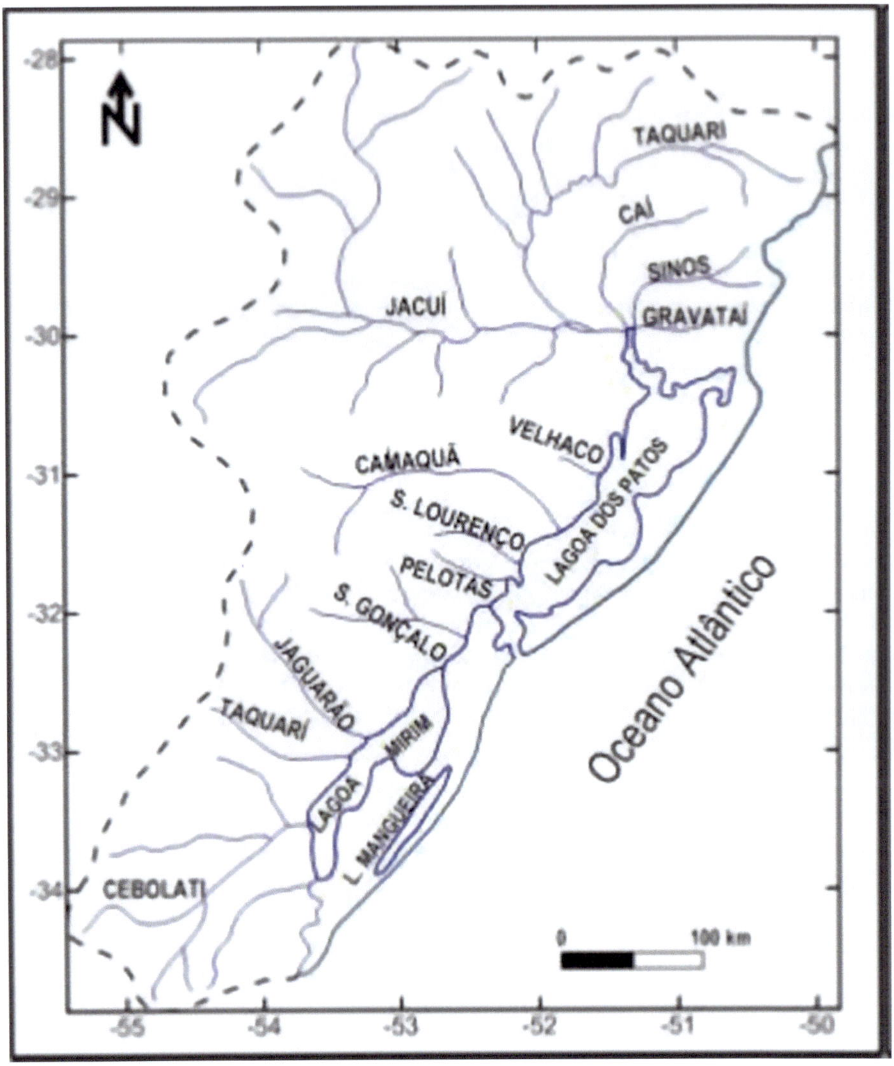

Fig. 18.2 The southeast hydrographic network of the Rio Grande do Sul State and northeast Uruguay

The coastal plain consists of unconsolidated Quaternary deposits that receives no bedload from the rivers, because it is trapped in lagoons and other coastal plain environments, e.g., Lagoa dos Patos and Lagoa Mirim (Toldo 1989), as exemplified by the multiple channel bars of the Delta do Jacuí River. Fine sediments, dominantly composed by silt and clay, are however transported by the Guaíba and Camaquã rivers into the lagoon (Fig. 18.2). The fine sediments that are not deposited in the lagoon, are transported through the Rio Grande inlet and deposited on the inner continental shelf.

The average annual discharge of the Jacuí, Taquari and Camaquã rivers is 1540 m^3s^{-1}; the Sinos, Caí and Gravataí rivers approximately 230 m^3s^{-1}; and the São Gonçalo channel 700 m^3s^{-1}; bringing the average annual freshwater discharge into Lagoa dos Patos to around 2400 m^3s^{-1} (Vaz et al. 2006). The average annual suspended solid sediment discharge of the river Guaíba into Lagoa dos Patos, between the years 2003 and 2006 was estimated in 1.0985 M t $year^{-1}$ (Andrade Neto et al. 2012).

Lagos dos Patos is the most extensive lagoon system in South America with an area of 10,000 km^2. It has a 180 km long main axis, aligned northeast-southwest, with a maximum width of 60 km and average depth of 6 m. The lagoon in the context of the southern Brazilian coast is a zone of convergence of the drainage network from the southeast basin of Rio Grande do Sul state and the northeast of Uruguay (Fig. 18.2).

18.1.3 Waves and Tides

The RS coast is regularly subject to storms associated with frontal systems and extratropical cyclones, which generate southerly waves and storm surges. Studies of the wave climate was initiated by Wainer (1963), Motta (1969), and Coli (2000). Strauch et al. (2009) using a Datawell waverider in a depth of 17 m, found similarity in wave height, period and power along the coast including a peak during the summer and autumn season. Overall heights, periods and energy show a progressive increase with the propagation of the waves from east to south, with south-southeast waves being the highest.

As alternative to the measured wave data, which usually represent short time series of measurements from the southern coast of Brazil, reanalysis models have allowed a detailed description of the wave climate in places where long-term measurements are not available or do not exist (see Chap. 2). The application of this resource has provided long continuous time series, with satisfactory spatial resolution data from the reanalysis about generation and waves propagation from deep water (Tolman 1997). This data has been compared with the results from waverider moored off Tramandaí beach (Sprovieri and Toldo 2015).

The wave time-series data obtained by the transfer method were statistically analyzed to characterize the wave climate in shallow waters. Table 18.1 shows the distribution of significant wave height (m) and direction (°) in shallow water. South-southeast waves have the highest frequency of occurrence (33.36 %), with significant height ranging between 0.5 and 1.5 m, representing 76.81 %. In addition 25.60 % arrive from the northeast, with significant height ranging between 0.5 and 1.5 m.

Table 18.2 shows the distribution of significant wave height (m) and peak period (s) of waves in shallow water. The most frequent significant wave height is 1–1.5 m, (41.87 %) and the most frequent peak period is from 6 to 8 s (33.24 %).

Waves in shallow water arrive predominantly from northeast and southwest (99.92 %), with significant wave height ranging mainly between 0 and 3 m (98.70 %)

Table 18.1 Distribution of significant wave height (m) and direction (°) from the reconstructed time series recorded in shallow water (17 m) (Sprovieri and Toldo 2015)

Hs (m) \ Dir	N	NNE	NE	ENE	E	ESE	SE	SSE	S	SSW	SW	WSW	W	WNW	NW	NNW	Events	Freq (%)
0 - 0.5		1	359	243	78	49	34	128	112	28	8	3	3	3			1049	2
0.5 - 1.0		2	1124	5055	663	337	289	4811	1861	31	24	10	2	3			14212	35
1.0 - 1.5				4429	2464	1502	1489	6339	797	5	2	1	3				17031	42
1.5 - 2.0				985	1157	1157	1237	1937	29	2	1						6505	16
2.0 - 2.5				95	270	325	465	308									1463	4
2.5 - 3.0				10	60	95	101	39									305	1
3.0 - 3.5				1	19	15	19	5									59	0
3.5 - 4.0					3	13	12	1									29	0
4.0 - 4.5						5	3										8	0
4.5 - 5.0						2	1	1									4	0
5.0 - 5.5						2	5										7	0
Events	0	3	1483	10818	4714	3502	3655	13569	2799	66	35	14	8	6	0	0		
Freq (%)	0	0	4	26	12	9	9	33	7	0	0	0	0	0	0	0		

Table 18.2 Distribution of significant wave height (m) and the peak period (s) from the reconstructed time-series recorded in shallow water (17 m)

Hs (m) \ Tp (s)	0 - 2	2 - 4	4 - 6	6 - 8	8 - 10	10 - 12	12 - 14	14 - 16	16 - 18	18 - 20	Events	Freq (%)
0 - 0.5		213	661	86	26	59	4				1049	2
0.5 - 1.0		78	2837	6945	2366	1718	257	11			14212	35
1.0 - 1.5			255	5947	6777	3389	610	50	1	2	17031	42
1.5 - 2.0				545	3495	1738	696	31			6505	16
2.0 - 2.5					587	630	242	4			1463	3
2.5 - 3.0					31	193	81				305	1
3.0 - 3.5					2	37	20				59	1
3.5 - 4.0						22	7				29	0
4.0 - 4.5						3	5				8	0
4.5 - 5.0							4				4	0
5.0 - 5.5						3	4				7	0
Events	0	291	3753	13523	13284	7792	1930	96	1	2		
Freq (%)	0	1	9	33	33	19	5	0	0	0		

Table 18.3 Seasonal statistics of wave parameters for the reconstructed time series (17 m depth)

Season	$H_{s(min)}$	$H_{s(max)}$	$H_{s(mean)\%}$	$H_{s(sd)}$
Summer	0.19	4.51	1.09	0.36
Autumn	0.14	5.46	1.19	0.49
Winter	0.17	4.12	1.20	0.46
Spring	0.21	3.98	1.25	0.40
Season	$T_{p(min)}$	$T_{p\ (max)}$	$T_{p\ (mean)\%}$	$T_{p\ (sd)}$
Summer	2.23	14.5	7.99	1.77
Autumn	2.16	18.75	8.68	2.19
Winter	2.04	15.02	8.71	2.27
Spring	2.11	15.25	8.47	1.78

Season	$Dir_{p(mean)}$	$Dir_{p\ (sd)}$
Summer	109.84	40.68
Autumn	125.16	39.72
Winter	121.88	44.20
Spring	110.93	41.58

and peak period from 4 to 14 s (99.73 %) (Tables 18.1 and 18.2). Extreme waves up to 5.66 m and T_p up to 18.75 s are also recorded in the Table.

Table 18.3 summarizes the seasonal wave climate. The data obtained with the reconstruction method of the time-series from shallow water (17 m) at Tramandaí beach are in accordance with previous studies (Wainer 1963; Strauch et al. 2009).

Table 18.4 Directional probabilities and statistics of H_s (m) for the reconstructed time-series in shallow water (17 m depth)

Direction	Probability (>1 %)	$Hs_{50\%}$	$Hs_{90\%}$	$Hs_{99\%}$	Hs_{12}
NE	3.6	0.60	0.77	0.87	0.92
ENE	26.6	1.01	1.51	2.00	2.41
E	11.6	1.33	1.91	2.71	3.33
ESE	8.6	1.44	2.11	3.05	4.49
SE	9.0	1.51	2.21	3.03	4.92
SSE	33.4	1.11	1.65	2.29	2.68
S	6.9	0.85	1.24	1.51	1.68

The determination of the waves (H_s) that exceed 50 %, 90 % and, 99 % of occurrence and for waves that exceed the probability of 12 h per year (H_{S12}) in shallow water are presented in Table 18.4.

18.1.4 Coastal Sediments

Beach sediment has uniform grain size, mainly represented by fine quartz sand (M_d=0.15 mm), with low carbonate contents (between 0.2 and 0.5 %). The concentration of heavy minerals varies widely, reaching values of up to 40 % at the backshore after storms north of the Lagoa dos Patos inlet. The shoreface surface consist mainly of sandy sediments, with well-sorted fine sand across the beach and the surf zone (Martins 1967; Siegle 1996; Nicolodi et al. 2002), except along the 60 km long southern coast where bimodal sediments dominate due to the presence of shells (Calliari and Klein 1993).

18.1.5 Coastal Provinces and Geomorphology

RS has an extensive sandy coast consisting of a 615 km long Holocene fine sand barrier with a uniform northeast-southwest orientation and a slight sinuosity. The nearshore zone is wide and shallow with seaward limit at 10–15 m depth, and the surf zone contains extensive shore parallel bars. The coast begins at Torres city (Praia Grande beach #1), in the north, and extends south to Chuí #32, at the border with Uruguay and is one of the longest sandy beach-barrier systems in the world.

The littoral can be divided into three sectors: northern, middle and southern. The north coast extends for 120 km between Dunas Altas slightly south of Cidreira beach and the northern border at Torres beach; the middle sector extends 275 km between Dunas Altas and Lagoa dos Patos inlet (Fig. 18.3); and the 220 km southern sector extends from Lagoa dos Patos inlet to Arroio Chuí on the border with the Uruguay.

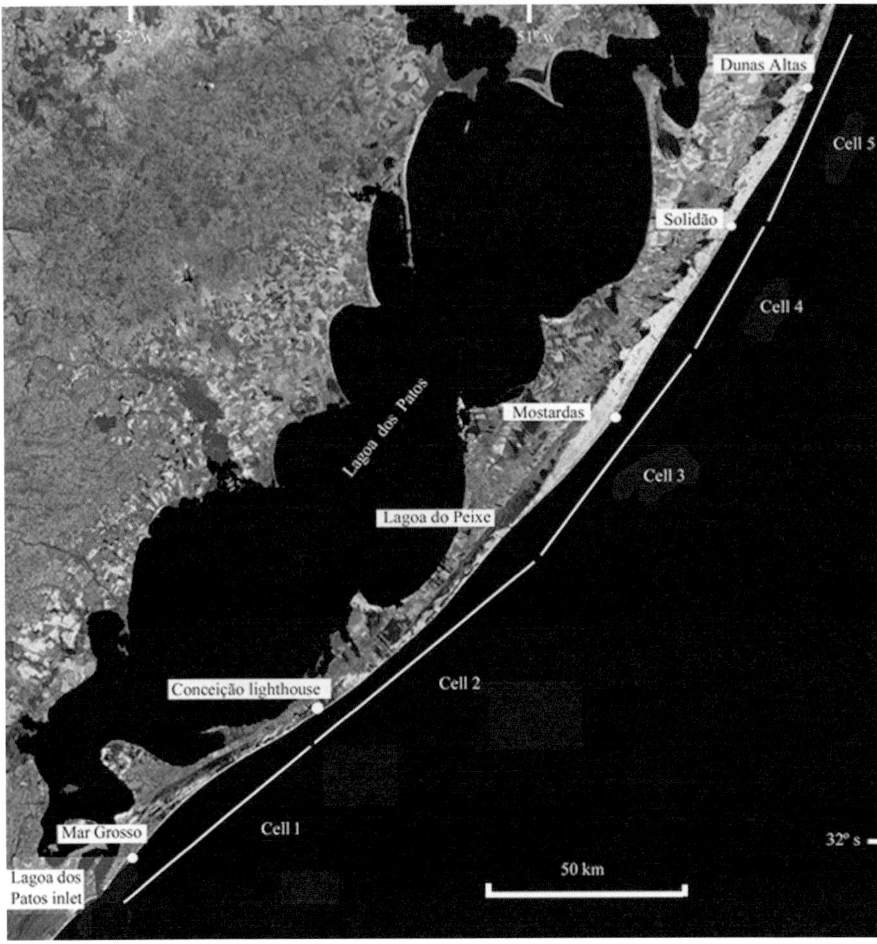

Fig. 18.3 Location of the midle sector and the five sediment cells along the middle coast of the Rio Grande do Sul, between Lagoa dos Patos inlet and Dunas Altas beach. The Landsat satellite image also shows the extent of the coastal dune field (*light gray color*) (Toldo et al. 2013)

18.2 Rio Grande do Sul Beach Systems

RS has a wave-dominated coast with wave energy flux controlling beach behavior (Tomazelli and Villwock 1992; Lima et al. 2001; Toldo et al. 2006a, 2013). Swell generated in the South Atlantic Ocean and sea generated by strong summer and spring local northeast winds dominate the coast. Except for the passage of cold fronts from the south or southeast, the oceanic waves are characterized by medium to high-energy with a significant height of 1.5 m and period of 7–9 s. The littoral has a small semi-diurnal tide with an average range of just 0.25 m. However, the meteorological tide can exceed the astronomical tide and reach 1.2 m. The coast is regularly subject to storms associated with frontal systems and extratropical cyclones, which can generate surges that cause anomalous high sea levels. Maximum water level elevation above the mean sea level and beach erosion are associated, respectively, with winter and summer storms. Longshore sediment transport is bidirectional from both the southwest and northeast with alternating direction occurring in most sectors. However, the net sediment transport is predominately towards the northeast as a result of the greater wave energy flux associated with the passages of cold fronts.

18.2.1 Coastal Processes and Parameters

The beach sediment budget represents the volume differences between the addition and removal of sediment in a given beach area (Rosati 2005), which can fluctuates over a given time period, due to storms or other processes. Although the sediment stock can be qualitatively evaluated, the quantity and rate of supply over a given time period is difficult to calculate, especially due to the inherent uncertainty in the calculation of the directional distribution of the energy flux and, consequently, the longshore transport.

Sand deposits are abundant on the Rio Grande do Sul inner shelf and the abundance of sand is represented by large coastal dune fields and the widening of the nearshore zone, along of the middle coast. The longshore transport however exhibit strong episodic behavior due to the wave forcing from south, closely related to the passage of cold fronts during winter (Calliari et al. 1998). Here we report an integrated analysis of the coastal morphology associated with longshore transport along the RS coast. In addition, zones of erosion and accretion were delimited by comparing a DGPS middle coast shoreline mapping and the beach line reproduced from the army chart collection. The results show extensive shore retreat along sections of the coast (Toldo et al. 2013).

The comparisons comprise the beach system, including the nearshore zone, surf zone, subaerial beach and adjacent coastal dune fields. The subaerial region between the backshore area and the dune field was delimited based on the high tide line, which is approximately at the base of the foredune. The submarine division between the inner continental shelf and the nearshore zone was set to a depth of 10 m.

The middle coast was divided into five beach cells using transverse boundaries (Rosati and Kraus 2001). The cells extend from the Lagos dos Patos inlet to Dunas Altas beach, a distance of 275 km (Fig. 18.3). The geomorphological criterion to distinguish southwest and northeast littoral cells was associated with the variations in the size of the coastal dune field, as well as significant changes in the shoreline alignment, which are used as topographic geoindicators.

Changes in the sediment budget along this coast were qualitatively evaluated by the following geo-indicators: (1) topographic measurements of the dune field height from a mosaic of Aster GDEM (Global Digital Elevation Model) image; (2) the

Table 18.5 Classification of transport mechanisms that lead to addition and loss in the RS coastal sediment budget

Addition (gain)	Removal (loss)
Littoral drift (into)	Littoral drift (out)
Aeolian transport (into)	Aeolian transport (inland)
Washout (into)	Storm surges (offshore)
	Storm surges (inland)
	Coastal jet (offshore)

Toldo et al. (2013)

Table 18.6 Length and classification of the five beach cells along the middle coast. The sediment budget ($\times 10^6$ m^3) was calculated from littoral drift rates throughout the period from 1998 to 2009

Cell	Length (km)	Sediment budget ($\times 10^6$ m^3)	Transport direction	Classification
(5) Dunas Altas	50	+0.10	North	Sink
(4) Solidão	35	+0.01	North	Sink
(3) Mostardas	50	+0.01	North	Sink
(2) Conceição	70	−7.60	North/South	Source
(1) Mar Grosso	70	+2.30	North	Sink

Toldo et al. (2013), Motta (2013), Motta and Toldo (2015)

width and slope of the nearshore zone obtained from Brazilian nautical charts surveyed in 1963; (3) mapping zones of erosion and accretion (shoreline change) using aerial photographs together with a GPS survey along of shoreline (Toldo and Almeida 2003); and (4) width and slope of the subaerial beach.

The mathematical estimation of the longshore sand transport along the middle coast was based on the CERC formula (U.S. Army Corps of Engineers 1984, p: 4–96), where the energy flux for computing longshore transport rate is based on the empirical relationship between the component of wave energy flux entering the surf zone and the immersed weight of sand transported. In addition, the parameters controlling the removal or addition of sediment for each cell were classified according to the transport mechanism including littoral drift, coastal jet, storm surges, washout and aeolian transport (Table 18.5).

The location of the shoreline determined in 1997 was compared to that determined using maps and aerial photographs from 1975 (Toldo et al. 2006a). The use of other geoindicators such as foredune height, bathymetry data points, width and slope of the subaerial beach allowed classification of the shoreline mobility. Segments where the coastline retreated were considered to be eroding, those where the coastline is moving seawards were classified as depositional (Table 18.6). The results of the 22 year temporal analysis (1975–1997) showed erosion was observed along much of the study area, particularly along the second cell.

Estimations of the littoral transport were used to determine the sediment balance in the beach system based on previous results (Motta and Toldo 2015). Mathematical calculations of the potential longshore transport, resulted in a large net northward annual volume. The application of this model on the middle coast predicts a substantial variation of the energy flux into the surf zone due to slight changes in the shoreline alignment, and consequently to the potential transport. The southernmost stretch, cell 1, has the highest rates of the littoral drift, decreasing to the north, in the cells 2, 3 and 4 (Table 18.6)

Aeolian sand transport is another important process. The strong and frequent northeast winds cause dunes to migrate towards southwest into the coastal plain (Tomazelli et al. 2000; Figueiredo and Calliari 2006).

In addition to these transport mechanisms other hydrodynamics processes are associated to shoreline changes. Analyses of satellite images reveal the presence of short-duration coastal currents that have not been studied in detail before. The results suggest that in some instances these currents can lead to the diffusion of suspended sediment seawards from the surf zone (Fig. 18.4). A strong coastal jet characterizes this coastal current. The occurrence and evolution of these jets are closely related to the passage of cold fronts, which form early in the winter. The characteristic circulation pattern over the shoreface consists of two segments, a northward coastal current intensified by the south and southeast winds, and a wide clockwise-rotating plume gyre. The conceptual model suggests that sediments are supplied to the surf zone and shoreface from the shoreline erosion. A fraction of the suspended load is then deposited offshore, developing large sand banks that can be seen at approximately the 10 m bathymetric contour. The physical process that control the formation and evolution of this coastal jet have not been fully explored, but good agreement was found between the shoreface accretion area that extends for more than 2 km offshore and the extension of the coastal jet over this area, as observed in the satellite image (Toldo et al. 2006a).

Sediment is transported longshore from the cell 2 to cells 1, 3, 4 and 5, respectively (Fig. 18.5). The aeolian transport adds sediment to all cells, which is deposited on the coastal plain. This onshore transport is particularly strong in the locations where there is a change in shoreline alignments, as seen between cells 2 and 3, 4 and 5 (Fig. 18.5). The transport of sediments into cells 3 and 5 results in a widening the upper and lower nearshore zone as well as the subaerial beach and dune field (Toldo et al. 2006a).

The eroding cell 2 shows significant short-term fluctuations. Shoreline changes along 3.17 km of beach in the cell 2 near the Conceição lighthouse (Fig. 18.6b) indicated that the shoreline retreated 87 m over 16 years (Lopes et al. 2008). This shoreline mapping was performed by comparing a mosaic of 1:20,000 scale aerial photographs from 1981 to the shoreline measured by DGPS surveying in 1997. The highest erosion rate of 5.4 m year^{-1} occurred along beach segment that includes the Conceição lighthouse (Lopes et al. 2008). The shoreline changes bordering Lagoa do Peixe (1987–2009), located between cells 2 and 3, averaged 1.6 m year^{-1} (Schossler et al. 2012) along a 24 km long section of beach located south of the lagoon inlet.

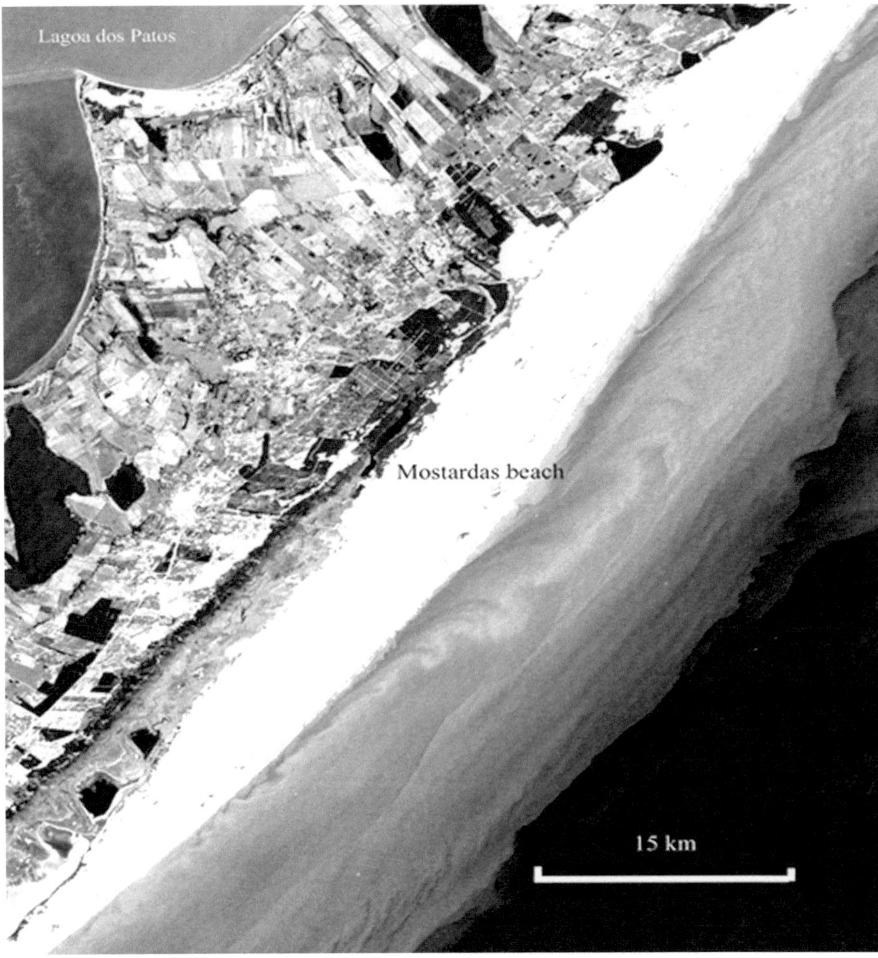

Fig. 18.4 Satellite image from May 20th 2000 shows subtle changes along the shoreline alignment in the Mostardas beach, the presence of coastal jets on the nearshore zone, and large dunes on the coastal plain (Modified from Toldo et al. 2006a)

The maximum erosion rates were distributed along the length of the beach studied for calculating the average rates, and indicate that large volumes of sediment were eroded and transported out of the cell 2. The estimated potential of sediment longshore transport based on the CERC formula predicts a substantial variation in energy flux in the surf zone, due to slight changes in shoreline alignments. The reduction in the sediment flux due to such changes produces a decrease in the longshore transport, meaning that part of the sediment arriving from the upstream stretch may be deposited or diverted offshore by the coastal jet (Dette 2001). Based on this, it is possible that decreases in the net longshore sand transport are responsible for accumulation and an increase in nearshore width from less than 1 km to

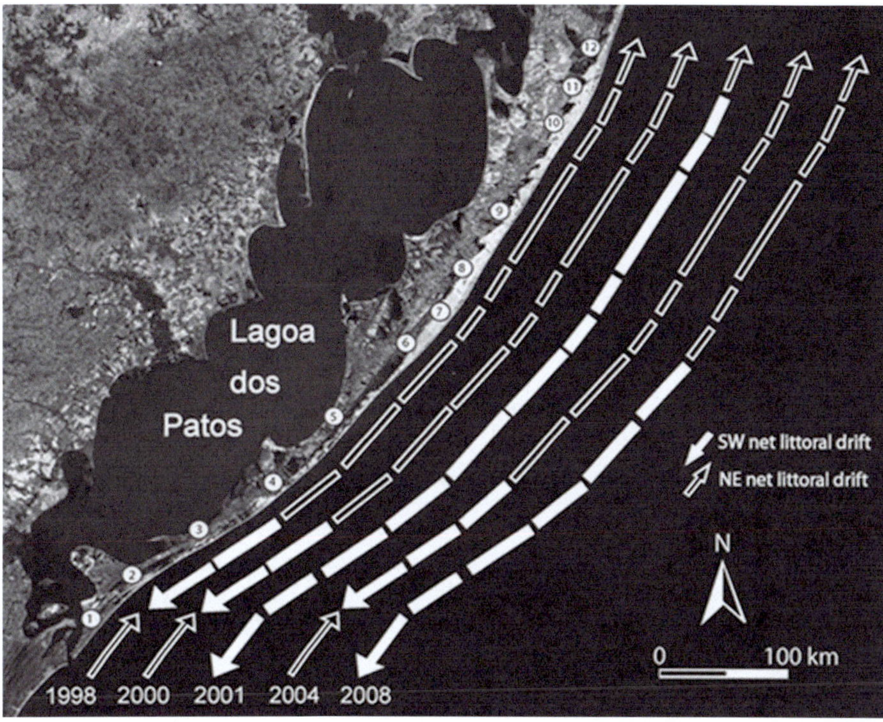

Fig. 18.5 The annual net littoral drift for each sub-cell (1–12, 1998–2008). Northeast transport dominated, except for 2001, while transport reversals were more common in the southern part of the middle coast (Motta 2013; Motta and Toldo 2015)

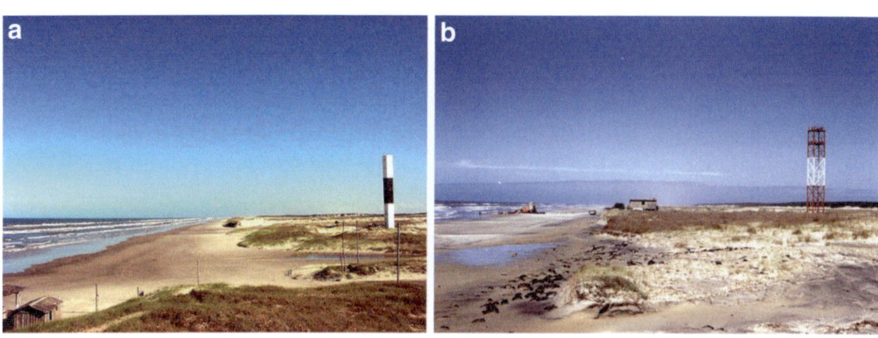

Fig. 18.6 (**a**) View to the south from Dunas Altas a stable beach with high foredune and a extensive dune field; and (**b**) the Conceição new and old lighthouses located along the cell 1, a section that has high erosion rates. The old lighthouse was built in the dunes approximately 200 m from the shoreline in 1929 and was destroyed in 1993 (note collapsed old lighthouse structure on the beach) (Toldo et al. 2013)

more than 3 km along Mostardas and Dunas Altas beaches (Fig. 18.5). It is interesting to note that wider dune fields are also associated with those beaches where the nearshore zone is wider (Toldo et al. 2006a). Therefore, based on the concept of sediment balance, the large sediment volumes eroded from cell 2 were transported to form the large bars found in the nearshore zone of Mostardas and Dunas Altas beaches as well as the extensive aeolian deposits (Fig. 18.5).

Also, Motta (2013) and Motta and Toldo (2015) analyzed the regional sediment budget on scale of sub-cells and observed another coastal process along southern half of the middle coast. This process is associated with reversal of the net littoral drift from northeast to southwest, in the opposite direction to that observed in the rest of the middle and north coasts (Fig. 18.5) (Motta 2013; Motta and Toldo 2015). This reversal was considered to be related to the position of extratropical cyclones in the Atlantic Ocean. Parise et al. (2009), identified three patterns of cyclone trajectories, one generated on the southern Uruguayan coast with displacement to the east propagating east and east-southeast waves (type 2). Littoral drift reversals have been associated with this cyclogenesis pattern. Large reversal of the wave pattern can also be proposed by the analysis of extreme wave events (Guimarães et al. 2014).

Other evidence of littoral drift reversals was also presented by Long and Paim (1987) who investigated the evolution of the Lagoa dos Patos inlet. They found that beach ridges developed to the southwest of the inlet, suggest a net littoral drift to the southwest along these two sub-cells. The genesis of the beach ridges along the cell 1, are also consistent with littoral drift reversals (Corrêa et al. 2004).

18.2.2 Ocean Beaches Types and States

18.2.2.1 Spatial and Variation and Beach Stages

Beaches along this coast are fully exposed with the exception of Torres where a rocky headland composed of sandstone, basalt and volcanic-clastic sequences provide a small degree of protection to coastal dynamics. The beaches generally consist of fine quartz sand with a low gradient (2°) swash zone usually containing low-relief beach cusps. Beaches range between intermediate and dissipative, with beach state controlled in part by grain-size variations through the contribution of shell fragments, gravel and coarse sand.

Morphodynamic studies of the RS coastline began with studies of the north and south coastal sectors (Calliari and Klein 1993, 1995; Calliari et al. 1996; Toldo et al. 1993). Using the Wright and Short (1984), classification and Short and Hesp (1982) morphometric parameters, three distinct beach sectors were identified along the southern Holocene barrier (Calliari and Klein 1993). Concurrently, Toldo et al. (1993), initiated the study of the northern sector. Several other studies on the RS beaches were used to develop a morphodynamic and sedimentology database called PraiaLOG (Pereira et al. 2010), which summarizes the information of the last 20 years (Table 18.7 and Fig. 18.7).

Table 18.7 Morphometric parameters: *N* number of profiles, *M* number of months, *Hb* breaker height (m), *T* wave period (s), *Mz* mean grain size (mm), *Am* medium sand percentage (%), *Ws* settling velocity (cm^{-s}), Ω Hb/T.Ws, *b* beach face slope (^{0}), *Yb* mean beach width (m), *σYb* standard deviation of Yb(m), *CV* backshore mobility index(%), *Vv* Volume change above datum (m^3/m), *σVv* standard deviation of Vv

Beaches	N	M	Hb	T	Mz	Am%	Ws	Ω	b	Yb	σYb	CV	Vv	σVv
Northern sector														
Praia Grande	8	15	1	10.8	0.212	7.0	2.11	4.2	1.56	134	19	15	39	17
Prainha	9	15	1.1	11.3	0.212	9.0	2.11	4.6	2.13	78	20	26	28	30
Praia da Cal	8	14	1.3	11.9	0.212	21.0	1.8	6.0	2.3	91	19	20	30	24
Guarita East	10	15	1.3	11.7	0.212	6.0	1.8	6.2	2.4	91	10	11	10	7
Guarita West	9	17	1.3	11.8	0.212	11	2.3	4.8	2.0	69	19	27	12	12
Capão sidewalk	10	24	1.6	12.3	0.212	27	2.2	5.9	2.6	121	13	10	38	19
Capão da Canoa lighthouse	9	23	1.3	11.7	0.212	27	2.2	5.0	2.6	140	13	9	29	4
Xangrilá	14	23	1.1	11.6	0.250	45	2.1	4.5	1.6	114	8	7	15	8
Imbé		3			0.212	30	2.2							
Tramandaí sidewalk	10	23	0.9	10.8	0.212	22	1.9	4.4	2.4	117	7	15	25	21
Tramandaí dunes	8	23	1	11.5	0.177	8	18	4.8	2.1	96	17	18	38	26
Cidreira	8	23	1.1	11.8	0.212	25	1.9	4.9	2.4	103	7	7	15	9
Middle sector														
Solidão lighthouse	10	38	1.3	12	0.212	35	1.9	5.7	2.1	89	15	17	21	19
São Simão	10	38	1.0	11	0.212	9.0	2.2	4.1	2.0	91	15	16	23	18
Mostardas lighthouse	9.0	34	1.1	10	0.212	25	2.2	5.0	2.3	75	9	12	19	15
Lagamarzinho	13	88	1.3	11	0.212	28	2.2	5.3	2.8	65	21	33	14	13
Conceição lighthouse	35	96	1.0	11	0.212	33	1.9	4.8	2.1	62	15	24	15	13
Estreito lighthouse A	8	15	0.9	10	0.180	36	2.0	4.5	1.7	88	12	12	14	11
Estreito lighthouse B	12	16	0.8	10	0.250	36	2.1	3.8	1.8	81	19	23	17	21
Mar Grosso	21	16	–	–	0.150	1.0	1.3	–	2.0	92	6	7	12	10

(continued)

Table 18.7 (continued)

Beaches	N	M	Hb	T	Mz	Am%	Ws	Ω	b	Yb	σYb	CV	Vv	σVv
Southern sector														
Terminal	52	3	0.8	9.8	0.125	0.0	1.6	6.3	2.0	77	6	8	9	2
Bahia street	38	43	0.9	11	0.125	4.0	1.3	6.2	1.25	126	6	4	3	2
Querência	15	11	0.6	8.7	0.177	2.0	1.8	3.8	1.9	134	7	5	4	3
Altair shipwreck														
Barcaça	23	14	–	–	0.180	14	2.0	3.1	4.4	108	11	10	11	8
Sarita lighthouse	9	8	0.7	8.5	0.177	7.0	1.8	5.0	2.5	70	10	14	4	3
Taim	8	7	0.8	8.4	0.180	6.0	2.0	5.0	3.2	85	6	7	8	3
Verga lighthouse	7	5	0.8	9.0	0.177	10	1.8	5.0	3.0	63	6	8	8	5
Albardão lighthouse	7	6	1.0	9.9	0.180	11	2.0	5.0	3.0	170	10	6	17	4
Concheiros	10	8	1.0	10.5	1.41/0.212	40	2.2/15	2.3	4.4	109	3	2	9	11
FronteiraAberta lighthouse	22	7	-	-	0.353	26	4.0	4.0	4.4	98	7	7	9	8
Hermenegildo	7	5	0.8	7.5	0.177	16	1.81	4.0	2.6	118	15	12	19	11
Chuí	5	5	0.9	7.5	0.177	12	1.81	4.8	2.2	147	12	8	20	12

See Fig. 18.7 for location

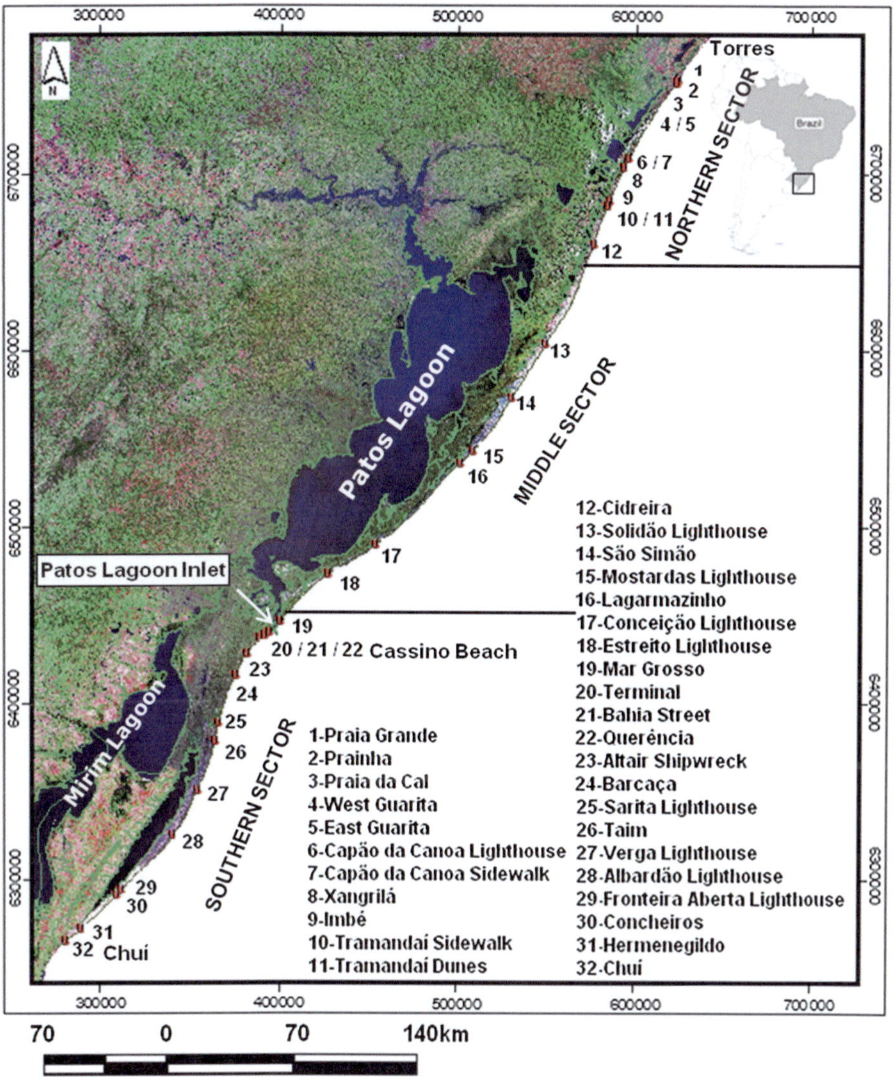

Fig. 18.7 The three coastal sectors and list of beaches along the RS coast (Modified from Pereira et al. 2010)

Northern Sector Beaches of Tôrres with the exception of Guarita East (#5) are relatively protected and display intermediate beach characteristics with high mobility coefficients and high vertical profile variation (Pivel 1997; Table 18.5). Using the Hegge et al. (1996) approach for sheltered beaches, Torres beach can be classified as a planar beach (Pereira et al. 2010). Tramandaí (#10,11) and Capão da Canoa (#6, 7) beaches are intermediate (Pereira et al. 2003). The beach from Arroio do Sal (south of Torres) to Imbé (#9), in northern RS, also was classified as intermediate (Weschenfelder et al. 1997). The high values found for backshore width and its standard deviation indicates the beach is mobile and therefore susceptible to significant episodes of erosion and accretion.

Middle Sector The beaches in the middle portion, between Estreito and the Solidão Lighthouse (#13–18), 228 km north of the inlet, were characterized as intermediate (Barletta and Calliari 2003; Barletta et al. 2006). Four profile surveyed north of the Patos Lagoon inlet for 12 km to Mar Grosso (#19) (De Oliveira and Calliari 2006) indicate that the beach is predominantly dissipative. The beach has three bars with waves breaking on the third only during higher wave events. During summer the RS beaches accrete with bars migrating toward the coast forming a ridge and runnel morphology (Wright and Short 1984). The profiles monitored also indicated that volumetric and morphologic changes increase toward the inlet as a function of the concentration of wave energy and associated circulations patterns. Most of the time, changes in dune morphology was minimum except during spring and summer, when winds blow onshore.

Southern Sector Although they have the same degree of exposure to coastal processes, the southern beaches have greater spatial variability due to considerable variations in beach sediment texture. The section between Patos Lagoon inlet and Cassino beach, subdivided in Terminal, Rua Bahia and Querência beaches (#20, #21, #22), consists of very fine sand, and are considered the most dissipative beaches along the RS coast (Tozzi and Calliari 2000). Along "Concheiros" sector (#29, #30, #31 on Fig. 18.6) polymodal sediments, consisting of gravel (shell fragments) coarse and fine sand (mainly quartz), produces both reflective and intermediate beaches (Calliari and Klein 1993; Klein 1997; Serau and Calliari 2003). Further south, towards the national border (205 and 217 km) the beaches are intermediate to dissipative. Two touristic beaches are located at this interval, Hermenegildo and Chuy, the first is experiencing intense erosion due to storm surges (Calliari et al. 1998; Klein and Calliari 1997; Esteves et al. 2003).

Pereira et al. (2010) applied a statistical approach to the beach profiles using clustering, multidimensional scaling and similarity analysis. It identified six beach groupings: one with dissipative tendencies, another with intermediate to reflective tendencies, and the other groups representing beaches with intermediate characteristics, mainly differentiated in terms of beach mobility (Table 18.7, Fig. 18.8). The six groups are in agreement with previous studies. Group 1 represents dissipative beaches (Cassino beach #20, #21, #22), while group 5 represents the intermediate beaches with reflective tendencies (Concheiros), and the other groups represent beaches with intermediate characteristics (Albardão, Tramandaí and Lagamarzinho). The application of statistical techniques allowed a quantitative differentiation among patterns, which had only previously suggested in the literature.

The results also demonstrate the presence of three intermediate beach stages (groups 2, 3 and 4) based on their morphology, beach mobility and sediment characteristics. Beach mobility (Short and Hesp 1982) which is an indicator of the

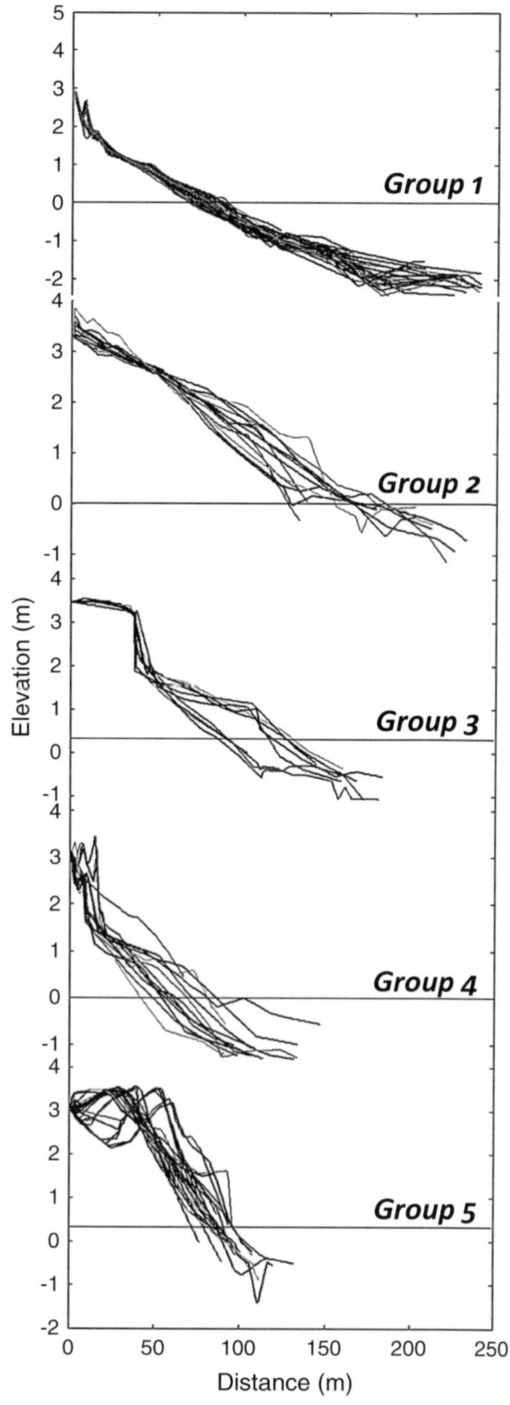

Fig. 18.8 Profiles from five of the six groups found using the multivariate analysis methods. Groups 1, 2, 3, 4 and 5 represented respectively by Terminal (#20), Albardão (#28), Tramandaí (#10), Lagamarzinho (#16) and Concheiros (#30) on Fig. 18.7 (Modified from Pereira et al. 2010)

net exchange of sediment between the surf zone and the sub-aerial beach profile can be defined as low for Group 2, moderate for Group 3 and high for Group 4. The presence of medium-sized sand is an important factor in the development of beach three-dimensional features and also in the differentiation among intermediate beach types (Pereira et al. 2003). In this context, the beaches from the groups 2, 3, and 4 (Albardão, Tramandaí and Lagamarzinho) show an increase in the precentage of medium sand (Table 18.7). Group 6 contains beaches that were expected to be inside the intermediate groups (2, 3 and 4) but were not, and they are considered as exceptions to the pattern. The association between morphometric and sedimentological parameters and aerial photographs indicated that beaches from Groups 2 and 3 may be classified as longshore bar-trough (LBT) and rhythmic bar and beach (RBB) (Pereira et al. 2003), which has respectively moderate-to-low and moderate-to-high mobility (Wright and Short 1984). Although the transverse bar and rip (TBR) stage was not found using this approach, it is believed that group 4 resembles this stage due to the high mobility observed (Pereira et al. 2010).

18.2.2.2 Bar Number and Multiple Bar Systems

The wave-dominated RS coast has a relatively gentle nearshore slope and is composed of fine sand. It is episodically exposed to storm waves, which favors the formation and maintenance of two to three shore parallel sand bars. However near the mouth of Lagoa dos Patos where very fine sand and lower gradients predominate cross shore profiles obtained using a sea sled and a total station allowed profiling to a depth of −5.5 m, 826 m from the shoreline, confirming the presence of four bars. The morphodynamic variables entered into the bar parameter defined by Short and Aagaard (1993), indicated a value of $B^* = 487$, confirming the applicability of this parameter in determining the number of bars on this beach. Temporal changes obtained over 71 daily profiles, simultaneously with an Argus video imaging database and hydrodynamic data installed on Cassino beach indicated the occurrence of three bars during an experiment when significant wave height reached 3 m (Guedes et al. 2009). The first corresponds to an intertidal bar, which is periodically exposed by wind setup, while the other two sub tidal bars are located at average distances of 99 and 237 m from the shoreline.

The inner bar displays high mobility over short time scales (hours), and appears to be controlled by swash and backwash processes. The second bar is more stable over intervals of hours, but display higher mobility over periods of days with offshore migration rates as high as 11.6 m d^{-1}. Shoreward migration rates reached 9.3 m d^{-1}, and were linked to low wave conditions, with significant wave height between 0.5 m and 0.75 m. During 13 days of small waves, the second bar migrating 45.5 m and attached to the inner bar and resulted in the development of rhythmic shoreline features.

A 2-year time series of video imaging (April 2005 to April 2007) also showed that Cassino beach has a three bars system (Fig. 18.9). During this period the formation of a new bar through the splitting of the second bar into two bars was observed

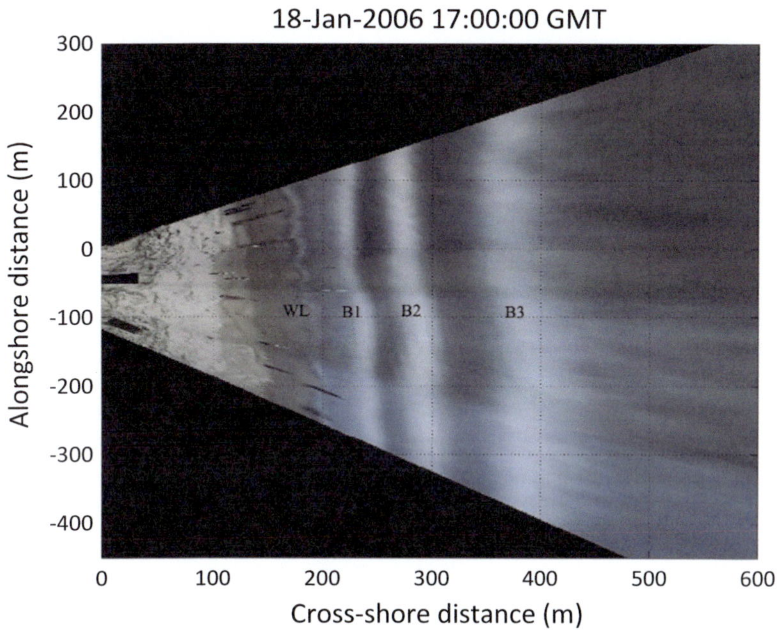

Fig. 18.9 Timex image mosaic for Cassino beach displaying the water line (*WL*), and the three Bars (B1, B2 and B3)

(Pereira et al. 2012). Similar processes were identified by Shand (2007) and was called Net Offshore bar Migration (NOM), except that in the Cassino beach the splitting happened at the bar instead of at the beach face. Before the total disappearance of the outer bar, the system had four bars. The analysis also indicated that bar variability ranges from weekly to seasonal, with the greatest variability explained by the weekly component. The total variance also has a significant correlation with the significant wave height.

18.2.2.3 Beach Mobility and States

Beach mobility relates to the exchange of sediments between the subaerial beach and subaqueous zones and the development of three-dimensional topography resulting from the associated temporal changes in beach states. Beach profiles and morphometric parameters obtained over the last two decades on some of the most populated beaches of the RS coast indicate that Praia Grande, Capão da Canoa and Tramandaí beaches in the northern littoral, and Hermenegildo beach in the southern littoral, experience high rates of mobility. This mobility is largely result of the medium grain size leading to more dynamic intermediate states (Fig. 18.10). In the case of Hermenegildo, high mobility is associated to the presence of high concentrations of shelly gravel sediments. This mobility is reflected in the behavior of

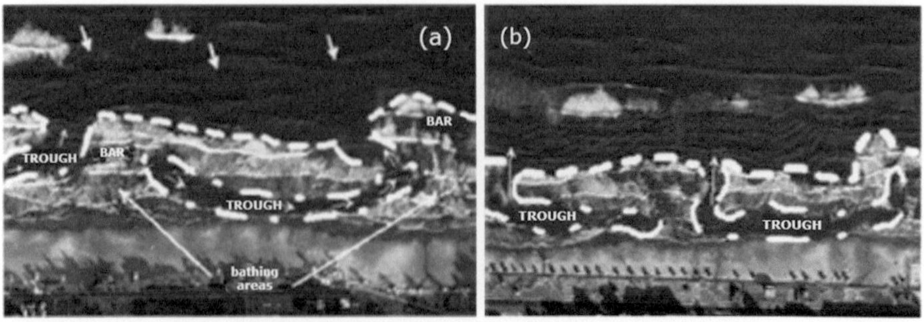

Fig. 18.10 (**a**) Capão da Canoa transverse bar and rip; and (**b**) Tramandaí rhythmic bar and beach

these beaches, which oscillates between the LBT and RBB states. Cassino and Chuí beaches in contrast, are characterized by dissipative and LBT beach states and a considerably lower mobility.

On the dissipative Cassino beach Argus images and high-frequency beach profiles surveyed since 2005 have shown that despite its predominantly dissipative state, other intermediate states can occur, particularly on the inner bars. Figure 18.11a shows Argus time-exposure mosaics indicating a dissipative system with three bars extending approximately 300 m offshore, whereas Fig. 18.11b shows low energy conditions without any breaker line; while Fig. 18.11c shows a LBT state; and Fig. 18.11d a RBB state; with a more chaotic bar patterns apparent in Fig. 18.11e. Finally, Fig. 18.11f shows rip currents morphology associated to intermediate states, which are occasionally identified in the video images.

During the summer of 2002 and 2003, daily monitoring of beach profiles were conducted at Cassino beach, in order to monitor changes across the inner and outer bar systems which could provide additional information to beach safety programs (see Sect. 18.3.5) (Pereira and Calliari 2005; Pereira et al. 2004). Bar crest distances and trough depth oscillated between 55 m and 0.89 m respectively for the first trough, and 160 m, and 1.78 m for the third trough. During low wave energy conditions, average migration rates of the first and second bar reached 4.8 m d^{-1}.

18.2.3 Beach and Dune Interactions

Foredunes along northern and middle littoral range are affected by changes in coastal orientation in relation to the prevailing northeast winds and beach morphodynamics and range are from well-developed to non-existent. The regional trend is therefore controlled by the coastal orientation, which influence the regional effectiveness of the aeolian transport processes. In the north between Torres and Mostardas beach the frontal dunes are well-developed. South of Mostardas to Mar Grosso beach there is a considerable reduction in the size and continuity. On the southern coast between Rio Grande and Chuí there are dunes (foredunes), hummocks and plains of sand (sand flats). The combination of coastal orientation

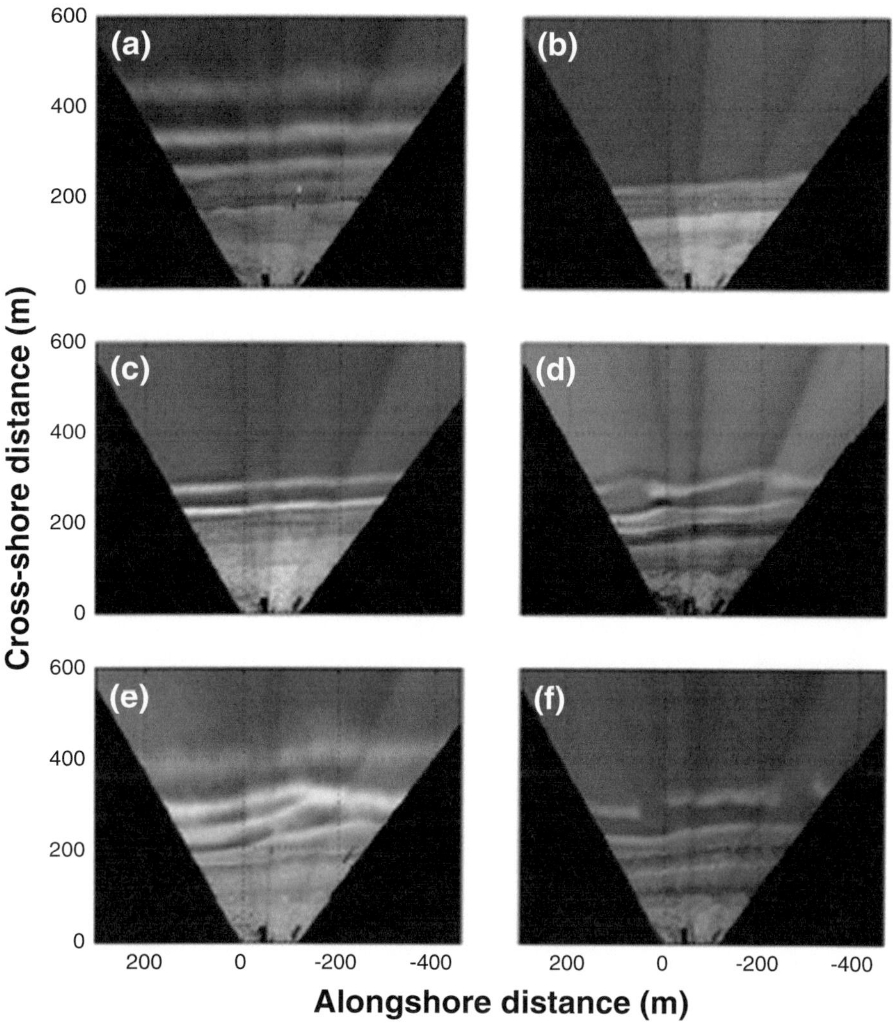

Fig. 18.11 Timex images from the Argus system at Cassino beach displaying (**a**) a dissipative system; (**b**) dissipative with low waves; (**c**) longshore bar and trough; (**d**) rhythmic bar and beach; (**e**) irregular bars; and (**f**) gaps where rip currents can be form

together with the frequency and intensity of the northeast winds favor the development of higher dunes close to Imbé, in the northern littoral. In this sector, the northeast winds approach obliquely onshore favoring sand transport to the foredunes. In the middle littoral, where the northeast winds are shore parallel or obliquely blowing offshore the foredunes are reduced or nonexistent.

Vegetation density and physiography also change laterally along the beach sectors. More vegetated stages with established dunes are associated with the northern littoral, while in the middle littoral less vegetated stages with eroded dunes dominate. Dune height however is not necessarily related to the beach width. There are beach sectors with wide beaches and low dunes as well as sectors with narrower beaches with high dunes. Foredunes ridge gaps are mainly related to the presence of washouts, however there is no relationship between dune height and washouts number, apart from alterations in the adjacent physiography and floristic composition.

The southern coast between Rio Grande and Chuí has a combination of well-developed foredunes, "hummocks" (non-coalescing and poorly developed frontal dunes) and "sand plains" (absence of dunes) (Seeliger 1992, 2003) associated with changes in grain size and coastline orientation. This pattern coincides with the variation in beach morphodynamics state along this sector showing the interrelationship between beach morphodynamic and the development of the frontal dunes (Calliari and Klein 1993; Tozzi 1999). Near Rio Grande where the beaches are more dissipative and the northeast wind component is most effective in transporting sand perpendicular to the coast, frontal dunes are well developed. Along the stretch between Sarita lighthouse and 15 km south of the Albardão Lighthouse, the combination of intermediate beaches with a more oblique northeast wind component contributes to the formation of "hummocks". Further south in the area called "concheiros do Albardão" where the beaches are composed by a mixture of shelly gravel and quartz sands with steeper beaches and shore parallel northeast winds the dunes consist of bare sand plains.

18.3 Beach Use

Approximately 80 % of the RS coastline is undeveloped. As a consequence coastal management is in its early stages focused on the northern sector where most of the developed beaches are located (Esteves 2004). However since a federal road located relatively close to the beach has been recently paved and now links the northern, middle and southern sectors of the coastline, these presently undeveloped sectors require special attention to ensure the implementation and regulation of an effective management plan.

18.3.1 Beach Development and Management

Based on rates of population growth, intensity of beachfront development, state of beach conservation, rates of shoreline change and Holocene coastal-evolution patterns, four classes of management strategies have been identified defined (Esteves 2004):

1. *critical areas* requiring corrective measures where the shoreline is experiencing significant changes (erosion and accretion) and either highly developed or experiencing population growth (Fig. 18.12);
2. *priority areas*, have low to moderate development but have a potential for future intensification of occupation and are also experiencing substantial shoreline changes. These require urgent regulation to restrict development and uses;

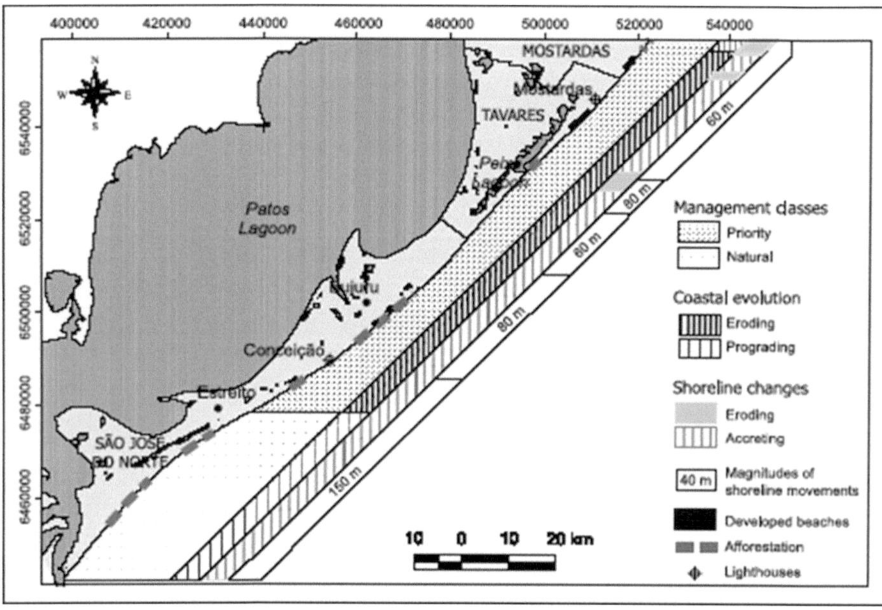

Fig. 18.12 Middle Sector: the shores from Mostardas to Estreito are classified as *priority* management areas. Beaches south of Estreito are classified as *natural* areas but impacted by jetties in the extreme south and afforestation and deserve special attention (Modifed from Esteves 2004)

3. *areas of future concern*, are not under pressure at present and consist of eroding shores located close to critical or priority areas; here regulation measures are recommended; and
4. *natural areas*, are mainly preserved, either eroding or accreting, with no signs of changes in the near future.

Using this approach Esteves (2004) found that 177 km (29 %) of the coast should be considered as *critical* particularly in the northern sector, at Cassino beach and between Cassino and Hermenegildo beaches. *Priority* areas comprises 198 km (32 %) and are mostly located in the middle sector; 178 km (29 %) are classified as *natural* areas and are located along the southern coast of São José do Norte and between Cassino and Albardão; and 65 km (10 %) between Hermenegildo to Albardão represent areas of *future concern*.

18.3.2 Impacts of Litter

Pianowski (1997) measured solid waste and plastic beads on four RS beaches, two in conservation units (Guarita and Taim beach located respectively at the northern and southern littoral) and two popular outside these areas (Praia Grande and Cassino beach). He found the popular beaches had substantially more than the beaches in the conservation units as shown in Table 18.8.

Table 18.8 Concentration of litter on four beaches

Beach	litter (items m^{-1})	(gm^{-1})
Cassino	8.87	67.6
Praia Grande	6.65	56.47
Guarita	3.81	11
Taim	2.84	53.3

The source of solid waste was predominantly tourists during the summer and marine origin between fall and spring where it comes from vessels or from the continent. This is especially true for Cassino, Praia Grande and Guarita, whereas for Taim it may come from the sea since there are no tourist activity at this beach.

Plastic was the main contaminant found in all the study areas representing 40–80 % of the total samples. It was dominant at Cassino (28 %) and Guarita (37 %), followed by building material Cassino (28 %) and Praia Grande (35 %), while wood predominated at Taim beach (49 %). Plastic beads showed low concentration and most of the beads were white. The beach which presented the highest concentration of plastic beads was the Taim beach, with an annual mean of 112.3 items m^{-1}.

An innovative study investigating the perception of beach users to solid waste pollution and quantifying the input of tourist-related litter to Cassino beach was undertaken by Santos et al. (2005). The study was conducted in areas occupied by beach users with different socio-economic characteristics. It found that daily litter input was higher in the region frequented by people with of low socio-economic standing, indicating that higher education levels raises awareness and behavior in relation to their own residues. Cigarette butts, followed by plastics are the main kind of litter generated. Strong correlation between visitor density and litter generation showed that tourism is the main source of marine debris and beach contamination is positively related to beach visitor density.

A systematic study on beach litter contamination was undertaken in a 12-month field investigation at Cassino beach (Wetzel et al. 2004). Estimates of litter accumulated around the most urbanized area of Cassino beach were among the highest recorded on a global basis. Although the main problem appeared to be the wastes left at beach by beach-users during the summer season especially in the urbanized area, year-round input from shipping, lagoon discharge as well as other sources cannot be neglected. Since the same characteristics of coastal circulation and urbanized patterns predominate along most of the developed RS coastline the same pattern of litter distribution is expected for the other coastal areas. Plastic was the main source of litter both in terms of numbers of items and weight. The study, which considered four stations, verified that plastic contributed 56.1 % of the total items counted and 40.3 % of the total weight of trash. Other important items in terms of weight were wood (10.7 %), fishing material (14.3 %) and a second category of debris (17.4 %) constituted by glass bottles and cloths. In terms of counts, fishing material, metal bottle tops and cigarette butts were the main debris. According Wetzel et al. (2004), management strategies to mitigate the problem include: increase in environmental education, recycling centers, anti-litter campaigns during the summer season, and control of food stands that occupy the shoreline. On a more general basis governmental integration, including other municipalities from the Lagoa dos Patos drainage basin appears to be fundamental, for reducing contributions by river discharge to the litter found along Cassino beach. At the same time the identification of litter from ocean sources suggests the need for integration at the international level.

18.3.3 Impact by Car Traffic

Impact of vehicles on Cassino beach was investigated via a survey (Vieira et al. 2004), that found cultural, geomorphological, sedimentological and political factors have been causing an unprecedented impact on the environment. The beach has been severely modified by intense traffic of vehicles (Fig. 18.13), which has been degrading the habitat of important species responsible for the ecosystem survival.

Fig. 18.13 Car traffic and parking at Cassino beach during the summer. The jetties of Lagoa dos Patos inlet are at the *top* (Modified from Vieira et al. 2004)

The study focused on determining the physical parameters related to very fine sand compaction using penetration resistance and hydraulic conductivity at two specific areas; one under intense impact and another on an undisturbed site. It was possible to identify physical impact across the beach, which affected pioneer vegetation, dune formation and species survival. The vehicular traffic on this beach began with the era of the automobile in Brazil and accelerated in 1970s and 1980s, becoming alarming since the 1990s. During the summer season 12,000 vehicles per day, are found along the 6 km beach between the center of the beach and the jetties. This stretch links with the main road system and helps to distribute the vehicle flow between the beach and downtown Rio Grande city. In addition there are incidents such as collisions, people being run over by cars, falls, and even traffic jam. The flow is generally on the backshore zone, due to better sand compaction. In this way the traffic flow occurs over a movable backshore strip that is dependent on mean sea elevation, which is highly variable since it is usually affected by the local wind (meteorological tide), causing the track to vary in a width from 9 to 18 m. Sand under the vehicle track is also compacted with a penetration depth four times less than the un-impacted area, making it impossible in many cases, to penetrate 10 cm (pressures up to 100 kgf cm^{-2}).

18.3.4 Beach Hazards and Safety

Beaches were analyzed in relation to safety and hazards that occur during the summer when there is a high concentration of bathers from the nearby urban centers. One of the most dangerous beach in RS is Torres where the presence of rocky headlands, turbid water obscuring the submerged rocks and strong currents against the northern headland of Guarita beach make it extremely hazardous. Most of the exposed beaches along the RS coast possess distinct hazard to bathers associated to their beach state and hazard rating (Short and Hogan 1994).

The LBT state, for example, is hazardous because of the troughs are usually deeper than the height of bathers, together with associated to rip currents. Deep troughs have been recorded at beaches such as Praia Grande in Torres (Pivel 1997), Tramandaí, Cassino (Pereira et al. 2004; Guedes et al. 2009), Hermenegildo (Tozzi 1999), and Chuí (Calliari and Klein 1993). The RBB state is also common on many RS beaches that have hazards related to the three-dimensionality of the bars which produces a complex surf-zone bathymetry with considerable spatial variations in water depth and the presence of rip currents which have also been documented on various beaches including Capão da Canoa and Tramandaí (Fig. 18.10). Rip currents can also occur against fixed structures such as the jetties located at the mouth of Mampituba River in Tôrres.

The combination of higher population during summer seasons and rip-dominated intermediate beaches result in the most hazardous beaches along the northern littoral and in the south at Hermenegildo beach.

At Cassino beach the total number of beach rescues during the 2002 summer, were equally associated to both dissipative and intermediate states, while during the 2003 summer 2003 a deep trough in the intermediate states resulted in a very hazardous morphology. Pereira and Calliari (2005) recorded a stable beach morphology with a 2.4 m deep trough located distance of 96 m from the water line.

18.3.5 Fluid Mud Hazards on the Beach and Surf Zone

Deposition of fluid mud on the Cassino beach and surf zone have been recorded since 1901. The process is associated to the transport and deposition of fluid mud delivered during dredging operations in the inlet and shoreface and then remobilized by storm waves that transport it to the beach. On at least two occasions, the fluid mud presented a potential risk to surfers owing to the lateral gradients in wave energy resulting from wave attenuation over muddy bottoms. This gradient changes wave set up along the shore generating oblique and longshore currents, which can transport surfers toward areas where they remain isolated from the beach by fluid mud banks in excess of 1.5 m thick (Calliari et al. 2001). This situation has caused the entrapment of the surfers, leading them to life-threatening hypothermia as the accidents occurred during fall and winter when the water and ambient temperature was low. Fortunately, in both cases local fishermen and surfers using cables and "bodyboards" rescued them.

The presence of mud between the beach face and backshore also represents an additional hazard for car traffic. The fact that Cassino beach is typically a wide highly dissipative beach composed of very fine compact sand encourages a steady flow of vehicles (see Sect. 18.3.4). However any mud deposited across the beach is rapidly covered by a thin layer of sand as soon as the sea level drops after the storms and forms a hidden trap for cars which can get stuck in the mud which is sometimes more than 0.3 m thick. It becomes impossible to remove the cars without the use of caterpillar machines. The accidents become more serious when cars hit the mud at high velocity, stopping abruptly and causing serious bodily injury to tens of drivers during the 14 month duration of the mud deposit.

18.3.6 Washouts and Flooding Beach Hazards

While there are few fluvial discharges along the RS coast there is considerable pluvial drainage from the dunes system to the beach in the form of washouts. These ephemeral watercourses play an important role in the drainage of pluvial waters from the swamps located behind the frontal dunes systems. The number and spatial distribution of these features varies along the coastline as a function of the seasonality, rainfall, geology and geomorphology of the coastal plain (Figueiredo and Calliari 2006). Urbanization visibly increases the washout concentration along the beach and dune field due to the reduction of the infiltration area and soil compaction, which is caused by road pavement and any kind of structure. These rills are generally active following several days of heavy rainfall and cause severe erosion of the frontal dunes and sub aerial beach. At most of the beach resorts along the RS coastline this can lead to flash flooding of the lower urbanized areas, which have serious consequences on basic sanitation for the urbanized areas as well as water quality in the surfzone since domestic wastes are not treated. These occurrences adversely affect tourism and related activities.

Washouts are also a beach hazard and a source of serious accidents and fatalities. The fine sand that predominates along the beaches provides good conditions for automobile traffic. However, after high precipitations events deep, difficult to see washouts, can form posing a severe hazard to cars and causing accidents to the unaware users. Many accidents have been reported during such conditions, including cars lost in quick sand in the ephemeral creeks. These washouts are also highly unstable with a potential risk for dune collapse. A fatality was reported in November of 2001, at Inhame district (S.J. do Norte city), when an 8 years old boy, who was playing on the edge of the scarp, was buried by the sand of the collapsing dune.

Guimarães et al. (2015), used a high resolution analysis of the interaction of irregular waves with natural and urban structures to predict extreme wave runup. Horizontal runup data, instantaneous flooding maps, and wave propagation beyond the coastline are numerically predicted to provide the information on risk conditions based on six storm events in the cities of Imbé and Tramandaí. This allowed for an identification of critical vulnerable areas subject to washing or flooding. The time analysis of the swash runup showed that the dune zones on the Tramandaí and Imbé beaches work as a natural protection structure to the wave impact, with no dune overtopping predicted in these zones. On the other hand, urban areas without dune protection in Imbé beach are exposed to wave flooding. These results confirmed that the urban occupation in the former estuary zone in Imbé resulted in this area having a high risk of marine flooding during periods of high run-up and storm surges.

18.4 Summary and Conclusions

1. *The Postglacial Marine Transgression* has been the main forcing function operating to translate RS barriers shoreward, across the glacially exposed continental shelf to their present location. Geological inheritance and shelf slope has been critical in determining where the barriers were geographically positioned as sea level roughly stabilized (+/− 1–3 m) or slowed down considerably about 6.5 ka. Shelf gradient would have also determined whether a barrier could form or not. The sediment supply, wave and wind energy and direction, and tidal range were critical in determining the ongoing barrier evolution and style (aggradation, progradation or retrogradation).

2. The coastline is characterized by a *long sandy barrier* with a main northeast-southwest orientation. Superimposed on this general trend there are several concave and convex sectors, which extend from the rocky headland of Torres to Chuí at the Brazilian-Uruguayan border. The barrier is wave-dominated with a mean significant height of 1.5 m and a microtidal regime, resulting in only five inlets along 615 m of shoreline.
3. Although the astronomic tide is reduced to a mean annual amplitude of 0.3 m, storms from the southern quadrant induce *storm surges* which can reach 1.5 m which inundate and erode the shoreline. Net littoral drift along the barrier is towards the northeast. As a reults of the northerly transport beaches located south of jetties (Praia Grande, Cassino and Chuí) are prograding while beach erosion occurs for a few kilometers north of the structures. Due to changes in coastline orientation and inner shelf gradient, beaches along the shoreline present differences in degree of exposure to the hydrodynamic factors.
4. *Grain size variations*, mainly related to different carbonate content and medium to fine quartz sand result in distinct morphodynamic behavior along several sectors. Beaches to the north of Albardão lighthouse are composed of unimodal fine to very fine quartz sand which favor dissipative to intermediate modal stages. To the south of Albardão beaches are composed by bimodal sediments (shells and fine to medium quartz sand) favoring intermediate to reflective stages.
5. The characterization and differentiation along the 615 km coast was undertaken using *beach profiles* and their resulting morphometric parameters. More recently the use of multivariatc statistical tools identified six groups of beaches: three with dissipative, intermediate to reflective characteristics, and three groups with intermediate characteristics. The three intermediate groups were differentiated in terms of beach mobility and percentage of medium-size sand, with higher mobility related to higher amounts of medium-size sand. This approach can be used to identify sectors with different morphodynamic behavior.

References

Andrade Neto JS, Rigon LT, Toldo EE Jr, Schettini CAF (2012) Descarga sólida em suspensão do sistema fluvial do Guaíba e sua variabilidade temporal. Pesquisas em Geociências 39:161–171

Barletta R, Calliari LJ (2003) An assessment of the atmospheric and wave aspects determining beach morphodynamic characteristics along the central coast of RS State, Southern Brazil. J Coast Res 35:300–308

Barletta RC, Franco DC, Mello E, Pereira OS, Calliari LJ (2006) Application of empirical orthogonal functions for the analysis of southern Brazilian beach profiles. J Coast Res 39:351–354

Calliari LJ, Klein AHF (1993) Características morfodinâmicas e sedimentológicas das praias oceânicas entre Rio Grande e Chuí, RS. Pesquisas 20(1):48–56

Calliari LJ, Klein AHF (1995) Variabilidade das praias oceânicas entre Cassino e Chui, RS. Pesquisas 11:55–63

Calliari LJ, Klein AHF, Barros FCR (1996) Beach differentiation along the Rio Grande do Sul coastline (Southern Brazil). Rev Chil Hist Nat 69:485–493

Calliari LJ, Tozzi H, Klein AHF (1998) Beach morphology and coastline erosion associated with storm surges in southern Brazil-Rio Grande to Chuí, RS. An Acad Bras Cienc 70(2):231–247

Calliari LJ, Speranski N, Torronteguy M, Oliveira MB (2001) The mud banks of Cassino beach, southern Brazil: characteristics, processes and effects. J Coast Res 34:318–325

Carraro CC, Gamermann N, Eick NC, Bortoluzzi CA, Jost H, Pinto JF (1974) Mapa Geológico do Estado do Rio Grande do Sul, Série Mapas, n 8. Instituto de Geociências, UFRGS, Porto Alegre

Coli AB (2000) Estudo sobre o clima de onda em Rio Grande, RS. Dissertação, Fundação Universidade de Rio Grande, 76 pp

Corrêa ICS (1987) Sedimentos superficiais da plataforma e talude continental do Rio Grande do Sul. Pesquisas 19:95–104

Corrêa ICS (1996) Les variations du niveau de la mer durant les derniers 17.500 ans BP: L'exemple de la plateforme continentale du Rio Grande do Sul, Brésil. Mar Geol 130:163–178

Corrêa ICS, Aliotta S, Weschenfelder J (2004) Estrutura e evolução dos cordões arenosos pleistocênicos no canal de acesso da Laguna dos Patos-RS, Brasil. Pesquisas em Geociências 31(2):69–78

Corrêa ICS, Toldo EE Jr, Weschenfelder J, Baitelli R, Ayup-Zouain RN, Dehnhardt BA, Martins LRS (2007) Plataforma e talude continental do Rio Grande do Sul: síntese dos conhecimentos. In: 50 Anos de Geologia do Instituto de Geociências. Editora, Comunicação e Identidade, Porto Alegre pp 341–353

De Oliveira AO, Calliari LJ (2006) Morfodinâmica da praia do Mar Grosso, São José do Norte/RS. Gravel 4:23–36

Dette HH (2001) Influence of shoreline alignment on offshore sediment transport. In: Proceedings of the coastal dynamics'91, Lund, Sweden, p 938

Dillenburg SR, Roy PS, Cowell PJ, Tomazelli LJ (2000) Influence of antecedent topography on coastal evolution as tested by the shoreface translation barrier model (STM). J Coast Res 16:71–81

Esteves LS (2004) Shoreline change and coastal evolution as parameters to identify priority areas for management in Rio Grande do Sul. Pesquisas em Geociências 31(2):15–30

Esteves LS, Silva ARP, Arejano TB, Pivel MAG, Vranjac MP, Barletta RC (2003) Coastal development and human impacts along the Rio Grande do Sul beaches, Brazil. J Coast Res (35):548–556

Figueiredo SA, Calliari LJ (2006) Washouts in the middle and northern littoral of Rio Grande do Sul state, Brazil. Distribution and implications. J Coast Res 39:366–370

Guedes RM, Calliari LJ, Pereira PS (2009) Morfodinâmica da praia e zona de arrebentação do Cassino, RS através de técnicas de vídeo imageamento e perfis de praia. Pesquisas em Geociências 36:165–180

Guimarães PV, Farina L, Toldo E Jr (2014) Analysis of extreme wave events in the southern coast of Brazil. Natural hazards and earth system sciences discussions. Nat Hazards Earth Syst Sci 14:3195–3205

Guimarães PV, Farina L, Toldo E Jr, Diaz-Hernandez G, Akhmatskaya E (2015) Numerical simulation of extreme wave runup during storm events in Tramandaí Beach, Rio Grande do Sul, Brazil. Coast Eng 95:171–180

Hegge B, Eliot I, Hsu J (1996) Sheltered sandy beaches of southwestern Australia. J Coast Res 12:748–760

Klein AHF (1997) Concheiros do Albardão: variações espaço-temporais dos sedimentos e da morfologia praial. Dissertação, Univ Fed do Rio Grande do Sul, 120 p

Klein AHF, Calliari LJ (1997) Praias oceânicas entre Cassino e Chuí (RS): Variações da permeabilidade e sua influência no processo de troca de sedimentos. In: VI Congresso da ABEQUA, Curitiba, PR, Brasil, p 407

Lima SF, Almeida LESB, Toldo EE Jr (2001) Estimate of longshore sediments transport from waves data to the Rio Grande do Sul coast. Pesquisas em Geociências 28(2):99–107
Long T, Paim OS (1987) Modelo de evolução histórica e holocênica da Lagoa dos Patos, RS.In: I Congresso da Assoc Brasileira de estudos do Quaternário. ABEQUA, Porto Alegre, RS, Brasil, p 227
Lopes CG, Zanatta A, Toldo EE Jr, Nunes JC (2008) Mobilidade de curto prazo da linha de praia do Litoral Norte e Médio do RS. In: 44 Congresso Brasileiro de Geologia, CBG, Curitiba, PR, CD
Martins LR (1967) Aspectos texturais e deposicionais dos sedimentos praiais e eólicos da Planície Costeira do Rio Grande do Sul. UFRGS Publicação Especial 13, 100p
Motta VF (1969) Relatório diagnóstico sobre a melhoria e o aprofundamento do acesso pela barra de Rio Grande. Relatório, Univ Fed do Rio Grande do Sul
Motta LM (2013) Sandy sediment budget in the mid-coast of Rio Grande do Sul, Brazil. Dissertação, Univ Fed do Rio Grande do Sul, Porto Alegre, 44 p
Motta LM, Toldo EE Jr (2015) Regional sediment budget in wave-dominated and dissipative-intermediate beaches. J Mar Res 73(3–4):49–69(21)
Nicolodi JL, Siegle E, Rechden FR, Corrêa ICS (2002) Análise sedimentológica do pós-praia no Rio Grande do Sul. In: Martins LR, Toldo EE Jr, Dillenburg SR (eds) Erosão Costeira: Causas, análise de risco e sua relação com a gênese de depósitos minerais. Univ Fed do Rio Grande do Sul, Porto Alegre
Parise CK, Calliari LJ, Krusche N (2009) Extreme storm surge in the south of Brazil: atmospheric conditions and shore erosion. Braz J Oceanogr 57(3):175–188
Pereira PS, Calliari LJ (2005) Variação morfodinâmica diária da Praia do Cassino, RS, durante os verões de 2002/2003 no setor do terminal turístico. Braz J Aquat Sci Technol 9(1):7–11
Pereira PS, Calliari LJ, Lélis RJF, Figueiredo AS (2003) Riscos associados ao banho de mar e sua relação com a heterogeneidade morfodinâmica das praiasdo Rio Grande do Sul, Brasil: Projeto Segurança nas Praias. In: IX Congresso da Associação Brasileira de Estudos do Quaternário. ABEQUA. Recife, PE, CD
Pereira PS, Calliari LJ, Lélis RJF (2004) High frequency beach profile monitoring: Implications in beach safety at Cassino beach, southern Brazil. J Coast Res (39):909–912
Pereira PS, Calliari LJ, Barletta RC (2010) Heterogeneity and homogeneity of Southern Brazilian beaches: a morphodynamic and statistical approach. Cont Shelf Res 30:270–280
Pereira PS, Calliari LJ, Guedes RM, Schetinni CA (2012) Variabilidade temporal dos bancos arenosos da praia do Cassino (RS): Uma análise através das imagens de video. Pesquisas em Geociências 39(3):195–211
Pianowski F (1997) Resíduos sólidos e esférulas plásticas nas praias do Rio Grande do Sul – Brasil. Trabalho de Graduação, Fundação Universidade Federal do Rio Grande, Rio Grande, 79 pp
Pivel MAG (1997) Caracterização morfodinâmica e sedimentológica das praias de Torres (RS). Trabalho de Graduação. Fundação Universidade Federal do Rio Grande, Rio Grande, 72 pp
Rosati JD (2005) Concepts in sediment budgets. J Coast Res 21(2):307–322
Rosati JD, Kraus NC (2001) Sediment budget analysis system (SBAS).ERDC/CHL CHETN-XIV-3, U.S. Army Engineer Res Dev Center, Vicksburg, MS, # pp
Santos IR, Friedrich AC, Wallner-Kersanach M, Fillmann G (2005) Influence of socio-economic characteristics of beach users on litter generation. Ocean Coast Manag 48(9–10):742–752
Schossler V, Toldo EE Jr, Gruber NLS, Aquino FE (2012) Dinâmica morfológica do canal livre da Lagoa do Peixe, Brasil. In: I Congreso Ibero americano de Gestión Integrada de Áreas Litorales, Cádiz, Espanha, p 563
Seeliger U (1992) Coastal foredunes of southern Brazil: physography, habitats, and vegetation. In: Seeliger U (ed) Coastal vegetation of Latin America. Academic, San Diego, p 3657
Seeliger U (2003) Response of southern Brazilian coastal foredunes to natural and human induced disturbance. J Coast Res (I36):51–55

Serau AK, Calliari LJ (2003) Changes of an oceanic beach face in the extreme south of Brazil, during the passage of a coastal storm. J Coast Res (35):194–201
Shand RD (2007) Bar splitting: system attribute and sediment budget implications for a net offshore migrating bar system. J Coast Res (50):721–730
Short AD, Aagaard T (1993) Single and multi-bar beach change models. J Coast Res 15:141–157
Short AD, Hesp P (1982) Wave beach and dune interactions southeastern Australia. Mar Geol 48:259–284
Short AD, Hogan CL (1994) Rips and beach hazards, their impact on public safety and implications for coastal management. J Coast Res 12:197–209
Siegle E (1996) Distribuição dos sedimentos litorâneos entre o farol da Conceição e farol do Chuí, RS, e fatores condicionantes. Trabalho de Graduação. Universidade Federal do Rio Grande 88 pp
Sprovieri F, Toldo EE Jr (2015) Reconstruction and analysis of wave climate in southern Brazil. (Unpublished results)
Strauch JC, Cuchiara DC, Toldo EE Jr, Almeida LE (2009) O Padrão das ondas de verão e outono no litoral sul e norte do Rio Grande do Sul. Rev Bras Recur Hidr 14(4):29–37
Toldo EE Jr (1989) Os efeitos do transporte sedimentar na distribuição dos tamanhos de grão e morfodinâmica da Lagoa dos Patos. Mestrado em Geociências. Universidade Federal do Rio Grande do Sul, 143 pp
Toldo EE Jr, Almeida LESB (2003) A linha d'água como indicadora da posição da linha de praia. In: IX Congresso da Associação Brasileira de Estudos do Quaternário. ABEQUA. Recife, PE, CD
Toldo EE Jr, Dillenburg SR, Cunha LOBP, Almeida LESB, Tabajara LL, Martins RR (1993) Parâmetros morfodinâmicos da praia de Imbé, RS. Pesquisas 19:27–32
Toldo EE Jr, Dillenburg SR, Corrêa ICS, Almeida LESB (2000) Holocene sedimentation in Lagoa dos Patos lagoon, Rio Grande do Sul, Brazil. J Coast Res 16(3):816–822
Toldo EE Jr, Nicolodi JL, Almeida LESB, Corrêa ICS, Esteves LS (2006a) Coastal dunes and shoreface width as a function of longshore transport. J Coast Res (SI39):390–394
Toldo EE Jr, Dillenburg SR, Corrêa ICS, Almeida LESB, Weschenfelder J, Gruber NLS (2006b) Sedimentação de longo e curto período na Lagoa dos Patos, Sul do Brasil. Pesquisas em Geociências 33(2):79–86
Toldo EE Jr, Motta LM, Almeida LESB, Nunes JC (2013) Large morphological change linked to the sediment budget in the Rio Grande do Sul Coast. In: Coastal Dynamics, Arcachon. Proceedings of the Coastal Dynamics. SHOM, p 1687
Tolman HL (1997) User manual and system documentation of WAVEWATCH-III version 1.15. NOAA/NWS/NCEP/OMB Tech Note 151 97 pp
Tomazelli LJ, Villwock JA (1992) Considerações sobre o ambiente praial e a deriva litorânea de sedimentos ao longo do litoral norte do RS. Pesquisas em Geociências 1:3–12
Tomazelli LJ, Dillenburg SR, Villwock JA (2000) Late Quaternary geological history of the Rio Grande do Sul coastal plain, Southern Brazil. Rev Bras Geosci 30(3):474–476
Tozzi HAM (1999) Influência das tempestades extratropicais sobre o estoque subaéreo das praias entre Rio Grande e Chuí, RS. Dissertação. Universidade Federal do Rio Grande do Sul, Porto Alegre, 115 pp
Tozzi HAM, Calliari LJ (2000) Morfodinâmica da praia do Cassino. Pesquisas em Geociencias 27:29–42
U.S. Army Corps of Engineers (1984) Shore protection manual, vol 1, 4th edn. U.S. Army Corps Engs, Washington, DC
Vaz AC, Möller OO Jr, Almeida TL (2006) Análise quantitativa da descarga dos rios afluentes da Lagoa dos Patos. Atlantica 28(1):13–23
Vieira H, Calliari LJ, Oliveira GP (2004) O estudo do impacto da circulação de veículos em praias arenosas através de parâmetros físicos: um estudo de caso. Engevista 6(3):54–63

Wainer IJ (1963) Análise e previsão das alturas de ondas em Tramandaí. Relatório Técnico, Universidade Federal do Rio Grande do Sul, Porto Alegre, 30 pp

Weschenfelder J, Ayup-Zouain RN, Zommer SLC, Souto RP (1997) Caracterização morfológica das praias oceânicas entre Imbé e Arroio do Sal, RS. Notas Técnicas 9:35–48

Wetzel L, Fillmann G, Niencheski LFH (2004) Litter contamination processes and management perspectives on the southern Brazilian coast. Int J Environ Pollut 21(2):153–165

Wright LD, Short AD (1984) Morphodynamic variability of surf zones and beaches: a synthesis. Mar Geol 56:93–118

Chapter 19
Sandy Beaches of Brazilian Oceanic Islands

Lauro Júlio Calliari, Pedro de Souza Pereira, Andrew D. Short, Fernando C. Sobral, Arthur A. Machado, Yuri G. Pinheiro, and Catherine Fitzpatrick

Abstract Brazil has four oceanic islands, three of which are discussed in this chapter: Rocas atoll, Fernando de Noronha and Trindade. Each of these islands is located on seafloor fracture zones extending east of the coast. Rocas and Fernando de Noronha lie 260 and 360 km respectively east of Natal. Rocas consists of a coral atoll, containing in its lagoon three sandy cays with dynamic beaches. The atoll is uninhabited and only visited by scientists. The Fernando de Noronha archipelago contains 21 islands the largest of which is inhabited and contains 15 sandy beaches, some considered the most beautiful in Brazil. It is a popular tourist and surfing destination. Trindade lies 1140 km east of Vitória and is a bare volcanic peak manned by a Naval base with a oceanographic research station and containing 12 sandy and cobble beaches, including the deadly "camel wave". The islands and their variety of beaches are presented in this chapter.

19.1 Introduction

The Brazilian coastline it is well known for its continental beaches, which are a major tourist attraction. On the other hand, there are several oceanic islands off the Brazilian coast that are not well known not only by the foreign tourists but also by Brazilians.

L.J. Calliari (✉) • F.C. Sobral • A.A. Machado • Y.G. Pinheiro • C. Fitzpatrick
Institute of Oceanography, Federal University of Rio Grande,
Campus Carreiros, Av. Italia km8 s/n, Caixa-Postal: 474,
CEP: 96201-900 Rio Grande, RS, Brazil
e-mail: lcalliari@log.furg.br; fcalmon.sobral@gmail.com;
oceaam@gmail.com; oceyuri@gmail.com; cath.e.fitz@gmail.com

P.d.S. Pereira
Geological Oceanography Laboratory, Center of Geoscience and Technology,
Oceanography Department, Federal University of Pernambuco,
Av. Arquitetura, s/n, Cidade Universitária, CEP: 50740-550 Recife, PE, Brazil
e-mail: pedro.pereira@pq.cnpq.br

A.D. Short
School of Geosciences, University of Sydney, Sydney, NSW 2006, Australia
e-mail: andrew.short@sydney.edu.au

A.D. Short, A.H. da F. Klein (eds.), *Brazilian Beach Systems*, Coastal Research Library 17, DOI 10.1007/978-3-319-30394-9_19

Fig. 19.1 Location of the Brazilian oceanic islands

The best known islands along the Brazilian coast are the coastal islands such as Arvoredo (SC), Ilha do Mel (PR), Ilha Bela (SP), Ilha Grande (RJ), Abrolhos (BA), Ilha de Itamaracá (PE), Ilha do Marajó (PA), as well as many others. However in terms of oceanic islands in the deep ocean, there are very few of them, just Archipelago of São Pedro and São Paulo, Rocas Atoll, Archipelago of Fernando de Noronha and Trindade and Martin Vaz (Fig. 19.1). All of these islands are volcanic islands with their origin related to hotspots along seafloor fracture zones.

Several beaches have been developed on these islands from sediment derived from island erosion and wave attack and also to biological production of calcium carbonate. Beaches on three oceanic islands are briefly reviewed in this chapter, together with the islands geology, climate, wave climate, sedimentology, beach behavior and hazards. It needs to be highlighted that due to the lack of beaches the Archipelago of São Pedro and São Paulo and Martin Vaz Island are not considered.

19.2 Geology

The Archipelago of Fernando de Noronha, Rocas Atoll and Trindade Island have a similar origin, with the islands emerging as part of the oriental portion of two oceanic volcanic chains orientated east-west (Almeida 1955, 2006).

The origin of the islands is related to the Atlantic mid-ocean ridge, which is crossed by transform faults. These faults are a result of plate spreading, and are called fractures zones, which are present along the entire mid-ocean ridge and may extend for hundred of kilometers. High relief can be found along these faults indicating regions of oceanic crust weakness. It is on those weakness points were the oceanic elevations, *guyots* and islands are formed, especially when they drift over hotspots or mantle plumes.

According to Almeida (2006) this process was responsible for the formation of the Brazilian oceanic islands, with the excetion of São Pedro and São Paulo, which emerged from an active transform fracture zone that bisects the rift-valley of the mid-Atlantic ridge.

Fernando de Noronha archipelago (Fig. 19.1) is located with Rocas Atoll (Fig. 19.1) along the Fernando de Noronha Chain at 4°S (Fig. 19.2). The Noronha archipelago is comprises 21 volcanic islands (area 26 km^2), located 360 km from Natal city, the closest continental city, while Rocas has an area of 6.56 km^2 and is located 267 km from the same city.

Fernando de Noronha Island is the largest island in the archipelago, with an area of 16.4 km^2 (Almeida 2006). This island is the top of a larger volcanic mountain that is approximately 75 km in diameter. The surrounding seabed around the island can reach almost 4000 m. The archipelago is composed of volcanic and subvolcanic rocks the result of two different volcanic episodes.

According to Ottmann (1963), Rocas Atoll evolved during the Holocene when a coral algae atoll was constructed on top of a reef platform (Kikuchi and Leão 1997). Initially, the island was a *guyot* as with many other topographic highs along the Fernando de Noronha chain. The atoll is surrounded by seabed up to 4000 m deep.

The island of Trindade (Fig. 19.1) is located at 21°S along the Vitória-Trindade Chain (Fig. 19.3), which is over 1000 km long. Trindade and Martin Vaz islands are the youngest ones along the east-west chain and have not been completely eroded by atmospheric and marine processes.

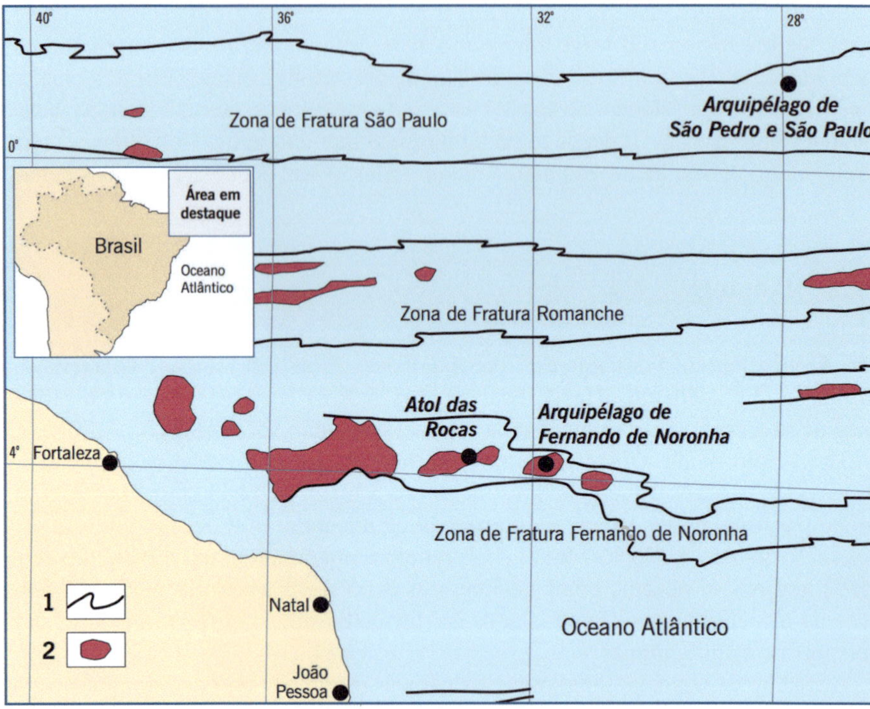

Fig. 19.2 Oceanic region where Fernando de Noronha archipelago and Rocas Atoll are located (Source: Almeida 2006)

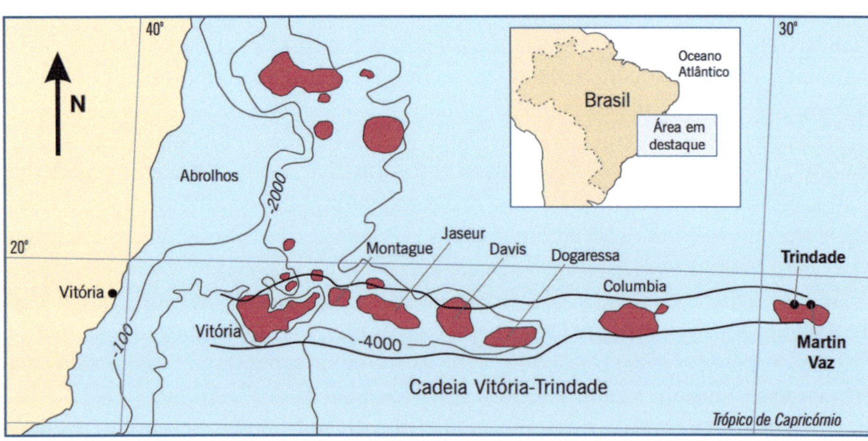

Fig. 19.3 Trindade island oceanic location

Trindade Island lies 1170 km east Vitoria (ES), the closest continental city. The island is approximately 9.28 km^2 in area with a diameter of 67 km and a surrounding depth of approximately 5000 m.

19.3 Climate

The three islands have tropical oceanic climates, with Frenando de Noronha and Rocas the more humid and Trindade more semi-arid. Fernando de Noronha has a humid tropical climate (Aw according to Köppen classification), with annual average temperature of 26.4 °C and range of just 25.7–27.1 °C, and annual precipitation of 1420 mm, with considerable interannual variability. There are two well-marked seasons, with the wet period from February to July and the drier period from August to January (Teixeira et al. 2003).

The prevailing winds in the equatorial south Atlantic are the southeast trade winds with average velocity of 6.6 m s^{-1} (Manso et al. 2011), with greater intensity from July to September (11 m s^{-1}). During the austral summer, winds from the southeast and east blow over 50 % and 35 % of the time, respectively (Hoflich 1984). The relative humidity is relatively constant, with an annual average of 81 %.

Rocas Atoll is located 145 km from Fernando de Noronha and has a similar climate. According to Kikuchi (1994) rainfall is distributed unevenly throughout the year, with a monthly average of 860 mm, ranging from 183 mm (April) to 2663 mm (August). Air temperature ranges from 17.5 °C (April) to 35.8 °C (February).

At Rocas the prevailing winds blow from east-southeast all year, with a frequency of 45 %. Between June and August (austral winter) southeast winds occur 35 % of the time. Between December and April (austral summer) southeast winds and east winds occur about 20 % of the time. Wind velocity ranges from 6 to 10 m s^{-1} throughout the year, but during the winter winds with a speed between 11 and 15 m s^{-1} are common. Speeds exceeding 20 m s^{-1} are recorded more often during the summer.

Trindade island has tropical semi-humid to semi-arid a climate, with a lower annual average rainfall (923 mm) The tropical climate is mitigated by its oceanic location and the easterly trade winds. The annual average temperature is 25.2 °C, with February the hottest month of the year (30.2 °C) and August the coldest (17.3 °C). The rainfall is both seasonal and variable (Clemente et al. 2009), with torrential rain called "Pirajá" occurring almost every day in summer. During winter (April to October) cold fronts periodically reach the island with intervals of 1 week (Alves 1998; Almeida 2002). The easterly trade winds (45 %) have an average velocity of 6.36 m s^{-1} in, with the maximum recorded wind velocity 19.62 m s^{-1} with a direction of 119.96° (east). The strongest winds come from extra-tropical cyclones that generate strong south and southeast winds.

19.4 Wave Climate and Tides

The wave climate of Fernando de Noronha and Rocas is dominated by southeast trade wind waves arriving most of the year, with higher energy northern hemisphere swells arriving during the austral summer.

On Rocas studies have shown that over 80 % of the waves come from the east quadrant and about 15 % of the northeast quadrant characterized by waves of short period from 4 to 7 s and between 1 and 2 m height. These values suggest that the waves are governed by the local climate, associated with the trade winds regime, together with waves from storms that occurs on the North Atlantic Ocean (Valentini and Rosman 1993).

Melo and Alves (1993) studying waves on the continent at the same latitude of Rocas and Fernando de Noronha noted that between December and March the wave climate behavior changes, with the occurrence of waves with 15 s and 18 s period and heights up to 2 m, coming from the northern hemisphere. According to Manso et al. (2011) during the summer the waves can reach up to 5 m on the northwest coast of Fernando de Noronha. The prevailing waves in the winter are from southeast, with an average height of 1.6 m (Hoflich 1984), and arrive along the southeast coast of Fernando de Noronha.

Based on 10 years of reanalysis wave data (2005–2014) from a grid point near the Trindade Island (20.5°S and 29.5°W) the WAVEWATCH III model simulations (NOAA) show that waves arrive from the south (33.7 %), southwest (23.4 %), east (18.1 %), north (10.3 %) and southeast (10.1 %), with the remaining (4.4 %) swell from the northwest, northeast and west (Fig. 19.4). The significant average wave height is 2.14 m, average period is 11.07 s and the maximum period is 21.46 s. The most frequent significant wave height for all directions is 1–3 m, representing 89.20 % of the total. Significant wave heights between 3 and 4 m are also common (9.45 %) for directions between east and southwest (75–245°). Significant wave height between 4 and 6 m are less frequent (1.1 %) and arrive mainly from south and are usually formed by extratropical cyclones generated in the southwest Atlantic.

Fernando de Noronha and Atol das Rocas have a micro-meso tidal semi-diurnal tide range. The tide forecast published by the Directory of Hydrography and

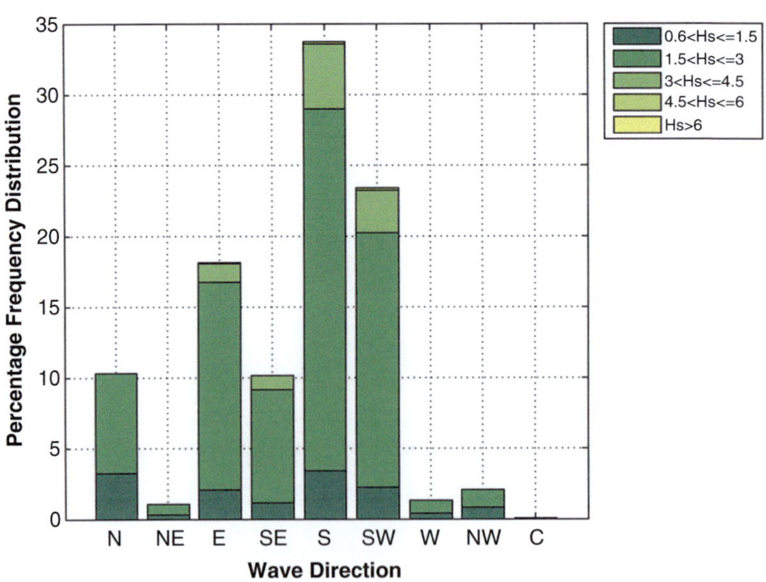

Fig. 19.4 Wave height and directions for Trindade island from NOAA WWII reanalysis data

Navigation from the Brazilian Navy (DHN) for Fernando de Noronha Archipelago has a spring tide range of 2.5 m and neap range of 1.3 m. The tidal amplitudes in Rocas Atoll are estimated from the navy forcast for Fernando de Noronha. Trindade island is semi-diurnal having a micro-tidal range of 1.3 m.

19.5 Rocas Atoll

19.5.1 Introduction

Rocas Atoll is the only atoll in the Southern Atlantic Ocean and is part of the Fernando de Noronha chain and located 145 km west of F. Noronha Archipelago (Almeida 2006). The uninhabited atoll was discovered in 1503 by the Portuguese sailor Gonçalo Coelho. It is an ellipsoid-shaped atoll with an internal area of about 7.5 km^2. Its largest axis (east-west) is 3.7 km long and the shortest (north-south) is 2.5 km long (Kikuchi 2002) (Fig. 19.5).

As with many other atolls, Rocas Atoll evolved as part of a complex interaction between physical, chemical and biological processes which resulted in the formation of the biogenic reef as well as the associated carbonate sand deposits atop a mid-ocean guyot (Lino et al. 2014).

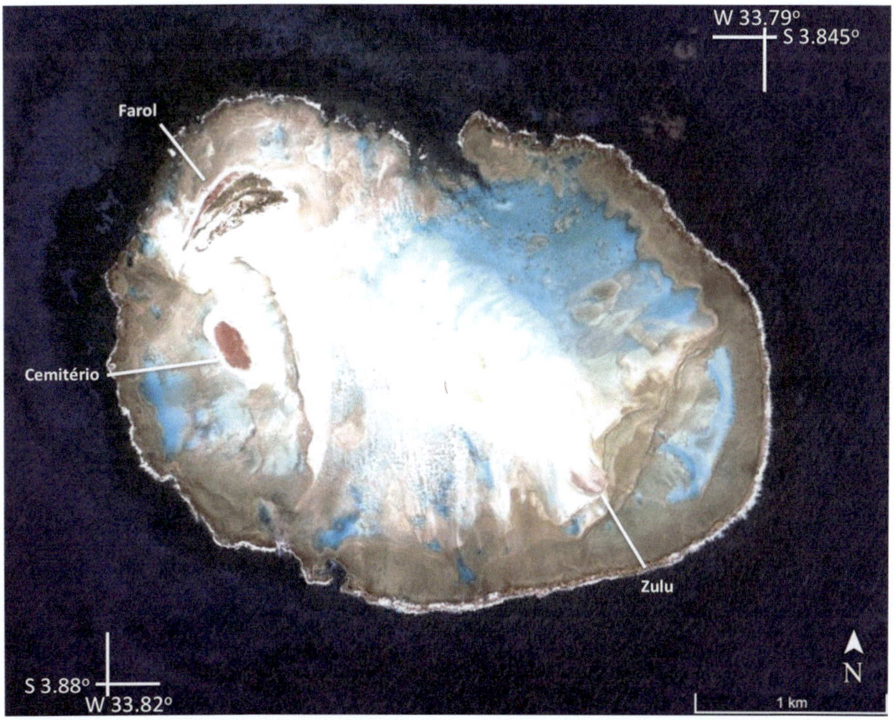

Fig. 19.5 Satellite image of Rocas Atoll with the name of the three main islets

Kikuchi (2002) drilled cores through the reef to a depth of 11.6 m without reaching the basement rock, on which the entire reef had grown, thereby suggesting the minimum carbonate depth is 11.7 m below the island surface.

As with Fernando de Noronha, Rocas is exposed to two wave climates, one with moderate waves generated by the southeasterly trade winds during the austral winter and higher waves during the boreal winter, coming from north and northwest directions. This pattern has controlled not only the reef development but also the sediment deposition pattern inside the lagoon.

19.5.2 Morphological and Sedimentological Aspects

Rocas's morphology has been studied by Andrade (1959), Ottmann (1963), Kikuchi (1994) and Pereira et al. (2010). According to Pereira et al. (2010) the reef complex is clearly distinguished in terms of its morphological and sedimentological features, with a reef front, an algal ridge, a reef flat, sediment deposits, a lagoon, small sand deposits and natural pools (Fig. 19.6).

Figure 19.6 indicates that most of the island area is the reef flat, followed by the lagoon and sand deposits, with most of the sediment composed of sand and gravel sized biogenic material (Pereira et al. 2013). The sand deposits cover the lagoon floor and also serves as sediment source for the formation of the three islets (Farol, Cemitério and Zulu), which have been formed by the internal lagoon circulation caused by the local winds and wave refraction.

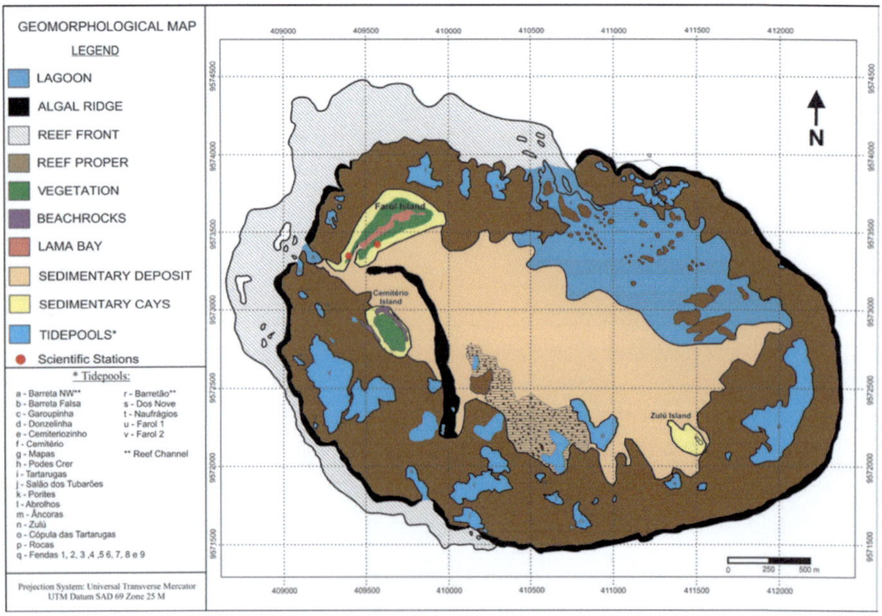

Fig. 19.6 Geomorphology of Rocas Atoll (Source: Pereira et al. 2013)

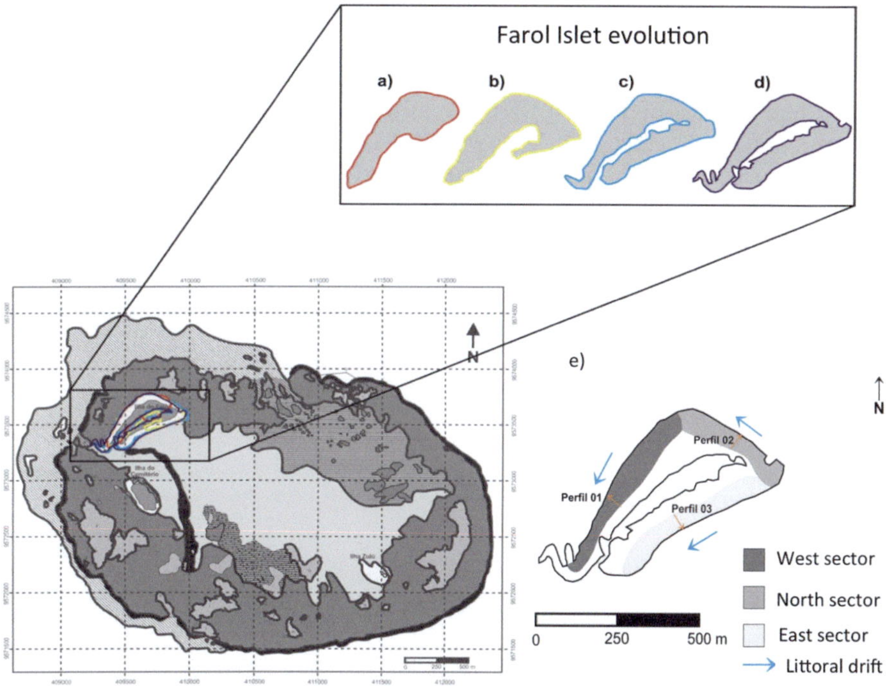

Fig. 19.7 Farol islet evolution: (**a**) 1959; (**b**) 1997; (**c**) 2010; (**d**) 2014; (**e**) shows the direction of sediment transport along the islet, illustrating its evolution and why it grows southwest (Adapted from Lino et al. 2014)

Fig. 19.8 View of the south of the Farol islet where the two spits shown in Fig. 19.3 can be seen separated by a channel (Image courtesy of Igor Pinheiro)

Farol islet is the largest and the most mobile. Its western side is accreting due to southerly transport of sediment. Figure 19.7 shows Farol's evolution from 1959 to 2012, with the islet growing and straightening. These changes appear to be caused by longshore drift generated by northerly wave refraction outside the reef and then, after crossing the algae ridge, inside the atoll. According to Costa et al. (2014) the Farol Islet has doubled it size between 1984 and 2014 eroding its northwest face (oceanward) and accreting its southwest face (lagoonward).

Beach profiles where carried out on Farol islet by Pereira et al. (2010) and Lino et al. (2014). The aim of the profiles was to record the evolution of the islet as shown in Fig. 19.7, with sediment normally been eroded from the northern part and deposited at the southern part of the islet (Fig. 19.8) along two spits. From the evolution shown in Figs. 19.7 and 19.8, it is expected that somewhere in the future the two spits will connect forming a small lagoon, which may be filled by sediments transported by the wind and wave overwash.

Sediment size inside the atoll varies from fine sand to gravel, with most of the sediment inside the lagoon been poorly to moderately sorted medium sand (Pereira et al. 2013). Pereira et al. (2013) also found the sediment has a high diversity of biogenic carbonate having at least eleven animal groups. The main particles found were coralline algae, bivalves, bryozoans, crustaceans, vertebrate bones as well as many other calcareous sediment types. While coralline algae is the main sediment component, one striking feature was the high diversity of foraminifera observed in the sediments. Lino et al. (2014) also examined sedimentological properties of the islet and found there it ranged from moderately sorted medium to coarse sand and was 100 % carbonate.

19.5.3 Beach System

Rocas atoll beaches are all associated with the islets and consist of biogenic sediment generated by the reef ecosystem and arranged by waves and currents. As they are located on the reef flats away from the exposed reef edge they are not exposed to high incoming waves, rather they receive lower ocean waves attenuated crossing the reef flats and shorter wind waves generated across the reef flat.

Since the beaches are not exposed to higher waves, they do not have any morphological striking features such as sandbars, beach cusps and rip currents. Figure 19.9 shows some of the Cemitério islet (Fig. 19.9a) and Farol islet beaches (Fig. 19.9b, c, d). Figure 19.9a, b, c show the sharp break in slope between the beach and underlying beachrock flats (Pereira et al. 2010), typical of this beach type (Short 2006). According to Lino (personal communication) the average incident wave height at Farol islet is under 0.5 m. This low wave height together with a spring tidal range of 2.5 m leads to a RTR equal or above 5, typical of tide-modified beaches.

19.5.4 Coastal Hazards

Rocas is a National Biological Reserve (Brasil 2007) and only managers, researchers and mariners can go there. Even though the number of people visiting the island is very limited, there are a large number of hazards in the island, all of them amplified by the remote location of the island and its absence of any facilities.

Among the hazards are the sharp reef topography that may cause several injuries; the deeper water inside the lagoon; the waves breaking heavily on the reef edge; and waves and currents moving a cross the reef flats and lagoon. This last one may be the greatest hazard on the island. During the boreal winter higher waves, between 3 and 5 m, coming from northwest, can break on the reef front making it unfeasible to land on or leave the island. Also, the higher waves drive stronger lagoon circulation, which results in hazardous conditions inside the lagoon.

Fig. 19.9 Images of Ilha do Cemitério (**a**) and Ilha do Farol (**b**, **c**, **d**). Note the presence of beach rock on the terrace in (**a**) and (**b**) (Images **a**, **c** and **d** are courtesy from Anderson P. Lino and **b** from Mirella B. Costa)

The atoll and the shallow reefs has also been a graveyard for many ships over the centuries. One of the very first shipwrecks was one of the ships on the Gonçalo Coelho expedition. Due to this fact, many people have died over the centuries, as a result, some of then where buried inside of the reef lagoon, on the Cemitério Islet, Cemitério being a Portuguese word for cemetery.

19.6 Fernando de Noronha

19.6.1 Introduction

Fernando de Noronha is an archipelago of small islands located in the South Atlantic 350 km north-northeast of Cape Calcanhar, the northeast tip of Brazil and 530 km northeast of Recife. It lies just south of the equator at 3°50′ S. The main island is 10 km long, averaging 2 km in width, with the widest point 3.2 km. It has an area of 18.4 km^{2}and comprises 91 % of the archipelago. Four smaller uninhabited islands extend another 3.5 km immediately east of the island, which together with another

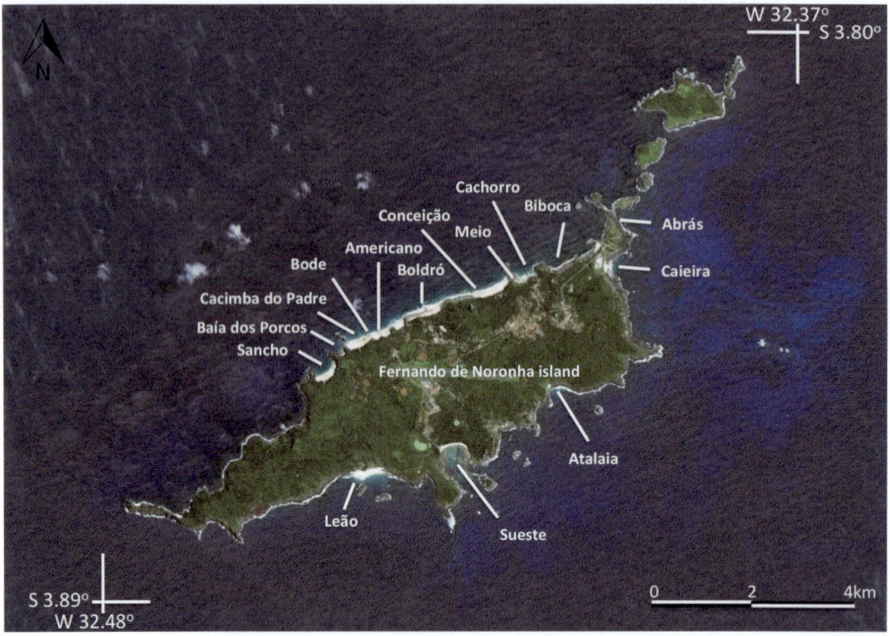

Fig. 19.10 Satellite image of Fernando de Noronha with location of the main beaches (Source: Google Earth)

20 small islets scattered off the north and south coasts form the archipelago (Fig. 19.10). The island is part of Pernambuco state and serviced from Recife. It has a population of 2930.

The island is a World Heritage Site and the Fernando de Noronha Marine National Park covers 70 % of the island including the western third of the island, the south coast and all the surrounding smaller islands. In addition the occupied central core of the island is an Environmental Protection Area.

The island is the peak of a volcanic mountain range the base of which lies 750 m below sea level (Fig. 19.11). The volcanic peaks rise to a maximum of 321 m at the towering Pico Mountain, with the island forming an undulating plateau with an average elevation of about 45 m.

19.6.2 The Coast and Beach Systems

The island has a generally rugged rocky coast of volcanic origin, with basalt cliffs, bluffs and boulders dominating much of the shore. The main island has a shoreline length of 37 km, of which 28.5 km (77 %) is rocky, with 32 beaches occupying just 8.4 km (23 %). All are bordered by rocky headlands and points and average only 260 m in length. Seventeen of the beaches are composed of boulders, with only 15 sandy beaches (Table 19.1). The boulder beaches are predominately of volcanic

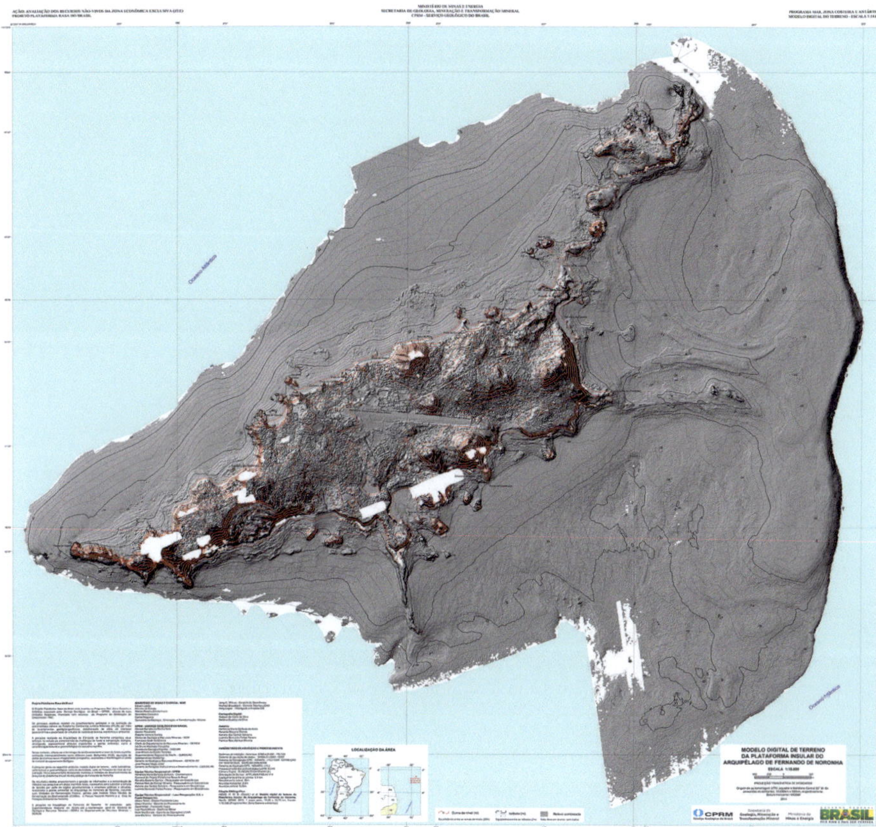

Fig. 19.11 Relief map of Fernando de Noronha showing the submarine volcanic platform and surrounding islands that make up the achipelago (Source: CPRM)

Table 19.1 Characteristics of Fernando de Noronha coast

	Rocky (km)	Sand beaches (number)	Sand beaches (km)	Boulder beaches (number)	Boulder beaches (km)
South coast	18.7	5	1.5	9	1.8
North coast	9.8	10	3.4	8	1.7
Total	28.5	15	4.9	17	3.5

origin, and largely composed of rounded basalt boulders. The sandy beaches are composed of calcareous sand (46–97 % according to Barcellos et al. 2011), which ranges in size from 0.1 to 0.5 mm (from fine to medium sand; Barcellos et al. 2011 and Manso et al. 2011). The carbonate sand is derived from the surrounding coral reefs, with the remainder from erodion of the volcanic rocks.

All the boulder beaches are steep and reflective, with some fronted by intertidal rock flats (Fig. 19.12a). They are predominately well exposed to waves and are extremely hazardous as a result, with some of the more embayed also fronted by

Fig. 19.12 Fernando de Noronha north coast beaches: (**a**) Biboca boulder beach; (**b**) the pocket beach of Baía dos Porcos is dominated by the towering basalt cliff, enclosed by basalt reefs and backed by basalt boulders; (**c**) the popular surfing beach at Cacimba do Padre; (**d**) the near continuous stretch of north coast reflective sandy beaches between Bode (foreground), Americano and Cacimba do Padre; (**e**) LTT at Conceicão, with Pico Mountain behind; and (**f**) the attached breakwater at St Antônio Harbor (Photos: AD Short)

strong topographic rips. Most of the sandy beaches are also bordered and underlain and/or backed by boulder beaches, which become exposed when the sand is eroded (Fig. 19.12b).

The north coast sandy beaches are more numerous and extensive, forming a near continuous 4 km long band of seven beaches between Cacimba do Padre and Cachorro (Fig. 19.12c, d). These beaches are predominately reflective to low a

b

Fig. 19.13 Boldró beach during (**a**) boreal summer conditions (September 2005) when lower waves dominate and the beach is reflective, wide and free of rips; and (**b**) at the end of the boreal winter conditions (April 2011), when high waves, a narrow eroded beach, with exposed beachrock and a strong central topographic rip (*red arrow*) prevail (Source: Google Earth)

tide terrace. Reflective conditions prevail on the slightly coarser sand beaches (Sancho, Porcos, Padre, Americano, Bode) and/or during the calmer boreal summer months. During the more energetic boreal winter the higher waves maintain LTT and even TBR conditions, particularly on the beaches with slightly finer sand (Boldró, Conceicao, Meio, Cachorro; Fig. 19.12e). The waves tend to break heavily on the LTT and at Padre produces a famous plunging surf break, called Cacimba do Padre, where surf contests are held (Fig. 19.12c). During periods of higher waves beach and topographic rips operate on many of the beaches, resulting in hazardous swimming conditions (Fig. 19.13). Good surf, particularly during the boreal winter months, is also found along the north coast at Cacimba do Padre, Boldró, Conceição,

Table 19.2 Volumetric data calculated to obtain the mobility coefficient in some beaches of Fernando de Noronha

Beach	Conceição			Boldró		Cacimba do Padre			Sancho
Beach profile	FNC1	FNC2	FNC3	FNB1	FNB2	FNCP1	FNCP2	FNCP3	FNS1
N	4	5	4	5	4	4	4	5	5
Vv (m^3/m)	98.2	124.7	131	116.3	104.9	327.1	107.4	178.0	124.0
σVv (m^3/m)	47.8	62.0	37.4	28.2	39.2	129.5	63.2	34.5	18.0
CVv	48.7	49.8	28.6	24.2	37.4	39.6	58.9	21.4	14.5

Source: Manso et al. (2011)

(N) Number of samples, (Vv) mean volume of the beach; (σVv) standard deviation from the mean volume; (CVv) coefficient of variation of the volume of the beach

Meio, Cachorro, Biboca and Abrás beaches. The harbor of St Antonio has a 350 m long attached breakwater that has trapped sand on its western side forming a curving 200 m long sheltered sandy beach, the easternmost on the north coast. Numerous boats anchor in the bay seaward of the harbor, with others docking in the harbor and smaller craft hauled up onto the berm (Fig. 19.12f).

Manso et al. (2011) measured beach sand volumes on four north coast beaches (Table 19.2) and found they ranged from a low of 98 $m^3 m^{-1}$, to a high of 327 $m^3 m^{-1}$, with a mobility of between 14 and 58 $m^3 m^{-1}$. The mobility is a product of the highly seasonal north coast wave climate, with higher waves and erosion dominating during the boreal winter, and calmer conditions and beach recovery during the boreal summer (Fig. 19.13). As the sign at Cachorro beach says: "*On times when Inner Sea is being submitted to rough waves, the sand is taken away from the beach letting volcanic stones appear. During still sea the sand returns to the beach through marine flow.*"

The south coast has 14 beaches only five of which are sandy: Caeira, Atalia, Baía do Sueste (2) and Praia do Leão. Caeira is a very accessible but exposed and hazardous beach fronted by a reef flats with a strong topographic rip dominating the centre of the surf zone (Fig. 19.14a). It has a sign that reads: *This beach does not have life-guards on duty and is sometime subject to strong marine currents and dangerous waves.* Atalia is a restricted national park beach. It has a small patch of sand fronted by coral reefs and rock pools, with waves breaking heavily on the outer reef. Baía do Sueste is deeply embayed and only receives low to zero waves, which spill aross a shallow LTT. Praia do Leão is fully exposed to southerly waves, which wrap around Viũva Hill to form a cuspate foreland. The waves break heavily on the LTT, with a strong rip currents against the eastern and western rocks (Fig. 19.14b). The eastern end also has a boulder beach fronted by a raised algal platform, which contains small potholes.

Fig. 19.14 Fernando de Noronha south coast beaches (**a**) Caeira is one of the few south coast sandy beaches. The embayed beach is backed by high tide boulders, with rock flats to either side and strong permanent rip flowing out the center (*red arrow*); and (**b**) the exposed Leão beach, with a strong rip against the eastern rocks (*red arrow*) and a well developed raised algal platform and boulders in foreground (Photos: AD Short)

19.6.3 Beach Hazards

The island beaches are predominately small, embayed and well exposed to their respective wave climates. Most are readily accessible, though those in the National Park require a permit to visit, with bookings required for Atalaia. On the popular north coast during the austral winter low waves and idyllic swimming conditions can prevail, with clear clean water and calm seas. However during the austral summer period higher waves from the northern hemisphere produce heavy plunging waves on most of the beaches, form beach rips on some of the longer sandy beaches and strong topographic rips on all the embayed beaches. As Fig. 19.13 illustrates how the beaches can switch from calm and reflective in winter (boreal summer) to hazardous with larger waves and rip currents in summer (boreal winter). The south coast beaches receive higher waves during the austral winter and can have strong rips, with Caeira (Fig. 19.14a) and Praia do Leão (Fig. 19.14b) being the most hazardous beach, while Sueste remains sheltered year round. In addition rocks and reefs are common on most of the beaches. The combination of higher waves, rips and rocks can result in very hazardous bathing conditions, which are exacerbated by the absence of lifeguards and first aid. In addition as most of the bathers are visiting tourists with little knowledge of the island's beaches, which increases the level of risk. As a result many of the beaches have signs warning of the hazards and lack of lifeguards.

19.7 Trindade Island

19.7.1 Introduction

Trindade Island is located in the South Atlantic Ocean 1140 km east of Vitória (ES) at 20°30′S. It is part of the Vitória-Trindade Chain an underwater volcanic chain oriented east-west which starts on the continental slope 175 km from the coast (Fig. 19.1). This chain forms a discontinuous linear series of guyots and submarines mountains (Almeida 2006). The most significant guyots are represented by the Victory, Montague, Jaseur, Davis, Dogaressa and Columbia banks (Figs. 19.3 and 19.15). The Victoria and Davis banks are at depths of 55–70 m and their top are respectively 48 km and 46 km in diameter. Trindade and Martin Vaz islands are the easternmost part of this chain (Gorini 1969). Trindade island is 80 Ma in age and is the emerged top of a large elongated volcanic mountain with a diameter of 67 km at its base at a depth of 5500 m.

The emerged part of the island together with its abrasion platform has an area of 13.5 km^2 with an extremely rugged relief as a result of weathering processes on a highly heterogeneous rock massive. Five volcanic episodes gave rise to five geological formations whose age from the oldest to the newest range from 3.6 Ma to less than 200 ka. The rocks are sodium-alkaline rich and silica undersatured lavas and intrusions, in addition to several pyroclastic rocks. Algal reefs, narrow beaches, localized dunes, limited fluvial deposits at the coastline, several cones and slope aprons compose the coast on this island which is predominately composed of volcanic and subvolcanic rocks. It is the only place in the Brazilian territory where a part of a volcanic cone is still recognizable (Almeida 2002).

The island is aligned northwest-southeast and is 5.9 km long and 2.72 km wide (Fig. 19.16). Its central peaks rise 550 and 600 m above sea level and consist of extrusions of phonolitic rocks (volcanic rocks sub-saturated in silica). The eastern region of the island is lower as a result of volcanic flows, which have formed more gentle slopes, together with beaches, which increase in width towards the southeast. In 1957 during the International Geophysical year, the Brazilian Navy, which has provided support for research and conservation activities in the island since 1916,

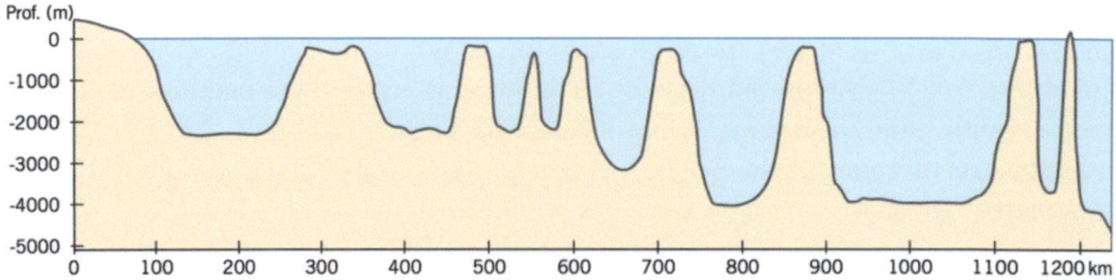

Fig. 19.15 Cross-section of underwater morphology of Vitória-Trindade volcanic chain (Source: Almeida 2006)

Fig. 19.16 Trindade Island with location of the better developed beaches and the places where the camel wave occurs (Source: Google Earth)

established the Oceanographic Station of the Trindade Island (POIT) (Alves 1998). Since then, the Navy has given logistical support for oceanographic research and also ensures the national sovereignty by maintaining military installations.

19.7.2 The Coast and Beach Systems

The island's predominantly rocky coast, headlands and protrusions of volcanic rocks have formed several embayment, which on the north, northwest, southwest and west coast are defined by several sharp and elongated headlands composed of phonolitic bodies and dykes. The geological framework has resulted in a morphological differentiation between east and west side of the island the former being straighter with better developed beaches. The total length of the coastline is 21.7 km of which 16.7 km (76.7 %) is rocky and 5.1 km (24 %) consists of 16 sand and boulder beaches. The predominantly sandy beaches occupy 1.2 km (24 %), sand and pebbles comprise 1.94 km (38 %) and pebbles and boulders occupy 1.9 km (38 %) (Table 19.3). The sandy beaches predominate on the east coast while mixed sand, pebbles and boulders beaches predominate on the western side of the island. On the

Table 19.3 Beach characteristics of the Trindade east and west coasts

	Rocky (km)	Sand beaches (number)	Sand beaches (km)	Sand and boulder beaches (number)	Sand and boulder beaches (km)	Boulder beaches (number)	Boulder beaches (km)
East coast	5.9	6	1.20	1	0.53	2	0.82
West coast	11.8	0	0	4	1.41	3	1.10
Total	16.7	6	1.20	5	1.94	5	1.92

southeast tip of the island is an abrasion platform containing two of the island most significant features: Ponta do Túnel (where a tunnel is located) and Paredão which is a rocky formation composed by volcanic tuffs.

The beach pebbles and boulders are predominantly of volcanic origin and mainly consist of various pyroclastic material with size ranging between 8 and 60 cm (Castro and Antonello 2006). On some beaches the material is delivered by debris slopes and as fluvial deposits delivered down small gullies and streams originating in the higher parts of the island.

The east and southwest sector of the island, where the better developed beaches are located, receives higher wave energy as waves coming from deep ocean undergo little energy dissipation over the narrow island platform which is 100 m deep just 1 km offshore. The differential erosion resulting from the variable island lithologies favors beach compartmentalization, as well as controlling the compartments grain size and mineralogy, and degree of wave exposure. Rock formations and coralline algae reefs in the surf zone represent additional topographic controls and source of sediments that influence the energy and hydrodynamic circulation. The beach sediments reflect the mineralogy of the adjacent rock formation, which are composed of tephras of phonolite and volcanic tuffs with high percentages of heavy minerals such as magnetite, hematite concretions as well as mineralogical assemblages of acid rocks due to different eruptive phases (Almeida 1961; Castro and Antonello 2006; Clemente et al. 2006).

The six beaches with greatest volume of sediments are from northeast to southwest: Cabritas, Calheta, Andrada, Tartarugas, Túnel (Vermelha) and Príncipe (Fig. 19.16). These are mainly sandy beaches composed of well-sorted medium to coarse sand, with calcareous algae fragments from the reef flat making up from 19 to 66 % of the sediments (Fig. 19.17). As a general pattern calcareous content increases toward the foreshore. The beach with the least amount of carbonate is Túnel beach (also known as Vermelha (Red beach)) where sediment from pyroclastic and ankaratrite (alkaline volcanic rocks) fragments (which gives a reddish color to this beach) predominate. High heavy minerals occur mainly on the beach face, with percentage ranging from 30 % (Calheta beach) and 63 % (Príncipe beach). On Tartarugas beach layers of 10 cm thick heavy mineral concentration (around 90 %) occur showing the hydraulic selection process on the beach face.

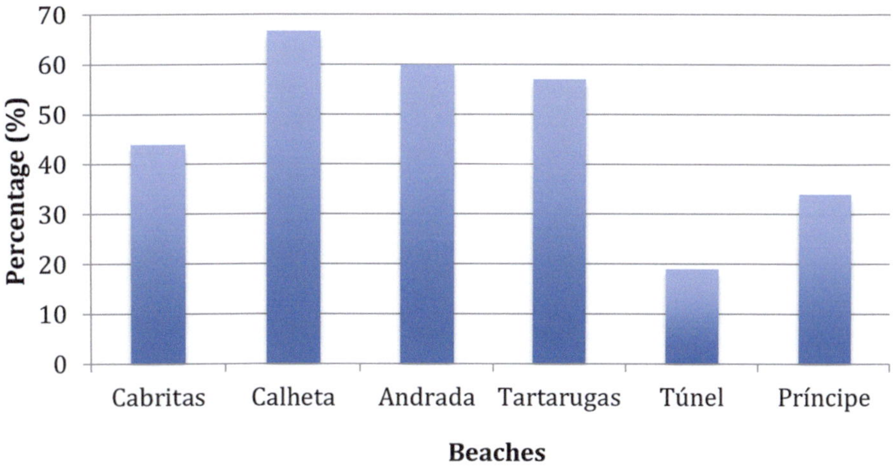

Fig. 19.17 Average percentage of calcium carbonate per beach

Table 19.4 Values of the embayment scaling parameter (δ) after Short and Masselink (1999) of the most developed beaches of Trindade Island. Si shoreline length, Cl width between headlands, H_b breaker height

Beach	Si (m)	Ci (m)	Hb (m)	$\delta = Si^2/100$ Ci Hb	Reef flat width (m)
Túnel or Vermelha	452.43	229.54	2.19	4.06	
Tartarugas	761.61	561.48	2.19	4.70	120
Andrada	409.5	249.11	2.19	3.06	117
Calheta	212	104.49	2.19	1.95	98
Cabritas	781.4	654.56	2.19	4.24	84
Príncipe	123.68	911.53	2.19	7.62	

Despite the presence of beachrock on the beach face, and reef flats of calcareous algae whose widths vary between 84 and 120 m, most beaches display cellular surf zone circulation. This is especially true for Túnel and Príncipe beaches where coralline-algae reefs are absent. On the Tartarugas and Cabritas beaches the reef flat is continuous, while discontinuous on Andrada and Calheta beaches. Table 19.4 shows that the non-dimensional embayment scaling parameter (δ) (Short and Masselink 1999), are between 2 and 7, which indicates cellular circulation, meaning that the circulation is controlled by the boundary topography.

Figure 19.18 illustrates the reflective-low tide terrace (R-LTT) profiles of Cabritas, Andrada, Tartarugas and Túnel beaches. Apart from Túnel Beach all are modified by coral calcareous algae and beach sandstones (beachrock) and intertidal reef flats. Cabritas (Fig. 19.18a) and Tartarugas beaches (Fig. 19.18d) do not show a significant exchange of sand between the beach face and the surf zone. These two beaches have scattered sand deposits that may be absent from the lower foreshore where beachrock and calcareous algae pavements are found. The beach profile

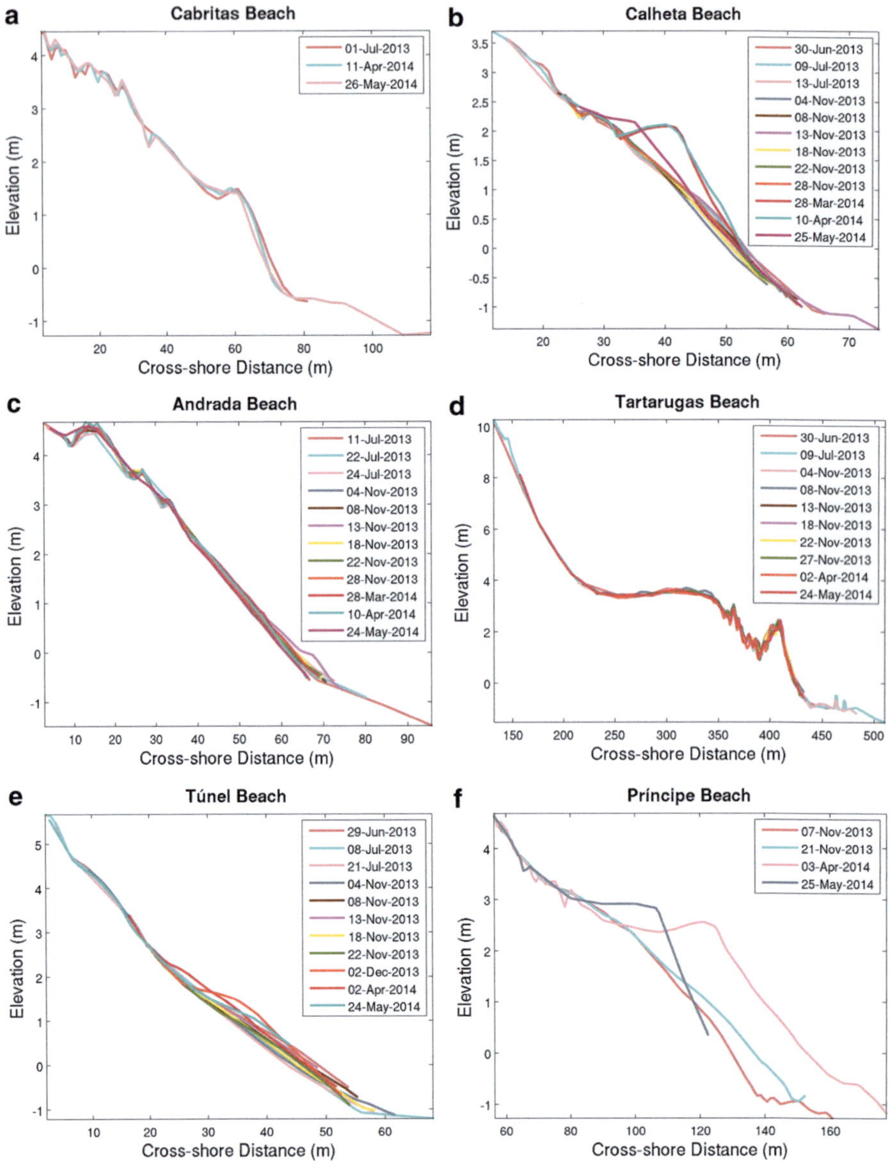

Fig. 19.18 Beach profile envelopes for the six more developed beaches at the Trindade Island: (**a**) Cabritas, (**b**) Calheta, (**c**) Andrada, (**d**) Tartarugas, (**e**) Túnel and (**f**) Príncipe

envelopes indicate that the reefs flats play an important role in protecting and maintaining the sediment on Cabritas and Tartarugas beaches. Data provided by researchers on the island indicate that seaward of the reef flat the depth suddenly increases making it virtually impossible for sand removed during storms to return. On the

Fig. 19.19 (**a**) General view of Cabritas beach; (**b**) Cabritas beach with pebbles covering the sand at lower foreshore and connected to the reef flat exposed during low tide

Fig. 19.20 Tartarugas beach (**a**) general view; (**b**) beachrock slabs and fragments; and (**c**) turtle excavation on the backshore during the nesting season

Cabritas beach the reflective lower foreshore consists of pebbles that cover the sand, which connects to the reef flat (Fig. 19.19a, b).

Tartarugas beach is the widest on the island (350 m) (Fig. 19.20a) and is the site of the second largest breeding area for the giant green turtle and one of the largest in the world (Clemente et al. 2011). The lower foreshore consists of sand, rock fragments and heavy minerals partially covered by slabs of beachrock that emerge at

Fig. 19.21 Calheta beach: (a) general view and; (b) sandy sector used for landing small boats

Fig. 19.22 Príncipe beach: (a) general view and; (b) southwest sector of the beach which is exposed to swell

low tide (Fig. 19.20b). At low tide waves break on the reef flats and maintain intermediate characteristics. Several beach profiles indicate that the most significant morphological changes on the wide backshore are produced by the turtles digging their nests (Fig. 19.20c). The southeast wind have also developed climbing dunes composed of calcareous sand (Fig. 19.20a) (Castro and Antonello 2006; Castro 2010).

The intermediate Calheta (Figs. 19.18b and 19.21a, b) and reflective Príncipe beaches (Figs. 19.18f and 19.22a, b) are located on opposite sides of the island. They tend to be eroded with low sediment volume between June and November, with accretion and higher volume between March and April. The subaerial beach profiles vary by up to 1–2 m and represent the exchange of sand between the beach and surf zone similar to seasonal behavior that occurs on many Brazilian mainland beaches. Unlike other beaches which are more sheltered both by rocky headlands and algae reefs, Príncipe beach does not have a reef and is exposed to the storm

Fig. 19.23 (**a**) Andrada beach bounded by a rocky phonolitic headland in the south and by surf zone, with outcrops of alkaline lava at the north; and (**b**) Pebbles of volcanic tuffs and phonolitic rock fragments in the south

Fig. 19.24 (**a**) Túnel or Vermelha beach with the túnel and Paredão in the extreme south and (**b**) detail of the volcanic tuff aprons reaching the backshore

waves from the south-southwest generated by extra-tropical cyclones in the southwest Atlantic. Calheta beach, despite being partially protected by algal reefs, has an opening where there is a sandy beach currently used in landing operations with inflatable boats (Fig. 19.21b).

The distinct seasonality of Calheta and Príncipe beaches are also subtly expressed at Andrada (Figs. 19.18c and 19.23a, b) and Túnel beaches (Figs. 19.18e and 19.24a, b). The stability of these beaches is due to their protection by surf zone outcrops of ankaratrite, submerged reef flats and the Túnel point headland. At Túnel beach volcanic tuff debris slopes deliver material directly to the backshore. Well developed beach cusps and a strong backswash are characteristics of this beach. On the Andrada beach debris of phonolitic rocks are spread at the southern end (Fig. 19.23b).

19.7.3 Beach and Coastal Hazards

All six sandy beaches have permanent hazards particularly rocks and submerged reef flats and consequently a high level of beach risk. The least hazardous beaches are Calheta, Andrada, Túnel and Príncipe, however as they are all reflective they have plunging breakers during high tide, with strong backwash and deep water off the base of the beach. Rip currents can also occur at the northern end of Andrada and Túnel beach and at the southern end of the Príncipe, with boundary rips located along the rocky headlands.

A hazard that caused seven deaths since 1963 is associated with a recurring event called 'Camel Wave'. This wave has a profile similar to the hump of a camel. According to report and videos made by sailors and other military persons, people who are fishing at certain places are surprised and swept away by the suddenly appearance of this wave. Our interpretation of videos, reports and data associated with the events is that waves with significant height around 2 m experience accentuated shoaling when propagating over the narrow insular rock platform. In certain situations the largest wave of the series is able to cross the platform without breaking until it reaches a steep slope where is produces a strong surge which causes an abnormal rise of the water level spreading over the rocks and carrying people away. Alternatively, videos of wave breaking across the platform at the southeast end of the island show the interaction of incident and reflected waves crests which reach greater heights as they break and dissipate their energy across the platform sweeping everything is on the surface. Data associated with events that caused deaths and a 'camel wave' experienced by a researcher from the Sea Turtle project (TAMAR) are listed in Table 19.5. The locations of the accidents are indicated in Fig. 19.16, with the highest occurrence (4 causalities) associated with the place called Pedra da Garoupa (Stone grouper) north of the Príncipe beach. Figures 19.25a, b show the situation experienced by TAMAR researcher at the EME beach before and after being hit by a Camel wave.

Table 19.5 Wave parameter for some of the days when the Camel wave occurred on the Trindade Island

Deaths (causalities) and accident at the Trindade Island due to Camel Wave.					
Local	Date	Hs (m)	Tp (s)	Dp	Observation
Ponta Norte	22/10/2005	1.96 m	12.84	349.85°	1 death
Praia do M	15/08/2007	3.49 m	13,15	197.75	Fig. 19.20a, b event
Pedra do Xaréu	03/06/2008	3.03 m	14.89	200.51°	1 death
Pedra da Garoupa	17/07/2010	3.35 m	12.33	210.09°	1 death

a b

Fig. 19.25 A TAMAR researcher fishing at EME beach before (**a**) and during (**b**) a Camel wave

19.8 Summary

The three Brazilian oceanic islands with sand beaches are presented in this chapter, each with very difference morphological and beach characteristics. Fernando de Noronha and Rocas lie just south of the equator east of Natal, while Trindade lies at 20°S 1140 km east of Vitoria. Fernando de Noronha and Rocas have a more humid tropical climate and distinctly bi-directional north and south wave climate, while Trindade is more arid with a predominance of southerly waves. Only Fernando is open to the public and is a popular tourist island, with special permission and some difficulty required to reach Rocas and Trindade with both however open for researchers.

Rocas is a low atoll, surrounded by fringing reef, with three small beaches located on sand cays formed on the reef flat, and only exposed to reformed and lagoonal wind waves. The beaches are tide-modified owing to the very low waves, 100 % carbonate and highly dynamic in planform.

Fernando de Noronha has a predominately rocky basalt shore with high cliffs, in amongst which are 32 beaches, just 15 of which are sandy, the sand derived predominately from adjacent coral reefs, as well as the bedrock. Most of the sand beaches are located along the popular north coast and range from R to LTT both spatially, and seasonally with the bi-modal wave climate. The south coast has just five sandy beaches, which range from very sheltered to exposed.

Trindade is a remote volcanic island with a more arid climate and greater exposure to larger southerly waves. The predominately rocky island has six small embayed sandy beaches and three boulder on the northeast coast, with four mixed sand-boulder beaches and three boulder beaches on the southwest coast. Most of these beaches are also fronted by beachrock and/or coral algal reef flats and all are hazardous owing to plunging waves, rips and rocks.

Acknowledgments To the Brazilian Navy by the 1st Naval District and the Secretariat of the Interministerial Commission for Sea Resources (SECIRM) which respectively provide logistical and operational support and coordination of the Programme for Research in the Trindade Island (PROTRINDADE). Also, to CNPq that through the process n[0] 404 799/2012-4 of the notice 39/2012 Program Archipelago and Oceanic Islands provided funding for the development of research.

References

Almeida FFM (1955) Geologia e Petrologia do Arquipélago de Fernando de Noronha. DNPM/DGM – Departamento Nacional de Produção Mineral – Diretoria de Geologia e Recursos Minerais, Rio de Janeiro

Almeida FFM (1961) Geologia e Petrologia da Ilha da Trindade. DGM/DNPM, Rio de Janeiro

Almeida FFM (2002) A ilha da Trindade. In: Schobbe nhaus C, Campos DA Queiroz ET, Winge M, Berbert-Born M (eds) Sítios geológicos e palentológicos do Brasil. CPRM, Brasília

Almeida FFM (2006) Ilhas oceânicas brasileiras e suas relações com a tectônica atlântica. Terræ Didatica 2(1):3–18

Alves RJV (1998) Ilha da Trindade & Arquipélago Martin Vaz:um ensaio geobotânico. Serviço de Documentação da Marinha, Rio de Janeiro

Andrade GO (1959) O recife anular das Rocas (Um registro das recentes variações eustáticas no Atlântico equatorial). Anais da Associação dos Geógrafos Brasileiros XI:29–61

Barcellos RL, Coelho-Júnior C, Lins SRRM, Silva MS, Camargo PB, Travassos PEPF (2011) Variações sedimentológicas e morfológicas de curta prazo em praias insulares: o caso do SE da ilha de Fernando Noronha, Atlântico Sul, Brasil. Journal of Integrated Coastal Zone Management 11(4):471–478

Brasil (2007) Plano de Manejo para a Reserva Biológica do Atol das Rocas. Brasília., Ministério do Meio Ambiente. Instituto Chico Mendes de Conservação da Biodiversidade. Brasília

Castro JWA (2010) Ilhas oceânicas da Trindade e Fernando de Noronha, Brasil: uma visão da geologia ambiental. Revista de Gestão Costeira Integrada 10(3):303–319

Castro JWA, Antonello LL (2006) Geologia das ilhas oceânicas brasileiras. In: Ilhas oceânicas Brasileiras-da pesquisa ao Manejo. MMA:SBF, Brasilia, pp 29–57

Clemente EP, Schaefer CE, Albuquerque Filho MR, Oliveira FS, Alves RJ, Melo VF (2006) Solos endêmicos da Ilha da Trindade. In: Ilhas oceânicas Brasileiras-da pesquisa ao Manejo. MMA:SBF, Brasilia, pp 59–82

Clemente EP, Schaefer CE, Oliveira FS, Albuquerque Filho MR, Alves RJ, Firme Sá MM, Melo VF, Corrêa GR (2009) Toposequência de solos na Ilha da Trindade. Atlântico Sul Rev Bras Ciên Solo 33(5):1357–1371

Clemente EP, Schaefer CER, Oliveira FS (2011) Proposta de Zoneamento Ambiental para a Ilha da Trindade (ES). Boletim de Pesquisa e Desenvolvimento 171. EMBRAPA

Costa MBSF, Macedo EC, Siegle E (2014) Shoreline changes analysis at Rocas Atoll as a tool to improve spatial planning strategies. 2014 Ocean Sciences Meeting, Honolulu, Hawaii

CPRM – Companhia de Pesquisa de Recursos Minerais. Accesed on 10 Oct 2015 – URL: http://geobank.sa.cprm.gov.br

Gorini MA (1969) Geologic observations on the "Comissão Oceanográfica Leste I" aboard the Research Vessel "Almirante Saldanha". An.Acad.brasil.Ciênc. Rio de Janeiro. 41: 642R–643R

Hoflich O (1984) Climate of the South Atlantic ocean. In: Van Loon H (ed) Climates of the oceans. Elsevier Science, Amsterdam, pp 1–192

Kikuchi RKP (1994) Geomorfologia, Estratigrafia e Sedimentologia do Rocas Atoll (Rebio-IBAMA/RN). Pós-Graduação em Geologia. Salvador, Universidade Federal da Bahia:144

Kikuchi RKP (2002) Rocas Atoll, Litoral do Nordeste do Brasil – Único atol do Atlântico Sul Equatorial Ocidental. In: Schobbenhaus C, Campos DA, Queiroz ET, Winge M, Berbert-Born MLC (eds) Sítios Geológicos e Paleontológicos do Brasil, vol 01, 1st edn. DNPM/CPRM – Comissão Brasileira de Sítios Geológicos e Paleobiológicos (SIGEP), Brasilia, pp 379–390

Kikuchi RKP, Leão ZMAN (1997) Rocas (Southwestern Equatorial Atlantic, Brazil): an atoll built primarily by coralline algae. In: Proceedings of the 8th international coral reef symposium, Panama, Allen Press

Lino AP, Silva AMC, Pereira NS, Ferreira Junior CV, Souza JPG (2014) Morfodinâmica da Ilha do Farol, Rocas Atoll – Atlântico Sul. Estudos Geológicos (UFPE) 24:65–81

Manso VA, Menor EA, Valença LMM, Neumann VHML, Pereira NS, Junior CFAS, Silva ERM (2011) Morfodinâmica de praias setentrionais da Ilha de Fernando de Noronha. Revista da Gestão Costeira Integrada 11(3):327–339

Melo EF, Alves JHGM (1993) Nota sobre a chegada de ondulações longínquas à costa brasileira. X Simpósio Brasileiro de Recursos Hídricos, Gramado, ABRH

NOAA WW3 – National Oceanic and Atmospheric Administration. Accesed on 10 Oct 2015 – URL: ftp://polar.ncep.noaa.gov/pub/history/waves

Ottmann F (1963) "L'Rocas Atoll" dans l'Atlantique sud tropical. Revue de Géographie Physique et de Géologie Dinamique 2:101–107

Pereira NS, Manso VAV, Silva AMC, Silva MB (2010) Mapeamento Geomorfológico e Morfodinâmica do Rocas Atoll, Atlântico Sul. Revista de Gestão Costeira Integrada 10(3):331–345

Pereira NS, Manso VAV, Macedo RJA, Dias JMA, Silva AMC (2013) Detrital carbonate sedimentation of the Rocas Atoll, South Atlantic. An Acad Bras Cienc 85(1):57–72

Short A (2006) Australian beach systems – nature and distribution. J Coast Res 22(1):11–27

Short AD, Masselink GM (1999) Embayed and structurally controlled beaches. In: Short AD (ed) Handbook of beach and shoreface morphodynamics. Wiley, Chichester, pp 230–249

Teixeira W, Cordani UG, Menor EA, Teixeira MG, Linsker R (2003) Arquipélago de Fernando de Noronha o paraíso do vulcão. Terra Virgem Editora, São Paulo, p 167

Valentini EMS, Rosman PCC (1993) Erosão costeira em Fortaleza. Revista Brasileira de Engenharia: Cadernos de Recursos Hídricos, vol 10

Chapter 20
Brazilian Beach Systems: Review and Overview

Andrew D. Short and Antonio Henrique da F. Klein

Abstract This chapter reviews the nature and status of the Brazilian coast and particularly its beach systems as detailed in the preceding nineteen chapters. It begins by reviewing the nature of the beaches in each of the seven coastal regions and seventeen coastal states, from the muddy Amazon Gulf in the north to the long sandy beaches of Rio Grande do Sul in the south. This is followed by a state by state review of beach erosion and inundation; beach development and impacts; and beach hazards, risk and safety. It finishes with a discussion of the range of beach research required in Brazil in order to obtain a better understanding of the coast and ts beach systems.

20.1 Introduction

In writing and editing this book we have be able to examine in detail the entire 9000 km long Brazilian coast and its 4000 open coast and bay beach systems. In this chapter we review the nature and status of the coast, and particularly its beach systems, as documented by the authors in the preceding 19 chapters. Table 20.1 provides a state-by-state overview of the coast in terms of its length and number of beaches, both on the open coast and in some of the larger bays and islands.

Brazilian beaches extend from the equatorial tropics at 4°N to temperate latitudes at 34°S. They range from the longest beaches in South America, more than 200 km long, to many small pockets of sand, hemmed in by protruding rocky headlands; they include the wide low gradient sand and mud flats of the Amazon; the beachrock and reef-dominated northeast; steep barless reflective beaches composed of coarser sand; and the wide fine grained multi-bar dissipative beaches in the

A.D. Short (✉)
School of Geosciences, University of Sydney, Sydney, NSW 2006, Australia
e-mail: andrew.short@sydney.edu.au

A.H. da F. Klein
Departamento de Geociências, Laboratório de Oceanografia Costeira, Universidade Federal de Santa Catarina, Florianópolis, SC 88040-900, Brazil
e-mail: antonio.klein@ufsc.br

A.D. Short, A.H. da F. Klein (eds.), *Brazilian Beach Systems*, Coastal Research Library 17, DOI 10.1007/978-3-319-30394-9_20

Table 20.1 The Brazilian coast by state: listing the length of each state's coastline and number of beaches, for both open and bay/island coasts[a]

	Coast length (km)			Beach number		
State	Open	Bay/Is	Total	Open	Bay/Is.	Total
1. Amapá	710	491	1200	10		10
2. Pará	790	1210	2000	130	284	414
3. Maranhão	1237		1237	223		223
4. Piauí	70		70	33		33
5. Ceará	595		595	210		210
6. Rio Grande Norte	435		435	247		247
7. Paraíba	145		145	79		79
8. Pernambuco	198		198	152		152
9. Alagoas	258		258	129		129
10. Sergipe	165		165	11		11
11. Bahia	1003	169	1172	436	123	559
12. Espiríto Santos	479		479	327		327
13. Rio de Janeiro	642	812	1454	242	637	879
14. São Paulo	670		670	242		242
15. Paraná	150		150	32		32
16. Santa Catarina	697	160	857	227	393	620
17. Rio Grande do Sul	615		615	7		7
	8859	**2842**	**11,701**	**2737**	**1437**	**4174**

[a]This table is based on measurements taken in Google earth and may differ from the numbers presented in some of the chapters

south. Some are located in undeveloped national parks while others are surrounded by heavily developed high-rise-lined shores such as of Fortaleza (CE), Recife (PE), Copacabana (RJ), Santos (SP) and Balneário Camboriú (SC). This book also cover three Brazil's oceanic island beaches which range from carbonate-rich sandy cays on atolls, to pocket-beaches backed by high volcanic cliffs. As such every beach type and state from wave to tide-dominated, from reflective to dissipative is represented along the Brazilian coast, together with every type of development.

All these beaches have however been formed by waves depositing sand at the shore. All respond daily to the changing waves, tides and winds. From a natural perspective the beaches can be classified by the interaction of waves and tide into three beach types and 15 beach states, as presented in Chap. 1 (Klein and Short 2016). This will be the subject of the first Sect. (20.2). No matter what the type of beach however, the stability of the shoreline, whether it is stable, eroding or prograding, will depend on additional parameters, particularly the sediment budget and accommodation space and in some locations human impact. The major areas of shoreline erosion and coastal inundation and their causes will be the subject of Sect. 20.3. These beaches have also been the focus of considerable development, especially since the 1970s. The level of development and particularly the adverse impact of some of this development is the subject of Sect. 20.4. The beaches and their development attract millions of people to the shore to relax, bathe and surf, and undertake business. However the beaches are inherently hazardous with breaking waves, rock and reefs and particularly rip currents. The impact of these hazards on the beach-going population will be examined in Sect. 20.5. Finally, this book provides an overview or stocktake of what we know about Brazil's beach systems in 2016 and in the final Sect. 20.6 it identifies the areas where research and monitoring is required in order to ensure that Brazil's greatest natural asset – its coast and beaches, are better understood, managed and maintained for future generation of Brazilians.

20.2 Beach Types and States

In Chap. 1 (Klein and Short 2016) the Brazilian coast was divided into seven regions (Fig. 20.1) based on previous classifications and the nature of the beach systems, as summarized in Table 1.6. This section will review the beach systems in each of these regions, the full details of which are provided in the foregoing chapters. As mentioned above the beaches cover the full specturm of beach types (WD, TM, TD) and states (D to R) (Fig. 20.2). This is to be expected given the range in tides from micro to mega and H_s from very low to >1.5 m.

20.2.1 Region1: The Tide-Dominated Amazon Delta Coast

The Amazon coast is one of the longest (~1200 km) and definitely the most complex and variable of Brazil's coastal regions. It extends either side of the world's greatest river, which flows into a gulf dominated by tides ranging up to 11 m. The massive water and sediment discharge, most of which is suspended mud, impacts not just the Amapá coast but the coast downdrift all the way to the Orinoco River in Venezuela. Region 1 extends from the northern Amapá border at Cape Orange, down through the Amapá and Pará Amazon river mouth coast, including its major North and South channels and numerous distributaries and islands, to the northeast tip of Marajó Island. Tides range from 4 m in the north to 11 m at the mouth. Easterly trade winds generate moderate waves, which after they cross the wide shallow shelf and inter and sub-tidal zones, are usually very low to zero at the shore. While mud makes up 99 % of the Amazon sediments, there is still sufficient sand (~6 M m^3 $year^{-1}$, Dunne et al. 1998) to be deposited at the shore to form sand beaches and sand flats.

Santos et al. (2016) divided the coast into four sub-provinces each of which contains tide-dominated beaches. The 550 km long northern Amapá coast, while dominated by mud, has low gradient tide-dominated beaches (B + SF to MF) occupying 20 % of the coast. The 500 km long southern Amapá Amazon river coast has tide-dominated sandy beaches fronted by wide tidal sand flats located on the eastern shores of the more exposed river mouth islands, while along the northern banks of the North Channel and some of the distributaries smaller pocket beaches occur usually hemmed in by mangroves and varea forest and fronted by river-truncated mud flats. These beaches occupy only 9 % and 1 % respectively of this river and mud-dominated sub-province. The 1500 km long Pará Amazon riverine-island

Fig. 20.1 Map of Brazil showing the seven coastal regions

Fig. 20.2 Plot of indicative regional beach states using Ω-RTR plot. RTR range from 0.1 to 100 across the full wave to tide-dominated beach types, and Ω ranges from 0.1 to 13 across the full reflective to dissipative range. See relevant state chapters for more detailed description of beach types and states

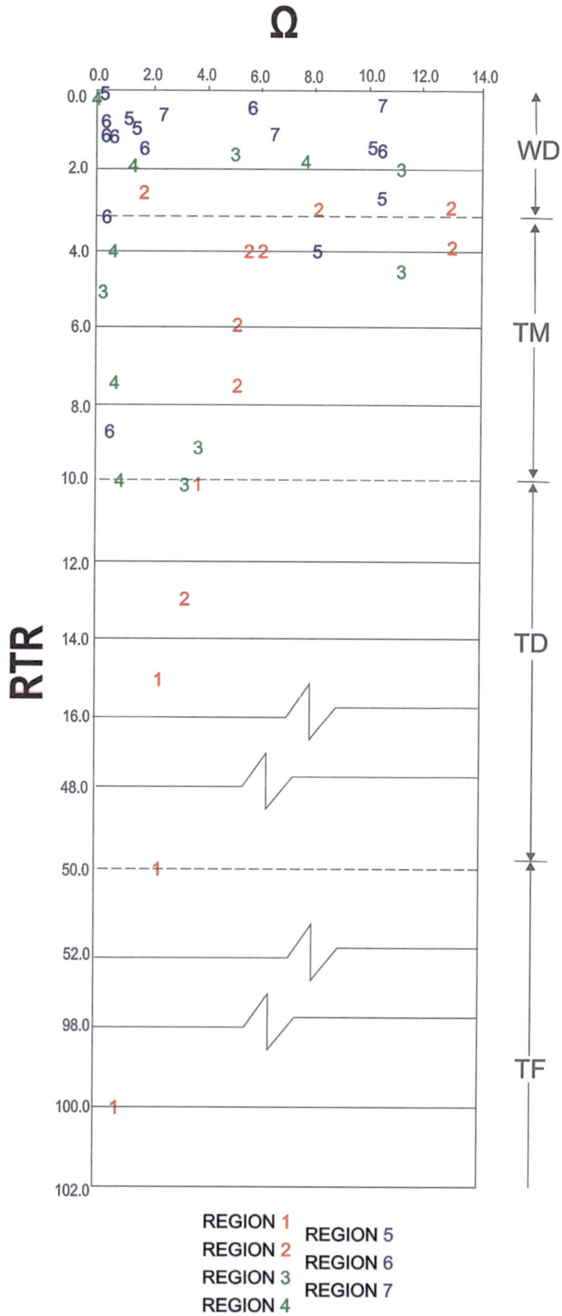

sub-province is dominated by the mouth of the Amazon, and includes the Amazon's South channel (Fig. 1.6). About 50 tide-dominated beaches and sand-mud flats, averaging 2 km in length, occupy about 7 % of the shore, all located along more exposed eastern sections of the major river mouth islands. The final sub-province is the 200 km long northern shore of Marajó Island, which faces north into the South channel. It contains 39 tide-dominated sandy beaches fronted by both sand flats and ridged sand flats (B + SF and RSF). The beaches average about 1 km in length and occupy about 20 % of the coast the remainder dominated by mangroves. All experience westerly longshore sediment transport. In total about 260 beaches occupy about 420 km (~7 %) of this predominately river-dominated mud and mangrove-lined province. As Fig. 20.2 indicates the Amapá beaches (Region 1) are all tide-dominated with an RTR between 10 and 100.

20.2.2 Region 2: Tide-Dominated Barriers and Estuaries of Pará-Maranhão

The large tides that dominate the Amazon Gulf extend eastwards along the east coast of Pará and Maranhão, decreasing from 10 m at the mouth of Marajó Bay to 3 m at Belém and on the eastern Maranhão coast (Fig. 1.7). At the same time on the open coast the easterly trade wind waves generate more energy at the shore particularly at high tide resulting in a combination of tide-dominated and tide-modified beaches (Fig. 20.2, region 2). This 1050 km long region begins at the northeast tip of Marajó island and includes the 285 beaches that lie either side of the elongate 180 km long Marajó bay (Pereira et al. 2016a). The bay beaches are all tide-dominated consisting both of B + SF, and on some western shore, B + RSF. The 'low' northern half of the bay is dominated more by tidal creeks and their associated ebb-tidal deltas and mangroves, while along the southern 'high' half of the bay the beaches are short and often bordered and backed by the bluffs of the Barreiras Formation. Mangroves increase southwards into the bay, finally dominating the shore of the Tocantins river mouth.

The open east Pará and western Maranhão coast commence at the mouth of Marajó Bay and extends 600 km east to São Marcos bay. This mega to macro-tidal coast consists of a series of 35 major promontories bordered by 5–10 km wide, up to 20 km long tide-dominated estuaries. Each of the promontories contains regressive barriers and 'drumstick' barrier islands that recurved into the bay mouths. Sediment transport is predominately to the west, though continually interrupted by the numerous estuaries. The easterly waves are modulated by the tides with moderate waves reaching the exposed beaches at high tide and little or no waves at low tide. The beaches range from tide-modified, including R + LTR on the most exposed tips of the promontories, to tide-dominated moving into the bay mouths and in lee of the bay mouth shoals (Pereira et al. 2016b). Ninety beaches occupy the eastern Pará section and another 42 the western Maranhão section. Figure 20.2 locates these beaches in the tide-modified to tide-dominated section of the Ω-RTR plot (Fig. 20.2).

20.2.3 Region 3: Northeastern Tide-Modified Barrier-Dune Coast

At the eastern mouth of São Marcos bay the coast changes dramatically from the highly irregular tide-dominated estuarine coast to the west, to a straighter more continuous, more wave-dominated coast to the east. This 1300 km long region includes the coasts of eastern Maranhão, Piauí, Ceará and the northern (or equatorial) coast of Rio Grande do Norte (Fig. 1.8). The coast has meso-tides averaging about 3 m and easterly trade wind waves between 1 and 2 m. The trade winds also have a profound impact on the coast forming extensive dune transgression along much of the coast and together with the easterly waves driving substantial westerly longshore sand transport, estimated at 100,000s m^3 $year^{-1}$ including headland overpassing and bypassing. Near the shore the waves are modulated by outcrops of the Barreiras Formation, which forms headlands, cliffs and scarps, and by shore parallel beachrock reefs, together with numerous inlets. Combined these produce a series of headland-bound embayed beaches, with wave height locally controlled by wave refraction and attenuation around and over the reefs.

The 230 m eastern **Maranhão** coast has 26 generally long sandy beaches, the longest 65 km (Pereira et al. 2016b). In the west the more exposed beaches and those with fine sand are multi-bar, dissipative wave-dominated beaches (Fig. 20.2), all backed by massive active transgressive dunes, all part of the large Lençôis Maranhenses national park. While towards the east and the Parnaíba river mouth, wave height decreases and tide-modified and some tide-dominated beaches predominate (Fig. 20.2).

The 66 km long **Piauí** coast consists of several larger headland-tied embayed beaches, together with some beachrock reefs. As wave energy increases from east to west within the embayments the full range of tide-modified reflective, intermediate and dissipative beach states are found (Fig. 20.2), with numerous well-developed low tide rips (Paula et al. 2016). Beachrock increases to the east with the shore-parallel rock usually backed by steep reflective beaches, which are only active at high tide. Sometimes there is a step (>1 m) between the beachrock and the nearshore. All the beaches are backed by massive active transgressive dunes, most of them without foredunes.

The **Ceará** coast extends east for 573 km and contains 216 beaches (Pinheiro et al. 2016). Tides remain meso, waves average 1.1–1.6 m and tide–modified beaches predominate. This section can be divided into a western sector dominated by barrier islands and lagoons and active transgressive dunes, and an eastern section with rocky coast, abrasion platforms, and tabular sandy-clay deposits derived from the eroding cliffs of the Barreiras Formation. Morais et al. (2006) further divided the coast into five sub-sectors, based on watershed boundaries, sediment delivery and the morphological characteristics of the beaches. Pinheiro et al. (2016) found the full range of wave-dominated and tide-modified beaches along the coast (Fig. 20.2). The tide-modified beaches predominate (48 %) of which reflective with low tidal bars and rips (R+LTR) are the most common (26 %), followed by reflective and low tide terrace (R+LTT) (21 %) with only 0.5 % ultradissipative (UD). Wave-dominated beaches make up 14 %, with LBT and TBR the most common beaches states, while reflective high tide beaches fronted by intertidal rocks flats comprise 38 % of the beaches.

The northern coast of **Rio Grande do Norte** extends east for 244 km to Touros where there is a dramatic change in the orientation of the coast as it swings south into the next region. Eighty percent of the northern coast consists of generally longer embayed sandy beaches and like the beaches to the west are tide-modified (predominately R+LTT), together with some tide-dominated beaches (Fig. 8.7). Dune transgression increases northward within each embayment (Vital et al. 2016). In addition there are outcrops of the Barreiras Formation and beachrock reefs.

20.2.4 Region 4: Eastern Beachrock Coast

The eastern beachrock coast extends for 760 km between between Touros and Alagoas and includes the coasts of eastern (or oriental) Rio Grande do Norte, Paraíba, Pernambuco and Alagoas (Fig. 1.9). This coast faces due east, then south-east, into both the easterly trade winds and their waves, plus longer swell arriving from the south, with wave averaging 1.5–2 m, while tides average 2.5 m. The coast consists of two sections a northern 220 km long more open coast and the 540 km central-southern sector, which is dominated by inshore beachrock and algal reefs.

The northern **Rio Grande do Norte** sector consists of a series of more than 20 headland-tied beaches, which occupy 61 % of the coast (Vital et al. 2016), with wave energy increasing northwards within each curving embayment and all backed by substantial dune transgression, dominated by longwalled parabolics. Beachrock, while present, is limited in extent and the beaches range from reflective to rip-dominated intermediate, with LTT and TBR being the most common beach states (Fig. 8.7). The waves and wind drive northerly sand transport along the beaches with headland bypassing and/or overpassing occurring at most headlands. The Barreiras Formation occupies 39 % of the coast and outcrops along the coast forming numerous headland and cliffs,

The central-southern sector extends from just inside **Paraíba** to Coruripe in southern **Alagoas**. This section is dominated by extensive shore-parallel beachrock reefs, together with nearshore and inner shelf beachrock and algal reefs, that both attenuate the breaking waves as well as cause wave refraction and diffraction which produces a more crenulate shoreline consisting of sandy bays and salients. Of interest is the total lack of transgressive dunes, a product of the lower inshore waves, lower rates of onshore sand transport and more stable foredune. The 145 km long Paraíba coast, south of Traição bay is dominated by near continuous exposed shore parallel and submerged inner shelf reefs, particularly around João Pessoa. The reefs lower wave energy and result in predominately tide-modified R+LTT beaches (Fig. 20.2).

Pereira et al. (2016c) divide the 200 km long **Pernambuco** coast into three sections based on the level of beachrock control. The 70 km long northern section has near continuous beachrock reefs backing by lower energy tide-modified beaches (R+LTT); the 60 km long central section, including Recife, has moderate beachrock control with both protected R+LTT beaches and more exposed beaches with

R+TBR. Along the 70 km long southern the reefs are discontinuous and wave energy is higher at the shore resulting in predominately rip-dominated R+TBR.

The 250 km long **Alagoas** coast consist of three sections. The 160 km northern-central sector is dominated by beachrock and algal reefs, lying between 1 and 4 km offshore. Like the Pernambuco coast, the reefs dominate the inshore waves climate lowering waves at the shore to maintain tide-modified predominately R+LTT beaches (Fig. 20.2) and generating sandy bays and salients in lee of the reefs. The exposed reefs essentially cease at Lagoa do Roterio beyond which is a 40 km long section of higher energy wave and rip-dominated, long near continuous beaches, down to Coruripe, when the next region begins with the regressive ridges of Sao Francisco deltaic plain, which includes the southern 50 km of the Alagoas coast.

20.2.5 Region 5: Eastern Wave-Dominated Deltaic Coast

Between the São Francisco delta and Cabo Frio lies the 2000 km long 'drip-feed wave-dominated deltaic coast' (Dominguez 2009), that includes the southern section of Alagoas and all of Sergipe, Bahia, Espírito Santo and the northern coast of Rio de Janeiro (Fig. 1.10). This coast is backed by a humid hinterland that feeds several rivers including the large São Francisco, together with the Jequitinhonha, Doce and Paraíba do Sul, which have deposited large protruding wave-dominated delta typified by extensive regressive protruding beach and foredune ridge plains. It is these deltas that have 'drip-feed' sediment to the coast. Between the deltas however the coast is sediment deficient and the coastal tableland (Barreiras Formation) and in places Cretaceous and Precambrian rocks reach the shore forming cliffs and rocky headlands with the Barreiras Formation actively retreating in many places. The coast is also dominated by a reef-covered shallow inner shelf which results in significant wave attenuation and refraction-diffraction over and around the reefs and generally low waves at the shore. While deepwater waves generated by the trades and southern swell average 1.5 m, they are significantly lowered in lee of the many reefs. Tides range from meso in the north (2.5 m) to micro in the south (1.6 m), with beaches switching between wave-dominated on exposed sections to tide-modified along sheltered sections.

The wave-dominated São Francisco river delta dominates the southern 50 km of the **Alagoas** coast and first 50 km of the Sergipe coast. The remaining 115 km of the **Sergipe** coast is dominated by three regressive barriers bordered by wide tidal inlets, with low gradient multi-bar dissipative beaches dominating the open shore.

The 1000 km long **Bahia** coast contains 436 open coast beaches. As Dominguez et al. (2016b) state "There are various coastal landforms along the Bahia coast, including cliffs and rivers, a wave-dominated delta, beach-ridge plains, coral reefs and large bays." They go on to divide the coast into four sections. The 230 km long Northern Littoral Coast extends down to Salvador. It trends north-northeast and contains Precambrian rock outcrops and beachrock reefs particularly in the south, generally narrow stable to slightly erosional foredune capped barriers and wave-dominated intermediate beaches. Its 90 beaches tend to be relatively short and bordered by reefs, rocks and tidal inlets. The Mezozoic Rifts Coast commences at Itaparica island and extends 230 km south to Itacaré and is characterized by exposures of eroding Mesozoic sedimentary rocks which result in a crenulate coastline fringed in places by coral reefs. The variation in both wave energy, sediment source (lithic to carbonate) and sediment size (0.1–0.9 mm) results in the 125 beaches ranging from reflective in lee of reefs, to higher energy intermediate on exposed sections (Fig. 20.2). The Jequitinhonha River Deltaic Coast occupies the next 240 km of the coast. This section has 70 generally longer beaches composed of fine sand and exposed to higher waves, which maintain higher energy intermediate to dissipative beaches. Near the river mouths coarser sand maintain more reflective beaches. The 300 km long southern Sediment Starved Southern Coast is dominated by cliffed exposures of the Barreiras Formation (Dominguez et al. 2016b), which supply most of the sediment to the adjoining beaches. Numerous coral reefs line the coast inducing wave attenuation and refraction and the formation of coastal salients, including the large Caravelas and Corumbau salients. Because of the lower wave energy the 165 beaches are typically R to LTT.

Espírito Santo has a 480 km long coast containing 327 beaches, which occupy 71 % of the coast, of which 25 % are embayed and pocket beaches, 22 % located between eroding cliffs and abrasion terraces and 24 % exposed sandy beaches (Albino et al. 2016). The coast contains two sedimentary plains totaling 180 km in length associated with the large Doce and Itabapoana river deltas. These contain exposed sandy beaches, which depending on grain size, are either higher energy intermediate or dissipative (Fig. 20.2). South of the Doce river the coast is more crenulate with outcrops of the Barreiras Formation, crystalline rocks and Neogene sedimentary deposits forming the numerous smaller embayments and cliffed-abrasion sections. The beaches here range from intermediate to reflective.

The northern **Rio de Janeiro** coast contains two sub-provinces (Muehe and Lins-de-Barro 2016), the northern protruding Paraiba do Sul coastal plain extends 180 km south of the border. Its beaches while well-exposed vary in beach state from intermediate to dissipative on the 90 km long northern coast where sands are finer (0.4 mm) to LTT to R along the southern coast where sands are coarser (1.2 mm) (Fig. 20.2). The remaining 160 km of coast down to Cabo Frio consists of two sectors, the Macaé-Búzios coastal plain, and outcrops of the rocky basement, which forms major promontories at Macaé, Cape Búzios and Cape Frio resulting is a series of smaller embayed beaches of variable energy, and ranging from reflective to intermediate.

20.2.6 Region 6: Southeast Wave-Dominated Rocky-Embayed Coast

At Cabo Frio there is a major change in coastal orientation, coastal processes and beach systems. The east-facing often sheltered beaches to the north give way to the initially south-facing more exposed beaches to the south contained within the São Paulo Bight. The moderate energy trade wind waves to the north are replaced by the prevailing higher southerly swell. In addition the crystalline rocks of the Serra do

Mar mountain range dominate much of the coast controlling coastal orientation and resulting in considerable rocky shores and numerous embayed and pocket beaches. This 1700 km long region includes the southern coast of Rio de Janeiro, the São Paulo and Paraná coasts and the coast of Santa Catarina down to Cape Santa Marta (Fig. 1.11).

The southern **Rio de Janeiro** coast is dominated by the Serra do Mar mountains. The south-facing coast is heavily indented containing the larger Guanabara, Sepetiba and Ilha Grande bays and numerous rocky bays and embayments of varying length. In amongst the rocky sections are 242 open coast beaches and 647 smaller bay beaches. The beaches range from small pockets of sand to several long barriers in the east and west including the 42 km long Marambaia beach. The open coast beaches are generally well exposed to the southerly waves and with micro-tides are all wave-dominated. The main control on beach state is sand size, which ranges from 0.15 to 1.0 mm (Muehe and Lins-de-Barros 2016). The coarser sand maintains steep reflective beaches, while the finer sand maintains intermediate beaches (Fig. 20.2).

São Paulo state has 670 m of coast containing 242 beaches. Mahiques et al. (2016) divide the coast into six sectors, all of which are dominated by the Serra do Mar crystalline massif. The highly indented Northern Scarps Coast contains a series of small bays and islands, with the 280 km of mainland shore containing 155 small wave-dominated beaches, which depending on exposure and sand size range from reflective to dissipative. In contrast the 50 km long Bertioga coastal plain contains just nine beaches, dominated by four higher energy intermediate to dissipative beaches (Fig. 20.2). The Santos sector is more heavily indented, with variable orientation, its 90 km of coast containing 35 embayed beaches which range from reflective to intermediate depending on exposure. South of Santos is 260 km of southeast facing coast containing two long coastal plains (Itanhaém-Peruíbe and Cananéia-Iguape) separated by rocky embayed sections. The longer beaches are intermediate to dissipative, while intermediate states tend to dominate the embayed beaches.

The **Paraná** coast is just 150 km long and dominated by a wide regressive coastal plain, with the Serra do Mar mountains located up to 50 km inland. The coast has just 32 beaches separated by inlets and some rocky headlands. The exposed beaches dominate the coast with beach type depending on grain size ranging from predominately fine sand dissipative to intermediate. The sandy beaches extend into the large tidal inlets where they become more tide-modified to tide-dominated (Angulo et al. 2016) (Fig. 20.2).

The Serra do Mar mountains returns to the coast along much of the **Santa Catarina** coast down to Cape Santa Marta, the southern boundary of this region. This 576 km section contains 240 beaches, which occupy 60 % of the coast, the remainder being rocky. Klein et al. (2016) divide the Santa Catarina coast into five sub-provinces. The first from the border to Farol Cabeçudas tends to have longer exposed intermediate to dissipative beaches. This is followed by a heavily indented rock-dominated section that extends down to the entrance to North Bay. This sector contains numerous smaller embayed reflective beaches with variable orientation.

The rock-dominated Santa Catarina island forms the next section which has a mix of sheltered reflective and exposed intermediate beaches. The backing South Bay contains the southernmost mangroves in South America. The final section down to Cape Santa Marta remains rock-dominated with 37 generally well-exposed embayed intermediate beaches (Fig. 17.11) occupying the 87 km of east facing coast.

20.2.7 Region 7: Wave-Dominated Beaches and Barrier Coast of Rio Grande do Sul

The southern-most region extends from Santa Catarina's Cape Santa Marta down to Chuí at the border between Rio Grande do Sul and Uruguay a distance of 735 km (Fig. 1.12). It contains 12 long beaches all composed of fine sand and facing east-southeast into the prevailing southerly swell, which drives substantial northerly longshore sand transport. They are all higher energy dissipative (Fig. 20.2), with bar number ranging from three to four depending on grain size. Rips can however dominate the inner bar (Calliari and Toldo 2016). The beaches are part of the longest and largest barrier systems in South America and are all backed by episodes of massive dune transgression which trends south under the prevailing northerly winds.

20.3 Beach Erosion and Coastal Inundation

The Brazilian coast is dominated by sandy beaches which in the north and more sheltered areas are fronted by sand and mud flats. All these systems are composed of unconsolidated sediments exposed to wind, waves and tidal current action, which can erode, transport and deposit the sediment, causing shoreline change and instability. As many of these same shores are the focus of human occupation and development, as well as recreation, housing, tourism, and industry, it is important to understand the nature and stability of these shores in order to plan, develop and manage them so as to avoid placing people, buildings and infrastructure at unnecessary risk.

This section will review what is presently known about the location and causes of coastal erosion and coastal inundation in Brazil, with the following Sect. 20.4 reviewing the hazards posed to bathers along the Brazilian coast.

The **Amapá** coast has very dynamic shoreline owing to the massive fluvial and sediment discharge from the Amazon, together with seasonal fluctuations in water level, mega-tides and northwesterly sediment transport. The coast experienced rapid accretion until about 1 ka. Many of the beaches are now unstable and experiencing erosion and/or associated longshore sand transport, while others are bordered by dynamic tidal channels and associated tidal shoals. Vegetation debris including large logs litters many of the upper beaches.

The coast between Cape Cassiporé and North Cape is presently eroding between 0.5 and1.0 m year^{-1}, south of Cape Cassiporé erosion at rates up to 78.5 m y^{-1} have been recorded, while on Maracá Island erosion has reached 5–10 m year^{-1} (Santos et al. 2016). The coast also has a very low gradient, most is low lying (<5 m) and is inundated during the seasonal wet season, with flooding posing a risk to Macapá and Santana, as well as the numerous villages scattered along the river channels and their distributaries. The remaining coast is largely undeveloped or used for cattle ranching and where settlement does occur, particularly along the river channels, it is well adapted to the seasonal inundation, with most structures built on stilts.

The **Pará** coast is both long and highly variable and dominated by macro to mega-tides and moderate tide-modulated waves. Sediment transport is to the west, which is manifest on the eastern barrier islands by highly dynamic barrier shores that feed west tending multiple recurved spits and islands. The large tide-dominated estuaries that separate the islands also induce shoreline change as their large tidal deltas respond to changing coastal processes, while inside the estuaries removal of mangroves and infilling of tidal channels is inducing erosion. As a consequence the islands and inlets have very dynamic shorelines with areas of both considerable erosion and accretion. In the Bragança region between 1972 and 1998 60.5 % of shoreline experienced erosion and 39.4 % progradation (Pereira et al. 2016a). On occupied parts of the coast, particularly the tourist beaches, this has resulted in beach and foredune erosion and destruction of property (Fig. 4.13; Pereira et al. 2016a). There is medium risk of coastal inundation in the cities of Belém and Bragança, and while much of the low-lying mangrove-barrier shore is exposed, it is considered medium to low risk owing to the low population density.

The **Maranhão** coast is largely undeveloped. The western barrier island coast is an extension of the eastern Pará barrier island coast and experiencing similar coastal processes and erosion-accretion cycles. The barrier islands are however largely undeveloped. Likewise much of the wave-dominated eastern coast is located in the large Lençois Maranhenses national park or undeveloped. The massive transgressive dune system lies within the park is the largest active dune system in Brazil. However in San Marcos bay along the highly developed Sao Luis coast beach and foredune erosion between 0.1 and 1.5 m year^{-1} has threatened and destroyed urban properties and infrastructure (Pereira et al. 2016b), while there is also exposure to inundation.

The **Piauí** coast is relatively short and predominatly stable. However beach monitoring by Paula et al. (2016) concluded that 50 % of coastal beaches are at high risk to erosion due to natural causes; 41 % at medium risk but have little development and only seasonal use; 4.5 % have very low risk of erosion; while 4.5 % more urbanized beaches have a very high erosion risk, associated with anthropogenic factors together with a contribution from natural factors. Some areas including the Parnaíba river mouth are also considered at a high risk of periodic river flooding.

Ceará has an exposed highly dynamic coast experiencing massive westerly longshore sand transport by both waves and wind. In addition there has been considerable infrastructure built along the coast that has interrupted this transport resulting in area of severe erosion, this is exacerbated by more than 3500 dams that are trapping sand and reducing the delivering of sand to the coast (Pinheiro et al. 2016). The worst erosion is occurring at the Port of Pecém and at Fortaleza where the Mucuripe Harbour construction in the 1960s, together with beach groins, interrupted the longshore sand transport leading to up to 200 m of beach erosion along the Fortaleza tourist beaches and the ensuing construction of kilometers of seawalls. Coastal inundation is also a problem at Fortaleza, particularly as storm surges up to 4.5 m can cause coastal flooding (Pinheiro et al. 2016).

Rio Grande do Norte has two coasts, the northern coast is a continuation of the Ceará coast and has similar issues with massive westerly longshore sand transport by waves and wind. The north coast is however largely undeveloped and there is little property at risk. The more exposed and more developed eastern coast experiences northerly longshore sand transport and Vital et al. (2006 and 2016) noted seven factors indicating shoreline erosion including: general shoreline erosion over the past 60 years: erosion of the Barreiras Formation forming sea cliffs; erosion and burial of mangroves; exposure of peat and lagoonal deposits at the shore; persistent destruction of coastal infrastructure; heavy mineral concentrations on the foreshore; and development of beach embayments. Construction of groins at Macau, Caiçara do Norte and Touros beaches has accelerated erosion, which at Macau and Guamaré is endangering oil pumping stations. Erosion is also occurring along most of the cliffed coasts where the Barreiras Formation is exposed. Coastal inundation is also a problem in low-lying areas adjacent to the Apodi and Mossoró rivers (Vital et al. 2016).

In **Paraíba**, 42 % of the 140 km coast is eroding, while coastal inundation is an issue in the major cities of João Pessoa and Recife both of which are built on low-lying land (Dominguez et al. 2016a). In **Pernambuco** coastal erosion was first documented at Olinda in 1914, and has been continuing ever since, particularly along the heavily developed and modified Recife-Olinda beaches, and in addition erosion has been recorded along one third of the 200 km long coast (Pereira et al. 2016c). The erosion is attributed to the unorganized urbanization processes that even occupy the backshore zone, together with changes in the river courses; removal of riparian vegetation; mangrove landfill; building of dams; deforestation of coastal vegetation; and inappropriate coastal structures.

In **Alagoas** coastal vulnerability has increased owing to river damming that has reduced the contribution of river sediments, with erosion is mainly concentrated in the more popular and heavily developed northern part of the state.

In **Sergipe** most of the coast (57 %) in stable owing to fluvial sediment input by rivers, while 21 % is eroding. The most severe erosion is at the mouth of São Francisco river, where Vila do Cabeço was completely eroded, and also at Atalaia Nova (north of Aracaju). This erosion is attributed to dams built for power stations from the 1950s, substantially decreasing the supply of sediments (Dominguez et al. 2016a). Coastal inundation is a problem in Aracaju where the high population density and low terrain between the Real and Vaza-Barris rivers results in a high risk of flooding.

The long coast of **Bahia** is divided by Dominguez et al. (2016b) into four sectors, three of which are experiencing coastal erosion, particularly the 'sediment starved southern coast' (see Fig. 12.12). This sector contains extensive areas of the Barreiras Formation, which is cliffed and eroding and contributing sediment to the beaches. They attribute the most severe erosion to sediment interuption by engineering works associated with port facilities (e.g. the Port of Ilhéus); sediment retention on unstable capes (e.g. the Caravelas Plain); negative long-term sediment budget (e.g. the sediment starved southern coast); reductions in fluvial and sediment river discharges resulting from natural processes or human intervention; and lateral migration of the mouths of small rivers. Localised flooding is a problem at Salvador, Ilhéus and Porto Seguro.

Erosion along the coast of **Espirito Santo** has been investigated by Albino et al. (2016) who found several erosion hot spots (see Fig. 13.16). On the north shoreline erosion between Baleia and Fruta is associated with the dynamic Doce river mouth. Erosion is also occurring in the center around Morro, and in the south between Ubu and Maimbá, where the Barreiras Formation is exposed and eroding at the coast. They found that the erosion could be attributed to both natural causes such as reduced sediment availability, as well as human occupation of the beach and foredunes, resulting in property and infrastructure being threatened by erosion. Inundation is a major issue in the Rio Doce region, Vitória and the inner areas of the Paraiba do Sul River due to the low terrain and high level of urbanization.

The **Rio de Janeiro** coast has as Muehe and Lins-de Barros (2016) state "remarkable resilience", with the most beaches recovering after erosion events. They do however note erosion on the north coast extending 10 km downdrift of the Barra do Furado jetties, which interrupted the sediment supply; and at Virgem beach which has retreated up to 12 m. Between the Espírito Santo border to Cabo Frio, erosion is occurring south of the Paraíba do Sul river in Atafona with a rate of 7.5 m yr^{-1}. The erosion is a result of sand is being held on the inner shelf by mud coverage coming from the river and by the dominant littoral drift from south. The highly urbanized coast of Macaé and Rio das Ostras is also eroding, while the Rio de Janeiro-Niterói metropolitan area with its high population density, is exposed to erosion, flooding and landslides. The expansion of urban areas into lagoonal wetlands (e.g. Barra da Tijuca) with limited drainage capacity, is a risk that will increase as sea level rises and storm intensity increases. In the south the long Marambaia barrier is at risk of breaching due to erosion of its lagoonal shore.

In its natural state the **São Paulo** coast is long, highly variable and in general stable, however the intense occupation of the coast for residential, tourism and industry has lead to considerable erosion such as at Massaguaçu (see Chap. 2). According to Mahiques et al. (2016) there is "intensive erosion in all of the (four) compartments" along the coast. Souza and Suguio (2003) found "that approximately 42 % of the state's sandy beaches present high to very high erosion risk, reaching up to 66 % of the state's coastline." However, the authors suggest that these risks are mainly associated with natural rather than anthropogenic activities. Coastal inundation is a problem between Santos (SP) and Itajaí (SC), owing to the presence of three major ports, relatively high population densities and the socio-economic importance of these centers.

Along the sandy **Paraná** coast Angulo et al. (2016) found that the exposed ocean beaches while dynamic were relatively stable, unlike the beaches associated with the large estuarine inlets (Paranaguá and Guaratuba bays) whose sandy entrances are highly dynamics over a period of decades to centuries resulting in 10–100s meter of shoreline change (erosion and accretion). Bay mouth beaches including Ilha do Mel are experiencing considerably dynamics as a result. In addition northerly longshore sand transport generates headland sand bypassing and shoreline mobility at Guaratuba beach.

A number of studies have investigated coastal erosion and vulnerability along the **Santa Catarina** coast. Perinotto et al. (2012) in an assessment of erosion 'hot spots' found that 60 % of the spots were in an 'erosional state' with 15 % exposed to 'severe erosion'. Klein et al. (2006) identified erosion hot spots in SC provinces 1 and 2 (Fig. 17.1) and found the most at risk were Barra do Sul, Barra Velha, Piçarras, Gravatá/Navegantes, Barra Sul/Balneário Camboriú. On the Santa Catarina Island erosion is occurring at Ingleses, Canasvieiras, Barra da Lagoa, Campeche, Armação, Pântano do Sul and Naufragado (Horn Filho 2006). They found that while negative sediment budget can be recognized in some of these spots (e.g. Piçarras), most erosion problems seem to be more related to episodic storm surges together with rigid structures, including river mouth training walls, enhancing the erosion. Coastal inundation caused by storm surges have affected 13 municipalities, with the ten most affect having high levels of urbanization and armoring of the coast. Finally the central-southern dune fields are migrating at rates between 5 and 14 m yr^{-1}, which in some areas are smothering road, houses and infrastructure.

The long sandy **Rio Grande do Sul** coast has the highest rates of northerly longshore sand transport in Brazil, and as might be expected any perturbations in the transport rate may affect shoreline stability. Dillenburg and Hesp (2009) divided the coast into five sectors and from north to south and found Torres-Tramanfaí was prograding slightly; Tranamdaí to Mostardas and Mostardas to Estrteito was experiencing long term regression with Hermenegildo beach retreating at 3.6 m^{-1} yr^{-1}; Estreito to Verga was prograding; and the southern Verga to Chui was undergoing long term recession. Toldo et al. (2013) found that sediment was being lost longshore, inland to aeolian transport and storm surges, and offshore to coastal jets. They found that along the middle coast (Dunas Atlas to Patos Inlet) the coastal could be divided into five sediment cells, which based on rates of longshore transport varied from erosional to accretionary, with the shoreline at Dunas Atlas having retreated more than 200 m during the past century (Fig. 18.7). RS has medium to very low inundation risk apart from places like of Hermenegildo, where the construction of houses extend to the waterline, where they are exposed to waves. At Rio Grande, the high season attracts almost 100,000 inhabitants and the combination of the occupation with a high-energy coastal dynamics, results in Rio Grande region having one of the highest flood risks on the whole Brazilian coast. Approximately 80 % of the RS coastline is undeveloped. As a consequence coastal management is in its early stages and focused on the northern sector where most of the developed beaches are located (Esteves 2004). However a federal road close to the coast now links the northern, middle and southern sectors of the coast. These presently undeveloped sectors will require special attention to ensure the implementation and regulations of an effective management plan (Calliari and Toldo 2016).

20.3.1 Summary

Coastal erosion along the Brazilian coast is both highly variable in location and intensity, and can be attributed to a number of causes, both natural and anthropogenic. Away from river mouths, much of the coast received a supply of sediment derived from the inner shelf and delivered during the early to mid-Holocene. This supply is now either exhausted or decreasing, resulting in a switch from a positive to negative sediment budget, with more sand being exported out of coastal cells to longshore, onshore and even offshore, than is being supplied from the shelf. This is usually manifest in a switch from regressive, to stable or even eroding shores. The other major natural cause of erosion is an in-balance in the longshore sand transport, leading to shoreline retreat in areas of negative balance (e.g. RGS).

Other natural causes of erosion are associated with the pulsative nature of longshore sand transport, particularly around headlands where both subaqueous bypassing and aeolian overpassing can occur, the latter also hindered and even stopped by urban development (e.g. Ingleses Beach, SC; Vieira da Silva et al. 2016). Also most river mouths, inlets and tidal creeks are highly dynamic leading to both shoreline erosion and accretion (e.g. Paraná). They also interrupt longshore sand transport and in places like eastern Pará-Maranhão lead to the formation of highly mobile barrier islands. The erosion along the Amapá coast can be attributed to both river mouth dynamics and pulsative longshore mud transport (see Anthony et al. 2010).

In most states coastal development is also contributing significantly the shoreline erosion and is usually the site of erosion hot spots. This development takes a number of forms. Port construction has interrupted or stopped longshore transport leading to downdrift erosion (e.g. Fortaleza, Suape, Ilhéus). Beachfront development of buildings and roads can place the structures in the beach hazard zone exposing them directly to erosion, undermining and destruction, at the same time interrupting or locking up the available sand leading to further erosion. The response to this erosion is often the construction of seawall and groynes, both of which further interrupt the natural beach sand system and can lead to downdrift erosion including end effects. These structures also degrade the beach amenity, in places replacing the sandy beach with a rock wall, as well as posing an additional hazard, in terms of rocks and debris, to the bathing public. As noted above other adverse anthropogenic impacts on the coast are also caused by dam construction decreasing fluvial sand supply; wetland infilling changing the tidal prism and tidal flows; mangrove removal leading to shoreline instability and exposure to flooding; beach mining removing sand from the system; port and sewerage pollution closing beaches; high buildings overshadowing tourist beaches; and general congestion along heavily developed tourist beaches.

Longshore sand transport (littoral drift) exerts a major control on shoreline stability, with negative sand budgets leading to erosion and positive budgets to accretion. Understanding the nature and rates of this transport is therefore critical to effective management of each littoral sediment cells. These cells may be 100s km long as in the case of Rio Grande do Sul and Amapá, or small pockets of sand in embayed beaches. The entire coast of Brazil is exposed to predominately southerly through easterly waves, which result in generally northerly-northwesterly longshore sand transport, though some permanent and seasonal reversals do occur (Dominguez 2009; Toldo et al. 2013). Most of this transport takes place in the surf zone, with breaking waves suspending the sediment, which is then transported northward by the net northerly currents. This transport is extremely critical to beach stability and long-term evolution, i.e. stable, accrete or erode. As a result there have been a number of studies to try and estimate the transport rates. As the rates vary considerably depending on wave climate, shoreline orientation and obstacles (headlands, estuaries, etc.), all studies have focused on local to regional estimates.

The longshore transport can be broken into a number of major cells. Starting in the south at region 7, the Rio Grande do Sul coast is expected to have the highest rates of longshore transport in Brazil, on the order of a few hundred thousand cubic meters per year. In region 6 the transport is largely interrupted by the presence of bedrock headland, islands and large bays, leading to numerous smaller sediment cells. In region 5 northerly transport resumes but is interrupted in places by river mouths, reefs, bedrock and some larger bays. Region 4 has a more continuous coast, but again interrupted by several river mouths and estuaries together with headlands. Region 3, like Region 1, experiences large rates (~ few 100,000 m^3 $year^{-1}$) of northwest sand transport both along the shore and in the massive dune fields. Bypassing and overpassing is common on most of the headlands. Region 2 experiences considerable local west to northwest transport, which is however continually interrupted by the numerous estuaries in the east and bays, headlands and mangroves in the west. The northern region 1 is dominated by fluvial and tidal transport in the Amazon River mouth, with waves, tides and the Brazilian current maintaining northerly sand and particularly mud transport along the northern Amapá coast, which continues for 1500 km to Venezuela.

Estimates of the actual rates of longshore transport are highly conjectural and largely based on empirical formulae, such as the CERC formulae (Shore Protection Manual 1984). They range from 500,000 m^3 yr^{-1} in Rio Grande do Sul (CPE 2009); 20,000 to 500,000 m^3 yr^{-1} in Santa Catarina (Klein et al. 2016); 10,000 to 100,000 m^3 yr^{-1} along the Paraná coast (Sayão 1989; Lessa et al. 2000; Lamour 2000; Lamour et al. 2006), to as much as 700,000 m^3 $year^{-1}$ and reaching a maximum of 1,400,000 m^3 yr^{-1} in Ceará (Pinheiro et al. 2001; Maia et al. 2005; Morais et al. 2006), which seems excessive. On the higher energy southeast Australian coast rates up to 500,000 m^3 yr^{-1} have been calculated and calibrated (Pattearson et al. 2012). Based on these rates we would expect maximum rates along the Brazilian coast to be less, but still on the order of a few hundred thousand cubic meter per year. Clearly considerable more research and monitoring of waves, currents and sand transport is required before acceptable estimates of longshore transport can be obtained.

20.4 Beach Development and Impacts

Much of the Brazilian coast has experienced a large increase in coastal population since the 1970s. This increase is associated with the development of ports and industry; the growth of coastal cities, towns and villages; the spread of second homes at the beach; and the growth in coastal tourism and development. Regrettably much of this development commenced at a time when coastal management was non-existent or in its infancy, and despite the subsequent Brazilian coastal management policy (Nicolodi and Zamboni 2008) and state coastal management plans, much of this development is poorly planned and/or sited, leading to a range of issues along the coast. The most critical are related to:

- development too close to the shore which often removes the foredune and back beach, and in many places has led to the armouring of the beach to protect the development and in worst case scenarios the destruction of the beach (Fig. 20.3a);
- high rise development too close to the beaches resulting in 'coastal squeeze', general congestion and sun shading in the afternoon (Fig. 20.3b);
- beach and water pollution from litter and non-existent or poorly constructed/sited water treatment/sewer, which has led to the closure of some popular tourist beaches;
- ribbon development of second homes extending for many kilometers along the beach (Fig. 20.3c);
- development in active coastal dune fields, which both places property at risk and interrupts or stops the natural dune movement and transport of sand (Fig. 20.3d)
- restricted beach access where private property blocks access along long sections of shore.

All this has resulted in a range of problems including destruction of property constructed in the beach hazard zone leading to costly coastal remediation and a general degradation of the beach, beach amenity and coastal environment. The following provide a snapshot of some of these issues along the Brazilian coast.

In the northern **Amazon Gulf states** (Regions 1 and 2: Amapá, Pará and Maranhão) coastal development is limited in extent, with the coast often too difficult or too hazardous to reach or develop. As a consequence while this is a highly dynamic coast with areas of substantial shoreline erosion and movement, there are relative few erosion hot spots. The main areas of coastal development, especially tourist development, occur on a few popular Pará beaches near Belém and in the north at Atalaia and Ajuruteua. In each of these areas development, including houses and bars have been built on the foredune and even the beach, resulting in erosion and destruction of some of the buildings (Fig. 20.3a; Pereira et al. 2016a). Other problems associated with this uncontrolled development and beach usage include sewer pollution, vehicle congestion, noise (loudspeakers), conflicts between difference beach users and hazardous beach conditions leading to rescues and drownings.

Fig. 20.3 Some issues facing the Brazilian coast. (**a**) Beachfront development and seawall protecting a road undermined by beach erosion, Ajuruteua, PA; (**b**) Balneário Camboriú, SC is a very popular and highly developed and congested beach, which is shaded by the buildings in the afternoon; (**c**) second home development occupies 100's km of the coast and in places extends down onto the beach and foredune, such as here at Balneário Rincão, SC; and (**d**) houses have been built in and in front of the active dune field at Ingleses SC, which has stopped the supply of sand to be beach via headland overpassing, leading to erosion of the beach and destruction of property

The **Region 3** coast is also largely undeveloped owing to the extensive beaches and backing massive transgressive dune systems, particularly in eastern Maranhão, Piauí, western Ceará and northern Rio Grande do Norte. In Piauí only 5 % of the coast is urbanized, however most is exposed to erosion owing to both anthropogenic and natural factors. Unplanned coastal development is a major problem including occupation of Pleistocene dunes that has led to their destabilisation and migrating and subsequent burial of property and river courses (Paula et al. 2016).

The eastern Ceará coast is also in places highly developed and has resulted in a range of major issues, which have been studied by Morais et al. (2006, 2008). They found urbanization, property speculation, second home construction and land subdivision has affected dunes, beaches and estuary margins and exacerbated coastal erosion and been responsible for village stress and coastal urbanization. They found 20 places with coastal erosion, some experiencing very high rates of erosion resulting in serious damage to urban infrastructure as at Bitupitá, Maceio, Taíba, Icaraí, Caponga, Iguape and Redonda beaches. The Fortaleza Mucuripe Harbour presents a classical example of the lack of knowledge of coastal processes. The jetty, built in 1939, changed the wave refraction and stopped the longshore transport, as a consequence the Fortaleza's beaches eroded about 200 m (Maia et al. 1998). At Icaraí, Pacheco, Caponga, Mundaú, Canto Verde and Requenguela there has been damage to roads, sidewalks, kiosks, sheds, and entertainment areas, in addition to problems like rocks, beach erosion, spoils, groins, garbage, and difficulty of beach access (Pinheiro and Rocha 2007; Medeiros 2012; Medeiros et al. 2014). Today, coastal erosion is perceived as the most significant threat to tourism and traditional economic activities.

Region 4 includes the major coastal cities of Natal, João Pessoa, Olinda-Recife and Maceió. Along the eastern coast of Rio Grande do Norte Vital et al. (2016) found that one of the main causes of beach erosion was the construction of hard engineering works, as along the beachfront at Ponta Negra, Natal's most popular beach. At Macau and Guamaré shoreline erosion is endangering oil pumping stations, with the erosion accelerated by the construction of groins at Macau, Caiçara do Norte and Touros beaches. There is extensive erosion along the Tertiary cliffs in all states except SE. Along the highly developed Recife coast Pereira et al. (2016c) found that the unplanned urbanisation allowed development on the beach and foredune, which has in turn led to 20 km of hard engineering works, which have increased the problem, leading to further works. All this was done "*without the proper knowledge of the coastal processes and without any technical support. This did not solve the problem but rather had magnified and transferred it to neighboring downdrift areas*" (Pereira et al. 2016c). They go on to state "*As an answer to (the erosion problems) the state government had started a new project in which most of the beaches of the area will be nourished and several hard engineering structures or will be covered by sand or will be reduced in size so they can not interfere as much with the longshore….*"

The long **Region 5** has a considerable mix of coast and coastal development, with Aracaju, Salvador and Vitoria being the most highly developed coast cities. In Bahia Dominguez et al. (2016b) "*… attribute the most severe erosion to sediment interruption by engineering works associated with port facilities (eg the Port of Ilhéus)..*" as well as other natural processes. In Vitoria bathing is frequently prohibited at the main city and tourist beach because of pollution (Fig. 20.4). Elsewhere in the state Albino et al. (2006, 2016) found "*The intense occupation of the beach, foredunes and ridges increases erosion vulnerability due to reduced sediment availability*". Resulting in "*Roads, parking places and kiosks are frequently threatened by erosion.*" The "*coast have been subjected … to anthropogenic impact, though port construction, urban occupation and uncontrolled touristic development. As a consequence beach sediment transport and morphology have been changed … erosion has been exacerbated by engineering projects …. and longshore transport is interrupted.*"

Sergipe has a relatively short (165 km) coast consisting of several long regressive-stable barriers separated by large dynamic inlets. Outside of Aracaju much of the coast is undeveloped. Aracaju has developed along the mouth of Rio Sergipe and there has been considerable port development, which is protected by seawalls and groynes. The open coast and particularly the popular Atalaia beach is well managed

Fig. 20.4 Example of sign prohibiting bathing owing to polluted waters at Vitoria, ES (Photo: A D Short)

with a broad buffer zone and set back. However further south at Mosqueiro and Caueira bars, roads and restaurants have been built too close to the shore and have been damaged by erosion. The biggest erosion problem is at Saco where development along the shore of the dynamic inlet mouth has led to the construction of 3 km of irregular seawalls.

The **Region 6** has the highest coastal population in Brazil with the major cities of Rio de Janeiro, São Paulo, Curitiba, Joinville, Itajaí and Florianópolis all located on or close to the coast. The result has been extensive and intensive coastal development for ports, industry, urbanization and tourism. In the Rio de Janeiro region Muehe and Lins-de-Barros (2016) found that "*Because they are located in heavily modified urban areas, the beaches of this sector have undergone many interventions including land fill, dredging and the building of seawalls. These modifications have changed the original condition of the beach morphodynamics especially of Copacabana Beach.*" Brazil's most famous beach was also substantially modified when in the 1960s coarser dredged sand was used to widen the beach, so as to increase the width of the road. This resulted in the development of a steep reflective beach face. In Guanabara Bay Muehe and Lins-de-Barros (2016) found the shoreline "*… has been modified by landfills including the …. center of the city and its port. Artificial coastlines are also found at Niterói and São Gonçalo …..At the back of the bay, …the … mangrove swamps have been highly impacted by the pollution coming from the rivers..*".

In **São Paulo** Mahiques et al. (2016) found "*Anthropogenic activity along the São Paulo coast is more prevelant along the center-north coast of the state, being most dominant in the Santos lowlands, where several economic activities and urbanization are strongly linked to Brazil's largest harbour. Santos has had severe environmental impacts related to pollution (Torres et al. 2009) and coastal erosion.*" They also used Massaguaçú beach to illustrate the rapid coastal development since the 1960s, which transformed the beach and backshore from a largely natural well-vegetated environment to a densely populated urban city (Fig. 15.9). As a consequence of this development including railways and highway too close to the shore the beach is experiencing erosion, with hard structures being built to protect the infrastructure. Santos has also experienced water pollution (Torres et al. 2009) and coastal erosion (Italiani 2014), much of it related to port installation and activities.

In **Parana** state laws have largely kept development off the beach and foredune and even removed a squatter settlement on the foredune. While most erosion in the state is due to natural causes Angulo et al. (2016) found that "*… the southern coastal sector is heavily occupied with erosion problems related to natural coastline shift induced by ebb-tidal delta dynamics as well as human destruction of foredune ridges and constructions over the beach and the beach dynamic fringe.*"

In **Santa Catarina** Bonetti et al. (2012) mapped the coast and found that infrastructure (ports, industry, tourism and housing) occupies 48 km (39 %) of the coast, particularly in the north and in Florianópolis bay. Klein et al. (2006) and Horn Filho (2006) also that "*.. coastal degradation, (was) mostly related to coastal expansion..*" and lack of management plans.

Region 7 includes the long barriers and beaches of southern Santa Catarina and Rio Grande do Sul. While there are no big cities there are a number of larger towns like Torres, Tramandaí and Rio Grande-Cassino on or near the coast. In addition there are extensive areas of second homes spread for many kilometers along the coast of both states, most of which extend down to the foredune and beach (Fig. 20.3c). Esteves (2004) assessed the vulnerability of the RS coast based on four levels of coastal development and shoreline stability. She found that 177 km (29 %) of the coast is in a critical conditions (high development and erosion) particularly in the north and round Cassino. Also at Cassino Lélis and Calliari (2006) and Calliari et al. (2010a) found that the long inlet jetties have both significantly modified the coast, while dredging of the inlet lead to massive mud deposition on Cassino beach, one of the most popular beaches in the state.

20.4.1 Summary

The Brazilian coast has experienced a large increase in population, occupation, tourism and development over the past 40 years. This increase is manifest in the rapid growth of coastal cities and towns, the spread of second homes along the coast and the boom in tourist trade and development along the coast. These same pressures are likely to increase into the future. Unfortunately much of the existing development has been un- or poorly-planned, and where management plans did exist they were often ignored. While integrated coastal management (through the Coastal Management-GERCO) was proposed three decades ago in 1987, not all coastal states have developed their State Plans for Coastal Management (DOU 2004).

In addition there was little or no knowledge of the impact of development on coastal processes. As a result there has been a series of adverse impacts along the developed parts of the coast as mentioned above. While most of the development cannot be undone or removed, there is an urgent need to ensure future coastal development is planned and regulated so as to minimize its impact on the coast and to optimize the use of the coast without degrading it. Unless changes take place soon, the Brazilian coast will continue to be degraded, coastal processes will continue to be interrupted, erosion will increase and the coast will become an increasingly expensive problem, rather than an economic potential.

As Pinheiro et al. (2016) state, in Ceará coastal erosion is the most significant threat to maintaining income in many areas that depend on tourism and traditional economic activities. As a result this is a challenge for both coastal dwellers and the coastal managers to find new ways of living with the coast, including redesigning coastal occupation as well as minimizing the impacts of the sea.

As is evident in this review there is need to break the nexus between coastal development and coastal degradation. It has started to happen in some states. Perhaps the best example is Paraná where there has been a combination of protecting parts of the coast in national parks, while regulating development in the remainder of the coast. The Guaraqueçaba Environmental Protected Area was established in 1985 and the Superagüi National Park, which covers the entire northern Paraná coastal zone, was established in 1989. Also during the 1980s State laws were passed to prevent environmental and urban degradation (Sampaio 2006). However there have been several cases, where the state regulations were not respected. Another good example of coastal tourist development is at Atalaia beach at Aracaju, SE where there is a 100 m wide protected foredune area, which provides an excellent buffer zone between the beach and backing infrastructure; planned wooden walkways cross the foredune to the beach; regulated beach amenities and kiosks are spaced along the beach; and considerable free parking for beach goers. Also in Santa Catarina a number of illegal houses built too close to the shore and in the foredune area were ordered to be demolished and removed by the state.

20.5 Beach Hazards, Risk and Safety

20.5.1 Introduction

Beach hazards are elements of the beach environment that expose the beach public to danger. These hazards may be physical, biological, chemical or anthropogenic. Beach risk occurs when the public is exposed to beach hazards, while beach safety involves programs and resources that are implemented to minimize the level of beach risk. The Brazilian coast, and particularly its 4000 beach systems, has a wide range of beach hazards that pose a risk to bathers, and there has been considerable effort in the form of lifeguards and signage to help minimize this risk. However much more needs to be done as Brazil still has a high level of beach risk resulting in numerous rescues and drowning.

The main hazards along the Brazilian coast are:

- Physical hazards include breaking waves; variable surf zone topography (bars, channels, troughs); surf zone currents that may move onshore, longshore and offshore, particularly rip currents; inlets and river mouths with deep channels and strong currents; beachrock reefs; coastal and algal reefs; rocks and headlands.
- Biological hazards include sharks and marine stingers.
- Chemical hazards relate to beach pollution from sewerage, shipping and ports;
- Anthropogenic hazards include structures such as seawalls and breakwaters; vehicles on beaches; pollution; beach congestion and conflicting usage when bathers, bodyboards, surfboards, jetskis and windsurfers all compete for space.

All of these hazards pose a risk to beach users, with the level of risk a function of the type and number of hazards and the type and number of beach users. Hazards can range from a low of 1 to extreme of 10 based on beach state, wave height and tide range, while local factors such as rocks, headlands, structures and pollution will increase this rating (Short and Brander 2015).

The presence of these hazards and the risk level is manifest in the number of first aids, rescues and fatalities that occur in the beach environment. In Brazil it is estimated there are 60,000 rescues each year and about 1000 drownings along its coast. This is an unacceptably high level. By comparison Australia with a coast three times the length has about 50 beach drownings a year.

Beach safety is aimed at identifying, reducing and minimizing the public exposure to these hazards, thereby reducing the level of risk. The Sociedade Brasileira de Salvamento Aquatico (Brazilian Society for Aquatic Rescue) (SOBRASA) is the lead agency in Brazil dedicated to reducing this level. Formed in 1995 SOBRASA has developed a range of programs and initiatives to help improve public awareness of the hazards and at the beach improve the safety resources, particularly lifeguards, as well as improving medical response to beach injuries and rescues (http://www.sobrasa.org).

The Brazil's 2700 open coast beaches are exposed to low to moderate to occasionally high waves. Many of these beaches have beach rips, and many also have topographic rips flowing out through or against beachrock reefs, rocks, reefs, headland, jetties and other man made structures. SOBRASA has found that rip currents account for between 60 and 70 % of beach rescues, a figure comparable with other countries (Short and Hogan 1994). They also found that the rips and rescues are most common when waves exceed 0.5 m. On a beach and rip-dominated coast like Brazil, rip currents, both beach rips and topographic rips, are therefore the major beach hazard.

Beach hazards only pose a risk to people when they enter the beach environment. Along the Brazilian coast the level of risk can therefore vary considerably between beaches, and is dependent on the number and type of people on the beach. Whereas a group of surfers can usually handle waves, rips and such hazards, a group of children from the interior with no beach knowledge would be at great risk. In addition many popular beaches experience major seasonal variation in beach population, with massive numbers coming to the beaches during the main vacation period. The level of beach risk will therefore varies considerably in both time and space depending on the prevailing level of beach hazards, which depend in turn on waves, tide, beach state and local factors and the prevailing beach population, which depends on season, weather and time and day of week. Assessing both beach hazards and risk therefore requires continual monitoring of the beach environment, which is best undertaken by lifeguards and remotely using video cameras which can view the beach and surf and estimate the beach population.

20.5.2 Regional Beach Hazards

This section briefly reviews beach hazards along the Brazilian coast.

Amapá is the most aquatic state in Brazil with an extensive network of rivers and channels and numerous people living along the channels. However while it was a long coastline, it has very low beach population, little beach usage and relatively low hazard beaches away from the river mouths. It therefore has a low level of beach risk.

In **Pará** the highest level of beach risk on popular beaches is caused by the high seasonal beach population, high waves particularly at mid to high tide, macro tides and intertidal rocks (Pereira et al. 2016a). They also found "*The lack of local coastal management plans together with the large number of beachgoers (and vehicles on the beach), many with low swimming skills and poor perception of physical hazards, results in a high level of beach risk and incidents.*"

While much of the **Maranhão** coast in undeveloped the sheltered São Luís beaches while free of rips, are closed to the public because of pollution from unauthorised sewerage outlets (Pereira et al. 2016b). Along the **Piauí** coast Paula et al. (2016) identified the following hazards to bathers: river mouth currents, channels and shoals; rips currents; beachrock reefs; submerged rocks; steep beaches adjacent to deep water; and sewer pollution.

In Ceará the popular **Fortaleza** tourist coast has a high risk level with 300–400 rescues each year and occasional drowning. Albuquerque et al. (2010) attribute most drowning and rescues to beach rips. They add that this problem is exacerbated by the poor siting of lifeguards towers and lack of signs and media to warn the public of the hazards.

In **Rio Grande do Norte** Vital et al. (2016) found "*The natural hazards include strong rip and longshore currents, breaking waves and variable topography associated with the many beachrock reefs and headlands*", with the level of risk increasing seasonally during the summer vacation period.

On the **Pernambuco** coast Pereira et al. (2016c) identified the greatest hazards were associated with the many beachrock reefs. These reefs are both a hazards in themselves as well as inducing variable bathymetry and strong topographic rips in

Fig. 20.5 Sign warning of sharks at Boa Viagem beach, Recife, PN (Photo A D Short)

gaps between the reefs, including along the popular Boa Viagem beach (Fig. 10.3). On the open coast beach rips and tidal sand shoals also trap people during rising tides, while sharks are a major biological hazards with an average of three attacks per year (Fig. 20.5).

In **Bahia** Carvalho (2002) found that along the popular Salvador coast rips currents were the major beach hazard. Dominguez et al. (2016b) extrapolated these findings to the rest of the coast and found that the more exposed rip-dominated northern and central sections posed the greatest threat to bathers, with lower risk on the more sheltered reflective beaches, apart from the southern coast where seacliff falls is also a hazard.

In **Espírito Santo** in addition to the usual hazards associated with beachrock, beach and topographic rips, the popular Camburi tourist beach has been frequentely closed to bathing owing to pollution from the adjacent major port (Fig. 20.4).

Along the exposed higher energy, micro-tidal **São Paulo Bight** (Rio de Janeiro, São Paulo, Parana and Santa Catarina coasts), the beach are wave-dominated and beach and topographic rips pose the greatest threat to bathers. In Santa Catarina Klein et al. (2016) found that of the 246 beaches in the state, just over half had 670 beach rips between them, together with 65 topographic rips (Figs. 17.5, 17.6, 17.7, 17.8 and 17.9). They added that the high number of rips is compounded by the lack of hazard awareness by the public. In Santa Catarina marine stingers can be a hazard as well as water pollution. In Parana Angelotti and Noernberg (2010) found the majority of bather swam in non-patrolled areas, half could not swim and most did not ask the lifeguard about bathing conditions. The most comprehensive beach safety study was undertaken by Klein et al. (2005) in Santa Catarina where they investigated beach safety management and public knowledge of beach hazards. They used the results to work with the Fire Department (responsible for lifeguards) to implement a range of measures (see below) to improve public awareness and lifeguard effectiveness.

Along the high energy **Rio Grande do Sul** coast Calliari and Toldo (2016) found "*The combination of higher population during summer seasons and rip-dominated intermediate beaches result in the most hazardous beaches along the northern littoral and Hermenegildo beach (in the south).*" Another interesting periodic hazards is the accumulation of fluid mud up to 1.5 m thick on the popular Cassino beach (Calliari et al. 2001). This mud can trap surfers exposing them in winter to hypothermia, while mud deposited on the beach and covered by sand is a hazard to cars driving on the beach and has resulted in serious accidents. They also found that water quality deteriorates after heavy rain owing to sewer contamination, and beach washouts caused by the rain are a hazard to vehicle traffic.

20.5.3 Mitigating Beach Risk

The only comprehensive study of beach risk followed by the development of a program to reduce that risk took place in Santa Caterina. As reported by Klein et al. (2003, 2005) and Mocellin (2006) the Santa Catarina beach safety program implemented the following measures to improve public beach awareness and safety: (1) publicity about the meaning of warning flags; (2) larger flags; (3) more pro-active lifeguards; (4) more lifeguards; (5) better trained lifeguards; (6) a civil lifeguard association; (7) water safety campaign; and (8) beach safety education of children and teenagers. As a consequence the number of drowning decreased by 80 % along the thirteen patrolled beaches in the study. In Fortaleza a pro-active program by lifeguards to warn people about beach hazards resulted in a 50 % reduction in beach rescues (Albuquerque et al. 2010). Calliari et al. (2010b) provide the most recent overview of beach hazards and safety in Brazil. They concluded that the Santa Catarina beach safety program had been successful and that similar pro-active programs should be implemented elsewhere, together with the use of video cameras to monitor beach conditions and hazard along the Brazilian beaches.

On a national scale the lead organization in implementing beach safety measures. SOBRASA (http://www.sobrasa.org) is a non-governmental organization and a full member of the International Lifesaving Federation (ILS), with lifesaving representatives from 24 states. Since its foundation in 1995 it has developed a wide range of preventative programs aimed at both educating the public, particular children about beach safety, as well as working to improve and expand the network of beach lifeguards together with beach safety signage and information.

Based on the above the following is evident about beach hazards, risk and safety in Brazil:

- Brazil has an inherently hazardous beaches and inlets, with the major physical hazards being beach rip currents, topographic rips, beachrock reefs, rocks and headlands; tidal inlets and river mouths, and the large tide ranges;
- The major biological hazard is sharks, particularly in northeast Brazil, but marine stingers can occur elsewhere;
- Pollution is a hazard in a number of developed locations where sewerage flows onto the beach and in some place is derived from shipping in adjacent ports;
- Using this environment is a Brazilian and tourist population that generally has a low level of hazards awareness and ability identify and deal with these hazards, e.g. what to do if caught in a rip current. As a consequence there is a need for beach safety education in general, and in particularly amongst the younger Brazilians;
- Finally, there is a need to increase, expand, improve and co-ordinate the life-guarding services as targeted by SOBRASA. This improvement should include better training, more lifeguards, better equipment (e.g. jet-skis, helicopters, medical facilities); and better signage to both educate and warn of hazards (Fig. 20.6). This should be coupled with beach management plans that provide a well-managed and safe environment for people to recreate and bathe at the beach.

Fig. 20.6 Lifeguards in their distinctive red and yellow uniforms placing red flags to identify and warn of rip currents along Boa Viagem beach, PN (Photo: A D Short)

Until such measures are implemented there will continue to be a high level of beach risk along the coast and a high level of rescues, accidents and drowning.

20.6 Future Beach Research

This book highlights what is known about the Brazilian coast and particularly its beaches at the present time. It shows that there is a good general understanding of the range of beach systems along the coast and to a degree of their behavior. It also examines the impact of coastal development on the beaches as well as assessing the extent of coastal erosion and inundation, together with the beach hazards and level of public risk along the coast. The book also highlights areas where we require more information about coastal processes, particular waves, beach response and longer-term beach behavior.

Brazil has a long and highly variable coastline ranging from a tide-dominated north to a wave-dominated south, together with sections dominated by reefs, rivers and bedrock geology. As a consequence there is a need to conduct research across the full spectrum of Brazilian beach and coastal environments in order to obtain both a comprehensive and at the same time a detailed understanding of the Brazilian coast and its numerous coastal systems. Such an understanding is required to effectively manage the coast, as well as prepare for the impacts of increasing coastal population and development and climate change. The wide range of coastal environments also provide an excellent natural laboratory for Brazilian and other researchers to study a tremendous range of coastal processes (waves, tides, currents, sediment transport, etc.) and systems (beaches, tidal inlets, deltas, regressive and transgressive barriers, beachrock and coral reefs, rocky coast, and their associated ecosystems) from tropical to temperate environments.

There are many issues facing the Brazilian coast and its future. In order to effectively manage the coast – its beaches, their development and the public who use them, a better understanding of the coast, its processes and its behavior, is required. Below are some of the areas where more information, monitoring and research is required and where national coastal programs should be implemented as required. It is important to note that the same recommendations were made by the Brazilian coastal community in the final document of First Brazilian Beach Symposium (Finkl and Klein 2003).

1. Wave climate – Brazil is exposed to waves from a range of tropical, temperate and higher latitude sources and consequently has a varied wave climate along its 9000 km of coast. Long term wave monitoring is required right along the coast to both understand the present wave climate as well as detect climate-induced changes in wave climate. In 2015 there were six operational buoys (Recife, Praia do Forte, Santos, Paranaguá, Tramandaí and Rio Grande) (see http://redeondas.herokuapp.com/), however additional buoys are required particularly for parts of the northeast and the north coast.

2. Tides – Brazilian tides range from 11 m to less than 1 m. The micro to mega-tides have a major impact on both beach behavior and sediment transport. A greater understanding of the role of tides and tidal currents is required, particularly in the higher tide ranges of northern Brazil. Also as rising sea level may affect changes in tide range, monitoring and modeling is required to detect and predict the impact of rising sea level on tides. Tides are presently monitored at eleven sites along the coast, but should be monitored along the entire coast by a national organisation.
3. Winds – Coastal dunes dominate much of the Brazilian coast and transport massive volumes of marine sand inland, as well as via headland overpassing to adjoining beach systems. The source, rates of transport and dune migration and their impact all require further study if the dune systems and the sand they transport are to be effective managed.
4. Sediment cells – Longshore sediment transport is predominately to the north along the Brazilian coast. However estimates on the rate of transport can vary by two orders of magnitude, and the identification of sediment cells boundaries both longshore and offshore remains unclear. In addition headland bypassing and overpassing is common. Research is required to obtain accurate measures of sediment transport, cell topography and boundaries, sources of beach nourishment and the impact of human occupation and climate change on the sediment cells and their sediment budget.
5. Coastal-barrier chronology – the chronology or evolution of coastal sectors is required to both understanding how and when the coast evolved, as well as using this information to predict present and future coastal evolution and changes. In addition there have been very few studies of the inner continental shelf, which can play a major role in shoreline behavior.
6. Coastal ecosystems – the Brazilian coast spans the tropics to temperate latitudes and has a wide range of associated sub-tidal (e.g. coral reefs), intertidal (e.g. mangroves), and subaerial (e.g. coastal dune vegetation) ecosystems, all of which are an integral part of the coastal environment and need to be fully understood in order to effectively manage the coast so as to ensure their viability.
7. Shoreline stability – most of the Brazilian coast is sand (and mud) and is inherently unstable responding to waves, tides and sediment budgets. Monitoring of representative coastal sites is required right around the coast to gauge shoreline stability (stable, accreting, eroding) and the drivers of these changes, particularly their response to climate indicators such as ENSO. This can be undertaken with traditional surveying and/or using video cameras, as it occurring at some sites (e.g. Boa Viagem, PN; Massaguaçu, SP; Cassino, RS) and Lidar.
8. Human impacts – the Brazilian coast has undergone massive changes in the past few decades owing to second homes, residential, tourist, urban and port development. Much of this development is poorly planned and located and having adverse impacts of the coast leading to coastal erosion, degraded water quality, pollution, and ground water seepage onto beaches. Where structures are threatened they are often protected with seawalls and groins, which further

exacerbate the problem as well as degrading beach amenity. Past poorly sited development needs to be removed or rectified wherever possible and all future development needs to be planned and managed so as to avoid further degradation of the beaches and placement of development in the coastal hazard zone.

9. Coastal management – integrated coastal management is required throughout the Brazilian coast. Coastal hazards need to be identified, coastal zoning undertaken, coastal management plan developed and the coast developed in an integrated, sustainable way. Most importantly these plans need to be enforced and illegal development prevented.
10. Beach hazards and safety – Brazil has inherently hazardous beaches which result in an unacceptably high level of beach risk as manifest in the high number of rescues and drowning. Beach safety measures including education, information, signage and emergency medical facilities and an improved and expanded lifeguard service are required.
11. Coastal impacts of climate change – climate change will impact the entire coast directly through rising sea level, changing wave climate, changing tide range and rising water temperature, and indirectly though a wide range of secondary and tertiary impacts, such as longshore sediment transport and coastal squeeze. The drivers and impacts of climate change and its impact along the Brazilian coast require ongoing study.
12. Coastal science and scientists – coastal processes do not stop at borders, and there is an urgent need for integrated, multi-disciplinary collaborative research between institutions throughout Brazil and its neighbors, in order to obtain a comprehensive understanding of the Brazilian coast, an understanding that is required for its effective management.

20.7 Conclusions

This chapter has reviewed the nature of Brazil's beach systems, which have tides range from 0.5 to 11 m and waves from near zero to averaging over 1.5 m. As a result the beaches range from wave-dominated to tide-modified and tide-dominated, and within these three types have the full range of beach states, including those fronted by intertidal rock and reef flats. This book has set out to assess the extent, nature and status of these beaches. It has also found that many beaches are unstable and threatened by erosion. The causes of erosion are both natural and a result of anthropogenic factors, the latter including interruption of longshore transport and development too close to the shore. A response to this erosion has been the construction of seawalls, which in turn can destroys the beach-foredune and exacerbate the erosion. Accompanying this development has been an increase in beach usage, which has lead to an increase in public risk on Brazil's naturally hazardous beaches. This risk is manifest in the high number of rescues and drowning which need to be addressed though an expansion in beach safety education and resources.

Brazil has a long, varied and magnificent coast, a coast that increasingly attracts millions of Brazilians as well as visitors. In accommodating this demand parts of the coast have been unnecessarily degraded. The future of Brazil's coasts and beaches requires greater scientific knowledge of the coast and its behavior; better coastal management that is regulated and enforced; and an expansion of beach safety resources to ensure Brazilians can enjoy the beach in greater safety.

References

Albino J, Girardi G, Nascimento KA (2006) Espírito Santo. In: Muehe D (ed) Progradação e Erosão do Litoral Brasileiro. Ministério de Meio Ambiente (MMA), Brasília, pp 227–263

Albino J, Neto NC, Oliveria TCA (2016) In: Short AD, Klein AHF (eds) Brazilian beach systems, Coastal research library. Springer, Dordrecht, pp 333–362

Albuquerque MG, Calliari LJ, Pinheiro LS (2010) Analysis of major risk associated with sea-bathing at Futuro, Fortaleza-Ceará. Braz J Aquat Sci Technol 14:1–8

Angelotti R, Noernberg MA (2010) Risk analysis of sea bathing in Pontal do Paraná county, PR – Season 2003/2004. Braz J Aquat Sci Technol 14(1):65–75, http://dx.doi.org/10.14210/bjast.v14n1.p65-75

Angulo RJ, Borozone CA, Noenberg MA, Quadros CJL, Souza MC, Rosa LC (2016) In: Short AD, Klein AHF (eds) Brazilian beach systems, Coastal research library. Springer, Dordrecht, pp 419–464

Anthony EJ, Gardel A, Gratiot N, Proisy C, Allison MA (2010) The Amazon-influenced muddy coast of South America: a review of mud-bank–shoreline interactions. Earth Sci Rev 103:99–121. doi:10.1016/j.earscirev.2010.09.008

Bonetti J, Klein AHF, Muler M, De Luca CB, Vieira da Silva G, Toldo EE Jr, González M (2012) Chapter 16: Spatial and numerical methodologies on coastal erosion and flooding risk assessement. In: Finkl C (ed) Coastal hazards, vol 16, Coastal research library series. Springer, Dordrecht, pp 423–442

Calliari LJ, Toldo EE Jr (2016) Ocean beaches of Rio Grande do Sul. In: Short AD, Klein AHF (eds) Brazilian beach systems, Coastal research library. Springer, Dordrecht, pp 507–542

Calliari LJ, Speranski N, Torronteguy M, Oliveira MB (2001) The mud banks of Cassino beach, southern Brazil: characteristics, processes and effects. J Coast Res 34:318–325

Calliari LJ, Guedes RMC, Lélis RF, Antiqueira JA, Figueiredo SA (2010a) Perggos e riscos associados a processos costeiros no littoral sul do Brasil (RS): uma sintese. Braz J Aquat Sci Technol 14(1):51–63

Calliari LJ, Klein AHF, Albuquerque MG, Mocellin O (2010b) Beach safety management in Brazil. In: Leatherman S, Fleremeyer J (eds) Rip currents. CRC Press, Boca Raton, pp 215–223

Carvalho MP (2002) Fatores Meteorológicos, Oceanográficos, Morfodinâmicos, Geológicos E Urbanos Relacionados À Incidência De Afogamentos Nas Praias Da Costa Atlântica De Salvador. IGEO-UFBA. Dissertação de Mestrado, Salvador, Bahia, Brazil, 167pp

CPE (Coastal Planning and Engineering do Brasil) (2009) Modelagem Numérica como Subsídio aos Estudos à Fixação da Barra do Rio Araranguá. Relatório Técnico, Florianópolis, 89p

Dillenburg SR, Hesp PA (2009) Geology and geomorphology of Holocene coastal barriers of Brazil. Springer, Berlin, p 380

Dominguez JML (2009) The coastal zone of Brazil. In: Dillenburg SR, Hesp PA (eds) Geology and geomorphology of holocene coastal barriers of Brazil. Springer, Berlin, pp 17–51

Dominguez JML, Bittencourt ACSP, Nascimento AS, Nascimento L (2016a) The sandy becahes of the states of Sergipe and Alagoas. In: Short AD, Klein AHF (eds) Brazilian beach systems, Coastal research library. Springer, Dordrecht, pp 281–305
Dominguez JML, Bittercourt ACS, Nascimento L, Nascimento AS (2016b) Beaches in the sate of Bahia- the importance of geological setting. In: Short AD, Klein AHF (eds) Brazilian beach systems, Coastal research library. Springer, Dordrecht, pp 307–332
DOU (2004) Decreto n° 5.300 de 7 de dezembro de 2004. Regulamenta a Lei no 7.661, de 16 de maio de 1988, que institui o Plano Nacional de Gerenciamento Costeiro – PNGC, dispõe sobre regras de uso e ocupação da zona costeira e estabelece critérios de gestão da orla marítima, e dá outras providências. Publicado no D.O.U. de 8.12.2004, Brasília, DF, Brasil. Disponível em, https://www.planalto.gov.br/ccivil_03/_Ato2004-2006/2004/Decreto/D5300.htm
Dunne T, Mertes LAK, Meade R, Richey JE, Forsberg BR (1998) Exchanges of sediment between the floodplain and channel of the Amazon in Brazil. Geol Soc Am Bull 110(4):450–467. doi:10.1130/0016-7606(1998)110<0450:EOSBTF>2.3.CO;2
Esteves LS (2004) Shoreline change and coastal evolution as parameters to identify priority areas for management in Rio Grande do Sul. Pesquisas em Geociências 31(2):15–30
Finkl CW, Klein AHF (2003) Foreword: proceedings of the Brazilian symposium on sandy beaches. J Coast Res SI 35:1–2
Horn Filho NO (2006) Ilha de Santa Catarina. In: Muehe D et al (eds) Erosão e progradação do litoral brasileiro. Ministério do Meio Ambiente, Brasília, pp 413–436
Italiani DM (2014) Resposta morfodinâmica à alimentação artificial da Ponta da Praia, Santos, SP. Dissertation, University of São Paulo
Klein AHF, Short AD (2016) Brazilian beach systems. In: Short AD, Klein AHF (eds) Brazilian beach systems, Coastal research library. Springer, Dordrecht, pp 1–36
Klein AHF, Santana GG, Diehl FL, Menezes JT (2003) Analysis of hazards associeted with sea bathing: results of five year work oceanic beaches of Santa Catarina State, Southern Brazil. J Coast Res SI(35):107–116
Klein AHF, Mocellim O, Menezes JT, Berribilli M, Vintém G, Dafferner G, Diehl FL, Sperb RM, Santana GG (2005) Beach safety management on the coast of Santa Catarina. Brazil Z Geomorph 141:47–58
Klein AHF, Menezes JT, Diehl FL, Abreu JGN, Polette M, Sperb RM, Sperb RC (2006) Litoral Centro Norte. In: Dieter M (ed) Erosão e Progradação do Litoral Brasileiro, 1st edn. Ministério do Meio Ambiente/Secretaria de Qualidade Ambiental nos Assentamentos Humanos, Brasilia, pp 402–412
Klein AHF, Short AD, Bonetti J (2016) Santa Caterina beach systems. In Short AD, Klein AHF (eds) Brazilian beach systems. Coastal Research Library, Springer, Dordrecht, pp 465–506
Lamour MR (2000) Dinâmica sedimentar do canal da Galheta via de acesso ao porto de Paranaguá, PR. MSc, dissertation, Universidade Federal do Paraná, Curitiba, p 100
Lamour MR, Angulo RJ, Soares CR (2006) Bathymetric evolution of critical silting sectors on Galheta channel, access way to Paranaguá port, Paraná state – Brazil. J Coast Res 23:49–58
Lélis RJ, Calliari LJ (2006) Historical shoreline changes near lagoonal and riverstabilized inlets in Rio Grande do Sul state, southern Brazil. J Coast Res 39:301–305
Lessa GC, Angulo RJ, Giannini PCF, Araújo AD (2000) Stratigraphy and Holocene evolution of a regressive barrier in south Brazil. Mar Geol 165:87–108
Mahiques MM, Siegle E, Alcántara-Carrió J, Silva FG, Sousa PHGO, Martins CC (2016) The beaches of the Sate of São Paulo. In: Short AD, Klein AHF (eds) Brazilian beach systems, Coastal research library. Springer, Dordrecht, pp 397–418
Maia LP, Jiménez JA, Serra J, Morais JO, Sánchez-Arcilla A (1998) The Fortaleza (NE Brazil) waterfront: port versus coastal management. J Coast Res 14(4):1284–1292
Maia LP, Freire GSS, Lacerda LD (2005) Accelerated dune migration and eolian transport during el niño events along the NE Brazilian Coast. J Coast Res 6:1121–1126. doi:10.2112/03-702A.1

Medeiros ECS (2012) Capacidade de carga e percepção ambiental da Praia do Icaraí – Caucaia – CE. MSc. dissertation, Federal University of Ceará

Medeiros ECS, Pantalena AF, Miola B, Lima RS, Soares MO (2014) Percepção ambiental da erosão costeira em uma praia no litoral do Nordeste do Brasil (Praia da Taíba, CE). Rev Gest Cost Integ 14:471–482. doi:10.5894/rgci488

Mocellin O (2006) Determinação do nível de risco público ao banho de mar das praias arenosas do litoral centro norte de Santa Catarina. Unpublished MSc. dissertation, Programa de Pós-graduação em Ciência e Tecnologia Ambiental, Universidade do Vale do Itajaí, Itajaí, SC, Brasil, 162p

Morais JO, Pinheiro LS, Medeiros C, Pitombeira ES (2006) The gabions for the protection of Caponga Beach, Ceará/Brazil: Hazards and management. J Coastal Res SI 39:848–851

Morais JO, Pinheiro LS, Cavalcante AA, Paula DP, Silva RL (2008) Erosão costeira em praias adjacentes as desembocaduras fluviais: O caso de Pontal de Maceió, Ceará, Brasil. Rev Gest Cost Integ 8:61–76. doi:10.5894/rgci132

Muehe D, Lins-de-Barros F (2016) The beaches of Rio de Janerio. In: Short AD, Klein AHF (eds) Brazilian beach systems, Coastal research library. Springer, Dordrecht, pp 363–396

Nicolodi JL, Zamboni A (2008) Gestão Costeira. In: Zamboni A, Nicolodi JL (eds) Macrodiagnóstico da Zona Costeira e Marinha do Brasil, vol 1. Ministério do Meio Ambiente, Brasilia, pp 213–241

Pattearson D, Boswood P, Elias G (2012) Tweed River entrance sand bypassing long term average sand transport rate. 20th NSW Coastal Conference, Tweed Heads, 16pp

Paula JEA, Morais JO, Baptista EMC, Gomes ML, Pinheriro LS (2016) Piauí beach systems. In: Short AD, Klein AHF (eds) Brazilian beach systems, Coastal research library. Springer, Dordrecht, pp 153–174

Pereira LCC, Vila-Concejo A, Short AD (2016a) Coastal morphodynamic processes on the macro-tidal beaches of Pará state under tidally modulated wave conditions. In Short AD, Klein AHF (eds) Brazilian beach systems, Coastal research library. Springer, Dordrecht, pp 95–124

Pereira LCC, Nascimento WT, Silva IR, Vila-Concejo A, Short AD (2016b) Maranhão beach systems, including the himan impact on São Luís beaches. In Short AD, Klien AHF (eds) Brazilian beach systems, Coastal research library. Springer, Dordrecht, pp 95–124

Pereira PS, Araújo TCM, Manso VA (2016c) Tropical sandy beaches of pernambuco state.In Short AD, Klien AHF (eds) Brazilian beach systems, Coastal research library. Springer, Dordrecht, pp 251–280

Perinotto RRC, Campos AV, Bonetti J, Rudorff FR (2012) Avaliação do estado erosional das praias de Santa Catarina com base na análise de geoindicadores. In: de Anais CD (ed) Congresso Brasileiro Sobre Desastres Naturais, Rio Claro, SP. UNESP, Rio Claro, pp 1–11

Pinheiro LS, Rocha GC (2007) Percepção Dos usuários quanto aos problemas relacionados à erosão na praia da Caponga-Cascavel-Ce. In: VII Encontro Nacional da Pós-Graduação em Geografia, Universidade Federal Fluminense, Niteroi

Pinheiro LS, Morais JO, Medeiros C (2001) Mudanças da linha de praia e feições morfológicas em Cascavel- Ceará. Arq Cienc Mar 34:117–130

Pinheiro LS, Morais JO, Maia LP (2016) The beaches of Ceará. In: Short AD, Klein AHF (eds) Brazilian beach systems, Coastal research library. Springer, Dordrecht, pp 175–200

Sampaio R (2006) Ocupação das orlas das praias paranaenses pelo uso balneário. Desenvolvimento e Meio Ambiente 13:169–186, http://dx.doi.org/10.5380/dma.v13i0.9850

Santos VF, Short AD, Mendes AC (2016) In: Short AD, Klein AHF (eds) Brazilian beach systems, Coastal research library. Springer, Dordrecht, pp 67–94

Sayão OJ (1989) Littoral drift along some beaches in Brazil. In: Magoon OT, Converse H, Miner D, Tobin LT, Clark D (eds) Proceedings of the 6th symposium on coastal and ocean management, Charleston, USA, American Society of Civil Engineers 4, pp 3638–3746

Shore Protection Manual (1984) 4th edn, U.S. Army Corps of Engineers, Coastal Engineering Research Center, U.S. Government Printing Office, Washington, DC

Short AD, Brander RW (2015) Beach hazards and risk assessment. In: Bierens J (ed) Drowning: prevention, rescue, treatment. Springer, Berlin, pp 287–292

Short AD, Hogan CL (1994) Rip currents and beach hazards: their impact on public safety and implications for coastal management. J Coast Res SI 12:197–209

Souza CRG, Suguio K (2003) The coastal erosion risk zoning and the State of São Paulo Plan for Coastal Management. J Coast Res 19(Special issue 35):530–547

Suguio K (2003b) Tópicos de geociências para o desenvolvimento sustentável: as regiões litorâneas. Rev Inst Geoc USP 1:1–40

Toldo Jr EE, Motta LM, Almeida LESB, Nunes JC (2013) Large morphological change linked to the sediment budget in the Rio Grande do Sul Coast. In: Coastal Dynamics, Arcachon. Proceedings of Coastal Dynamics. SHOM, p 1687

Torres JR, Abessa DMS, Santos FC, Maranhão LA, Davanso MB, Nascimento MRL, Mozeto A (2009) Effects of dredging operations on sediment quality: contaminant mobilization in dredged sediments from the Porto of Santos, SP, Brazil. J Soils Sediments 9:420–432

Vieira da Silva G, Muler M, Prado MFV, Short AD, Klein AHF, Toldo Jr EE (2016) Shoreline change analysis and insight into the sediment transport path along Santa Catarina Island north shore, Brazil. J Coast Res 32. doi:10.2112/JCOASTRES-D-15-00164.1

Vital H, Amaro VE, Silveira IM (2006) Coastal Erosion on the Rio Grande doNorte State (Northeastern Brazil): causes and factors versus effects and associated processes. J Coast Res SI 39:1307–1310

Vital H, Silveira IM, Tabosa WF, Lima ZMC, Pinheiro Lima Filho F, Souza FES, Chaves MS, Pimenta FM, Gomes MP (2016) The Beaches of Rio Grande do Norte. In: Short AD, Klein AHF (eds) Brazilian beach systems, Coastal research library. Springer, Dordrecht, pp 201–230

Website: http://www.sobrasa.org

Index

A.D. Short, A.H. da F. Klein (eds.), *Brazilian Beach Systems*, Coastal Research Library 17, DOI 10.1007/978-3-319-30394-9

MIX
Papier aus verantwortungsvollen Quellen
Paper from responsible sources
FSC® C105338

If you have any concerns about our products,
you can contact us on
ProductSafety@springernature.com

In case Publisher is established outside the EU,
the EU authorized representative is:
Springer Nature Customer Service Center GmbH
Europaplatz 3, 69115 Heidelberg, Germany

Printed by Libri Plureos GmbH
in Hamburg, Germany